STUDY GUIDE AND SOLUTIONS MANUAL FOR

ORGANIC CHEMISTRY

Susan McMurry

BROOKS/COLE PUBLISHING COMPANY
MONTEREY, CALIFORNIA

Brooks/Cole Publishing Company
A Division of Wadsworth, Inc.

Printed in the United States of America

10 9 8 7 6

QD251.2.M43 1984 547 83-7744

ISBN 0-534-02675-3

Sponsoring Editor: Michael Needham
Production Coordinator: Louise Rixey
Cover Design: Stan Rice

PREFACE

What enters your mind when you hear the words "organic chemistry"? Some of you may think, "the chemistry of life," or "the chemistry of carbon." Other responses might include, "pre-med," "pressure," "difficult," or "memorization." Although organic chemistry is, formally, the study of the compounds of carbon, the discipline of organic chemistry encompasses many skills that are common to other areas of study. Organic chemistry is as much a liberal art as a science, and mastery of the concepts and techniques of organic chemistry can lead to an enhanced competence in other fields.

As you proceed to solve the problems that accompany the text, you will bring to the task many problem-solving techniques. For example, planning an organic synthesis requires the skills of a chess player; you must plan your moves while looking several steps ahead, and you must keep your plan flexible. Structure determination problems are detective problems, in which many clues must be assembled to yield the most likely solution. Naming organic compounds is similar to the systematic naming of biological specimens; in both cases, a set of rules must first be learned and then applied to the specimen or compound under study.

The problems in the text fall into two categories—simple, or drill, problems, and complex problems. Simple problems, which appear throughout the text and in the end of chapter problems, test your knowledge of one fact or technique at a time. You may need to rely on memorization to solve these problems, which you should work on first. More complicated problems require you to recall facts from several parts of the text and involve using some of the problem-solving techniques mentioned above. As each major type of problem—synthesis, nomenclature, or structure determination—is introduced in the text, one solution to that type of problem is extensively worked out in the Solutions Manual.

I would like to offer several suggestions that may help you with problem-solving.

1. The text is organized, for the most part, into chapters that describe individual functional groups of organic compounds. As you study each functional group, *make sure that you understand the structure and reactivity of that group*. In case your memory fails you, you can rely on your knowledge of functional groups to help you.
2. *Use molecular models*. It is difficult to imagine the three-dimensional architecture of an organic molecule when looking at a two-dimensional drawing. Models will help you to appreciate the structural aspects of organic chemistry and are indispensible tools for understanding stereochemistry.
3. I have gone to great lengths to make this Solutions Manual as clear, attractive, and error-free as possible. Nevertheless, I advise you to *use the Solutions Manual in moderation*. The principal use of this book should be to

check answers to problems you have worked out. Also, this Manual can give you assistance if you are unable to solve a problem. The Solutions Manual should not be used as a substitute for mental effort; at times, struggling with a difficult problem is the only way to teach yourself.

4. *Read through the appendices at the end of the Solutions Manual.* Some of these appendices contain tables that may help you in working problems; others present information related to the history of organic chemistry. The nomenclature appendix should be helpful for naming complex molecules.

Acknowledgements. First, I would like to thank my husband, John McMurry, for suggesting this project and for supporting my efforts while this book was being written. Virginia Severn Goodman and Terry Auld were responsible for typing the preliminary manuscript. Finally, my sincere appreciation goes to Sonja Erion, who typed the final copy of the Solutions Manual. I thank her not only for the beautiful job she did in layout and typing, but also for her scientific editing and her many suggestions, which have added to the quality of this book.

CONTENTS

Solutions to Problems

Appendices

CHAPTER 1 STRUCTURE AND BONDING

1.1 The first step in finding the ground state electronic configuration of an element is to locate its atomic number. For boron, the atomic number is 5; boron thus has 5 protons and *5 electrons*. Next, we assign the electrons to the proper energy levels, starting with lowest level:

boron
$2p$ ↑ __ __
$2s$ ↑↓
$1s$ ↑↓

Remember that only two electrons can occupy the same orbital, and that they must be of opposite spin.

A different way to represent the ground state electron configuration is to simply write down the occupied orbitals and to indicate the number of electrons in each orbital. For example, the electron configuration for boron is $1s^22s^22p$.

Often, we are interested only in the electrons in the outermost shell. We can then represent all filled levels by the symbol for the inert gas having the same levels filled. In the case of boron the filled $1s$ energy level is represented by [He], and the *valence shell electron configuration* is symbolized by $[He]2s^22p$.

Let's consider an element with many electrons. Phosphorus, with an atomic number of 15, has *15 electrons*. Assigning these to energy levels:

phosphorus
$3p$ ↑ ↑ ↑
$3s$ ↑↓
$2p$ ↑↓ ↑↓ ↑↓
$2s$ ↑↓
$1s$ ↑↓

Notice that the $3p$ electrons are all in different orbitals. According to *Hund's rule*, we must place one electron into each orbital of the same energy level until all orbitals are half-filled.

The more concise way to represent ground state electron configuration for phosphorus: $1s^22s^22p^63s^23p^3$

Valence shell electron configuration: $[Ne]3s^23p^3$

Iron (atomic number 26)
$3d$ ↑↓ ↑ ↑ ↑ ↑
$4s$ ↑↓
$3p$ ↑↓ ↑↓ ↑↓
$3s$ ↑↓
$2p$ ↑↓ ↑↓ ↑↓
$2s$ ↑↓
$1s$ ↑↓

$1s^22s^22p^63s^23p^64s^23d^6$

$[Ar]4s^23d^6$

Selenium (atomic number 34)
$4p$ ↑↓ ↑ ↑
$3d$ ↑↓ ↑↓ ↑↓ ↑↓ ↑↓
$4s$ ↑↓
$3p$ ↑↓ ↑↓ ↑↓
$3s$ ↑↓
$2p$ ↑↓ ↑↓ ↑↓
$2s$ ↑↓
$1s$ ↑↓

$1s^22s^22p^63s^23p^64s^23d^{10}4p^4$

$[Ar]4s^23d^{10}4p^4$

1.2 Writing Lewis dot structures requires that you first determine the number of valence, or outer-shell, electrons for each element. For chloroform, we know that carbon has four valence electrons, hydrogen has one, and each chlorine has seven.

```
 ·
·C·    4 x 1 =  4
 ·
H·     1 x 1 =  1
 ··
:Cl·   7 x 3 = 21
 ··           ----
               26 total valence electrons
```

Next, use two electrons for each single bond.

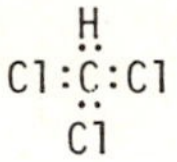

```
    H
    ··
Cl:C:Cl
    ··
    Cl
```

Finally, use the remaining electrons to achieve an inert gas configuration for all atoms.

a) $CHCl_3$

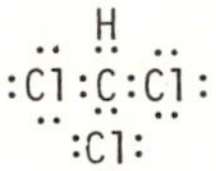

```
      H
  ··  ··  ··
 :Cl:C:Cl:
  ··  ··  ··
     :Cl:
      ··
```

b) H_2S Total valence electrons = 8

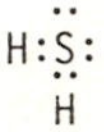

```
   ··
 H:S:
   ··
   H
```

c) CH_3NH_2 Total valence electrons = 14

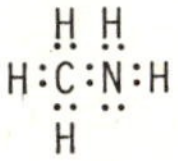

```
   H H
   ·· ··
 H:C:N:H
   ·· ··
   H
```

d) BH_3 Total valence electrons = 6

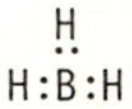

```
   H
   ··
 H:B:H
```

Note that it is not possible for boron to achieve an inert gas configuration because there are only six valence electrons.

e) NaH Total valence electrons = 2 Na:H

f) CH_3OCH_3 Total valence electrons = 20

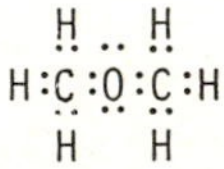

```
   H  ··  H
   ··     ··
 H:C:O:C:H
   ·· ·· ··
   H      H
```

1.3 Bonds formed between an element with low ionization potential and an element with high electron affinity are ionic. Bonds formed between an element in the middle of the periodic table and another element are most often covalent, but there are exceptions.

ionic bonds: $CaCl_2$, KF

covalent bonds: CH_3-Cl, CH_3-NH_2, BF_3, H_2N-NH_2

1.4

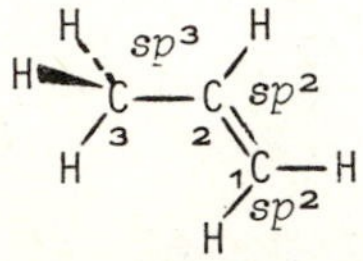

```
  H   sp³   H
   \       /
H—C₃———C₂ sp²
  /       \\
 H         C₁—H
          /sp²
         H
```

propene

The *C3-H bonds* are sigma bonds formed by overlap of an sp^3 orbital of carbon 3 with an s orbital of hydrogen.

The *C2-H and C1-H bonds* are sigma bonds formed by overlap of an sp^2 orbital of carbon with an s orbital of hydrogen.

The *C2-C3 bond* is a sigma bond formed by overlap of an sp^3 orbital of carbon 3 with an sp^2 orbital of carbon 2.

There are two *C1-C2 bonds*. One is a sigma bond formed by overlap of an sp^2 orbital of carbon 1 with an sp^2 orbital of carbon 2. The other is a pi bond formed by overlap of a *p* orbital of carbon 1 with a *p* orbital of carbon 2. All four atoms attached to the carbon-carbon double bond lie in the same plane. The C-C-C bond angle is approximately 120°.

1.5 $H_3\overset{sp^3}{C}_3-\overset{sp}{C}_2\equiv\overset{sp}{C}_1-H$ propyne

The *C3-H bonds* are sigma bonds formed by overlap of an sp^3 orbital of carbon 3 with an *s* orbital of hydrogen.

The *C1-H bond* is a sigma bond formed by overlap of an *sp* orbital of carbon 1 with an *s* orbital of hydrogen.

The *C2-C3 bond* is a sigma bond formed by overlap of an *sp* orbital of carbon 2 with an sp^3 orbital of carbon 3.

There are three *C1-C2 bonds*. One is a sigma bond formed by overlap of an *sp* orbital of carbon 1 with an *sp* orbital of carbon 2. The other two bonds are pi bonds formed by overlap of two *p* orbitals of carbon 1 with two *p* orbitals of carbon 2.

The three carbon atoms of propyne are approximately linear.

1.6 $H_2\overset{sp^2}{C}_1=\overset{sp^2}{C}_2H-\overset{sp^2}{C}_3H=\overset{sp^2}{C}_4H_2$ 1,3-butadiene

All atoms lie in the same plane, and all bond angles are approximately 120°.

1.7

The two nitrogen atoms are triple-bonded. The sigma bond is formed by the overlap of one *sp* orbital from each nitrogen, and the two pi bonds are formed by the overlap of two *p* orbitals from one nitrogen atom with two *p* orbitals from the other nitrogen atom. A lone pair of electrons occupies the remaining *sp* orbital of each nitrogen.

1.8 a)

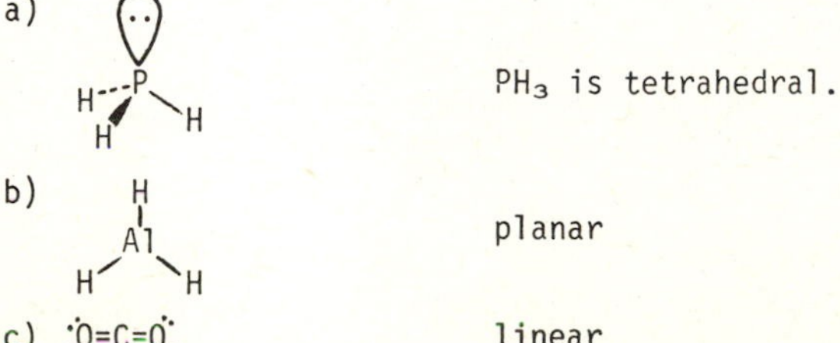

a) PH_3 is tetrahedral.

b) AlH_3 planar

c) :Ö=C=Ö: linear

d) 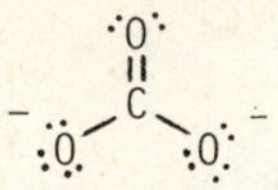 planar

e) 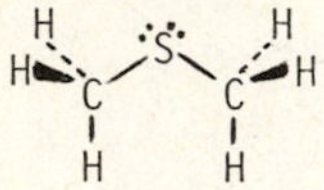 The C-S-C bond angle is approximately 109°.

f) 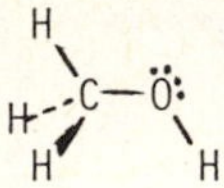 The C-O-H bond angle is approximately 109°.

1.9

	element	atomic number	ground state electronic configuration
a)	Na	11	$1s^2 2s^2 2p^6 3s$
b)	Al	13	$1s^2 2s^2 2p^6 3s^2 3p$
c)	As	33	$1s^2 2s^2 2p^6 3s^2 3p^6 4s^2 3d^{10} 4p^3$
d)	Si	14	$1s^2 2s^2 2p^6 3s^2 3p^2$
e)	Cr	24	$1s^2 2s^2 2p^6 3s^2 3p^6 4s^2 3d^4$

1.10 a) H:C:::C:H 10 valence electrons

b) 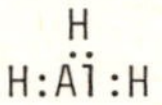 6 valence electrons

c) 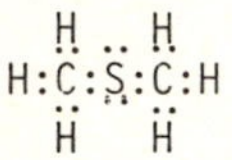 20 valence electrons

d) 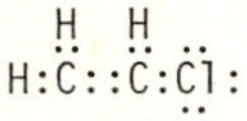 18 valence electrons

e) 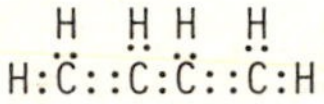 22 valence electrons

f) 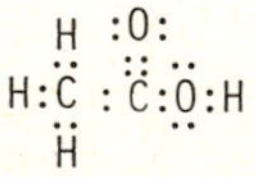 24 valence electrons

1.11 In problems of this sort, carbon is listed first and hydrogen second. All other elements are listed in alphabetical order after C and H.

a) C_6H_6O phenol

b) $C_9H_8O_4$ aspirin

c) $C_6H_8O_6$ vitamin C

d) $C_{10}H_{14}N_2$ nicotine

e) $C_{13}H_{21}ClN_2O_2$ Novocain

f) $C_6H_{12}O_6$ glucose

1.12 To work a problem of this sort requires that you examine *all* the possibilities consistent with the rules of valency. You must systematically consider all possible attachments, including those that have branches, rings and multiple bonds.

a) 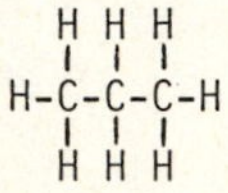 b) 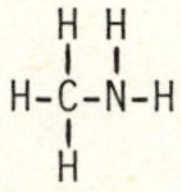 c) 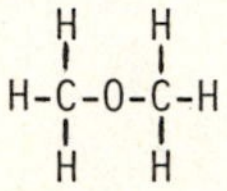 and 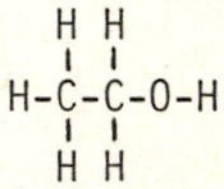

d) $CH_3CH_2CH_2Br$ and $CH_3CHBrCH_3$ (structural formulas)

e) CH_3CHO (H-C-C-H with C=O), $H_2C{=}CH(OH)$, and ethylene oxide (three-membered ring: O, C, C, each C bearing two H)

f) $CH_3CH_2CH_2NH_2$, $CH_3CH_2NHCH_3$, $CH_3CH(NH_2)CH_3$, $(CH_3)_3N$ (structural formulas)

1.13 a) H–C–C–C–H propane: all three carbons sp^3

b) $(CH_3)_2C{=}CH_2$: CH_3 carbons sp^3; sp^2C=C sp^2

c) $H_2C{=}CH{-}C{\equiv}CH$ (drawn H–C sp^2 = sp^2 C–C–C≡H, with "sp sp" on the last two carbons)

d) Cyclobutene: C=C carbons sp^2, sp^2; CH_2 carbons sp^3, sp^3

e) H–C–O–C–H: both carbons sp^3

f) Toluene: ring carbons all sp^2; CH_3 carbon sp^3

1.14 a) sp^3 b) sp^3 c) sp^2

1.15 All angles are approximate.

a) 109° b) 109° c) 109° d) 120°

1.16 a) NH_4^+ (tetrahedral drawing)

The ammonium ion is tetrahedral because nitrogen is sp^3 hybridized.

b) $B(CH_3)_3$ (three-dimensional drawing)

The boron-carbon bonded portion of the molecule is planar because of sp^2 hybridization at boron. The $-CH_3$ portions are tetrahedral.

c) Benzene, C_6H_6 (structural drawing)

Benzene is planar because of sp^2 hybridization of all six carbon atoms.

d) $P(CH_3)_3$ (three-dimensional drawing with lone pair on P)

The phosphorus atom is tetrahedral because of sp^3 hybridization. (Don't forget the lone pair.) The $-CH_3$ portions are tetrahedral.

1.17 a) SO_2 has eighteen valence electrons (six from sulfur and six from each oxygen). The oxygen-sulfur-oxygen bond angle is approximately 120°.

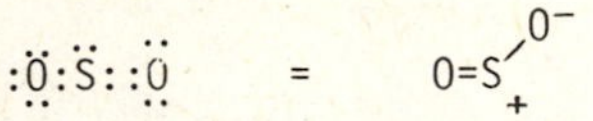

b) SO_3 has 24 valence electrons. SO_3 is a planar molecule.

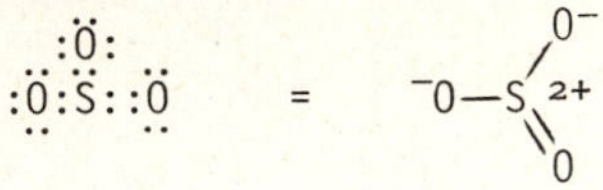

c) Four oxygens and one sulfur contribute 30 valence electrons. In addition, there are two electrons that give SO_4^{2-} its negative charge. The total number of electrons used to draw the Lewis structure is 32. SO_4^{2-} is a tetrahedral anion.

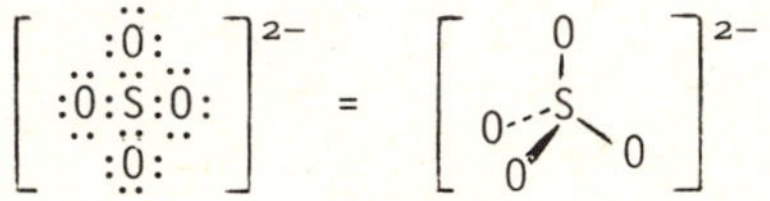

1.18 a) Each Ti has four valence electrons; each Cl has seven. Total number of valence electrons is 32.

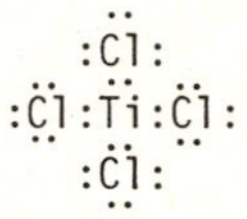

b) There are 16 valence electrons.

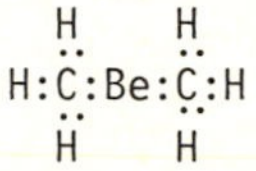

c) There are 26 valence electrons (five from phosphorus, four from each carbon, and one from each hydrogen.

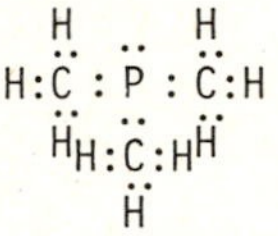

1.19 a)

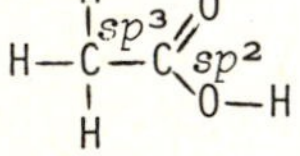

b)

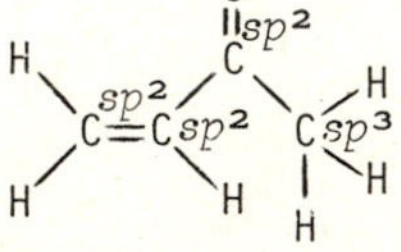

c)

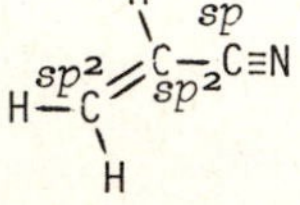

d)

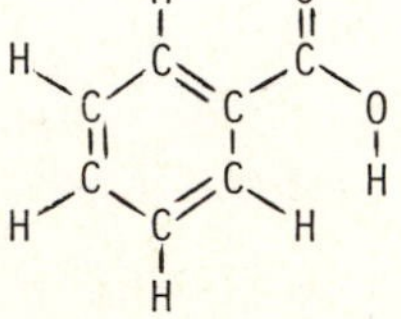

all carbons are sp^2 hybridized

1.20 a) sp^3 b) sp^2 c) sp d) sp^3

1.21 All angles are approximate.

a) 109° b) 180° c) 109°

1.22 Covalent: CH_3O--H, H--ONO_2, $CH_3\overset{O}{\overset{\|}{C}}$--Cl, CH_3O--Cl, $CH_3\overset{O}{\overset{\|}{C}}O$--$CH_3$, F--F

Ionic: CH_3O--Na, Na--NH_2, $CH_3\overset{O}{\overset{\|}{C}}O$--$NH_4$

1.23 least ionic → most ionic

a) Cl--Cl, H_3C--Cl, NH_4--Cl

b) CH_3NH--$NHCH_3$, CH_3O--NH_2, $CH_3\overset{O}{\overset{\|}{C}}O$--$NH_4$

c) H_3C--H, $CH_3\overset{O}{\overset{\|}{C}}O$--H, H--Cl

d) $(CH_3)_3Si$--CH_3, Br--CH_3, Li--CH_3

1.24 The central carbon of $R_3C^{\oplus}$, which has three valence shell electrons, is sp^2 hybridized. The carbocation is planar and is isoelectronic with (has the same number of electrons as) a trivalent boron compound.

1.25 Both $R_3C{:}^{\ominus}$ and $R_3C\cdot$ have tetrahedral geometries and are sp^3 hybridized. Note that $R_3C{:}^{\ominus}$ has the same electronic configuration as $:NH_3$.

1.26 To answer this problem correctly we must first remember the Pauli Exclusion Principle, which states that two electrons in the same orbital must have opposite spins. Thus, the two electrons of triplet (spin-unpaired) methylene must occupy different orbitals. Because carbon uses two other orbitals to bond with hydrogen, there are a total of four occupied orbitals. We therefore predict sp^3 hybridization and tetrahedral geometry for triplet methylene. In singlet (spin-paired) methylene the two electrons can occupy the same orbital because they have opposite spins. Including the two C-H bonds, there are a total of three occupied orbitals. We predict sp^2 hybridization and planar geometry for singlet methylene.

109° vacant *p* orbital 120°

triplet methylene (tetrahedral) singlet methylene (planar)

What you should know:

After doing these problems you should be able to :

1. Predict ground state electron configuration of elements;
2. Draw Lewis dot structures of compounds;
3. Predict whether bonds are covalent or ionic;
4. Understand sigma bonds, pi bonds, and hybridization. You should be able to determine the type of hybridization of an atom and tell whether it is bonded to another atom by sigma or pi bonds;
5. Predict bond angles and shapes of molecules;
6. Convert Kekulé structures into molecular formulas, and *vice versa*;
7. Convert Lewis dot structures into Kekulé structures and *vice versa*.

CHAPTER 2 BONDING AND MOLECULAR PROPERTIES

2.1 a) This problem will be worked out in detail, using the rules in the text.

```
        *                           H
                                    |
   * /‾‾‾\ *            H\        C         /H
    |     |         =      C=/      \C
   * \   / *               |         ||
      N                  H/ C=\    /C \H
      ..                        N
                                ..
```

Remember: (1) carbon atoms occur at the intersection of two line-bonds (indicated by * above); (2) to satisfy valency, a hydrogen is implicitly understood to be bonded to each of the above carbons. The molecular formula of pyridine is C_5H_5N.

b) Cyclohexanone; $C_6H_{10}O$

c) Indole; C_8H_7N

d) The Kekulé structure of BHT is:

BHT

The molecular formula of BHT is $C_{15}H_{24}O$.

2.2 There are several possible Kekulé structures for each molecular formula.

a)

```
  H H H H H          H H H H              H
  | | | | |          | | | |              |
H-C-C-C-C-C-H      H-C-C-C-C-H         H H-C-H H
  | | | | |          | | | |            |  |  |
  H H H H H          H H | H          H-C--C--C-H
                       H-C-H            |  |  |
                         |             H H-C-H H
                         H                |
                                          H
```

C_5H_{12}

b)

```
  H H H            H H H
  | | |            | | |
H-C-C-N-H        H-C-N-C-H
  | |              |   |
  H H              H   H
```

C_2H_7N

c)

```
  H H O          H H H            H O H          H
  | | ||         | | |            | || |         |
H-C-C-C-H      H-C=C-C-OH       H-C-C-C-H      H-C-O
  | |                |            |   |          | |
  H H                H            H   H        H-C-C-H
                                                 | |
                                                 H H

    O   H          H   H          H H H          H OH H
   / \  |           \ /           | | |          | |  |
H-C---C-C-H          C          H-C-C=C-OH     H-C=C--C-H
  |   | |           / \           |                   |
  H   H H        H-C---C-OH       H                   H
                   |   |
                   H   H
```

C_3H_6O

d)

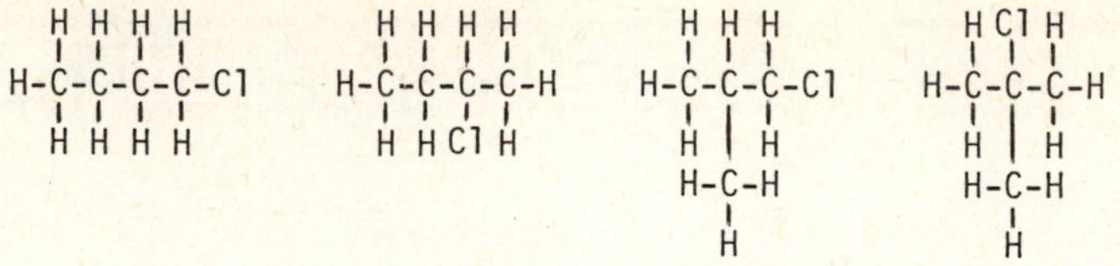

```
  H H H H        H H H H        H H H          H Cl H
  | | | |        | | | |        | | |          | |  |
H-C-C-C-C-Cl   H-C-C-C-C-H    H-C-C-C-Cl     H-C-C--C-H      C4H9Cl
  | | | |        | | | |        | | |          | |  |
  H H H H        H H Cl H       H | H          H |  H
                                H-C-H          H-C-H
                                  |              |
                                  H              H
```

C_4H_9Cl

2.3 Formal charge (FC) = # of outer shell electrons − $\frac{\text{\# bonding electrons}}{2}$ − # non-bonding electrons

a) $H_2C{=}N{=}\ddot{N}:$ The Lewis dot structure is

```
   H  1  2
   ..   ..
 H:C::N::N:
```

Hydrogen	$FC = 1 - \frac{2}{2} - 0 = 0$
Carbon	$FC = 4 - \frac{8}{2} - 0 = 0$
Nitrogen 1	$FC = 5 - \frac{8}{2} - 0 = +1$
Nitrogen 2	$FC = 5 - \frac{4}{2} - 4 = -1$

Remember: Outer shell electrons are the valence electrons particular to a specific element. Bonding electrons are those electrons involved in bonding to other elements. Non-bonding electrons are those electrons in lone pairs.

b)

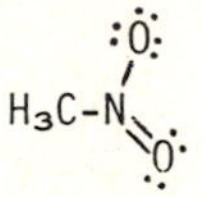

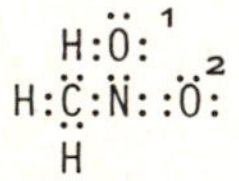

```
       ..                      ..  1
      :O:                    H:O:
      /                        ..    2
 H3C-N             ≡        H:C:N::O:
      \\                      ..    ..
       O:                     H
       ..
```

Hydrogen	$FC = 1 - \frac{2}{2} - 0 = 0$
Carbon	$FC = 4 - \frac{8}{2} - 0 = 0$
Nitrogen	$FC = 5 - \frac{8}{2} - 0 = +1$
Oxygen 1	$FC = 6 - \frac{4}{2} - 4 = 0$
Oxygen 2	$FC = 6 - \frac{2}{2} - 6 = -1$

c)

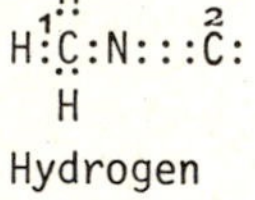

```
   H
  1..     2
 H:C:N:::C:
   ..
   H
```

Hydrogen	$FC = 1 - \frac{2}{2} - 0 = 0$
Carbon 1	$FC = 4 - \frac{8}{2} - 0 = 0$
Carbon 2	$FC = 4 - \frac{6}{2} - 2 = -1$
Nitrogen	$FC = 5 - \frac{8}{2} - 0 = +1$

2.4 We must look at the polarization of the individual bonds of a molecule to account for an observed dipole moment. In the case of methanol, CH_3OH, the individual bond polarities can be estimated from Table 2.1. A bond is polarized in the direction of the more electronegative element (the larger numbers in Table 2.1). In addition, we must take into account the contribution of the two lone pairs of oxygen. Indicating the individual bond polarities by arrows, we can predict the direction of the dipole moment.

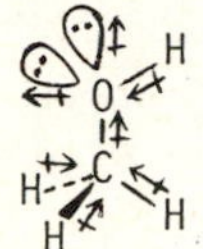

↑ net dipole moment

If we were to calculate the dipole moment from the individual bond moments, we would have to use vector analysis. It is usually possible, however, to estimate qualitatively the direction and relative magnitude of the dipole moment by estimating the net direction of the bond polarities.

b) 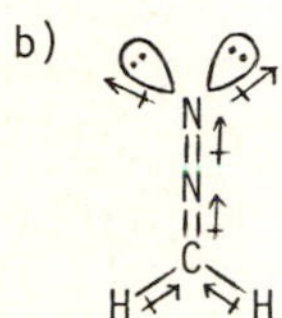

↑ net dipole moment

c) 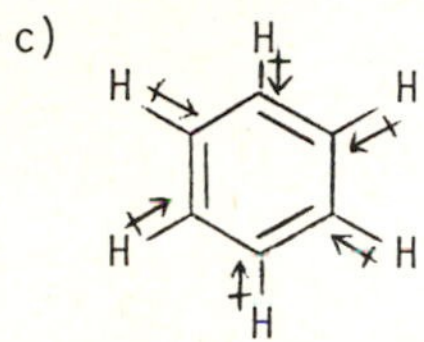

0 dipole moment

Notice that all individual bond moments cancel each other to give zero net dipole moment.

2.5 a)

```
          H
          ..
     F  H:C:H
     ..   ..
   F:B  : O:
     ..   ..
     F  H:C:H
          ..
          H
```

boron: $FC = 3 - \frac{8}{2} - 0 = -1$

oxygen: $FC = 6 - \frac{6}{2} - 2 = +1$

The formal charge of -1 for boron indicates that boron has a net negative charge; oxygen has a net positive charge.

b)

```
             H
             ..
     Cl   H:C:H  H
     ..     ..   ..
  Cl:Al  :  N  : C:H
     ..     ..   ..
     Cl   H:C:H  H
             ..
             H
```

aluminum: $FC = 3 - \frac{8}{2} - 0 = -1$

nitrogen: $FC = 5 - \frac{8}{2} - 0 = +1$

The formal charge of -1 for aluminum indicates that it has a net negative charge; the formal charge +1 for nitrogen indicates that it has a net positive charge.

2.6 A Lewis base has a non-bonding electron pair to share; a Lewis acid has a vacant orbital to accept an electron pair.

Lewis acids: $MgBr_2$, $B(CH_3)_3$

Lewis bases: $:P(CH_3)_3$, $:\ddot{S}(CH_3)_2$, $CH_3C{\equiv}N:$, $C_5H_5N:$

2.7 a) benzene, C_6H_6

molecular weight = 6(12.011) + 6(1.008) = 78.11 g/mol

$$\% \text{ carbon} = \frac{6(12.011)}{78.11} \times 100 = \frac{72.06}{78.11} \times 100 = 92.25\%$$

$$\% \text{ hydrogen} = \frac{6(1.008)}{78.11} \times 100 = \frac{6.048}{78.11} \times 100 = 7.75\%$$

b) $C_{14}H_{15}NO_7$

molecular weight = 309.28 g/mol

$$\% \text{ carbon} = \frac{168.15}{309.28} \times 100 = 54.37\%$$

$$\% \text{ hydrogen} = \frac{15.12}{309.28} \times 100 = 4.89\%$$

$$\% \text{ nitrogen} = \frac{14.01}{309.28} \times 100 = 4.53\%$$

$$\% \text{ oxygen} = \frac{112.00}{309.28} \times 100 = 36.21\%$$

c) $C_{20}H_{24}N_2O_2$

molecular weight = 324.42 g/mol

$$\% \text{ carbon} = \frac{240.22}{324.42} \times 100 = 74.05\%$$

$$\% \text{ hydrogen} = \frac{24.19}{324.42} \times 100 = 7.46\%$$

$$\% \text{ nitrogen} = \frac{28.01}{324.42} \times 100 = 8.63\%$$

$$\% \text{ oxygen} = \frac{32.00}{324.42} \times 100 = 9.86\%$$

d) $C_{18}H_{20}O_2$

molecular weight = 268.36 g/mol

$$\% \text{ carbon} = \frac{216.20}{268.36} \times 100 = 80.56\%$$

$$\% \text{ hydrogen} = \frac{20.16}{268.36} \times 100 = 7.51\%$$

$$\% \text{ oxygen} = \frac{32.00}{268.36} \times 100 = 11.92\%$$

2.8 Citral has a percent composition: 78.9% carbon, 10.6% hydrogen, 10.5% oxygen. Assume that we have a 100 g sample of citral. It contains: 78.9 g carbon, 10.6 g hydrogen, 10.5 g oxygen. If we now divide the weight of each element in the 100 g sample by the atomic weight of the element, we get the *relative* number of moles of each element in the sample.

$$\text{for C} \quad \frac{78.9 \text{ g}}{12.0 \text{ g/mol}} = 6.56 \text{ moles C}$$

$$\text{for H} \quad \frac{10.6 \text{ g}}{12.0 \text{ g/mol}} = 10.51 \text{ moles H}$$

$$\text{for O} \quad \frac{10.5 \text{ g}}{12.0 \text{ g/mol}} = 0.656 \text{ moles O}$$

Next, it is necessary to find the ratio of the numbers of moles of the elements of citral in terms of the smallest whole numbers:

C : H : O = 6.56 : 10.51 : 0.656 = 10 : 16 : 1

So, the empirical formula of citral is $C_{10}H_{16}O$. The formula weight corresponding to $C_{10}H_{16}O$ is 152; since the actual molecular weight given is 152 g/mol, we know that the molecular formula of citral is $C_{10}H_{16}O$.

2.9 The first step in solving this problem is to determine how many mg of carbon and hydrogen are in an 8.00 mg sample of squalene.

$$\text{\# mg carbon} = \text{\# mg } CO_2 \times \frac{\text{at.wt. C}}{\text{m.w. } CO_2} = 25.6 \text{ mg} \times \frac{12.0}{44.0} = 6.98 \text{ mg C}$$

$$\text{\# mg hydrogen} = \text{\# mg } H_2O \times \frac{2(\text{at.wt. H})}{\text{m.w. } H_2O} = 8.75 \text{ mg} \times \frac{2.02}{18.0} = 0.98 \text{ mg H}$$

6.98 mg + 0.98 mg = 7.96 mg (C + H). Because this number is so close to 8.00 mg, we can assume that squalene contains only C and H.

Next, it is necessary to *determine the percent* of carbon and hydrogen in the sample.

$$\frac{6.98 \text{ mg}}{8.00 \text{ mg}} \times 100 = 87.2\% \text{ carbon}$$

$$\frac{0.98 \text{ mg}}{8.00 \text{ mg}} \times 100 = 12.2\% \text{ hydrogen}$$

Now, as in problem 2.8, we must *find the relative number of moles* of carbon and hydrogen.

$$\frac{87.2 \text{ g}}{12.0 \text{ g/mol}} = 7.26 \text{ moles carbon}$$

$$\frac{12.2 \text{ g}}{1.01 \text{ g/mol}} = 12.1 \text{ moles hydrogen}$$

Finding the nearest whole number ratio, we obtain:

C : H = 7.26 : 12.1 = 3 : 5

Thus, the empirical formula of squalene is C_3H_5. The formula weight of a compound having the formula C_3H_5 is 3(12.0) + 5(1.01) = 41.0. Since the actual molecular weight of squalene is 410 g/mol, the molecular formula = 10 x (empirical formula), or $C_{30}H_{50}$.

2.10 a) 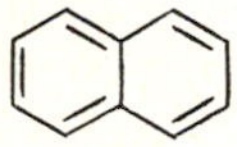 b)  c) 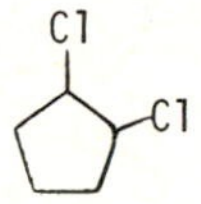 d)  e)

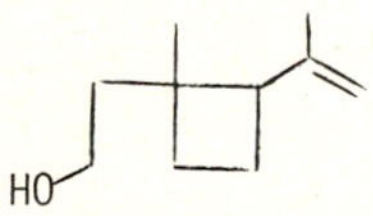

2.11 a) 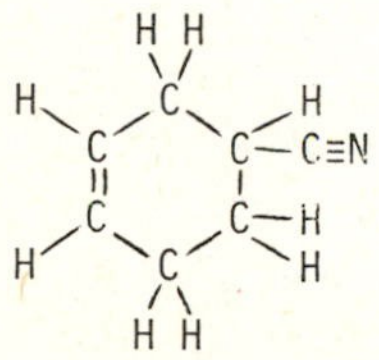 b) 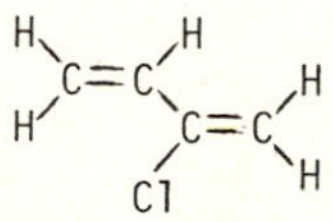 c)

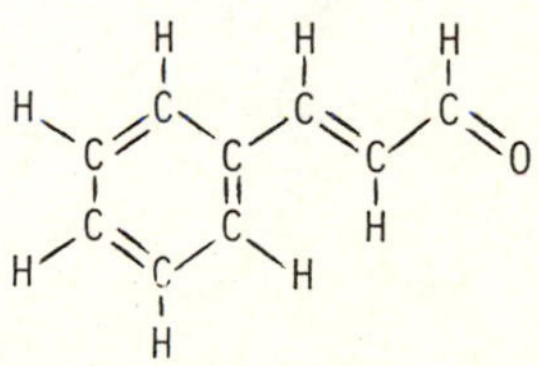

d) [structure: $(CH_3)_2C{=}CH$–cyclopropane ring bearing $C(CH_3)_2$ and $COOH$]

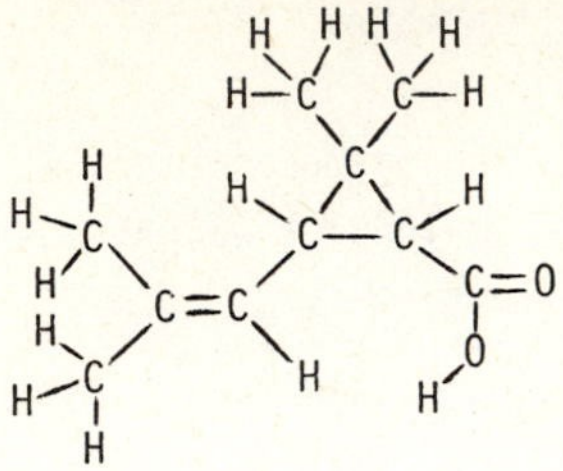

2.12 a) $(CH_3)_3O{:}BF_4$

oxygen	$FC = 6 - \frac{6}{2} - 2 = +1$
boron	$FC = 3 - \frac{8}{2} - 0 = -1$

b) $:CH_2-\overset{1}{N}\equiv\overset{2}{N}:$

carbon	$FC = 4 - \frac{6}{2} - 2 = -1$
nitrogen 1	$FC = 5 - \frac{8}{2} - 0 = +1$
nitrogen 2	$FC = 5 - \frac{6}{2} - 2 = \ \ 0$

c) $CH_2=\overset{1}{N}=\overset{2}{\ddot{N}}:$

carbon	$FC = 4 - \frac{8}{2} - 0 = \ \ 0$
nitrogen 1	$FC = 5 - \frac{8}{2} - 0 = +1$
nitrogen 2	$FC = 5 - \frac{4}{2} - 6 = -1$

d) $:\overset{1}{\ddot{O}}=\overset{2}{\ddot{O}}-\overset{3}{\ddot{O}}:$

oxygen 1	$FC = 6 - \frac{4}{2} - 4 = \ \ 0$
oxygen 2	$FC = 6 - \frac{6}{2} - 2 = +1$
oxygen 3	$FC = 6 - \frac{2}{2} - 6 = -1$

e) $:CH_2-P(C_6H_5)_3$

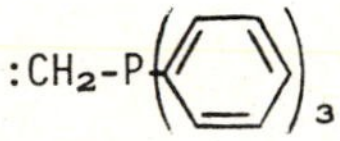

carbon	$FC = 4 - \frac{6}{2} - 2 = -1$
phosphorus	$FC = 5 - \frac{8}{2} - 0 = +1$

f) [pyridine ring]$N-\ddot{O}:$

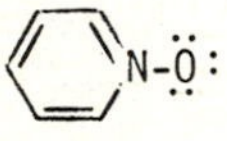

nitrogen	$FC = 5 - \frac{8}{2} - 0 = +1$
oxygen	$FC = 6 - \frac{2}{2} - 6 = -1$

2.13 a) 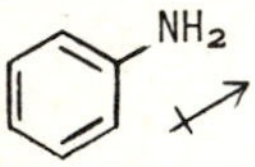 [benzene ring]–NH_2 (dipole arrow)

b) no dipole moment

c) 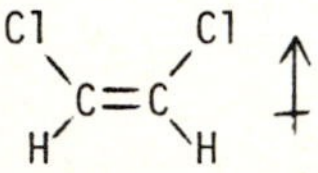 Cl(H)C=C(H)Cl (cis) (dipole arrow ↑)

d) no dipole moment

e) Li-H

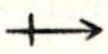

⟼

f) 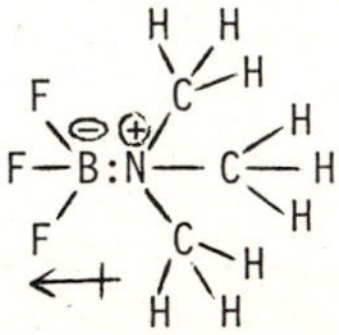 $F_3B^{\ominus}{:}N^{\oplus}(CH_3)_3$ (dipole arrow ⟻)

g) 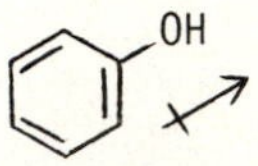 [benzene ring]–OH (dipole arrow)

h) 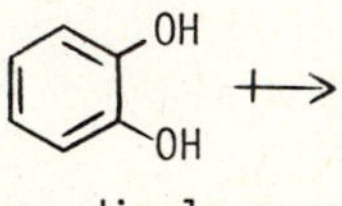 [benzene ring with two OH groups] ⟼

i) no dipole moment

2.14 a) $CH_3\ddot{O}H + H^{\oplus} \longrightarrow CH_3\overset{\oplus}{\ddot{O}}H_2$
base acid

b) $CH_3\ddot{O}H + :\overset{\ominus}{\ddot{N}}H_2 \longrightarrow CH_3\ddot{O}:^{\ominus} + :NH_3$
acid base

c) 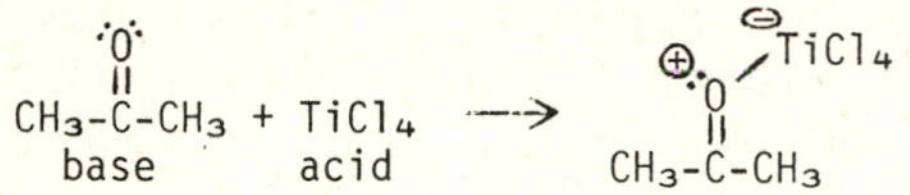 $CH_3-\overset{\overset{\cdot\cdot}{O}}{\overset{\|}{C}}-CH_3 + TiCl_4 \longrightarrow CH_3-\overset{\overset{\oplus}{O}-\overset{\ominus}{Ti}Cl_4}{\overset{\|}{C}}-CH_3$
base acid

d) 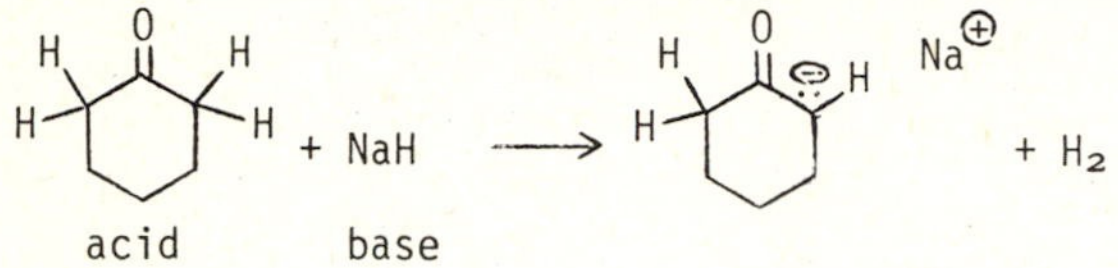 cyclohexanone + NaH ⟶ enolate anion $Na^{\oplus}$ + H_2
acid base

e) 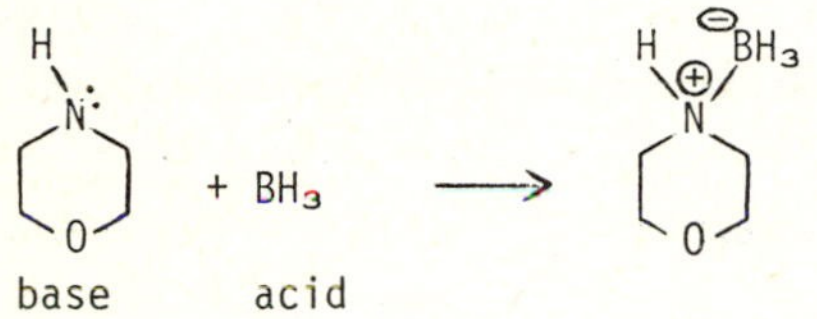 morpholine + BH_3 ⟶ morpholine–$\overset{\ominus}{B}H_3$ adduct ($\overset{\oplus}{N}$)
base acid

f) 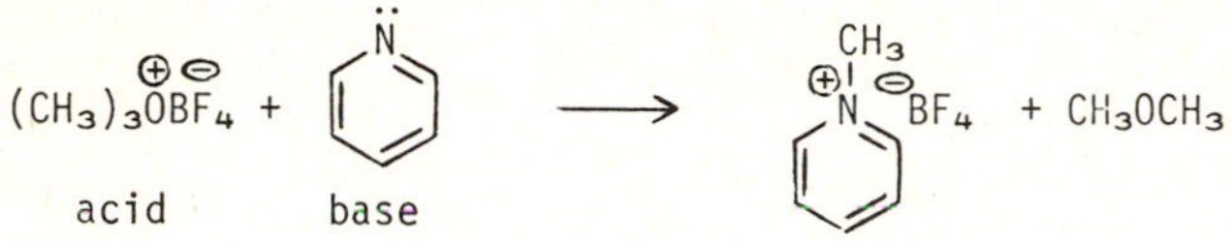 $(CH_3)_3\overset{\oplus}{O}\overset{\ominus}{B}F_4$ + pyridine ⟶ N-methylpyridinium $\overset{\ominus}{B}F_4$ + CH_3OCH_3
acid base

2.15 a) $C_9H_8O_4$ 60.00% C; 4.47% H; 35.52% O
b) $C_{15}H_{26}O$ 81.02% C; 11.79% H; 7.20% O
c) $C_{16}H_{30}O$ 80.61% C; 12.68% H; 6.71% O
d) $C_{17}H_{19}NO_3$ 71.56% C; 6.71% H; 4.91% N; 16.82% O
e) $C_{21}H_{22}N_2O_2$ 75.42% C; 6.63% H; 8.38% N; 9.57% O

2.16 The molecular formula of α-pinene is $C_{10}H_{16}$; see problem 2.8 for method of solution.

2.17 Because the percentages of carbon and hydrogen do not add up to 100%, another element is present in jasmone. The other element is oxygen, present as 9.6% by weight.

As in problem 2.8, assume that we have a 100 g sample. Then we would have

80.7 g carbon $\frac{80.7\text{ g}}{12.0\text{ g/mol}} = 6.72$ mol carbon

9.7 g hydrogen $\frac{9.7\text{ g}}{1.01\text{ g/mol}} = 9.62$ mol hydrogen

9.6 g oxygen $\frac{9.6\text{ g}}{16.0\text{ g/mol}} = 0.60$ mol oxygen

C : H : O = 6.72 : 9.62 : 0.60 = 11.2 : 16.0 : 1 or $C_{11}H_{16}O$.
A compound of this empirical formula has a formula weight of 164. Since this weight agrees with the observed molecular weight, the molecular formula of jasmone is $C_{11}H_{16}O$.

2.18 The empirical formula of myrcene is C_5H_8; its molecular formula is $C_{10}H_{16}$. See problem 2.9 for the method of solution.

2.19 Weight of carbon in sample = $0.0147 \text{ g} \times \frac{12.0}{44.0} = 0.00401 \text{ g}$

Weight of hydrogen in sample = $0.0043 \text{ g} \times \frac{2.02}{18.0} = 0.00048 \text{ g}$

Since these weights do not add up to 0.0050 g, we assume that oxygen is also present.

Weight of oxygen = 0.0050 g - 0.00401 g - 0.00048 g = 0.00051 g oxygen

% C = 80.2% relative number of moles C = $\frac{80.2}{12.0} = 6.68$

% H = 9.6% H = $\frac{9.6}{1.01} = 9.50$

% O = 10.2% O = $\frac{10.2}{16.0} = 0.64$

C : H : O = 6.68 : 9.50 : 0.64 = 10.4 : 14.8 : 1 ≃ 21 : 30 : 2

A compound of empirical formula $C_{21}H_{30}O_2$ has a formula weight of 314, which is identical to the given molecular weight of progesterone. Thus, the molecular formula of progesterone is $C_{21}H_{30}O_2$.

2.20 Weight of carbon = $0.086 \text{ g } CO_2 \times \frac{12.0}{44.0} = 0.0235 \text{ g C}$

Weight of hydrogen = $0.051 \text{ g } H_2O \times \frac{2.02}{18.0} = 0.0057 \text{ g H}$

Weight of nitrogen = $\frac{8.6 \text{ mL}}{22.4 \text{ mL/mmol}} \times 0.028 \text{ g/mmol} = 0.0108 \text{ g N}$

Since these weights add up to 0.040 g, no other elements are present.

$\% \text{C} = \frac{0.0235 \text{ g}}{0.040 \text{ g}} \times 100 = 58.8\%$

$\% \text{H} = \frac{0.0057 \text{ g}}{0.040 \text{ g}} \times 100 = 14.2\%$

$\% \text{N} = \frac{0.0108 \text{ g}}{0.040 \text{ g}} \times 100 = 27.0\%$

$\frac{58.8 \text{ g}}{12.0 \text{ g/mol}} = 4.90 \text{ mol C}$

$\frac{14.2 \text{ g}}{1.01 \text{ g/mol}} = 14.1 \text{ mol H}$

$\frac{27.0 \text{ g}}{14.0 \text{ g/mol}} = 1.93 \text{ mol N}$

C : H : N = 4.90 : 14.1 : 1.93 = 2.5 : 7.3 : 1 or 5 : 14-15 : 2.

The formula weight for $C_5H_{14}N_2$ = 102. Since this weight is equal to the given molecular weight, $C_5H_{14}N_2$ is the molecular formula of cadaverine. Cadaverine has a vile stench.

2.21 $H_3C-S(O^-)_2-CH_3$ (S bearing 2+ charge; each O with three lone pairs and − charge)

FC of each oxygen = $6 - \frac{2}{2} - 6 = -1$

FC of sulfur = $6 - \frac{8}{2} - 0 = +2$

The presence of formal charges indicates that electrons are not shared equally between S and O; they are strongly attracted to oxygen. The S-O bonds, therefore, are strongly polar. If the geometry of the molecule were planar (as drawn above), the dipole moments of the individual S-O bonds would cancel, resulting in a net dipole moment of zero. Tetrahedral geometry, however, predicts a large dipole moment.

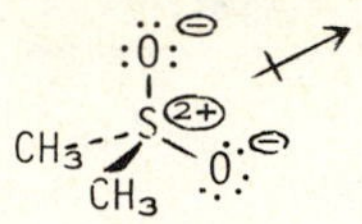

$$\ddot{O}^{\ominus}-S^{2+}(CH_3)_2-\ddot{O}^{\ominus}$$

What you should know:

After doing these problems you should be able to:

1. Draw chemical structures from molecular formulas;
2. Calculate formal charges for atoms in chemical compounds;
3. Predict the direction of dipole moment for simple compounds;
4. Recognize Lewis acids and bases;
5. Calculate
   a) percent composition;
   b) empirical formula.

3.1 a) $Cl-C_6H_4-CH(CCl_3)-C_6H_4-Cl$ — halide; aromatic ring

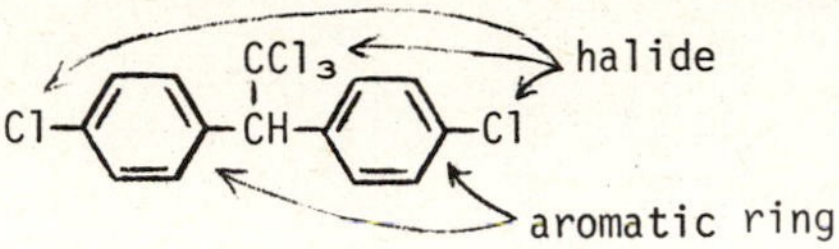

b) $C_6H_5CH_2CH(NH_2)-COOH$ — amine; carboxylic acid; aromatic ring

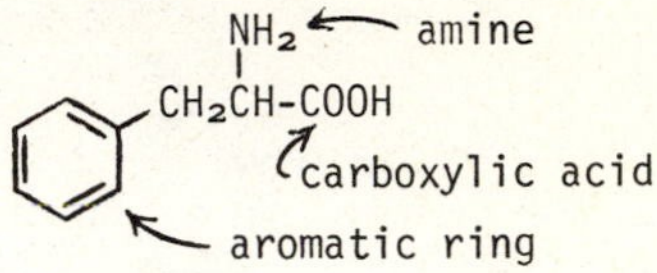

c) $CH_2=CH-CHO$ — aldehyde; double bond

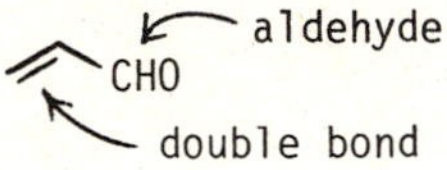

d) $C_6H_5CH=CH_2$ — double bond; aromatic ring

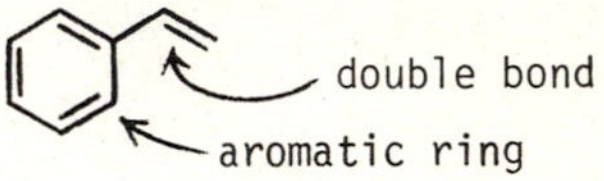

3.2 a) Using the shorthand we learned in Chapter 2 we can draw the isomers of C_8H_{18} in a less tedious fashion. There are 18 possible isomers of C_8H_{18}.

octane

heptanes

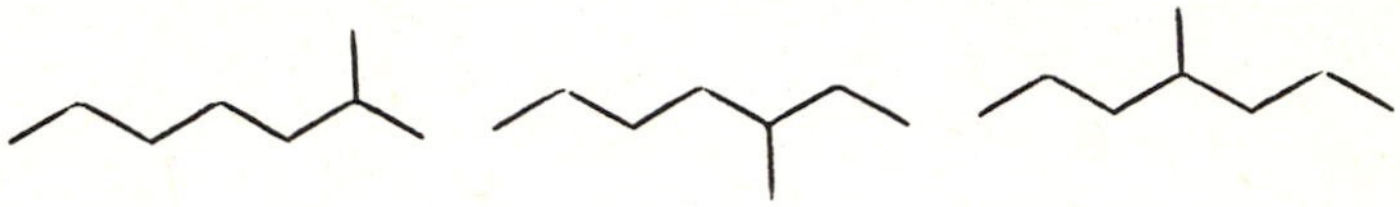

hexanes

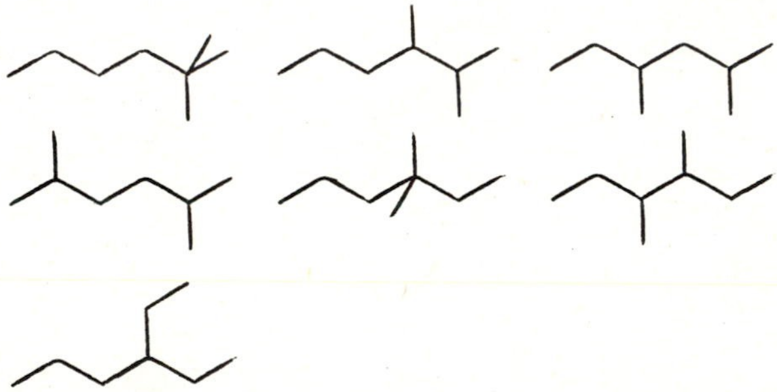

pentanes

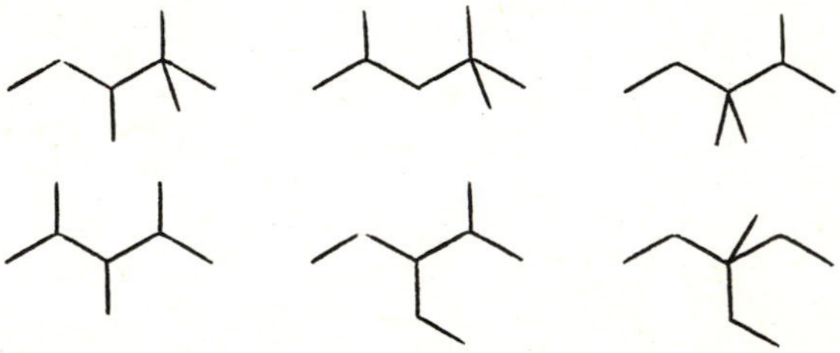

butane

b) $CH_3CH_2CH_2COCH_3$ (C=O), $CH_3CH(CH_3)COCH_3$, $CH_3CH_2COCH_2CH_3$, $CH_3COCH_2CH_2CH_3$

$CH_3COCH(CH_3)_2$, $HCOCH_2CH_2CH_2CH_3$, $HCOCH(CH_3)CH_2CH_3$, $HCOCH_2CH(CH_3)CH_3$, $HCOC(CH_3)_2CH_3$

c)

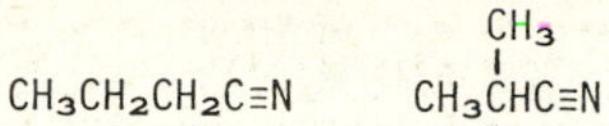

<u>3.3</u> Working a problem of this sort requires that you examine *all* the possibilities in a systematic way.

a)

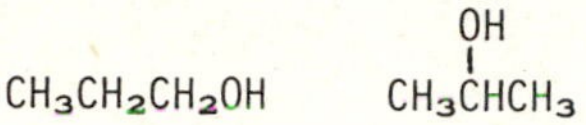

b) This problem will be used to show a systematic approach:

1. Draw the simplest long-chain parent *alkane*. $CH_3CH_2CH_2CH_3$
2. Find the number of different sites for attaching a substituent (here the substituent is bromine). For this alkane two different sites for attachment of bromine are possible ($-CH_3$ and $-CH_2-$).
3. At each different site, replace an -H by a -Br and draw the isomer.

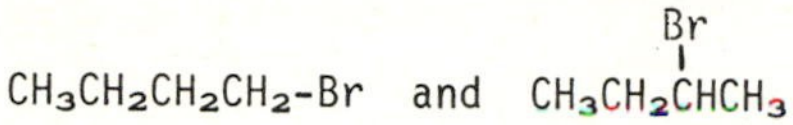

4. Draw the simplest *branched* alkane.

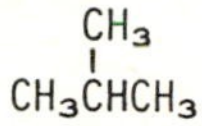

5. Find the number of *different* sites. (There are two for the above alkane.)
6. Replace an -H with a -Br and draw the isomers:

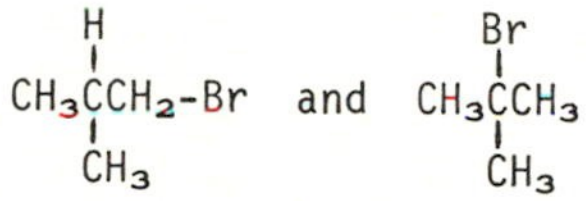

7. Proceed with the next simplest branched alkane. In this problem we have already drawn all isomers.

c)

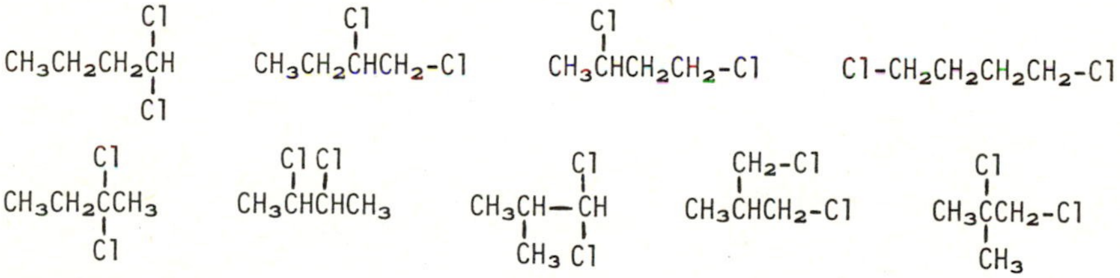

(9 isomers)

<u>3.4</u> a)

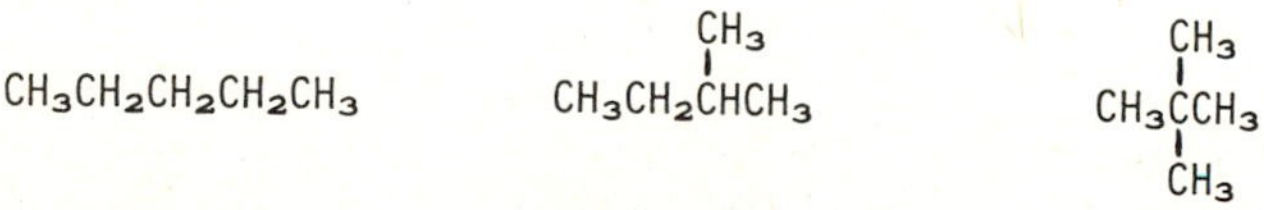

pentane　　2-methylbutane　　2,2-dimethylpropane

b)

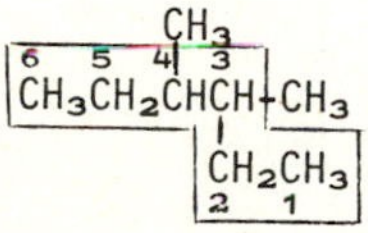

The longest chain is a *hexane*.
The substituents are: 3-methyl, 4-methyl.
The IUPAC name is 3,4-dimethylhexane.

c)

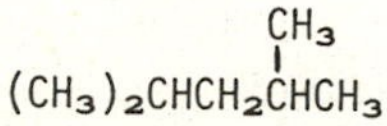

2,4-dimethylpentane

d) $\begin{array}{l} \qquad\qquad\quad CH_3 \\ (CH_3)_3CCH_2CH_2CH \\ \qquad\qquad\quad CH_2CH_3 \end{array}$ 2,2,5-trimethylheptane

3.5

a) 3,4-dimethylnonane $\quad \begin{array}{l} \qquad\qquad\qquad\qquad\quad CH_3 \\ CH_3CH_2CH_2CH_2CH_2CHCHCH_2CH_3 \\ \qquad\qquad\qquad\qquad\qquad CH_3 \end{array}$

b) 3-ethyl-4,4-dimethylheptane $\quad \begin{array}{l} \qquad\qquad\quad CH_3 \\ CH_3CH_2CH_2C{-}CHCH_2CH_3 \\ \qquad\qquad\quad CH_3\ CH_2CH_3 \end{array}$

c) 2,2-dimethyl-4-propyloctane $\quad \begin{array}{l} CH_3CH_2CH_2CH_2CHCH_2C(CH_3)_3 \\ \qquad\qquad\qquad\quad CH_2CH_2CH_3 \end{array}$

d) 2,2,4-trimethylpentane $\quad \begin{array}{l} \quad CH_3\ \ CH_3 \\ CH_3CHCH_2CCH_3 \\ \qquad\qquad CH_3 \end{array}$

3.6

a) $\begin{array}{l} \qquad\qquad\qquad CH_3 \\ CH_3CH_2CH_2CH_2CH \\ \qquad\qquad\qquad CH_3 \end{array}$ The longest chain is a *hexane*. The correct name is 2-methylhexane.

b) $\begin{array}{l} \qquad\qquad\quad CH_3 \\ CH_3CH_2CH_2CHCH{-}CH_3 \\ \qquad\qquad\qquad CH_2CH_2CH_3 \end{array}$ The longest chain is an *octane*. The correct name is 4,5-dimethyloctane.

c) $\begin{array}{l} \quad CH_3 \\ CH_3C{-}CHCH_2CH_3 \\ \quad CH_3\ CH_2CH_3 \end{array}$ The numbering of substituents should start from the other end of the chain. Also, substituents should be cited in alphabetical order. The correct name is 3-ethyl-2,2-dimethylpentane.

d) $\begin{array}{l} \qquad\quad CH_3 \\ CH_3{-}CHCHCH_2CH_2CH_3 \\ \qquad CH_2CH_3 \end{array}$ The longest chain is a *heptane*; substituents are numbered starting from wrong end. The correct name is 3,4-dimethylheptane.

e) $\begin{array}{l} \qquad\qquad\qquad CH_3 \\ CH_3CH_2CH_2CHCHCH_3 \\ \qquad\qquad\qquad\quad CH_3 \end{array}$ The name must include "di-" when two substituents are the same. The correct name is 2,3-dimethylhexane.

f) $\begin{array}{l} \qquad\quad CH_3 \\ CH_3CH_2CCH_2CH_3 \\ \qquad\quad CH_3 \end{array}$ The number "3" must be repeated in the name. The correct name is 3,3-dimethylpentane.

3.7

$CH_3CH_2CH_2CH_2CH_2{-}$	$CH_3CH_2CH_2CHCH_3$	$CH_3CH_2CHCH_2CH_3$	$\begin{array}{l} CH_3CH_2CHCH_2{-} \\ \qquad\quad CH_3 \end{array}$
pentyl	1-methylbutyl	1-ethylpropyl	2-methylbutyl
$\begin{array}{l} CH_3CHCH_2CH_2{-} \\ \quad CH_3 \end{array}$	$\begin{array}{l} CH_3CH_2CCH_3 \\ \qquad\quad CH_3 \end{array}$	$\begin{array}{l} CH_3CHCHCH_3 \\ \qquad\ CH_3 \end{array}$	$\begin{array}{l} \quad CH_3 \\ CH_3CCH_2{-} \\ \quad CH_3 \end{array}$
3-methylbutyl	1,1-dimethylpropyl	1,2-dimethylpropyl	2,2-dimethylpropyl

3.8 Many other answers to these problems are acceptable.

a) $CH_3CH-CH-CHCH_3$ with CH_3 on each of the three CH carbons (arrows mark these three CH carbons) — 2,3,4-trimethylpentane

b) $CH_3CH-CH_2-CHCH_3$ with CH_3 on C-2 and C-4 (each $CH_3CH(CH_3)$ end group boxed) — 2,4-dimethylpentane

c) $CH_3C(CH_3)_2CH_2CH_3$ (the quaternary carbon labelled q, the CH_2 labelled s) — 2,2-dimethylbutane

d) $CH_3CH(Cl)CH_3$ — 2-chloropropane

3.9 a) $(CH_3)_2CHCH_2C(CH_3)_3$ (carbons labelled p, t, s, q, p)

b) $(CH_3)_2CHCH(CH_3)CH_2CH_2CH_3$ (carbons labelled p, t, t, p, s, s, p)

c) $CH_3CH_2C(CH_3)_2C(CH_3)_2CH_2CH_3$ (carbons labelled p, s, q, p, q, p, s, p)

3.10

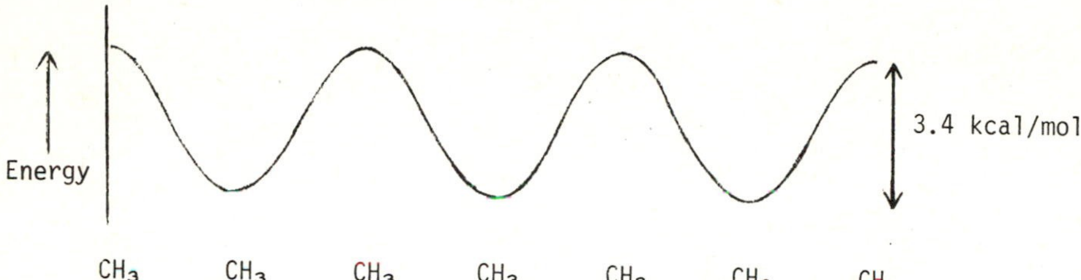

Newman projections at 0°, 60°, 120°, 180°, 240°, 300°, 360°; axis labels: 0° 60° 120° 180° 240° 300° 360°

3.11 This conformer of 2,3-dimethylbutane is the most stable since it is staggered and has the fewest CH_3-CH_3 *gauche* interactions.

3.12 a) The *most stable* conformer occurs at 60°, 180°, and 300°.

b) The *least stable* conformer occurs at 0°, 120°, 240°, and 360°.

Newman projection labels: 1.4 kcal/mol (CH_3–H); 1.4 kcal/mol (CH_3–H); 1.0 kcal/mol (H–H)

c), d)

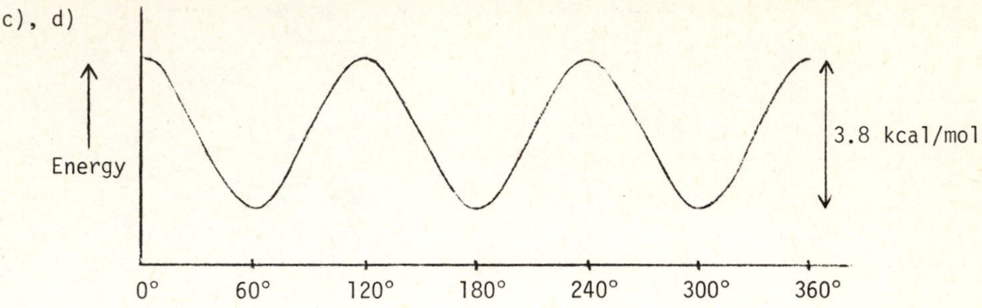

3.13 a) 3.8 kcal/mol

b) The data in Table 3.7 show that 99.9% of molecules are in the more stable conformation if the energy difference between conformations is 4.1 kcal/mol. Since the energy difference here is 3.8 kcal/mol, approximately 99.8% of molecules adopt the more stable conformation.

c) 0.2% of molecules adopt the less stable conformation.

3.14 a) b)

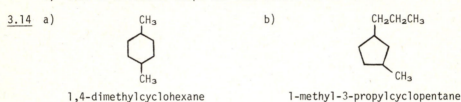

1,4-dimethylcyclohexane 1-methyl-3-propylcyclopentane

c) d)

CH_2CH_3

Br

3-cyclobutylpentane 1-bromo-4-ethylcyclodecane

3.15 a) b) c)

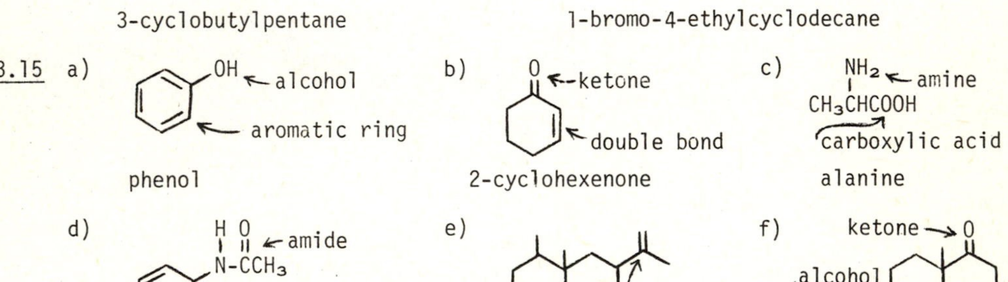

phenol 2-cyclohexenone alanine

d) H O ← amide, N-CCH_3, aromatic ring — acetanilide

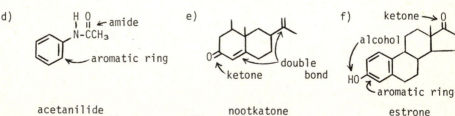

e) O, ketone, double bond — nootkatone

f) ketone → O, alcohol, HO, aromatic ring — estrone

g) 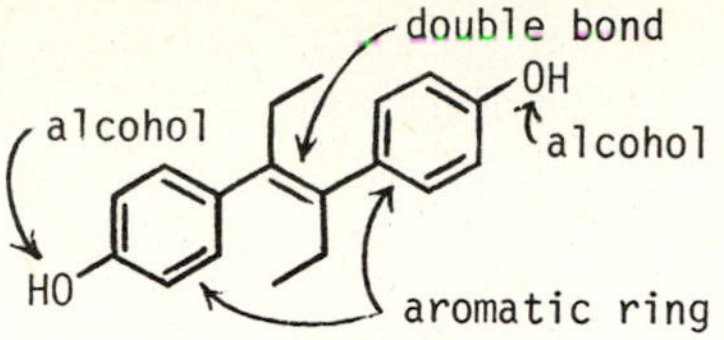 diethylstilbesterol — functional groups: alcohol, double bond, OH alcohol, aromatic ring (HO)

h) 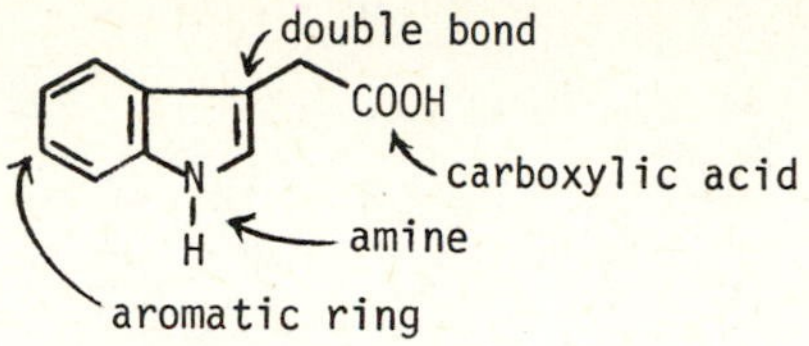 3-indoleacetic acid — functional groups: double bond, COOH carboxylic acid, N–H amine, aromatic ring

i) 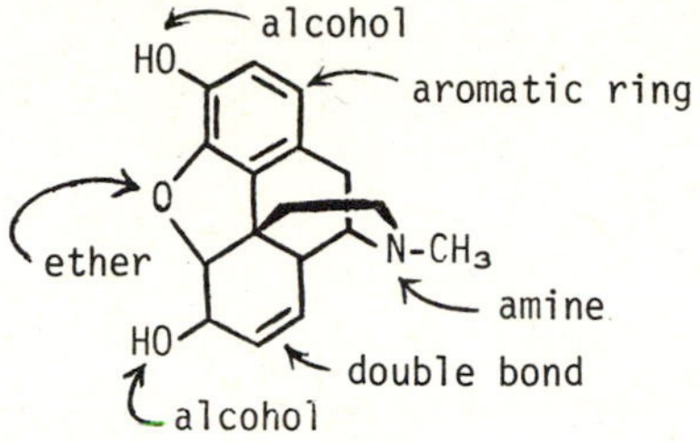 morphine — functional groups: HO alcohol, aromatic ring, O ether, N-CH_3 amine, double bond, HO alcohol

<u>3.16</u> Other answers to this problem and to problem 3.17 are acceptable.

a) $CH_3CH_2\overset{O}{\overset{\|}{C}}CH_2CH_3$ b) $CH_3\overset{O}{\overset{\|}{C}}NHCH_2CH_3$ c) $CH_3\overset{O}{\overset{\|}{C}}OCH_2CH_2CH_3$ d) 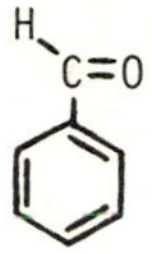 H–C=O attached to a benzene ring

e) $CH_3\overset{O}{\overset{\|}{C}}CH_2\overset{O}{\overset{\|}{C}}OCH_2CH_3$ f) $H_2NCH_2CH_2OH$ g) $CH_3C{\equiv}CCH_2C{\equiv}N$

<u>3.17</u> a) $CH_3\overset{O}{\overset{\|}{C}}CH_2CH_3$ C_4H_8O b) $CH_3CH_2CH_2CH_2C{\equiv}N$ C_5H_9N c) $H\overset{O}{\overset{\|}{C}}CH_2CH_2\overset{O}{\overset{\|}{C}}H$ $C_4H_6O_2$

d) $CH_3CH_2CH{=}CHCH_2CH_2Br$ $C_6H_{11}Br$ e) $CH_3CH_2CH_2CH_2CH_2CH_3$ C_6H_{14}

f)  (cyclohexane) An <u>alkane</u> of formula C_6H_{12} must contain one ring.

g) 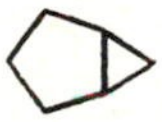 (bicyclo[3.1.0]hexane) C_6H_{10} contains two rings. h) $CH_3CH{=}CHCH{=}CH_2$ C_5H_8

i)  (cyclopentenone) C_5H_6O

<u>3.18</u> First, draw all straight-chain isomers. Then proceed to the simplest branched structure.

a) $CH_3CH_2CH_2CH_2\text{-}OH$ $CH_3CH_2\overset{OH}{\overset{|}{C}}HCH_3$ $CH_3\overset{CH_3}{\overset{|}{C}}HCH_2\text{-}OH$ $CH_3\underset{CH_3}{\underset{|}{\overset{OH}{\overset{|}{C}}}}CH_3$

There are *4 isomers* of $C_4H_{10}O$.

b)

```
                                       NH2                  NH2
                                       |                    |
CH3CH2CH2CH2CH2-NH2      CH3CH2CH2CHCH3       CH3CH2CHCH2CH3

                         NH2              NH2
                         |                |
CH3CH2CHCH2-NH2     CH3CH2CCH3      CH3CHCHCH3        H2N-CH2CH2CHCH3
      |                  |                |                     |
      CH3                CH3              CH3                   CH3

   CH3
   |
CH3CCH2NH2      CH3CH2CH2CH2-NHCH3        CH3CH2CH2-NHCH2CH3
   |
   CH3

                                         CH3                  CH3
                                         |                    |
CH3CH2CH-NHCH3      CH3CHCH2-NHCH3      CH3C-NHCH3      CH3CH2NHCH
      |                 |                |                    |
      CH3               CH3              CH3                  CH3

CH3CH2CH2-NCH3      CH3CH2-NCH2CH3      CH3CH— NCH3
          |                |               |     |
          CH3              CH3             CH3   CH3
```

There are *17 isomers* of $C_5H_{13}N$. The nitrogen can be bonded to one, two, or three alkyl groups.

c)

```
          O                  O                  O
          ||                 ||                 ||
CH3CH2CH2CCH3       CH3CH2CCH2CH3       CH3CHCCH3
                                           |
                                           CH3
```

There are *3 ketone isomers* with the formula $C_5H_{10}O$.

d)

```
              O                 O                 O           CH3 O
              ||                ||                ||          |   ||
CH3CH2CH2CH2CH      CH3CHCH2CH       CH3CH2CHCH       CH3C —— CH
                       |                   |              |
                       CH3                 CH3            CH3
```

There are *4 isomeric aldehydes* with the formula $C_5H_{10}O$. Remember that the aldehyde functional group can occur only at the end of a chain.

e)

```
                                          N
                                          |||
                       C≡N                C
                      /                   |
  (benzene ring)   (benzene ring)   (benzene ring)   (benzene ring)—CH2C≡N
       \                |                 |
        C≡N             |                 |
   |
   CH3                  CH3               CH3
```

There are *4 isomeric aromatic nitriles* with the formula C_8H_7N.

f)

```
       O            O              O                 O CH3
       ||           ||             ||                || |
CH3CH2COCH3    CH3COCH2CH3    HCOCH2CH2CH3      HCOCH
                                                    |
                                                    CH3
```

There are *4 esters* with the formula $C_4H_8O_2$.

g)

```
                                                  CH3
                                                  |
CH3CH2OCH2CH3        CH3OCH2CH2CH3         CH3OCH
                                                  |
                                                  CH3
```

There are *3 ethers* with the formula $C_4H_{10}O$.

3.19

```
a) CH3CH2OH     b)    CH3        c) CH3CHCH3     d) CH3CHCH2OH
                      |                 |               |
                   CH3C-C≡N             Br              OH
                      |
                      CH3
```

e) $CH_3CH(CH_3)C(=O)CH_3$

f) CH_3 CH_3 CH_3 CH_3

g) $H_3C-C(CH_3)_2-CH_3$

3.20 Try to attack this problem in a systematic way. Don't forget to examine *all* the possibilities.

$CH_3CH_2CH_2CH_2CH_2Br$ $CH_3CH_2CH_2CH(Br)CH_3$ $CH_3CH_2CH(Br)CH_2CH_3$ $CH_3CH_2CH(CH_3)CH_2Br$

$CH_3CH_2C(Br)(CH_3)CH_3$ $CH_3CH(Br)CH(CH_3)CH_3$ $BrCH_2CH_2CH(CH_3)CH_3$ $CH_3C(CH_3)_2CH_2Br$

There are eight monobromopentanes.

To find the dibromopentanes, work with one monobromopentane at a time. Find all the dibromopentanes for each monopentane, but take care to avoid duplicates.

$CH_3CH_2CH_2CH_2CHBr_2$ $CH_3CH_2CH_2CHBrCH_2Br$ $CH_3CH_2CHBrCH_2CH_2Br$

$CH_3CHBrCH_2CH_2CH_2Br$ $CH_2BrCH_2CH_2CH_2CH_2Br$ $CH_3CH_2CH_2CBr_2CH_3$

$CH_3CH_2CHBrCHBrCH_3$ $CH_3CHBrCH_2CHBrCH_3$ $CH_3CH_2CBr_2CH_2CH_3$

$CH_3CH_2CH(CH_3)CHBr_2$ $CH_3CH_2CH(CH_2Br)CH_2Br$ $CH_3CH_2C(Br)(CH_3)CH_2Br$ $CH_3CHBrCH(CH_3)CH_2Br$

$CH_2BrCH_2CH(CH_3)CH_2Br$ $CH_3CHBrC(Br)(CH_3)CH_3$ $CH_2BrCH_2C(Br)(CH_3)CH_3$ $CH_2BrCHBrCH(CH_3)CH_3$

$CH_3CBr_2CH(CH_3)CH_3$ $CHBr_2CH_2CH(CH_3)CH_3$ $CH_3C(CH_3)_2CHBr_2$ $CH_3C(CH_2Br)(CH_3)CH_2Br$

There are 21 dibromopentanes!

3.21 a) O, C *sp*²   b) -C≡N *sp*   c) O, C, OH *sp*²

d) $H_3C-O-CH_3$ *sp*³   e) C-M *sp*³

3.22 The purpose of this problem is to teach you to recognize identical structures when they are drawn slightly differently.

a) 1 and 2 are the same.   b) All structures are the same.

c) 1 and 2 are the same.   d) 1 and 3 are the same.

e) 1 and 3 are the same.   f) All structures are the same.

3.23 a) 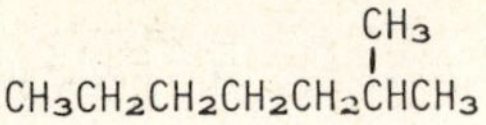 $CH_3CH_2CH_2CH_2CH_2CH(CH_3)CH_3$

b) 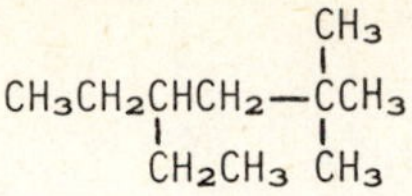 $CH_3CH_2CH(CH_2CH_3)CH_2—C(CH_3)_2CH_3$

c) $CH_3CH_2CH_2CH_2C(CH_3)(CH_2CH_3)—CH(CH_3)CH_2CH_3$

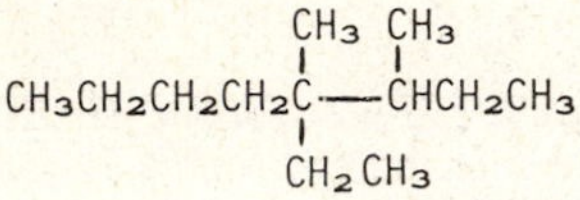

d) 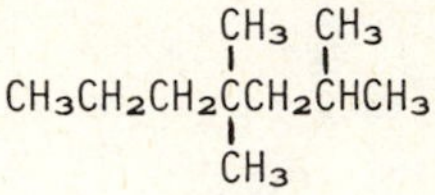 $CH_3CH_2CH_2C(CH_3)_2CH_2CH(CH_3)CH_3$

e) 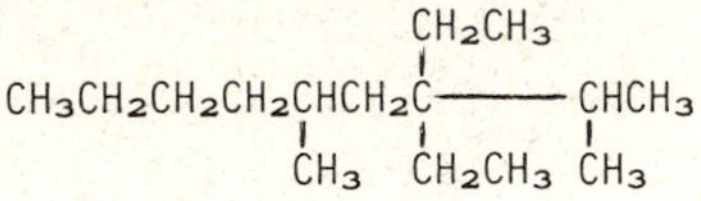 $CH_3CH_2CH_2CH_2CH(CH_3)CH_2C(CH_2CH_3)(CH_2CH_3)—CH(CH_3)CH_3$

f) 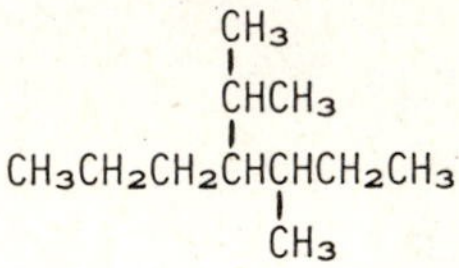 $CH_3CH_2CH_2CH(CH(CH_3)CH_3)CH(CH_3)CH_2CH_3$

g) 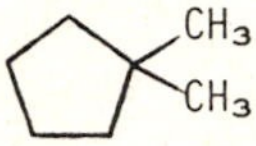 1,1-dimethylcyclopentane (cyclopentane ring with two CH_3 on one carbon)

3.24 Isomers are compounds of the same molecular formula. In problem 3.24a, you should draw a structure having the molecular formula $C_5H_{11}Br$. Other correct solutions to many parts of this question are possible.

a) $CH_3CH_2CH_2CH_2CH_2Br$ b) 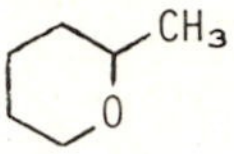 2-methyltetrahydropyran (six-membered ring containing O, with CH_3 on the carbon next to O) c) 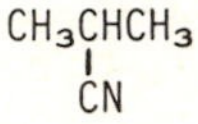 $CH_3CH(CN)CH_3$ d) 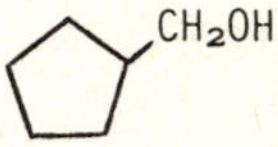 cyclopentyl–CH_2OH

e) There are no aldehyde isomers. However, CH_3COCH_3 is a ketone isomer. f) 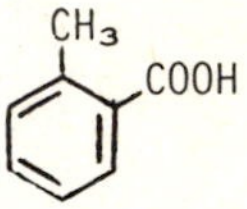 benzene ring bearing CH_3 and COOH on adjacent carbons

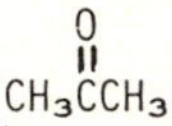

3.25 a) 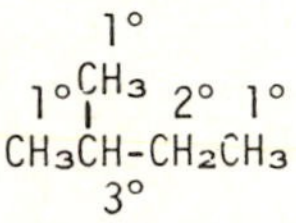 $CH_3CH(CH_3)-CH_2CH_3$ — the three CH_3 carbons 1°, CH 3°, CH_2 2°

b) 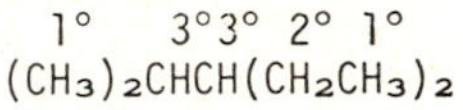 $(CH_3)_2CHCH(CH_2CH_3)_2$ — 1° 3° 3° 2° 1°

c) 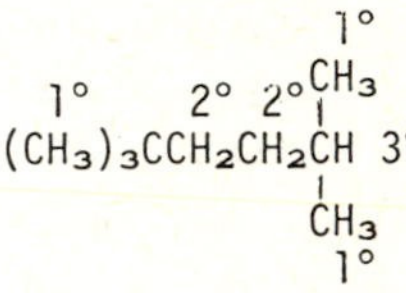 $(CH_3)_3CCH_2CH_2CH(CH_3)CH_3$ — 1° 2° 2° 3°; the CH_3 groups are 1°

d) 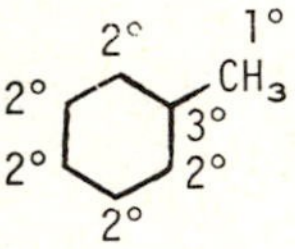 methylcyclohexane — CH_3 1°; ring carbon bearing CH_3 3°; the other ring carbons 2°

e) 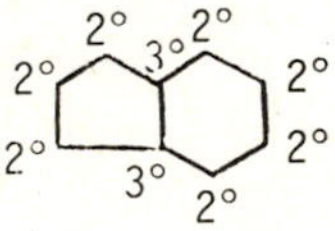 bicyclic structure (five-membered ring fused to six-membered ring) — two ring-fusion carbons 3°; all other ring carbons 2°

f) 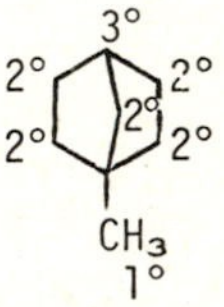 bridged bicyclic structure bearing CH_3 — 3° at the top bridgehead; ring carbons 2°; CH_3 1°

3.26 a) CH_3CH_3 b) 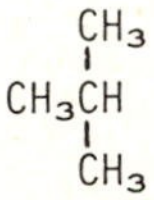 $CH_3CH(CH_3)CH_3$ c)  cyclohexane d) $CH_3CH_2CH_2CH_3$

3.27 a) 2-methylpentane b) 2,2-dimethylbutane c) 2,3,3-trimethylhexane
d) 5-ethyl-2-methylheptane e) 3,3,5-trimethyloctane
f) 2,2,3,3-tetramethylhexane g) 5-ethyl-3,5-dimethyloctane

3.28 $CH_3CH_2CH_2CH_2CH_2CH_3$ 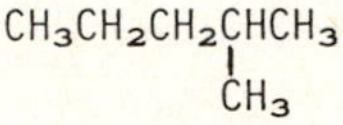 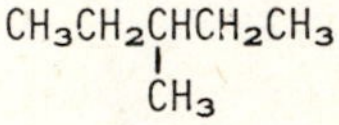 hexane

$CH_3CH_2CH_2CH(CH_3)CH_3$ 2-methylpentane

$CH_3CH_2CH(CH_3)CH_2CH_3$ 3-methylpentane

CH_3
$CH_3CH_2CCH_3$
CH_3

2,2-dimethylbutane

CH_3
$CH_3CHCHCH_3$
CH_3

2,3-dimethylbutane

3.29

$CH_3CH_2CH_2CH_2CH_2CH_2CH_3$

heptane

CH_3
$CH_3CH_2CH_2CH_2CHCH_3$

2-methylhexane

CH_3
$CH_3CH_2CH_2CHCH_2CH_3$

3-methylhexane

CH_3
$CH_3CH_2CH_2CCH_3$
CH_3

2,2-dimethylpentane

CH_3
$CH_3CH_2CHCHCH_3$
CH_3

2,3-dimethylpentane

CH_3
$CH_3CHCH_2CHCH_3$
CH_3

2,4-dimethylpentane

CH_3
$CH_3CH_2CCH_2CH_3$
CH_3

3,3-dimethylpentane

$CH_3CH_2CHCH_2CH_3$
CH_2CH_3

3-ethylpentane

CH_3 CH_3
$CH_3CH—CCH_3$
CH_3

2,2,3-trimethylbutane

3.30 a)

CH_2CH_3 CH_3
$CH_3CHCH_2CH_2CH_2CCH_3$
CH_3

The longest chain is an octane.

correct name: 2,2,6-trimethyloctane

b)

CH_3
$CH_3CHCHCH_2CH_2CH_3$
CH_2CH_3

The longest chain is a hexane; numbering should start from the other end.

correct name: 3-ethyl-2-methylhexane

c)

CH_3
$CH_3CH_2C—CHCH_2CH_3$
CH_3 CH_2CH_3

Numbering should start from the other end.

correct name: 4-ethyl-3,3-dimethylhexane

d)

CH_3 CH_3
$CH_3CH_2CH—CCH_2CH_2CH_2CH_3$
CH_3

Numbering should start from the other end.

correct name: 3,4,4-trimethyloctane

e)

CH_3
$CH_3CH_2CH_2CHCH_2CHCH_3$
CH_3CHCH_3

The longest chain is an octane.

correct name: 2,3,5-trimethyloctane

f)

CH_3
CH_3

The subsituents should have the lowest possible numbers.

correct name: 1,3-dimethylcyclohexane

3.31 a)

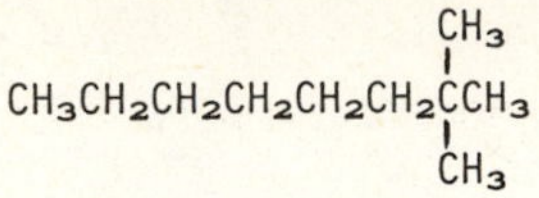

$$CH_3CH_2CH_2CH_2CH_2CH_2C(CH_3)_2CH_3$$

2,2-dimethyloctane

b)

$$CH_3C(CH_3)_2CH_2C(CH_2CH_3)_2CH_2CH_3$$

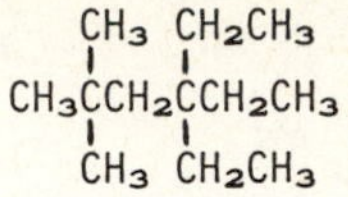

4,4-diethyl-2,2-dimethylhexane

c)

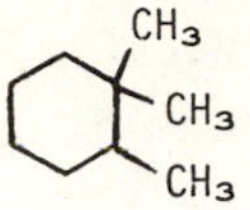

1,1,2-trimethylcyclohexane

d)

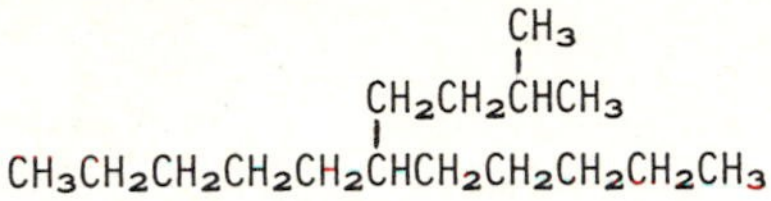

$$CH_3CH_2CH_2CH_2CH_2CH(CH_2CH_2CH(CH_3)CH_3)CH_2CH_2CH_2CH_2CH_3$$

5-(3-methylbutyl)-undecane

Remember that you must choose an alkane whose principal chain is long enough so that the substituent does not become part of the principal chain.

3.32

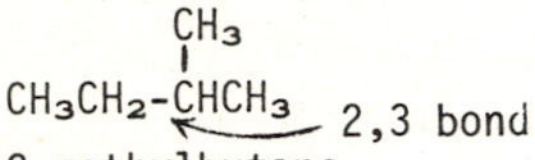

CH_3CH_2-$CH(CH_3)CH_3$ ← 2,3 bond

2-methylbutane

CH3 ← 0.9 kcal/mol

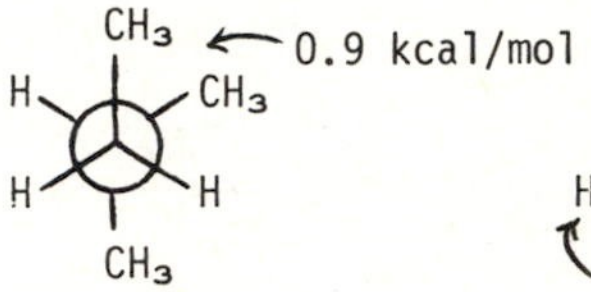

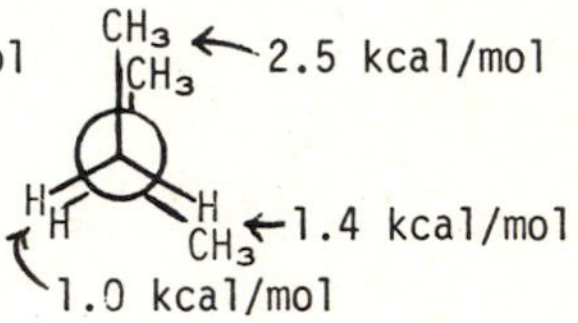

Most stable conformation

CH3/CH3 ← 2.5 kcal/mol; H/CH3 ← 1.4 kcal/mol; H/H ← 1.0 kcal/mol

Least stable conformation

The energy difference between the two conformations is $(2.5 + 1.4 + 1.0) - 0.9 = 4.0$ kcal/mol.

Consider the least stable conformation to be at zero degrees. Keeping the "front" of the projection unchanged, rotate the "back" by 60° to obtain each conformation.

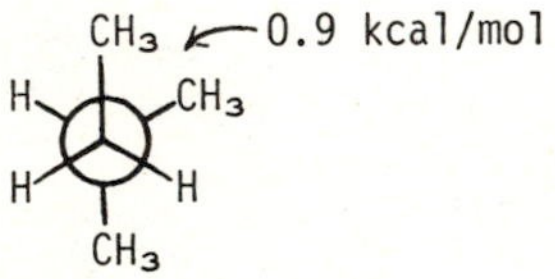

CH3 ← 0.9 kcal/mol

at 60° energy = 0.9 kcal/mol

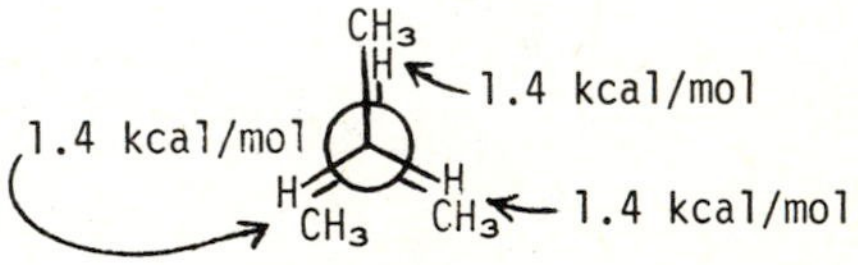

CH3/H ← 1.4 kcal/mol; 1.4 kcal/mol; CH3/H ← 1.4 kcal/mol

at 120° energy = 4.2 kcal/mol

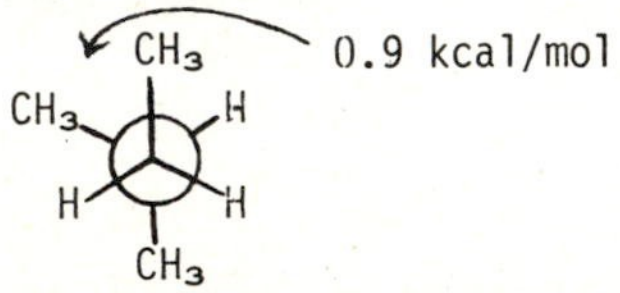

CH3 ← 0.9 kcal/mol

at 180° energy = 0.9 kcal/mol

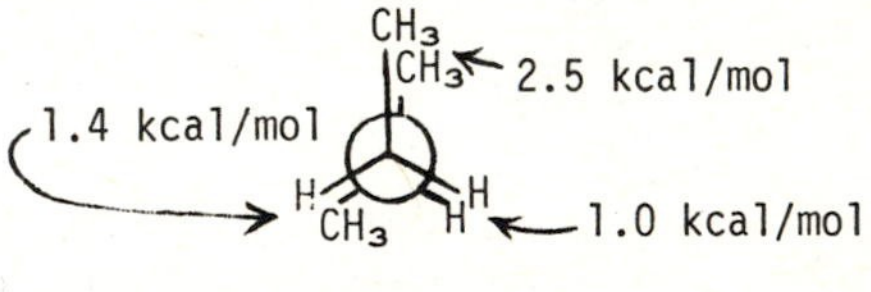

CH3/CH3 ← 2.5 kcal/mol; 1.4 kcal/mol; H/H ← 1.0 kcal/mol

at 240° energy = 4.9 kcal/mol

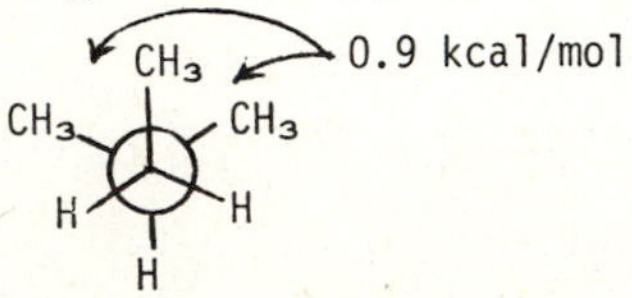

CH3 ← 0.9 kcal/mol; 0.9 kcal/mol

at 300° energy = 1.8 kcal/mol

Use the lowest energy conformation as the energy minimum. The highest energy conformation is 4.0 kcal/mol higher in energy than the lowest energy conformation.

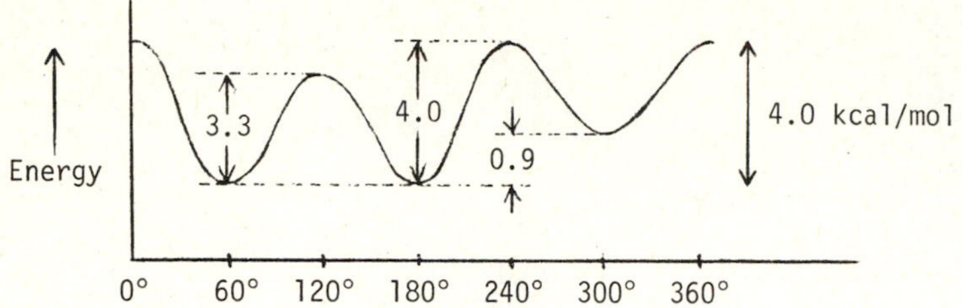

3.33

2 CH_3-CH_3 *gauche*
= 2(0.9 kcal/mol)
= 1.8 kcal/mol

3 CH_3-CH_3 *gauche*
= 3(0.9 kcal/mol)
= 2.7 kcal/mol

3 CH_3-CH_3 *gauche*
= 3(0.9 kcal/mol)
= 2.7 kcal/mol

3.34 Since we are not told the values of the interactions for 1,2-dibromobutane, the diagram can only be qualitative.

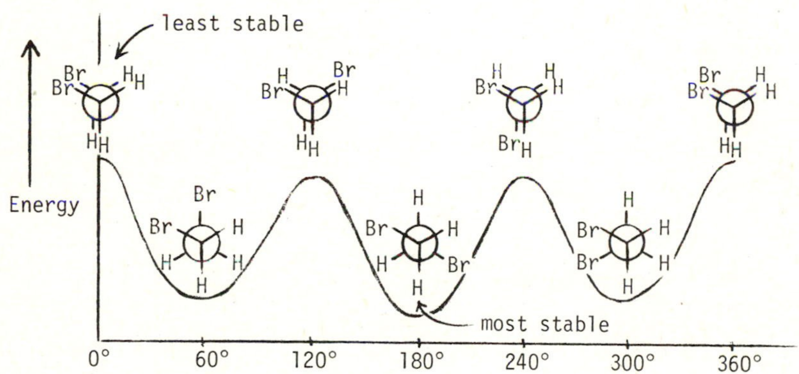

The *anti* conformer is at 180°.

The *gauche* conformers are at 60°, 300°.

The *anti* conformer has no dipole moment because the net dipole moments of the individual bonds cancel. The *gauche* conformer, however, has a dipole moment. Because the observed dipole moment is 1.0 D at room temperature, a mixture of conformers must be present.

3.35 The highest energy conformation of bromoethane is 3.6 kcal/mol. Since this includes two H-H eclipsing interactions of 1.0 kcal/mol each, the value of an H-Br interaction is 3.6 - 2(1.0) = 1.6 kcal/mol.

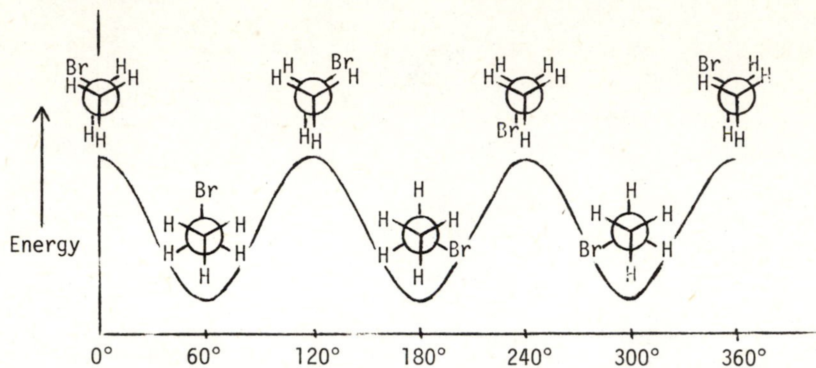

3.36 Because malic acid has 2-COOH groups, the formula for the rest of the molecule is C_2H_4O. Possible structures for malic acid are:

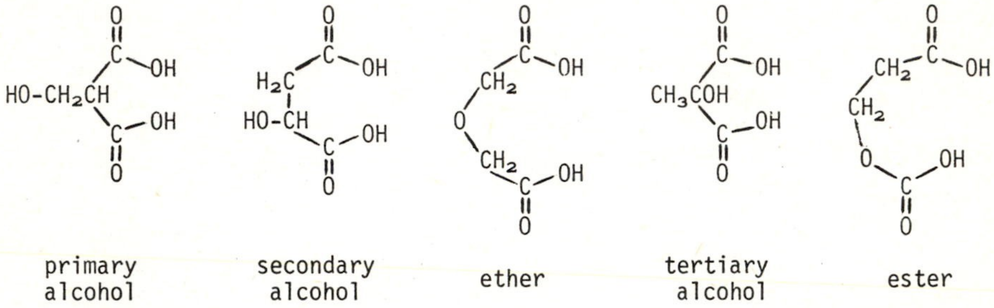

primary alcohol	secondary alcohol	ether	tertiary alcohol	ester

Because only one of these compounds (the second one) is a secondary alcohol, it must be malic acid.

3.37 ester; double bond

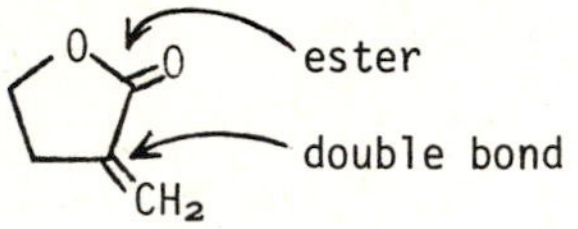

3.38 Many students do not know where to begin when they are confronted with this type of problem. To start, read the problem carefully, word for word. Then try to interpret parts of the problem. For example:

1) Formaldehyde is an aldehyde, H-C(=O)-H.
2) It trimerizes -- that is, 3 formaldehydes come together to form a compound C_3H_6O. No atoms are eliminated, so all of the original atoms are still present.
3) There are no carbonyls. This means that trioxan cannot contain any

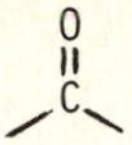

$\overset{O}{\underset{/\ \ \backslash}{\overset{\|}{C}}}$ functional groups. If you look back to Table 3.1, you can see that the only oxygen functional groups that can be present are either ethers or alcohols.

4) A monobromo derivative is a compound in which one of the H's has been replaced by a Br. Because only one monobromo derivative is possible, we know that there can only be one type of hydrogen in trioxan. The only possibility for trioxan is:

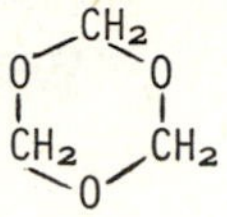

trioxan

3.39 a)

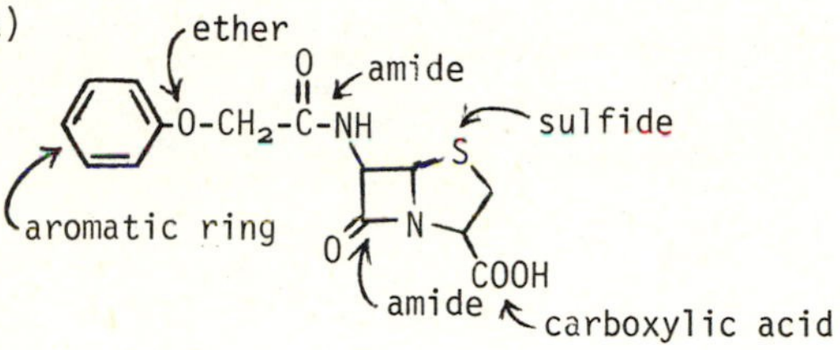

Penicillin V

ether, amide, sulfide, aromatic ring, amide, carboxylic acid

b)

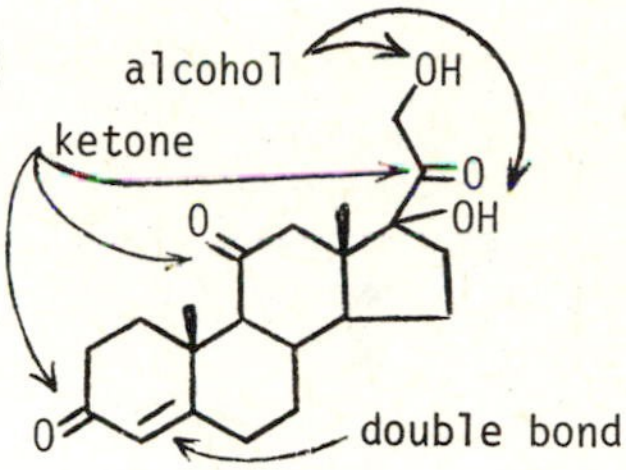

cortisone

alcohol, ketone, double bond

c)

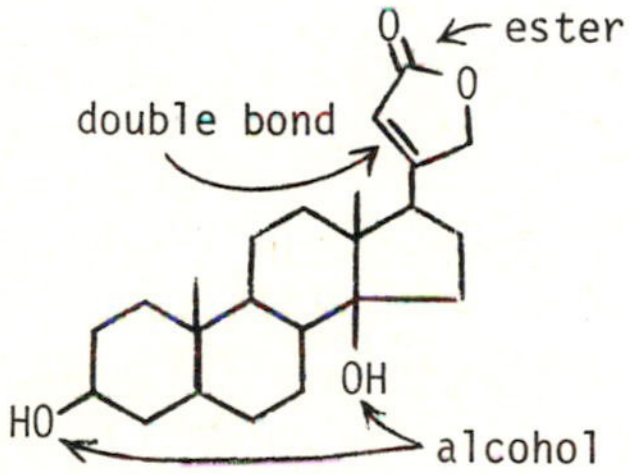

digitoxigenin

ester, double bond, alcohol

d)

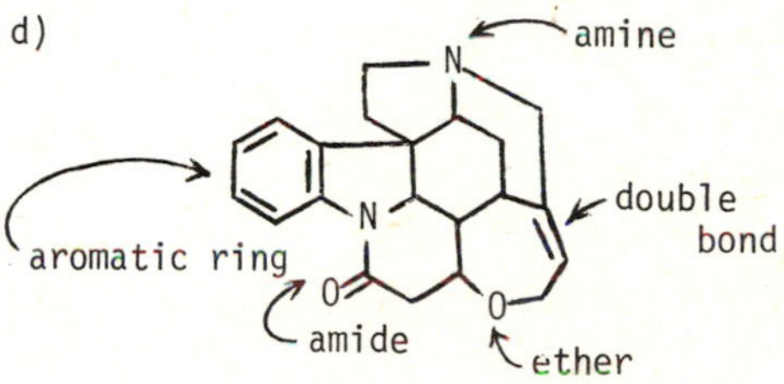

strychnine

amine, aromatic ring, double bond, amide, ether

3.40 If $BrCH_2CH_2CH_2Br \xrightarrow{Na}$ cyclopropane (CH_2)$_3$

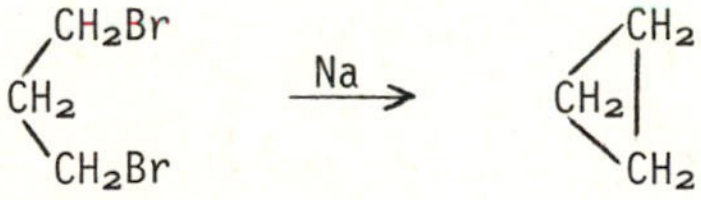

Then $C(CH_2Br)_4 \xrightarrow{Na}$ spiropentane

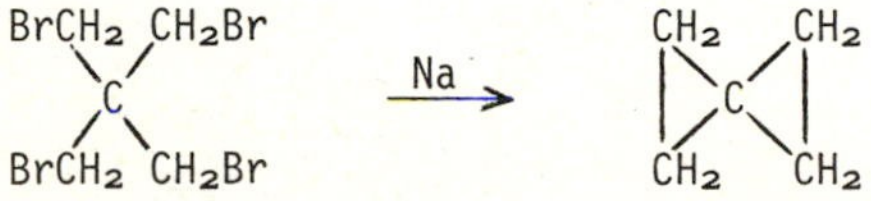

The two rings are perpendicular in order to keep the geometry of the central carbon as close to tetrahedral as possible.

What you should know:

After doing these problems you should be able to:

1. Recognize and identify functional groups in complex molecules; draw molecules containing a given functional group;
2. Systematically draw all possible isomers of a given structure;
3. Correctly name complicated alkanes, including cycloalkanes; correctly draw alkanes, given their IUPAC name;
4. Identify carbon and hydrogen as being primary, secondary, or tertiary;
5. Draw Newman projections of conformations about a single bond and draw energy *vs.* rotation graphs.

CHAPTER 4 ORGANIC REACTIONS

4.1 a) $CH_3\overset{O}{\overset{\|}{C}}CH_3$ — O $\delta-$; C $\delta+$

ketone

b) $CH_2{=}CH{-}Cl$ — Cl $\delta-$ ← halide; CH $\delta+$ ← double bond

c) $CH_2{=}CH\overset{O}{\overset{\|}{C}}OCH_2CH_3$ — O $\delta-$ ← ester; C $\delta+$, O $\delta-$; ← double bond

d) $(CH_3CH_2)_4Pb$ — each CH_2 $\delta-$; Pb $\delta+$

organometallic

4.2 nucleophiles $:NH_3$, $^-:\ddot{\underset{..}{I}}:$, $^-:C{\equiv}N:$

electrophiles H^+, Mg^{2+}, CH_3COCl

4.3 Irradiation initiates the chlorination reaction by producing chlorine radicals. Although these radicals are consumed in the propagation steps, new Cl· radicals are formed to carry on the reaction. After irradiation stops, chlorine radicals are still present to carry on the propagation steps, but, as time goes on, radicals combine with each other in termination reactions that remove radicals from the reaction mixture. Because the number of radicals is decreasing, fewer propagation cycles occur, and the reaction gradually slows down and stops.

4.4 Pentane has three types of hydrogen atoms, $\overset{a}{CH_3}\overset{b}{CH_2}\overset{c}{CH_2}\overset{b}{CH_2}\overset{a}{CH_3}$. Although monochlorination produces $CH_3CH_2CH_2CH_2CH_2Cl$, it is not possible to avoid producing $CH_3CH_2CH_2CHClCH_3$ and $CH_3CH_2CHClCH_2CH_3$ as well. Since neopentane has only one type of hydrogen, monochlorination yields a single product.

4.5 a) $Cl_2 \xrightarrow{h\nu} 2\ Cl\cdot$

Bond broken	$\Delta H°$
Cl-Cl	58 kcal/mol

$\Delta H°_a$ = +58 kcal/mol

b) $C_6H_5CH_2{-}H + Cl\cdot \longrightarrow C_6H_5\dot{C}H_2 + H{-}Cl$

Bond broken	$\Delta H°$	Bond formed	$\Delta H°$
$C_6H_5CH_2$-H	85 kcal/mol	H-Cl	103 kcal/mol

$\Delta H°_b$ = 85 kcal/mol - 103 kcal/mol = -18 kcal/mol

c) (benzyl radical, $C_6H_5\dot{C}H_2$) + Cl-Cl ⟶ $C_6H_5CH_2$-Cl + Cl·

Bond broken	$\Delta H°$	Bond formed	$\Delta H°$
Cl-Cl	58 kcal/mol	$C_6H_5CH_2$-Cl	70 kcal/mol

$\Delta H°_c$ = 58 kcal/mol - 70 kcal/mol = -12 kcal/mol

$\Delta H°_{b+c} = \Delta H°_b + \Delta H°_c$ = -18 kcal/mol + (-12 kcal/mol) = -30 kcal/mol

4.6 a) $CH_3CH_2OCH_3$ + HI ⟶ CH_3CH_2OH + CH_3I

Bonds broken	$\Delta H°$	Bonds formed	$\Delta H°$
CH_3CH_2O-CH_3	81 kcal/mol	CH_3CH_2O-H	103 kcal/mol
H-I	71 kcal/mol	CH_3-I	56 kcal/mol
$\Delta H°_{\text{bonds broken}}$	= 152 kcal/mol	$\Delta H°_{\text{bonds formed}}$	= 159 kcal/mol

$\Delta H°_{\text{overall}} = \Delta H°_{\text{bonds broken}} - \Delta H°_{\text{bonds formed}}$ = 152 kcal/mol - 159 kcal/mol = -7 kcal/mol

b) CH_3Cl + NH_3 ⟶ CH_3NH_2 + HCl

Bonds broken	$\Delta H°$	Bonds formed	$\Delta H°$
CH_3-Cl	84 kcal/mol	CH_3-NH_2	80 kcal/mol
NH_2-H	103 kcal/mol	H-Cl	103 kcal/mol
$\Delta H°_{\text{bonds broken}}$	= 187 kcal/mol	$\Delta H°_{\text{bonds formed}}$	= 183 kcal/mol

$\Delta H°_{\text{overall}} = \Delta H°_{\text{bonds broken}} - \Delta H°_{\text{bonds formed}}$ = 187 kcal/mol - 183 kcal/mol = 4 kcal/mol

c) C_6H_5-OH + HBr ⟶ C_6H_5-Br + H_2O

Bonds broken	$\Delta H°$	Bonds formed	$\Delta H°$
C_6H_5O-H	112 kcal/mol	C_6H_5-Br	82 kcal/mol
H-Br	88 kcal/mol	H-OH	119 kcal/mol
$\Delta H°_{\text{bonds broken}}$	= 200 kcal/mol	$\Delta H°_{\text{bonds formed}}$	= 201 kcal/mol

$\Delta H°_{\text{overall}} = \Delta H°_{\text{bonds broken}} - \Delta H°_{\text{bonds formed}}$ = 200 kcal/mol - 201 kcal/mol = -1 kcal/mol

4.7 Transition state‡

Possible transition state

Energy

Reaction Progress ⟶

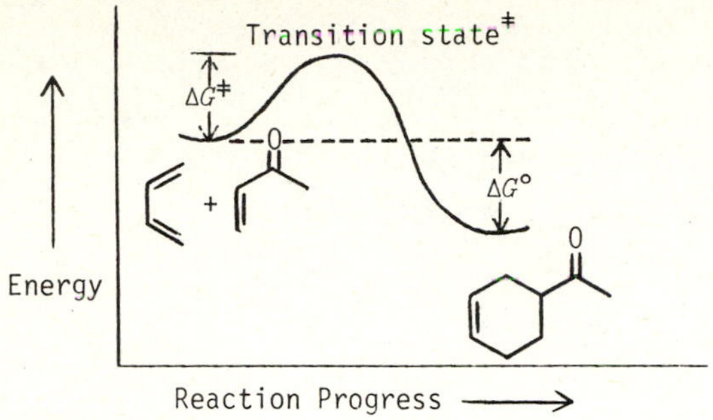

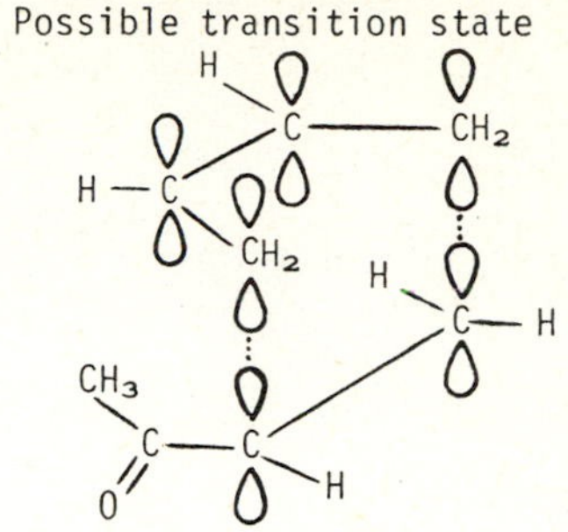

The Diels-Alder reaction is a single-step process.

4.8

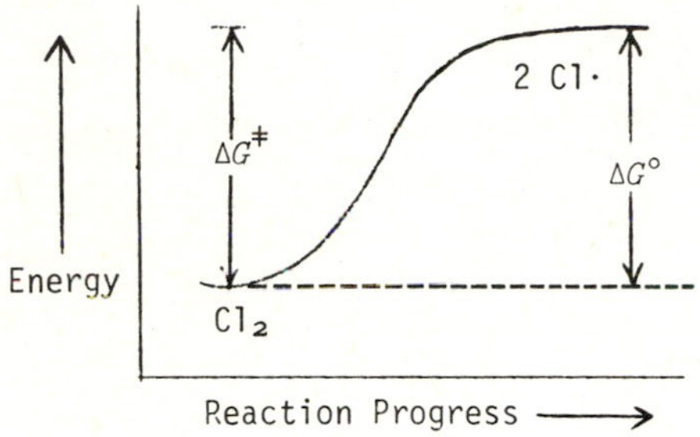

Energy

Reaction Progress ⟶

This reaction is highly endothermic, and $\Delta G°$ is therefore positive. One interesting feature of this uni-molecular reaction is that $\Delta G^{\ddagger}$, the activation energy, is equal to $\Delta G°$. As soon as enough energy has been put into the Cl-Cl bond to break it, reaction occurs.

4.9 a) $\overset{\delta+}{C}\overset{\delta-}{N}$: $CH_3C{\equiv}N$ nitrile b) ether c) ketone, ester d) ketone, double bonds

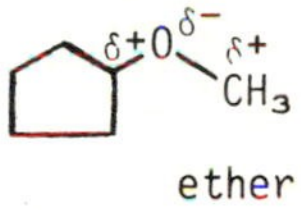

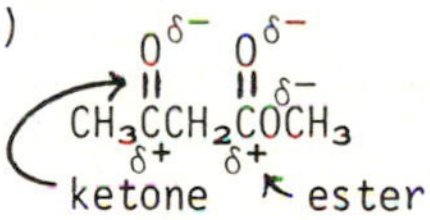

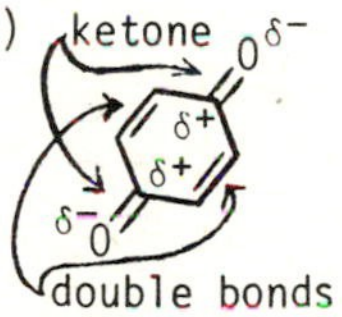

4.10 a) polar reaction b) pericyclic reaction c) radical reaction

4.11 a) In a *polar reaction*, electron-rich sites in the functional groups of one molecule react with electron-poor sites in the functional groups of another molecule.

b) *Heterolytic breakage* of a bond occurs when both bonding electrons leave with one fragment.

c) *Homolytic breakage* of a bond occurs when one bonding electron leaves with each fragment.

d) A *radical reaction* is a reaction in which odd-electron species are produced or consumed.

e) A *functional group* is a group of atoms that has a characteristic reactivity.

f) *Polarization* is the temporary change in the electron distribution of atoms or functional groups due to interactions with reagents or solvent.

4.12 There are many acceptable answers to each part of this question.

a) $^{-}$:B̤̈r: b) H^+ c), d) CH_3Br + $^{-}$:Ö̤H ⟶ CH_3OH + $^{-}$:B̤̈r:

e) H–OH ⟶ H^+ + $^{-}$:Ö̤H f) Cl–Cl ⟶ 2 :C̤̈l·

4.13 a) The carbonyl carbon of this aldehyde acts as an *electrophile*; the oxygen, however, behaves as a *nucleophile*.

b) $^{-}$:B̤̈r: is a *nucleophile*.

c) :B̤̈r$^+$ is an *electrophile*.

d) $CH_3CO_2^-$ behaves as a *nucleophile*.

e) The bond polarity of the C-Br bond causes carbon to be relatively electron-poor and to react with nucleophilic reagents. CH_3Br is thus an *electrophile*.

f) The carbonyl carbon behaves as an *electrophile*; oxygen can behave as a *nucleophile*.

g) $CH_3CH_2\ddot{N}H_2$ is a *nucleophile* because of the lone electron pair of nitrogen.

h) BF_3 is an *electrophile*.

Neutral molecules may behave either as electrophiles or nucleophiles, depending on their structure and the site of reaction.

4.14

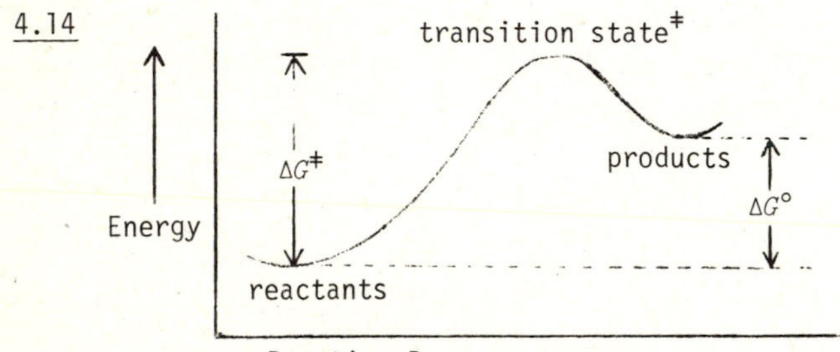

$\Delta G°$ is positive.

4.15

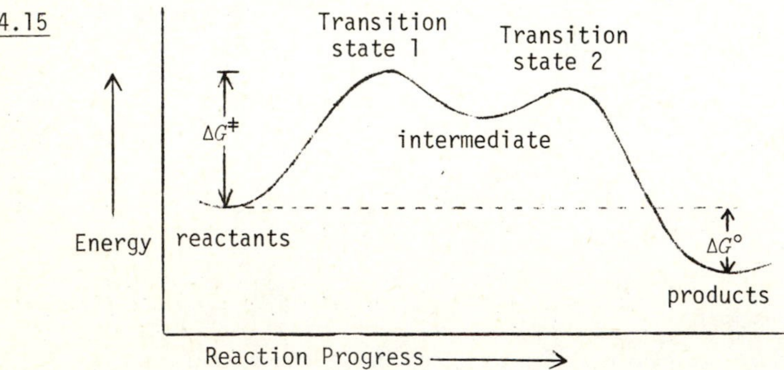

$\Delta G°$ is negative.

4.16 Transition states and intermediates are both relatively unstable species in a reaction. A transition state is a maximum energy species that occurs during a reaction; an intermediate is the species that occurs at an energy minimum between two transition states. Even though an intermediate may be of such high energy that it cannot be isolated, it is still of lower energy than a transition state.

4.17

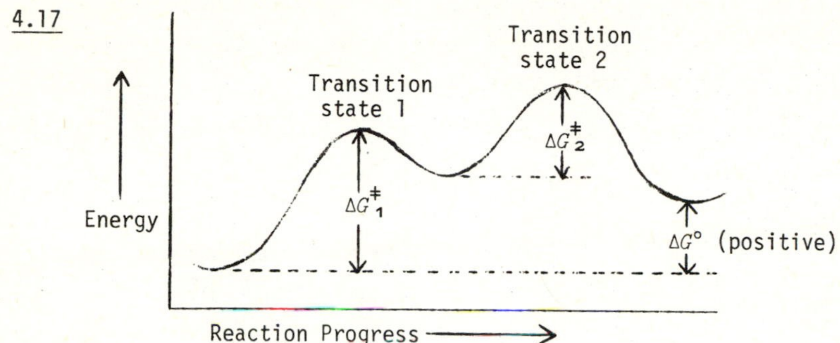

There are two steps in the reaction; step 2 is faster since $\Delta G_2^{\ddagger} < \Delta G_1^{\ddagger}$.

4.18 Refer to problems 4.5 and 4.6 if you need help with this problem.

a) CH_3-OH + H-Br ⟶ CH_3-Br + H-OH
91 kcal/mol 88 kcal/mol 70 kcal/mol 119 kcal/mol
$\Delta H° = -10$ kcal/mol

b) CH_3CH_2O-H + CH_3-Cl ⟶ CH_3CH_2O-CH_3 + H-Cl
103 kcal/mol 84 kcal/mol 81 kcal/mol 103 kcal/mol
$\Delta H° = +3$ kcal/mol

c) t-C_4H_9Br + H-I ⟶ t-C_4H_9I + H-Br
65 kcal/mol 71 kcal/mol 50 kcal/mol 88 kcal/mol
$\Delta H° = -2$ kcal/mol

4.19 a) CH_3CH_2-H + Cl_2 ⟶ CH_3CH_2-Cl + H-Cl
98 kcal/mol 58 kcal/mol 81 kcal/mol 103 kcal/mol
$\Delta H° = -28$ kcal/mol

b) CH_3CH_2-H + Br_2 ⟶ CH_3CH_2-Br + H-Br
98 kcal/mol 46 kcal/mol 68 kcal/mol 88 kcal/mol
$\Delta H° = -12$ kcal/mol

c) CH_3CH_2-H + I_2 ⟶ CH_3CH_2-I + H-I
98 kcal/mol 36 kcal/mol 53 kcal/mol 71 kcal/mol
$\Delta H° = +10$ kcal/mol

Of the three halogenation reactions, chlorination proceeds the most readily; iodination is the most difficult reaction.

4.20

Bonds broken	$\Delta H°$	Bonds formed	$\Delta H°$
CH_3-CH_3	88 kcal/mol	CH_3-Br	70 kcal/mol
Br_2	46 kcal/mol	CH_3-Br	70 kcal/mol
	$\Delta H°$ = 134 kcal/mol		$\Delta H°$ = 140 kcal/mol

$$\Delta H°_{overall} = \Delta H°_{\substack{bonds\\broken}} - \Delta H°_{\substack{bonds\\formed}} = 134 \text{ kcal/mol} - 140 \text{ kcal/mol} = \underline{-6 \text{ kcal/mol}}$$

$\Delta H°$ for bromoethane formation is -12 kcal/mol; $\Delta H°$ for bromomethane formation is -6 kcal/mol. Although both of these reactions have negative $\Delta H°$, the reaction that forms bromoethane is more favorable.

4.21 To predict the most stable radical, first look in Table 4.5 to find the bond dissociation energy of the bond in the hydrocarbon that is broken to form the radical.

For example, for $CH_3\cdot$, the bond dissociation energy for CH_3-H is 104 kcal/mol.

CH_3CH_2-H ⟶ $CH_3\dot{C}H_2$ + H· $\Delta H°$ = 98 kcal/mol

C_6H_5-H ⟶ $C_6H_5\cdot$ + H· $\Delta H°$ = 112 kcal/mol

CH_2=$CHCH_2$-H ⟶ CH_2=$CH\dot{C}H_2$ + H· $\Delta H°$ = 87 kcal/mol

$(CH_3)_3C$-H ⟶ $(CH_3)_3C\cdot$ + H· $\Delta H°$ = 91 kcal/mol

Remember that the largest *positive* value for $\Delta H°$ represents the bond that is hardest to break; the radical resulting from the bond having the largest $\Delta H°$ is the least stable.

least stable ⟶ most stable

$C_6H_5\cdot$ < $\cdot CH_3$ < $CH_3\dot{C}H_2$ < $(CH_3)_3C\cdot$ < CH_2=$CH\dot{C}H_2$

4.22 The following bond dissociation energies can be found in Table 4.5.

bond	$\Delta H°$
H-Cl	103 kcal/mol
Cl-Cl	58 kcal/mol
C_2H_5-H	98 kcal/mol
$(CH_3)_2CH$-H	95 kcal/mol
$(CH_3)_3C$-H	91 kcal/mol
C_2H_5-Cl	81 kcal/mol
$(CH_3)_2CH$-Cl	80 kcal/mol
$(CH_3)_3C$-Cl	79 kcal/mol

Using these values in the equation R-H + Cl_2 ⟶ R-Cl + HCl we can determine the values of $\Delta H°$ for chlorination at each position of 2-methylbutane.

$$CH_3CH_2CH(CH_3)CH_3 + Cl_2 \longrightarrow HCl + CH_2ClCH_2CH(CH_3)CH_3 \text{ and } CH_3CH_2CH(CH_3)CH_2Cl + CH_3CHClCH(CH_3)CH_3 + CH_3CH_2CCl(CH_3)CH_3$$

$\Delta H°$ -28 kcal/mol -30 kcal/mol -33 kcal/mol

Formation of the tertiary chloride is favored, but all three $\Delta H°$ values are quite close to one another. Consequently, a mixture of products is expected.

4.23 The following compounds yield single monohalogenation products since each has only one kind of hydrogen atom.

C_2H_6, (cyclohexane), $CH_3C{\equiv}CCH_3$

4.24 For the first series of steps:

a) $\Delta H° = +58$ kcal/mol
b) $\Delta H° = +1$ kcal/mol
c) $\Delta H° = -26$ kcal/mol

(b + c) $\Delta H°_{overall} = -25$ kcal/mol

For the alternate series:

a) $\Delta H° = +58$ kcal/mol
b) $\Delta H° = +20$ kcal/mol
c) $\Delta H° = -45$ kcal/mol

(b + c) $\Delta H°_{overall} = -25$ kcal/mol

For both series of reactions, the radical-producing initiation step has $\Delta H° = +58$ kcal/mol.

In both series, $\Delta H°$ for the propagation steps b + c is -25 kcal/mol. In series 2, however, one step has $\Delta H° = +20$ kcal/mol; this step is energetically much less favorable than any step in series 1 and disfavors the second series as a whole. The first route for chlorination of methane is thus more likely to occur.

What you should know:

After doing these problems you should be able to:

1. Understand the meaning of bond polarity and be able to determine the polarity of a given bond;
2. Recognize polar reactions, radical reactions, pericyclic reactions, electrophiles and nucleophiles;
3. Understand the concepts of equilibrium and rate;
4. Know the meaning of bond dissociation energy; use bond dissociation energies to calculate $\Delta H°$ of simple reactions;
5. Draw reaction energy diagrams and label them properly.

CHAPTER 5 STRUCTURE AND ELECTROPHILIC ADDITION REACTIONS

5.1 a) The formula of a C_8 *alkane* is C_8H_{18}. C_8H_{14}, which contains four fewer (or two pairs fewer) hydrogens than C_8H_{18}, may have two double bonds, or two rings, or one of each, or one triple bond.

b) C_5H_6 -- three double bonds and/or rings

c) $C_{12}H_{20}$ -- three double bonds and/or rings

d) An alkane with 20 carbons has the formula $C_{20}H_{42}$. $C_{20}H_{32}$ has ten hydrogens, or five hydrogen pairs, fewer than $C_{20}H_{42}$; $C_{20}H_{32}$ thus contains five double bonds and/or rings.

e) $C_{40}H_{56}$ -- 13 double bonds and/or rings

f) $C_{30}H_{50}$ -- six double bonds and/or rings

5.2 a) C_4H_8 contains one double bond or one ring. The ring compounds are:

```
H2C—CH2                 CH2
 |   |                  / \
H2C—CH2            H2C———CHCH3
```

The double bond compounds: $CH_3CH_2CH{=}CH_2$, $CH_3CH{=}CHCH_3$, $(CH_3)_2C{=}CH_2$

b) C_4H_6 has two double bonds, two rings, or one of each (or one triple bond).

Triple-bond compounds: $CH_3CH_2C{\equiv}CH$, $CH_3C{\equiv}CCH_3$

Two double bonds: $CH_2{=}CHCH{=}CH_2$, $CH_3CH{=}C{=}CH_2$

Two rings:

```
 HC—CH2
 | \ |
H2C—CH
```

One ring and one double bond:

```
H2C—CH         CH2           CH3          CH3
 |   ||        ||             |            |
H2C—CH         C             CH            C
              / \            / \          // \
          H2C———CH2      HC===CH      HC———CH2
```

c) C_3H_4 contains two double bonds, or one ring and one double bond, or one triple bond. (It is not possible to form two rings with three carbon atoms.)

Triple bond: $CH_3C{\equiv}CH$

Two double bonds: $CH_2{=}C{=}CH_2$

One ring and one double bond:

```
  CH2
  / \
HC===CH
```

d) C_5H_8 contains two double bonds, or two rings, or one double bond and one ring, or one triple bond. 23 structures are possible for C_5H_8. Included are a few on them.

Triple bond: $CH_3CH_2CH_2C{\equiv}CH$, $CH_3CH_2C{\equiv}CCH_3$, $(CH_3)_2CHC{\equiv}CH$

Two double bonds: $CH_3CH{=}CHCH{=}CH_2$, $CH_3CH_2CH{=}C{=}CH_2$, $CH_2{=}C(CH_3)CH{=}CH_2$

Two rings:

```
    CH-CH2           CH2
   /  |  |          /   \
H2C   |  |       HC<————>C-CH3
   \  |  |          \   /
    CH-CH2           CH2
```

One ring and one double bond:

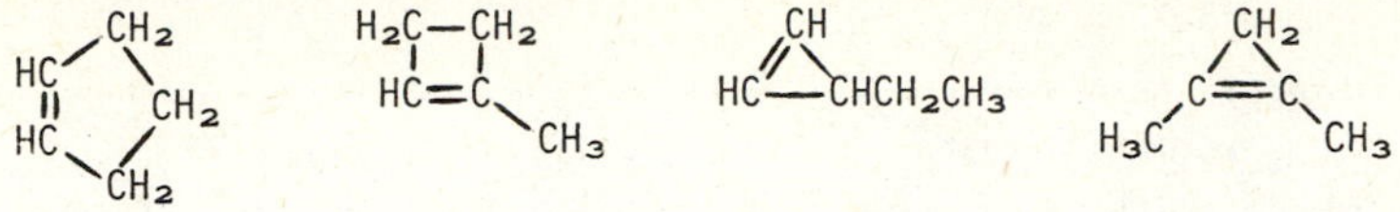

```
   ,CH2          H2C—CH2          CH              CH2
HC'   \           |   |          //\             /  \
 ||    CH2       HC=C          HC——CHCH2CH3    ,C==C,
HC.   /              \                      H3C'      `CH3
   `CH2               CH3
```

5.3 a) Subtract one hydrogen for each nitrogen present to find the base formula C_6H_4. Compared to the alkane C_6H_{14}, the compound of formula C_6H_5N has ten fewer hydrogens, or five fewer hydrogen pairs, and contains five double bonds and/or rings.

b) $C_6H_5NO_2$ also contains five double bonds and/or rings because oxygen does not affect the base formula of a compound.

c) A halogen atom is equivalent to a hydrogen atom in the calculation of the base formula. Here, the "base formula" is C_8H_{12}. $C_8H_9Cl_3$ has three double bonds and/or rings.

d) $C_9H_{16}Br_2$ -- one double bond or ring

e) $C_{10}H_{12}N_2O_3$ -- six double bonds and/or rings

f) $C_{20}H_{32}O_2$ -- five double bonds and/or rings

5.4 Compounds (c), (e), and (f) can exist as pairs of *cis-trans* isomers.

```
                              cis                  trans

                        CH3CH2   CH3         CH3CH2    H
c) CH3CH2CH=CHCH3             C=C                  C=C
                             H   H                H   CH3

                         Cl     Cl            Cl     H
e) ClCH=CHCl               C=C                  C=C
                          H    H               H    Cl

                         Br     Cl            Br     H
f) BrCH=CHCl               C=C                  C=C
                          H    H               H    Cl
```

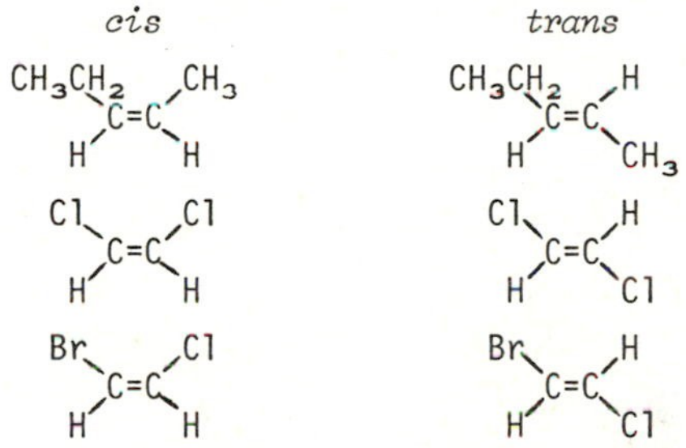

5.5 In all of these examples, ΔH_{hydrog} for the *trans* cycloalkenes is higher than ΔH_{hydrog} for the *cis* compounds, indicating that the *trans*-cycloalkenes are less stable than *cis*-cycloalkenes. Build models of the two cyclooctenes and notice the large amount of strain in *trans*-cyclooctene relative to *cis*-cyclooctene. This strain causes the *trans* isomer to be of higher energy and to have a ΔH_{hydrog} larger than *cis*-cyclooctene. Use models to construct the other four cycloalkenes. As ring size increases, the problem of strain for the *trans* rings becomes less severe, and ΔH_{hydrog} decreases.

5.6 a)

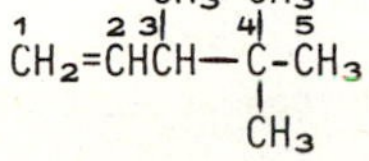

```
        CH3 CH3
 1    2 3|  4| 5
CH2=CHCH—C-CH3
             |
            CH3
```

1) Find the longest carbon chain containing the double bond, and name the parent compound. Here, the longest chain contains five carbons, and the compound is a *pentene*.

2) Number the carbon atoms, giving to the double bond the lowest possible number.

3) Name the compound: 3,4,4-trimethyl-1-pentene.

b) $CH_3CH_2CH{=}C(CH_3)CH_2CH_3$ 3-methyl-3-hexene

c) $CH_3CH{=}CHCH(CH_3)CH{=}CHCH(CH_3)_2$ 4,7-dimethyl-2,5-octadiene

5.7 a)

```
            CH3
            |
CH2=CHCH2CH2C=CH2
```

2-methyl-1,5-hexadiene

b)

```
             CH2CH3
             |
CH3CH2CH2CH=CC(CH3)3
```

3-ethyl-2,2-dimethyl-3-heptene

c)

```
              CH3
              |
CH3CH=CHCH=CHC—— C=CH2
              |   |
              CH3 CH3
```

2,3,3-trimethyl-1,4,6-octatriene

d)

```
  CH3      CH3
  |        |
CH3CH      CHCH3
     \    /
      C=C
     /    \
CH3CH      CHCH3
  |        |
  CH3      CH3
```

3,4-diisopropyl-2,5-dimethyl-3-hexene

5.8 a)

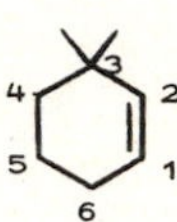

3,3-dimethylcyclohexene

b)

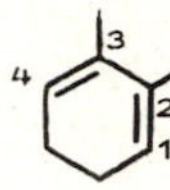

2,3-dimethyl-1,3-cyclohexadiene

c)

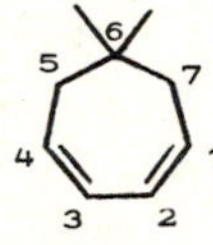

6,6-dimethyl-1,3-cycloheptadiene

d)

1,2,5,5-tetramethyl-1,3-cyclopentadiene

5.9 highest priority ⟶ lowest priority

a) $-Cl$, $-OH$, $-CH_3$, $-H$

b) $-CH_2OH$, $-CH{=}CH_2$, $-CH_2CH_3$, $-CH_3$

c) $-COOH$, $-CH_2OH$, $-CN$, $-CH_2NH_2$

d) $-CH_2OCH_3$, $-CN$, $-C{\equiv}CH$, $-CH_2CH_3$

5.10 a)

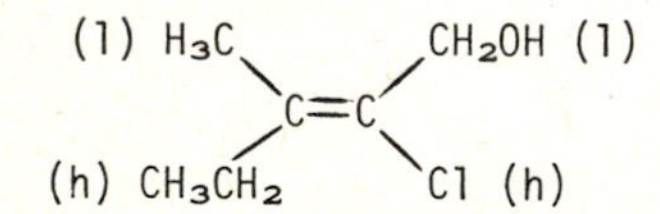

Z

First, consider substituents on the right side of the double bond. $-Cl$ ranks higher than $-CH_2OH$ by Cahn-Ingold-Prelog rules. On the left side, $-CH_2CH_3$ ranks higher than $-CH_3$. The isomer has *Z* configuration because the higher priority groups are on the same side of the double bond.

b)

```
(h) Cl            CH2CH3 (1)
      \          /
       C=C
      /          \
(1) CH3O          CH2CH2CH3 (h)
```

E

c) (h) CH_3 ... C=C ... COOH (h); (l) ... CH_2OH (l)   Z

Notice that in ranking substituents on the left side of the bond, the upper substituent is of higher priority because of the methyl group attached to the ring.

d) (l) H, (h) CH_3 ... C=C ... CN (h), CH_2NH_2 (l)   E

5.11 a) cyclohexene + HCl ⟶ chlorocyclohexane

Since the starting material is symmetrical, only one product is possible.

b) $(CH_3)_2C{=}CHCH_2CH_3 \xrightarrow{H_2SO_4} (CH_3)_2C(OSO_3H)CH_2CH_2CH_3$

c) $CH_3CH_2CH_2CH{=}CH_2 \xrightarrow{H_3PO_4,\ KI} CH_3CH_2CH_2CH(I)CH_3$

2-iodopentane

d) methylenecyclohexane ($=CH_2$) + HBr ⟶ (CH_3, Br)

1-bromo-1-methylcyclohexane

5.12 The second step in the electrophilic addition of HCl to alkenes is exothermic. According to the Hammond postulate, the transition state should resemble the carbocation intermediate.

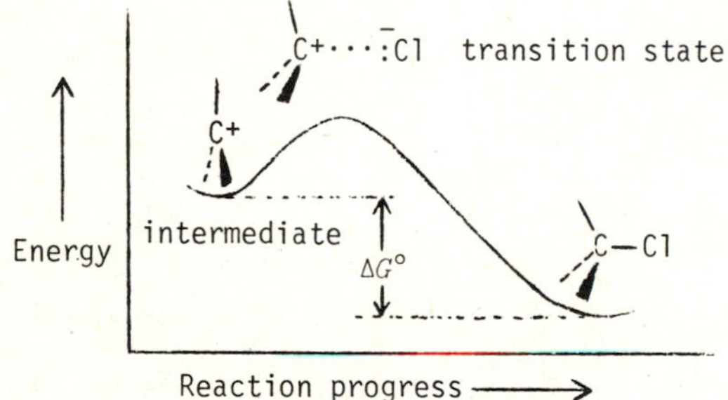

5.13 See problems 5.1, 5.3 for method of solution.

a) C_6H_6 -- four degrees of unsaturation

b) C_6H_{10} -- two degrees of unsaturation

c) $C_{10}H_{16}$ -- three degrees of unsaturation

d) $C_6H_6Cl_6$ -- one degree of unsaturation

e) C_5H_5N -- four degrees of unsaturation

f) $C_{10}H_{10}O_2$ -- six degrees of unsaturation

5.14 The purpose of this problem is to give you experience in calculating the number of double bonds and/or rings in a formula. Additionally, you will gain practice in writing structures containing various functional groups. Remember that any formulas that satisfy the rules of valency are acceptable. Try to identify functional groups in the formulas you draw.

a) $C_{10}H_{16}$ -- three degrees of unsaturation. Examples:

$CH_3CH_2CH{=}CHCH{=}CHCH{=}CHCH_2CH_3$

CH_3, $C{=}CH_2$, CH_3, $C{=}C$, $H_2C{=}C$, CH_3, CH_3

b) C_8H_8O. The parent hydrocarbon is C_8H_8; each structure contains five degrees of unsaturation. Examples:

O, CCH_3 ← ketone; ← aromatic ring

O, CH_2CH ← aldehyde; ← aromatic ring

double bonds; ketone; O

O, CH ← aldehyde; ← aromatic ring; CH_3

O ← ketone; $CCH{=}CH_2$; double bonds

c) This compound has C_7H_{12} as its base formula. $C_7H_{10}Cl_2$ has two degrees of unsaturation. Examples:

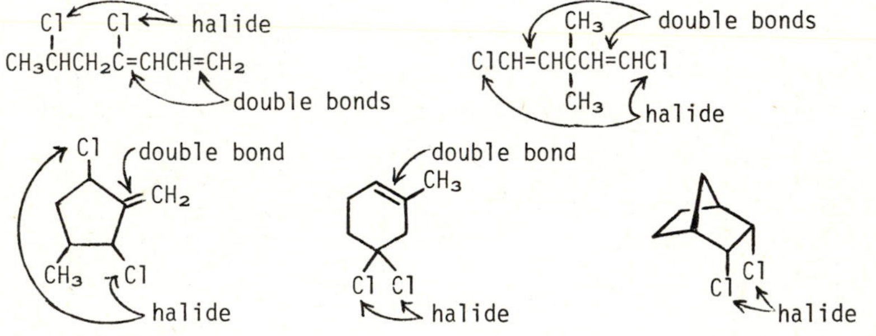

d) $C_{10}H_{16}O_2$ -- three degrees of unsaturation

$CH_2{=}CHCH{=}CHCH_2CH_2CH_2CH_2CH_2COH$ (C=O) ← carboxylic acid; double bonds

O, O ← ketone

HO, O ← ketone; alcohol

H, CH_3 ketone; O, O; H_3C, CH_3, CH_3 CH_3

HO ← alcohol → OH

e) $C_5H_9NO_2$ -- two degrees of unsaturation

ketone, amide: $CH_3CH_2\overset{O}{\overset{\|}{C}}CH_2\overset{O}{\overset{\|}{C}}NH_2$

double bond, nitro group: $CH_2{=}CHCH_2CH_2CH_2\overset{O}{\overset{\|}{N^+}}{-}O^-$

double bond, amine, carboxylic acid: $CH_2{=}CHCHNHCH_2\overset{O}{\overset{\|}{C}}OH$

alcohol, amide

ketone, amine, ether

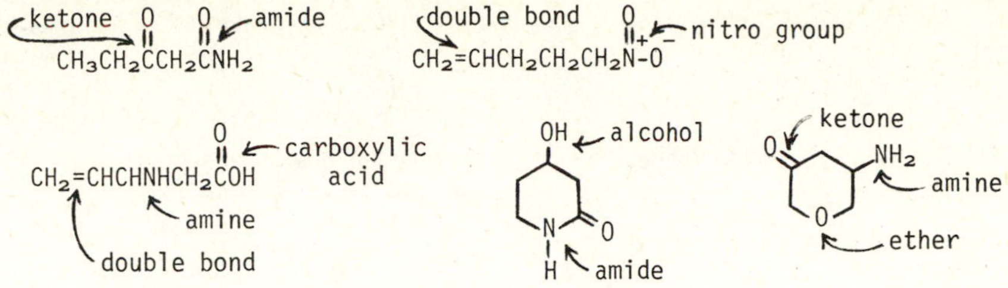

f) $C_8H_{10}ClNO$ -- four degrees of unsaturation

halide, double bonds, amide: $ClCH{=}CHCH{=}CHCH{=}CHCH_2\overset{O}{\overset{\|}{C}}NH_2$

double bonds, ether, halide, amine

alcohol, aromatic ring, amine, halide

halide, double bonds, ketone, amine

ketone, amine, halide

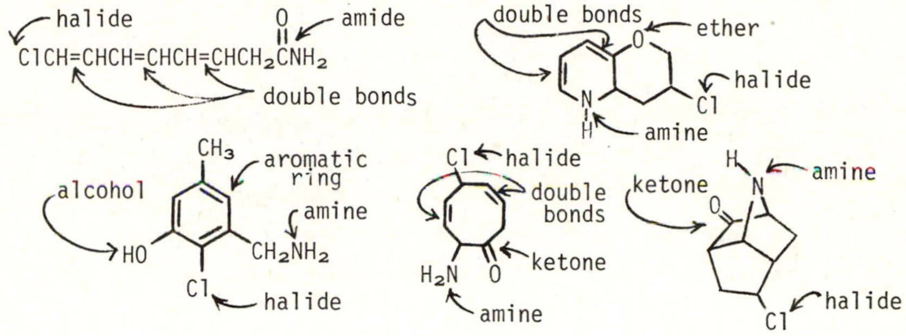

5.15 Interpreting problems of this sort is often difficult. To start, you should train yourself to *read every word* of the problem. Then you should try to solve the problem phrase by phrase. For example, "A compound of formula $C_{10}H_{14}$" describes a compound having four degrees of unsaturation ($C_{10}H_{14}$ has four fewer hydrogen pairs than a C_{10} acyclic alkane). The phrase "undergoes catalytic hydrogenation" means that H_2 is added to the double bonds. "Absorbs only two equivalents of H_2" means that only *two* of the degrees of unsaturation are double bonds (or a triple bond). The other two must be rings.

5.16
a) 4-methyl-2-hexene
b) 4-butyl-7-methyl-2-octene
c) 2-ethyl-1-butene
d) 3,4-dimethyl-1,5-heptadiene
e) 4-methyl-1,3-hexadiene
f) 1,2-butadiene
g) 3,3-dimethyl-1-butene
h) 2,2,5,5-tetramethyl-3-hexene

5.17 Because the longest carbon chain contains 8 carbons, and because there are three double bonds present, ocimene is an octatriene. Start numbering at the end that will give the lowest number to the first double bond (1,3,6 is lower than 2,5,7). Number the methyl substituents and, finally, name the compound.

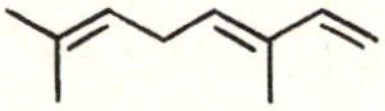

(3*E*)-3,7-dimethyl-1,3,6-octatriene

5.18 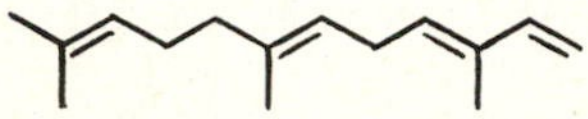 (3*E*,6*E*)-3,7,11-trimethyl-1,3,6,10-dodecatetraene

5.19 a)
$$\begin{array}{c} CH_3 \\ | \\ CH_3CH{=}CCH_2C{=}CH_2 \\ | \\ CH_3 \end{array}$$

b)
$$\begin{array}{c} CH_2CH_2CH_3 \\ | \\ CH_3CH_2CH{=}CHCHC(CH_3)_2CH{=}CH_2 \end{array}$$

c)
$$\begin{array}{c} CH_3 \\ | \\ CH_3CHCH{=}C{=}CH_2 \end{array}$$

d)
$$\begin{array}{c} CH_3CH_3 \\ || \\ CH_3C{=}CHCH_2CH{=}CCH{=}CH_2 \end{array}$$

e)
$$\begin{array}{c} CH_2CH_2CH_2CH_3 \\ | \\ CH_3CH_2CH_2CH_2C{=}CHCH_3 \end{array}$$

f) $(CH_3)_3CCH{=}CHC(CH_3)_3$

5.20 a) $CH_2{=}CHCH{=}C(CH_3)_2$ — Correct name: 4-methyl-1,3-pentadiene
Numbering must start at the other end.

b)
$$\begin{array}{c} CH_2 \\ \| \\ CH_3CH_2CCH{=}CH_2 \end{array}$$
Correct name: 2-ethyl-1,3-butadiene
The parent chain must contain *both* double bonds.

c) $CH_3CH{=}CHCH_2CH{=}CHCH_2CH_3$ — Correct name: 2,5-octadiene
Numbering must start at the other end.

d)
$$\begin{array}{c} CH_2CH_3 \\ | \\ CH_3CH_2CH_2C{=}CHCH_2CH_2CH_3 \end{array}$$
Correct name: 4-ethyl-4-octene
Numbering must start at the other end.

e)
$$\begin{array}{c} CH_2CH_2CH_3 \\ | \\ CH_3CH_2CH_2CH{=}CCH_2CH_3 \end{array}$$
Correct name: 4-ethyl-4-octene
The longest chain containing the double bond is an *octene*.

f) $H_2C{=}CHCH_2CH{=}CH_2$ — Correct name: 1,4-pentadiene
The longest chain must contain both double bonds.

5.21 The central carbon of allene forms two sigma bonds and two pi bonds. The central carbon is *sp* hybridized, and the carbon-carbon bond angle is 180°, indicating linear geometry for the carbons of allene.

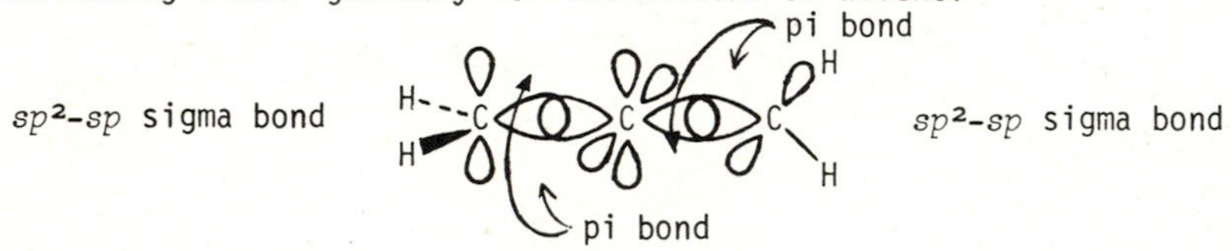

pi bond

*sp*²-*sp* sigma bond — H, H, C, C, C, H, H — *sp*²-*sp* sigma bond

pi bond

5.22 As expected, the two *trans* compounds are more stable than their *cis* counterparts. The *cis-trans* difference is much more extreme for the tetramethyl compound, however. Build a model of *cis*-2,2,5,5-tetramethyl-3-hexene and notice the extreme crowding of the methyl groups. Steric interference makes the *cis* isomer much less stable than the *trans* isomer and causes *cis* ΔH_{hydrog} to be much larger than *trans* ΔH_{hydrog}.

5.23 Because its heat of hydrogenation is so much larger, 1,2-pentadiene must be much less stable than 1,4-pentadiene. This instability may be due to strain encountered when one carbon must form a double bond to each of two different carbons.

5.24 higher priority ⟶ lower priority

a) $-I$, $-Br$, $-CH_3$, $-H$

b) $-OCH_3$, $-OH$, $-COOH$, $-H$

c) $-COOCH_3$, $-COOH$, $-CH_2OH$, $-CH_3$

d) $-COCH_3$, $-CH_2CH_2OH$, $-CH_2CH_3$, $-CH_3$

e) $-CH_2Br$, $-C{\equiv}N$, $-CH_2NH_2$, $-CH{=}CH_2$

f) $-CH_2OCH_3$, $-CH_2OH$, $-CH{=}CH_2$, $-CH_2CH_3$

5.25 a) 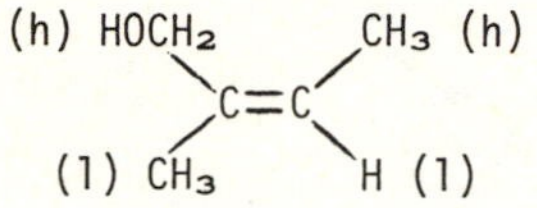 (h) $HOCH_2$, CH_3 (h) / C=C / (l) CH_3, H (l) — *Z*

b) 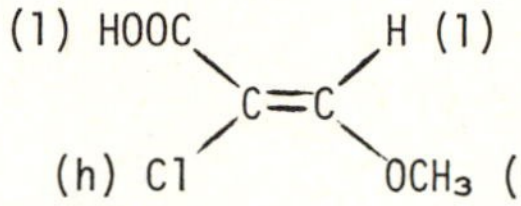 (l) HOOC, H (l) / C=C / (h) Cl, OCH_3 (h) — *Z*

c) 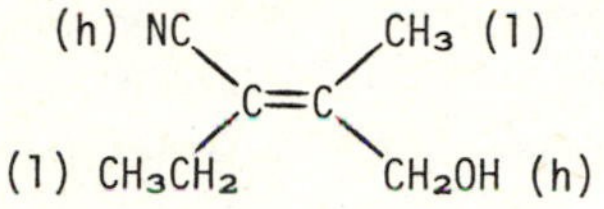 (h) NC, CH_3 (l) / C=C / (l) CH_3CH_2, CH_2OH (h) — *E*

d) 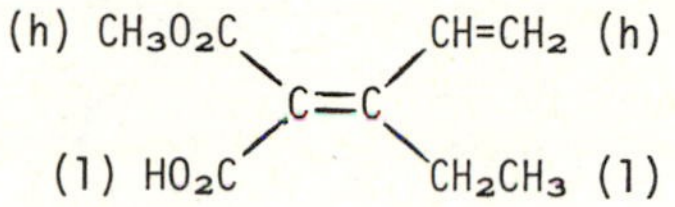 (h) CH_3O_2C, $CH{=}CH_2$ (h) / C=C / (l) HO_2C, CH_2CH_3 (l) — *Z*

5.26 a) 3-methylcyclohexene

b) 2,3-dimethylcyclopentene

c) ethylcyclobutadiene

d) 1,2-dimethyl-1,4-cyclohexadiene

e) 5-methyl-1,3-cyclohexadiene

f) 1,5-cyclooctadiene

5.27 a) 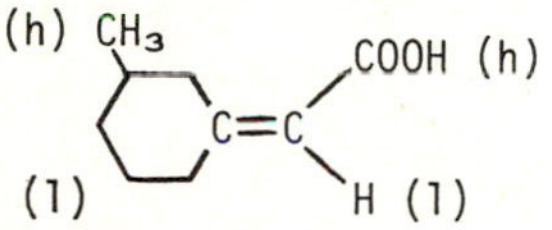 (h) CH_3 (on cyclohexylidene ring), COOH (h) / C=C / (l) ring, H (l) — *Z* (correct)

b) 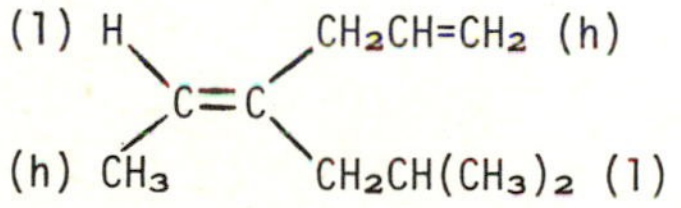 (l) H, $CH_2CH{=}CH_2$ (h) / C=C / (h) CH_3, $CH_2CH(CH_3)_2$ (l) — *E* (correct)

c) 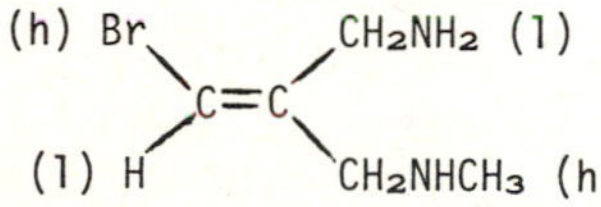 (h) Br, CH_2NH_2 (l) / C=C / (l) H, CH_2NHCH_3 (h) — *E* (incorrect)

d) 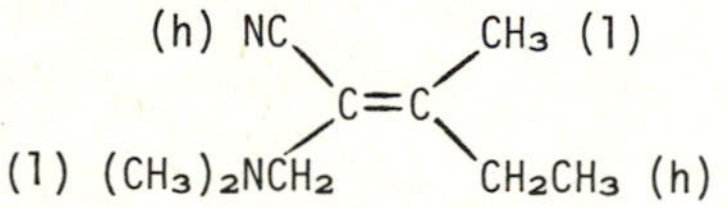 (h) NC, CH_3 (l) / C=C / (l) $(CH_3)_2NCH_2$, CH_2CH_3 (h) — *E* (correct)

e) 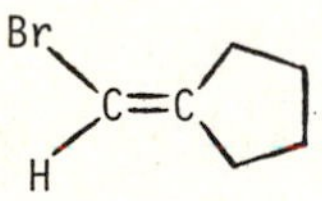 Br, H / C=C (cyclopentylidene)

This compound does not exhibit *E-Z* isomerism.

f) 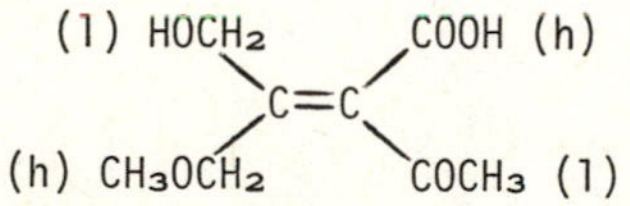 (l) $HOCH_2$, COOH (h) / C=C / (h) CH_3OCH_2, $COCH_3$ (l) — *E* (correct)

All structures except c and e are designated correctly.

5.28

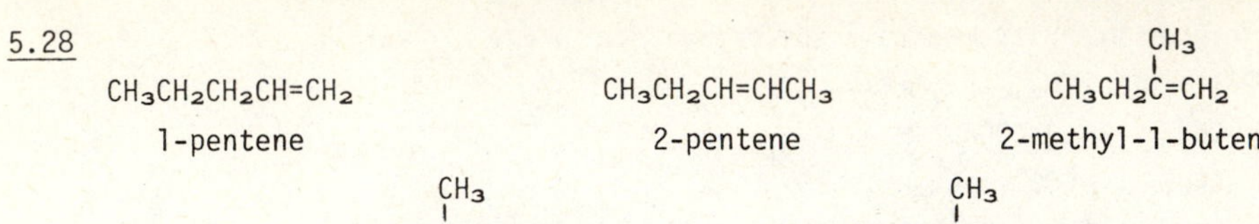

$CH_3CH_2CH_2CH{=}CH_2$ 1-pentene

$CH_3CH_2CH{=}CHCH_3$ 2-pentene

$CH_3CH_2C(CH_3){=}CH_2$ 2-methyl-1-butene

$CH_3CH(CH_3)CH{=}CH_2$ 3-methyl-1-butene

$CH_3CH{=}C(CH_3)CH_3$ 2-methyl-2-butene

5.29

$CH_3CH_2CH_2CH_2CH{=}CH_2$ 1-hexene

$CH_3CH_2CH_2CH{=}CHCH_3$ 2-hexene

$CH_3CH_2CH{=}CHCH_2CH_3$ 3-hexene

$CH_3CH_2CH_2C(CH_3){=}CH_2$ 2-methyl-1-pentene

$CH_3CH_2CH(CH_3)CH{=}CH_2$ 3-methyl-1-pentene

$CH_3CH(CH_3)CH_2CH{=}CH_2$ 4-methyl-1-pentene

$CH_3CH_2CH{=}C(CH_3)CH_3$ 2-methyl-2-pentene

$CH_3CH_2C(CH_3){=}CHCH_3$ 3-methyl-2-pentene

$CH_3CH(CH_3)CH{=}CHCH_3$ 4-methyl-2-pentene

$CH_3CH(CH_3)C(CH_3){=}CH_2$ 2,3-dimethyl-1-butene

$CH_3C(CH_3)_2CH{=}CH_2$ 3,3-dimethyl-1-butene

$(CH_3)_2C{=}C(CH_3)_2$ 2,3-dimethyl-2-butene

$CH_3CH_2C(CH_2CH_3){=}CH_2$ 2-ethyl-1-butene

5.30 See problems 5.1-5.3 for a method of solution.

a) $C_{27}H_{46}O$ five degrees of unsaturation

b) $C_{14}H_9Cl_5$ eight degrees of unsaturation

c) $C_{20}H_{34}O_5$ four degrees of unsaturation

d) $C_8H_{10}N_4O_2$ six degrees of unsaturation

e) $C_{21}H_{28}O_5$ eight degrees of unsaturation

f) $C_{17}H_{23}NO_3$ seven degrees of unsaturation

g) $C_6H_8O_6$ three degrees of unsaturation

5.31

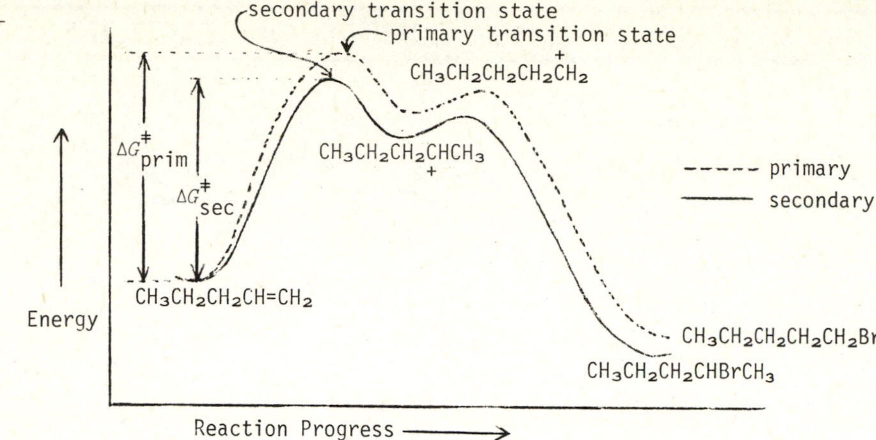

5.32

	Transition State #1	Transition State #2
2-bromopentane path	H, n-C_3H_7, δ^+, H, H, H	H_3C, H, n-C_3H_7, $\bar{:}Br$
1-bromopentane path	H, H, n-C_3H_7, δ^+, H, H	H, H, n-C_4H_9, $\bar{:}Br$

The first step (protonation) for both reaction paths is endothermic, and both transition states resemble the carbocation intermediate. Transition states for the exothermic second step also resemble the carbocation intermediate. Transition state #1 for 1-bromopentane is more like the carbocation intermediate than is transition state #1 for 2-bromopentane.

What you should know:

After doing these problems, you should be able to:

1. Calculate the number of double bonds and/or rings (degrees of unsaturation) of any compound, including those containing N,O, and halogen;
2. Use the concept of heat of hydrogenation to predict stability of alkenes;
3. Name cyclic and acyclic alkenes;
4. Use E,Z nomenclature to designate the arrangement of groups about a double bond;
5. Understand the types of reactions that alkenes undergo (electrophilic addition);
6. Understand the Hammond Postulate.

CHAPTER 6 ALKENES. ADDITION/ELIMINATION REACTIONS

6.1

CH_3 CH_3 — Cl_2 → [$\overset{+}{Cl}$; CH_3 CH_3 ; $:\ddot{Cl}:^-$] → Cl CH_3 ; CH_3 Cl

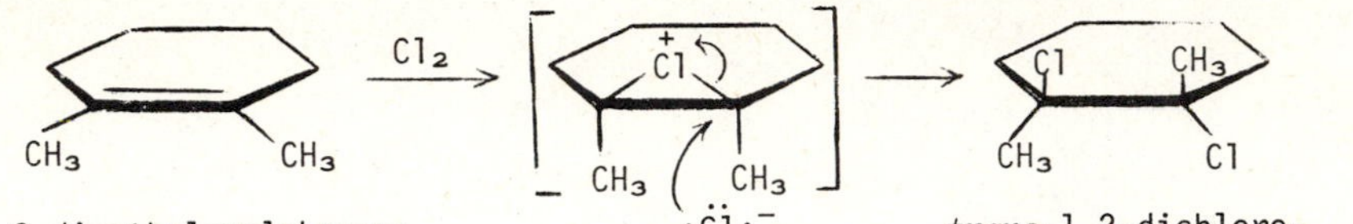

1,2-dimethylcyclohexene — *trans*-1,2-dichloro-1,2-dimethylcyclohexane

The chlorines are *trans* to one another in the product, as are the methyl groups.

6.2

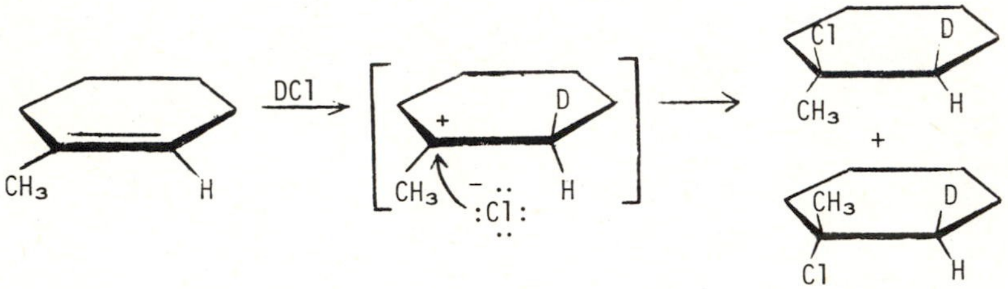

CH_3 H — DCl → [+ D ; CH_3 $:\ddot{Cl}:^-$ H] → Cl D ; CH_3 H + CH_3 D ; Cl H

Addition of hydrogen (or deuterium) halides involves formation of an open carbocation, not a cyclic halonium ion intermediate. The carbocation, which is sp^2 hybridized and planar, can be attacked by chloride from either top or bottom, yielding products in which deuterium and chlorine can be either *cis* or *trans* to each other.

6.3 NBS is the source of the electrophilic Br^+ ion in bromohydrin formation. Attack of the alkene pi electrons on Br^+ forms a cyclic bromonium ion. When this bromonium ion is opened by water, a partial positive charge develops at the carbon whose bond to bromine is being cleaved.

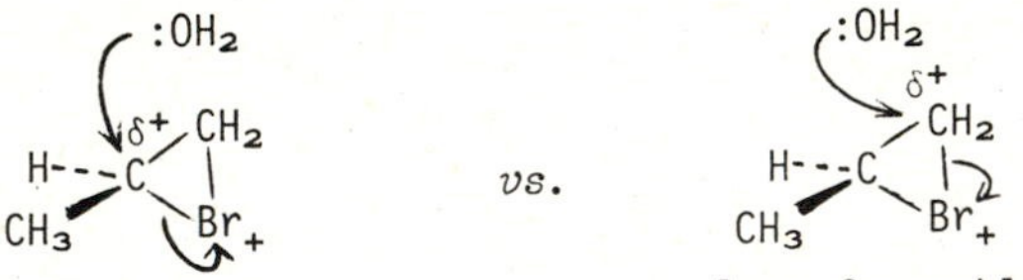

$:OH_2$; H---C($\delta+$) CH_2 ; CH_3 Br_+ *vs.* $:OH_2$; $\delta+$ CH_2 ; H---C ; CH_3 Br_+

less favorable

Since a secondary carbon can stabilize this charge better than a primary carbon, opening of the bromonium ion occurs at the secondary carbon to yield Markovnikov product.

6.4 Addition of IN_3 to the alkene yields a product in which I is bonded to the primary carbon and N_3 is bonded to the secondary carbon. If addition occurs with Markovnikov orientation, I^+ must be the electrophile, and the reaction must proceed through an iodonium ion intermediate. Opening of the iodonium ion gives Markovnikov product for the reasons discussed in Problem 6.3. The bond polarity of iodine azide is $I \overset{\leftrightarrow}{—} N_3$.

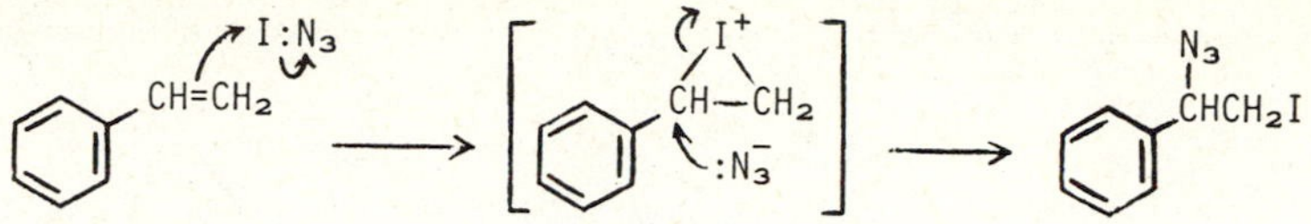

$$C_6H_5CH{=}CH_2 \xrightarrow{I:N_3} \left[C_6H_5\overset{I^+}{CH{-}CH_2} \;\; :N_3^- \right] \longrightarrow C_6H_5CH(N_3)CH_2I$$

6.5 Recall the mechanism of hydroboration and note that the hydrogen added to the double bond comes from borane. The product of hydroboration with BD_3 has deuterium bonded to the more substituted carbon; -D and -OH are *cis* to one another.

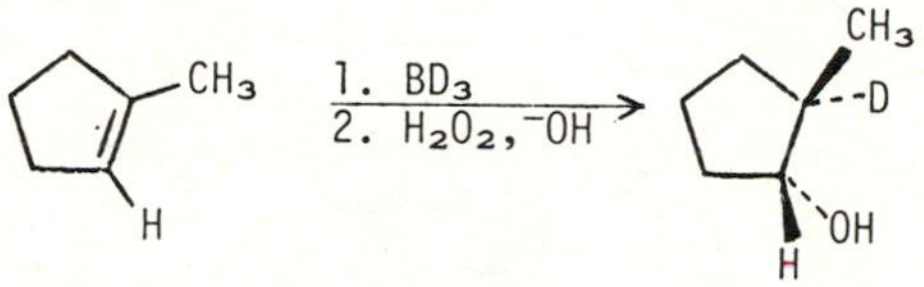

1-methylcyclopentene $\xrightarrow[2.\ H_2O_2,\ ^-OH]{1.\ BD_3}$ product with CH_3, -D, -OH and H (-D and -OH *cis*)

6.6

Energy

$\Delta G_1^{\ddagger}$ $\Delta G_2^{\ddagger}$ ΔG_2° ΔG_1°

$CH_3CH{=}C(CH_3)CH_3$

$CH_3\dot{C}HCBr(CH_3)CH_3$

$CH_3CHBr\dot{C}(CH_3)CH_3$

$CH_3CH_2CBr(CH_3)CH_3$

$CH_3CHBrCH(CH_3)CH_3$

Reaction Progress ⟶

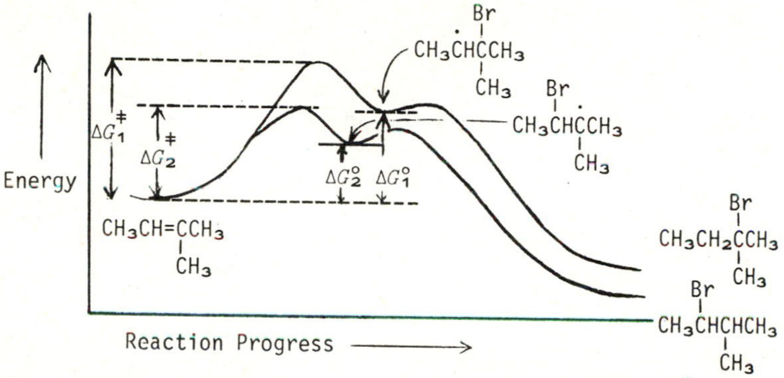

The reaction producing the more stable radical ($CH_3CHBr\dot{C}(CH_3)_2$) has a lower ΔG° for the first step. According to the Hammond Postulate, the more stable radical also forms faster, and $\Delta G^\ddagger$ for its formation is also lower.

6.7 a) $(CH_3)(CH_3CH_2)C{=}CH_2$ or $CH_3CH{=}C(CH_3)_2$ $\xrightarrow{HBr}$ $CH_3CH_2CBr(CH_3)_2$

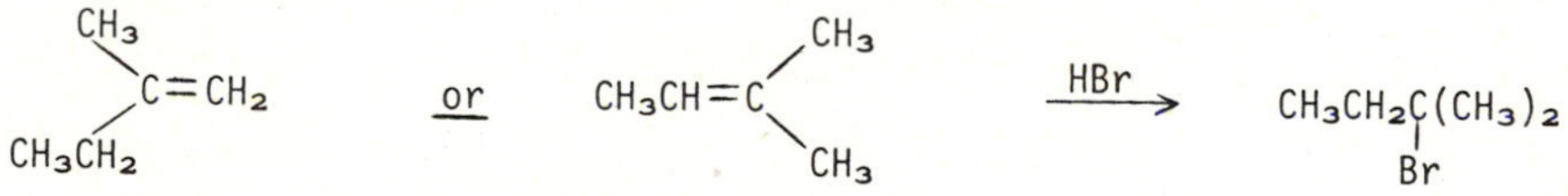

2-methyl-1-butene 2-methyl-2-butene

b) $CH_3CH_2CH{=}CH_2 \xrightarrow[\text{peroxides}]{HBr} CH_3CH_2CH_2CH_2Br$

1-butene

c) This product cannot be synthesized cleanly. Either hydroboration/oxidation or oxymercuaration of $CH_3CH{=}CHCH_2CH_3$ produces a mixture of two alcohols.

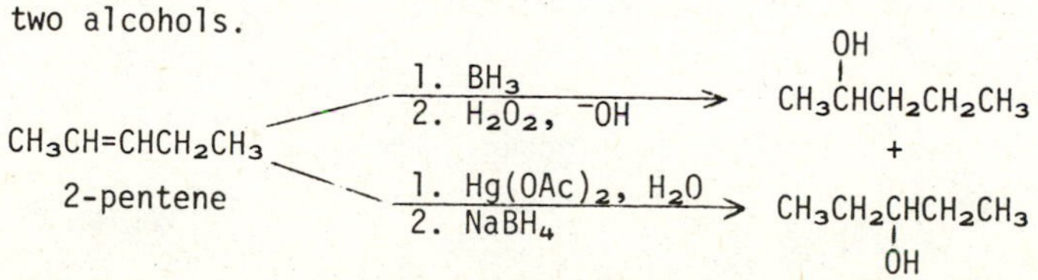

$CH_3CH{=}CHCH_2CH_3$ (2-pentene) $\xrightarrow[2.\ H_2O_2,\ ^-OH]{1.\ BH_3}$ or $\xrightarrow[2.\ NaBH_4]{1.\ Hg(OAc)_2,\ H_2O}$ $CH_3CH(OH)CH_2CH_2CH_3$ + $CH_3CH_2CH(OH)CH_2CH_3$

d) 3-methyl-3-hexene $\xrightarrow[\text{peroxides}]{\text{HBr}}$ $CH_3CH_2CH(Br)CH(CH_3)CH_2CH_3$

$$CH_3(CH_3CH_2)C{=}CHCH_2CH_3 \xrightarrow[\text{peroxides}]{HBr} CH_3CH_2\underset{Br}{CH}\overset{CH_3}{CH}CH_2CH_3$$

e)

$$CH_3CH{=}C(CH_3)CH_2CH_2CH_3 \quad or \quad (CH_3)(CH_3CH_2)C{=}CHCH_2CH_3 \quad or \quad CH_3CH_2CH_2C({=}CH_2)CH_2CH_3$$

3-methyl-2-hexene or 3-methyl-3-hexene or 2-ethyl-1-pentene

$$\downarrow \text{1. } Hg(OAc)_2,\ H_2O \quad \text{2. } NaBH_4$$

$$CH_3CH_2\overset{CH_3}{\underset{OH}{C}}CH_2CH_2CH_3$$

f) methylenecyclohexane ($=CH_2$) $\xrightarrow[\text{2. } H_2O_2,\ ^-OH]{\text{1. } BH_3}$ cyclohexyl–CH_2OH

6.8 a) 1-methylcyclohexene $\xrightarrow[\text{pyridine}]{OsO_4}$ [cyclic osmate ester, CH_3 and H cis on ring carbons, O_2Os bridging the two O] $\xrightarrow[H_2O]{NaHSO_3}$ cis-1-methylcyclohexane-1,2-diol (CH_3, OH, OH, H)

b) $CH_3CH_2CH{=}C(CH_3)_2$ $\xrightarrow[\text{2. } NaHSO_3,\ H_2O]{\text{1. } OsO_4,\ \text{pyridine}}$ $CH_3CH_2CH(OH)C(OH)(CH_3)_2$

2-methyl-2-pentene

c) $CH_2{=}CHCH{=}CH_2$ $\xrightarrow[\text{2. } NaHSO_3,\ H_2O]{\text{1. 2 } OsO_4,\ \text{pyridine}}$ $CH_2(OH)CH(OH)CH(OH)CH_2OH$

1,3-butadiene

Alternatively, aqueous alkaline $KMnO_4$ can be used to carry out hydroxylation of the above alkenes.

6.9

cis-2-butene (H, H / CH_3, CH_3 on C=C) $\xrightarrow[\text{pyridine}]{OsO_4}$ [cyclic osmate: O_2Os, O–C(H)(CH_3)–C(H)(CH_3)–O] $\xrightarrow[H_2O]{NaHSO_3}$ HO, OH; H–C–C–H; CH_3, CH_3

trans-2-butene (H, CH_3 / CH_3, H on C=C) $\xrightarrow[\text{pyridine}]{OsO_4}$ [cyclic osmate: O_2Os, O–C(H)(CH_3)–C(CH_3)(H)–O] $\xrightarrow[H_2O]{NaHSO_3}$ HO, OH; H–C–C–CH_3; CH_3, H

Formation of the cyclic osmate, which occurs with *syn* stereochemistry, retains the *cis-trans* stereochemistry of the double bond, since osmate formation is a single-step reaction. Treatment of the osmate ester with $NaHSO_3$ does not affect the stereochemistry of the carbon-oxygen bond. The diol produced from *cis*-2-butene is isomeric with the diol produced from *trans*-2-butene.

6.10 1,2-dimethylcyclohexene $\xrightarrow[\text{2. Zn, } H_3O^+]{\text{1. } O_3}$ $CH_3CO(CH_2)_4COCH_3$ a diketone

6.11 Make models of the *cis* and *trans* diols. Notice that it is much easier to form a five-membered cyclic periodate from the *cis* diol than from the *trans* diol. We therefore predict that the *cis* periodate intermediate will be of lower energy than the *trans* periodate intermediate because of the lack of strain in the *cis* periodate ring.

Remembering that any factor that lowers the energy of a transition state or intermediate should also lower $\Delta G^{\ddagger}$ and increase the rate of reaction, we predict that diol cleavage should proceed much more rapidly for *cis* diols than for *trans* diols.

6.12 a) $(CH_3)_2CBrCH_2CH_2CH_3$ $\xrightarrow[\Delta]{\text{DBN}}$ $(CH_3)_2C{=}CHCH_2CH_3$ + $CH_2{=}C(CH_3)CH_2CH_2CH_3$

2-bromo-2-methylpentane → 2-methyl-2-pentene (major) + 2-methyl-1-pentene (minor)

b) $CH_3CH_2CH(OH)CH(CH_3)_2$ $\xrightarrow[\text{pyridine}]{POCl_3}$ $CH_3CH_2CH{=}C(CH_3)_2$ + $CH_3CH{=}CHCH(CH_3)_2$

2-methyl-3-pentanol → 2-methyl-2-pentene (major) + 4-methyl-2-pentene (minor)

c) cyclohexyl–$CBr(CH_3)_2$ $\xrightarrow[\Delta]{\text{DBN}}$ cyclohexylidene$=C(CH_3)_2$ (major) + cyclohexyl–$C(=CH_2)CH_3$ (minor)

6.13 Try to solve this problem phrase by phrase.

1) $C_{10}H_{18}O$ has *two* double bonds and/or rings.
2) $C_{10}H_{18}O$ must be an alcohol because it undergoes reaction with H_2SO_4.
3) When $C_{10}H_{18}O$ is treated with dilute H_2SO_4, a mixture of alkenes of the formula $C_{10}H_{16}$ is produced.
4) Since the major alkene product B yields only one product, C_5H_8O, on ozonolysis, B and A contain two rings. A therefore has no double bonds.

A (1-cyclopentylcyclopentanol, OH) $\xrightarrow[H_2SO_4]{\text{dilute}}$ 1-cyclopentylcyclopentene + B (cyclopentylidenecyclopentane)

B $\xrightarrow[\text{2. Zn, } H_3O^+]{\text{1. } O_3}$ 2 cyclopentanone (=O)

6.14 a) $C_6H_5CH{=}CH_2 \xrightarrow{H_2/Pd} C_6H_5CH_2CH_3$

b) $C_6H_5CH{=}CH_2 \xrightarrow{Br_2} C_6H_5CHBrCH_2Br$

c) $C_6H_5CH{=}CH_2 \xrightarrow{HBr} C_6H_5CHBrCH_3$

d) $C_6H_5CH{=}CH_2 \xrightarrow[2.\ NaHSO_3]{1.\ OsO_4} C_6H_5CH(OH)CH_2OH$

e) $C_6H_5CH{=}CH_2 \xrightarrow{D_2/Pd} C_6H_5CHDCH_2D$

6.15 a)

$CH_3CH_2CH_2CH_2C(CH_3){=}CH_2$ 2-methyl-1-hexene

$CH_3CH_2CH_2CH{=}C(CH_3)_2$ 2-methyl-2-hexene

$CH_3CH_2CH{=}CHCH(CH_3)_2$ 2-methyl-3-hexene

$CH_3CH{=}CHCH_2CH(CH_3)_2$ 5-methyl-2-hexene

$CH_2{=}CHCH_2CH_2CH(CH_3)_2$ 5-methyl-1-hexene

$\xrightarrow{H_2/Pd}$ $CH_3CH_2CH_2CH_2CH(CH_3)_2$ 2-methylhexane

b)

3,3-dimethylcyclohexene

4,4-dimethylcyclohexene

$\xrightarrow{H_2/Pd}$ 1,1-dimethylcyclohexane

c) $CH_3CH{=}CHCH_2CH(CH_3)_2 \xrightarrow{Br_2/CCl_4} CH_3CHBrCHBrCH_2CH(CH_3)_2$

5-methyl-2-hexene → 2,3-dibromo-5-methylhexane

d) $CH_3CH_2CH_2CH{=}CH_2 \xrightarrow[2.\ NaBH_4]{1.\ Hg(OAc)_2,\ H_2O} CH_3CH_2CH_2CH(OH)CH_3$

1-pentene → 2-pentanol

e) $CH_3CH_2CH_2CH_2C(CH_3){=}CHCH_3 \xrightarrow[peroxides]{HBr} CH_3CH_2CH_2CH_2CH(CH_3)CHBrCH_3$

3-methyl-2-heptene → 2-bromo-3-methylheptane

f) $CH_3CH_2CH_2CH_2CH(CH_3)CH{=}CH_2 \xrightarrow[ether]{HCl} CH_3CH_2CH_2CH_2CH(CH_3)CHClCH_3$

3-methyl-1-heptene → 2-chloro-3-methylheptane

6.16 a) 1-methylcyclohexene $\xrightarrow[\text{2. Zn, } H_3O^+]{\text{1. } O_3}$ (keto aldehyde: CH_3, H)

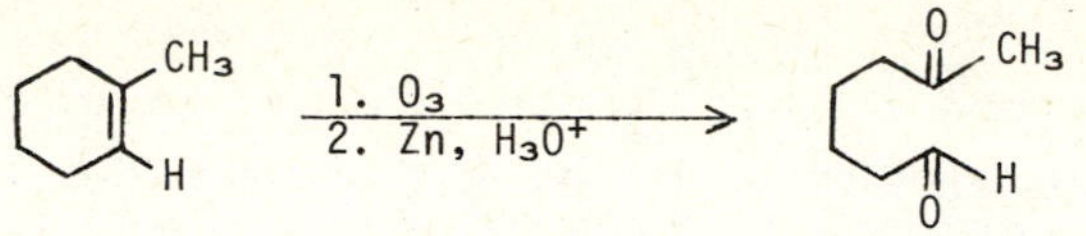

b) cyclohexene $\xrightarrow[H_3O^+]{KMnO_4}$ (CO_2H, CO_2H)

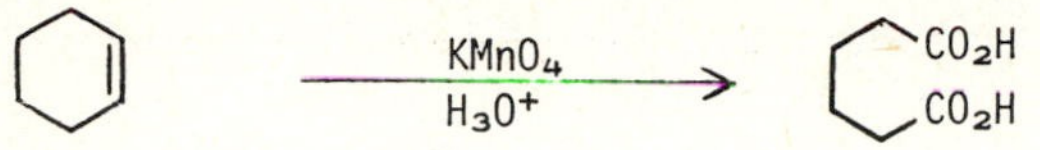

c) 1-methylcyclohexene $\xrightarrow[\text{2. } H_2O_2,\ ^-OH]{\text{1. } BH_3}$ (H, CH_3, H, OH)

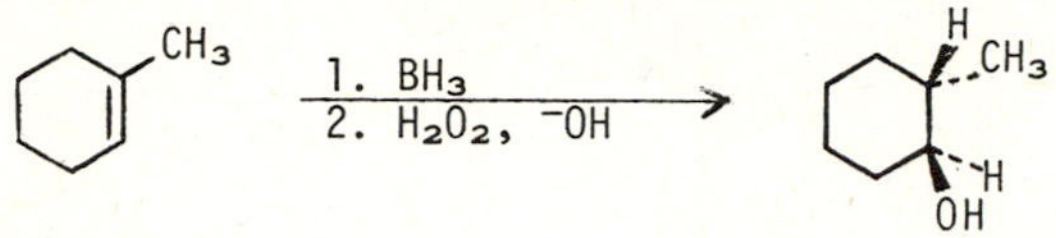

Remember that -H and -OH add *cis* across the double bond.

d) 1-methylcyclohexene $\xrightarrow[\text{2. } NaBH_4]{\text{1. } Hg(OAc)_2,\ H_2O}$ (OH, CH_3)

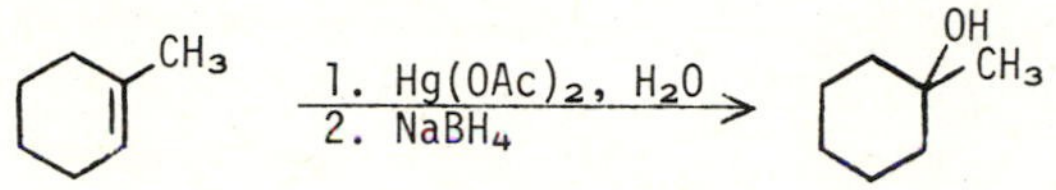

e) (CH_3, H, H, H) $\xrightarrow[\text{2. } NaHSO_3,\ H_2O]{\text{1. } OsO_4\text{, pyridine}}$ (CH_3, H, H, OH, OH, H) + (CH_3, H, OH, H, H, OH)

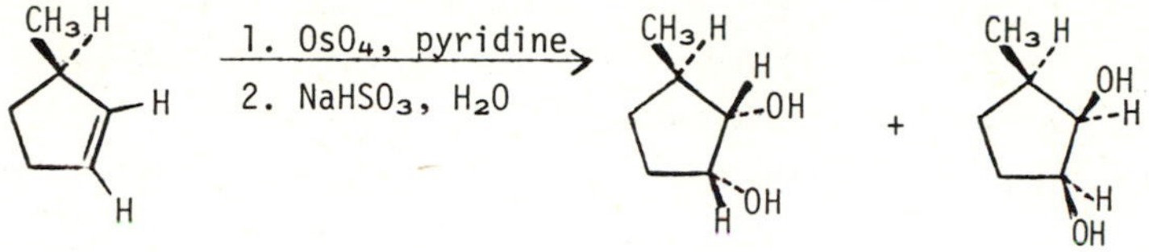

major minor

6.17 a) cyclopentene $\xrightarrow[\text{2. } NaHSO_3,\ H_2O]{\text{1. } OsO_4\text{, pyridine}}$ (H, OH, OH, H)

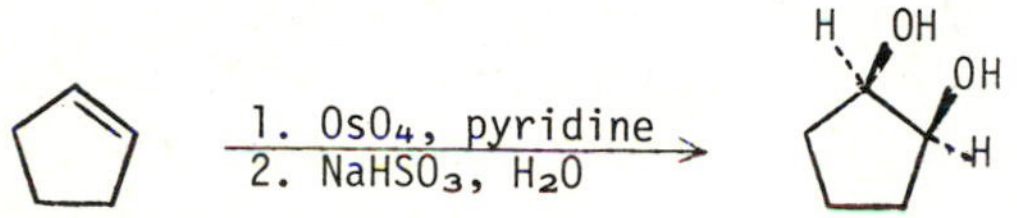

b) cyclopentene $\xrightarrow[\text{2. } NaBH_4]{\text{1. } Hg(OAc)_2,\ H_2O}$ (H, OH)

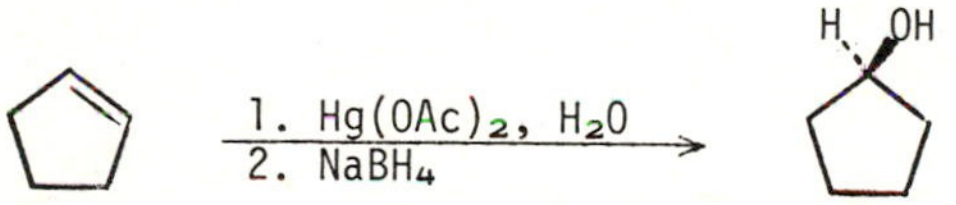

c) cyclopentene $\xrightarrow[\text{2. } NaBH_4]{\text{1. } Hg(OAc)_2,\ CH_3OH}$ (H, OCH_3)

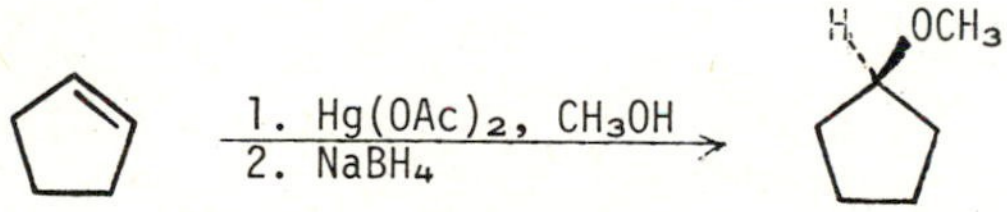

d) $CH_3CH{=}CHCH(CH_3)_2 \xrightarrow[\text{2. } NaBH_4,\ H_2O]{\text{1. } O_3} CH_3CH_2OH + (CH_3)_2CHCH_2OH$

e) $(CH_3)_2C{=}CH_2 \xrightarrow[\text{2. } H_2O_2,\ ^-OH]{\text{1. } BH_3} (CH_3)_2CHCH_2OH$

f) (CH_3, OH) $\xrightarrow[\text{pyridine, or dilute } H_2SO_4]{POCl_3 \text{ in}}$ (CH_3)

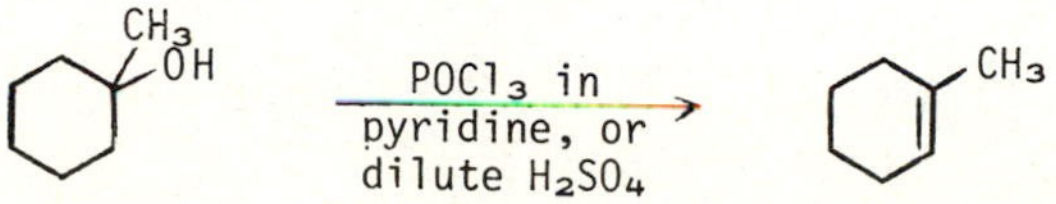

6.18 a) (H^+, $CHCH_3$) ⇌ [(+, CH_2CH_3, H)] $\underset{}{\overset{-H^+}{\rightleftharpoons}}$ (CH_2CH_3)

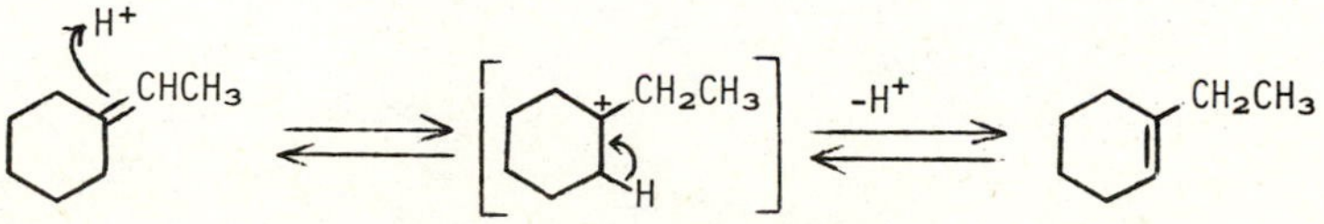

most stable carbocation

b) Protonation occurs to produce the most stable cation, which can then lose $-H^+$ to form either of two alkenes. Because 1-ethylcyclohexene is the major product of this equilibrium, it must be the more stable product.

6.19 Because ozonolysis gives only one product, we can assume that the alkene is symmetrical.

$$(H_3C)_2C{=}O \quad O{=}C(CH_3)_2 \xleftarrow[2.\ Zn,\ H_3O^+]{1.\ O_3} (H_3C)_2C{=}C(CH_3)_2$$

6.20 Do not get discouraged by the amount of information in this problem. Read slowly and interpret piece by piece. We know the following:

1) Hydrocarbon A (C_6H_{12}) has one double bond or ring.
2) Because A reacts with one equivalent of H_2, it has one double bond and no ring.
3) Compound A forms a diol when reacted with OsO_4.
4) When diols are oxidized with $KMnO_4$ they give either carboxylic acids or ketones, depending on the substitution pattern of the diol.
   a) A ketone is produced from what was originally a disubstituted carbon in the double bond.
   b) A carboxylic acid is produced from what was originally a monosubstituted carbon in the double bond.
5) One fragment from $KMnO_4$ oxidation is a carboxylic acid, CH_3CH_2COOH.
   a) This fragment was $CH_3CH_2CH{=}$ (a monosubstituted double bond) in compound A.
   b) It contains three of the six carbons of compound A.
6) a) The other fragment contains three carbons.
   b) It is a ketone.
   c) The only three carbon ketone is acetone, $O{=}C(CH_3)_2$.
   d) This fragment was $=C(CH_3)_2$ in compound A.
7) If we join the fragment in 5a with the one in 6d, we get:

$$CH_3CH_2CH{=}C(CH_3)_2 \qquad \underline{A} \qquad C_6H_{12}$$

The complete scheme:

$$CH_3CH_2CH_2CH(CH_3)_2 \xleftarrow{H_2} \underset{\underline{A}}{CH_3CH_2CH{=}C(CH_3)_2} \xrightarrow[2.\ NaHSO_3,\ H_2O]{1.\ OsO_4,\ pyridine} \underset{\underline{B}}{CH_3CH_2CH(OH){-}C(OH)(CH_3)CH_3}$$

$$\underline{B} \xrightarrow{KMnO_4,\ H^+} \underset{\underline{C}}{CH_3CH_2COOH + O{=}C(CH_3)_2}$$

6.21 The oxidative cleavage reaction of alkenes with O_3, followed by Zn in acid, produces aldehyde and ketone functional groups at sites where double bonds used to be. On ozonolysis, these two dienes yield only aldehydes because all double bonds are monosubstituted.

$$\text{cyclohexa-1,3-diene} \xrightarrow[\text{2. Zn, } H_3O^+]{\text{1. } O_3} OHCCH_2CH_2CHO + OHCCHO$$

Because the other diene is symmetrical, only one dialdehyde, $OCHCH_2CHO$, is produced.

6.22 a) If the hydrocarbon reacts with only one equivalent of hydrogen, it has only one double bond.

b) If only one type of aldehyde is produced on ozonolysis, the alkene must be symmetrical.

Putting these two facts together allows us to deduce that $CH_3CH_2CH_2CH_2CH{=}CHCH_2CH_2CH_2CH_3$ is the unknown hydrocarbon.

$$\underset{\text{5-decene}}{CH_3CH_2CH_2CH_2CH{=}CHCH_2CH_2CH_2CH_3} \xrightarrow[\text{2. Zn, } H_3O^+]{\text{1. } O_3} \underset{\text{pentanal}}{2\ CH_3CH_2CH_2CH_2\overset{O}{\overset{\|}{C}}H}$$

$$\text{5-decene} \xrightarrow{H_2/Pd} \underset{\text{decane}}{CH_3CH_2CH_2CH_2CH_2CH_2CH_2CH_2CH_2CH_3}$$

6.23 a)

$$\underset{\text{propene}}{CH_3CH{=}CH_2} \xrightarrow[\text{ether}]{HI} CH_3\overset{I}{\overset{|}{C}}HCH_3 \qquad \underset{\text{2-methylpropene}}{CH_3\underset{CH_3}{\underset{|}{C}}{=}CH_2} \xrightarrow[\text{ether}]{HI} CH_3\overset{I}{\overset{|}{\underset{CH_3}{\underset{|}{C}}}}CH_3$$

Addition of HI to 2-methyl propene occurs at a faster rate. The intermediate carbocation resulting from electrophilic addition of HI is more stable (tertiary) than the carbocation produced from propene (secondary).

b)

$$\text{cyclohexene} \xrightarrow[CCl_4]{Br_2} \textit{trans}\text{-1,2-dibromocyclohexane} \qquad \text{1-methylcyclohexene} \xrightarrow[CCl_4]{Br_2} \textit{trans}\text{-1,2-dibromo-1-methylcyclohexane}$$

Addition of Br_2 to 1-methylcyclohexene occurs at a faster rate for reasons similar to those given in part a.

c)

$$\underset{\text{ethylene}}{H_2C{=}CH_2} \xrightarrow[\text{ether}]{HBr} CH_3CH_2Br \qquad \underset{\text{vinyl bromide}}{H_2C{=}CHBr} \xrightarrow[\text{ether}]{HBr} CH_3CHBr_2$$

Addition of HBr to ethylene occurs at a faster rate. The electronegativity of bromine in vinyl bromide makes the double bond less electron-rich and less reactive toward electrophilic reagents.

d)

$$\underset{\text{propene}}{CH_3CH{=}CH_2} \xrightarrow[\text{ether}]{HCl} CH_3\overset{Cl}{\overset{|}{C}}HCH_3 \qquad \underset{\text{propenoic acid}}{CH_3CH{=}CHCOOH} \xrightarrow{HCl} CH_3CHClCH_2COOH$$

Addition of HCl to propene occurs at a faster rate for reasons similar to those given in part c.

6.24 a)

$$\underset{\text{2-butene}}{CH_3CH{=}CHCH_3} \xrightarrow{HBr} CH_3CH_2\overset{Br}{\overset{|}{C}}HCH_3$$

b)

$$\underset{\text{2-butene}}{3\ CH_3CH{=}CHCH_3} \xrightarrow[THF]{BH_3} (CH_3CH_2\overset{CH_3}{\overset{|}{C}}H)_3B \xrightarrow[^-OH]{H_2O_2} 3\ CH_3CH_2\overset{OH}{\overset{|}{C}}HCH_3$$

c)

$$\underset{\text{2-methylpropene}}{(CH_3)_2C{=}CH_2} \xrightarrow[\text{peroxides}]{HBr} (CH_3)_2CHCH_2Br$$

d)

$$\underset{\text{2-methyl-2-butene}}{CH_3CH{=}C(CH_3)_2} \xrightarrow[\text{peroxides}]{HI} CH_3CH_2\overset{I}{\overset{|}{C}}(CH_3)_2$$

Addition of HI always gives Markovnikov products.

6.25 What do we know about the hydrocarbon whose structure we want to determine?

a) It contains two double bonds (because it absorbs two equivalents of H_2);

b) It is symmetrical (because it only produces one kind of fragment);

c) A double bond is present at each end of the fragment (because a dialdehyde is produced). $OHCCH_2CH_2CHO$ must have come from

H\C(=)/CH₂CH₂\C(=)/H

d) If the hydrocarbon contains only two double bonds and yields only one type of ozonolysis product, the hydrocarbon must be 1,5-cyclooctadiene.

1,5-cyclooctadiene

6.26 a) This reaction is an alkene-forming dehydrohalogenation using a strong base. The product that will form in greatest abundance is the disubstituted alkene $CH_3CH{=}CHCH_3$.

b) Non-Markovnikov addition using peroxides succeeds only with HBr. Other hydrogen halides yield products of Markovnikov addition.

c) Hydroxylation of double bonds produces *cis*, not *trans*, diols.

d) Dehydration of a primary alcohol with H_2SO_4 occurs only under very severe conditions.

e) Ozone reacts with <u>both</u> double bonds of 1,4-cyclohexadiene.

f) Because hydroboration is a *cis* addition, the -H and the -OH added to the double bond must be *cis* to each other.

6.27

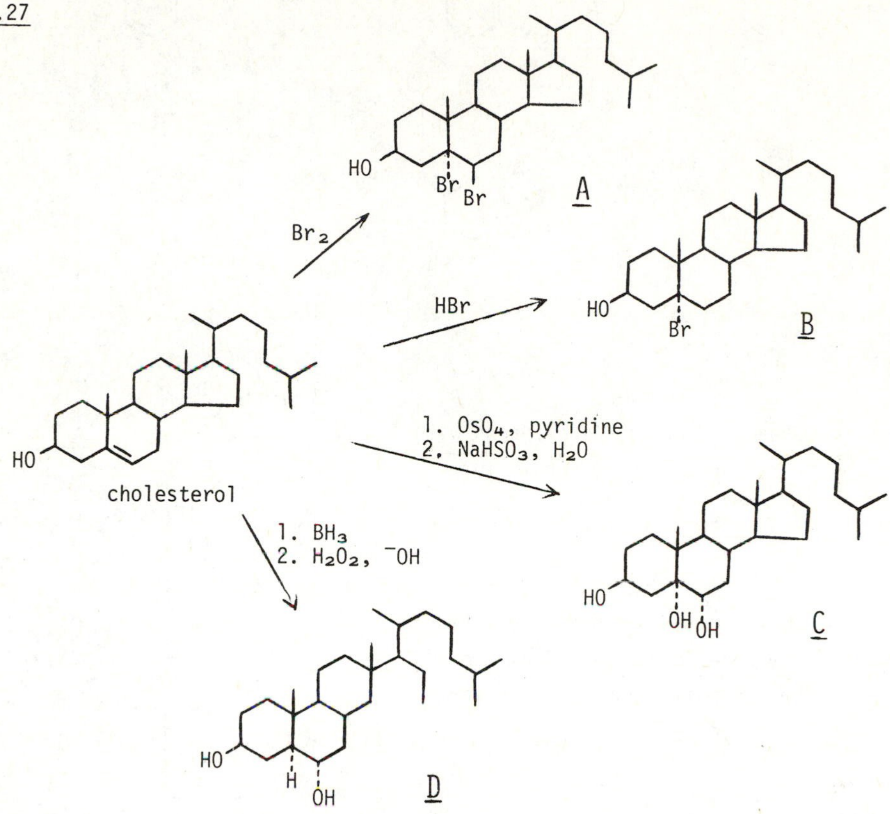

HO Br Br A

Br_2

HBr

HO Br B

HO

cholesterol

1. OsO_4, pyridine
2. $NaHSO_3$, H_2O

HO OH OH C

1. BH_3
2. H_2O_2, ^-OH

HO H OH D

6.28 Compound A has three degrees of unsaturation. Because compound A contains only one double bond, the other two degrees of unsaturation must be rings.

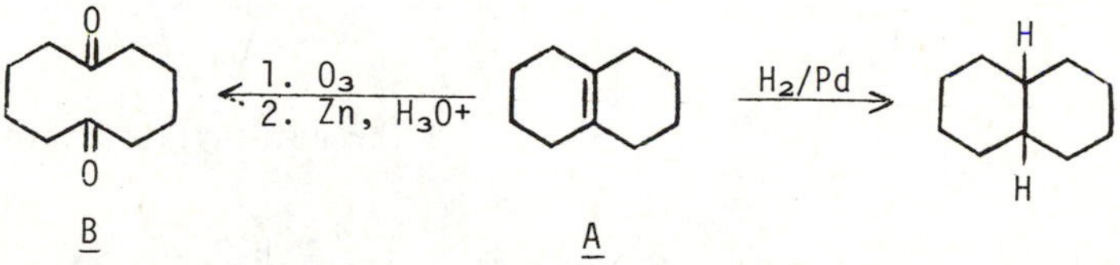

O O B

1. O_3
2. Zn, H_3O+

A

H_2/Pd

H H

6.29 C_8H_8 has five double bonds and/or rings. One of these double bonds reacts with H_2/Pd. Stronger conditions (H_2/Rh) cause the uptake of four equivalents of H_2. C_8H_8 thus contains four double bonds, three of which are in an aromatic ring, and one C=C double bond. A good guess for C_8H_8 at this point is:

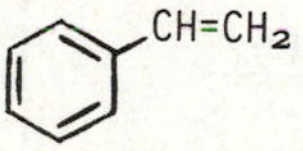

$CH=CH_2$

Reaction of a double bond with $KMnO_4$ yields cleavage products of the highest possible degree of oxidation. In this case, the products are $CO_2 + C_6H_5CO_2H$.

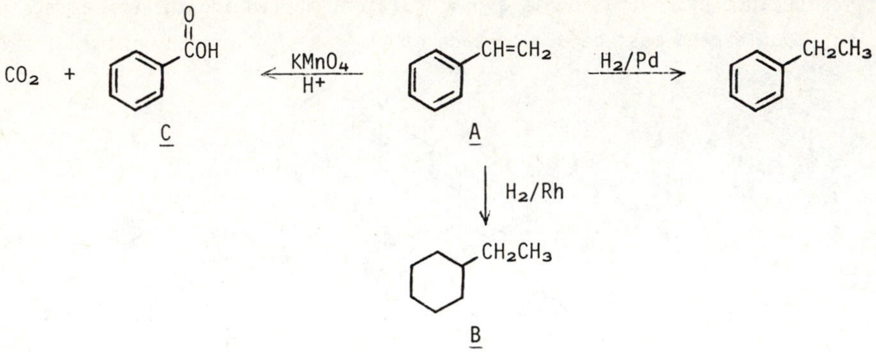

CO_2 + C (C_6H_5COOH) ← ($KMnO_4$ / H^+) — A ($C_6H_5CH{=}CH_2$) — (H_2/Pd) → $C_6H_5CH_2CH_3$

A — (H_2/Rh) → B (cyclohexyl–CH_2CH_3)

6.30

Reaction of the double bond with Br_2 forms a cyclic bromonium ion.

The bromonium ion can be attacked by an electron pair from the nucleophilic -OH group to form the cyclic bromo ether.

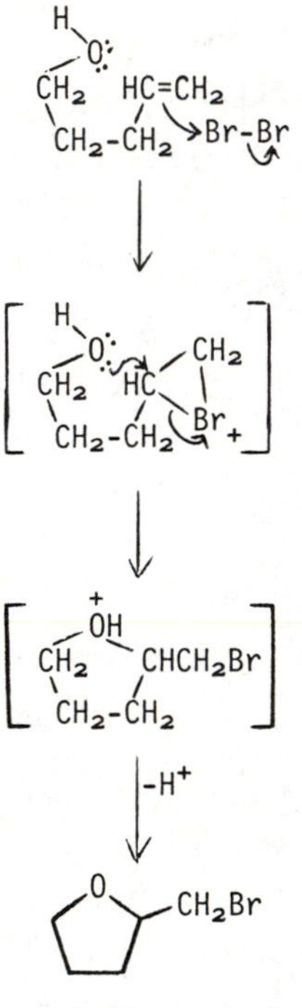

$-H^+$

2-(bromomethyl)tetrahydrofuran

6.31

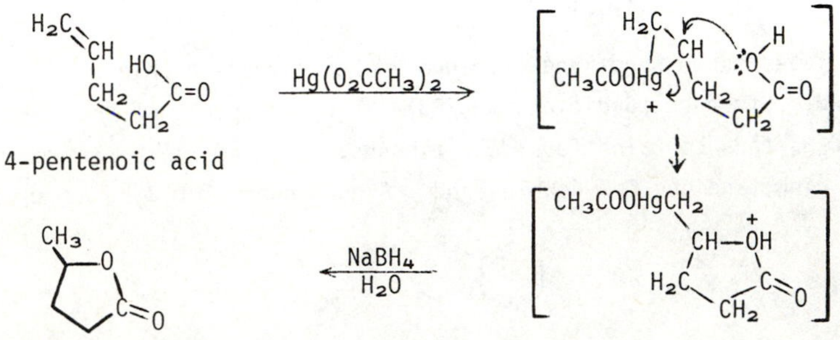

4-pentenoic acid — $Hg(O_2CCH_3)_2$ →

CH_3COOHg

$CH_3COOHgCH_2$

$NaBH_4$ / H_2O

4-methylbutanolide

Reaction with mercuric acetate forms a mercurinium ion intermediate. An electron pair from the nucleophilic carboxylic acid oxygen attacks the cyclic ion to form an organomercury intermediate. Reaction of the intermediate with $NaBH_4$ removes mercury.

6.32 α-Terpinene, $C_{10}H_{16}$, has three degrees of unsaturation -- two double bonds and one ring.

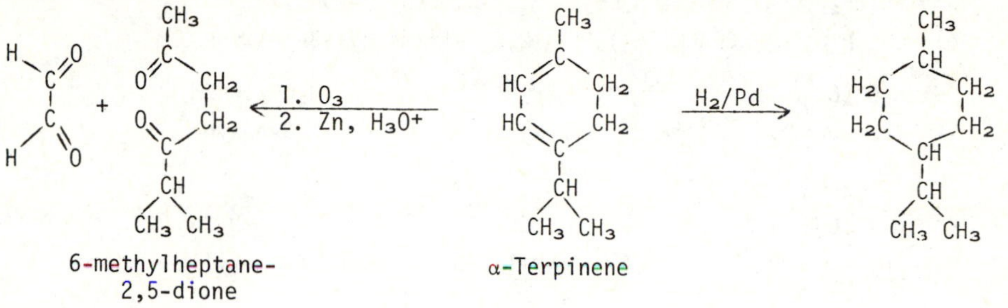

CH₃ structures: glyoxal (OHC–CHO) + 6-methylheptane-2,5-dione ← 1. O_3 2. Zn, H_3O^+ — α-Terpinene — H_2/Pd →

6-methylheptane-2,5-dione

α-Terpinene

What you should know:

After doing these problems you should be able to:

1. Explain, using mechanistic arguments, the reactions of alkenes and the products observed;
2. Predict the major product of alkene-forming reactions;
3. Know how to carry out transformations of alkenes;
4. Deduce the structure of alkenes, given the molecular formula and the products of cleavage.

CHAPTER 7 ALKYNES

7.1 a) $(CH_3)_2CHC{\equiv}CCH(CH_3)_2$ — 2,5-dimethyl-3-hexyne

b) $HC{\equiv}CC(CH_3)_3$ — 3,3-dimethyl-1-butyne

c) $CH_3CH{=}CHCH{=}CHC{\equiv}CCH_3$ — 2,4-octadien-6-yne (not 4,6-octadiene-2-yne)

d) $CH_3CH_2C(CH_3)_2C{\equiv}CCH_2CH_2CH_3$ — 3,3-dimethyl-4-octyne

e) $CH_3CH_2C(CH_3)_2C{\equiv}CCH(CH_3)_2$ — 2,5,5-trimethyl-3-heptyne

7.2 $CH_3CH_2CH_2CH_2C{\equiv}CH$ 1-hexyne; $CH_3CH_2CH_2C{\equiv}CCH_3$ 2-hexyne; $CH_3CH_2C{\equiv}CCH_2CH_3$ 3-hexyne

$CH_3CH_2CH(CH_3)C{\equiv}CH$ 3-methyl-1-pentyne; $CH_3CH(CH_3)CH_2C{\equiv}CH$ 4-methyl-1-pentyne; $CH_3CH(CH_3)C{\equiv}CCH_3$ 4-methyl-2-pentyne

$CH_3C(CH_3)_2C{\equiv}CH$ 3,3-dimethyl-1-butyne

7.3 Hydroboration of an alkyne is a *syn* addition that yields an intermediate trialkenylborane. When the trialkenylborane is treated with CH_3COOD, the boron-carbon bond is replaced by a deuterium-carbon bond with retention of stereochemistry.

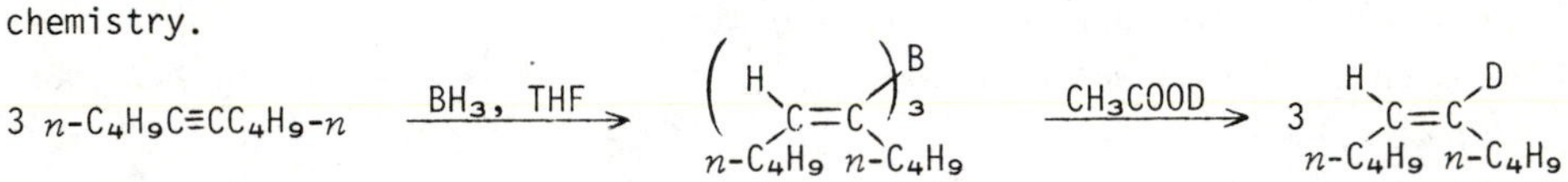

$$3\ n\text{-}C_4H_9C{\equiv}CC_4H_9\text{-}n \xrightarrow{BH_3,\ THF} \left(\begin{matrix} H & & B \\ & C{=}C & \\ n\text{-}C_4H_9 & & n\text{-}C_4H_9 \end{matrix}\right)_3 \xrightarrow{CH_3COOD} 3\ \begin{matrix} H & & D \\ & C{=}C & \\ n\text{-}C_4H_9 & & n\text{-}C_4H_9 \end{matrix}$$

7.4 Boron becomes attached to the less substituted carbon when BH_3 reacts with an alkene. Of the two alkenes that might be used to form disiamylborane ($(CH_3)_2C{=}CHCH_3$ or $(CH_3)_2CHCH{=}CH_2$), only the first yields the desired borane. Reaction stops after BH_3 adds to two equivalents of alkene because addition to a third equivalent is sterically hindered.

$$\underset{\text{2-methyl-2-butene}}{2\ (CH_3)_2C{=}CHCH_3} \xrightarrow{BH_3,\ THF} \underset{\text{disiamylborane}}{H\text{-}B\left[\begin{matrix}CH_3\\ |\\ CHCH(CH_3)_2\end{matrix}\right]_2}$$

7.5 Disiamylborane is prepared by addition of BH_3 to two equivalents of 2-methyl-2-butene (problem 7.4). When disiamylborane adds to an alkyne, an alkenylborane is produced. Oxidation of this alkenylborane with $H_2O_2/^-OH$ then yields a carbonyl compound, along with two equivalents of 3-methyl-2-butanol.

$$\left[(CH_3)_2CH\overset{CH_3}{\overset{|}{C}}H\right]_2BH + R\text{-}C{\equiv}CH \xrightarrow{THF} \left[(CH_3)_2CH\overset{CH_3}{\overset{|}{C}}H\right]_2B\text{-}CH{=}CHR$$

$$\downarrow H_2O_2/^{-}OH$$

$$R\text{-}CH_2CHO + 2\ (CH_3)_2CH\overset{OH}{\overset{|}{C}}HCH_3$$

3-methyl-2-butanol

7.6 $RC{\equiv}CH \xrightarrow[NH_3]{NaNH_2} [RC{\equiv}\bar{C}{:}\overset{+}{Na}] \xrightarrow[THF]{R'X} RC{\equiv}CR'$

Alkyne		R'X (X=Br or I)	Product
a) $CH_3CH_2CH_2C{\equiv}CH$		CH_3X	
	or		$CH_3CH_2CH_2C{\equiv}CCH_3$
$HC{\equiv}CCH_3$		$CH_3CH_2CH_2X$	2-hexyne
b) $(CH_3)_2CHC{\equiv}CH$		CH_3CH_2X	$(CH_3)_2CHC{\equiv}CCH_2CH_3$ 2-methyl-3-hexyne
c) cyclohexyl-$C{\equiv}CH$		CH_3X	cyclohexyl-$C{\equiv}CCH_3$
d) $(CH_3)_2CHCH_2C{\equiv}CH$		CH_3X	
	or		$(CH_3)_2CHCH_2C{\equiv}CCH_3$
$HC{\equiv}CCH_3$		$(CH_3)_2CHCH_2X$	5-methyl-2-hexyne
e) $HC{\equiv}CC(CH_3)_3$		CH_3CH_2X	$CH_3CH_2C{\equiv}CC(CH_3)_3$ 2,2-dimethyl-3-hexyne

Products (b), (c), and (e) can be synthesized by only one route because only primary halides can be used for acetylide alkylations.

7.7 $CH_3C{\equiv}CH \xrightarrow[2.\ CH_3Br,\ THF]{1.\ NaNH_2,\ NH_3} CH_3C{\equiv}CCH_3 \xrightarrow[\text{Lindlar catalyst}]{H_2}$ (H)(CH₃)C=C(H)(CH₃)

cis-2-butene

7.8 a) $HC{\equiv}CH \xrightarrow[2.\ CH_3CH_2CH_2Br,\ THF]{1.\ NaNH_2,\ NH_3} CH_3CH_2CH_2C{\equiv}CH \xrightarrow[THF]{2\ BuLi} [CH_3CH_2\ddot{\bar{C}}HC{\equiv}\bar{C}{:}]$

$$\downarrow CH_3Br$$

$$CH_3CH_2\overset{CH_3}{\overset{|}{C}}HC{\equiv}CH \xleftarrow{H_3O^+} \left[CH_3CH_2\overset{CH_3}{\overset{|}{C}}HC{\equiv}\bar{C}{:}\right]$$

3-methyl-1-pentyne

b) $HC{\equiv}CH \xrightarrow[2.\ CH_3CH_2Br,\ THF]{1.\ NaNH_2,\ NH_3} CH_3CH_2C{\equiv}CH \xrightarrow[THF]{2\ BuLi} [CH_3\ddot{\bar{C}}HC{\equiv}\bar{C}{:}]$

$$\downarrow CH_3Br$$

$$CH_3\overset{CH_3}{\overset{|}{C}}HC{\equiv}CCH_3 \xleftarrow{CH_3Br} \left[CH_3\overset{CH_3}{\overset{|}{C}}HC{\equiv}\bar{C}{:}\right]$$

4-methyl-2-pentyne

7.9 We know from Table 7.2 that the pK_a of ethylene is 44 and that the pK_a of methane is 49. The pK_a of butane is close to that of methane -- a reasonable estimate is 50. Thus, butane is an extremely weak acid, and its conjugate base, butyllithium, is an extremely strong base.

7.10 a) Phenylacetylene ($C_6H_5C{\equiv}CH$) $\xrightarrow[H_2O]{KMnO_4}$ Benzoic acid (C_6H_5COOH) + CO_2

b) $CH_3(CH_2)_7C{\equiv}C(CH_2)_7C{\equiv}C(CH_2)_7CH_3 \xrightarrow[H_2O]{KMnO_4} 2\ CH_3(CH_2)_7COOH + HOOC(CH_2)_7COOH$

c) $CH_3(CH_2)_5C{\equiv}C(CH_2)_5CH_3 \xrightarrow[H_2O]{KMnO_4} 2\ CH_3(CH_2)_5COOH$

7.11 The starting material is $CH_3CH_2CH_2C{\equiv}CCH_2CH_2CH_3$.

a) Either O_3 or $KMnO_4$ cleaves 4-octyne into two four-carbon fragments.

$$CH_3CH_2CH_2C{\equiv}CCH_2CH_2CH_3 \xrightarrow[H_2O]{KMnO_4} 2\ CH_3CH_2CH_2COOH$$

butanoic acid

b) To reduce a triple bond to a double bond with *cis* stereochemistry, use either H_2/Pd with Lindlar catalyst, or BH_3, followed by CH_3COOH.

$$CH_3CH_2CH_2C{\equiv}CCH_2CH_2CH_3 \xrightarrow[\text{Lindlar}]{H_2} (CH_3CH_2CH_2)(H)C{=}C(H)(CH_2CH_2CH_3)$$

cis-4-octene

c) Addition of HBr to *cis*-4-octene (part b) yields 4-bromooctane.

$$(CH_3CH_2CH_2)(H)C{=}C(H)(CH_2CH_2CH_3) \xrightarrow{HBr} CH_3CH_2CH_2CHBrCH_2CH_2CH_2CH_3$$

4-bromooctane

Alternatively, lithium/ammonia reduction of 4-octyne, followed by addition of HBr, gives 4-bromooctane.

d) Hydration of *cis*-4-octene (part b) yields 4-hydroxyoctane (4-octanol).

$$(CH_3CH_2CH_2)(H)C{=}C(H)(CH_2CH_2CH_3) \xrightarrow[2.\ NaBH_4]{1.\ Hg(OAc)_2,\ H_2O} CH_3CH_2CH_2CH(OH)CH_2CH_2CH_2CH_3$$

4-hydroxyoctane

e) Addition of Cl_2 to 4-octene (part b) yields 4,5-dichlorooctane.

$$(CH_3CH_2CH_2)(H)C{=}C(H)(CH_2CH_2CH_3) \xrightarrow{Cl_2} CH_3CH_2CH_2CHClCHClCH_2CH_2CH_3$$

4,5-dichlorooctane

7.12 The following syntheses are explained in detail in order to illustrate "retrosynthetic" logic -- the system of planning syntheses by working backwards.

a) 1) $CH_3CH_2CH_2CH_2CH_2CH_2CH_2CH_2CH_2CH_3$. An immediate precursor might be an alkene or alkyne. Try n-$C_8H_{17}C{\equiv}CH$, which can be reduced to decane by H_2/Pd.

2) The alkyne n-$C_8H_{17}C{\equiv}CH$ can be formed by alkylation of $\overset{+}{Na}{:}\overset{-}{C}{\equiv}CH$ by n-$C_8H_{17}Br$.

3) $\overset{+}{Na}{:}\overset{-}{C}{\equiv}CH$ can be formed by treatment of $HC{\equiv}CH$ with $NaNH_2$, NH_3.

The complete sequence:

$$HC{\equiv}CH \xrightarrow[NH_3]{NaNH_2} [\overset{+}{Na}:\overset{-}{C}{\equiv}CH] \xrightarrow[THF]{n\text{-}C_8H_{17}Br} n\text{-}C_8H_{17}C{\equiv}CH \xrightarrow{H_2/Pd} n\text{-}C_{10}H_{22}$$

b) 1) $CH_3CH_2CH_2C(CH_3)_2CH_2CH_3$. An immediate precursor is $CH_3CH_2CH_2C(CH_3)_2C{\equiv}CH$. This might be formed by alkylation of $:\overset{-}{C}{\equiv}CH$ with $CH_3CH_2CH_2C(CH_3)_2Br$, *but* acetylide alkylation only works with primary halides.

2) It *is* possible, however, to form $CH_3CH_2CH_2C(CH_3)_2C{\equiv}CH$ via

$$CH_3CH_2CH_2CH(CH_3)C{\equiv}CH \xrightarrow[\text{2. } CH_3I \text{ 3. } H_3O^+]{\text{1. 2 BuLi, THF}} CH_3CH_2CH_2C(CH_3)_2C{\equiv}CH.$$

This is an example of a dianion alkylation.

3) $CH_3CH_2CH_2CH_2(CH_3)C{\equiv}CH$ can be formed by another dianion alkylation.

$$CH_3CH_2CH_2CH_2CH_2C{\equiv}CH \xrightarrow[\text{2. } CH_3I \text{ 3. } H_3O^+]{\text{1. 2 BuLi, THF}} CH_3CH_2CH_2CH(CH_3)C{\equiv}CH$$

4) The alkyne necessary for step 3 can be synthesized by an acetylide alkylation.

$$HC{\equiv}CH \xrightarrow[NH_3]{NaNH_2} [\overset{+}{Na}:\overset{-}{C}{\equiv}CH] \xrightarrow[THF]{CH_3CH_2CH_2CH_2Br} CH_3CH_2CH_2CH_2C{\equiv}CH$$

The complete sequence:

$$HC{\equiv}CH \xrightarrow[NH_3]{NaNH_2} [\overset{+}{Na}:\overset{-}{C}{\equiv}CH] \xrightarrow[THF]{CH_3CH_2CH_2CH_2Br} CH_3CH_2CH_2CH_2C{\equiv}CH$$

$$\downarrow \text{1. 2 BuLi, THF; 2. } CH_3I\text{; 3. } H_3O^+$$

$$CH_3CH_2CH_2\overset{CH_3}{\overset{|}{C}}HC{\equiv}CH \xrightarrow[\text{2. } CH_3I \text{ 3. } H_3O^+]{\text{1. 2 BuLi, THF}} CH_3CH_2CH_2\underset{CH_3}{\underset{|}{\overset{CH_3}{\overset{|}{C}}}}C{\equiv}CH$$

$$\downarrow H_2/Pd$$

$$CH_3CH_2CH_2\underset{CH_3}{\underset{|}{\overset{CH_3}{\overset{|}{C}}}}CH_2CH_3$$

c) 1) $CH_3CH_2CH_2CH_2CH_2CHO$ can be made by treating $CH_3CH_2CH_2CH_2C{\equiv}CH$ with disiamylborane followed by H_2O_2, HO^-.

2) $CH_3CH_2CH_2CH_2C{\equiv}CH$ can be synthesized from $CH_3CH_2CH_2CH_2Br$ and $\overset{+}{Na}:\overset{-}{C}{\equiv}CH$.

The complete sequence:

$$HC{\equiv}CH \xrightarrow[NH_3]{NaNH_2} [\overset{+}{Na}:\overset{-}{C}{\equiv}CH] \xrightarrow[THF]{CH_3CH_2CH_2CH_2Br} CH_3CH_2CH_2CH_2C{\equiv}CH$$

$$\downarrow \text{1. } H\text{-}BR_2\text{; 2. } H_2O_2/^-OH$$

$$CH_3CH_2CH_2CH_2CH_2\overset{O}{\overset{\|}{C}}H$$

d) 1) $CH_3CH_2CH_2CH_2CH_2\overset{O}{\overset{\|}{C}}CH_3$. This ketone is the product of hydration of $CH_3CH_2CH_2CH_2CH_2C{\equiv}CH$ with H_2SO_4, H_3O, Hg^{2+}.

2) $CH_3CH_2CH_2CH_2CH_2Br + \overset{+}{Na}:\overset{-}{C}{\equiv}CH \xrightarrow{THF} CH_3CH_2CH_2CH_2CH_2C{\equiv}CH$

The complete sequence:

$$HC{\equiv}CH \xrightarrow[NH_3]{NaNH_2} [\overset{+}{Na}:\overset{-}{C}{\equiv}CH] \xrightarrow[THF]{CH_3CH_2CH_2CH_2CH_2Br} CH_3CH_2CH_2CH_2CH_2C{\equiv}CH \xrightarrow{H_2SO_4,\ H_2O,\ Hg^{2+}} CH_3CH_2CH_2CH_2CH_2\overset{O}{\overset{\|}{C}}CH_3$$

7.13 a) 2,2-dimethyl-3-hexyne
b) 2,5-octadiyne
c) 3,6-dimethyl-2-hepten-4-yne
d) 3,3-dimethyl-1,5-hexadiyne
e) 1,3-hexadien-5-yne
e) 3,6-diethyl-2-methyl-4-octyne

7.14 a)

```
            CH3
            |
CH3CH2CH2C≡CCCH2CH3
            |
            CH3
```

b)

```
         CH3
         |
CH3C≡CC≡CCHCH2CHC≡CH
               |
               CH2CH3
```

c) $(CH_3)_3CC{\equiv}CC(CH_3)_3$

d)

```
              CH3
             /
      CH2-C≡C-CH
     /          \
  CH2            CHCH3
     \           |
      CH2        CH2
         \      /
          CH2-CH2
```

e) $HC{\equiv}CCH{=}CHCH{=}CHCH_3$

f) $CH_3CH_2C{\equiv}CCH_2C(CH_3)_2CHClCH{=}CH_2$

7.15 a) $(CH_3)_3CCH_2CH_2C{\equiv}CCH_2CH_3$. correct name; 7,7-dimethyl-3-octyne
This compound is an *octyne*.

b) $HC{\equiv}CC(CH_3)_2CH_2CH_2CH(CH_3)_2$. correct name; 3,3,6-trimethyl-1-heptyne
Start numbering from the opposite end.

c) $CH_3CH{=}CHCH_2CH(CH_3)C{\equiv}CH$. correct name; 3-methyl-5-hepten-1-yne
Try not to break up the name of the compound more than necessary.

d)

```
                 CH3
                 |
        CH3      CHCH3
        |        |
HC≡CCH2CHCH2CH2CHCH3.
```

correct name; 4,7,8-trimethyl-1-nonyne
Choose the longest chain, and number from the opposite end.

e) $CH_3CH_2CH{=}CHC{\equiv}CH$. correct name; 3-hexen-1-yne

f) (cyclohexane ring bearing C≡CH and CH_3 substituents) correct name; 1-ethynyl-3-methylcyclohexane
The ring positions are numbered incorrectly.

7.16 a) $CH_3CH{=}CHC{\equiv}CC{\equiv}CCH{=}CHCH{=}CHCH{=}CH_2$. 1,3,5,11-tridecatetraen-7,9-diyne
Using *E-Z* notation: (3*E*,5*E*,11*E*)-1,3,5,11-tridecatetraen-7,9-diyne
The parent alkane of this hydrocarbon is tridecane.

b) $CH_3C{\equiv}CC{\equiv}CC{\equiv}CC{\equiv}CC{\equiv}CCH{=}CH_2$. 1-tridecen-3,5,7,9,11-pentayne
This hydrocarbon is also of the tridecane family.

7.17

B ← (H_2, Lindlar) — PhCH=CH–C≡H — (H_2/Pd) → A

B: Ph–CH=CH–CH=CH_2

A: Ph–$CH_2CH_2CH_2CH_3$

7.18 a) An acyclic alkane with eight carbons has the formula C_8H_{18}. C_8H_{10} has eight fewer hydrogens, or four fewer pairs of hydrogens, than C_8H_{18}. Thus, C_8H_{10} contains four degrees of unsaturation.

b) Because only one equivalent of H_2 is absorbed over the Lindlar catalyst, *one* triple bond is present.

c) Three equivalents of H_2 are absorbed when reduction is done over a palladium catalyst; two of them hydrogenate the triple bond already found to be present. Therefore, one *double* bond must also be present.

d) C_8H_{10} must contain one ring.

Many structures for C_8H_{10} are possible.

(cyclohexenyl)–C≡CH; dimethylcyclobutenyl–C≡CH (CH_3, CH_3, C≡CH); CH_2=CHCH(cyclopropyl)C≡CH

7.19

$CH_3CH_2CH_2CH_2C{\equiv}CH$

→ (1 equiv. HBr, CCl_4) → $CH_3CH_2CH_2CH_2C(Br){=}CH_2$ (a)

→ (1 equiv. Cl_2, CCl_4) → $CH_3CH_2CH_2CH_2$(Cl)C=C(Cl)H (b)

→ (H_2, Lindlar catalyst) → $CH_3CH_2CH_2CH_2CH{=}CH_2$ (c)

→ (1. $NaNH_2$, NH_3; 2. CH_3Br, THF) → $CH_3CH_2CH_2CH_2C{\equiv}CCH_3$ (d)

→ (H_2O, H_2SO_4, Hg^{2+}) → $CH_3CH_2CH_2CH_2C(=O)CH_3$ (e)

7.20 a) n-$C_4H_9C{\equiv}CC_4H_9$-n → (H_2, Lindlar) → (cis) n-C_4H_9(H)C=C(H)C_4H_9-n

b) → (Li/NH_3) → (trans) n-C_4H_9(H)C=C(H)C_4H_9-n

c) $n\text{-}C_4H_9C{\equiv}CC_4H_9\text{-}n \xrightarrow[CCl_4]{\text{1 equiv. }Br_2}$ $(n\text{-}C_4H_9)(Br)C{=}C(Br)(C_4H_9\text{-}n)$

n-C₄H₉ and Br on one carbon, Br and C₄H₉-n on the other (trans Br):

$$\begin{matrix} n\text{-}C_4H_9 & & Br \\ & C{=}C & \\ Br & & C_4H_9\text{-}n \end{matrix}$$

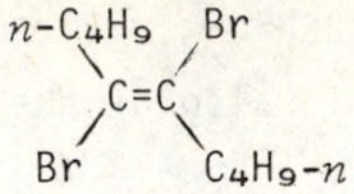

d) $\xrightarrow[2.\ H_2O_2,\ ^-OH]{1.\ BH_3,\ THF}$ $n\text{-}C_4H_9\overset{O}{\overset{\|}{C}}CH_2C_4H_9\text{-}n$

e) $\xrightarrow[2.\ CH_3COOH]{1.\ BH_3,\ THF}$

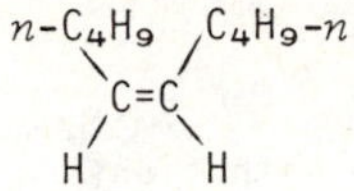

$$\begin{matrix} n\text{-}C_4H_9 & & C_4H_9\text{-}n \\ & C{=}C & \\ H & & H \end{matrix}$$

f) $\xrightarrow[2.\ CH_3COOD]{1.\ BH_3,\ THF}$

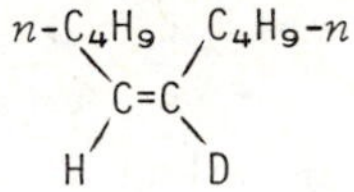

$$\begin{matrix} n\text{-}C_4H_9 & & C_4H_9\text{-}n \\ & C{=}C & \\ H & & D \end{matrix}$$

g) $\xrightarrow[2.\ CH_3COOH]{1.\ BD_3,\ THF}$

$$\begin{matrix} n\text{-}C_4H_9 & & C_4H_9\text{-}n \\ & C{=}C & \\ H & & D \end{matrix}$$

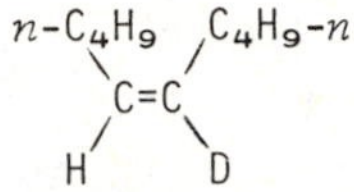

h) $\xrightarrow[Hg^{2+}]{H_2O,\ H_2SO_4}$ $n\text{-}C_4H_9\overset{O}{\overset{\|}{C}}CH_2C_4H_9\text{-}n$

7.21 a) $CH_3C{\equiv}CC_3H_7\text{-}n \xrightarrow[CCl_4]{\text{2 equiv. }Br_2} CH_3CBr_2CBr_2C_3H_7\text{-}n$

b) $\xrightarrow[2.\ CH_3COOH]{1.\ BH_3,\ THF}$

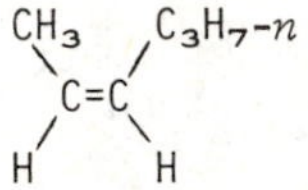

$$\begin{matrix} CH_3 & & C_3H_7\text{-}n \\ & C{=}C & \\ H & & H \end{matrix}$$

c) $\xrightarrow[2.\ H_2O_2,\ ^-OH]{1.\ BH_3,\ THF}$ $CH_3\overset{O}{\overset{\|}{C}}CH_2C_3H_7\text{-}n \ + \ CH_3CH_2\overset{O}{\overset{\|}{C}}C_3H_7\text{-}n$

d) $\xrightarrow[ether]{HBr}$

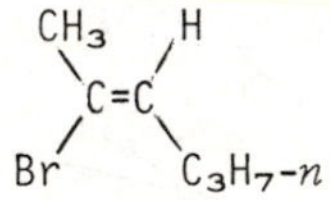

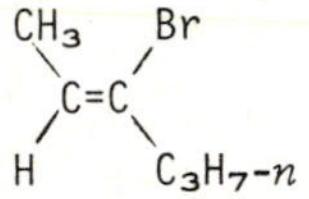

$$\begin{matrix} CH_3 & & H \\ & C{=}C & \\ Br & & C_3H_7\text{-}n \end{matrix} \quad + \quad \begin{matrix} CH_3 & & Br \\ & C{=}C & \\ H & & C_3H_7\text{-}n \end{matrix}$$

e) $\xrightarrow{Li/NH_3}$

$$\begin{matrix} CH_3 & & H \\ & C{=}C & \\ H & & C_3H_7\text{-}n \end{matrix}$$

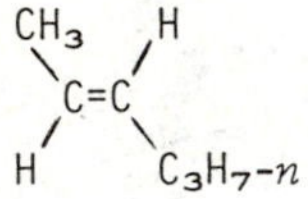

f) $\xrightarrow[Hg^{2+}]{H_2O,\ H_2SO_4}$ $CH_3\overset{O}{\overset{\|}{C}}CH_2C_3H_7\text{-}n \ + \ CH_3CH_2\overset{O}{\overset{\|}{C}}C_3H_7\text{-}n$

7.22 sp^3 sp

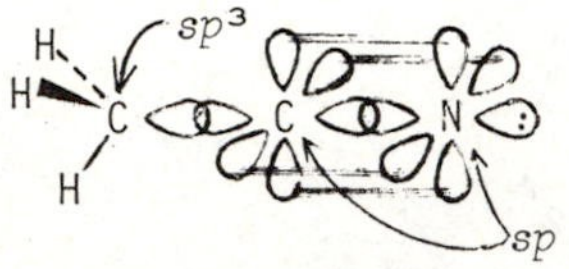

Each of the two pi bonds between carbon and nitrogen is formed by overlap of one *p* orbital of carbon with one *p* orbital of nitrogen.

7.23 a) $CH_3CH_2C{\equiv}CH \xrightarrow[\substack{2.\ CH_3I \\ 3.\ H_3O^+}]{1.\ 2\ BuLi,\ THF} CH_3\overset{CH_3}{\overset{|}{C}}HC{\equiv}CH$

b) $CH_3CH_2C{\equiv}CH \xrightarrow[Hg^{2+}]{H_2O,\ H_2SO_4} CH_3CH_2\overset{O}{\overset{\|}{C}}CH_3$

c) $CH_3CH_2C{\equiv}CH \xrightarrow[\text{2. } H_2O_2,\ ^-OH]{\text{1. Disiamylborane}} CH_3CH_2CH_2CHO$

d) $C_6H_5C{\equiv}CH \xrightarrow[\text{2. } CH_3I,\ THF]{\text{1. } NaNH_2,\ NH_3} C_6H_5C{\equiv}CCH_3$

e) $C_6H_5C{\equiv}CCH_3 \xrightarrow[\text{Lindlar}]{H_2} C_6H_5(H)C{=}C(H)CH_3$

f) $CH_3CH_2C{\equiv}CH \xrightarrow[H_2O]{KMnO_4} CH_3CH_2COOH + CO_2$

g) $CH_3CH_2CH_2CH_2CH{=}CH_2 \xrightarrow[CCl_4]{Br_2} CH_3CH_2CH_2CH_2CHBrCH_2Br$

$\downarrow$ 1. 2 $NaNH_2$; 2. H_3O+

$CH_3CH_2CH_2CH_2C{\equiv}CH$

7.24 a) *trans*-5-decene (n-C_4H_9(H)C=C(H)C_4H_9-n) $\xrightarrow[CCl_4]{Br_2}$ n-$C_4H_9CHBrCHBrC_4H_9$-n $\xrightarrow[\text{2. } H_3O+]{\text{1. 2 } NaNH_2}$ n-$C_4H_9C{\equiv}CC_4H_9$-n $\xrightarrow[\text{Lindlar}]{H_2}$ *cis*-5-decene (n-C_4H_9, C_4H_9-n on same side; H, H)

b) *cis*-5-decene (n-C_4H_9, C_4H_9-n on same side; H, H) $\xrightarrow[CCl_4]{Br_2}$ n-$C_4H_9CHBrCHBrC_4H_9$-n $\xrightarrow[\text{2. } H_3O+]{\text{1. 2 } NaNH_2}$ n-$C_4H_9C{\equiv}CC_4H_9$-n $\xrightarrow{Li/NH_3}$ *trans*-5-decene (n-C_4H_9(H)C=C(H)C_4H_9-n)

7.25 Both $KMnO_4$ and O_3 oxidation of alkynes yield carboxylic acids; terminal alkynes give CO_2, also.

a) $CH_3(CH_2)_5C{\equiv}CH \xrightarrow[H_2O]{KMnO_4} CH_3(CH_2)_5COOH + CO_2$

b) $C_6H_5C{\equiv}CCH_3 \xrightarrow[H_2O]{KMnO_4} C_6H_5COOH + CH_3COOH$

c) In this case, the existence of only one product, a diacid, indicates the presence of a ring.

cyclodecyne $\xrightarrow[H_2O]{KMnO_4}$ $HOOC(CH_3)_8COOH$

d) Notice that the products of this ozonolysis contain aldehyde and ketone functional groups, as well as a carboxylic acid and CO_2. The parent hydrocarbon must therefore contain a double and a triple bond.

$$CH_3CH{=}\overset{CH_3}{\overset{|}{C}}CH_2CH_2C{\equiv}CH \xrightarrow[\text{2. Zn, } H_3O^+]{\text{1. } O_3} CH_3CHO + CH_3\overset{O}{\overset{\|}{C}}CH_2CH_2COOH + CO_2$$

e) (1-ethynylcyclohexene, ring with C≡CH substituent)

$$\xrightarrow[\text{2. Zn, } H_3O^+]{\text{1. } O_3} OHCCH_2CH_2CH_2CH_2\overset{O}{\overset{\|}{C}}COOH + CO_2$$

7.26 a)

$$HC{\equiv}CH \xrightarrow[\text{2. } BrCH_2(CH_2)_6CH_2Br\text{, THF}]{\text{1. 2 } NaNH_2\text{, } NH_3} \text{cyclodecyne} \xrightarrow[\text{Lindlar}]{H_2} \text{cis-cyclodecene}$$

(cyclodecyne drawn as CH_2–CH_2–CH_2–C≡C–CH_2–CH_2–CH_2–CH_2–CH_2 ring; product ring with H, H on the double bond)

b)

$$CH_3CH_2C{\equiv}CH \xrightarrow[\substack{\text{2. } CH_3Br \\ \text{3. } H_3O^+}]{\text{1. 2 BuLi, THF}} CH_3\overset{CH_3}{\overset{|}{C}}HC{\equiv}CH \xrightarrow[\substack{\text{2. } CH_3Br \\ \text{3. } H_3O^+}]{\text{1. 2 BuLi, THF}} (CH_3)_3CC{\equiv}CH$$

$$(CH_3)_3CC{\equiv}CH \xrightarrow[\text{2. } CH_3CH_2Br\text{, THF}]{\text{1. } NaNH_2\text{, } NH_3} (CH_3)_3CC{\equiv}CCH_2CH_3$$

c)

$$C_6H_5\overset{H}{\overset{|}{C}}{=}CH_2 \xrightarrow[CCl_4]{Br_2} C_6H_5CHBrCH_2Br \xrightarrow[\text{2. } H_3O^+]{\text{1. 2 } NaNH_2} C_6H_5C{\equiv}CH$$

A

$$C_6H_5\overset{H}{\overset{|}{C}}{=}CH_2 \xrightarrow[\text{peroxides}]{HBr} C_6H_5CH_2CH_2Br$$

B

$$\underset{A}{C_6H_5C{\equiv}CH} \xrightarrow[\text{2. product B, THF}]{\text{1. } NaNH_2\text{, } NH_3} C_6H_5C{\equiv}CCH_2CH_2C_6H_5$$

$$\xrightarrow{H_2\text{, Pd/C}} C_6H_5CH_2CH_2CH_2CH_2C_6H_5$$

7.27

$$CH_3\overset{O}{\overset{\|}{C}}CH_3 \xrightarrow[\text{2. } H_3O^+]{\text{1. } \overset{+}{Na}\,\overset{-}{:}C{\equiv}CH} CH_3\overset{OH}{\overset{|}{C}}CH_3(C{\equiv}CH) \xrightarrow{H_3O^+} CH_2{=}\overset{CH_3}{\overset{|}{C}}C{\equiv}CH \xrightarrow[\text{Lindlar}]{H_2} CH_2{=}\overset{CH_3}{\overset{|}{C}}CH{=}CH_2$$

2-methyl-1,3-butadiene

7.28 a) $CH_3CH_2C{\equiv}CH \xrightarrow{2\ Cl_2} CH_3CH_2CCl_2CHCl_2$

1,1,2,2-tetrachlorobutane

b) $CH_3CH_2C{\equiv}CH \xrightarrow[\text{Lindlar}]{H_2} CH_3CH_2CH{=}CH_2 \xrightarrow[\text{peroxides}]{HBr} CH_3CH_2CH_2CH_2Br$

$CH_3CH_2C{\equiv}CH \xrightarrow[\text{2. } CH_3CH_2CH_2CH_2Br,\ THF]{\text{1. } NaNH_2,\ NH_3} CH_3CH_2C{\equiv}CCH_2CH_2CH_2CH_3 \xrightarrow[Pd/C]{H_2}$ octane

c) $CH_3CH_2C{\equiv}CH \xrightarrow[\text{2. } H_2O_2,\ ^-OH]{\text{1. Disiamylborane}} CH_3CH_2CH_2CHO$

butanal

7.29 a) $HC{\equiv}CH \xrightarrow[NH_3]{NaNH_2} [\overset{+}{Na}:\overset{-}{C}{\equiv}CH] \xrightarrow[THF]{CH_3CH_2CH_2Br} CH_3CH_2CH_2C{\equiv}CH$

b) $HC{\equiv}CH \xrightarrow[NH_3]{NaNH_2} [\overset{+}{Na}:\overset{-}{C}{\equiv}CH] \xrightarrow[THF]{CH_3CH_2Br} CH_3CH_2C{\equiv}CH$

$\downarrow NaNH_2,\ NH_3$

$CH_3CH_2C{\equiv}CCH_2CH_3 \xleftarrow[THF]{CH_3CH_2Br} [CH_3CH_2C{\equiv}\overset{-}{C}:\overset{+}{Na}]$

c) Product from a) $\xrightarrow[\text{3. } H_3O^+]{\text{1. 2 BuLi, THF; 2. } CH_3CH_2Br}$ $CH_3CH_2\overset{CH_2CH_3}{\overset{|}{C}}HC{\equiv}CH$

d) Product from a) $\xrightarrow[Hg^{2+}]{H_2SO_4,\ H_2O}$ $CH_3CH_2CH_2\overset{O}{\overset{\|}{C}}CH_3$

e) $HC{\equiv}CH \xrightarrow[NH_3]{NaNH_2} [\overset{+}{Na}:\overset{-}{C}{\equiv}CH] \xrightarrow[THF]{CH_3CH_2Br} CH_3CH_2C{\equiv}CH \xrightarrow[\text{3. } H_3O^+]{\text{1. 2 BuLi; 2. } CH_3I}$ $CH_3\overset{CH_3}{\overset{|}{C}}HC{\equiv}CH$

$\downarrow H_2$, Lindlar

$CH_3\overset{CH_3}{\overset{|}{C}}HCH{=}CH_2$

f) Product from a) $\xrightarrow[NH_3]{NaNH_2} [\overset{+}{Na}:\overset{-}{C}{\equiv}CCH_2CH_2CH_3]$

$\downarrow CH_3CH_2CH_2Br$, THF

$CH_3CH_2CH_2C{\equiv}CCH_2CH_2CH_3 \xrightarrow[Hg^{2+}]{H_2SO_4,\ H_2O}$ $CH_3CH_2CH_2\overset{O}{\overset{\|}{C}}CH_2CH_2CH_2CH_3$

g) $HC{\equiv}CH \xrightarrow[NH_3]{NaNH_2} [\overset{+}{Na}:\overset{-}{C}{\equiv}CH] \xrightarrow[THF]{CH_3CH_2CH_2CH_2Br} CH_3CH_2CH_2CH_2C{\equiv}CH$

$\downarrow$ 1. Disiamylborane 2. $H_2O_2,\ ^-OH$

$CH_3CH_2CH_2CH_2CH_2CHO$

7.30 a) $CH_3CH_2C{\equiv}CCH_2CH_3 \xrightarrow[\text{2. } CH_3COOD]{\text{1. } BH_3}$ (H)(CH₃CH₂)C=C(D)(CH₂CH₃)

or $\xrightarrow[\text{2. } CH_3COOH]{\text{1. } BD_3}$

b)

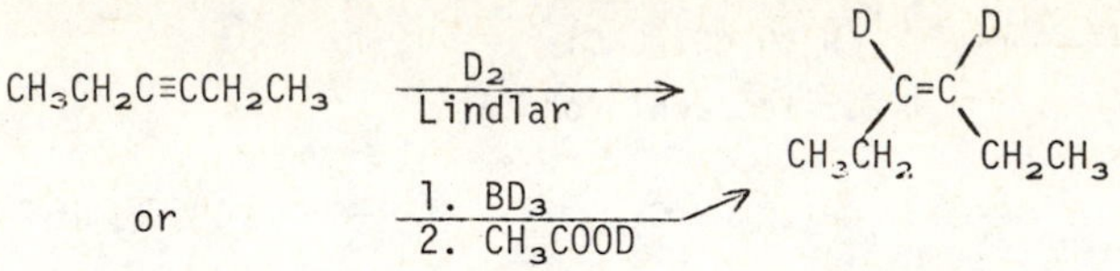

$$CH_3CH_2C{\equiv}CCH_2CH_3 \xrightarrow[\text{Lindlar}]{D_2} \text{(cis) } (CH_3CH_2)DC{=}CD(CH_2CH_3)$$

or $\xrightarrow[2.\ CH_3COOD]{1.\ BD_3}$

c)

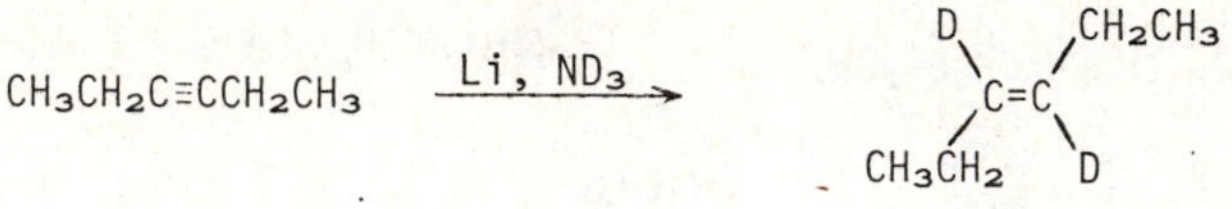

$$CH_3CH_2C{\equiv}CCH_2CH_3 \xrightarrow{Li,\ ND_3} \text{(trans) } (CH_3CH_2)DC{=}CD(CH_2CH_3)$$

d) $$CH_3CH_2CH_2C{\equiv}CH \xrightarrow[NH_3]{NaNH_2} [CH_3CH_2CH_2C{\equiv}\bar{C}{:}\overset{+}{Na}] \xrightarrow{D_3O^+} CH_3CH_2CH_2C{\equiv}CD$$

e) $$C_6H_5C{\equiv}CH \xrightarrow[NH_3]{NaNH_2} [C_6H_5C{\equiv}\bar{C}{:}\overset{+}{Na}] \xrightarrow{D_3O^+} C_6H_5C{\equiv}CD \xrightarrow[\text{Lindlar}]{D_2} C_6H_5CD{=}CD_2$$

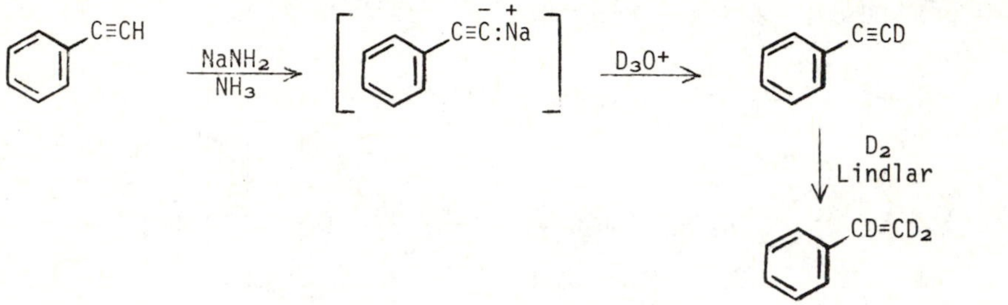

7.31 a) $$CH_3(CH_2)_4C{\equiv}CH \xrightarrow[2.\ H_2O_2,\ ^-OH]{1.\ \text{Disiamylborane}} CH_3(CH_2)_5CHO$$

b) $$C_6H_5C(Br){=}CH_2 \xrightarrow[2.\ H_3O^+]{1.\ NaNH_2} C_6H_5C{\equiv}CH$$

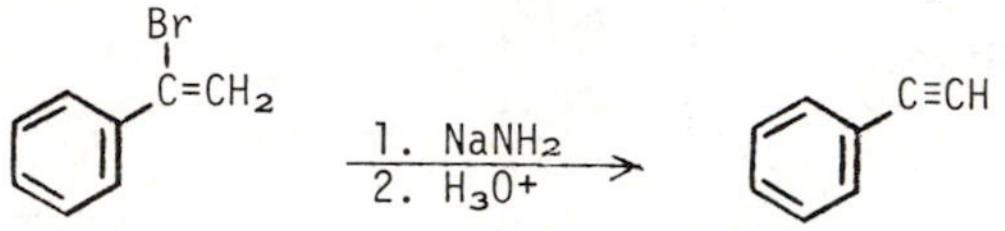

c) $$C_6H_5CBr_2CH_3 \xrightarrow[2.\ H_3O^+]{1.\ 2\ NaNH_2} C_6H_5C{\equiv}CH$$

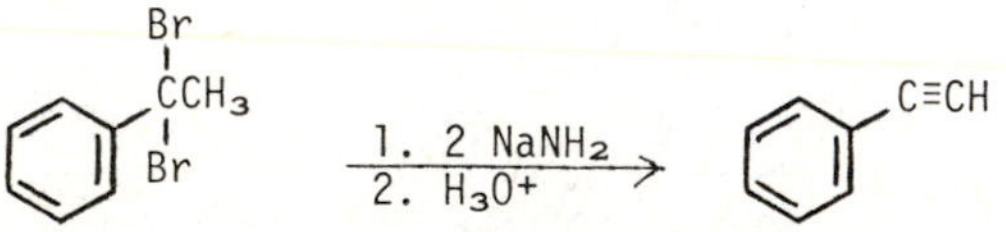

7.32

R1, R2, C, C, C, C, R3, R4

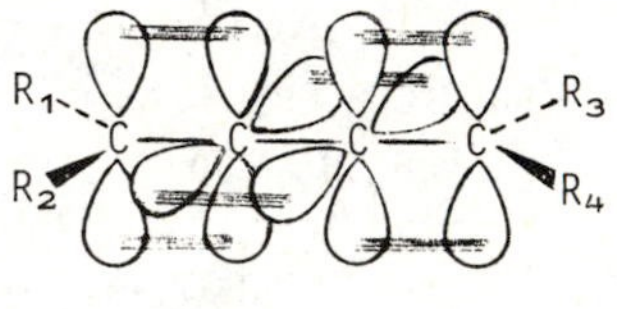

The simplest cumulene is pictured above. The carbons at the end of the cumulated double bonds are sp^2 hybridized and form one pi bond to the "interior" carbons. The "interior" carbons are sp hybridized; each carbon forms two pi bonds -- one to an "exterior" carbon and one to the other "interior" carbon. If you build a model of this cumulene, you can see that the substituents all lie in the same plane. This cumulene can thus exhibit *cis-trans* isomerism, just as simple alkenes can.

In general, the substituents at the ends of any cumulene having an odd number of adjacent double bonds lie in a plane; these cumulenes can exhibit *cis-trans* isomerism. The relationahip of substituents at the ends of any even-numbered cumulene is more complex and will be explained in problem 8.44.

7.33 Muscalure is a C_{30}(!) alkene. The only functional group present is the double bond between C_9 and C_{10}. Since our synthesis begins with acetylene, we can assume that the double bond can be produced by hydrogenation of a triple bond.

$$HC{\equiv}CH \xrightarrow[NH_3]{NaNH_2} [\overset{+}{Na}:\overset{-}{C}{\equiv}CH] \xrightarrow[THF]{CH_3(CH_2)_6CH_2Br} CH_3(CH_2)_7C{\equiv}CH$$

$$\xrightarrow{NaNH_2,\ NH_3} [CH_3(CH_2)_7C{\equiv}\overset{-}{C}:\overset{+}{Na}] \xrightarrow[THF]{CH_3(CH_2)_{18}CH_2Br} CH_3(CH_2)_7C{\equiv}C(CH_2)_{19}CH_3$$

$$\xrightarrow{H_2,\ Lindlar}$$

$CH_3(CH_2)_6CH_2$ and $CH_2(CH_2)_{18}CH_3$ on C=C, each carbon with H (cis)

(Z)-9-triacontene

7.34

A (cyclohexylidene with H–C–C≡CH) $\xrightarrow{3\ H_2/Pd}$ B (cyclohexyl–$CH_2CH_2CH_3$)

A $\xrightarrow{1.\ O_3;\ 2.\ Zn,\ H_3O^+}$ cyclohexanone (=O) plus other fragments

A $\xrightarrow{1.\ NaNH_2,\ NH_3;\ 2.\ CH_3I,\ THF}$ C (cyclohexylidene with H–C–C≡CCH_3)

7.35

$$\begin{array}{l} CH_2\text{-}C{\equiv}C\text{-}C{\equiv}C\text{-}CH_2 \\ | \qquad\qquad\qquad\quad | \\ CH_2\text{-}C{\equiv}C\text{-}C{\equiv}C\text{-}CH_2 \end{array}$$

H_2/Pd → bicyclic (two fused rings) product

1. O_3 2. Zn, H_3O^+ → 2 $HOOCCH_2CH_2COOH$ + 2 HOOC-COOH

7.36 1) Erythrogenic acid contains six degrees of unsaturation (see Sec. 5.1 for method of calculating unsaturation equivalents for compounds containing elements other than C and H).

2) One of these double bonds is contained in the carboxylic acid functional group -COOH; thus, five other degrees of unsaturation are present.

3) Because five equivalents of H_2 are absorbed on catalytic hydrogenation, erythrogenic acid contains no rings.

4) The presence of both aldehyde and carboxylic acid products of ozonolysis

indicates that both double and triple bonds are present in erythrogenic acid.

5) Only two ozonolysis products contain aldehyde functional groups; these fragments must have been double-bonded to each other in erythrogenic acid. $H_2C{=}CH(CH_2)_4C{\equiv}$

6) The other ozonolysis products result from cleavage of triple bonds. However, not enough information is available to tell in which order the fragments were attached. The two possible structures:

   A $H_2C{=}CH(CH_2)_4C{\equiv}C{-}C{\equiv}C(CH_2)_7COOH$

   B $H_2C{=}CH(CH_2)_4C{\equiv}C(CH_2)_7C{\equiv}CCOOH$

   One method of distinguishing between the two possible structures is to treat erythrogenic acid with two equivalents of H_2, using Lindlar catalyst. The resulting trialkene can then be ozonized. The fragment that originally contained the carboxylic acid will still have it intact and can be identified as the terminal carboxylic acid fragment.

What you should know:

After doing these problems, you should be able to:

1. Name alkynes according to IUPAC conventions, and draw any alkyne, given its IUPAC name;
2. Predict the products of reactions of alkynes;
3. Determine the structures of alkynes, given the structural formula and data from hydrogenation and/or cleavage reactions;
4. Specify reagents necessary to carry out reactions of alkynes;
5. Plan syntheses of alkynes;
6. Explain the hybridization of the triple bond and account for the weak acidity of terminal alkynes.

CHAPTER 8 INTRODUCTION TO STEREOCHEMISTRY

8.1 Use the formula $[\alpha]_D = \frac{\alpha}{\ell \times C}$ where

$[\alpha]_D$ = specific rotation

α = observed rotation

ℓ = path length of cell (in dm)

C = concentration (in g/mL)

in this problem:

α = +1.2°

ℓ = 5 cm = 0.5 dm

C = 1.5 g/10 mL = 0.15 g/mL

$$[\alpha]_D = \frac{+1.2°}{0.5 \text{ dm} \times 0.15 \text{ g/mL}} = +16°$$

8.2 The following rules may be used to locate the centers that are <u>not</u> chiral.

1. All $-CH_3$ carbons are achiral.
2. All $-CH_2$ carbons are achiral
   <u>corollary</u>; All $-CX_2$ carbons are achiral.
3. All C=C and C≡C carbons are achiral
   <u>corollary</u>; All benzene ring carbons are achiral because the ring is planar.

a) Toluene is achiral. Rules 1 and 3 indicate all carbons are achiral.

b) (structure: piperidine ring with N–H, starred carbon bearing $CH_2CH_2CH_3$)

coniine

Coniine is chiral. Rules 1 and 2 indicate that all carbons are achiral except for the starred carbon, which is bonded to four different groups.

c) Ethynylcyclohexanol is achiral.

d) $HOCH_2\overset{*}{C}H(NH_2)COOH$ Serine is chiral.

e) (structure: pyridine ring bonded to starred carbon of N-CH_3 pyrrolidine ring)

Nicotine is chiral.

f) Aminoadamantane is achiral; a plane of symmetry passes through the circled carbons. This plane is easier to see with a molecular model.

Aminoadamantane

8.3 Use the rules in problem 8.2 to determine the number of chiral centers.

a) Phenobarbital is achiral.

b) 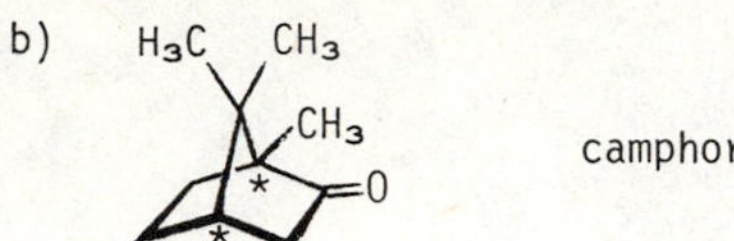

camphor

c) menthol

d) dextromethorphan

8.4 a) By rule 1, -H is of lowest priority, and -Br is of highest priority. By rule 2, $-CH_2CH_2OH$ is of higher priority than $-CH_2CH_3$.

highest ⟶ lowest

$-Br$, $-CH_2CH_2OH$, $-CH_2CH_3$, $-H$

b) By rule 3, -COOH can be considered as -C(O-)(O-)-OH. Since three "oxygens" are attached to a -COOH carbon and only one oxygen is attached to $-CH_2OH$, -COOH is of higher priority than $-CH_2OH$. $-CO_2CH_3$ is of higher priority than -COOH by rule 2, and -OH is of highest priority by rule 1.

highest ⟶ lowest

$-OH$, $-CO_2CH_3$, $-COOH$, $-CH_2OH$

c) $-NH_2$, $-CN$, $-CH_2NHCH_3$, $-CH_2NH_2$

d) $-Br$, $-Cl$, $-CH_2Br$, $-CH_2Cl$

8.5 The following scheme may be used to assign *R*,*S* configurations to chiral centers;

Step 1. For each chiral center, rank substituents by the Cahn-Ingold-Prelog system; give the number 4 to the lowest priority substituent. For part (a)

substituent	priority
-Br	1
-COOH	2
$-CH_3$	3
-H	4

Step 2. Manipulate the molecule so that the lowest priority group is oriented toward the rear. To avoid errors, use a molecular model of the compound.

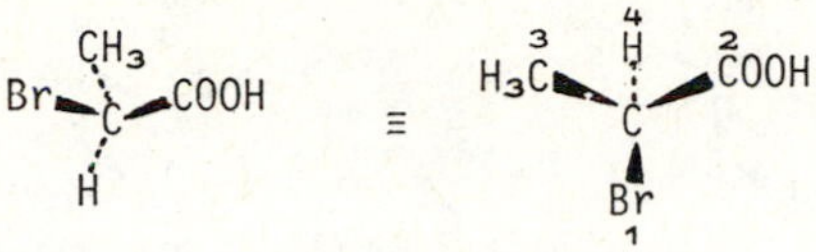

Step 3. Find the direction of rotation of the arrows that go from group 1 to group 2 to group 3. If the arrows have a clockwise rotation, the configuration is *R*; if the arrows have a counterclockwise rotation, the configuration is *S*. Here, the configuration is *S*.

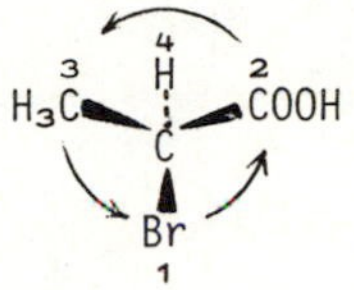

S

b) 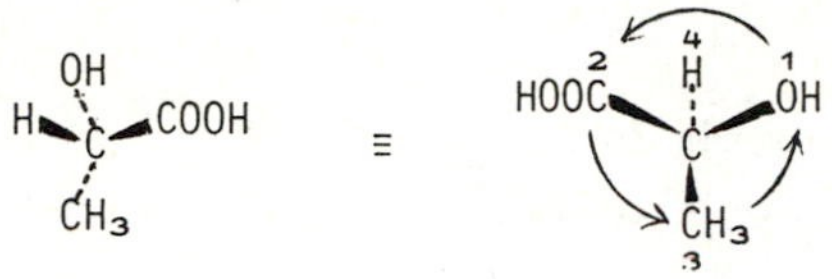 ≡ *S*

c) 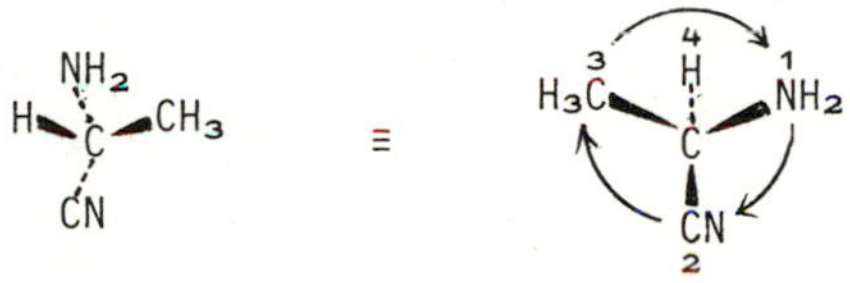 ≡ *R*

d) 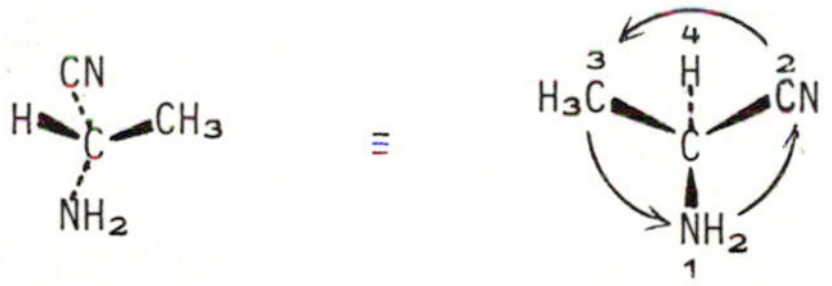 ≡ *S*

8.6 *R*,*S* assignments for more complicated molecules can be made by using a slight modification of the rules in problem 8.5. It is especially important to use molecular models when a compound has more than one chiral center.

Step 1. Assign priorities to groups on the *first* chiral center.

substituent	priority
-Br	1
-CH(OH)CH_3	2
-CH_3	3
-H	4

Step 2. Orient the model so that -H bonded to the first chiral center points to the back.

Step 3. Find the direction of the arrows that travel from 1 to 2 to 3 and assign an *R*,*S* configuration to the first chiral center.

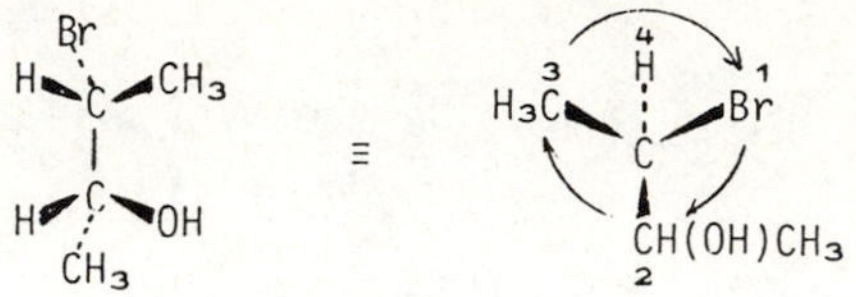

R

Step 4. Repeat steps 1-3 for the next chiral center.

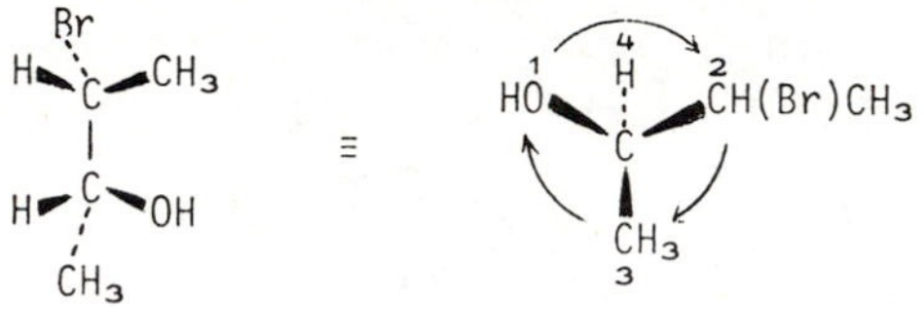

R

b) *S*,*R*

c) *R*,*S*

d) *S*,*S*

a, d are enantiomers and are diastereomeric to b, c.

b, c are enantiomers and are diastereomeric to a, d.

8.7

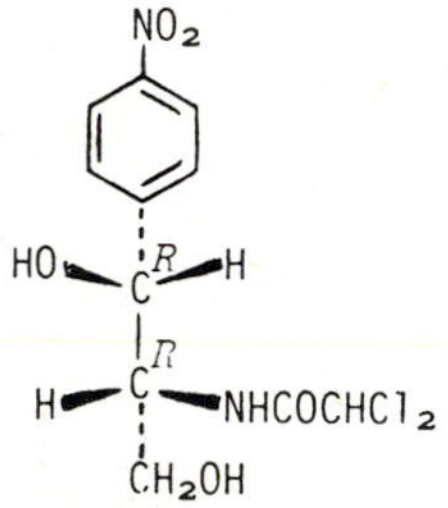

chloramphenicol

8.8 To decide if a structure represents a *meso* compound, try to locate a plane of symmetry that divides the molecule into two halves that are mirror images. Molecular models may be helpful.

a) plane of symmetry

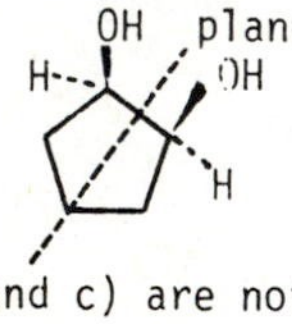

meso

b) and c) are not *meso* structures.

d) plane of symmetry

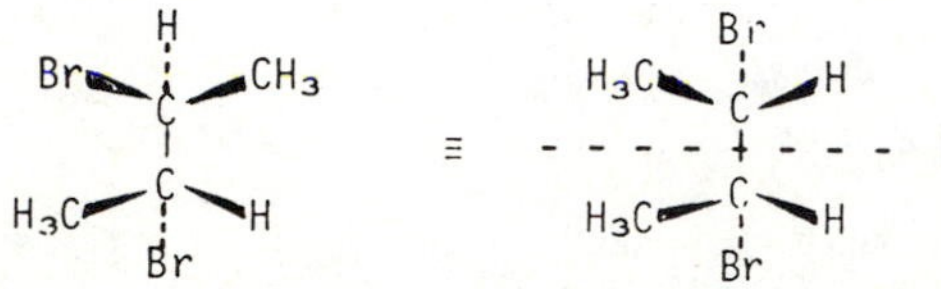

meso

8.10 Two manipulations for Fischer projections are allowable.

1. A Fischer projection may be rotated by 180°, but not by 90° or 270°.
2. Holding one group of Fischer projection, we may rotate the other three either clockwise or counterclockwise.

If projection (a) is rotated by 180°, the resulting projection is superimposable on projection (b). Thus, the first two structures are identical.

```
180°  CO2H              CH3
   H ─┼─ Br    =   Br ─┼─ H
      CH3              CO2H
     (a)               (b)
```

Notice that in projection (c) -Br and -H occupy the same position as they do in projection (b). $-CH_3$ and -COOH, however, are 180° different from their position in projection (b). Projections (b) and (c) are enantiomers.

Holding the $-CH_3$ group of projection (c) steady, rotate the other groups clockwise.

```
      CO2H                Br
  Br ─┼─ H       =    H ─┼─ CO2H
     [CH3] steady         CH3
      (c)                 (d)
```

The structure obtained is identical with projection (d). Structures (a) and (b) are identical and are enantiomeric with structures (c) and (d).

8.11 Manipulate two groups in the first structure to the positions they occupy in the second structure.

For example:

```
       CH3  180°           CHO
  Cl ─┼─ H        =    H ─┼─ Cl
       CHO                 CH3
```

Now, holding $-CH_3$ steady, rotate the groups clockwise.

```
       CHO                 H
   H ─┼─ Cl       =   Cl ─┼─ CHO
      [CH3] steady         CH3
```

Although this projection resembles the second one in placement of $-CH_3$ and -H, -Cl and -CHO are interchanged. The two projections thus represent enantiomers.

b) As above, rotate the first projection by 180°.

```
      CH2OH               CHO
   H ─┼─ OH       =   HO ─┼─ H
      CHO                 CH2OH
```

If we now compare this structure to the second structure of the pair, two of the groups occupy the same relative position.

CHO / HO—+—H / CH_2OH *vs.* CHO / HO—+—CH_2OH / H

There is, however, no rotation that can make these two structures superimposable; they are enantiomers.

c) The projections represent enantiomers.

8.12 One of the hardest spatial problems in organic chemistry is to visualize two-dimensional drawings as three-dimensional chemical structures. This difficulty becomes particularly troublesome when it is necessary to assign *R,S* configurations to structures, especially if they are drawn as Fischer projections. Working out these assignments is easier when models are used, but it is still possible to determine a configuration from a two-dimensional drawing.

The following system may be used for assigning *R,S* configurations. Problem (a) is used as an example.

Step 1. Rank substituents in priority order.

substituent	priority	
-Br	1	high
-COOH	2	
$-CH_3$	3	
-H	4	low

Step 2. Use either of the allowable manipulations of Fischer projections to bring the group of lowest priority to the top of the projection. In this case, hold -Br steady and rotate the other three groups clockwise.

COOH (2) / H (4)—+—CH_3 (3) / Br (1) steady ≡ H (4) / H_3C (3)—+—COOH (2) / Br (1)

Step 3. Indicate on the rotated projection the direction of the arrows that proceed from group 1 to group 2 to group 3. If the direction of the arrows is clockwise the configuration is *R*; if the direction is counterclockwise, the configuration is *S*. Here the configuration is *S*.

H (4) / H_3C (3)—+—COOH (2) / Br (1) *S*

b) CH_3 (3) 180° / HO (1)—+—CH_2CH_3 (2) / H (4) ≡ H (4) / CH_3CH_2 (2)—+—OH (1) / CH_3 (3) *S*

c) 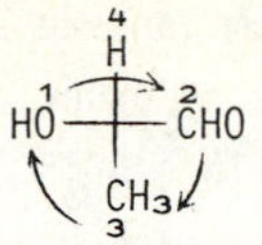 Fischer projection: H (4) top, HO (1) left, CHO (2) right, CH_3 (3) bottom. *R*

d) For the first chiral center:

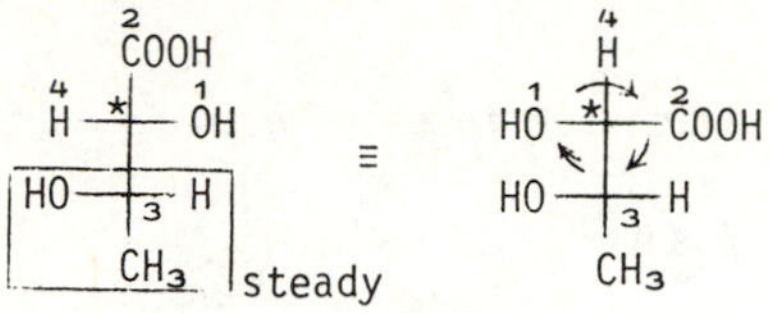

COOH (2), H (4), OH (1), HO, H (3), CH_3 steady ≡ H (4), HO (1), COOH (2), HO, H (3), CH_3

Chiral center one is *R*.

For the second chiral center:

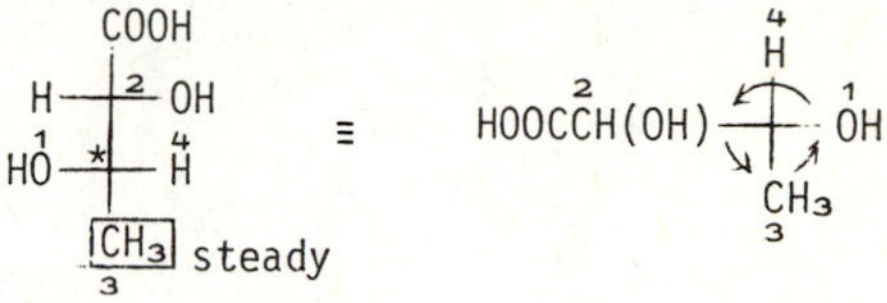

COOH, H, OH (2), HO (1), H (4), CH_3 (3) steady ≡ H (4), HOOCCH(OH) (2), OH (1), CH_3 (3)

Chiral center two is *S*.

8.13 If carbocations were tetrahedral, attack by :Br⁻ would take place on only one side of the 1-methylpropyl cation to yield only one steroisomer of 2-bromobutane. However, two enantiomeric tetrahedral carbocations would be present as a 50:50 mixture. Attack by :Br⁻ on each of them would produce a 50:50 mixture of the enantiomers of 2-bromobutane -- a racemic mixture.

8.14 Draw the bromonium ion resulting from bottom-side attack.

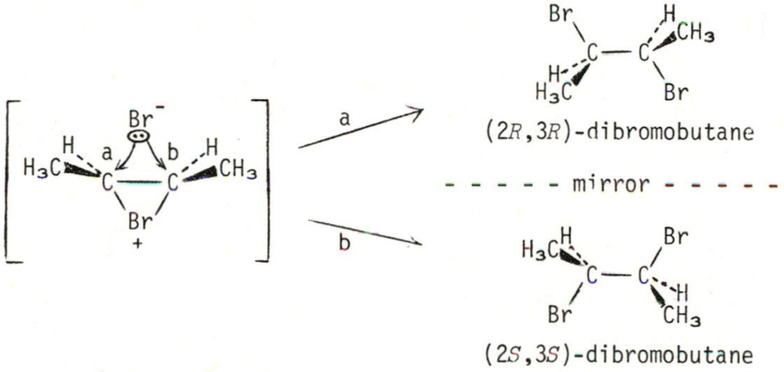

Bromonium ion (H_3C, H, C—C, H, CH_3; Br⁺ bridging; Br⁻ attacking by paths a and b)

a → (2*R*,3*R*)-dibromobutane

- - - - - mirror - - - - -

b → (2*S*,3*S*)-dibromobutane

Since attack on the bromonium ion by :Br⁻ is equally likely to occur by path (a) as to occur by path (b), the product is racemic.

8.15 Possible bromonium ion intermediates are shown below. For the sake of this problem, assume that attack of :Br⁻ is somewhat more likely at carbon 2.

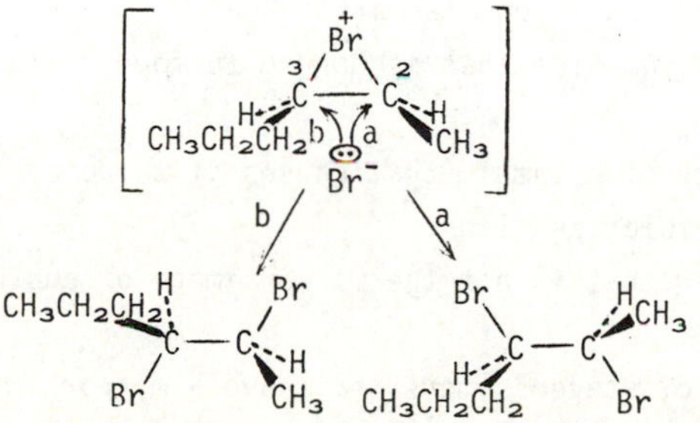

Bromonium ion 1 ($CH_3CH_2CH_2$, H on C3; H, CH_3 on C2; Br⁺ on top, Br⁻ attacking from below by paths a and b)

b → 2*S*,3*S* minor

a → 2*R*,3*R* major

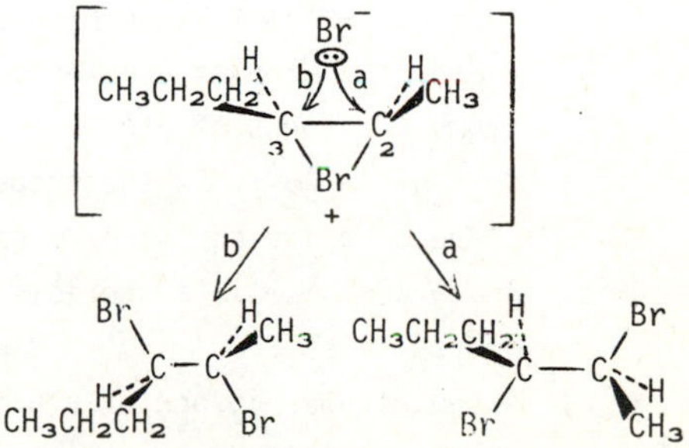

Bromonium ion 2 ($CH_3CH_2CH_2$, H on C3; H, CH_3 on C2; Br⁺ on bottom, Br⁻ attacking from above by paths a and b)

b → 2*R*,3*R* minor

a → 2*S*,3*S* major

The products of attack of bromide ion on each bromonium ion are shown above. Notice that the major products are enantiomers of each other, as are the minor products. Because the bromonium ions are formed in a 50:50 mixture, and because the percent attack at carbon two is the same for each bromonium ion, the amount of (2*R*,3*R*) and (2*S*,3*S*)-dibromohexanes will be equal, and the product will be a racemic mixture.

8.16 Remember from Problem 8.15 that attack at carbon is assumed to be more likely.

2*S*,3*R* minor — 2*R*,3*S* major — 2*R*,3*S* minor — 2*S*,3*R* major

The product of bromination of *trans*-2-hexene is a racemic mixture of (2*S*,3*R*)-2,3-dibromohexane and (2*R*,3*S*)-2,3-dibromohexane. The reasoning is explained in Problem 8.15.

8.17 for cholic acid; $[\alpha]_D = \dfrac{+2.22°}{0.1 \text{ dm} \times 0.6 \text{ g/mL}} = \dfrac{+2.22°}{0.06} = +37.0°$

8.18 for ecdysone; $[\alpha]_D = \dfrac{+0.087°}{0.2 \text{ dm} \times 0.007 \text{ g/mL}} = +62°$

8.19 a) *Chirality* is the property of "handedness" -- the property of an object that causes it to be non-superimposable on its mirror image.

b) A *chiral center* of a molecule is an atom that is bonded to four different atoms or groups of atoms.

c) *Optical activity* is the property of a sample that causes it to rotate the plane of polarization of plane-polarized light.

d) A *diastereomer* is a stereoisomer that is not the mirror image of another stereoisomer.

e) An *enantiomer* is one of a pair of stereoisomers that have a mirror image relationship.

f) A *racemate* is a 50:50 mixture of (+) and (-) enantiomers that behaves as if it were a pure compound and that is optically inactive.

8.20 The rules for determining chiral centers were listed in Problem 8.2.

a) $CH_3CH_2C(CH_3)_2\overset{*}{C}H(CH_3)CH_2CH_3$. This compound is *chiral.*

b) Aspirin is *achiral.*

c) 1,4-Dimethylcyclohexane is *achiral.*

d) $HOCH_2\overset{*}{C}H(OH)CHO$ is *chiral.*

e) $CH_3\overset{*}{C}H(C_6H_5)\overset{*}{C}Br(C_6H_5)CH_2Br$. *Two chiral* carbons are present in 1,2-dibromo-2,3-diphenylbutane.

f) It is surprising, considering its large number of carbons, that vitamin A is *achiral.*

8.21 a) $CH_3CH_2CH_2\overset{*}{C}H(CH_3)CH_2CH(CH_3)_2$. 2,4-Dimethylheptane has one chiral center.

b) $CH_3CH_2C(CH_3)_2CH_2CH(CH_2CH_3)_2$. 3-Ethyl-5,5-dimethylheptane is *achiral.*

c) Cl

Cl

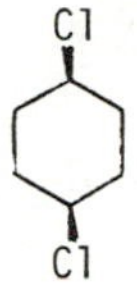

cis-1,4-Dichlorocyclohexane is *achiral.* Notice the plane of symmetry that passes through the -Cl groups.

d) $CH_3C{\equiv}C\overset{*}{C}H(CH_3)\overset{*}{C}H(CH_3)C{\equiv}CCH$. 4,5-Dimethyl-2,6-octadiyne has two chiral centers. The chirality of the compound depends upon the configuration at each of the chiral centers; a pair of enantiomers (each of which is chiral) and an achiral *meso* compound correspond to this structure.

8.22 If you have trouble with this sort of problem, use the following scheme:

1. Draw all alkanes having the formula C_5H_{12}.

$CH_3CH_2CH_2CH_2CH_3$	$CH_3CH_2CH(CH_3)_2$	$C(CH_3)_4$
3 kinds of -H	4 kinds of -H	1 kind of -H

2. Find the number of different kinds of hydrogen for each alkane.

3. Replace one of each different kind of -H with an -OH.

$CH_3CH_2CH_2CH_2CH_2OH$	$CH_3CH_2CH_2\overset{*}{C}H(OH)CH_3$	$CH_3CH_2CH(OH)CH_2CH_3$
achiral	*chiral*	*achiral*
$CH_3CH_2\overset{*}{C}H(CH_3)CH_2OH$	$CH_3CH_2C(CH_3)_2OH$	$CH_3\overset{*}{C}H(OH)CH(CH_3)_2$
chiral	*achiral*	*chiral*

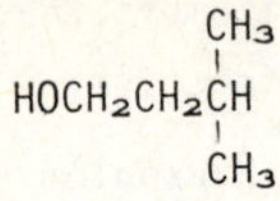

$$\begin{array}{c} CH_3 \\ | \\ HOCH_2CH_2CH \\ | \\ CH_3 \end{array} \qquad (CH_3)_3CCH_2OH$$

achiral *achiral*

4. Using the rules in Problem 8.2, determine if chiral centers are present.

8.23 Draw the five C_6H_{14} hexanes.

$CH_3CH_2CH_2CH_2CH_2CH_3$	$CH_3CH_2CH_2CH(CH_3)_2$	$CH_3CH_2CH(CH_3)CH_2CH_3$
3 kinds of -H	5 kinds of -H	4 kinds of -H

$(CH_3)_2CHCH(CH_3)_2$	$CH_3CH_2C(CH_3)_3$
2 kinds of -H	3 kinds of -H

17 monobromohexanes can be formed from the hexane isomers. You may need to draw all the bromohexanes to find those that are chiral; you may, however, prefer to make the substitution of -Br for -H in your mind.

The nine chiral bromhexanes:

$CH_3CH_2CH_2CH_2\overset{*}{C}HBrCH_3$	$CH_3CH_2CH_2\overset{*}{C}HBrCH_2CH_3$	$CH_3CH_2CH_2\overset{*}{C}H(CH_3)CH_2Br$
$CH_3CH_2\overset{*}{C}HBrCH(CH_3)_2$	$CH_3\overset{*}{C}HBrCH_2CH(CH_3)_2$	$CH_3CH_2\overset{*}{C}H(CH_3)CH_2CH_2Br$
$CH_3CH_2\overset{*}{C}H(CH_3)\overset{*}{C}HBrCH_3$	$(CH_3)_2CH\overset{*}{C}H(CH_3)CH_2Br$	$CH_3\overset{*}{C}HBrC(CH_3)_2$

8.24 a) $CH_3CH_2\overset{*}{C}H(OH)CH_3$

b) This carboxylic acid has no rings or carbon-carbon double or triple bonds. (The formula $C_5H_{10}O_2$ indicates that there is one double bond present, but it is in the carboxylic acid functional group.)

$CH_3CH_2\overset{*}{C}H(COOH)CH_3$

c) $CH_3\overset{*}{C}HBr\overset{*}{C}H(OH)CH_3$

d) $CH_3\overset{*}{C}HBrCHO$

8.25 chiral: golf club, monkey wrench, screw, beanstalk, ear, coin

achiral: basketball, fork, wine glass, snowflake

8.26 a)

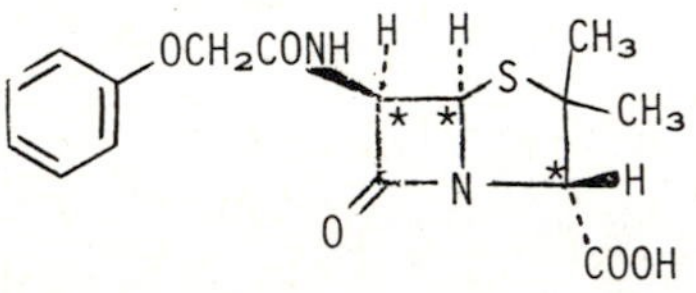

penicillin V

three chiral carbons

b)

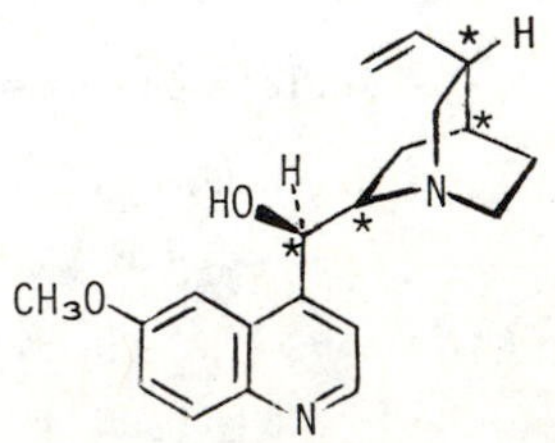

quinine

four chiral carbons

c)

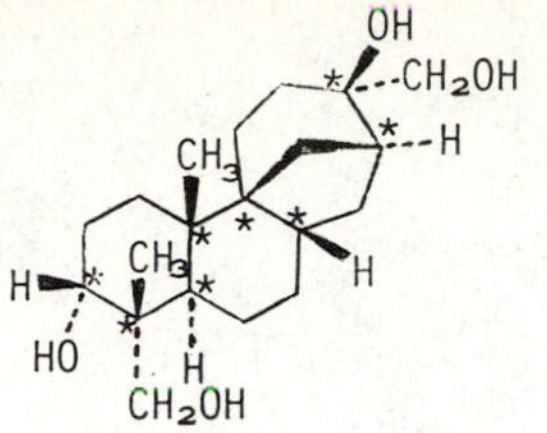

aphidicolin

eight chiral carbons

d)

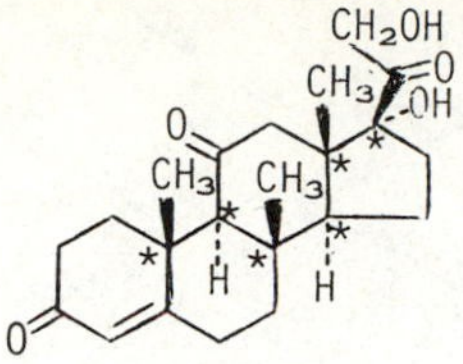

cortisone

six chiral carbons

8.27 a)

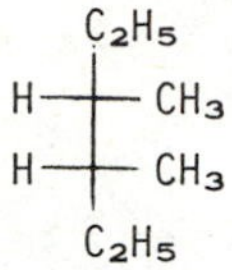

b)

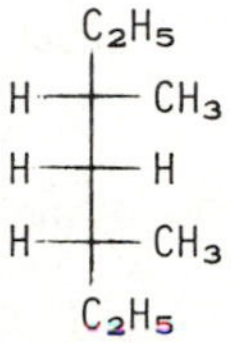

c)

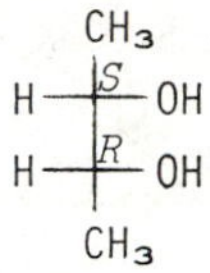

8.28 See Problem 8.10 for the method of solution.

b, c, d, and f represent pairs of identical molecules.

a and e represent pairs of enantiomers.

8.29 See Problem 8.4 for the method of solution.

highest priority ⟶ lowest priority

a) $-C(CH_3)_3$, $-CH{=}CH_2$, $-CH(CH_3)_2$, $-CH_2CH_3$

b) 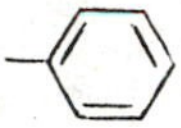, $-C{\equiv}CH$, $-C(CH_3)_3$, $-CH{=}CH_2$

c) $-COOCH_3$, $-COCH_3$, $-CH_2OCH_3$, $-CH_2CH_3$

d) $-Br$, $-CH_2Br$, $-CN$, $-CH_2CH_2Br$

8.30 a)

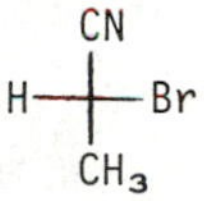

R

b)

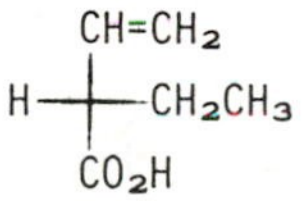

S

c)

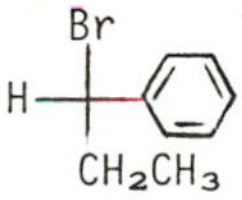

S

d)

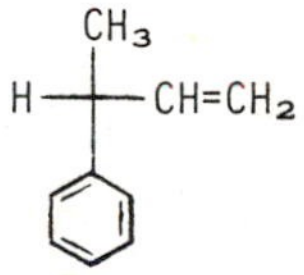

R

8.31 See Problem 8.12 for method of solution.

a)

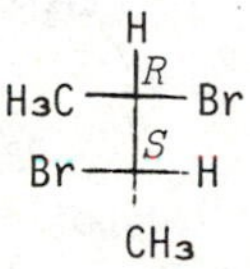

b)

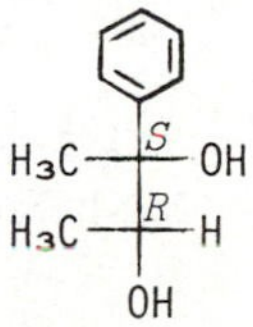

c)

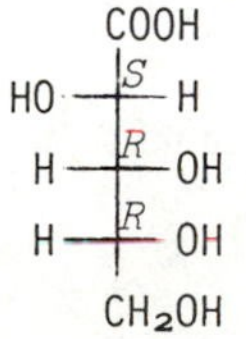

d)

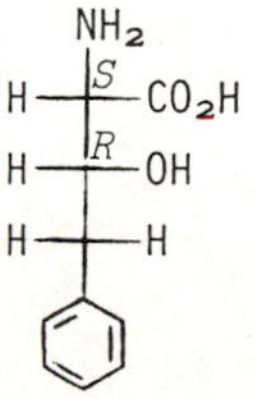

8.32 a)

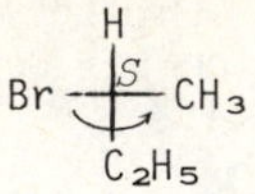

H, Br, CH_3, C_2H_5 (*S*)

(*S*)-2-bromobutane

$CH_3CH_2CHBrCH_3$

b)

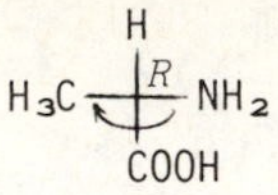

H, H_3C, NH_2, COOH (*R*)

(*R*)-alanine

$CH_3CH(NH_2)CO_2H$

c)

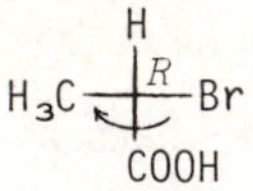

H, H_3C, Br, COOH (*R*)

(*R*)-2-bromopropanoic acid

$CH_3CHBrCOOH$

d)

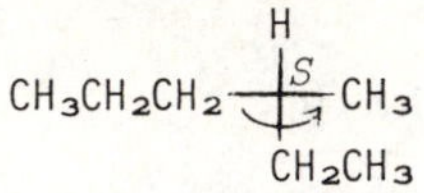

H, $CH_3CH_2CH_2$, CH_3, CH_2CH_3 (*S*)

(*S*)-3-methylhexane

$CH_3CH_2CH_2CH(CH_3)CH_2CH_3$

8.33

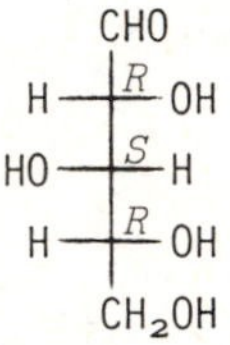

CHO; H—(*R*)—OH; HO—(*S*)—H; H—(*R*)—OH; CH_2OH

(+)-xylose

8.34 The intial product of hydroxylation of a double bond is a cyclic *osmate*. Drawings of the osmates for *cis*-2-butene and *trans*-2-butene are shown below. No carbon-oxygen bonds are broken in the cleavage step; cleavage occurs at osmium-oxygen bonds. The final stereochemistry, therefore, is the same as that of the initial adduct.

cis-2-butene $\xrightarrow{OsO_4}$ osmate + osmate $\xrightarrow{NaHSO_3}$ *meso*-2,3-butanediol

trans-2-butene $\xrightarrow{OsO_4}$ osmate + osmate $\xrightarrow{NaHSO_3}$ (2*R*,3*R*)-2,3-butanediol and (2*S*,3*S*)-2,3-butanediol

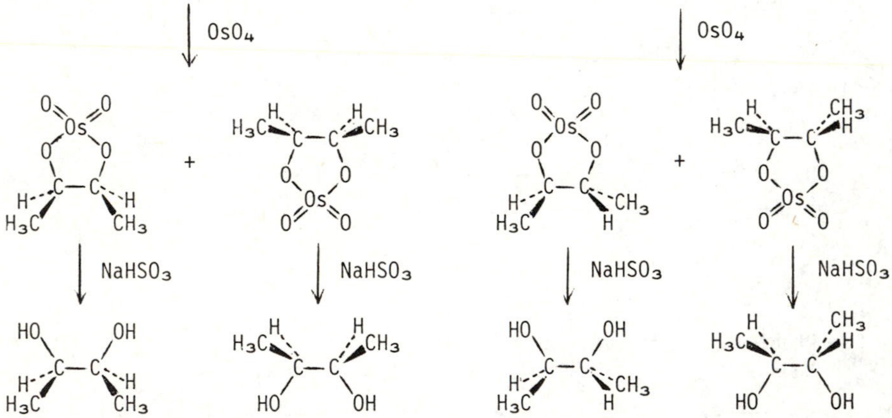

8.35

(*R*,*S*) epoxide: $CH_3CH_2CH_2$, H, O, H, $CH_2CH_2CH_3$ — (*S*,*R*) epoxide: $CH_3CH_2CH_2$, H, H, $CH_2CH_2CH_3$, O

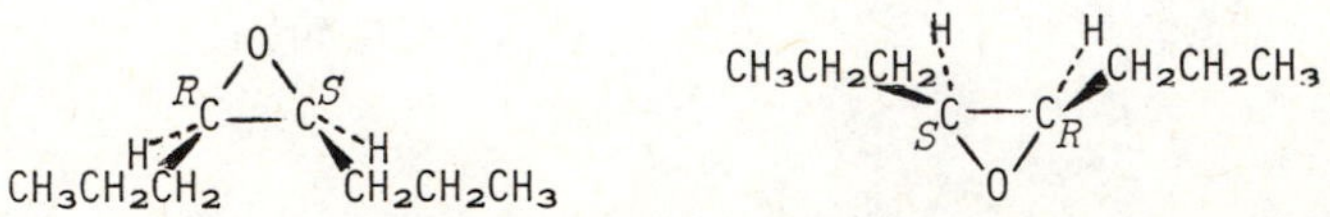

Peroxycarboxylic acids can attack either the "top" side or the "bottom" side of a double bond. The epoxide resulting from "top" side attack has two chiral

centers, but because it has a plane of symmetry it is a *meso* compound. The two epoxides are identical.

8.36

O
R C—C *R*
H CH₂CH₂CH₃
CH₃CH₂CH₂ H

The epoxide formed by "top-side" attack of a peroxyacid on *trans*-4-octene is pictured. This epoxide has two chiral centers of *R* configuration. The epoxide formed by "bottom side" attack has *S*,*S* configuration. The two epoxide enantiomers are formed in equal amounts and constitute a racemic mixture.

8.37 a)

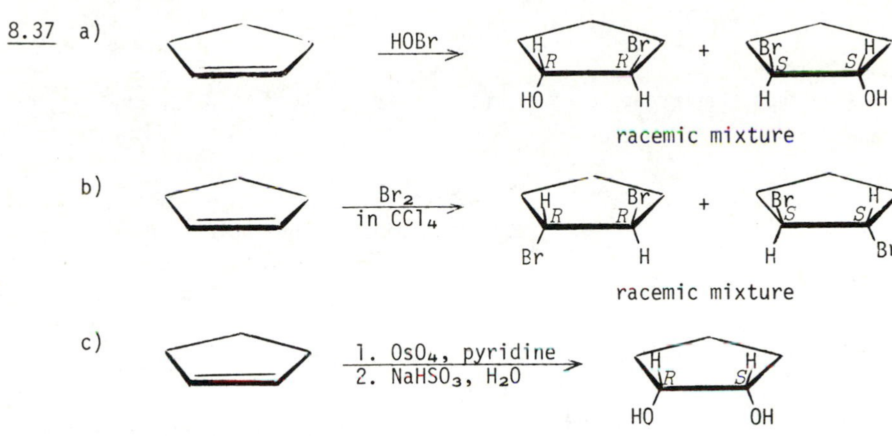

8.38 1. Write the structural formula and identify chiral centers.

$CH_3\overset{*}{C}H(OH)\overset{*}{C}H(OH)COOH$ (2*R*,3*S*)-2,3-dihydroxybutanoic acid

2. For the first chiral center assign priorities to the four groups.

For $CH_3\overset{*}{C}H(OH)\overset{*}{C}H(OH)COOH$
↑

-OH	1
-COOH	2
-[CH(OH)CH₃]	3
-H	4

3. Draw the Fischer "framework" and put the group of lowest priority on top.

H
HO—+—COOH
[CH(OH)CH₃]

4. Arrange the other 3 groups so that rotation from 1 to 2 to 3 proceeds in a clockwise direction (because carbon 2 has an *R* configuration).

5. For the next chiral center assign priorities to the 4 groups.

For $CH_3\overset{*}{C}H(OH)\overset{*}{C}H(OH)COOH$
↑

Group	Priority
-OH	1
-[CH(OH)COOH]	2
$-CH_3$	3
-H	4

6. Put the group of lowest priority at the bottom.

H
HO —+— COOH
—+—
H

7. Arrange the other groups so that rotation from 1 to 2 to 3 proceeds in a counterclockwise direction.

H
HO —+[*R*]— COOH
H_3C —+[*S*]— OH
H

(2*R*,3*S*)-2,3-dihydroxybutanoic acid

b) $CH_3CH_2CH_2\overset{*}{C}HBr\overset{*}{C}HBrCH_2CH_2CH_3$

H
$CH_3CH_2CH_2$ —+— Br
$CH_3CH_2CH_2$ —+— Br
H

=

$CH_2CH_2CH_3$
Br —+— H
Br —+— H
$CH_2CH_2CH_3$

meso-4,5-dibromooctane

c) $CH_3CH_2CH_2\overset{*}{C}HCl\overset{*}{C}HClCH_2CH_2CH_3$

H
$CH_3CH_2CH_2$ —+[*R*]— Cl
Cl —+[*R*]— $CH_2CH_2CH_3$
H

=

$CH_2CH_2CH_3$
Cl —+[*R*]— H
H —+[*R*]— Cl
$CH_2CH_2CH_3$

(4*R*,5*R*)-4,5-dibromooctane

8.39 $CH_2OH\overset{*}{C}H(OH)\overset{*}{C}H(OH)\overset{*}{C}H(OH)\overset{*}{C}H(OH)CHO$

Remember that the aldehyde carbon is carbon one.

^{1}CHO
H —+[*R*]— OH
HO —+[*S*]— H
H —+[*R*]— OH
H —+[*R*]— OH
$_6CH_2OH$

glucose

The same procedures used in the previous problem also work for glucose.

8.40 A: cyclobutane ring with COOH (S) and H; COOH (R) and H — B: H (R), COOH; COOH (R), H — C: COOH (S), H; H (S), COOH

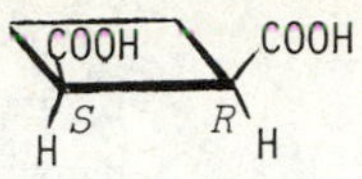

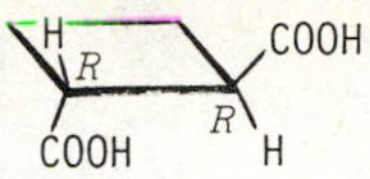

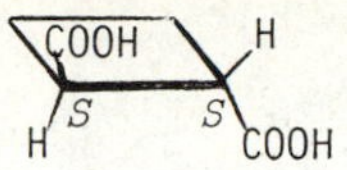

A B C

B and C are enantiomers. Compound A is their diastereomer and is a *meso* compound. B and C are optically active.

The two isomeric cyclobutane-1,3-dicarboxylic acids are diastereomers and are optically inactive; both are *meso* compounds.

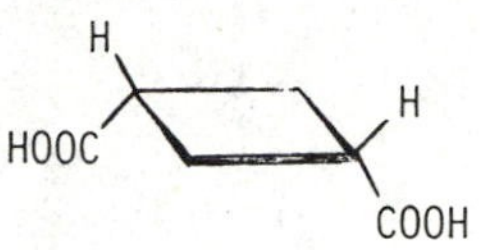

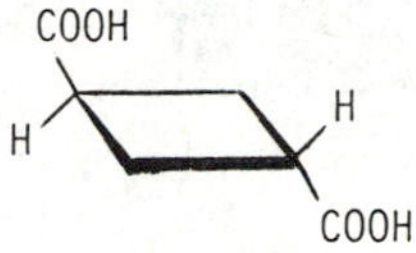

8.41 $CH_3C{\equiv}C\overset{*}{C}H(CH_3)CH_2CH_3$ (A) $\xrightarrow{2\ H_2}$ $CH_3CH_2CH_2CH(CH_3)CH_2CH_3$ (B)

A $\xrightarrow{1.\ O_3;\ 2.\ Zn,\ H_3O^+}$ $HOOC\overset{*}{C}H(CH_3)CH_2CH_3$ (C) + CH_3COOH

8.42 A has four multiple bonds and/or rings.

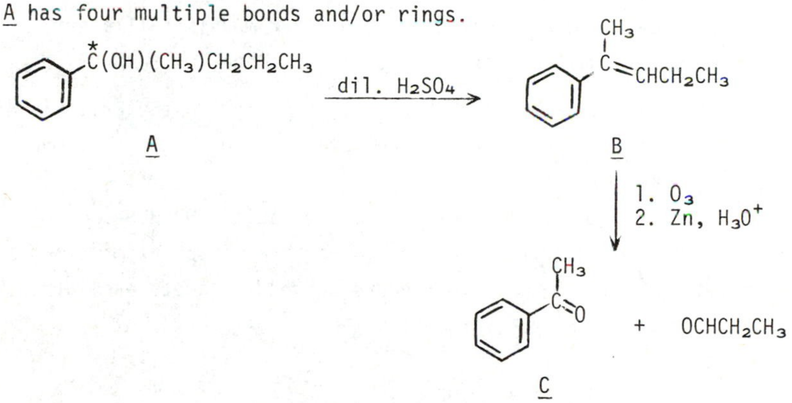

$C_6H_5\overset{*}{C}(OH)(CH_3)CH_2CH_2CH_3$ (A) $\xrightarrow{\text{dil. } H_2SO_4}$ $C_6H_5C(CH_3){=}CHCH_2CH_3$ (B)

B $\xrightarrow{1.\ O_3;\ 2.\ Zn,\ H_3O^+}$ $C_6H_5C(CH_3){=}O$ (C) + $OCHCH_2CH_3$

8.43 A molecule containing *n* chiral centers can give rise to a maximum of 2^n stereoisomers. On first inspection, we would predict eight stereoisomers for 2,4-dibromo-3-chloropentane ($CH_3\overset{*}{C}HBr\overset{*}{C}HCl\overset{*}{C}HBrCH_3$), which has three chiral carbons. After drawing the eight possible stereoisomers, it is apparent that only four 2,4-dibromo-3-chloropentanes are unique.

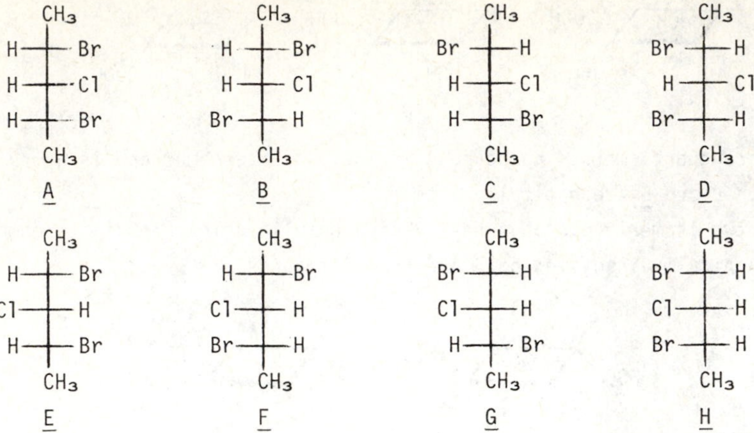

Identical: A and H, B and F, D and E, C and G

(A,H) and (D,E) are optically inactive *meso* compounds and are diastereomers.
(B,F) and (C,G) are enantiomers and should be optically active.
(A,H) and (D,E) are diastereomeric with (B,F) and (C,G).

8.44 A tetrahedrane can be chiral. Notice that the orientations of the four substituents in space are the same as the orientations of the four substituents of a tetrasubstituted carbon atom. If you were able to make a model of a tetrasubstituted tetrahedrane (without having your models fall apart), you would also be able to make a model of its mirror image.

8.45 If we analyze all carbon atoms for chirality according to the criteria of Problem 8.2, we find that no chiral carbon atoms are present. Yet mycomycin is chiral. How can this be?

Make a model of mycomycin. For simplicity, call -CH=CHCH=CHCH$_2$COOH "A" and -C≡CC≡CH "B". Remember from Chapter 5 that the carbon atoms of an allene are linear and that the pi bonds formed are perpendicular to each other. Attach substituents at the *sp*^{2} carbons.

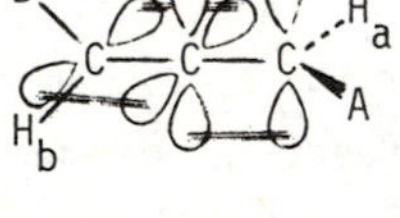

Notice that the substituents A, H$_a$, and all carbon atoms lie in a plane that is perpendicular to the plane that contains B, H$_b$, and all carbon atoms.

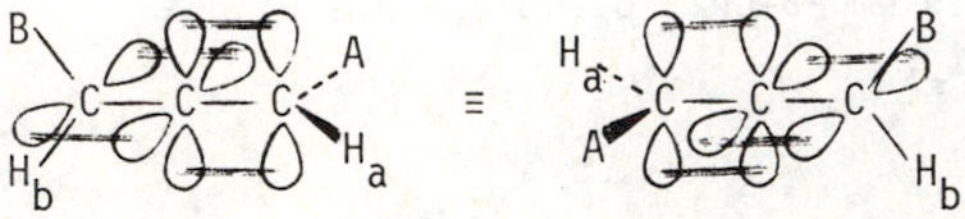

Now, make another model identical to the first, except for an exchange of A and H$_a$. This new allene is not superimposable on the original allene; in fact, the two allenes are mirror images. The two allenes are enantiomers and are chiral for the same reason that tetrasubstituted carbon atoms are chiral -- they possess no plane of symmetry.

8.46 4-Methylcyclohexylideneacetic acid is chiral for the same reason as an allene; it possesses no plane of symmetry and is not superimposable on its mirror image. As in the case of allenes, the two functional groups at one end lie in a plane perpendicular to the plane that contains the two functional groups at the other end.

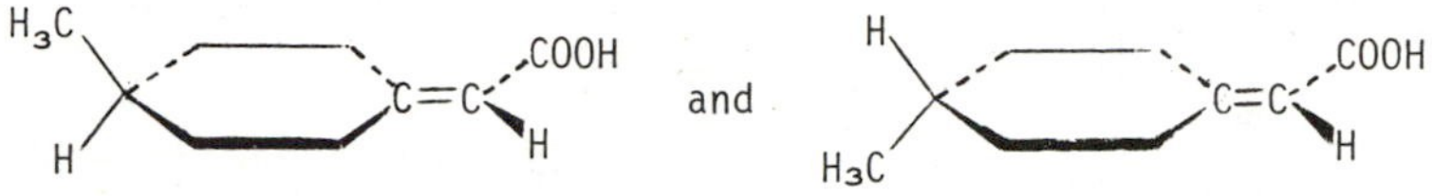

What you should know:

After doing these problems you should be able to:

1. Calculate the specific rotation of an optically active compound;
2. Determine if a compound is chiral;
3. Manipulate Fischer projections to determine if they are identical;
4. Assign *R* and *S* configurations to chiral centers;
5. Recognize *meso* compounds and find their plane of symmetry;
6. Predict the stereochemical outcome of reactions involving achiral starting materials.

CHAPTER 9 ALKYL HALIDES

9.1 a) $(CH_3)_2CClCH_2CH_2Cl$ 1,3-dichloro-3-methylbutane

b) $BrCH_2CH_2CH_2C(CH_3)_2CH_2Br$ 1,5-dibromo-2,2-dimethylpentane

c) $CH_3CHICH(CH_2CH_2Cl)CH_2CH_3$ 1-chloro-3-ethyl-4-iodopentane

9.2

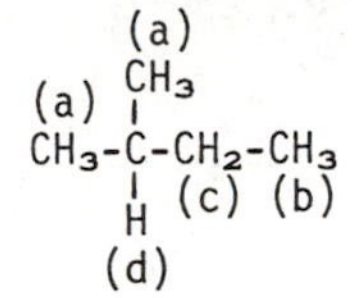

```
          (a)
      (a) CH3
          |
      CH3-C-CH2-CH3
          |  (c) (b)
          H
         (d)
```

2-methylbutane

Type of H	a	b	c	d
Number of H of each type	6	3	2	1
Relative reactivity	1.0	1.0	3.5	5.0
Number times reactivity	6.0	3.0	7.0	5.0
Percent chlorination	29%	14%	33%	24%

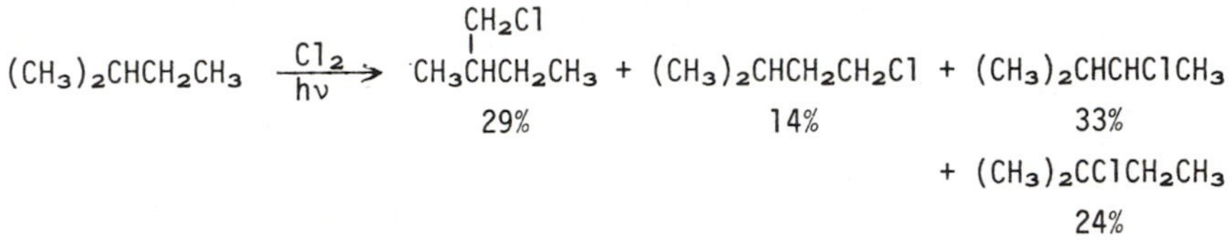

$$(CH_3)_2CHCH_2CH_3 \xrightarrow[h\nu]{Cl_2} CH_3CH(CH_2Cl)CH_2CH_3 + (CH_3)_2CHCH_2CH_2Cl + (CH_3)_2CHCHClCH_3 + (CH_3)_2CClCH_2CH_3$$

29% 14% 33% 24%

9.3

Abstraction of hydrogen by a bromine radical yields an allylic radical.

The allylic radical reacts with Br_2 to produce A and B.

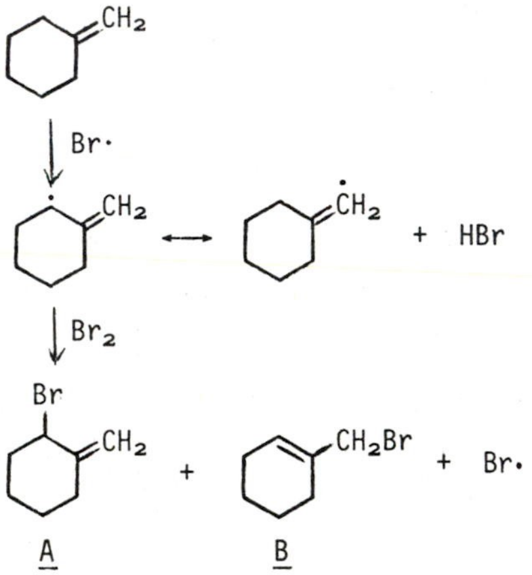

Methylenecyclohexane (CH_2) $\xrightarrow{Br\cdot}$ allylic radical ↔ resonance form ($\dot{C}H_2$) + HBr

$\xrightarrow{Br_2}$ A (2-bromo-1-methylenecyclohexane: Br, CH_2) + B (1-(bromomethyl)cyclohexene: CH_2Br) + Br·

Product B, which has a trisubstituted double bond, forms in preference to product A, which has a disubstituted double bond.

9.4 Table 4.5 shows that the bond dissociation energy for $C_6H_5CH_2$-H is 85 kcal/mol. This value is even smaller than the bond dissociation energy for allylic hydrogens, and thus it is relatively easy to form the $C_6H_5CH_2\cdot$ radical. The high bond dissociation energy for formation of $C_6H_5\cdot$, 112 kcal/mol, indicates that bromination on the benzene ring will not occur; the only product of reaction with NBS is $C_6H_5CH_2Br$.

9.5 a) [carbonate ion resonance structures: $^{-}$:O–C(=O)–O:$^{-}$ ⟷ O=C(O:$^{-}$)–O:$^{-}$ ⟷ $^{-}$:O–C(O:$^{-}$)=O]

b) $CH_3N^+(=O)–O^-$ ⟷ $CH_3N^+(–O^-)=O$

c) $\overset{+}{O}$: $O=\overset{+}{O}-\overset{-}{O}$: ⟷ $:\overset{-}{O}-\overset{+}{O}=O$

9.6 The acetate anion can be described as a resonance hybrid of two structures:

$CH_3C(=O)–O^-$ ⟷ $CH_3C(–O^-)=O$

These resonance structures indicate that both oxygen atoms are equivalent and have the same carbon-oxygen bond length, which is between the length of a C-O single bond (1.34 Å) and a C=O double bond (1.20 Å). Spectroscopic measurements confirm this interpretation.

9.7 Recall that the pK_a of CH_3-H is 49. Since methane is a very weak acid, $:\bar{C}H_3$ is a very strong base. Alkyl Grignard reagents have pK_b's close to that of $:\bar{C}H_3$, and alkenyl Grignards are somewhat weaker bases.

9.8 If Grignard reagents react with proton donors to convert R-MgX into R-H, they will also react with *deuterium* donors to convert R-MgX into R-D. In this case:

$CH_3CH(Br)CH_2CH_3 \xrightarrow{Mg} CH_3CH(MgBr)CH_2CH_3 \xrightarrow{D_2O} CH_3CH(D)CH_2CH_3$

9.9 It is not possible to prepare a Grignard reagent from a compound containing a functional group that is a good proton donor. As the Grignard reagent starts to form, it is immediately quenched by the proton source. The -COOH, -OH, and $-NH_2$ functional groups are too acidic to be compatible with preparation of a Grignard reagent. Terminal alkynes are also too acidic; for example, $CH_3CHBrC{\equiv}CH$ does not form a Grignard reagent.

9.10 a) cyclohexene $\xrightarrow[CCl_4]{NBS}$ 3-bromocyclohexene $\xrightarrow{(CH_3)_2\overset{-}{Cu}\overset{+}{Li}}$ 3-methylcyclohexene

3-methylcyclohexene

b) 2 $CH_3CH_2CH_2CH_2Br \xrightarrow[pentane]{4\ Li}$ 2 $CH_3CH_2CH_2CH_2Li$ + 2 LiBr $\xrightarrow[ether]{CuI}$ $(CH_3CH_2CH_2CH_2)_2\overset{-}{Cu}\overset{+}{Li}$ $\xrightarrow[ether]{CH_3CH_2CH_2CH_2Br}$ $CH_3CH_2CH_2CH_2CH_2CH_2CH_2CH_3$

octane

c) $CH_3CH_2CH_2CH{=}CH_2 \xrightarrow[\text{peroxides}]{\text{HBr}} CH_3CH_2CH_2CH_2CH_2Br$

$2\ CH_3CH_2CH_2CH_2CH_2Br \xrightarrow[\text{pentane}]{\text{4 Li}} 2\ CH_3(CH_2)_3CH_2Li + 2\ LiBr$

$\downarrow$ CuI, ether

$[CH_3(CH_2)_3CH_2]_2\overset{-}{Cu}\overset{+}{Li}$

$\downarrow$ $CH_3(CH_2)_3CH_2Br$, ether

$CH_3(CH_2)_8CH_3$

decane

9.11 a) 3,4-dibromo-2,6-dimethylheptane b) 5-iodo-2-hexene

c) 2-bromo-4-chloro-2,5-dimethylhexane d) 3-(bromomethyl)hexane

e) 1-bromo-6-chloro-2-hexyne

9.12 Abstraction of hydrogen by Br· can produce either of two allylic radicals. The first radical, resulting from abstraction of a secondary hydrogen, is more likely to be formed.

$CH_3\dot{C}HCH{=}CHCH_3 \longleftrightarrow CH_3CH{=}CH\dot{C}HCH_3$ (identical)

and

$CH_3CH_2CH{=}CH\dot{C}H_2 \longleftrightarrow CH_3CH_2\dot{C}HCH{=}CH_2$

Reaction of the radical intermediates with a bromine source leads to a mixture of products:

$CH_3CH_2CHBrCH{=}CH_2$ 3-bromo-1-pentene

and

$CH_3CHBrCH{=}CHCH_3$ *cis*- and *trans*-4-bromo-2-pentene

and

$CH_3CH_2CH{=}CHCH_2Br$ *cis*- and *trans*-1-bromo-2-pentene

The major product is 4-bromo-2-pentene instead of the desired product, 1-bromo-2-pentene.

9.13 Three different allylic radical intermediates can be formed. Bromination of these intermediates can yield as many as five bromoalkenes. This is definitely not a good reaction to use in a synthesis.

[structure: 1-methylcyclohexene allylic radical] CH_3 $\longleftrightarrow$ [resonance structure] CH_3 $\xrightarrow{\text{NBS}}$ [structure] Br, CH_3 3-bromo-2-methylcyclohexene

(allylic; secondary hydrogen abstracted)

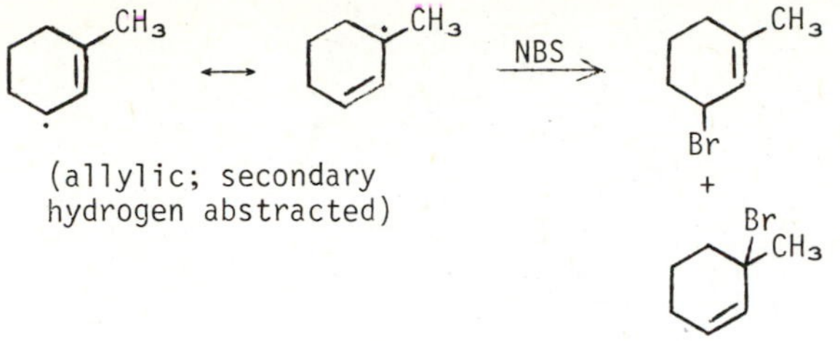

(allylic; secondary hydrogen abstracted)

NBS →

3-bromo-1-methylcyclohexene

+

3-bromo-3-methylcyclohexene

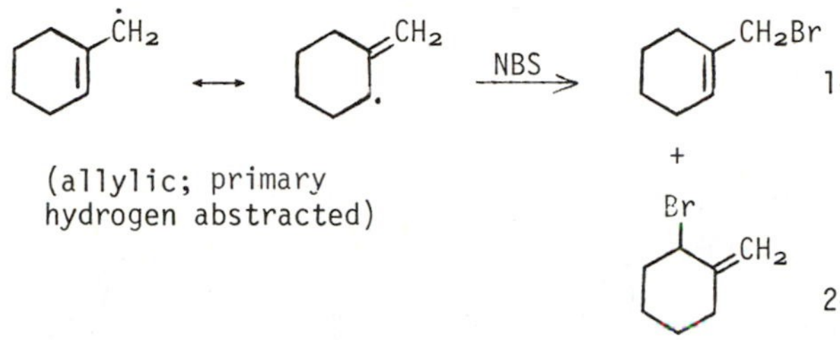

(allylic; primary hydrogen abstracted)

NBS →

1-(bromomethyl)cyclohexene

+

2-bromomethylenecyclohexane

9.14

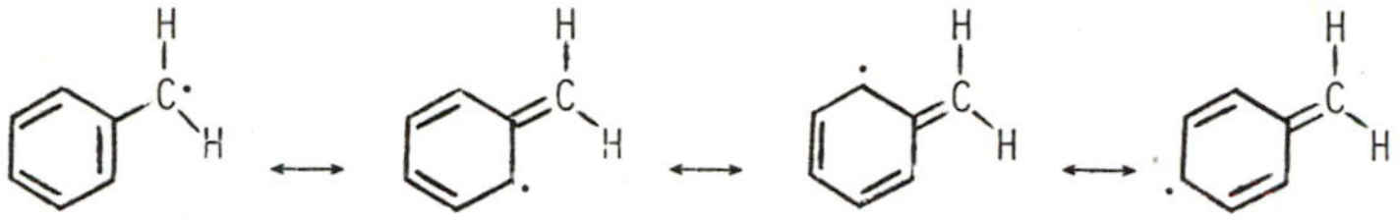

9.15 Two allylic radicals can form:

and

The second radical is much more likely to form because it is both allylic and benzylic, and it yields the following products:

9.16 a)

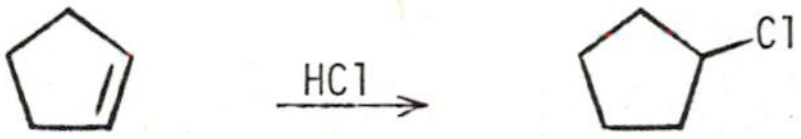

$\xrightarrow{HCl}$

chlorocyclopentane

b)

$\xrightarrow{HBr}$ $\xrightarrow{(CH_3)_2\overset{-}{Cu}\overset{+}{Li}}$

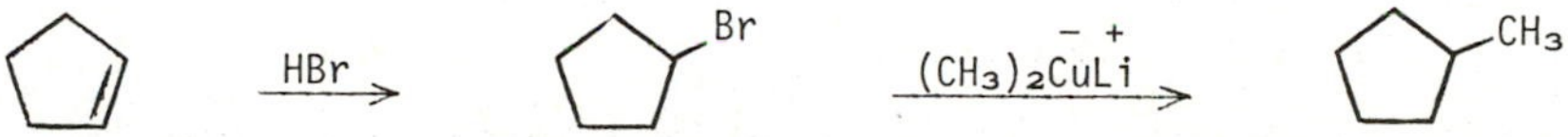

methylcyclopentane

c) Cyclopentene $\xrightarrow[CCl_4]{NBS}$ 3-bromocyclopentene

Br

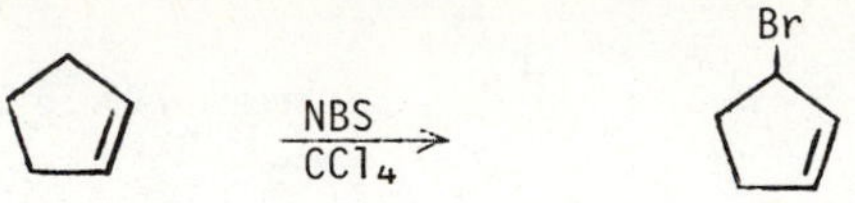

3-bromocyclopentene

d) Cyclopentene $\xrightarrow[2.\ NaBH_4]{1.\ Hg(OAc)_2,\ H_2O}$ cyclopentanol

OH

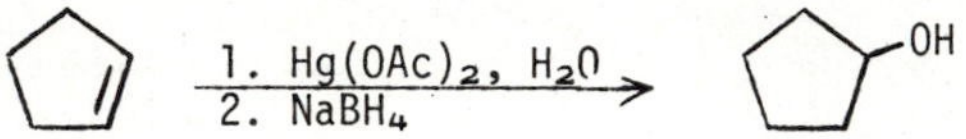

cyclopentanol

e) Cyclopentene $\xrightarrow{HBr}$ bromocyclopentane $\xrightarrow[pentane]{2\ Li}$ cyclopentyllithium

2 cyclopentyllithium $\xrightarrow[ether]{CuI}$ $(C_5H_9)_2\overset{-}{Cu}\overset{+}{Li}$ $\xrightarrow[ether]{C_5H_9Br}$ cyclopentylcyclopentane

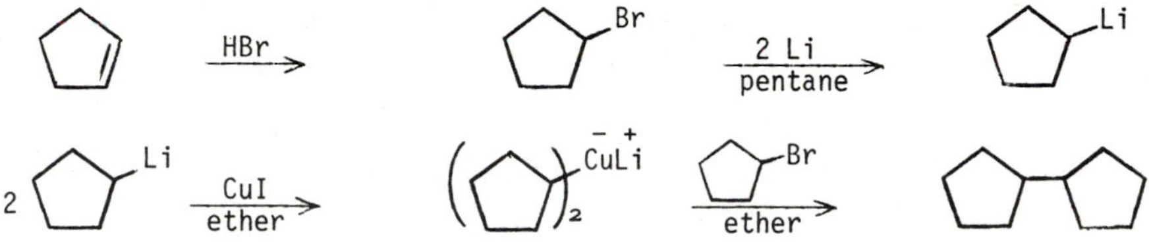

cyclopentyl-
cyclopentane

f) Cyclopentene $\xrightarrow{HBr}$ bromocyclopentane (Br) $\xrightarrow[ether]{Mg}$ cyclopentyl-MgBr $\xrightarrow{D_2O}$ cyclopentane-D

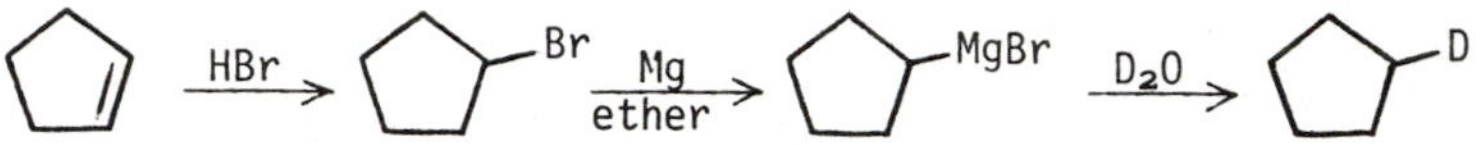

monodeutero-
cyclopentane

9.17 a) 1-methylcyclohexanol (CH_3, OH) $\xrightarrow{HBr}$ 1-bromo-1-methylcyclohexane (CH_3, Br)

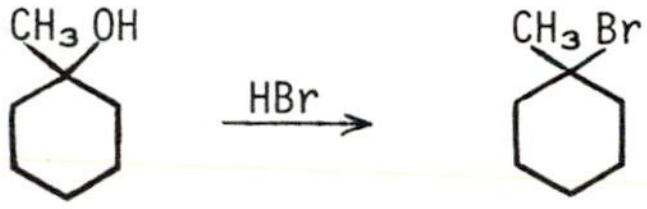

This is a good method for converting a tertiary alcohol into a bromide.

b) $CH_3CH_2CH_2CH_2OH \xrightarrow[pyridine]{SOCl_2} CH_3CH_2CH_2CH_2Cl$

c) Octahydronaphthalene $\xrightarrow{NBS}$ bromo product (Br) + bromo product (Br)

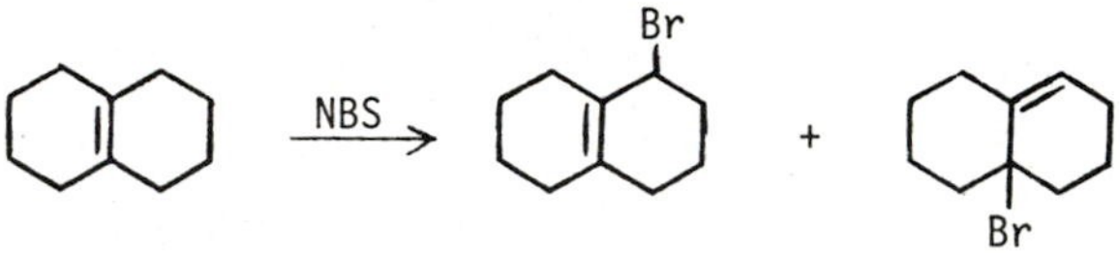

major minor

The major product contains a tetrasubstituted double bond; the minor product contains a trisubstituted double bond.

d) Cyclohexanol (OH) $\xrightarrow{PBr_3}$ bromocyclohexane (Br)

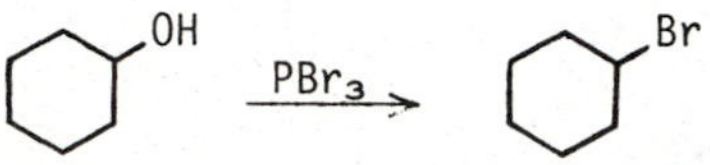

This is a good method for converting a primary or secondary alcohol to a bromide.

e)

$$CH_3CH_2CHBrCH_3 \xrightarrow[\text{ether}]{Mg} \underset{\underline{A}}{CH_3CH_2\overset{MgBr}{\overset{|}{C}}HCH_3} \xrightarrow{H_2O} \underset{\underline{B}}{CH_3CH_2CH_2CH_3}$$

f)

$$2\ CH_3CH_2CH_2CH_2Br \xrightarrow[\text{pentane}]{4\ Li} \underset{\underline{A}}{2\ CH_3CH_2CH_2CH_2Li} \xrightarrow{CuI} \underset{\underline{B}}{(CH_3CH_2CH_2CH_2)_2\overset{-}{Cu}\overset{+}{Li}}$$

g)

$$CH_3CH_2CH_2CH_2Br + (CH_3)_2\overset{-}{Cu}\overset{+}{Li} \xrightarrow[\text{ether}]{} CH_3CH_2CH_2CH_2CH_3 + CH_3Cu + LiBr$$

9.18 Resonance forms cannot differ in the position of nuclei. The two structure in (a) are not resonance forms; the carbon and hydrogen atoms outside the ring occupy different positions in each structure.

[benzocyclobutene structure with CH_2–CH_2] ⇹ [ring with two exocyclic $=CH_2$ groups] not resonance structures

The pairs of structures in parts (b), (c), and (d) are resonance forms.

9.19 a) [CH_3–C(=Ö:)–$\ddot{C}H^-$–H] ⟷ [CH_3–C(–Ö:$^-$)=CH–H]

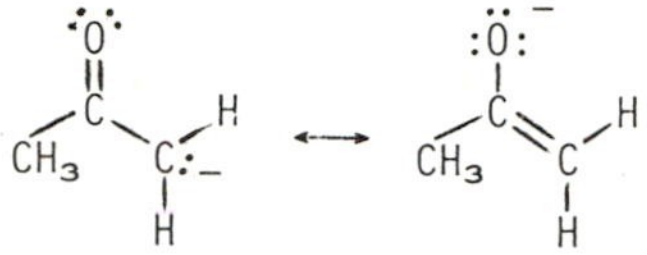

b) $CH_2=\overset{+}{N}=\overset{-}{\underset{\cdot\cdot}{N}}:$ $\qquad :\overset{-}{C}H_2-\overset{+}{N}\equiv N: \qquad :\overset{-}{C}H_2-\overset{\cdot\cdot}{N}=\overset{+}{N}:$

c) [$H_2\ddot{N}$–C(=$\overset{+}{N}H_2$)–$\ddot{N}H_2$] ⟷ [$H_2\ddot{N}$–C(–$:NH_2$)=$\overset{+}{N}H_2$] ⟷ [$H_2\overset{+}{N}$=C(–$:NH_2$)–$\ddot{N}H_2$] ⟷ [$H_2\ddot{N}$–C^+(–$:NH_2$)–$\ddot{N}H_2$]

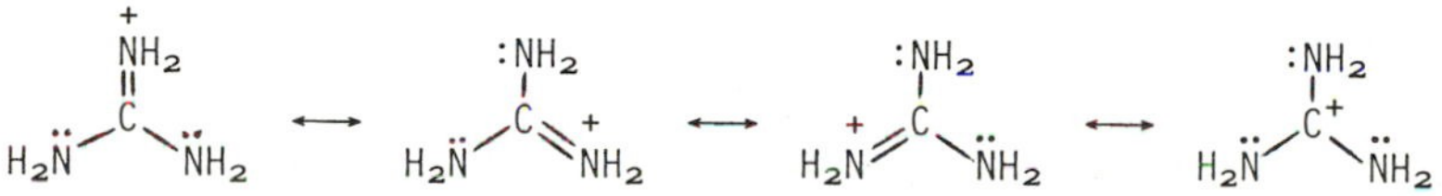

9.20 All of these reactions involve addition of a dialkylcopper reagent, $(CH_3CH_2CH_2CH_2)_2\overset{-}{Cu}\overset{+}{Li}$, to an alkyl halide. The dialkylcopper is prepared by treating 1-bromobutane with lithium, followed by addition of CuI:

$$2\ CH_3CH_2CH_2CH_2Br \xrightarrow[\text{pentane}]{2\ Li} 2\ CH_3CH_2CH_2CH_2Li \xrightarrow[\text{ether}]{CuI} (CH_3CH_2CH_2CH_2)_2\overset{-}{Cu}\overset{+}{Li}$$

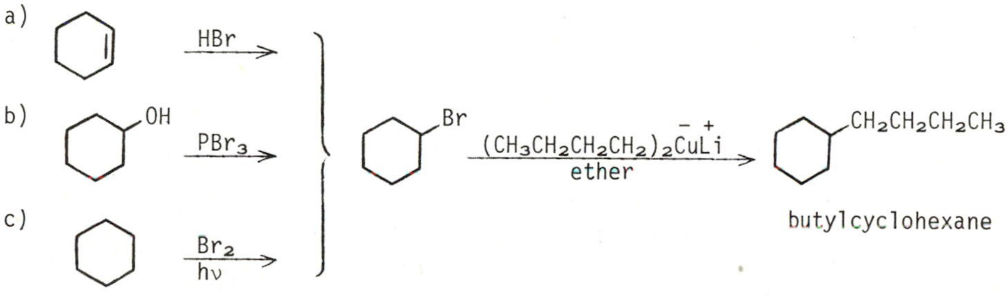

a) cyclohexene $\xrightarrow{HBr}$

b) cyclohexanol (OH) $\xrightarrow{PBr_3}$

c) cyclohexane $\xrightarrow[h\nu]{Br_2}$

} bromocyclohexane (Br) $\xrightarrow[\text{ether}]{(CH_3CH_2CH_2CH_2)_2\overset{-}{Cu}\overset{+}{Li}}$ cyclohexyl–$CH_2CH_2CH_2CH_3$

butylcyclohexane

9.21 a) Displacement of halides by $LiAlH_4$ is limited to primary and secondary halides; in addition, fluorine is unreactive to displacement.

b) Two allylic radicals can be produced.

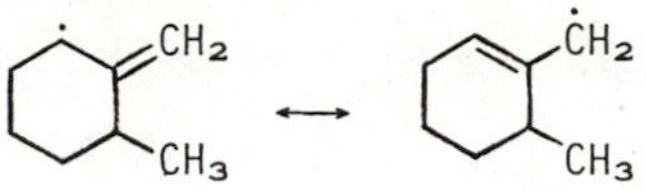

(1) ⟷ (2)

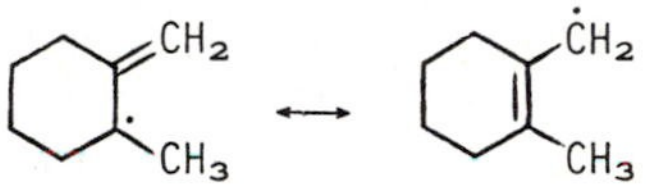

(3) ⟷ (4)

Instead of a single product, as many as four bromide products may result.

c) Dialkylcopper reagents do not react with fluoroalkanes.

9.22 Pairs (a) and (d) represent resonance structures; pairs (b) and (c) do not. For (b) and (c), a proton differs in position in each structure of the pair. The two structures of each pair represent different chemical species, rather than resonance forms of the same species.

What you should know:

After doing these problems, you should be able to:

1. Give IUPAC names of alkyl halides and draw structures corresponding to IUPAC names;
2. Know the mechanism of the radical halogenation of alkanes;
3. Be able to explain the stability order of radicals; including the allyl radical;
4. Be able to recognize and to draw simple resonance structures;
5. Know several methods for preparation of alkyl halides from alkanes, alkenes, and alcohols;
6. Know how to prepare Grignard reagents and dialkycopper reagents from alkyl halides, and be able to use them in planning syntheses.

CHAPTER 10 REACTIONS OF ALKYL HALIDES. NUCLEOPHILIC SUBSTITUTION REACTIONS

10.1 If back-side attack were necessary for S_N2 reaction, a molecule with a hindered "back-side" would not be able to react by an S_N2 mechanism. In this problem approach by the ethoxide ion from the back side of the bromoalkane is blocked by the rigid bicyclic ring system, and displacement can't occur.

10.2 Since the pK_b's of triethylamine (10.75) and quinuclidine (10.95) are similar, the difference in reaction rate must be due to a factor other than basicity.

Build molecular models of triethylamine and quinuclidine. A model of the most stable conformation of triethylamine shows that the ethyl groups interfere with approach of the nitrogen lone pair electrons to methyl iodide. In quinuclidine, however, the hydrocarbon framework is rigidly held back from the nitrogen lone pair. It is sterically easier for quinuclidine to approach methyl iodide, and reaction occurs at a faster rate.

10.3 a) $(CH_3)_2\ddot{N}:^-$ is more nucleophilic. A negatively charged reagent is more nucleophilic than its conjugate acid.

b) $(CH_3)_3B$ is a Lewis acid; $(CH_3)_3N:$ is a Lewis base. Since nucleophilicity roughly parallels basicity, $(CH_3)_3N:$ is more nucleophilic. $(CH_3)_3B$ is non-nucleophilic because it has no lone electron pair.

c) Since nucleophilicity increases in going down a column of the periodic table, $H_2\ddot{S}:$ is more nucleophilic than $H_2\ddot{O}:$.

10.4 a) The difference in this pair of reactions is in the *leaving group*. Since -OTos is a better leaving group than -Cl (see Table 10.5), S_N2 displacement by iodide on CH_3-OTos proceeds faster.

b) The *substrates* in these two reactions are different. Bromoethane is a primary bromoalkane and bromocyclohexane is a secondary bromoalkane. Since S_N2 reactions proceed faster at primary, rather than secondary, carbon atoms, S_N2 displacement on bromoethane is a faster reaction.

c) Ethoxide ion and cyanide ion are different *nucleophiles*. Table 10.3 shows that $:CN^-$ is more reactive than $CH_3CH_2\ddot{O}:^-$ in S_N2 reactions. S_N2 displacement on 2-bromopropane by $:CN^-$ thus proceeds at a faster rate.

d) The *solvent* in each reaction is different. Table 10.6 shows that S_N2 reactions run in hexamethylphosphoramide (HMPA) proceed faster than those run in other solvents. Thus, S_N2 displacement by acetylide ion on bromomethane proceeds faster in HMPA than in ether.

10.5 If solvolysis had proceeded with complete inversion, the product would have had a specific rotation of +53.6°. If complete racemization had occurred, $[\alpha]_D$ would have been zero.

The observed rotation was +5.3°. Since $\frac{+5.3°}{+53.6°} = 0.099$, 9.9% of the original tosylate was inverted. The remaining 90.1% of product must have been racemized.

10.6 The first step in an S_N1 displacement is ionization of the substrate to form a planar, *sp*² hybridized carbocation and a leaving group. The carbocation that would form from ionization of this haloalkane is prevented from achieving planarity by the rigid structure of the rest of the molecule. Because it is not possible to form the necessary carbocation, the reaction does not proceed by S_N1 substitution.

10.7

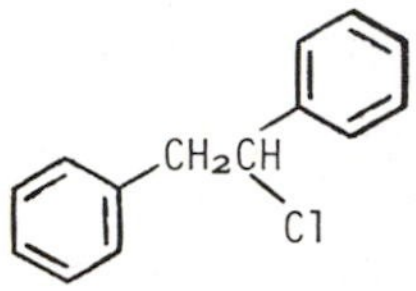

$C_6H_5CH_2CH(Cl)C_6H_5$ 1-chloro-1,2-diphenylethane

Nucleophilic substitution of 1-chloro-1,2-diphenylethane proceeds via an S_N1 mechanism because of stabilization of the carbocation intermediate by the phenyl group at C-1. In an S_N1 reaction, the rate-limiting step is carbocation formation; all subsequent steps occur at a faster rate and do not affect the rate of reaction. After carbocation formation, the rate does not depend on the identity of the nucleophile that combines with the carbocation. In this example, F:$^-$ and $(CH_3CH_2)_3N$: react at the same rate.

10.8

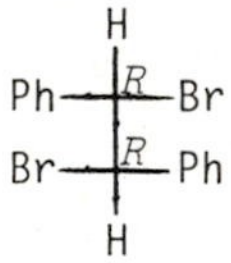

Fischer projection: H (top); Ph—*R*—Br; Br—*R*—Ph; H (bottom)

(1*R*,2*R*)-1,2-dibromo-1,2-diphenylethane

Convert this Fischer projection into a Newman projection and draw the conformation having *anti*-periplanar geometry for -H and -Br.

Newman projections: Br, Ph, H, H, Ph, Br = Br, Ph, Br, H, Ph, H

The alkene resulting from dehyrohalogenation is (*Z*)-1-bromo-1,2-diphenylethylene.

Ph, Br, H, Ph = H(Ph)C=C(Ph)Br

10.9

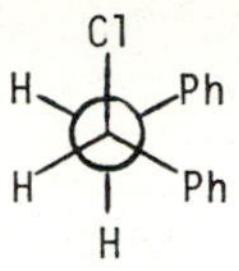

A: Newman projection — front: Cl (up), H, H; back: Ph, Ph, H

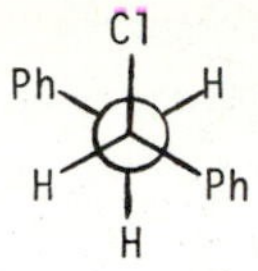

B: Newman projection — front: Cl (up), Ph, H; back: H, Ph, H

These two Newman projections place -H and -Cl in the correct *anti*-periplanar geometry for E_2 elimination.

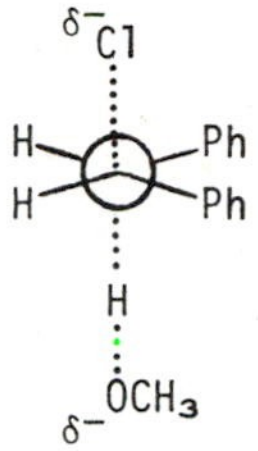

T.S. $\underline{A}^{\ddagger}$ — $^{\delta-}$Cl ⋯ C(H,H)–C(Ph,Ph) ⋯ H ⋯ OCH_3 $^{\delta-}$

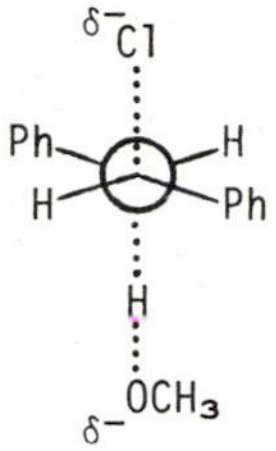

T.S. $\underline{B}^{\ddagger}$ — $^{\delta-}$Cl ⋯ C(Ph,H)–C(H,Ph) ⋯ H ⋯ OCH_3 $^{\delta-}$

Either transition state $\underline{A}^{\ddagger}$ or $\underline{B}^{\ddagger}$ can form when 1-chloro-1,2-diphenylethane undergoes E_2 elimination. Steric interactions of the two phenyl groups in T.S. $\underline{A}^{\ddagger}$ make this transition state (and the product resulting from it) of higher energy than transition state $\underline{B}^{\ddagger}$. Formation of the product from $\underline{B}^{\ddagger}$ is therefore favored, and *trans*-1,2-diphenylethylene is the major product.

10.10 a) 1-Bromobutane is a primary haloalkane, which reacts by either an S_N2 or an E_2 mechanism. Since 1-azidobutane is a substitution product, the reaction is an example of an *S_N2 reaction*.

b) Bromocyclohexane, a secondary haloalkane, can undergo elimination by either an E_1 or E_2 mechanism. Since $:OH^-$ is a strong base, elimination proceeds by the E_2 route.

c) 2-Methylbutanenitrile is a substitution product. Because $CH_3CH_2CHBrCH_3$ is a secondary bromoalkane, substitution could occur by either an S_N2 or S_N1 mechanism. However, $:CN^-$ is a good nucleophile and HMPA is a polar aprotic solvent. This reaction, therefore, is an example of an *S_N2* reaction.

d) This substitution reaction occurs by an S_N1 mechanism. Even though the starting material is a secondary halide, it can form a stabilized benzylic carbocation intermediate. Ionization of the halide is the rate-limiting step.

10.11 a) CH_3I reacts faster than CH_3Br because $:I^-$ is a better leaving group than $:Br^-$.

b) CH_3CH_2I reacts faster with $:OH^-$ in dimethylsulfoxide (DMSO) than in ethanol. Ethanol, a protic solvent, hydrogen-bonds with hydroxide ion and decreases its reactivity.

c) Under the S_N2 conditions of this reaction, CH_3Cl reacts faster than $(CH_3)_3CCl$. Approach of the nucleophile to the bulky $(CH_3)_3CCl$ molecule is hindered.

d) $H_2C{=}CHCH_2Br$ reacts faster because vinylic halides such as $H_2C{=}CHBr$ are unreactive to displacement reactions.

10.12 In this problem you are to imagine that you have been provided with the reagents necessary to synthesize the desired compounds, using just one substitution reaction.

a) $CH_3Br + \overset{+}{Na}\!:\!\overset{-}{C}{\equiv}CCH(CH_3)_2 \longrightarrow CH_3C{\equiv}CCH(CH_3)_2$

NOT $CH_3C{\equiv}\overset{-}{C}\!:\!\overset{+}{Na} + BrCH(CH_3)_2$. The strong base $CH_3C{\equiv}\overset{-}{C}\!:$ brings about E_2 elimination, producing $CH_3C{\equiv}CH$ and $H_2C{=}CHCH_3$.

b) $CH_3CH_2CH_2CH_2Br + NaCN \longrightarrow CH_3CH_2CH_2CH_2CN$

c) $(CH_3)_3C\overset{-}{O}\!: + CH_3I \longrightarrow (CH_3)_3COCH_3$. The reaction of $CH_3\overset{-}{O}\!:$ with $(CH_3)_3CI$ results in elimination, not substitution.

d) $(C_6H_5)_3P\!: + CH_3Br \longrightarrow (C_6H_5)_3\overset{+}{P}CH_3\ Br^-$

Triphenylphosphine is not different from other nucleophiles, even though it contains the unfamiliar phosphorus atom.

e) 1-methylcyclohexanol (CH_3, OH) + HBr $\longrightarrow$ 1-bromo-1-methylcyclohexane (CH_3, Br)

f) $CH_3CH_2CH(Br)CH_3 + NaN_3 \longrightarrow CH_3CH_2CH(N_3)CH_3$

Azide ($:N_3^-$) is a very good nucleophile that reacts with secondary halides.

10.13 Because 1-bromopropane is a primary haloalkane, the mode of reaction is either S_N2 or E_2, depending on the basicity and the amount of steric hindrance in the nucleophile.

a) $CH_3CH_2CH_2Br + NaNH_2 \longrightarrow CH_3CH_2CH_2NH_2 + CH_3CH{=}CH_2$

b) $CH_3CH_2CH_2Br + \overset{+}{K}\!:\!\overset{-}{O}C(CH_3)_3 \longrightarrow CH_3CH{=}CH_2$

$\overset{+}{K}\!:\!\overset{-}{O}$-*tert*-Butyl is a hindered, strong base that brings about elimination not substitution.

c) $CH_3CH_2CH_2Br + NaI \longrightarrow CH_3CH_2CH_2I$

d) $CH_3CH_2CH_2Br + NaCN \longrightarrow CH_3CH_2CH_2CN$

e) $CH_3CH_2CH_2Br + \overset{+}{Na}\!:\!\overset{-}{C}{\equiv}CH \longrightarrow CH_3CH_2CH_2C{\equiv}CH$

f) $CH_3CH_2CH_2Br + LiAlH_4 \longrightarrow CH_3CH_2CH_3$

g) $CH_3CH_2CH_2Br + Mg \longrightarrow CH_3CH_2CH_2MgBr \xrightarrow{H_2O} CH_3CH_2CH_3$

h) $CH_3CH_2CH_2Br + CH_3\overset{-}{S}\!: \longrightarrow CH_3CH_2CH_2SCH_3$

10.14 Remember two rules used to predict nucleophilicity:

1) In comparing nucleophiles that have the same attacking atom, nucleophilicity parallels basicity. (In other words, a more basic nucleophile is a more effective nucleophile.)
2) Nucleophilicity increases in going down a column of the periodic table.

Use Table 10.3 if necessary.

a) $N_3:^-$ is more nucleophilic than $H_3N:$.
b) $H_2O:$ is a better nucleophile because it is more basic (see Table 10.4).
c) BF_3 is a Lewis acid and is neither basic nor nucleophilic. $:F^-$ is a better nucleophile.
d) $(CH_3)_3P:$ is a better nucleophile than $(CH_3)_3N:$ by rule 2 (above).
e) $:Cl^-$ is a better nucleophile than ClO_4^-.
f) $:\bar{C}N$ is a better nucleophile than $:\bar{O}CH_3$.

10.15 An alcohol is converted to an ether by two different routes in this series of reactions. The two resulting ethers have identical structural formulas but differ in sign of specific rotation. Therefore, at some step or steps in these reaction sequences, inversion of configuration at the chiral carbon must have occurred. Let's study each step of the Phillips and Kenyon series to find where inversion is occurring.

$CH_3CH(OH)CH_2C_6H_5$ $[\alpha]_D = +33.0°$

(1) K → $CH_3CH(\bar{O}:\overset{+}{K})CH_2C_6H_5$ (2) CH_3CH_2Br → $CH_3CH(OCH_2CH_3)CH_2C_6H_5$ $[\alpha]_D = +23.5°$

(3) TosCl → $CH_3CH(OTos)CH_2C_6H_5$ $[\alpha]_D = +31.10°$ (4) CH_3CH_2OH, Δ → $CH_3CH(OCH_2CH_3)CH_2C_6H_5$ $[\alpha]_D = -19.9°$

In *step 1*, the relatively acidic hydroxyl proton reacts with potassium metal to produce a potassium alkoxide. Since the bond between carbon and oxygen has not been broken, no inversion has occurred in this step.

The potassium alkoxide acts as a nucleophile in the S_N2 displacement on CH_3CH_2Br in *step 2*. It is the C-Br bond of bromoethane, however, not the C-O bond of the alkoxide, that is broken; no inversion at the chiral carbon occurs in step 2.

The starting alcohol reacts with tosyl chloride in *step 3*. Again, since the O-H, and not the C-O, bond of the alcohol is broken, no inversion can occur at this step.

Inversion does occur at *step 4*. The -OTos group is displaced by CH_3CH_2OH. The C-O bond of the tosylate (OTos) is broken, and a new C-O bond is formed in this displacement.

Notice the specific rotations of the two enantiomeric products. The product of steps 1 and 2 should be enantiomerically pure because neither reaction has affected the C-O bond. Reaction 4 proceeds with some racemization at the chiral carbon to give a smaller value of $[\alpha]_D$.

10.16 a) Reaction with $Li(CH_3CH_2)_3BH$ proceeds by an S_N2 mechanism. Since the substrate is a tertiary haloalkane, which does not undergo reaction under S_N2 conditions, this reaction is unlikely to occur.

b) It is not possible to effect a substitution on a secondary haloalkane using a strong, bulky base. Elimination occurs instead, and produces $H_2C{=}CHCH_2CH_3$ and $CH_3CH{=}CHCH_3$.

c) Reaction of this secondary fluoroalkane with hydroxide yields both elimination and substitution products.

d) SO_2Cl converts alcohols to chlorides by an S_N2 mechanism. 1-Methyl-1-cyclohexanol is a tertiary alcohol, and does not undergo S_N2 substitution. Instead, E_2 elimination occurs to give 1-methylcyclohexene.

10.17 S_N1 reactivity:

most reactive ⟶ least reactive

a) C_6H_5—$C(CH_3)_2Cl$ > $(CH_3)_3C{-}Cl$ > $CH_3CH_2CH(NH_2)CH_3$

(most stable carbocation)

b) $(CH_3)_3C{-}Br$ > $(CH_3)_3C{-}F$ > $(CH_3)_3C{-}OH$

(best leaving group)

c) $(C_6H_5)_3CBr$ > C_6H_5—$CH(CH_3)Br$ > C_6H_5—CH_2Br

(most stable carbocation)

10.18 S_N2 reactivity:

most reactive ⟶ least reactive

a) $H_2C{=}CHCH_2Cl$ (primary allylic carbon atom) > $CH_3CH_2CH_2Cl$ > $CH_3CH_2CHClCH_3$ (secondary carbon atom)

b) $(CH_3)_2CHCH_2Br$ (least sterically hindered carbon atom) > $(CH_3)_2CHCHBrCH_3$ > $(CH_3)_3CCH_2Br$

c) $CH_3CH_2CH_2OTos$ (best leaving group) > $CH_3CH_2CH_2Br$ > $CH_3CH_2CH_2OCH_3$

10.19

H, H_3C, Br, n-C_6H_{13} = H_3C, H, C—Br, n-C_6H_{13}

(*R*)-2-bromooctane

(*R*)-2-Bromooctane is a secondary bromoalkane, which can theoretically undergo both S_N1 and S_N2 substitution. All of the nucleophiles listed are very reactive, however, and all reactions proceed by an S_N2 mechanism. Since S_N2 reactions proceed with inversion of configuration, the configuration at the chiral carbon atom is inverted. (This does not necessarily mean that all *R* isomers become *S* isomers after an S_N2 reaction; the *R*-*S* designation refers to the priorities of groups, and priorities may change when the nucleophile is varied.)

Nu: + H_3C(H)C—Br(n-C_6H_{13}) ⟶ Nu—C(H)(CH_3)(n-C_6H_{13}) + :Br^-

	nucleophile	product	
a)	$:\bar{C}N$	NC—C(H)(CH_3)(n-C_6H_{13})	*S*
b)	$CH_3CO:^-$ (C=O)	$CH_3C(=O)O$—C(H)(CH_3)(n-C_6H_{13})	*S*
c)	$CH_3\bar{S}:$	CH_3S—C(H)(CH_3)(n-C_6H_{13})	*S*
d)	Br^-:	Br—C(H)(CH_3)(n-C_6H_{13}) *S* + H_3C(H)C—Br(n-C_6H_{13})	*R*

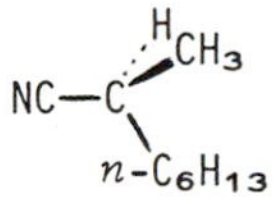

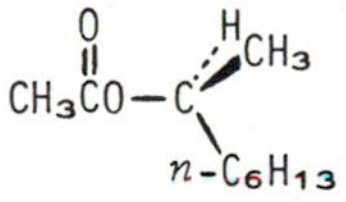

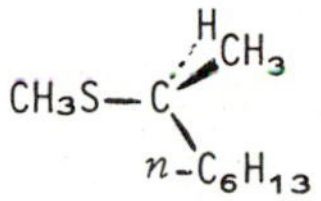

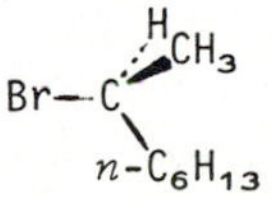

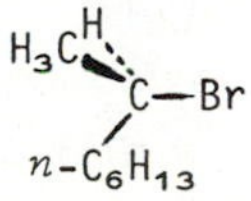

2-Bromooctane is 100% racemized after 50% of the original

(*R*)-2-bromooctane has reacted with $Br\!:^-$.

10.20 a) The rates of both S_N1 and S_N2 reactions are affected by the use of polar solvents. S_N1 reactions are accelerated because polar solvents stabilize developing charges in the transition state. Most S_N2 reactions, however, are slowed down by polar *protic* solvents because these solvents hydrogen-bond to the nucleophile and decrease its reactivity. Polar aprotic solvents solvate nucleophiles without hydrogen bonding and increase nucleophile reactivity in S_N2 reactions.

b) Good leaving groups (weak bases whose negative charge can be dispersed) increase the rates of S_N1 and S_N2 reactions.

c) A good attacking nucleophile accelerates the rate of an S_N2 reaction. Since the nucleophile is involved in the rate-limiting step of an S_N2 reaction, a good attacking nucleophile lowers the energy of the transition state and increases the rate of reaction. Choice of nucleophile has no effect on the rate of an S_N1 reaction because attack of the nucleophile occurs after the rate-limiting step.

d) Because the rate-limiting step in an S_N2 reaction involves attack of the nucleophile on the substrate, any factor that makes approach of the nucleophile more difficult slows down the rate of reaction. Especially important is the degree of crowding at the reacting carbon atom. Tertiary carbon atoms are too crowded to allow S_N2 substitution to occur. Even steric hindrance one carbon atom away from the reacting site causes a drastic slowdown in rate of reaction.

The rate-limiting step in an S_N1 reaction involves formation of a carbocation. Any structural factor in the substrate that stabilizes carbocations will increase the rate of reaction. Substrates that are tertiary, allylic, or benzylic react the fastest.

10.21

$$\text{cyclohexyl-}\ddot{\underset{\cdot\cdot}{O}}\!:^- + CH_3I \longrightarrow \text{cyclohexyl-}OCH_3$$

This is an excellent method of ether preparation since iodomethane is very reactive in S_N2 displacements.

$$CH_3\ddot{\underset{\cdot\cdot}{O}}\!:^- + \text{cyclohexyl-}I \longrightarrow \text{cyclohexyl-}OCH_3 + \text{cyclohexene}$$

Reaction of a secondary haloalkane with a basic nucleophile yields both substitution and elimination products. This is obviously a less satisfactory method of ether preparation.

10.22

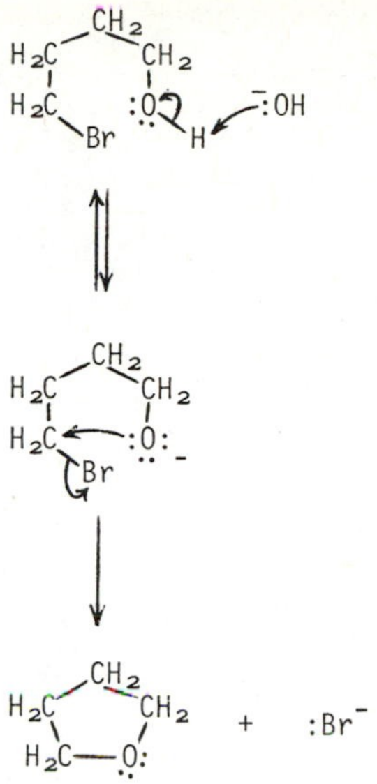

$$H_2C(-CH_2-CH_2-\ddot{O}-H)(-H_2C-Br) \quad \curvearrowleft {}^{-}{:}OH$$

Hydroxide removes a proton from the hydroxyl group of 4-bromo-1-butanol.

$$\rightleftarrows$$

$$H_2C(-CH_2-CH_2-{:}\ddot{O}{:}^-)(-H_2C-Br)$$

S_N2 displacement of :Br$^-$ by the alkoxide oxygen yields the cyclic ether tetrahydrofuran.

$$\downarrow$$

$$\text{(ring: } H_2C-CH_2-CH_2-\ddot{O}{:}-H_2C) \quad + \quad {:}Br^-$$

tetrahydrofuran

1,4-butanediol, $HOCH_2CH_2CH_2CH_2OH$ is also produced.

10.23 $BrCH_2CH_2Br \; + \; 2\ NaOH \longrightarrow HOCH_2CH_2OH$

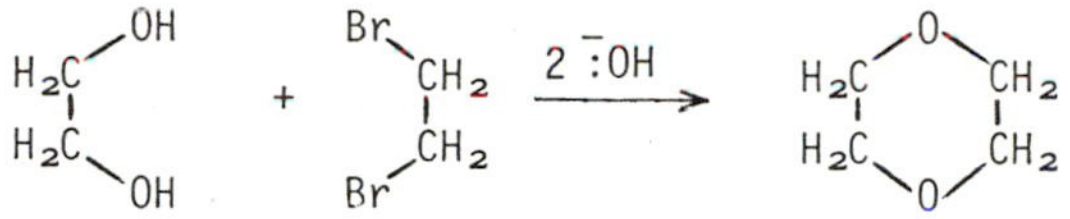

$$HOCH_2CH_2OH \; + \; BrCH_2CH_2Br \xrightarrow{2\ {}^{-}{:}OH} \text{(ring: } O-CH_2-CH_2-O-CH_2-CH_2)$$

10.24 a)

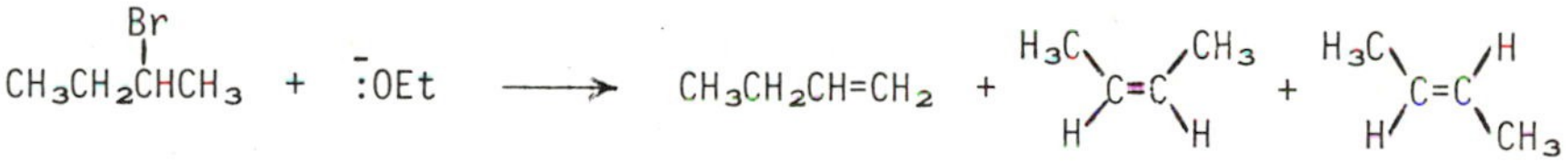

$$CH_3CH_2CH(Br)CH_3 \; + \; {}^{-}{:}OEt \longrightarrow CH_3CH_2CH{=}CH_2 \; + \; (H_3C)(H)C{=}C(CH_3)(H) \; + \; (H_3C)(H)C{=}C(H)(CH_3)$$

b) [bicyclic bromide with H_3C, CH_3 and Br] does not undergo nucleophilic substitution (see Problems 10.1 and 10.6).

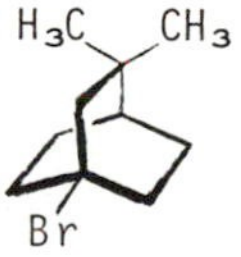

c) [chloro-methylcyclohexane chair with H, H, H, Cl, CH_3, H] $\xrightarrow{E_2}$ [methylcyclohexene with H, H, H, CH_3]

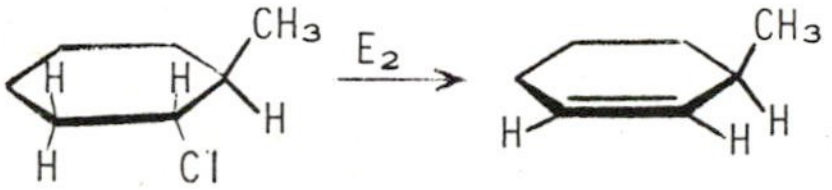

This haloalkane gives the less substituted cycloalkene (non-Zaitsev product). Elimination to form Zaitsev product is not likely to occur because the -Cl and -H involved cannot assume the *anti*-periplanar geometry preferred for E_2 elimination.

d) $(CH_3)_3C\text{-}OH \; + \; HCl \xrightarrow{0°} (CH_3)_3C\text{-}Cl$

10.25

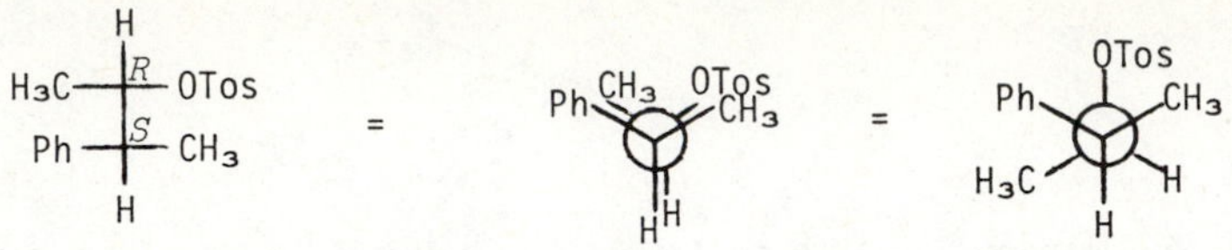

H_3C—[R]—OTos, Ph—[S]—CH_3 (Fischer projection, H top and bottom) = Newman projections

Draw the Fischer projection of the tosylate of (2*R*,3*S*)-3-phenyl-2-butanol. Now, draw the Newman projection that corresponds to the Fischer projection. The Newman projection can be rotated until the -OTos and the -H on the adjoining carbon atom are *anti*-periplanar. Even though this conformation has several *gauche* interactions, it is the only conformation in which -OTos and -H are 180° apart.

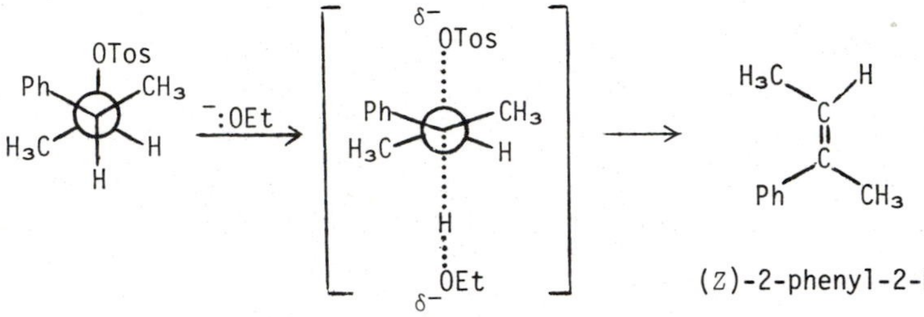

$\xrightarrow{^-:OEt}$ [δ^-OTos ⋯ H ⋯ OEtδ^-] $\longrightarrow$

(*Z*)-2-phenyl-2-butene

Elimination yields the *Z* isomer of 2-phenyl-2-butene. Refer to Chapter 5 for the method of assigning *E*,*Z* designation.

10.26 By the same arguments used in Problem 10.25, you can show that elimination from (2*R*,3*R*)-3-phenyl-2-butyl tosylate gives the *E*-alkene.

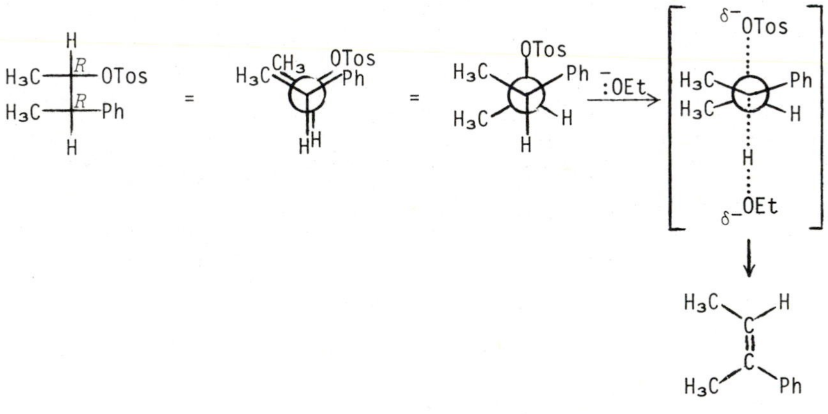

H_3C—[R]—OTos, H_3C—[R]—Ph (Fischer projection, H top and bottom) = Newman projections $\xrightarrow{^-:OEt}$ [δ^-OTos ⋯ H ⋯ OEtδ^-] ↓

(*E*)-2-phenyl-2-butene

The 2*S*,3*S* isomer also forms the *E*-alkene; the 2*S*,3*R* isomer leads to *Z*-alkene.

10.27 (*E*)-2-chloro-2-butene-1,4-dioic acid (HOOC–C(H)=C(Cl)–COOH)   (*Z*)-2-chloro-2-butene-1,4-dioic acid (HOOC–C(H)=C(Cl)–COOH)

Hydrogen and chlorine are *anti* to each other in the *Z* isomer and are *syn* in the *E* isomer. Since the *Z* isomer reacts fifty times faster than the *E* isomer, elimination must proceed more favorably when the substituents to be eliminated are *anti* to one another. This is the same stereochemical result as occurs in eliminations of alkyl halides.

10.28 a) 1-bromo-2-methylcyclohexane $\xrightarrow{^-:OC_2H_5}$ 3-methylcyclohexene

Only this alkene can result from *anti*-periplanar E_2 elimination.

b) $(CH_3)_2CHC(CH_3)(Br)CH_2CH_3 \xrightarrow[\text{heat}]{\text{HOAc}} (CH_3)_2C{=}C(CH_3)CH_2CH_3$

This alkene has the most substituted double bond.

c) $C_6H_{11}C(CH_3)_2Br \xrightarrow{^-:OC(CH_3)_3} C_6H_{10}{=}C(CH_3)_2$

This product contains the most substituted double bond.

10.29 Since 2-butanol is a secondary alcohol, substitution can occur by either an S_N1 or an S_N2 route, depending on reaction conditions. Two factors favor an S_N1 mechanism in this case. (1) The reaction is run under solvolysis (solvent as nucleophile) conditions in a polar, protic solvent. (2) Dilute acid converts a poor leaving group (-OH) into a good leaving group ($-\overset{+}{O}H_2$), which dissociates from 2-butanol more easily. The mechanism is given on the following page.

Protonation of oxygen...

...is followed by loss of water to form a planar carbocation.

Attack of water from either side of the planar cation yields racemic product.

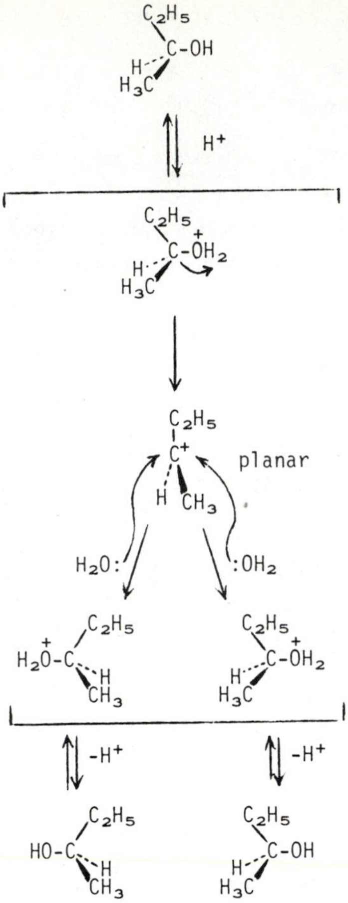

10.30 The chiral tertiary alcohol (*R*)-3-methyl-3-hexanol reacts with HBr by an S_N1 pathway. H^+ protonates the hydroxyl group, which dissociates to yield a planar, achiral carbocation. Attack by the nucleophilic bromide anion can occur from either side of the carbocation to produce (±)3-bromo-3-methylhexane.

10.31 Since carbon-deuterium bonds are slightly stronger than carbon-hydrogen bonds, more energy is required to break a C-D bond than to break a C-H bond. In a reaction where either a carbon-deuterium or a carbon-hydrogen bond is broken in the rate-limiting step, a higher percentage of C-H bond-breaking will occur because the energy of activation for C-H breakage is lower.

In the E_2 reaction of this problem, the transition state involves breaking either a C-H or C-D bond.

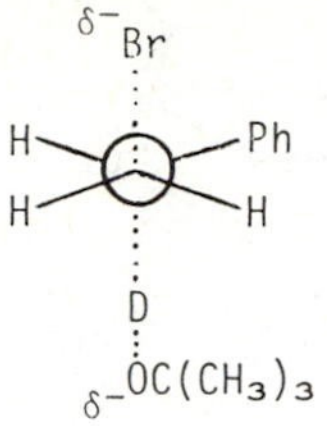

$A^{\ddagger}$ 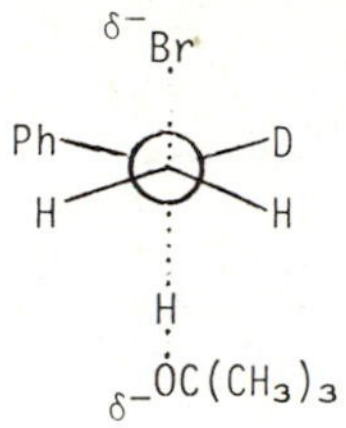 $B^{\ddagger}$

Transition state $A^{\ddagger}$ is of higher energy than transition state $B^{\ddagger}$ because more energy is required to break the C-D bond. The product that results from transition state $B^{\ddagger}$ is thus formed in greater abundance.

10.32 One of the steric requirements of E_2 elimination is the need for periplanar geometry, which optimizes orbital overlap in the transition state leading to alkene product. Two types of periplanar arrangements of substituents -- *syn* and *anti* -- are possible.

A model of the deuterated bromo compound shows that the deuterium, bromine, and the two carbon atoms that will constitute the double bond all lie in a plane. This arrangement of atoms leads to *syn* elimination. Even though *anti* elimination is usually preferred, it does not occur for this compound. Models show that *anti*-periplanar arrangement of bromine, hydrogen, and the two carbons can't occur because of the rigidity of the molecule.

10.33 We concluded in Problem 10.32 that E_2 elimination in compounds of this bicyclic structure occurs with *syn*-periplanar geometry.

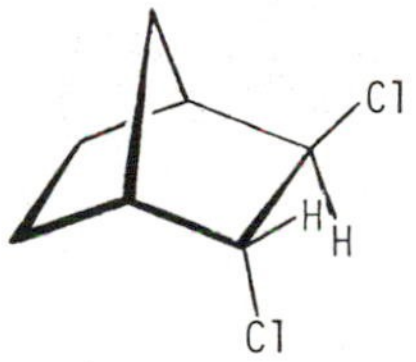

A 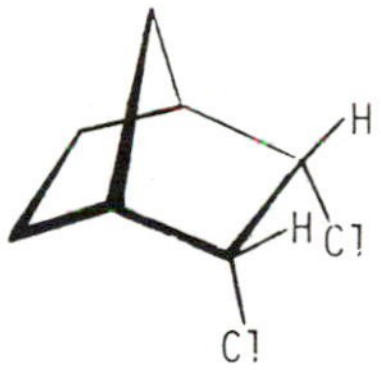 B

In compound A, -H and -Cl can be eliminated via the *syn*-periplanar route. Since neither *syn* nor *anti*-periplanar elimination is possible for B, elimination occurs by a slower E_1 route.

10.34

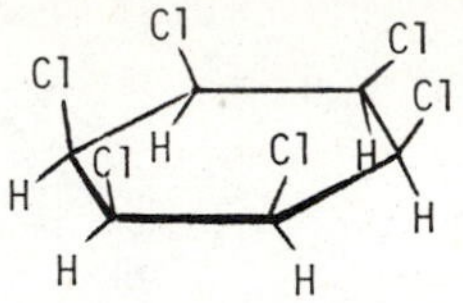

Diastereomer *8* reacts much more slowly in an E_2 reaction. No pair of hydrogen and chlorine atoms can assume the *anti*-periplanar orientation preferred for E_2 elimination.

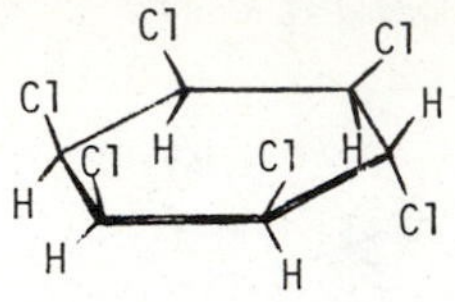

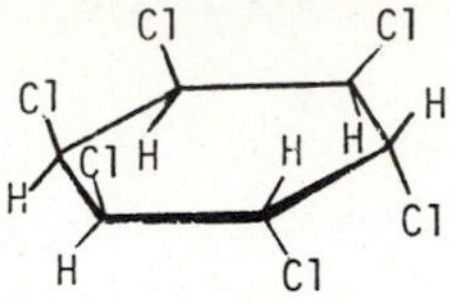

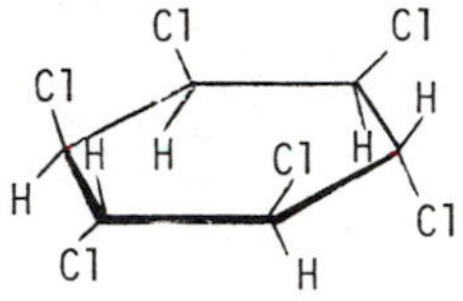

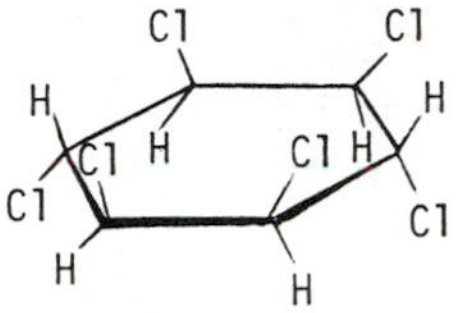

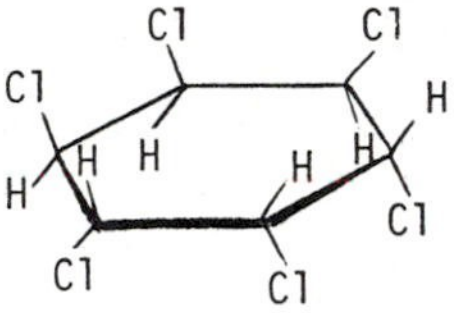

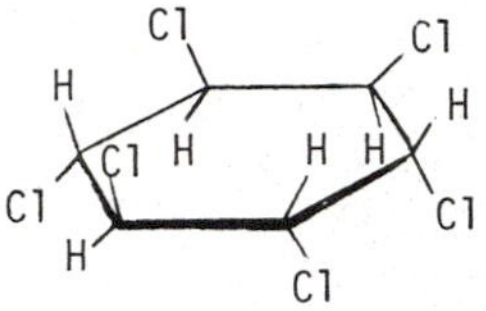

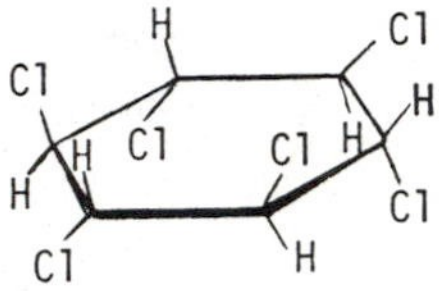

10.35 The two pieces of evidence indicate that the reaction proceeds by an S_N2 mechanism. S_N2 reactions proceed much faster in polar aprotic solvents such as DMF, and methyl esters react faster than ethyl esters. This reaction is an S_N2 displacement on a methyl ester by iodide.

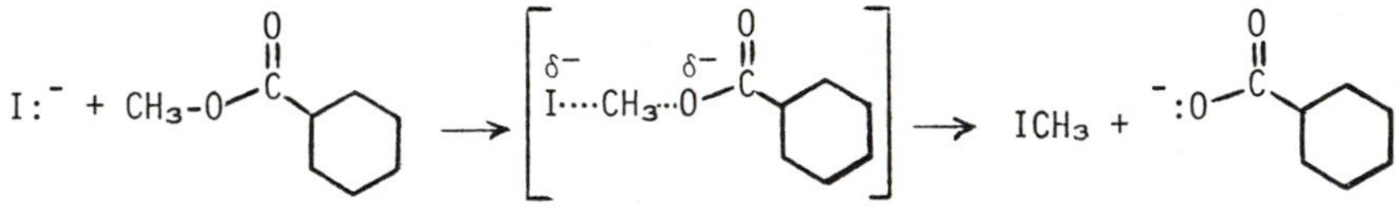

$$I:^- + CH_3\text{-}O\text{-}C(=O)\text{-}C_6H_{11} \rightarrow \left[\overset{\delta-}{I}\cdots CH_3\cdots\overset{\delta-}{O}\text{-}C(=O)\text{-}C_6H_{11}\right] \rightarrow ICH_3 + {}^-:O\text{-}C(=O)\text{-}C_6H_{11}$$

Other experiments can provide additional evidence for an S_N2 mechanism. We can determine if the reaction is second-order by varying the concentration of LiI. We can also vary the type of nucleophile to distinguish the mechanism from an S_N1 mechanism, which does not depend on the type of nucleophile.

10.36 $Cl:^-$ is a relatively poor leaving group and acetate is a relatively poor nucleophile; a displacement involving these two groups proceeds at a very slow rate. $I:^-$, however, is both a good nucleophile and a good leaving group. 1-Chlorooctane therefore reacts preferentially with iodide to form

1-iodooctane. Only a small amount of 1-iodooctane is formed (because of the low concentration of iodide ion) but 1-iodooctane is more reactive than 1-chlorooctane toward substitution by acetate. Reaction with acetate produces 1-octyl acetate and regenerates iodide ion. The whole process can now be repeated with another molecule of 1-chlorooctane. The net result is production of 1-octyl acetate; no iodide is consumed.

10.37 Two optically inactive structures are possible for compound A. Any other structure consistent with the series of reactions is optically active.

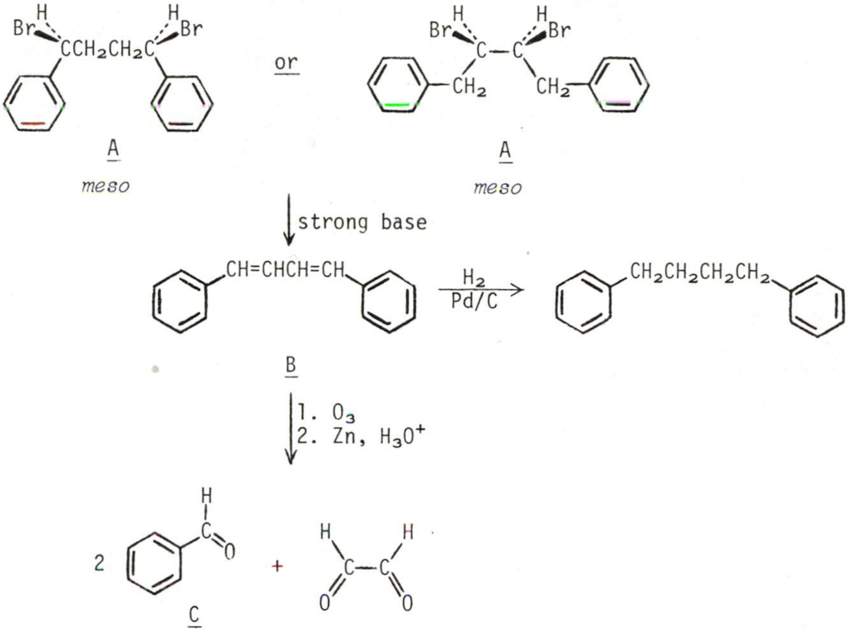

H Br CCH₂CH₂C H Br (with two phenyl rings)   or   H Br C—C H Br, CH₂ CH₂ (with two phenyl rings)

A meso   A meso

strong base

CH=CHCH=CH (phenyl at each end) $\xrightarrow[\text{Pd/C}]{H_2}$ $CH_2CH_2CH_2CH_2$ (phenyl at each end)

B

1. O_3
2. Zn, H_3O^+

2 C (benzaldehyde, H–C=O) + H–C(=O)–C(=O)–H

10.38

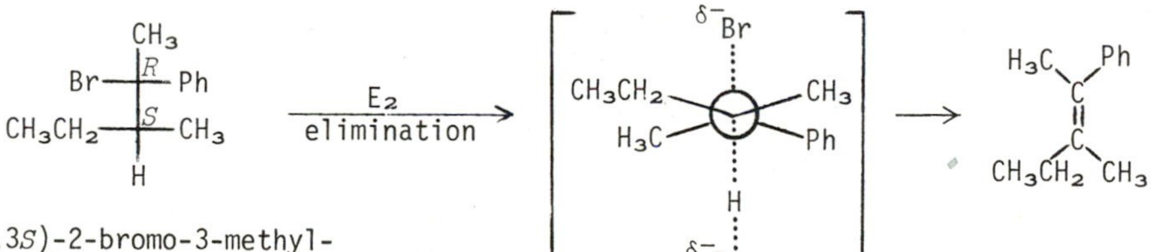

(2*R*,3*S*)-2-bromo-3-methyl-2-phenylpentane: CH_3, Br, Ph, CH_3CH_2, CH_3, H $\xrightarrow[\text{elimination}]{E2}$ [δ^-Br; CH_3CH_2, CH_3, H_3C, Ph; H; δ^-Base]   ⟶ H_3C, Ph, C=C, CH_3CH_2, CH_3

(*E*)-3-methyl-2-phenyl-2-pentene

The 2*S*,3*R* isomer also yields *E* product.

What you should know:

After doing these problems you should be able to:

1. Understand the nature of S_N2, S_N1, E_2, and E_1 reactions;
2. Classify reactions as S_N2, S_N1, E_2, or E_1, and predict their products;
3. Understand the effects of substrate structure, nucleophile, leaving group, and of solvent on substitutions and elimination;
4. Use substitution and elimination reactions in synthetic sequences;
5. Understand the stereochemistry of substitutions and eliminations, and predict the stereochemically correct products of these reactions.

CHAPTER 11 STRUCTURE DETERMINATION. MASS SPECTROSCOPY AND INFRARED SPECTROSCOPY

11.1 The following systematic approach may be helpful.

a) $M^{+\cdot} = 86$

1. Find the compound of molecular weight 86 that contains only -C and -H. Remember that a hydrocarbon with n carbon atoms can contain no more than $2n+2$ hydrogen atoms. Here, C_6H_{14} is the correct formula.
2. Find the formula corresponding to $M^{+\cdot} = 86$ that contains carbon, hydrogen, and one oxygen atom. If one oxygen atom (atomic weight = 16) is added to the base formula from step one, one carbon atom (atomic weight 12) and four hydrogen atoms (atomic weight 4) must be removed. This formula is $C_5H_{10}O$.
3. Proceed to find the remaining molecular formulas. Each time one oxygen is added, one carbon and four hydrogens must be removed. The remaining formulas for $M^{+\cdot} = 86$ are $C_4H_6O_2$, and $C_3H_2O_3$.

b) $M^{+\cdot} = 128$. The procedure is the same as in part a). The hydrocarbon having $M^{+\cdot} = 128$ is C_9H_{20}. The formula containing one oxygen is $C_8H_{16}O$. The remaining formulas are $C_7H_{12}O_2$, $C_6H_8O_3$, $C_5H_4O_4$.

c) $M^{+\cdot} = 156$. Possible formulas are $C_{11}H_{24}$, $C_{10}H_{20}O$, $C_9H_{16}O_2$, $C_8H_{12}O_3$, $C_7H_8O_4$, $C_6H_4O_5$.

d) $M^{+\cdot} = 180$. Possible formulas are $C_{13}H_{24}$, $C_{12}H_{20}O$, $C_{11}H_{16}O_2$, $C_{10}H_{12}O_3$, $C_9H_8O_4$, $C_8H_4O_5$.

11.2 Use the method described in 11.1a to solve this problem. The hydrocarbon (containing only C and H) having $M^{+\cdot} = 218$ is $C_{16}H_{26}$. Since nootkatone also contains oxygen, we must consider only those formulas that include oxygen. Using the previous procedure, we can determine that $C_{15}H_{22}O$, $C_{14}H_{18}O_2$, $C_{13}H_{14}O_3$, $C_{12}H_{10}O_4$, $C_{11}H_6O_5$ are possible formulas for nootkatone. The actual formula of nootkatone is $C_{15}H_{22}O$.

11.3 Each carbon atom has a 1.11% probability of being ^{13}C and a 98.89% probability of being ^{12}C. The ratio of the height of the ^{13}C peak to the height of the ^{12}C peak for a one-carbon compound is $\frac{1.11}{98.9} \times 100\% = 1.12\%$. For a six-carbon compound the contribution to $(M+1)^{+\cdot}$ from ^{13}C is;

$6 \times \frac{1.11}{98.9} \times 100\% = 6.72\%$. For benzene, the relative height of $(M+1)^{+\cdot}$ is 6.72% of the height of $M^{+\cdot}$.

(A similar line of reasoning can be used to calculate the contribution to $(M+1)^{+\cdot}$ from ^2H. The natural abundance of ^2H is 0.015%; the ratio of a ^2H peak to a ^1H peak for a one-hydrogen compound is 0.015%. For a six-hydrogen compound the contribution to $(M+1)^{+\cdot}$ from ^2H is 6 x 0.015% = 0.09%.)

For benzene, $(M+1)^{+\cdot}$ is 6.72% + 0.09% = 6.81% of $M^{+\cdot}$. Notice that ^2H contributes very little to the size of $(M+1)^{+\cdot}$.

11.4 Carbon is a tetravalent element; nitrogen is a trivalent element. If the structural unit $-CH_2-$ (formula weight 14) is replaced by the structural unit -NH- (formula weight 15), the molecular weight of the resulting compound increases by one. Since all neutral hydrocarbons have even molecular weights (C_nH_{2n+2}, C_nH_{2n}, etc.) the resulting nitrogen-containing compound has an odd-numbered molecular weight and molecular ion. If two $-CH_2-$ units are replaced by two -NH- units, the molecular weight of the resulting compound increases by two and remains an even number. You can continue this argument to prove that an odd number of -NH- groups results in an odd-numbered molecular weight and molecular ion. You can also prove that an odd number of $-NH_2$ or -N- groups results in an odd-numbered molecular ion.

11.5 Because $M^{+\cdot}$ is an odd number, pyridine contains an odd number of nitrogen atoms.

If pyridine contained one nitrogen atom (atomic weight 14) the remaining atoms would have a formula weight of 65, corresponding to $-C_5H_5$. C_5H_5N is, in fact, the molecular formula of pyridine.

11.6 The structural formula of neopentane and of its molecular ion are given below.

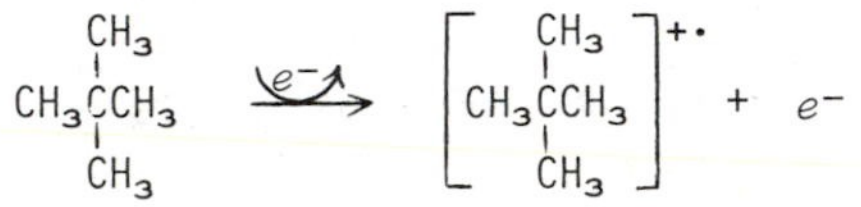

$$C(CH_3)_4 \xrightarrow{e^-} [C(CH_3)_4]^{+\cdot} + e^-$$

When the molecular ion fragments, neutral and positively charged species are produced. The fragment of m/z = 57 corresponds to $C_4H_9^+$.

The base peak usually represents the cation best able to stabilize positive charge. Since tertiary carbocations are relatively stable, $C_4H_9^+$ is most likely to be the *tert*-butyl cation.

11.7

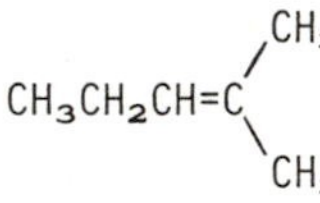

$CH_3CH_2CH{=}C(CH_3)_2$ — 2-methyl-2-pentene

$CH_3CH_2CH_2CH{=}CHCH_3$ — 2-hexene

Fragmentation occurs to a greater extent at the weakest carbon-carbon bonds; the positive charge remains with the fragment that is more able to stabilize it. A table of bond dissociation energies (Table 4.6) shows that allylic bonds have lower bond dissociation energies than the other bonds in these two compounds. The principal fragmentations of these compounds yield allylic cations.

$$^{+}CH_2CH{=}C(CH_3)_2 \qquad m/z = 69$$

$$^{+}CH_2CH{=}CHCH_3 \qquad m/z = 55$$

Spectrum b), which has 55 as its base peak, corresponds to 2-hexene. Although spectrum a) has $m/z = 41$ as its base peak, the peak at $m/z = 69$ is almost as abundant. Spectrum a) corresponds to 2-methyl-2-pentene.

11.8 a) $E = \dfrac{2.86 \times 10^{-3} \text{ kcal/mol}}{\lambda \text{ (in cm)}}$; in this part, $\lambda = 5 \times 10^{-9}$ cm

$$= \frac{2.86 \times 10^{-3} \text{ kcal/mol}}{5 \times 10^{-9}}$$

$$= 0.57 \times 10^{6} \text{ kcal/mol}$$

$$= 5.7 \times 10^{5} \text{ kcal/mol for gamma rays}$$

b) $E = 9.5 \times 10^{3}$ kcal/mol for X-rays

c) $\nu = \dfrac{c}{\lambda}$; $\lambda = \dfrac{c}{\nu} = \dfrac{3 \times 10^{10} \text{ cm/sec}}{6 \times 10^{15} \text{ Hz}} = 5 \times 10^{-6}$ cm

$$E = \frac{2.86 \times 10^{-3} \text{ kcal/mol}}{5 \times 10^{-6}} = 5.7 \times 10^{2} \text{ kcal/mol for ultraviolet light}$$

d) $E = 68$ kcal/mol for visible light

e) $E = 1.4$ kcal/mol for infrared radiation

f) $E = 9.5 \times 10^{-3}$ kcal/mol for microwave radiation

g) Imagine that your favorite AM radio station is at 100 kHz. This frequency is equivalent to 100×10^{3} Hz, or 1×10^{5} Hz.

$$\lambda = \frac{c}{\nu} = \frac{3 \times 10^{10} \text{ cm/sec}}{1 \times 10^{5} \text{ Hz}} = 3 \times 10^{5} \text{ cm}$$

$$E = \frac{2.86 \times 10^{-3} \text{ kcal/mol}}{3 \times 10^{5}} = 9.5 \times 10^{-9} \text{ kcal/mol for AM radio waves at 100 kHz.}$$

11.9 Wavenumber $= \dfrac{1}{\text{wavelength}}$; wavenumber has units of cm^{-1}. 1 μm = 10^{-4} cm

a) 3.10 μm = 3.10×10^{-4} cm; $\dfrac{1}{3.1 \times 10^{-4} \text{ cm}} = 3225 \text{ cm}^{-1}$

b) 5.85 μm; 1710 cm^{-1}

c) 6.75 μm; 1480 cm^{-1}

d) $\dfrac{1}{2250 \text{ cm}^{-1}} = 4.44 \times 10^{-4} \text{ cm} = 4.44$ μm

e) 970 cm^{-1}; 10.3 μm

f) 1560 cm^{-1}; 6.41 μm

11.10 a) A compound with a strong absorption at 1710 cm^{-1} contains a carbonyl group; it is either a ketone or aldehyde.

b) A nitro compound has a strong absorption at 1540 cm^{-1}.

c) A compound showing both carbonyl (1720 cm^{-1}) and -OH (2500-3000 cm^{-1} broad) absorptions is a carboxylic acid.

d) This compound contains the carbonyl functional group (1735 cm^{-1}). The absorption at 3500 cm^{-1} either could be due to an -OH group, in which case the compound would have carbonyl and alcohol functional groups (*not* a carboxylic acid), or it could be due to an N-H group, in which case the compound would be an amide.

11.11 Based on what we know at present, we can identify three absorptions in this spectrum.

a) The absorption at 2100 cm^{-1} is due to a -C≡C- stretch.

b) The absorption at 2900 cm^{-1} is due to a C-H stretch.

c) The absorption at 3300 cm^{-1} is due to a ≡C-H stretch.

11.12 In this problem, all formulas must represent hydrocarbons.

a) For $M^{+\cdot}$ = 64, the only possible molecular formula is C_5H_4.

b) For $M^{+\cdot}$ = 186, possible formulas are $C_{14}H_{18}$ and $C_{15}H_6$.

c) For $M^{+\cdot}$ = 158, the only reasonable formula is $C_{12}H_{14}$.

d) Three formulas for $M^{+\cdot}$ = 220 are possible: $C_{16}H_{28}$, $C_{17}H_{16}$, and $C_{18}H_4$.

11.13

	$M^{+\cdot}$	molecular formula	degree of unsaturation
a)	86	C_6H_{14}	0
b)	110	C_8H_{14}	2
c)	146	$C_{11}H_{14}$	5
d)	190	$C_{14}H_{22}$	4
		$C_{15}H_{10}$	11

11.14

	$M^{+\cdot}$	molecular formula	degree of unsaturation	possible structure
a)	132	$C_{10}H_{12}$	5	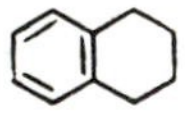
b)	166	$C_{13}H_{10}$	9	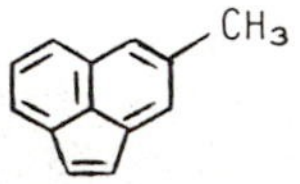 CH_3
		$C_{12}H_{22}$	2	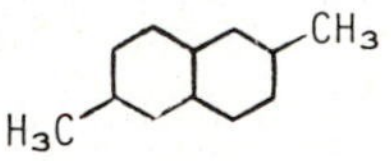 CH_3, H_3C
c)	84	C_6H_{12}	1	

11.15 Remember that compounds in this problem may contain carbon, hydrogen, oxygen, and nitrogen. In addition, the molecular ions of many compounds may have the same value of $M^{+\cdot}$. Some of the less likely molecular formulas -- those with few carbon or hydrogen atoms -- have been omitted.

a) $M^{+\cdot} = 74$. Any nitrogen-containing compound that shows a molecular ion at $M^{+\cdot} = 74$ must have an even number of nitrogen atoms.

Compounds Containing:

C, H; C_6H_2

C, H, O; $C_4H_{10}O$, $C_3H_6O_2$, $C_2H_2O_3$

C, H, N; $C_3H_{10}N_2$, CH_6N_4

C, H, N, O; $C_2H_6N_2O$, $CH_2N_2O_2$

b) $M^{+\cdot} = 131$ has an odd number of nitrogen atoms; no hydrocarbons correspond to this molecular ion.

Compounds containing:

C, H, N; C_9H_9N, $C_7H_5N_3$, $C_6H_{17}N_3$, $C_4H_{13}N_5$

C, H, N, O; $C_7H_{17}NO$, $C_6H_{13}NO_2$, $C_5H_9NO_3$, $C_4H_5NO_4$, $C_5H_{13}N_3O$, $C_4H_9N_3O_2$, $C_3H_5N_3O_3$, $C_3H_9N_5O$, C_8H_5NO

11.16 Reasonable molecular formulas for camphor are $C_{10}H_{16}O$, $C_9H_{12}O_2$, $C_8H_8O_3$. The actual formula, $C_{10}H_{16}O$, corresponds to three degrees of unsaturation. The ketone functional group accounts for one of these. Since camphor is a saturated compound, the other two degrees of unsaturation are due to two rings.

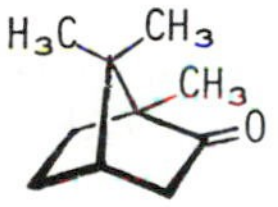

H_3C CH_3 CH_3 O

camphor

11.17 The molecular formula of nicotine is $C_{10}H_{14}N_2$. To find the base formula, subtract the number of nitrogens from the number of hydrogens. The base formula of nicotine, $C_{10}H_{12}$, indicates the presence of five degrees of unsaturation.

N N CH_3

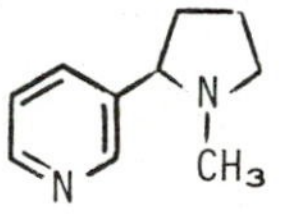

nicotine

11.18 In order to simplify this problem, neglect the ^{13}C and ^{2}H isotopes in determining the molecular ions of these compounds.

a) The formula weight of $-CH_3$ is 15; the atomic masses of the two bromine isotopes are 79 and 81. The two molecular ions of bromoethane occur at $M^{+\cdot} = 96$ (49.5%) and $M^{+\cdot} = 94$ (50.5%).

b) The formula weight of $-C_6H_{13}$ is 85; atomic masses of the two chlorine isotopes are 35 and 37. The two molecular ions of chlorohexane occur at

$M^{+\cdot}$ = 122 (24.5%) and $M^{+\cdot}$ = 120 (75.5%).

c) $M^{+\cdot}$ = 64 (24.5%) and $M^{+\cdot}$ = 62 (75.5%) are the molecular ions for vinyl chloride, C_2H_3Cl.

<u>11.19</u> Again, neglect ^{13}C and ^{2}H in these calculations.

a) Finding the molecular ions of chloroform is a statistics problem.

1) The probability that all three chlorine atoms are ^{37}Cl is $(0.245)^3 = 0.015$.

2) The probability that two chlorine atoms are ^{37}Cl and one is ^{35}Cl is $3(0.245)(0.245)(0.755) = 3(0.045) = 0.135$. The factor *3* enters the calculations because three permutations of two ^{37}Cl's and one ^{35}Cl are possible.

3) The probability that one chlorine atom is ^{37}Cl and two are ^{35}Cl is $3(0.245)(0.755)(0.755) = 3(0.140) = 0.420$.

4) The probability that all chlorine atoms are ^{35}Cl is $(0.755)^3 = 0.430$.

5) The mass of:
$CH^{37}Cl^{37}Cl^{37}Cl = 124$
$CH^{37}Cl^{37}Cl^{35}Cl = 122$
$CH^{37}Cl^{35}Cl^{35}Cl = 120$
$CH^{35}Cl^{35}Cl^{35}Cl = 118$

6) Therefore, molecular ions for chloroform occur at:

$M^{+\cdot}$	124	122	120	118
abundance	1.5%	13.5%	42.0%	43.0%

b) 1) The probability that bromochloromethane contains ^{81}Br and ^{37}Cl is $(0.495)(0.245) = 0.121$.

2) The probability of ^{81}Br and ^{35}Cl is $(0.495)(0.755) = 0.374$.

3) The probability of ^{79}Br and ^{37}Cl is $(0.505)(0.245) = 0.124$.

4) The probability of ^{79}Br and ^{35}Cl is $(0.505)(0.755) = 0.381$.

5) The mass of:
$CH_2{}^{81}Br^{37}Cl = 132$
$CH_2{}^{81}Br^{35}Cl = 130$
$CH_2{}^{79}Br^{37}Cl = 130$
$CH_2{}^{79}Br^{35}Cl = 128$

6) The molecular ions for bromochloromethane:

$M^{+\cdot}$	132	130	128
abundance	12.1%	49.8%	38.1%

c) The molecular ions for freon 12:

$M^{+\cdot}$	124	122	120
abundance	6.0%	37.0%	57.0%

11.20

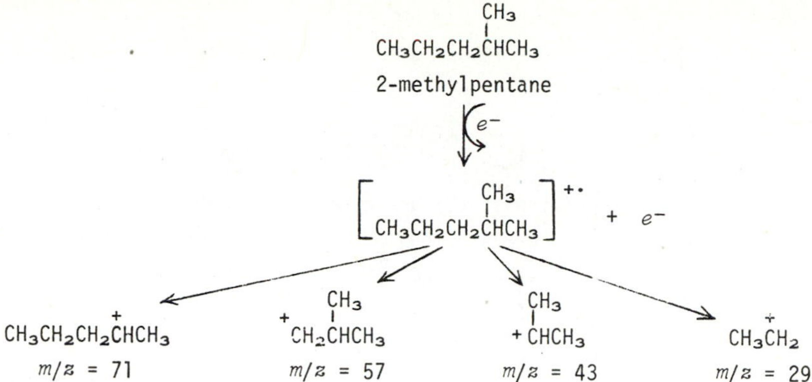

The molecular ion, at m/z = 86, is present in very low abundance. The base peak, at m/z = 43, represents a stable secondary carbocation.

11.21 Before doing the hydrogenation, familiarize yourself with the mass spectra of cyclohexene and cyclohexane. Note that $M^{+\cdot}$ is different for each compound.

After the reaction is underway, inject a sample from the reaction mixture onto the GC/MS. There should be two obvious peaks that can be unambiguously identified by their mass spectra. As the reaction proceeds, one of these peaks will increase in size and one will decrease; by this point you should be able to identify these peaks without running additional mass spectra. When the reaction is complete the gas chromatogram should show one peak whose mass spectrum is superimposable with that of cyclohexane.

11.22 See Problem 11.9 for the method of solution.

a) 3355 cm^{-1} b) 1720 cm^{-1} c) 2030 cm^{-1}

11.23 a) 5.70 μm b) 3.08 μm c) 5.80 μm

11.24 Spectrum b) differs from spectrum a) in several respects; note in particular the absorbances at 715 cm^{-1} (s), 1140 cm^{-1} (s), 1650 cm^{-1} (m), and 3000 cm^{-1} (m) in spectrum b). The absorbances at 1650 cm^{-1} (C=C stretch) and 3000 cm^{-1} (=C-H stretch) can be found in Table 11.1. They are sufficient to allow us to assign spectrum b) to cyclohexene and spectrum a) to cyclohexane.

11.25 a) $(CH_3)_3N$, a tertiary amine, does not exhibit a N-H stretching vibration at 3300-3500 cm^{-1}, as $CH_3CH_2NHCH_3$ does.

b) $CH_3CH_2COCH_3$ shows a strong ketone absorbance at 1710 cm^{-1}; $CH_3CH{=}CHCH_2OH$ shows the broad band at 3400-3640 cm^{-1} characteristic of alcohols, as well as alkene absorbances at 3000-3100 cm^{-1} and 1645-1670 cm^{-1}.

c) CH_3CH_2CHO exhibits an aldehyde band at 1725 cm^{-1}; $H_2C{=}CHOCH_3$ shows characteristic alkene absorbances, as well as a C-O stretch near 1200 cm^{-1}.

11.26

$$\text{1-methylcyclohexanol} \xrightarrow[\text{pyridine}]{POCl_3} \text{1-methylcyclohexene}$$

1-methylcyclohexanol 1-methylcyclohexene

The infrared spectrum of the starting alcohol shows a broad absorption at 3400-3640 cm^{-1}, due to an O-H stretch, and another strong absorption at 1050-1100 cm^{-1}, due to a C-O stretch. The alkene product exhibits medium intensity absorbances at 1645-1670 cm^{-1} and at 3000-3100 cm^{-1}. Monitoring the *disappearance* of one of the alcohol absorptions allows one to decide when the alcohol is totally dehydrated. It is also possible to monitor the *appearance* of one of the alkene absorbances.

11.27 The peak of maximum intensity (base peak) in the mass spectrum occurs at m/z = 67. This peak does *not* represent the molecular ion, since $M^{+\cdot}$ of a hydrocarbon must be an even number. Careful inspection reveals the molecular ion peak at m/z = 68. $M^{+\cdot}$ = 68 corresponds to a hydrocarbon of molecular formula C_5H_8 with a degree of unsaturation of two.

Fairly intense peaks in the mass spectrum occur at m/z = 67, 53, 40, 39, and 27. The peak at m/z = 67 correspondes to loss of one hydrogen atom, and the peak at m/z = 53 represents loss of a methyl group. The unknown hydrocarbon thus contains a methyl group.

Significant IR absorptions occur at 2130 cm^{-1} (-C≡C- stretch) and at 3320 cm^{-1} (≡C-H stretch). These bands indicate that the unknown hydrocarbon is a terminal alkyne. Possible structures for C_5H_8 are $CH_3CH_2CH_2C{\equiv}CH$ and $(CH_3)_2CHC{\equiv}CH$. [In fact, 1-pentyne is correct.]

11.28 The molecular ion, $M^{+\cdot}$ = 70, corresponds to the molecular formula C_5H_{10}. This compound has one double bond or ring.

The base peak in the mass spectrum occurs at m/z = 55. This peak represents loss of a methyl group from the molecular ion and indicates the presence of a methyl group in the unknown hydrocarbon. All other peaks occur with low intensity.

In the IR spectrum, it is possible to distinguish absorbances at 1660 cm^{-1} and at 3000 cm^{-1}; the 2960 cm^{-1} absorption is rather hard to detect because it occurs as a shoulder on the alkane C-H stretch at 2850-2960 cm^{-1}. These two absorptions are due to a double bond.

Possible structures for C_5H_{10} alkenes are $CH_3CH_2CH_2CH{=}CH_2$, $CH_3CH_2CH{=}CHCH_3$, $(CH_3)_2CHCH{=}CH_2$, $CH_3CH_2C(CH_3){=}CH_2$, and $(CH_3)_2C{=}CHCH_3$. [In fact, 2-methyl-2-butene is correct.]

11.29 Possible molecular formulas containing carbon, hydrogen, and oxygen and having $M^{+\bullet}$ = 150 are $C_{10}H_{14}O$, $C_9H_{10}O_2$, and $C_8H_6O_3$. The first formula has four degrees of unsaturation, the second has five degrees of unsaturation, and the third has six degrees of unsaturation. Since carvone has three double bonds (including the ketone) and one ring, $C_{10}H_{14}O$ is the correct molecular formula for carvone.

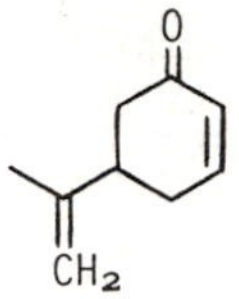

O

CH_2

carvone

What you should know:

After doing these problems, you should be able to:

1. Write molecular formulas containing carbon, hydrogen, nitrogen, and oxygen for a given value of $M^{+\bullet}$;
2. Use mass spectra to determine molecular weights and base peaks and to distinguish between hydrocarbons;
3. Calculate the energy of electromagnetic radiation; convert from wavelength to wavenumber, and *vice versa*;
4. Identify functional groups by their infrared absorptions.

CHAPTER 12 STRUCTURE DETERMINATION. NUCLEAR MAGNETIC RESONANCE SPECTROSCOPY

12.1 $$E = \frac{2.86 \times 10^{-3} \text{ kcal/mol}}{\lambda \text{ (in cm)}}$$

$$\lambda = \frac{c}{\nu} = \frac{3 \times 10^{10} \text{ cm/sec}}{\nu}$$

here ν = 56 MHz, or 5.6×10^7 Hz

so $$\lambda = \frac{3 \times 10^{10} \text{ cm/sec}}{5.6 \times 10^7 \text{ Hz}} = 0.54 \times 10^3 \text{ cm}$$

$$E = \frac{2.86 \times 10^{-3} \text{ kcal/mol}}{0.54 \times 10^3} = 5.3 \times 10^{-6} \text{ kcal/mol}$$

Compare this value with $E = 5.7 \times 10^{-6}$ kcal/mol for ^1H. It takes less energy to spin-flip a ^{19}F nucleus than to spin-flip a ^1H nucleus.

12.2 $$\lambda = \frac{c}{\nu} = \frac{3 \times 10^{10} \text{ cm/sec}}{\nu}$$

here ν = 100 MHz = 100×10^6 Hz, or 10^8 Hz

so $$\lambda = \frac{3 \times 10^{10} \text{ cm/sec}}{10^8 \text{ Hz}} = 3 \times 10^2 \text{ cm}$$

$$E = \frac{2.86 \times 10^{-3} \text{ kcal/mol}}{3 \times 10^2} = 9.5 \times 10^{-6} \text{ kcal/mol}$$

Increasing the spectrometer frequency increases the amount of energy needed for resonance.

12.3 a) $(H_3C)_2C{=}C(CH_3)_2$

This alkene has two different types of carbon atoms and shows two signals in the ^{13}C NMR. Since all protons are equivalent; only one ^1H NMR signal appears.

b) cyclohexane ring of six CH_2 groups (H_2C, CH_2, CH_2, CH_2, CH_2, H_2C)

All carbon atoms are equivalent, as are all hydrogen atoms. Consequently, the ^{13}C NMR spectrum and the ^1H NMR spectrum of cyclohexane each show one signal.

c) CH_3OCH_3

Dimethyl ether shows one ^{13}C NMR signal and one ^1H NMR signal.

d) $(CH_3)_3CC(=O)OCH_3$ — labels: a = $(CH_3)_3$, b = quaternary C, c = C=O, d = OCH_3

Four signals appear in the ^{13}C NMR spectrum of this ester because four different kinds of carbon atoms are present. The ^1H NMR spectrum shows two signals.

e)

^{13}C: one signal

^{1}H: one signal

f)

```
     H  CH3
     |  |
  H—C—C—CH3
     |  |
H3C—C—C—H
     |  |
   H3C  H
```

^{13}C: three signals

^{1}H: two signals

12.4 a) $\delta = \dfrac{\text{observed chemical shift (\# Hz away from TMS)}}{\text{spectrometer frequency (in Hz)}/10^6}$

We want to find the position of resonance (in Hz) when we know δ.

$2.1\ \delta = \dfrac{\text{observed chemical shift}}{60\ \text{MHz}/10^6} = \dfrac{\text{observed chemical shift}}{60\ \text{Hz}}$

126 Hz = observed chemical shift

b) If the ^{1}H NMR spectrum of acetone were recorded at 100 MHz, the position of the absorption would still be 2.1 δ because measurements expressed in δ (# of p.p.m.) are independent of the operating frequency of the NMR spectrometer.

c) $2.1\ \delta = \dfrac{\text{observed chemical shift}}{100\ \text{Hz}}$; observed chemical shift = 210 Hz

12.5 $\delta = \dfrac{\text{observed chemical shift (in Hz)}}{(60\ \text{MHz}/10^6)}$

a) $\delta = \dfrac{436\ \text{Hz}}{60\ \text{Hz}} = 7.27$ ppm for $C\mathit{H}Cl_3$

b) $\delta = \dfrac{183\ \text{Hz}}{60\ \text{Hz}} = 3.05$ ppm for $C\mathit{H}_3Cl$

c) $\delta = \dfrac{208\ \text{Hz}}{60\ \text{Hz}} = 3.47$ ppm for $C\mathit{H}_3OH$

d) $\delta = \dfrac{318\ \text{Hz}}{60\ \text{Hz}} = 5.30$ ppm for $C\mathit{H}_2Cl_2$

12.6 $\overset{4}{C}H_3\overset{3}{C}H_2\overset{2}{C}O_2\overset{1}{C}H_3$

δ (ppm)	assignment
9.3	4
27.6	3
51.4	1
174.6	2

12.7 a) $-\,-\,$[methylcyclopentane, carbons numbered 1–4] $-CH_3-\,-$ plane of symmetry

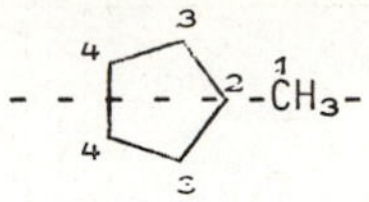

Four resonance lines are observed in the proton noise-decoupled ^{13}C NMR spectrum of methylcyclopentane.

b) [1-methylcyclohexene, carbons numbered 1–7]

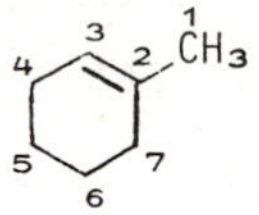

Seven resonance lines are seen. No two carbon atoms in 1-methylcyclohexene are equivalent because no plane of symmetry is present.

c) [o-xylene, carbons numbered 1–4] plane of symmetry

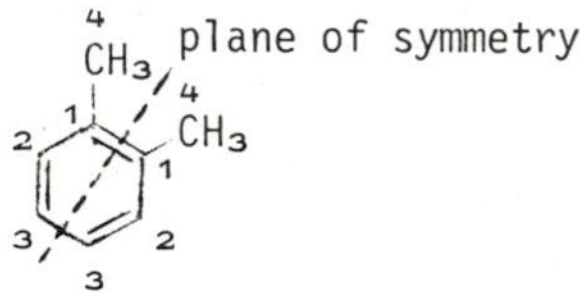

Four resonance lines are observed. A plane of symmetry causes one half of the carbon atoms to be equivalent to the other half.

d) $(CH_3)(H)C{=}C(CH_3)_2$ [carbons numbered 1–5]

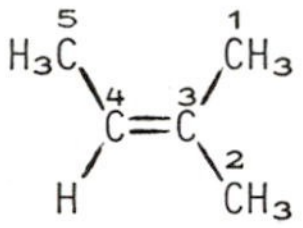

Five resonance lines are observed. Carbons 1 and 2 are non-equivalent because of the double bond stereochemistry.

12.8 Each part of this problem has several correct answers.

a) 1-Methylcyclohexene (see Prob. 12.7a) and 1,3-dimethylcyclopentene show seven resonance lines.

b) $(H_3C)_2CHCH_2CH_2CH_3$ Two of the six carbons are magnetically equivalent.

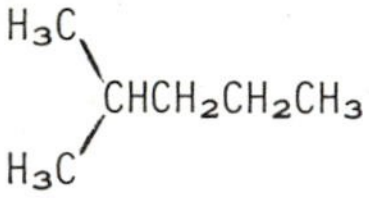

c) $(H_3C)_2CHCH_2Cl$ The two methyl groups are equivalent.

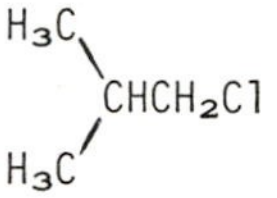

12.9 $\overset{1}{H_2C}=\overset{2}{C}H\overset{3}{C}H_3$

propene

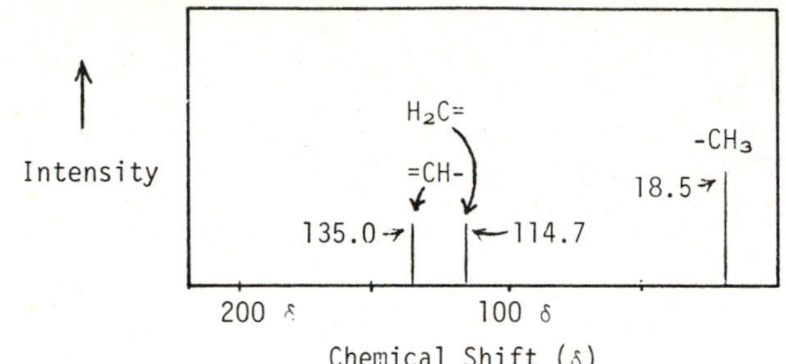

Proton noise-decoupled spectrum of propene

a) Propene has three different carbon atoms and shows three carbon resonances.

b) In an off-resonance spectrum:

The *methyl* carbon resonance splits into four peaks having a 1:3:3:1 ratio of intensity.

The $H_2\mathit{C}=$ carbon resonance splits into three peaks having a 1:2:1 ratio of intensity.

The $=\mathit{C}H-$ carbon resonance splits into two peaks having a 1:1 ratio of intensity.

12.10

- - - - - - - plane of symmetry

cyclopentene

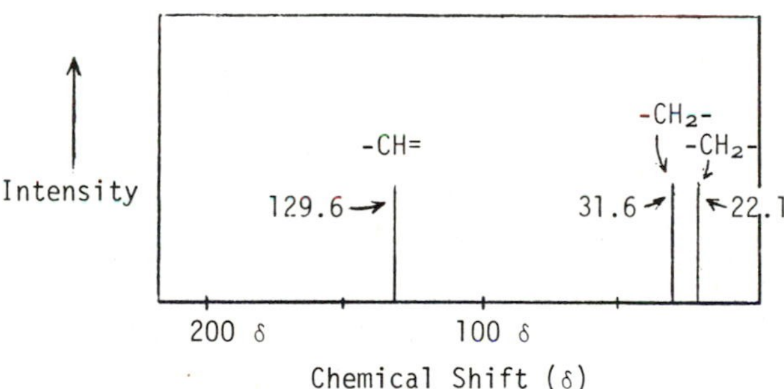

Proton noise-decoupled spectrum of cyclopentene

a) Symmetry causes two pairs of carbons to be equivalent. Three ^{13}C resonances are present.

b)

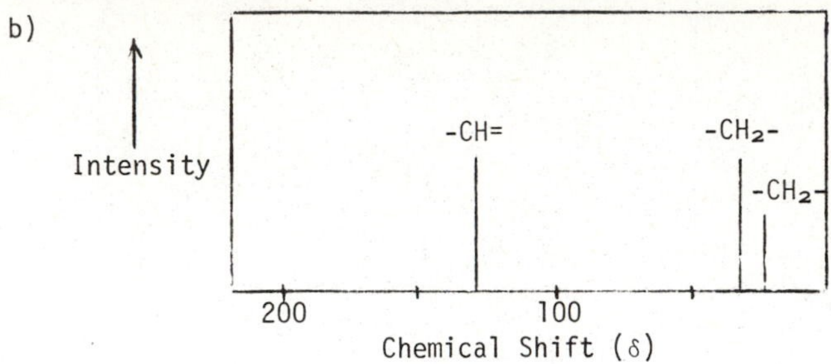

In a gated-decoupled spectrum, peak areas are proportional to the relative number of carbon atoms each peak represents.

c)

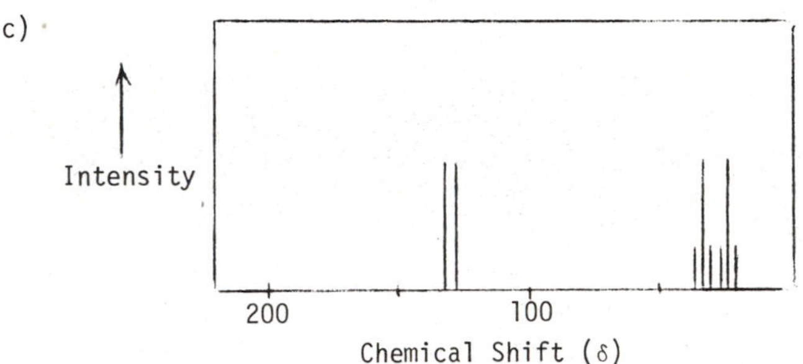

Off-resonance spectrum of cyclopentene.

12.11 Either of two products may result from addition of HBr to 2-methylpropene: $(CH_3)_2CHCH_2Br$ (non-Markovnikov product) and $(CH_3)_3CBr$ (Markovnikov product). ^{13}C NMR can easily distinguish between them.

	$(CH_3)_3CBr$	$(CH_3)_2CHCH_2Br$
number of non-equivalent carbon atoms	2	3
ratio of peak area	3:1	2:1:1
spin-spin splitting	one quartet one singlet	one quartet one triplet one doublet
approximate position of resonance	8-30 δ (q) 25-65 δ (s)	8-30 δ (q),(d) 25-65 δ (t)

The product of HBr addition to 2-methylpropene in the presence of peroxides has a ^{13}C NMR spectrum identical to that of $(CH_3)_2CHCH_2Br$.

12.12

	Compound	Kinds of non-equivalent protons
a)	$\overset{1}{C}H_3\overset{2}{C}H_2Br$	two
b)	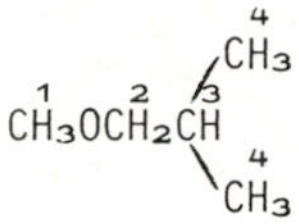 $\overset{1}{C}H_3O\overset{2}{C}H_2\overset{3}{C}H(\overset{4}{C}H_3)_2$	four
c)	$\overset{1}{C}H_3\overset{2}{C}H_2\overset{3}{C}H_2NO_2$	three
d)	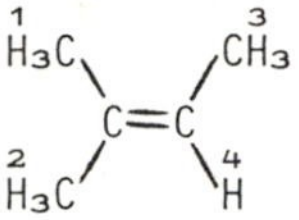 $(\overset{1}{H_3C})(\overset{2}{H_3C})C{=}C(\overset{3}{C}H_3)(\overset{4}{H})$	four

This is tricky! Methyl groups 1 and 2 are not equivalent, and the protons of any one methyl group are not necessarily equivalent to those of any other methyl group. If you try the "mental substitution" test, you can see that substitution for a proton on one methyl group yields a *Z* isomer, and substitution on the other methyl group yields an *E* isomer.

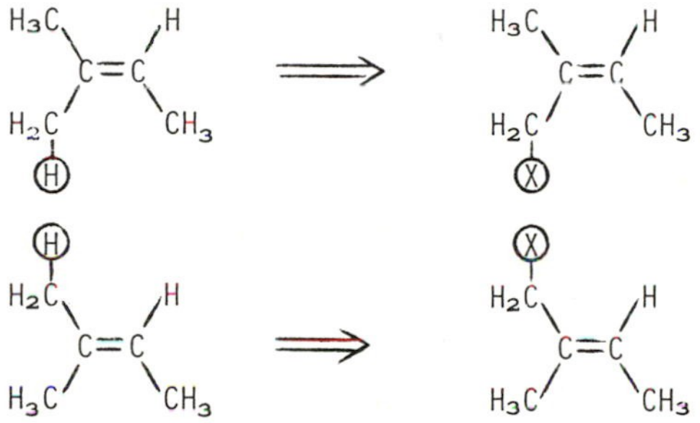

$(H_3C)(H\text{-}H_2C)C{=}C(H)(CH_3) \Longrightarrow (H_3C)(X\text{-}H_2C)C{=}C(H)(CH_3)$ Z

$(H\text{-}H_2C)(H_3C)C{=}C(H)(CH_3) \Longrightarrow (X\text{-}H_2C)(H_3C)C{=}C(H)(CH_3)$ E

e) 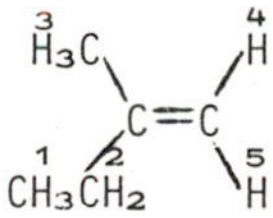 $(\overset{3}{H_3C})(\overset{1}{C}H_3\overset{2}{C}H_2)C{=}C(\overset{4}{H})(\overset{5}{H})$ five resonance lines

The two protons attached to the double bond are non-equivalent. See part (d) for explanation.

f) 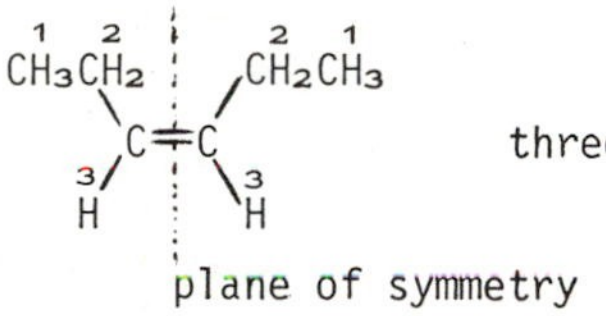 $(\overset{1}{C}H_3\overset{2}{C}H_2)(\overset{3}{H})C{=}C(\overset{2}{C}H_2\overset{1}{C}H_3)(\overset{3}{H})$ three resonance lines

plane of symmetry

12.13

	compound	δ	kind of proton
a)	cyclohexane	1.43	secondary alkyl
b)	CH_3COCH_3	2.17	methyl ketone
c)	C_6H_6	7.37	aromatic
d)	glyoxal	9.70	aldehyde
e)	CH_2Cl_2	5.30	protons adjacent to two halogens
f)	$(CH_3)_3N$	2.12	methyl protons adjacent to nitrogen
g)	*p*-dioxane	3.70	protons adjacent to an ether oxygen

12.14

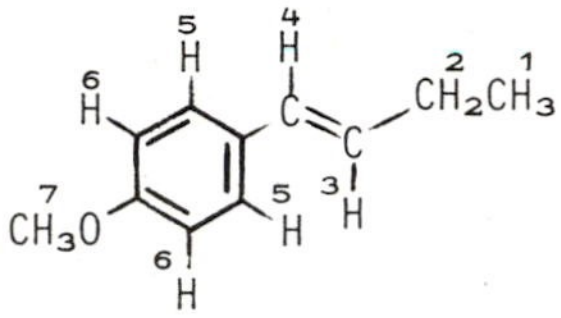

Seven different kinds of protons are present in the above structure.

proton	δ	kind of proton
1	1.0	primary alkyl
2	1.8	allylic
3	6.1	vinylic
4	6.2 (different from proton 3)	vinylic
5	7.2	aromatic
6	6.7	aromatic
7	3.8	ether

Note: The two "5" protons are equivalent to each other, as are the two "6" protons, because of free rotation around the bond joining the aromatic ring and the alkenyl functional group.

12.15

	Compound	Proton	Number of Adjacent Protons	Splitting
a)	$\overset{1}{CHBr_2}\overset{2}{CH_3}$	1	3	quartet
		2	1	doublet
b)	$\overset{1}{CH_3}O\overset{2}{CH_2}\overset{3}{CH_2}Br$	1	0	singlet
		2	2	triplet
		3	2	triplet
c)	$Cl\overset{1}{CH_2}\overset{2}{CH_2}\overset{1}{CH_2}Cl$	1	2	triplet
		2	4	quintet

Compound	Proton	Number of Adjacent protons	Splitting
d) $\overset{1}{CH_3}\overset{2}{CH_2}O\overset{O}{\overset{\|}{C}}\overset{3}{CH}(\overset{4}{CH_3})_2$	1	2	triplet
	2	3	quartet
	3	6	septet
	4	1	doublet

12.16 a) CH_3OCH_3 b) $CH_3CHClCH_3$ c) $ClCH_2CH_2OCH_2CH_2Cl$ d) $CH_3CH_2\overset{O}{\overset{\|}{C}}OCH_3$

12.17 The ^{1}H NMR spectrum shows two signals, corresponding to two types of hydrogens. These signals are in the ratio 9:14, or 2:3. Since the unknown contains 10 hydrogens, four protons are of one type and six are of the other type.

The upfield signal at 1.10 δ is due to saturated primary protons. The downfield signal at 3.50 δ is due to protons on carbon adjacent to an electronegative atom -- in this case, oxygen.

The signal at 1.1 δ is a *triplet*, and indicates two neighboring protons. The signal at 3.5 δ is a *quartet*, and indicates three neighboring protons. The unknown compound is diethyl ether, $CH_3CH_2OCH_2CH_3$.

12.18

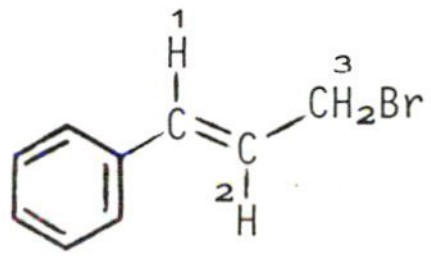

3-bromo-1-phenyl-1-propene

Coupling of the C2 proton to the C1 vinylic proton occurs with J = 16 Hz and causes the signal of the C2 proton to be split into a doublet. The C2 proton is also coupled to the two C3 protons with J = 8 Hz. This splitting causes each leg of the C2 proton doublet to be split into a triplet, producing six lines in all. Because of the size of the coupling constants, two of the lines coincide, and a quintet is observed.

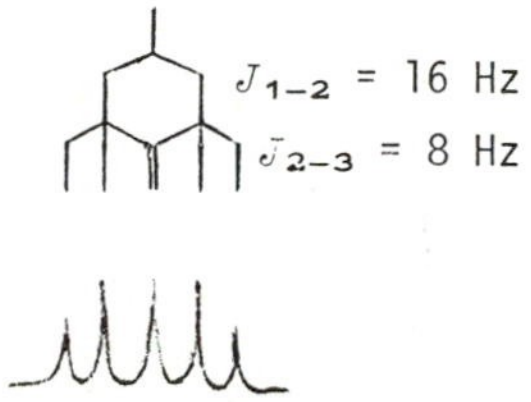

J_{1-2} = 16 Hz
J_{2-3} = 8 Hz

12.19 See Problem 12.5 for the method of solution.

a) 2.18 δ b) 4.78 δ c) 7.52 δ d) 9.05 δ

12.20 See Problem 12.4 for the method of solution.

a) 168 Hz b) 276 Hz c) 504 Hz d) 616 Hz

12.21 a) Since the symbol "δ" indicates ppm downfield from TMS, chloroform absorbs at 7.3 ppm.

b) $$\delta = \frac{\text{observed chemical shift (\# Hz from TMS)}}{\text{spectrometer frequency (in Hz)}/10^6}$$

$$7.3 = \frac{\text{chemical shift}}{360\ \text{MHz}/10^6} = \frac{\text{chemical shift}}{360\ \text{Hz}}$$

chemical shift = 2600 Hz

c) δ is still 7.3 because the chemical shift measured in δ is independent of the operating frequency of the spectrometer.

12.22 and 12.23

	Compound	Number of ^{13}C Absorptions	Splitting in off-resonance spectrum: Singlets	Doublets	Triplets	Quartets
a)	1,1-dimethylcyclohexane (H_3C CH$_3$ on carbon 2; carbons numbered 1–5)	5	1 (carbon 2)	0	3 (carbons 3, 4, 5)	1 (carbon 1)
b)	$\overset{1}{C}H_3\overset{2}{C}H_2O\overset{3}{C}H_3$	3	0	0	1 (carbon 2)	2 (carbons 1, 3)
c)	naphthalene (carbons numbered 1–3)	3	1 (carbon 1)	2 (carbons 2, 3)	0	0
d)	2-acetylbenzoic acid (C(=O)CH$_3$, COOH; carbons numbered 1–9)	9	4 (carbons 2, 3, 8, 9)	4 (carbons 4, 5, 6, 7)	0	1 (carbon 1)

12.24 ^{13}C NMR absorptions occur over a range of 250 ppm; ^{1}H NMR absorptions generally occur over a range of 10 ppm. The spread of peaks in ^{13}C NMR is much greater and accidental overlap is less likely. In addition, proton-noise decoupled ^{13}C NMR spectra are uncomplicated by spin-spin splitting. The total number of lines is smaller and, again, overlap is less common.

12.25 a) The *chemical shift* is the exact position at which a nucleus absorbs rf energy in an NMR spectrum.

b) *Spin-spin splitting* is the splitting of a single NMR resonance into multiple lines. Spin-spin splitting occurs when the effective magnetic field felt by a nucleus is influenced by the small magnetic moments of adjacent nuclei. In ^{13}C NMR the signal of a carbon bonded to n protons is split into $n+1$ peaks. In ^{1}H NMR the signal of a proton with n

neighboring protons is split into *n+1* peaks. The magnitude of spin-spin splitting is given by the coupling constant *J*.

c) The *applied magnetic field* is the magnetic field that is externally applied to a sample by an NMR spectrometer.

d) The *spectrometer operating frequency* is the frequency of applied rf energy employed by the spectrometer to bring a magnetic nucleus into resonance. The rf energy required depends on the magnetic field strength and on the nature of the nucleus being observed.

e) If the NMR signal of nucleus A is split by the spin of adjacent nucleus B, there is reciprocal splitting of the signal of nucleus B by the spin of nucleus A. The spins of the two nuclei are said to be coupled. The distance between two individual peaks within the multiplet of A is the same as the distance between two individual peaks within the multiplet of B. This distance, measured in Hz, is called the *coupling constant*.

12.26

	Compound	Non-equivalent protons
a)	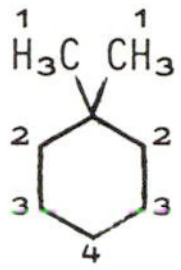 H_3C CH_3 (1, 1) on cyclohexane ring (2, 2, 3, 3, 4)	4
b)	$\overset{1}{CH_3}\overset{2}{CH_2}\overset{3}{CH_2}O\overset{4}{CH_3}$	4
c)	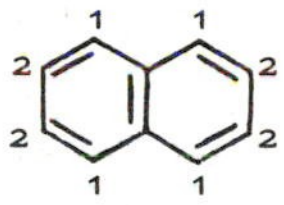 naphthalene (1, 1, 2, 2, 2, 2, 1, 1)	2
d)	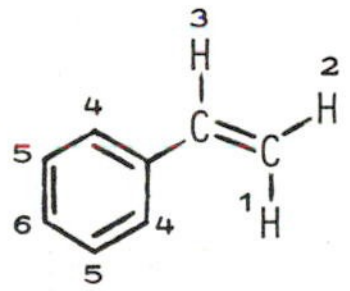 styrene: H (3), H (2), H (1); ring 4, 5, 6, 4, 5	6
e)	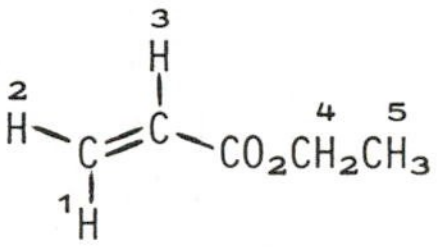 H (2), H (3), H (1): $H{-}C{=}C{-}CO_2\overset{4}{CH_2}\overset{5}{CH_3}$	5

12.27 The two possible products of hydroboration of 1-heptene are very similar. Each product has 16 protons of eight different types, including an alcohol proton; many of the proton resonances have similar chemical shifts. It is necessary to find a structural difference that causes an ^1H NMR signal to occur in a region of the spectrum where it will be distinguishable from other resonances.

The chemical shift of protons attached to a carbon bonded to oxygen (H-C-OH) occurs downfield from the resonances of the other protons. For 1-heptanol, this signal is a two-proton triplet at 3.6 δ; for 2-heptanol, the signal is a one-proton multiplet at 3.75 δ. The actual spectrum, which shows a triplet of relative area two, indicates that 1-heptanol is the hydroboration product.

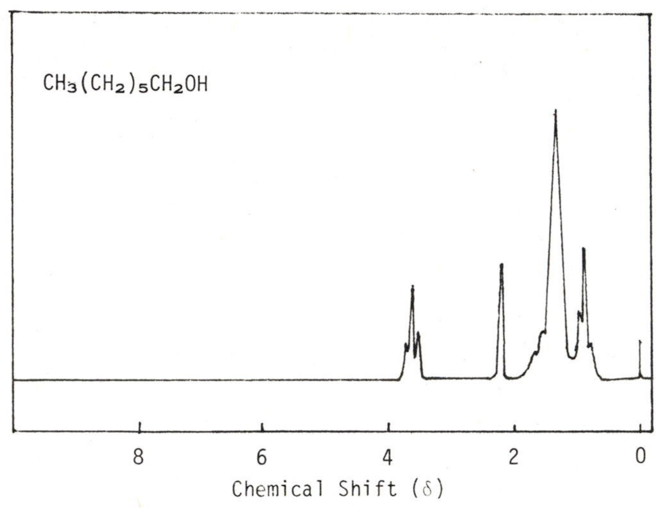

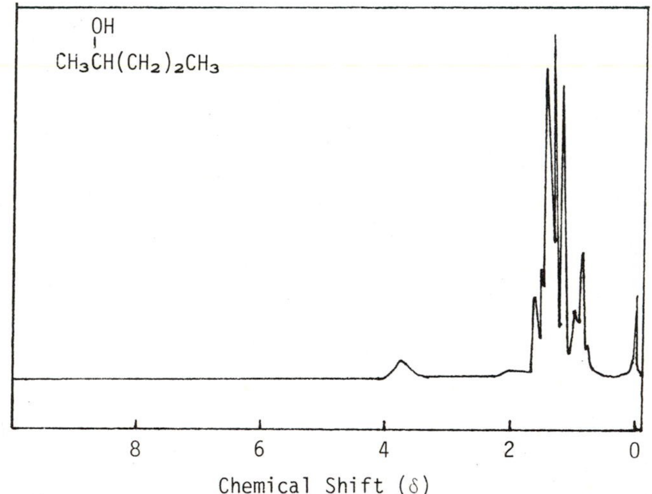

^{13}C NMR can also be used to distinguish between the two isomers. 1-Heptanol has one methyl and six different methylene carbons; 2-heptanol has two methyl, four methylene, and one methine carbon.

12.28 Use of ^{13}C NMR to distinguish between the two isomers has been described in the text. ^{1}H NMR can also be useful in this case.

Isomer A has only four kinds of protons because of symmetry. Its vinylic proton absorption (4.5-6.5 δ) represents two hydrogens. Isomer B contains six different kinds of protons. Its ^{1}H NMR shows an unsplit methyl group signal and one vinylic proton signal of relative area one. These differences make it possible to distinguish between A and B.

12.29 First, examine each isomer for structural differences that are obviously recognizable in the NMR spectrum. If it is not possible to pick out distinguishing features immediately, it may be necessary to sketch an approximate spectrum of each isomer for comparison.

a) $CH_3CH{=}CHCH_2CH_3$ has two vinylic protons with chemical shifts at 5.4-5.5 δ. Because ethylcyclopropane shows no signal in this region, it should be easy to distinguish one isomer from the other.

b) $CH_3CH_2OCH_2CH_3$ has two kinds of protons; its ^{1}H NMR spectrum consists of two resonances -- a triplet and a quartet. $CH_3OCH_2CH_2CH_3$ has four different types of protons, and its spectrum is more complex. In particular, the methyl group bonded to oxygen shows an unsplit singlet absorption.

c) Because these isomers have the same number and types of resonances, it is necessary to examine the ^{1}H NMR spectrum of each isomer for identification.

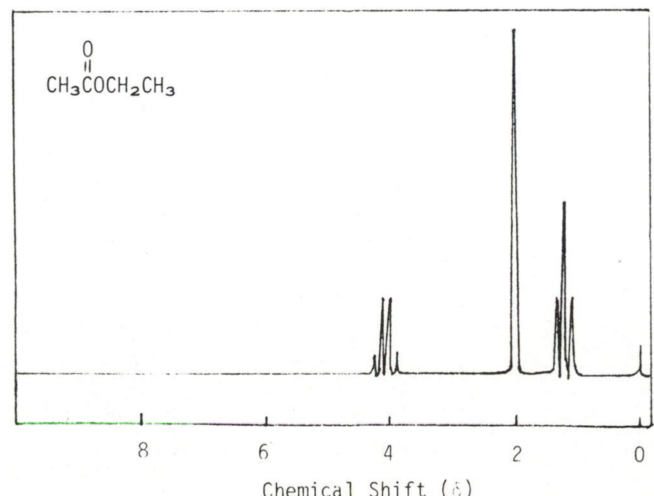

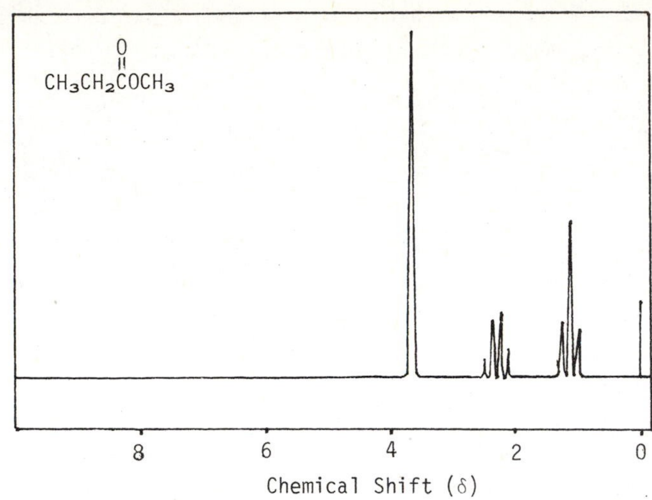

The absorptions at either 2.1-2.4 δ (protons adjacent to a carbonyl group) or at 3.3-4.0 δ (protons adjacent to oxygen) can be used to identify the isomers.

d) Each isomer contains four different kinds of protons -- two kinds of methyl protons and two kinds of vinylic protons. For the first isomer, the methyl resonances are both singlets, whereas for the second isomer, one resonance is a singlet and one is a doublet.

12.30 a) C_3H_6O contains one double bond or ring.

b) Possible structures for C_3H_6O include:

H_2C-CH_2 / H_2C-O (four-membered ring)	$H_2C-CHCH_3$ bridged by O (three-membered ring)	$H_2C{=}CHOCH_3$
cyclic ether	cyclic ether	ether, double bond
$H_2C{=}CHCH_2OH$	$H_2C{=}C(OH)CH_3$	H_2C, H_2C ring with C(H)(OH)
alcohol, double bond	alcohol, double bond	cyclic alcohol
$CH_3C(=O)CH_3$	$CH_3CH_2C(=O)H$	
ketone (acetone)	aldehyde	

c) Carbonyl functional groups absorb at 1710 cm^{-1} in the infrared. Only the last two compounds above show an infrared absorption in this region.

d) Because the aldehyde from part b) contains three different kinds of protons, its ^1H NMR spectrum exhibits three resonances. The ketone shows only one resonance. Since the unknown compound of this problem shows only one ^1H NMR absorption (in the methyl ketone region), it must be acetone.

12.31 Either ^1H NMR or ^{13}C NMR can be used to distinguish among these isomers. In either case, it is first necessary to find the number of different kinds of protons or carbon atoms.

Compound	H_2C-CH_2 / H_2C-CH_2 (cyclobutane)	$H_2C=CHCH_2CH_3$	$CH_3CH=CHCH_3$	$(CH_3)_2C=CH_2$
Kinds of protons	1	5	2	2
Kinds of carbon atoms	1	4	2	3
Number of ^1H NMR peaks	1	5	2	2
Number of ^{13}C NMR peaks	1	4	2	3

^{13}C NMR is the preferred method for identifying these compounds; each isomer differs in the number of absorptions in it ^{13}C NMR spectrum.

^1H NMR can also be used to distinguish among the isomers. The two isomers that show two ^1H NMR peaks differ in their splitting patterns.

12.32 $BrCH_2CH_2CH_2Br$

12.33 Possible structures for $C_4H_7ClO_2$ are $CH_3CH_2CO_2CH_2Cl$ and $ClCH_2CO_2CH_2CH_3$. Chemical shift data can distinguish between these two possible structures.

$CH_3CH_2C(=O)OCH_2Cl$ (I) $ClCH_2C(=O)OCH_2CH_3$ (II)

In I, the protons attached to the carbon bonded to both oxygen and chlorine ($-OCH_2Cl$) absorb far downfield (5.0-6.0 δ). Because no signal is present in this region of the ^1H NMR spectrum given, the unknown must be II.

12.34 a) $(CH_3)_2CHC(=O)CH_3$

b) $H_3C(Br)C=CH_2$ (H₃C and H on top; Br and H on bottom)

c) $H_3C(Cl)C=C(H)CH_2Cl$ (H₃C and H on top; Cl and CH_2Cl on bottom)

or the *E* isomer

d) $CH_3C(CH_3)(C_6H_5)CH_3$

e) $CH_3C(=O)CH_2CH_2Br$

f) $C_6H_5CH_2CH_2CH_2Br$

12.35 The molecular formula of the isomers, $C_{12}H_{16}$, corresponds to five multiple bonds and/or rings. We will try to differentiate between the isomers by using both ^1H NMR and ^{13}C NMR.

^1H NMR. Isomer A has two kinds of protons -- those in the four-membered rings and those in the eight-membered ring. Each group of eight protons absorbs in the allylic region of the spectrum (C=C-C-H; 1.5-2.5 δ). Two kinds of protons are also present in isomer B, but you will have to build a model to see the difference between them. One group of eight protons points toward a double bond; the other group of eight protons points away from the double bond. Both groups of protons are also allylic and absorb in the region of 1.5-2.5 δ. ^1H NMR thus cannot be used to distinguish between A and B.

^{13}C NMR. Isomer A has three different kinds of carbon atoms:

carbon atom	quantity	chemical shift
=C-	four	100-150 δ
$-CH_2-$ (in 4-membered rings)	four	15-55 δ
$-CH_2-$ (in 8-membered ring)	four	15-55 δ

Isomer B contains only two different kinds of carbon atoms:

carbon atom	quantity	chemical shift
=C-	four	100-150 δ
$-CH_2-$	eight	15-55 δ

Three resonances should appear in the ^{13}C NMR spectrum of A; only two resonances should appear in the ^{13}C NMR spectrum of B. Although ^1H NMR cannot distinguish between A and B, ^{13}C NMR will solve the problem.

12.36

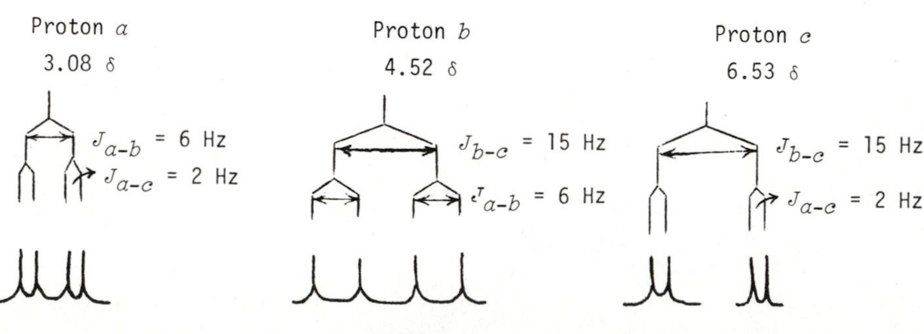

The absorptions in the ^1H NMR spectrum can be identified by comparison with the tree diagrams. H_a absorbs at 3.08 δ, H_b absorbs at 4.52 δ, and H_c absorbs at 6.35 δ.

12.37

$$C_6H_5C(=O)OCH_2CH_3$$

ethyl benzoate

Carbon	δ (ppm)
1	14
2	61
3	166
4, 5, 6, 7	127-133 (four absorptions)

12.38 Compound A (4 multiple bonds and/or rings) must be symmetrical because it exhibits only six peaks in its ^{13}C NMR spectrum. Saturated carbons account for two of these peaks (δ = 15, 28 ppm), and unsaturated carbons account for the other four (δ = 119, 129, 131, 143 ppm).

^1H NMR shows a triplet (3H at 1.1 δ), and a quartet (2H at 2.5 δ), indicating the presence of an ethyl group. The other signals (4H at 6.9-7.3 δ) are due to aromatic protons.

$$p\text{-}BrC_6H_4CH_2CH_3$$

Compound A

12.39 The peak in the mass spectrum at m/z = 84 is probably the molecular ion of the unknown compound and corresponds to a molecular weight of 84 (C_6H_{12} - one double bond or ring).

^{13}C NMR shows three different kinds of carbons and indicates a symmetrical hydrocarbon. The absorption at 132 δ is due to a vinylic carbon atom. A reasonable structure for the unknown is:

$$CH_3CH_2CH{=}CHCH_2CH_3$$

3-hexene

What you should know:

After doing these problems you should be able to:

1. Understand the phenomenon of Nuclear Magnetic Resonance, and be able to define terms associated with NMR;
2. Find the value of a chemical shift in Hz, given its value in δ, and *vice versa*;
3. Find the number of non-equivalent kinds of carbon atoms or protons in a structural formula;
4. Interpret ^{13}C NMR spectra using (a) proton noise-decoupled spectra, (b) gated decoupled spectra, and (c) off-resonance spectra;
5. Interpret ^{1}H NMR spectra, using (a) chemical shifts; (b) values of integration, (c) spin-spin splitting, and (d) coupling constants;
6. Construct tree diagrams to interpret complex splitting patterns;
7. Use ^{13}C NMR and ^{1}H NMR to determine the identity of reaction products.

CHAPTER 13 CONJUGATED DIENES. ULTRAVIOLET SPECTROSCOPY

13.1

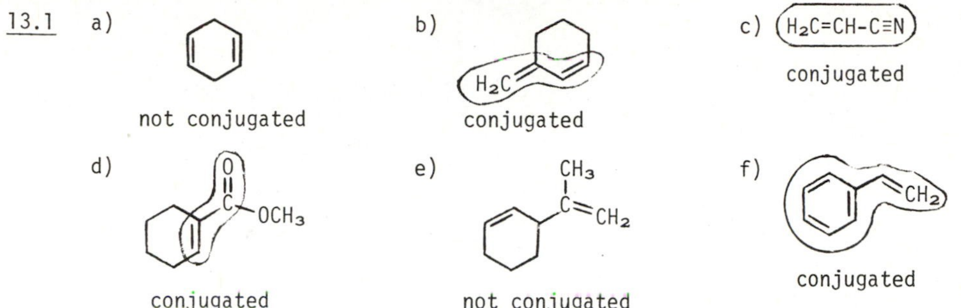

13.2 We would expect ΔH_{hydrog} = 30.3 + 30.3 = 60.6 kcal/mol for both double bonds of allene if the heat of hydrogenation for each double bond of allene were the same as that for an isolated double bond. The observed ΔH_{hydrog} = 71.3 kcal/mol is 10.7 kcal/mol *higher* than the expected value. Allene is higher in energy (less stable) than a non-conjugated diene, which in turn is less stable than a conjugated diene.

13.3 $CH_3CH{=}CHCH{=}CH_2$ 1,3-pentadiene

	product	name	results from
(1)	$CH_3CH{=}CHCHClCH_3$	4-chloro-2-pentene	1,2 addition 1,4 addition
(2)	$CH_3CH_2CHClCH{=}CH_2$	3-chloro-1-pentene	1,2 addition
(3)	$CH_3CHClCH_2CH{=}CH_2$	4-chloro-1-pentene	1,2 addition
(4)	$CH_3CH_2CH{=}CHCH_2Cl$	1-chloro-2-pentene	1,4 addition
(5)	$CH_3CH{=}CHCH_2CH_2Cl$	5-chloro-2-pentene	1,2 addition (non-Markovnikov addition)

13.4

$CH_3CH_2\overset{\delta+}{CH}\cdots CH \cdots \overset{\delta+}{CH_2}$ D $\xleftarrow{H^+}$ protonation on carbon 4

$CH_3\overset{\delta+}{CH}\cdots CH \cdots \overset{\delta+}{CH}CH_3$ A $\xleftarrow{H^+}$ protonation on carbon 1

$CH_3CH{=}CHCH{=}CH_2$

$\xrightarrow{H^+}$ (blocked) protonation on carbon 3 → $CH_3\overset{+}{C}HCH_2CH{=}CH_2$ C

$\xrightarrow{H^+}$ (blocked) protonation on carbon 2 → $CH_3CH{=}CHCH_2\overset{+}{C}H_2$ B

The positive charge of allylic carbocation A is delocalized over three secondary carbons; the positive charge of carbocation D is delocalized over two secondary carbons and one primary carbon; the positive charge of carbocations B and C is not delocalized. We therefore predict that carbocation A is the major intermediate formed, and that product (1) in problem 13.3 will predominate.

13.5

$$\underset{\substack{\textbf{B}\\ \text{secondary/primary}\\ \text{allylic carbocation}}}{\overset{\delta+}{C}H_2\text{=}\overset{\overset{CH_3}{|}\,\delta+}{C}\text{=}CHCH_2Br} \xleftarrow[\text{addition to carbon 4}]{Br^+} H_2C\text{=}\overset{\overset{CH_3}{|}}{C}\text{-}CH\text{=}CH_2 \xrightarrow[\text{addition to carbon 1}]{Br^+} \underset{\substack{\textbf{A}\\ \text{tertiary/primary}\\ \text{allylic carbocation}}}{CH_2Br\text{-}\overset{\overset{CH_3}{|}\,\delta+}{C}\text{=}CH\text{=}\overset{\delta+}{C}H_2}$$

Tertiary / primary allylic carbocation A is more stable than secondary/ primary allylic carbocation B. Since the products formed from the more stable intermediate predominate, 3,4-dibromo-3-methyl-1-butene is the major product of 1,2 addition of bromine to isoprene.

13.6 Figure 13.4 shows the three pi molecular orbitals of an allylic pi system. An allyl radical has three pi electrons. Two of them occupy the bonding molecular orbital, and the third electron occupies the non-bonding orbital.

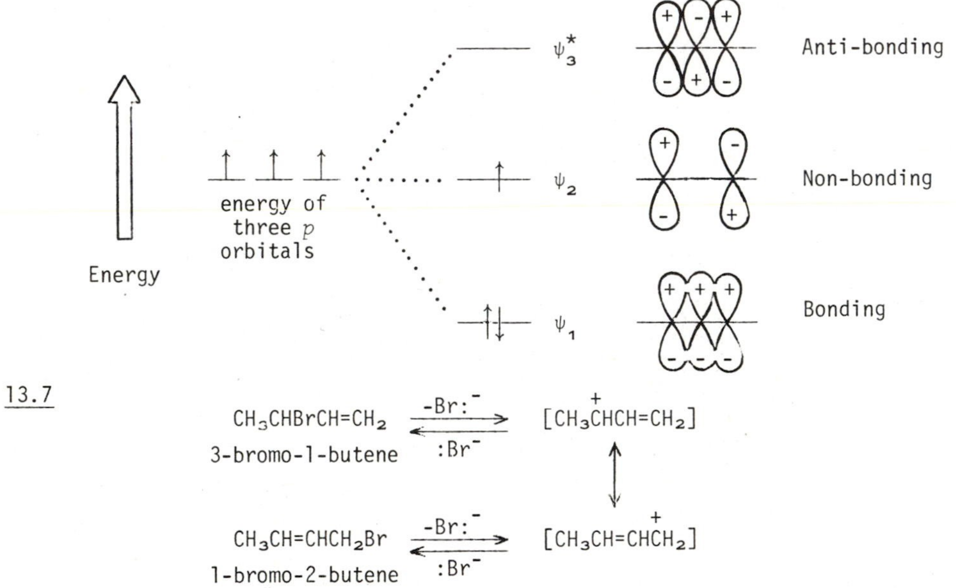

Allylic halides can undergo slow ionization to form stabilized carbocations. Both 3-bromo-1-butene and 1-bromo-2-butene form the same allylic carbocation,pictured above, upon ionization. Addition of bromide ion to the allylic carbocation then occurs to form a mixture of bromobutenes. Since the reaction is run under equilibrium conditions, the thermodynamically more stable 1- bromo-2-butene predominates.

13.8 *good dienophiles* *poor dienophiles*

a) $H_2C{=}CHNO_2$

b) 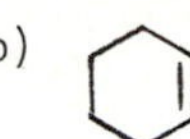

c)

d)

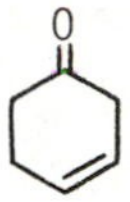

Since both a) and c) have an electron-withdrawing functional group conjugated with the double bond, these two compounds are reactive in the Diels-Alder reaction. Alkene b) has no electron-withdrawing functional group. Compound d) is not a good dienophile because its electron-withdrawing functional group is not conjugated with the double bond.

13.9 a) This diene has an *s-cis* conformation and should undergo Diels-Alder cycloaddition.

b) This diene has an *s-trans* conformation. Because the double bonds are in a fused ring system, it is not possible for them to rotate to an *s-cis* conformation.

c) Although this diene has an *s-trans* conformation as drawn, rotation to an *s-cis* diene can easily occur.

d) Rotation can also occur about the single bond of this *s-trans* diene. The resulting *s-cis* conformation, however, has an unfavorable steric interaction of the interior methyl group with a hydrogen at carbon 1. Rotation to the *s-cis* conformation is therefore disfavored.

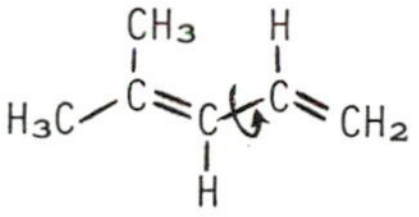

s-trans (more stable) *s-cis* (less stable)

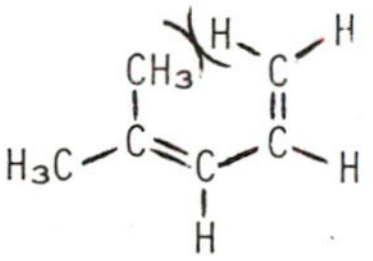

13.10 The difference in reactivity of the three cyclic dienes is due to steric factors. As the "non-diene" part of the molecule becomes larger, the carbon atoms at the end of the diene portion of the ring are forced farther apart. Overlap with the pi system of the dienophile in the pericyclic transition state is poorer, and reaction is slower.

13.11 200 nm = 200×10^{-7} cm = 2×10^{-5} cm

400 nm = 400×10^{-7} cm = 4×10^{-5} cm

$$E = \frac{2.86 \times 10^{-3}\ \text{kcal/mol}}{\lambda\ (\text{in cm})}$$

for $\lambda = 2 \times 10^{-5}$ cm

$$E = \frac{2.86 \times 10^{-3}\ \text{kcal/mol}}{2 \times 10^{-5}} = 1.4 \times 10^{2}\ \text{kcal/mol}$$

for $\lambda = 4 \times 10^{-5}$ cm

$$E = \frac{2.86 \times 10^{-3}\ \text{kcal/mol}}{4 \times 10^{-5}} = 0.72 \times 10^{2}\ \text{kcal/mol} = 72\ \text{kcal/mol}$$

The energy of electromagnetic radiation occurs over the range 72-140 kcal/mol.

13.12

	UV	IR	^{1}H NMR (at 60 MHz)
Energy (in kcal/mol)	72-140	1.1-11	5.7×10^{-6}

The energy required for UV transitions is greater than the energy required for IR or ^{1}H NMR transitions.

13.13 All compounds having *alternating* single and multiple bonds should show ultraviolet absorption in the range 200-400 nm. Only compound a) of those pictured is not UV-active. All of the compounds pictured below are UV active.

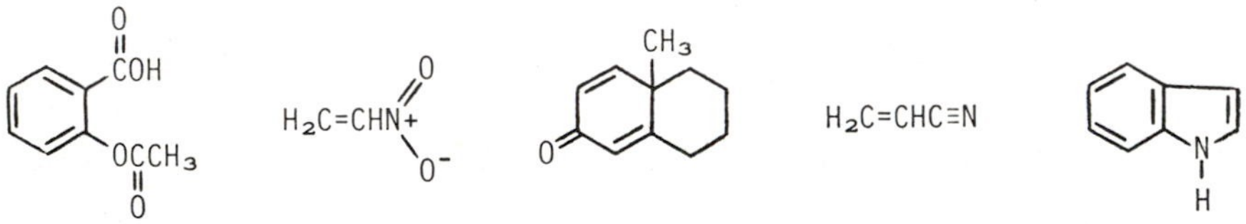

$H_2C{=}CHN^{+}(=O)O^{-}$ $H_2C{=}CHC{\equiv}N$

13.14 a) 3-methyl-2,4-hexadiene b) 1,3,5-heptatriene
c) 2,3,5-heptatriene d) 3-propyl-1,3-pentadiene

13.15 a) $CH_3CH{=}C{=}CHCH{=}C(H)CH_3$ b) 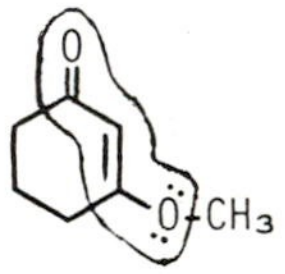 c)

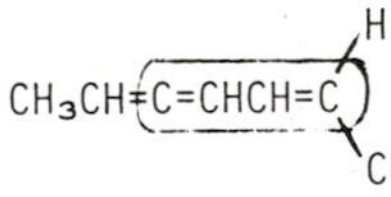

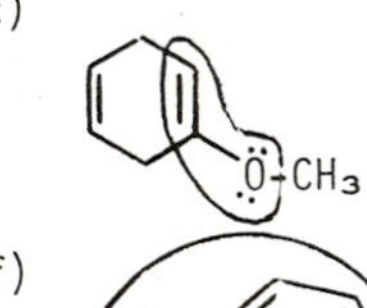

d) 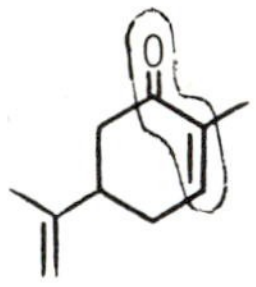 e) 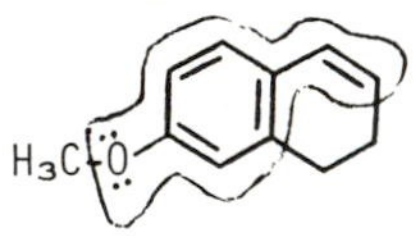 f)

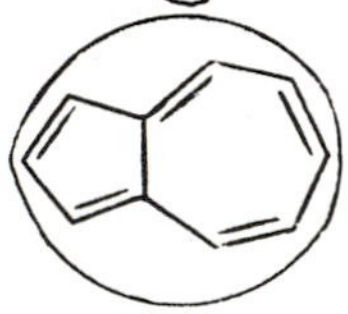

13.16 a) 1 mole Br_2 / CCl_4

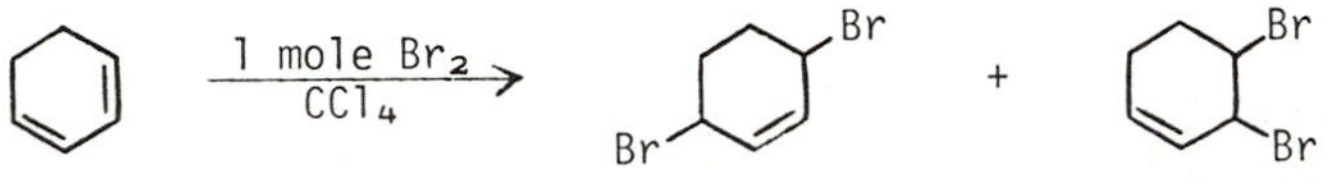

b) 1. O_3 2. Zn, H_3O^+

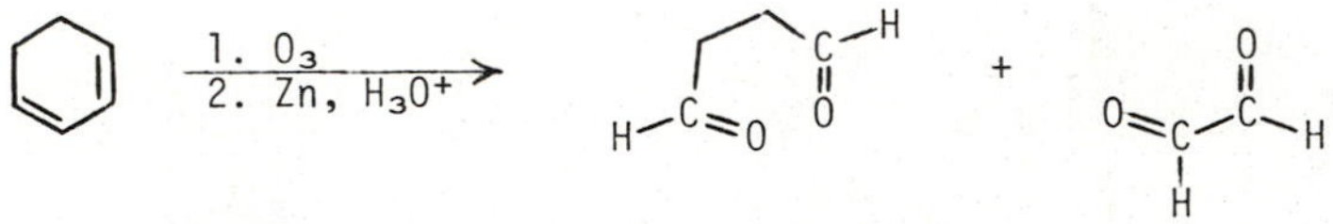

c) cyclohexadiene $\xrightarrow{\text{1 mole HCl}}$ 3-chlorocyclohexene (Cl)

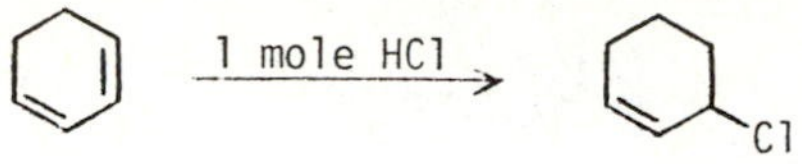

d) cyclohexadiene $\xrightarrow{\text{1 mole DCl}}$ (D, Cl) + (Cl, D)

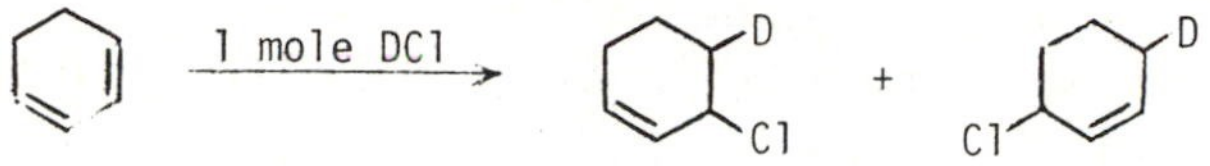

e) cyclohexadiene + H–C(=CH₂)–C(=O)–CH₃ ⟶ (H, COCH₃)

f) cyclohexadiene $\xrightarrow[\text{2. } NaHSO_3]{\text{1. } OsO_4}$ (HO, OH, HO, OH, H, H, H, H) + (H, OH, H, HO, HO, H, HO, H) + (HO, H, OH, H, H, OH, H, OH)

enantiomers

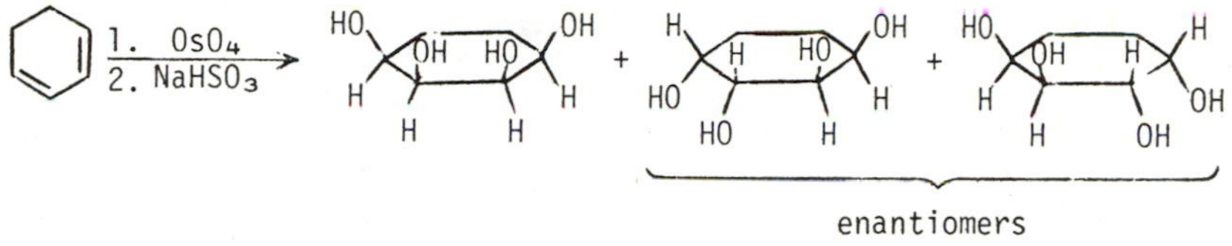

13.17

conjugated dienes: $CH_3CH{=}CHCH{=}CH_2$ 1,3-pentadiene; $H_2C{=}C(CH_3)CH{=}CH_2$ 2-methyl-1,3-butadiene

cumulated dienes: $CH_3CH_2CH{=}C{=}CH_2$ 1,2-pentadiene; $CH_3CH{=}C{=}CHCH_3$ 2,3-pentadiene; $H_2C{=}C{=}C(CH_3)_2$ 3-methyl-1,2-butadiene

non-conjugated diene: $H_2C{=}CHCH_2CH{=}CH_2$ 1,4-pentadiene

13.18 $CH_3CH_2C{\equiv}CCH_2CH_3$ (3-hexyne) has two different kinds of protons. They are visible as a triplet and a quartet in the saturated hydrocarbon region of the ^1H NMR spectrum.

The ^1H NMR spectrum of 2,4-hexadiene ($CH_3CH{=}CHCH{=}CHCH_3$) is more complex. Four protons absorb in the vinylic region (4.5-6.5 δ) and serve to distinguish the NMR spectrum of 2,4-hexadiene from that of 3-hexyne. In addition, 2,4-hexadiene shows ultraviolet absorption because it is conjugated.

13.19 a) cyclopentadiene + $H_2C{=}CH{-}C({=}O){-}H$ ⟶ (H, C=O, H)

b) cyclopentadiene + maleic anhydride (O, O, O) ⟶ (H, H, O, O, O)

c)

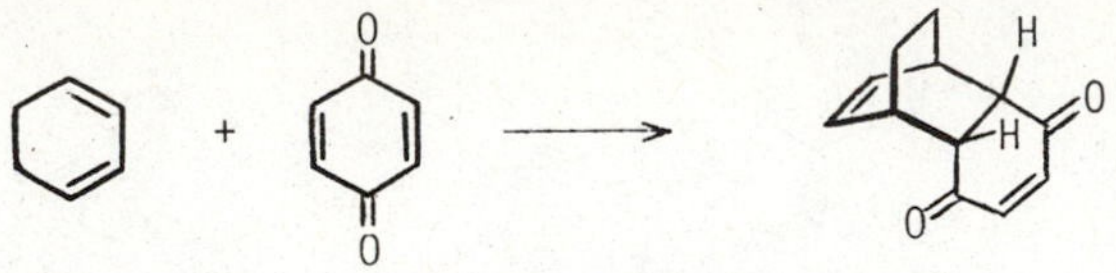

If two moles of cyclohexadiene are present for each mole of dienophile, you can also obtain a second product:

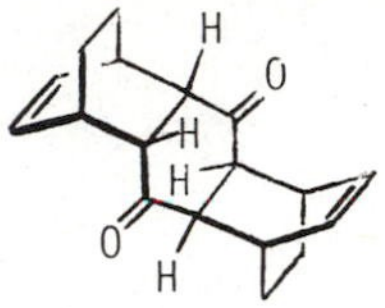

d)

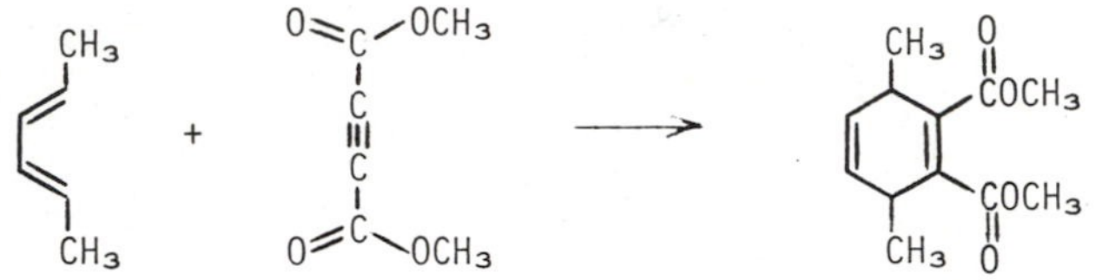

13.20

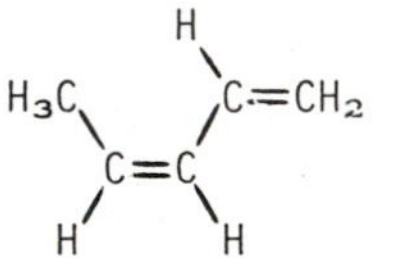

cis-1,3-pentadiene 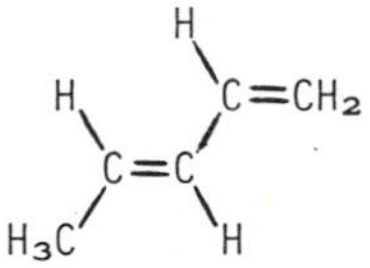 *trans*-1,3-pentadiene

Both pentadienes are more stable in *s-trans* conformations. To undergo Diels-Alder reactions, they must rotate about the single bond between the double bonds to assume *s-cis* conformations.

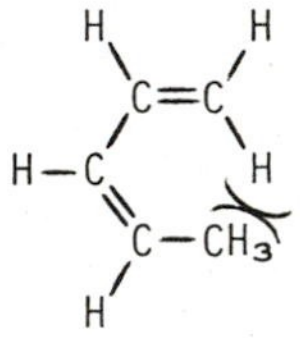

cis-1,3-pentadiene 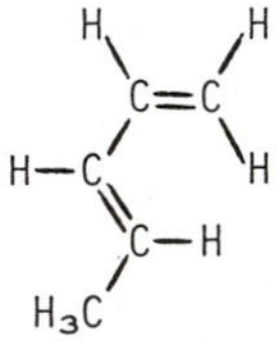 *trans*-1,3-pentadiene

It is not difficult for the *trans* isomer to assume the *s-cis* conformation. When *cis*-1,3-pentadiene rotates to the *s-cis* conformation, steric interaction occurs between the methyl-group protons and a proton on carbon 1. Since it is more difficult for *cis*-1,3-pentadiene to assume the *s-cis* conformation, it is less reactive in the Diels-Alder reaction.

13.21 Only compounds having alternating multiple bonds show π-π^* ultraviolet absorptions in the 200-400 nm range. The only two compounds that show UV absorptions in this range are:

cyclopentadiene and pyridine

13.22 Protonation on carbon 1:

1-phenyl-1,3-butadiene (C_6H_5–CH=CHCH=CH_2, carbons numbered 1 2 3 4) $\xrightarrow{H^+}$ [C_6H_5–$CH_2\overset{\delta+}{C}H$⋯CH⋯$\overset{\delta+}{C}H_2$] A, allylic

$\xrightarrow{:Cl^-}$ C_6H_5–CH_2CHClCH=CH_2 3-chloro-4-phenyl-1-butene

$\xrightarrow{:Cl^-}$ C_6H_5–CH_2CH=CHCH_2Cl 1-chloro-4-phenyl-2-butene

Protonation on carbon 2:

C_6H_5–CH=CHCH=CH_2 $\xrightarrow{H^+}$ [C_6H_5–$\overset{+}{C}HCH_2CH=CH_2$] B $\xrightarrow{:Cl^-}$ C_6H_5–CHClCH_2CH=CH_2 4-chloro-4-phenyl-1-butene

Protonation on carbon 3:

C_6H_5–CH=CHCH=CH_2 $\xrightarrow{H^+}$ [C_6H_5–CH=CH$CH_2\overset{+}{C}H_2$] C $\xrightarrow{:Cl^-}$ C_6H_5–CH=CHCH_2CH_2Cl 4-chloro-1-phenyl-1-butene

Protonation on carbon 4:

C_6H_5–CH=CHCH=CH_2 $\xrightarrow{H^+}$ [C_6H_5–$\overset{\delta+}{C}H$⋯CH⋯$\overset{\delta+}{C}HCH_3$] D, allylic

$\xrightarrow{:Cl^-}$ C_6H_5–CH=CHCHClCH_3 3-chloro-1-phenyl-1-butene

$\xrightarrow{:Cl^-}$ C_6H_5–CHClCH=CHCH_3 1-chloro-1-phenyl-2-butene

Carbocation D is most stable because it can use the pi systems of both the benzene ring and the side chain to further delocalize positive charge. 3-Chloro-1-phenyl-1-butene is the major product because it results from cation D and because its double bond can conjugate with the benzene ring to provide extra stability.

13.23 First, find the six-membered ring formed by the Diels-Alder reaction. Locate the new bonds; you should then be able to identify the diene and the dienophile.

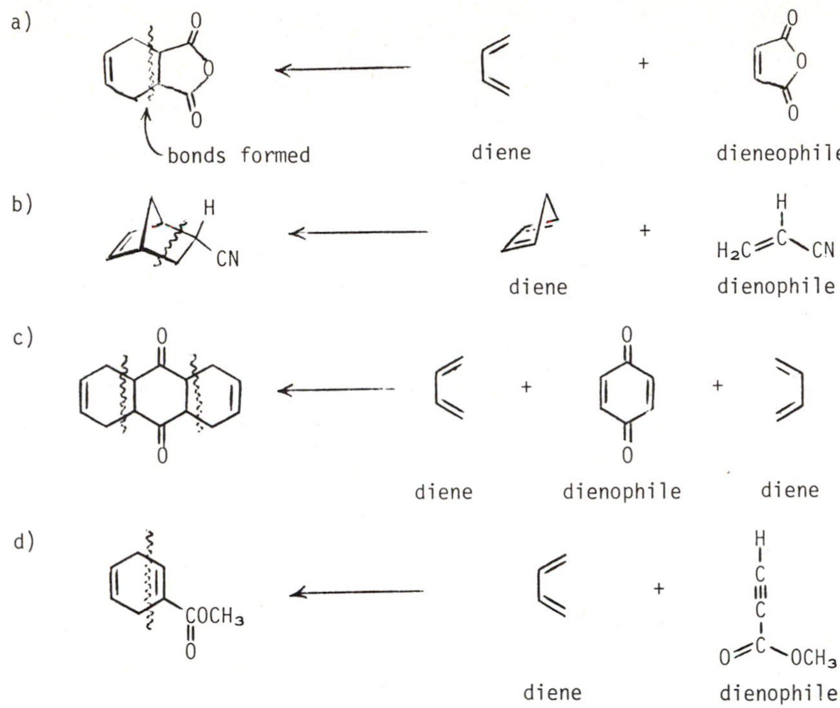

13.24 HC≡C-C≡CH cannot be used as a Diels-Alder diene because it is linear. The end carbons are too far apart to be able to react with a dienophile in a cyclic transition state.

13.25 In the instances when it is possible to make the Diels-Alder reaction reversible, the products are much more stable (of lower energy) than the reactants. In this case, the reactant is a non-conjugated diene, and the products are benzene, a stable conjugated molecule, and ethylene.

$\xrightarrow{\Delta}$ + $H_2C{=}CH_2$

13.26 A Diels-Alder reaction between α-pyrone (diene) and the alkyne dienophile yields the following product:

+ CO_2CH_3–C≡C–CO_2CH_3 $\xrightarrow{\Delta}$ CO_2CH_3, CO_2CH_3

The double bonds in this product are not conjugated. A more stable product is formed by loss of CO_2.

CO_2CH_3, CO_2CH_3 $\xrightarrow{\Delta}$ $C_6H_4(CO_2CH_3)_2$ + CO_2

This process can occur in a manner similar to the reverse Diels-Alder reaction of the previous problem.

13.27 The value of λ_{max} in the ultraviolet spectrum of dienes becomes larger with increasing methyl substitution. Since energy is inversely related to λ_{max}, the energy needed to produce ultraviolet absorption decreases with increasing methyl substitution.

diene	# of $-CH_3$ groups	λ_{max}	λ_{max} - λ_{max} (butadiene)
$H_2C{=}C(H){-}C(H){=}CH_2$	0	217 nm	0
$H_2C{=}C(H){-}C(H_3C){=}CH_2$	1	220	3
$H_3C(H)C{=}C(H){-}C(H){=}CH_2$	1	223	6
$H_2C{=}C(CH_3){-}C(H_3C){=}CH_2$	2	226	9
$H_3C(H)C{=}C(H){-}C(H){=}C(H)CH_3$	2	227	10
$(H_3C)_2C{=}C(H){-}C(H_3C){=}CH_2$	3	232	15
$(H_3C)_2C{=}C(H){-}C(H){=}C(CH_3)_2$	4	240	23

The average increase in λ_{max} is 5 nm per methyl group.

13.28 a) β-Ocimene, $C_{10}H_{16}$, has three degrees of unsaturation. Catalytic hydrogenation yields a hydrocarbon of formula $C_{10}H_{22}$. β-Ocimene thus contains three double bonds, and no rings.

b) The ultraviolet absorption at 232 nm indicates that β-ocimene is conjugated.

c) The carbon skeleton, as determined from hydrogenation, is:

$$\underset{\displaystyle\text{2,6-dimethyloctane}}{CH_3CH_2\overset{\displaystyle CH_3}{\overset{|}{C}}HCH_2CH_2CH_2\overset{\displaystyle CH_3}{\overset{|}{C}}HCH_3}$$

Ozonolysis data is used to determine the location of the double bonds.

The acetone fragment, which comes from carbon atoms 1 and 2 of 2,6-dimethyloctane, fixes the position of one double bond. Formic acid results from ozonolysis of a double bond at the other end of β-ocimene. Placement of the other fragments to comply with the carbon skeleton yields the following structural formula for β-ocimene.

$$\underset{\displaystyle\beta\text{-ocimene}}{H_2C{=}CH\overset{\displaystyle CH_3}{\overset{|}{C}}{=}CHCH_2CH{=}\overset{\displaystyle CH_3}{\overset{|}{C}}CH_3}$$

d)

$$H_2C{=}CH\overset{\displaystyle CH_3}{\overset{|}{C}}{=}CHCH_2CH{=}\overset{\displaystyle CH_3}{\overset{|}{C}}CH_3 \xrightarrow{1.\,O_3\;\; 2.\,H_2O_2} \underset{\text{formic acid}}{HCOOH} + \underset{\text{pyruvic acid}}{HOOC\overset{\displaystyle CH_3}{\overset{|}{C}}{=}O} + \underset{\text{malonic acid}}{HOOCCH_2COOH} + \underset{\text{acetone}}{O{=}\overset{\displaystyle CH_3}{\overset{|}{C}}CH_3}$$

$$H_2C{=}CH\overset{\displaystyle CH_3}{\overset{|}{C}}{=}CHCH_2CH{=}\overset{\displaystyle CH_3}{\overset{|}{C}}CH_3 \xrightarrow{H_2/Pd} CH_3CH_2\overset{\displaystyle CH_3}{\overset{|}{C}}HCH_2CH_2CH_2\overset{\displaystyle CH_3}{\overset{|}{C}}HCH_3$$

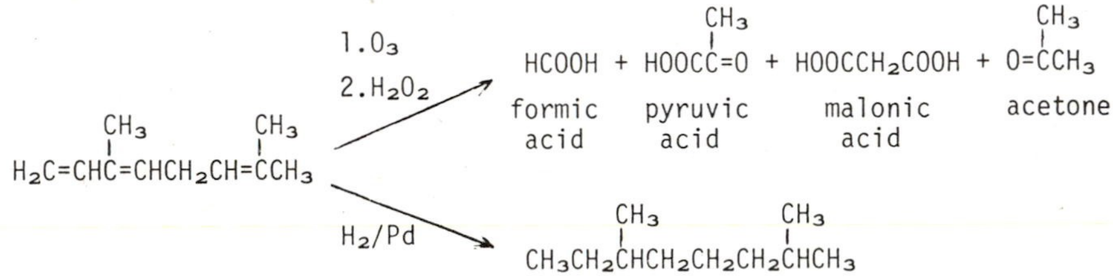

Note: ozonolysis followed by H_2O_2 yields ketones and *carboxylic acids*.

13.29 Much of what was proven for β-ocimene is also true for myrcene, since both hydrocarbons have the same carbon skeleton and contain conjugated double bonds. The difference between the two isomers is in the placement of double bonds.

The ozonolysis fragments are 2-oxopentanedioic acid (five carbon atoms), acetone (three carbon atoms), and two equivalents of formic acid (one carbon atom each). Putting these fragments together in a manner consistent with the data gives the following structural formula for myrcene:

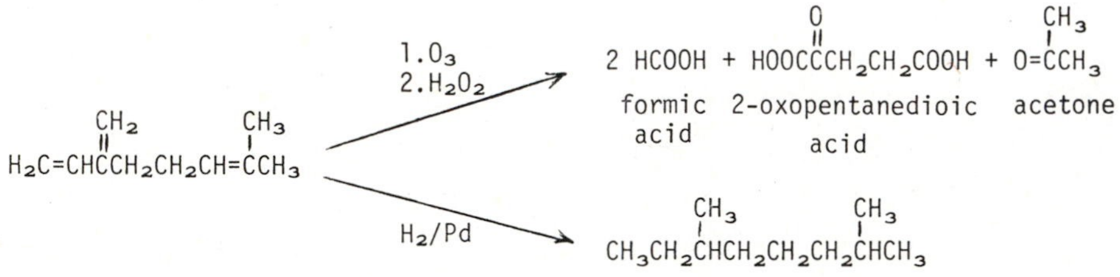

$$H_2C{=}CH\overset{\displaystyle CH_2}{\overset{\|}{C}}CH_2CH_2CH{=}\overset{\displaystyle CH_3}{\overset{|}{C}}CH_3 \xrightarrow{1.\,O_3\;\; 2.\,H_2O_2} \underset{\text{formic acid}}{2\ HCOOH} + \underset{\text{2-oxopentanedioic acid}}{HOOC\overset{\displaystyle O}{\overset{\|}{C}}CH_2CH_2COOH} + \underset{\text{acetone}}{O{=}\overset{\displaystyle CH_3}{\overset{|}{C}}CH_3}$$

$$H_2C{=}CH\overset{\displaystyle CH_2}{\overset{\|}{C}}CH_2CH_2CH{=}\overset{\displaystyle CH_3}{\overset{|}{C}}CH_3 \xrightarrow{H_2/Pd} CH_3CH_2\overset{\displaystyle CH_3}{\overset{|}{C}}HCH_2CH_2CH_2\overset{\displaystyle CH_3}{\overset{|}{C}}HCH_3$$

13.30 Double bonds can be conjugated not only with other multiple bonds but also with the lone-pair electrons of atoms such as oxygen and nitrogen. *p*-Toluidine has the same number of double bonds as benzene, yet its λ_{max} is 31 nm greater. The electron pair of the nitrogen atom can conjugate with the pi electrons of the three double bonds of the ring, extending the pi system and increasing λ_{max}.

13.31 Hydrogen ion protonates the nitrogen atom of *p*-toluidine and prevents its lone pair of electrons from conjugating with the ring double bonds. λ_{max} is therefore lowered to a value very close to that of benzene.

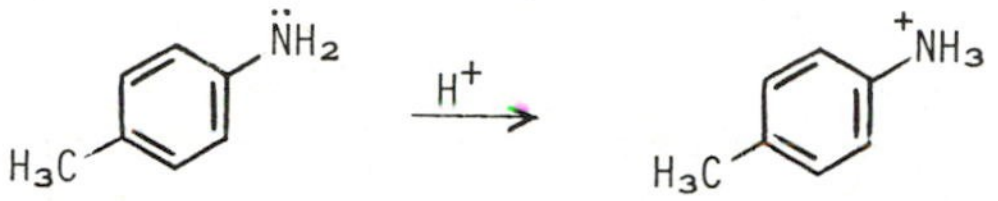

H_3C–C_6H_4–$\ddot{N}H_2$ $\xrightarrow{H^+}$ H_3C–C_6H_4–$\overset{+}{N}H_3$

13.32 Dilute NaOH serves to remove the proton from the -OH group, leaving the phenoxide anion.

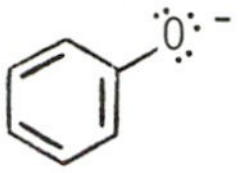

C_6H_5–$\ddot{O}$:$^-$

The increased electron density at oxygen makes it easier for an oxygen electron pair to conjugate with the pi electrons of the ring double bonds. The extended conjugation increases λ_{max} in a manner similar to *p*-toluidine (Problem 13.31).

13.33 a) Hydrocarbon A (4 multiple bonds and/or rings):

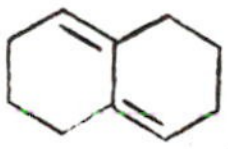

or

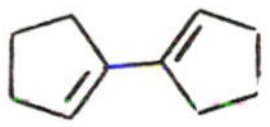

I II

b) Rotation about the central single bond of II allows the double bonds to assume the *s-cis* conformation necessary for a Diels-Alder reaction. Rotation is not possible for I.

c)

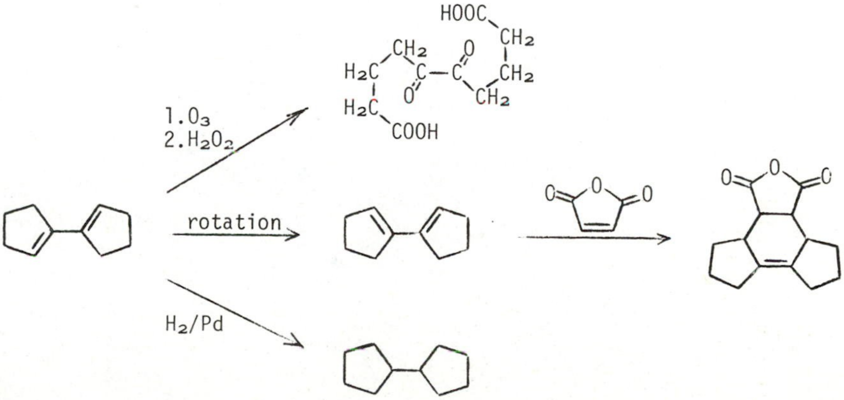

HOOC, CH_2, CH_2, O, H_2C, C, C, CH_2, O, CH_2, H_2C, COOH

1. O_3
2. H_2O_2

rotation

O, O, O

O, O, O

H_2/Pd

What you should know:

After doing these problems you should be able to:

1. Understand the reasons for the stability of conjugated dienes and recognize conjugated dienes;
2. Predict the products of electrophilic addition to conjugated dienes;
3. Predict the products of Diels-Alder reactions. You should be able to identify compounds as good dienes or dienophiles;
4. Predict if a compound absorbs radiation in the ultraviolet region;
5. Understand the concepts of kinetic control and thermodynamic control of reactions.

CHAPTER 14 BENZENE AND AROMATICITY

14.1 a)

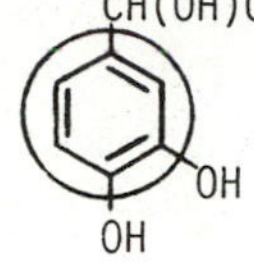

adrenaline

b) vitamin E

c) penicillin V

14.2 a) *m*-bromochlorobenzene

b) (3-methylbutyl)benzene

c) *p*-bromoaniline

d) 2,4-dichlorotoluene

e) 1-ethyl-2,4-dinitrobenzene

f) 1,2,3,5-tetramethylbenzene

14.3 a) The correct name is 1-bromo-2-chlorobenzene. You must use the lowest possible combination of numbers. You may also use the *ortho* designation.

b) The correct name is 2,4-dinitrotoluene. This is the same mistake as in a).

c) 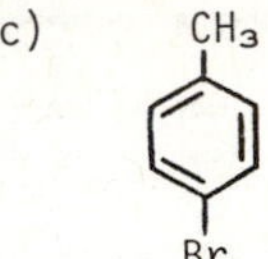

The correct name is *p*-bromotoluene. This compound should be named as a substituted toluene.

d) 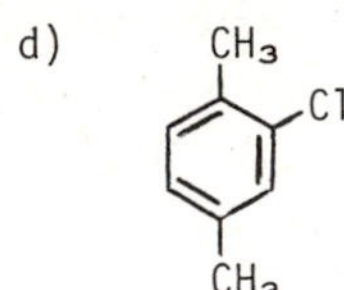

The correct name is 2-chloro-1,4-dimethylbenzene. Compounds with more than two substituents must be named by numbering the position of each substituent on the ring.

14.4

1,2,4-tribromobenzene 1,2,3-tribromobenzene 1,3,5-tribromobenzene

According to Kekulé theory, four tribromobenzenes are possible. Kekulé would say that the two 1,2,4-tribromobenzenes rapidly interconvert, and that only three isomers can be isolated.

14.5 Since all carbon atoms are equivalent in Ladenburg "benzene", only one monobromo derivative is possible. Four dibromo derivatives, including a pair of enantiomers, are possible.

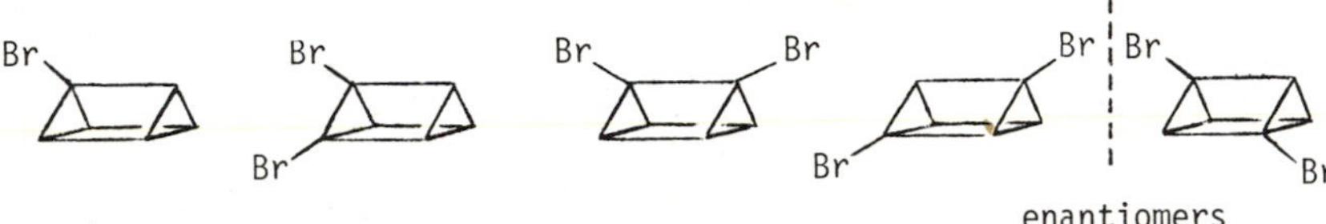

monobromo derivative

dibromo derivatives

enantiomers

Dewar "benzene", which is a bent molecule, has two different kinds of carbons. Three monobromo derivatives, including a pair of enantiomers, are possible.

enantiomers

The dibromo derivatives include three pairs of enantiomers and three other dibromo Dewar "benzenes".

\+ enantiomer + enantiomer + enantiomer

14.6

A $\xrightarrow[\text{2. Zn, } H_3O^+]{\text{1. } O_3}$

B $\xrightarrow[\text{2. Zn, } H_3O^+]{\text{1. } O_3}$

If *o*-xylene existed only as structure A, ozonolysis would cause cleavage at the bonds indicated and would yield two moles of pyruvaldehyde and one mole of glyoxal for each mole of A consumed. If *o*-xylene existed only as structure B, ozonolysis would yield one mole of 2,3-butanedione and two moles of glyoxal. If *o*-xylene existed as a resonance hybrid of A and B, the ratio of ozonolysis products would be; glyoxal : pyruvaldehyde : 2,3-butanedione = 3:2:1. Since this ratio is identical to the experimentally determined ratio, we know that A and B contribute equally to the structure of *o*-xylene.

14.7

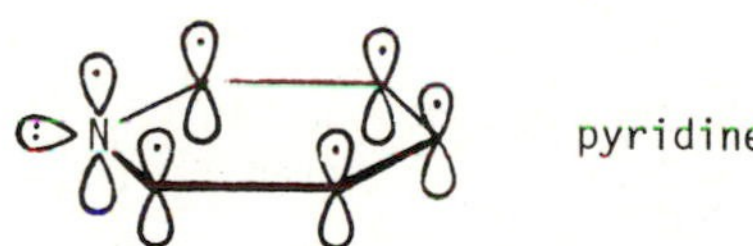

The electronic descriptions of pyridine and benzene are very similar. The pyridine ring is formed by the sigma overlap of carbon and nitrogen sp^2 orbitals. In addition, six p orbitals, perpendicular to the plane of the ring, hold six electrons. These six p orbitals form six molecular orbitals that allow electrons to be delocalized over the pi system of the pyridine ring. The lone pair of nitrogen electrons occupies an sp^2 orbital that lies in the plane of the ring.

14.8

Cyclodecapentaene has $4n + 2$ pi electrons ($n = 2$), but is is not flat. If cyclodecapentaene were flat, the hydrogen atoms circled would crowd each other across the ring. To avoid this interaction, cyclodecapentaene twists so that it is neither planar nor aromatic.

14.9

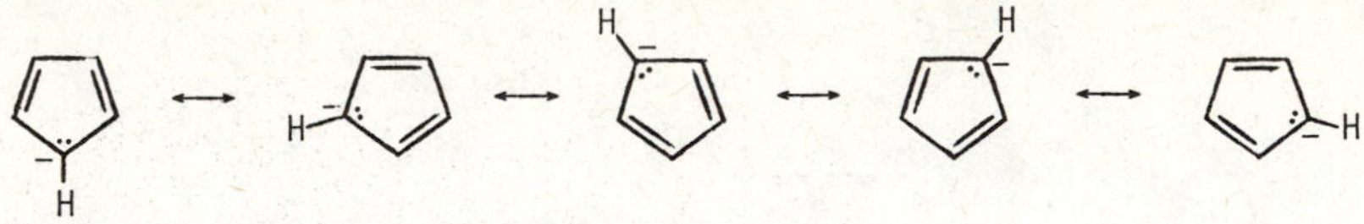

A compound that can be described by several resonance forms has a structure that can be represented by no one form. The structure of the cyclopentadienyl anion is a combination of all of the above structures and contains only one kind of carbon atom and one kind of hydrogen atom. All carbon-carbon bond lengths are equivalent; all carbon-hydrogen bond lengths are equivalent. Both the ^1H NMR and ^{13}C NMR spectra show only one absorption.

14.10 When cyclooctatetraene accepts two electrons, it becomes a $(4n + 2)$ pi electron aromatic ion. Cyclooctatetraenyl dianion is planar with a carbon-carbon bond angle of 135° (a regular octagon).

14.11

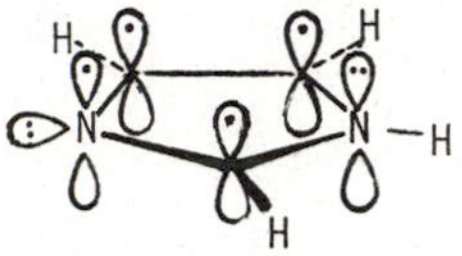

imidazole

The aromatic heterocycle imidazole contains six pi electrons. Each carbon contributes one electron, the nitrogen bonded to hydrogen contributes two electrons, and the remaining nitrogen contributes one electron. Each nitrogen is sp^2 hybridized.

14.12

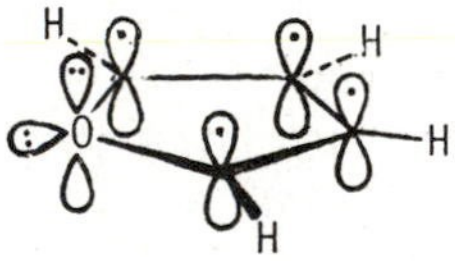

furan

The furan oxygen is sp^2 hybridized.

14.13

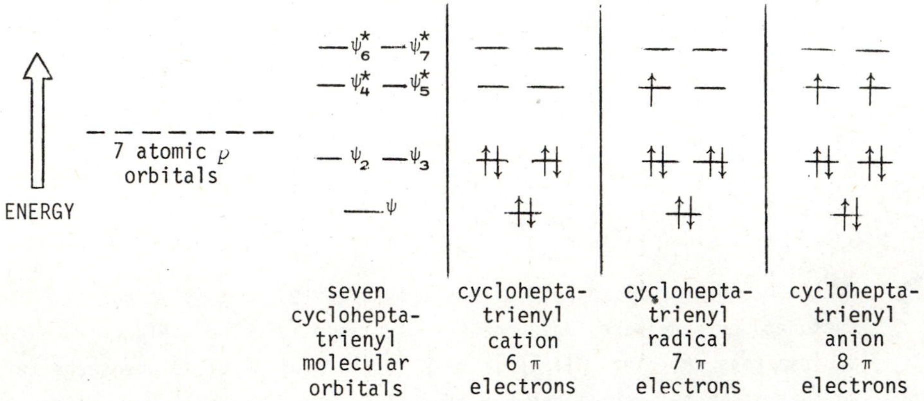

ENERGY

7 atomic p orbitals

ψ_6^* ψ_7^*

ψ_4^* ψ_5^*

ψ_2 ψ_3

ψ_1

seven cycloheptatrienyl molecular orbitals	cycloheptatrienyl cation 6 π electrons	cycloheptatrienyl radical 7 π electrons	cycloheptatrienyl anion 8 π electrons

The cycloheptatrienyl cation has six pi electrons and is aromatic.

14.14

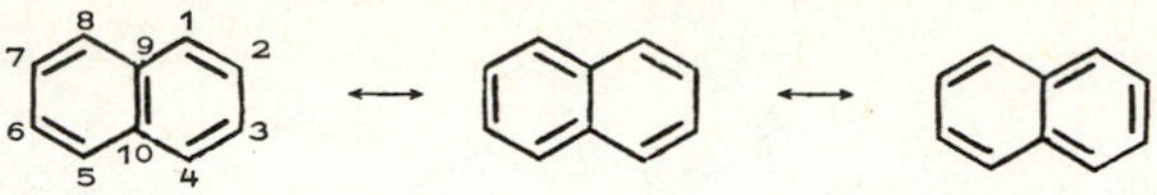

Notice that the bond between carbons 1 and 2 is represented as a double bond in two of the three resonance structures. The bond between carbons 2 and 3 is represented as a double bond in only one resonance structure. The C1-C2 bond has more double bond character in the resonance hybrid, and its carbon-carbon bond length is shorter than the C2-C3 bond length.

14.15 Naphthalene is a ten pi electron compound; the circle in each ring represents five electrons.

14.16 a) 2-methyl-5-phenylhexane b) *m*-bromobenzoic acid

c) 1-bromo-3,5-dimethylbenzene d) *o*-bromopropylbenzene

e) 1-fluoro-2,4-dinitrobenzene f) *p*-chloroaniline

14.17 a) 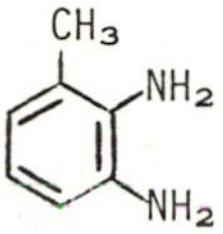 b) 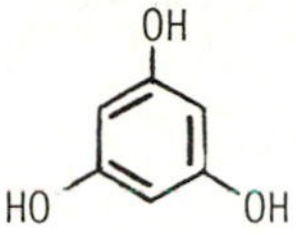 c) 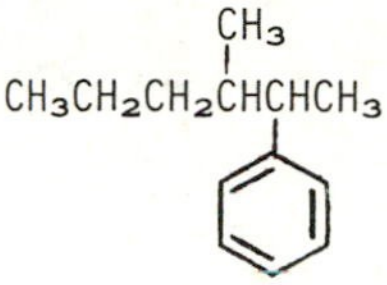 $CH_3CH_2CH_2CH(CH_3)CHCH_3$

d) 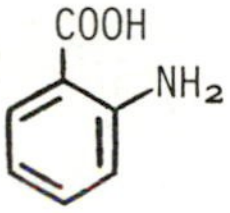 e) 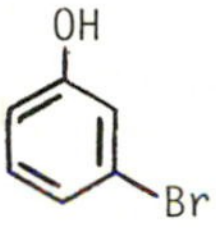 f) 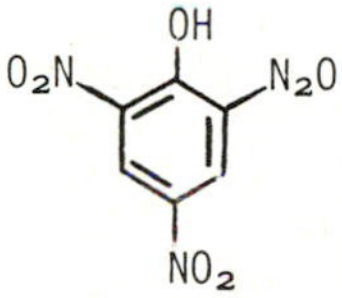 g)

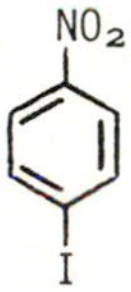

14.18 a)

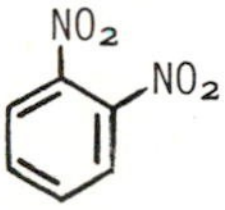

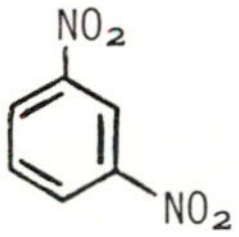

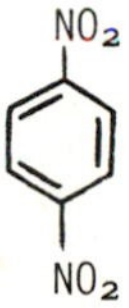

o-dinitrobenzene *m*-dinitrobenzene *p*-dinitrobenzene

b)

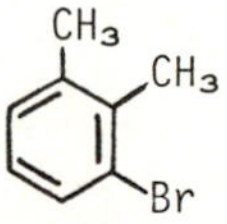

1-bromo-2,3-dimethylbenzene 4-bromo-1,2-dimethylbenzene 2-bromo-1,3-dimethylbenzene

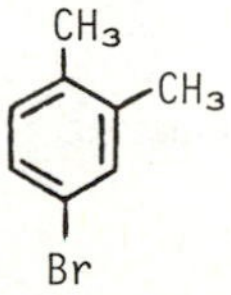

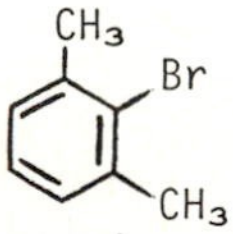

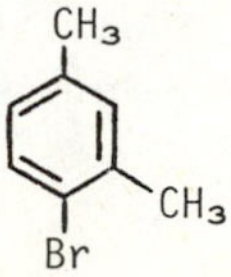

1-bromo-2,4-dimethylbenzene 1-bromo-3,5-dimethylbenzene 2-bromo-1,4-dimethylbenzene

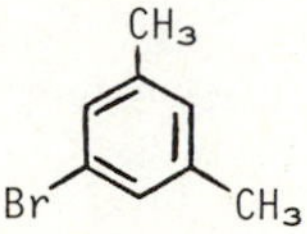

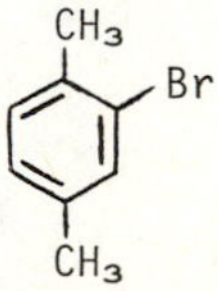

c)

2,3,4-trinitrophenol | 2,3,5-trinitrophenol | 2,3,6-trinitrophenol

2,4,5-trinitrophenol | 2,4,6-trinitrophenol | 3,4,5-trinitrophenol

<u>14.19</u> All aromatic compounds of formula C_7H_7Cl have one ring and three double bonds.

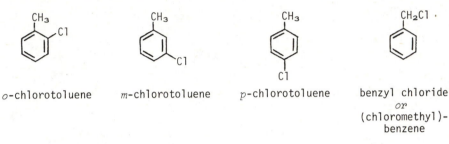

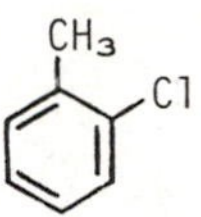

o-chlorotoluene

m-chlorotoluene

p-chlorotoluene

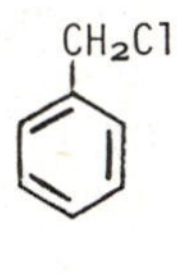

benzyl chloride
or
(chloromethyl)-
benzene

<u>14.20</u> Six of these compounds are illustrated and named in Problem 14.18 b).

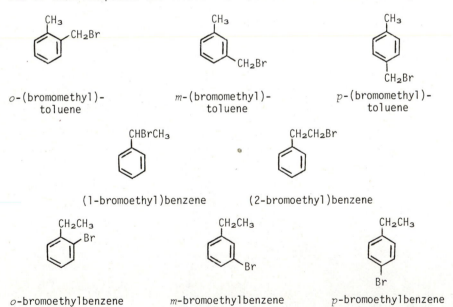

o-(bromomethyl)-
toluene

m-(bromomethyl)-
toluene

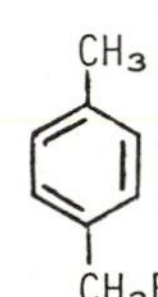

p-(bromomethyl)-
toluene

(1-bromoethyl)benzene

(2-bromoethyl)benzene

o-bromoethylbenzene

m-bromoethylbenzene

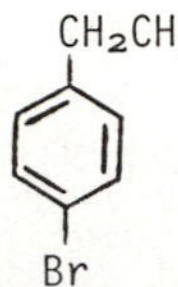

p-bromoethylbenzene

14.21 All compounds in this problem have four double bonds and/or rings and must be substituted benzenes if they are to be aromatic. They may be substituted by methyl, ethyl, propyl or butyl groups.

a) 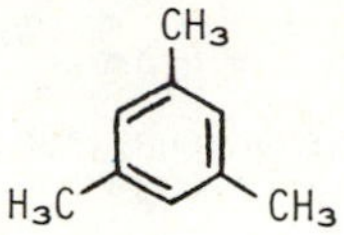 CH_3, H_3C, CH_3

b) 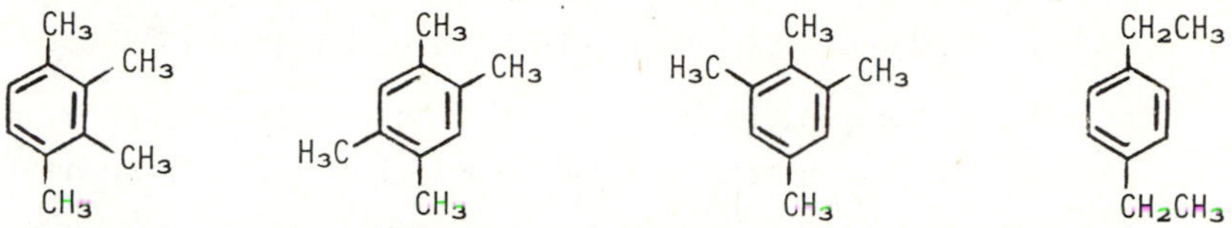 CH_3, CH_3, CH_3, CH_3 CH_3, CH_3, H_3C, CH_3 CH_3, H_3C, CH_3, CH_3 CH_2CH_3, CH_2CH_3

c) CH_3, CH_3 CH_2CH_3

d) 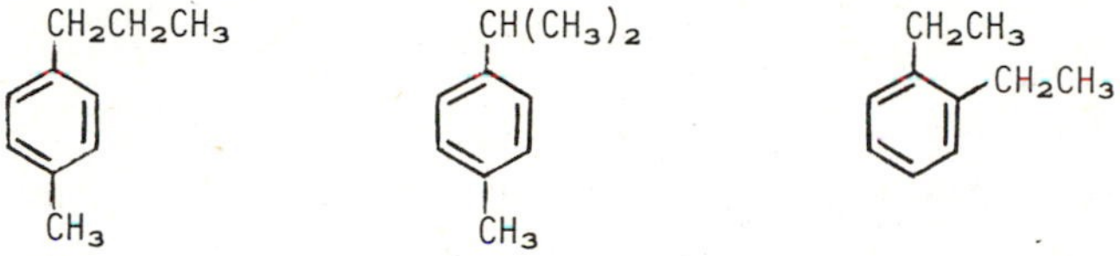 $CH_2CH_2CH_3$, CH_3 $CH(CH_3)_2$, CH_3 CH_2CH_3, CH_2CH_3

14.22

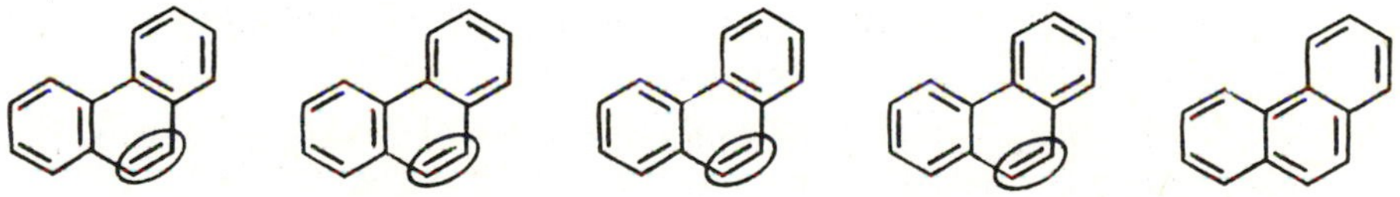

The circled bond is represented as a double bond in four of the five resonance forms of phenanthrene. This bond has more double-bond character and thus is shorter than the other carbon-carbon bonds of phenanthrene.

14.23 a) *Aromaticity* is a property of cyclic conjugated compounds having $(4n + 2)$ pi electrons. Aromatic compounds are unusually stable and unreactive, and undergo substitution, rather than addition, reactions.

b) A system of alternating single and multiple bonds having overlapping p orbitals is said to be *conjugated*. Compounds having conjugated bonds are stabilized by their ability to delocalize pi electrons.

c) The *Hückel 4n + 2 rule* predicts that compounds having $(4n + 2)$ pi electrons (where n = 0, 1, 2 ...) in a cyclic conjugated system will be aromatic.

d) For some compounds it is possible to draw two or more Kekulé structures that differ only in the placement of electrons and not of atoms. These structures are called resonance forms. The actual structure of the compound cannot be represented by any one of the resonance forms, but is a *resonance hybrid* of all of them.

14.24 The heat of hydrogenation is the amount of heat liberated when a compound reacts with hydrogen.

(1) benzene + 3 H_2 ⟶ cyclohexane ΔH_{hydrog} = 49.8 kcal/mol

(2) 1,3-cyclohexadiene + 2 H_2 ⟶ cyclohexane ΔH_{hydrog} = 55.4 kcal/mol

To find ΔH_{hydrog} for benzene + H_2 ⟶ 1,3-cyclohexadiene, subtract the ΔH_{hydrog} of reaction (2) from that of reaction (1):

49.8 kcal/mol - 55.4 kcal/mol = -5.6 kcal/mol

The heat of hydrogenation for this reaction is *negative*; the reaction requires heat (is endothermic) and is not likely to be a favorable process in the laboratory.

14.25

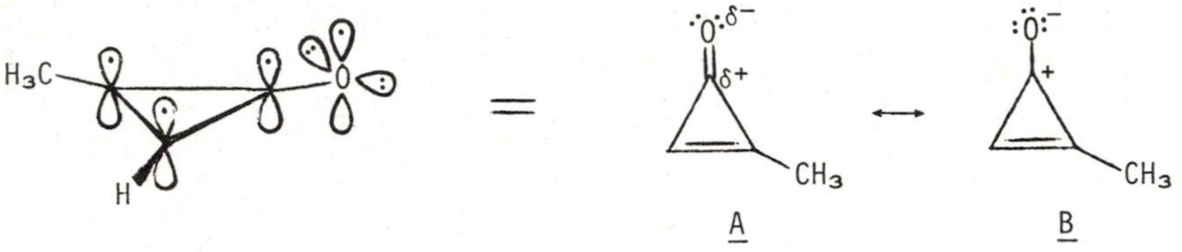

Methylcyclopropenone is a cyclic conjugated compound with a four electron pi system. The electronegative oxygen attracts the pi electrons of the carbon-oxygen pi bond.

In resonance structure B both carbonyl pi electrons are located on oxygen; the other two pi electrons remain in the ring. Since two is a Hückel number, the methylcyclopropanone ring fulfills the criteria of aromaticity and has the added stability of an aromatic ring.

14.26

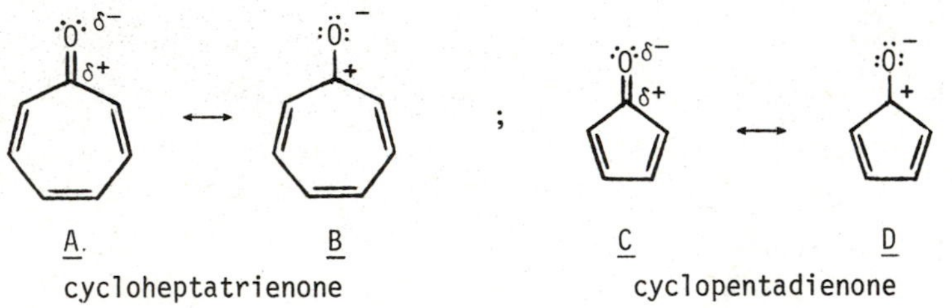

As in the previous problem, we can draw resonance forms in which both carbonyl pi electrons are located on oxygen. The cycloheptatrienone ring in B contains six pi electrons and should be stable, according to Hückel's rule. The cyclopentadienone ring in D contains four pi electrons and is not aromatic.

14.27 Check the number of electrons in the pi system of each compound. The species with a Hückel ($4n + 2$) number of pi electrons is the most stable.

a)

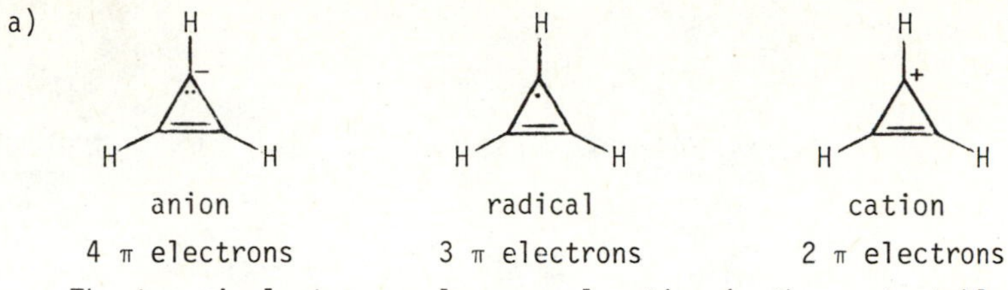

The two pi electron cyclopropenyl cation is the most stable.

b)

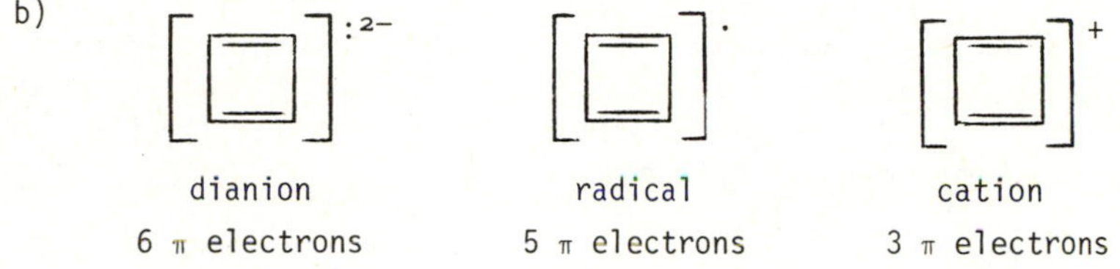

The six pi electron cyclobutadienyl dianion is the most stable.

c)

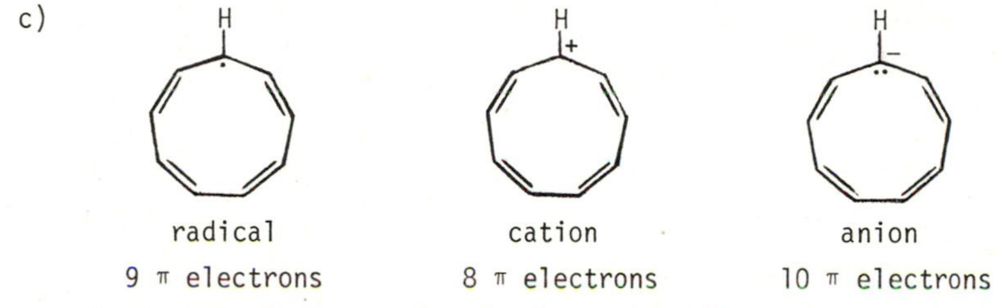

The ten pi electron anion is the most stable.

14.28

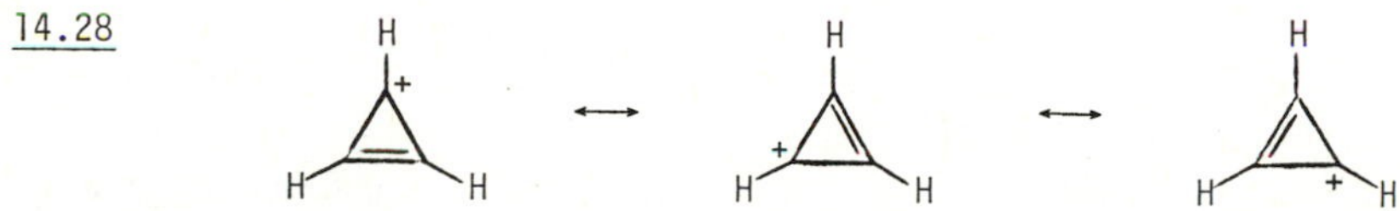

The product of the reaction of 3-chlorocyclopropene with $AgBF_4$ is the cyclopropenyl cation $C_3H_3^+$. The resonance structures of the cation indicate that all hydrogen atoms are equivalent; the ^1H NMR spectrum, which shows only one type of hydrogen atom, confirms this equivalence. The cyclopropenyl cation contains two pi electrons and should be aromatic, according to Hückel's rule.

14.29 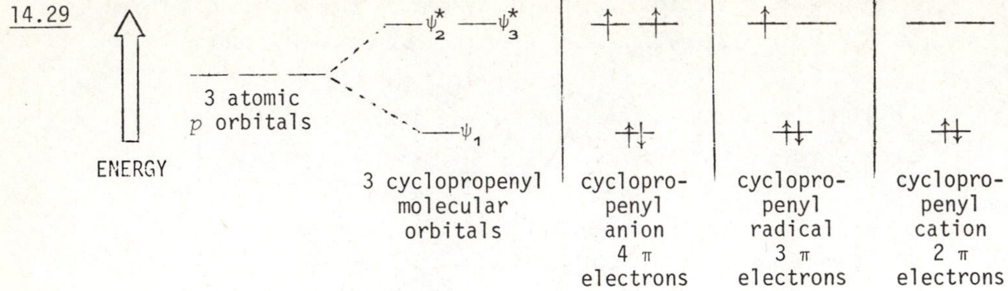

The cyclopropenyl cation is aromatic, according to Hückel's rule.

14.30 The circle in the cyclopropenyl cation represents two pi electrons.

14.31 Compound A has four multiple bonds and/or rings. Possible structures that yield three monobromo substitution products are:

Only structure I shows a six-proton singlet at 2.30 δ; I contains two identical methyl groups unsplit by other protons. The presence of four protons in the aromatic region of the ^{1}H NMR spectrum confirms that I is the correct structure.

14.32 The molecular weight of the hydrocarbon (120) corresponds to the structural formula C_9H_{12}, which indicates four double bonds and/or rings. The ^{1}H NMR singlet at 7.25 δ indicates the presence of five aromatic ring protons. The septet at 2.90 δ is due to a benzylic proton that has six neighboring protons.

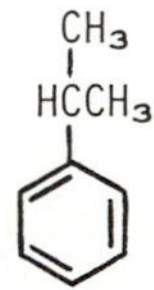

isopropylbenzene

14.33 a) Azulene is a planar, cyclic, conjugated hydrocarbon that has ten pi electrons. According to Hückel's rule it should be aromatic.

b) 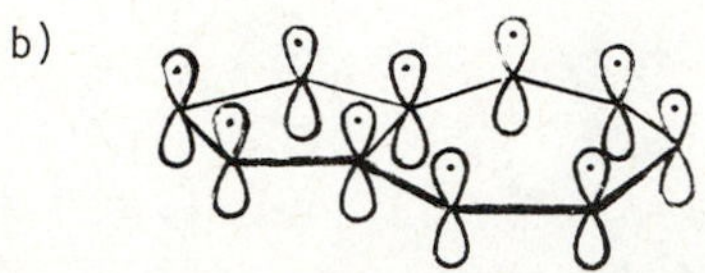

c)

Molecules with dipole moments are polar; electron density is drawn from one part of the molecule to another. In azulene, electron density is greater in the five-membered ring. The five-membered ring resembles the cyclopentadienyl anion in having six pi electrons and in satisfying Hückel's rule. The seven-membered ring resembles the cycloheptatrienyl cation, which also has six pi electrons. Electron density is drawn from the seven-membered ring to the five-membered ring, satisfying Hückel's rule for both rings and producing a dipole moment.

14.34

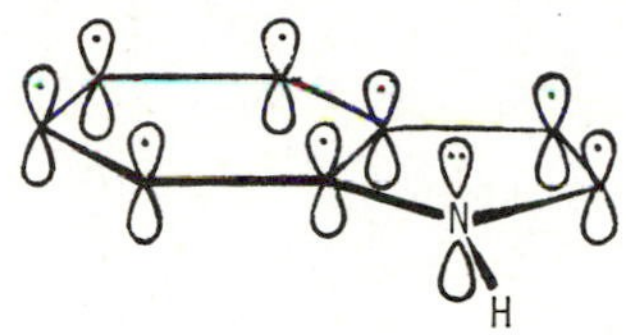

Indole has ten pi electrons and is aromatic. Two pi electrons come from the nitrogen atom. Indole and naphthalene both possess ten pi electrons in two rings.

14.35

A B C

When 4-pyrone is protonated, structure A is produced. B and C are resonance forms of A. In C, a lone pair of electrons of the ring oxygen is delocalized into the ring to produce a six pi electron system, which should be stable, according to Hückel's rule.

14.36 Pentalene has eight pi electrons; Hückel's rule predicts it to be unstable. Pentalene dianion, however, has ten pi electrons and should be stable.

14.37

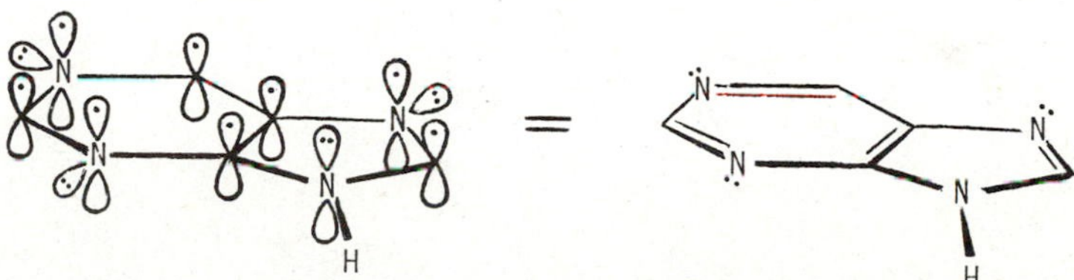

Purine is a ten pi electron aromatic molecule. The N-H nitrogen atom in the five-membered ring donates *both* electrons of its lone pair to the pi electron system, and each of the other three nitrogens donates one electron to the pi electron system.

What you should know:

After doing these problems you should be able to:

1. Recognize and name substituted benzene compounds;
2. Understand the resonance and molecular orbital descriptions of benzene;
3. Use Hückel's rule to predict aromaticity;
4. Draw orbital pictures of cyclic conjugated molecules.

CHAPTER 15 CHEMISTRY OF BENZENE. ELECTROPHILIC AROMATIC SUBSTITUTION

15.1

Benzene + $\overset{\delta+}{I}-\overset{\delta-}{Cl}$ → [intermediate: ring with I and H, +] → iodobenzene + HCl

ICl can be represented as $\overset{\delta+}{I}-\overset{\delta-}{Cl}$ because chlorine is more electronegative element than iodine. Iodine can act as an electrophile in electrophilic aromatic substitution.

15.2

Benzene + $Tl(OCOCF_3)_3$ → [ring with $Tl(OCOCF_3)_2$ and H, + ; + $^-OCOCF_3$] → $C_6H_5Tl(OCOCF_3)_2$ + H^+

Thallium acts as an electrophile and substitutes on the aromatic ring.

15.3

$C_6H_5SO_3H$ $\underset{H^+}{\overset{-H^+}{\rightleftarrows}}$ $C_6H_5SO_3^-$ $\underset{-H^+}{\overset{H^+}{\rightleftarrows}}$ [ring with H and SO_3^-, +] $\underset{SO_3}{\overset{-SO_3}{\rightleftarrows}}$ C_6H_5H

This mechanism is the reverse of the sulfonation mechanism illustrated in the text. H^+ is the attacking electrophile in the desulfonylation reaction.

15.4

Benzene + D^+ ⇄ [ring with D and H, +] ⇄ C_6H_5D + H^+

Benzene can be protonated by strong acids. Although either the deuterium-carbon or the hydrogen-carbon bond in the intermediate can be broken, preferential breaking of the hydrogen-carbon bond occurs because of the deuterium isotope effect. Attack by D^+ can occur at all positions of the ring and leads to eventual replacement of all hydrogen by deuterium.

15.5 Use Table 15.2 in the text to find the activating and deactivating effects of groups.

most reactive ⟶ least reactive

a) phenol > toluene > nitrobenzene

b) phenol > benzene > chlorobenzene > benzoic acid

c) aniline > benzene > bromobenzene > benzaldehyde

15.6 a) *para* attack

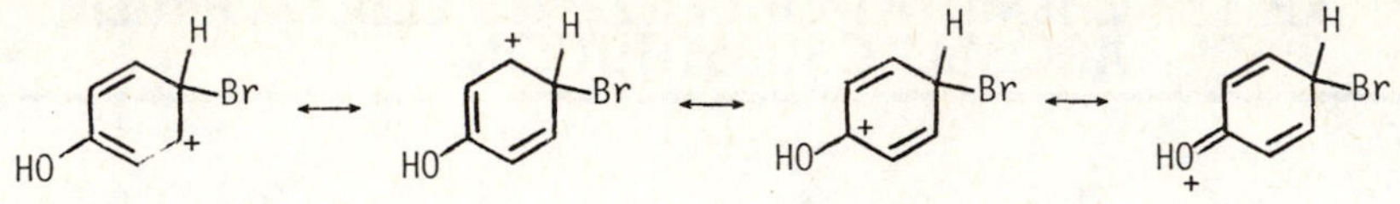

b) *meta* attack

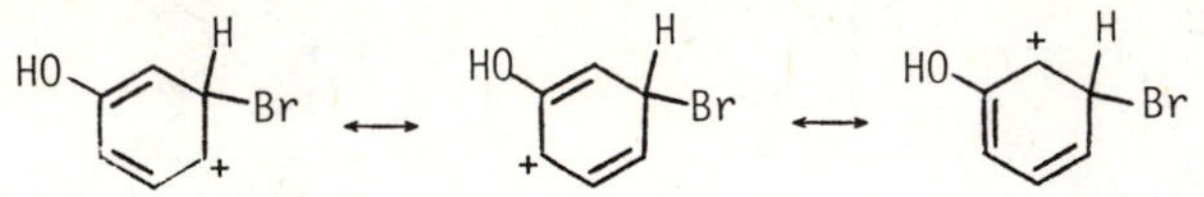

c) *ortho* attack

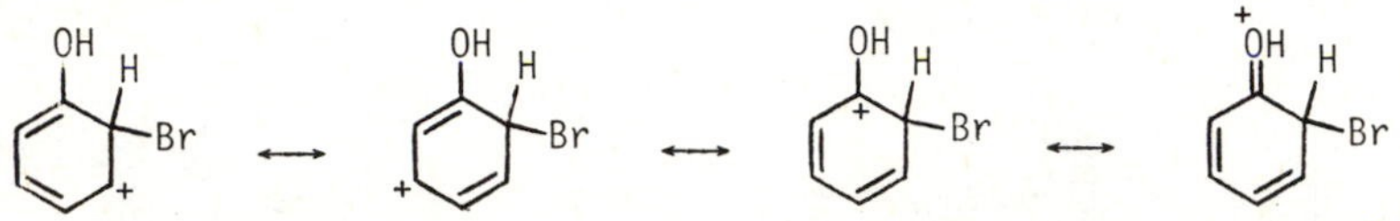

The more resonance forms that can be drawn, the greater is the extent of charge delocalization. Since the intermediates from *ortho* and *para* attack can be written in four resonance forms each, these intermediates are of lower energy. Reaction at the *ortho* and *para* positions is thus favored over reaction at the *meta* position.

15.7 *ortho* attack

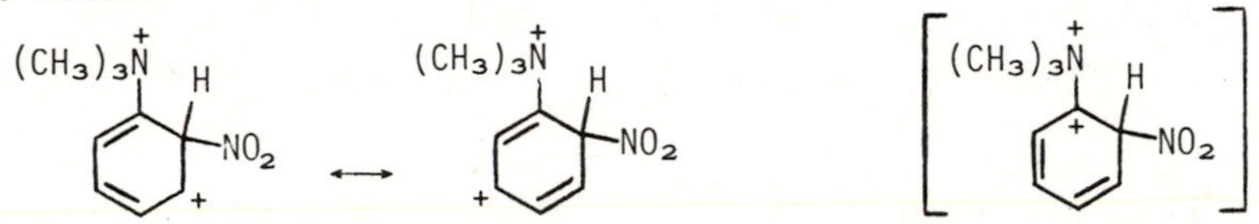

This form is not an important resonance form because two positive charges are next to each other.

meta attack - *MOST FAVORED*

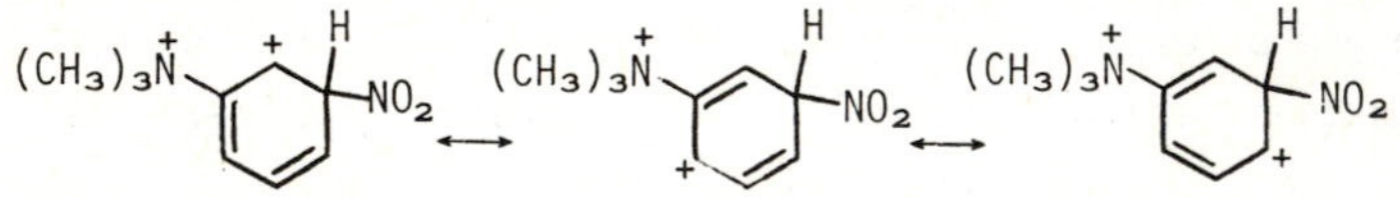

para attack

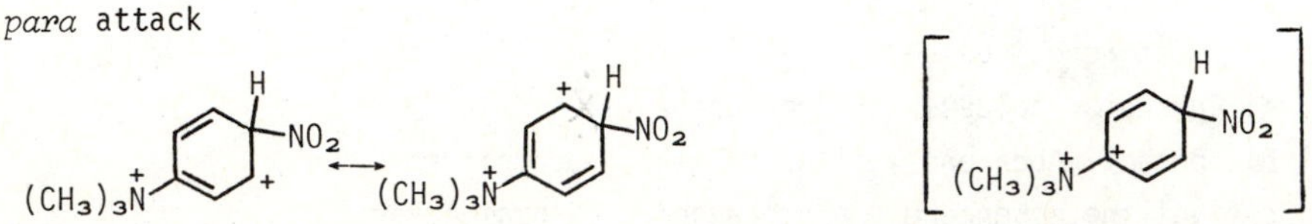

This form is not significant.

15.8 Refer to Table 15.3 in the text for the directing effects of substituents. You should memorize the effects of the most important groups.

a) Bromobenzene $\xrightarrow[H_2SO_4]{HNO_3}$ o-bromonitrobenzene (NO_2) + p-bromonitrobenzene (NO_2)

Even though bromine is a deactivator, it is an *ortho-para* director.

b) Nitrobenzene (NO_2) $\xrightarrow[FeBr_3]{Br_2}$ m-bromonitrobenzene (Br)

The $-NO_2$ group is a *m*-director.

c) Phenol (OH) $\xrightarrow{Cl_2}$ o-chlorophenol (Cl) + p-chlorophenol (Cl)

d) Chlorobenzene (Cl) $\xrightarrow[HSO_3F]{H_2O_2}$ o-chlorophenol (OH) + p-chlorophenol (OH)

e) Bromobenzene (Br) $\xrightarrow[H_2SO_4]{HNO_3}$ o-bromonitrobenzene (NO_2) + p-bromonitrobenzene (NO_2) $\xrightarrow[2.\ ^{-}OH]{1.\ SnCl_2,\ H_3O^+}$ o-bromoaniline (NH_2) + p-bromoaniline (NH_2)

f) Aniline (NH_2) $\xrightarrow{Br_2}$ o-bromoaniline (Br) + p-bromoaniline (Br)

No catalyst is necessary since aniline is highly activated.

15.9 Toluene is more reactive toward electrophilic substitution than trifluoromethylbenzene. The electronegativity of the three fluorine atoms causes the trifluoromethyl group to be electron-withdrawing and deactivating toward electrophilic substitution.

15.10

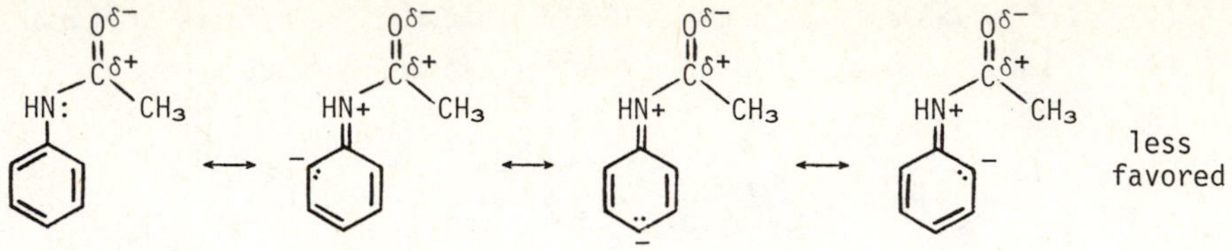

less favored

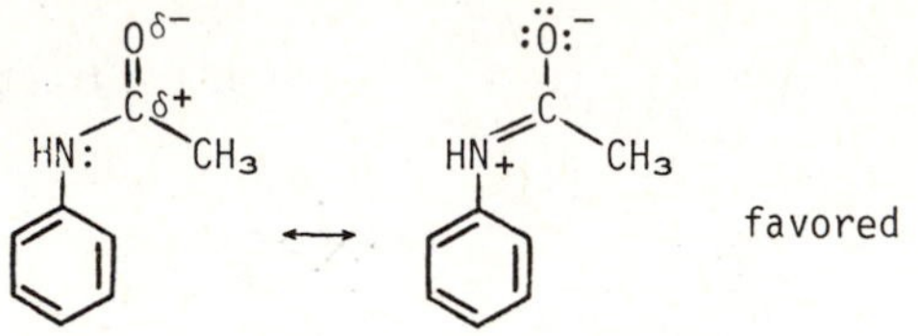

favored

For acetanilide, resonance donation of the nitrogen lone pair electrons to the aromatic ring is less favored because the resulting positive charge is next to the positively polarized carbonyl group. Resonance donation to the carbonyl oxygen is favored because of the electronegativity of oxygen. Since the nitrogen lone pair electrons are less available to the ring, the reactivity of the ring toward electrophilic substitution is decreased, and acetanilide is less reactive than aniline toward electrophilic substitution.

15.11

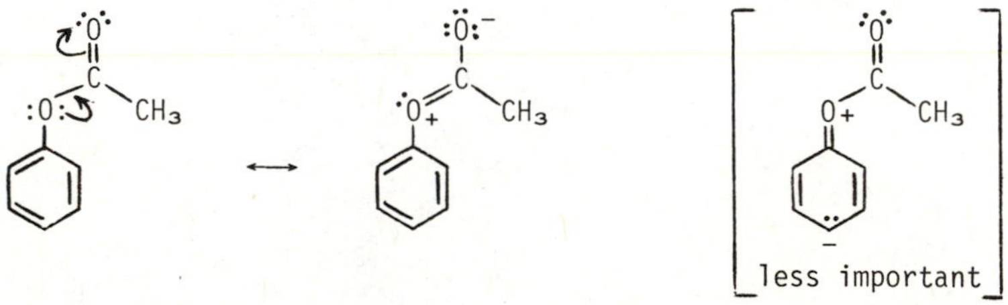

less important

Phenoxide ion is the most reactive and phenyl acetate is the least reactive towards electrophilic substitution. The full negative charge of the phenoxide anion can be delocalized into the ring, which becomes strongly activated toward electrophilic aromatic substitution. For phenyl acetate, delocalization of an electron pair onto the electronegative carbonyl oxygen makes the ring less reactive.

15.12 The aromatic ring is deactivated toward electrophilic aromatic substitution by the combined electron-withdrawing inductive effect of electronegative nitrogen and oxygen. The lone pair of electrons of nitrogen can, however, stabilize by resonance the *ortho* and *para* substituted intermediates.

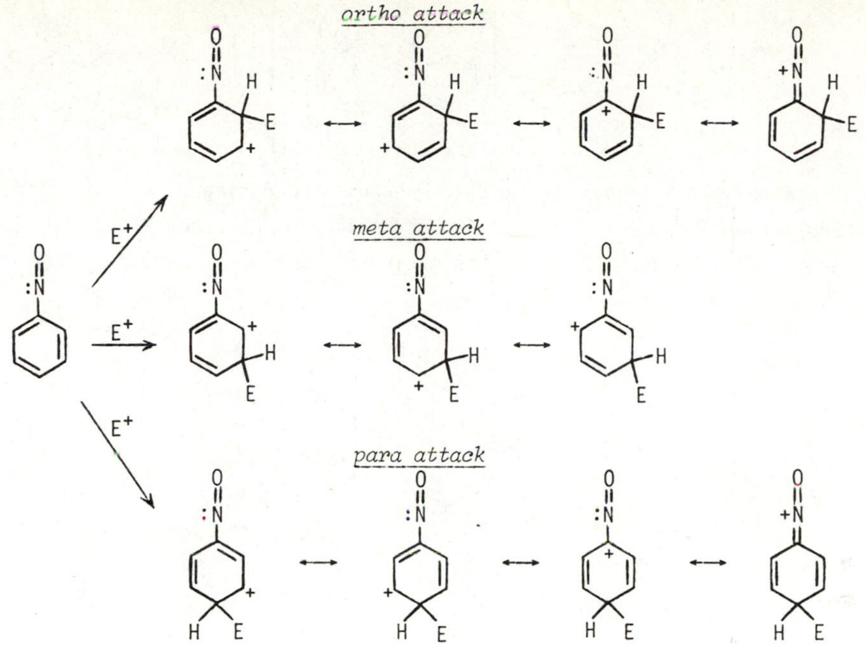

ortho attack

meta attack

para attack

This stabilization is more effective for intermediates of *ortho* and *para* attack than for *meta* attack.

15.13 a)

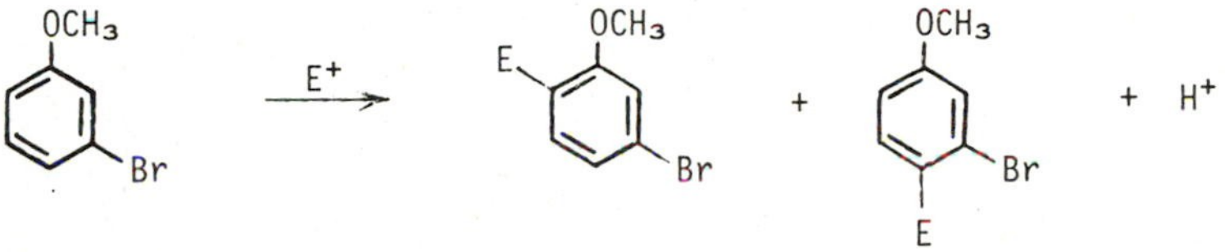

Both groups are *o-p* directors and direct substitution to the same positions. Attack does not occur between the two groups for steric reasons.

b)

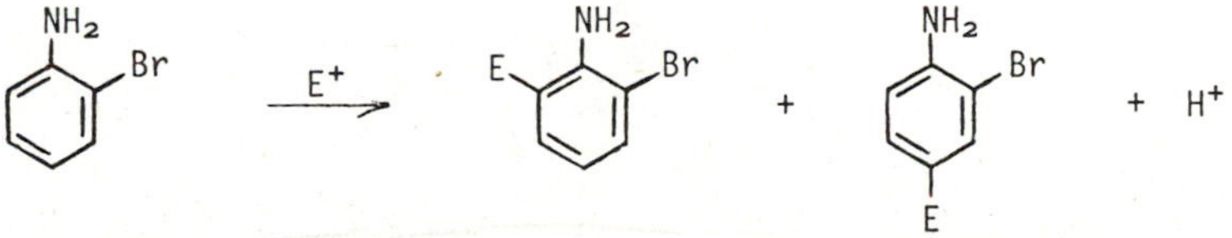

Both groups are *o-p* directors but direct substitution to different positions. Because the $-NH_2$ group is a more powerful activator, substitution occurs *ortho* and *para* to it.

c)

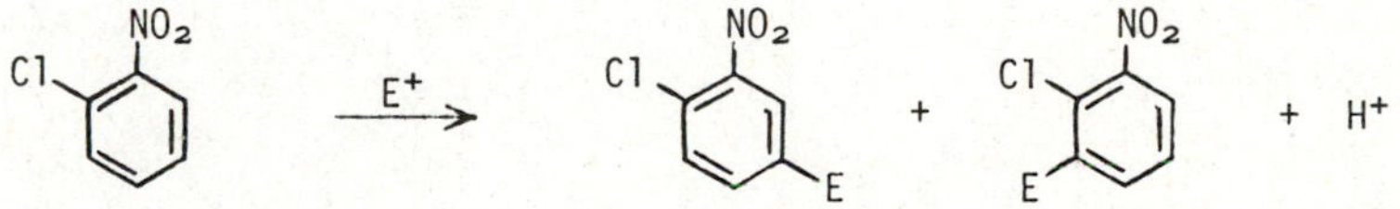

Both groups are deactivating, but they orient substitution toward the same positions.

15.14

The carbonyl oxygens make the chlorine-containing ring electron-deficient and susceptible to attack by the nucleophile $:\bar{O}CH_3$. They also stabilize the negatively charged complex. This nucleophilic aromatic substitution occurs by an addition-elimination pathway.

15.15

Energy

intermediate

$+ :OH^-$

$+ :Cl^-$

Reaction Progress ⟶

15.16

$\xrightarrow{:\bar{O}H}$ $+ :Br^-$ $\xrightarrow{H_2O}$

p-bromotoluene *p*-methylphenol *m*-methylphenol

m-bromotoluene $\xrightarrow{:\bar{O}H}$ $+ :Br^-$ $\xrightarrow{H_2O}$ *o*-methylphenol *m*-methylphenol *p*-methylphenol

Treatment of *m*-bromotoluene with NaOH leads to two benzyne intermediates, which react with water to yield three methylphenol products.

15.17 a) Bromobenzene $\xrightarrow[H_2SO_4]{HNO_3}$ p-bromonitrobenzene + o-bromonitrobenzene

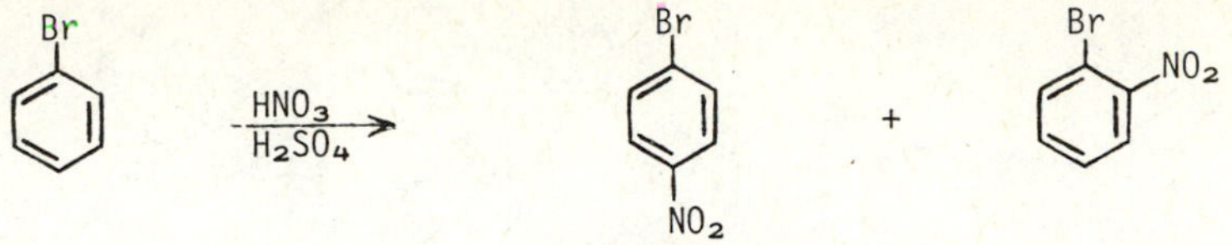

p-bromonitrobenzene *o*-bromonitrobenzene

b) Benzonitrile $\xrightarrow[H_2SO_4]{HNO_3}$

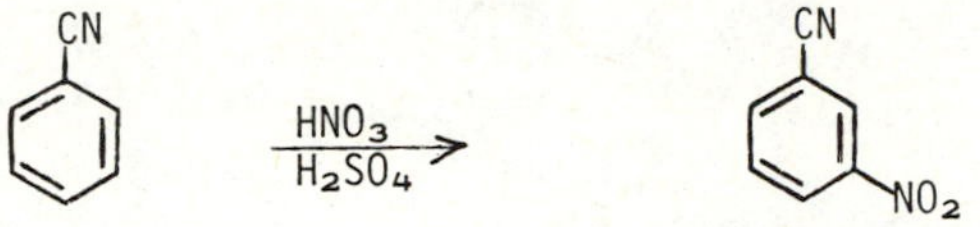

m-nitrobenzonitrile

c) Benzoic acid $\xrightarrow[H_2SO_4]{HNO_3}$

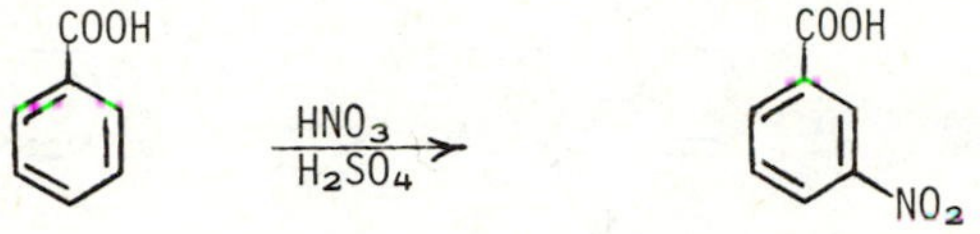

m-nitrobenzoic acid

d) Nitrobenzene $\xrightarrow[H_2SO_4]{HNO_3}$

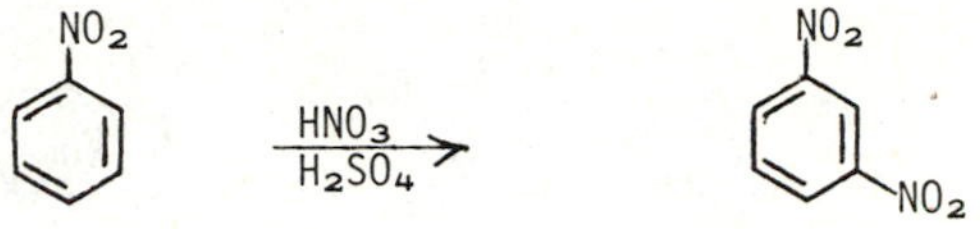

m-dinitrobenzene

e) Benzenesulfonic acid $\xrightarrow[H_2SO_4]{HNO_3}$

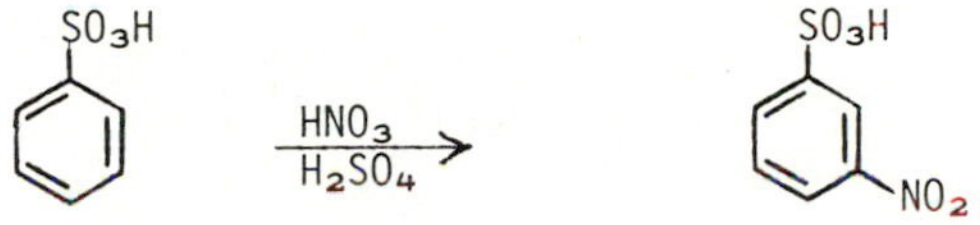

m-nitrobenzenesulfonic acid

f) Methoxybenzene $\xrightarrow[H_2SO_4]{HNO_3}$

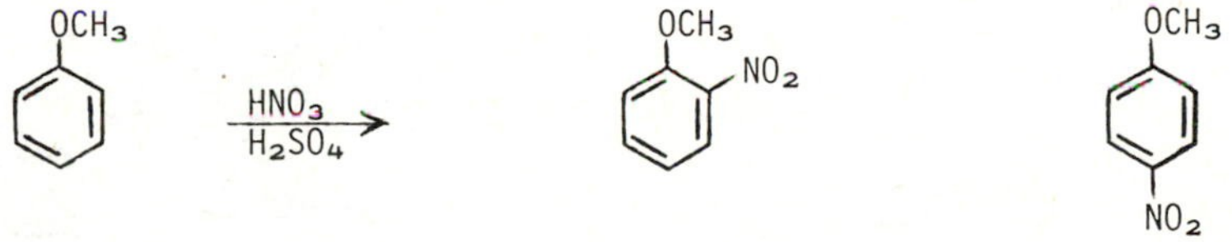

o-methoxynitrobenzene *p*-methoxynitrobenzene

Only methoxybenzene reacts faster than benzene.

15.18 most reactive ⟶ least reactive

a) benzene > chlorobenzene > *o*-dichlorobenzene
b) phenol > nitrobenzene > *p*-bromonitrobenzene
c) *o*-xylene > fluorobenzene > benzaldehyde
d) *p*-methoxybenzonitrile > *p*-methylbenzonitrile > benzonitrile

15.19

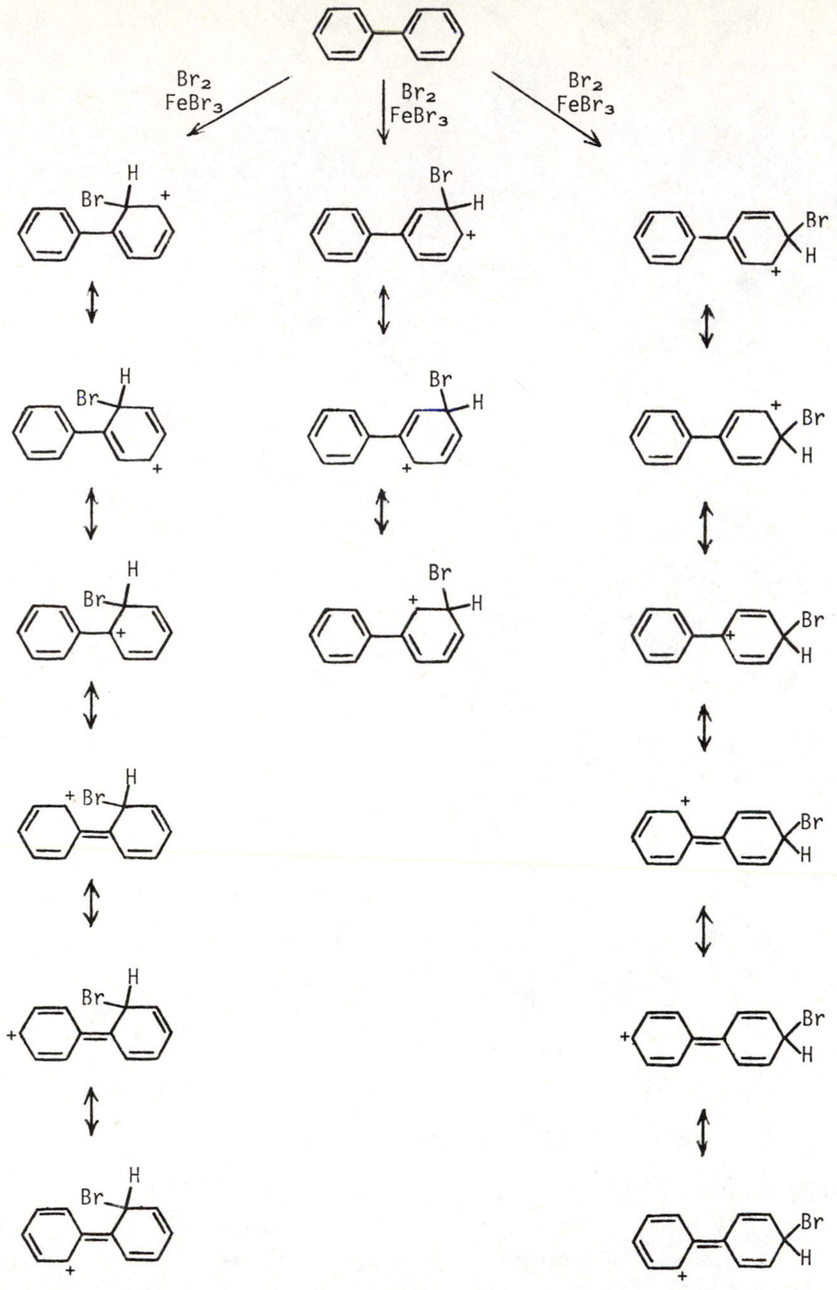

Resonance structures show that bromination occurs in the *ortho* and *para* positions of the rings. The positively charged intermediate formed from *ortho* or *para* attack can be stabilized by resonance contributions from the other ring of biphenyl. This stabilization is not possible for *meta* attack.

15.20 Attack occurs on the unsubstituted ring because bromine is a deactivating group. Attack occurs at the *ortho* and *para* positions of the ring because the positively charged intermediate can be stabilized by resonance contributions from bromine and from the second ring (Problem 15.19).

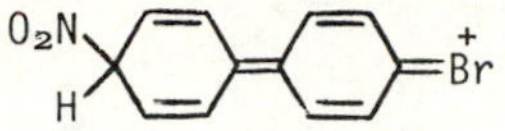

15.21 The -CN group is a *meta*-directing deactivator for both inductive and resonance reasons. In 3-phenylpropanenitrile, the saturated side chain does not allow resonance interactions of -CN with the aromatic ring. In addition, the -CN group is too far from the ring for its inductive effect to be strongly felt. The side chain acts as an alkyl substituent, and *ortho-para* substitution is observed.

In 3-phenylpropenenitrile, the -CN group interacts with the ring through the pi electrons of the side chain. Drawings of resonance forms show that -CN deactivates the ring toward electrophilic substitution. Substitution occurs at the *meta* position.

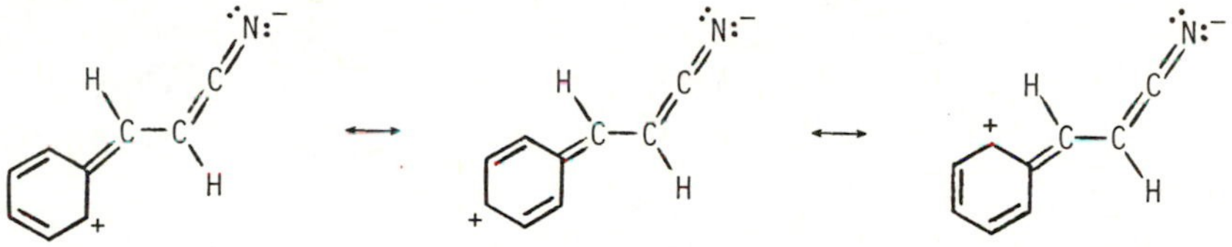

15.22 a)

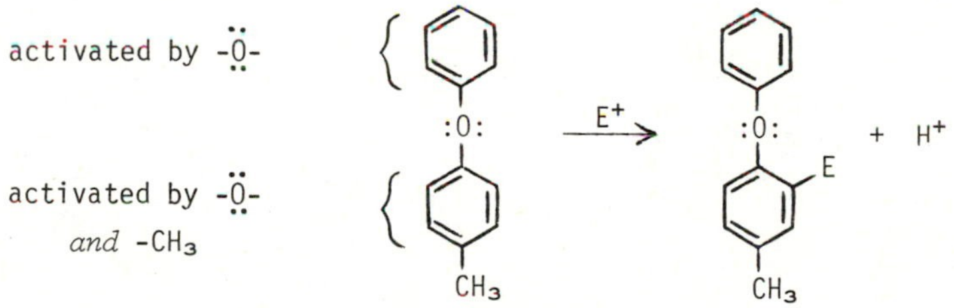

activated by -Ö-

activated by -Ö- *and* $-CH_3$

$\xrightarrow{E^+}$ + H^+

Substitution occurs in the more activated ring. The position of substitution is determined by the more powerful activating group -- in this case, by -Ö-.

b)

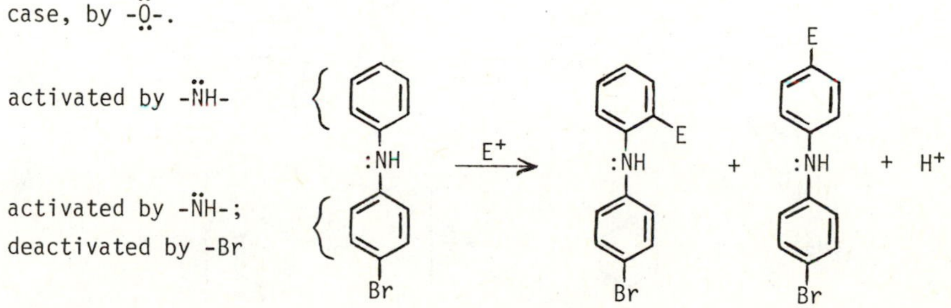

activated by -N̈H-

activated by -N̈H-; deactivated by -Br

$\xrightarrow{E^+}$ + + H^+

The upper ring is more activated than the lower ring. -N̈H- is an *ortho-para* director.

c)

activated by $-C_6H_4CH_3$

activated by $-CH_3$ *and* $-C_6H_5$

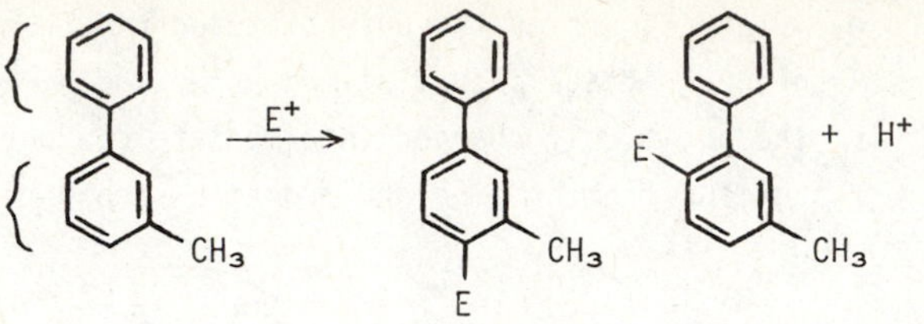

E^+ → + H^+

Substitution occurs in the *ortho* and *para* positions of the more activated ring. Substitution does not occur between $-C_6H_5$ and $-CH_3$ for steric reasons.

15.23 deactivated by C=O

activated by -N̈H-

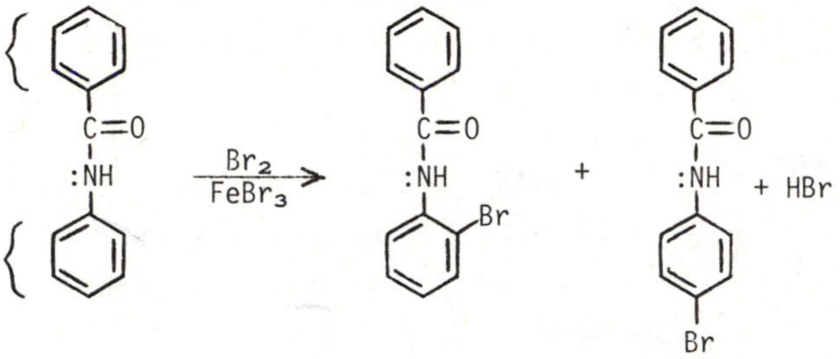

$\xrightarrow[FeBr_3]{Br_2}$ + + HBr

Attack occurs in the activated ring and yields *ortho* and *para* bromination products. The intermediate is resonance-stabilized by overlap of the nitrogen lone pair electrons with the pi electrons of the substituted ring.

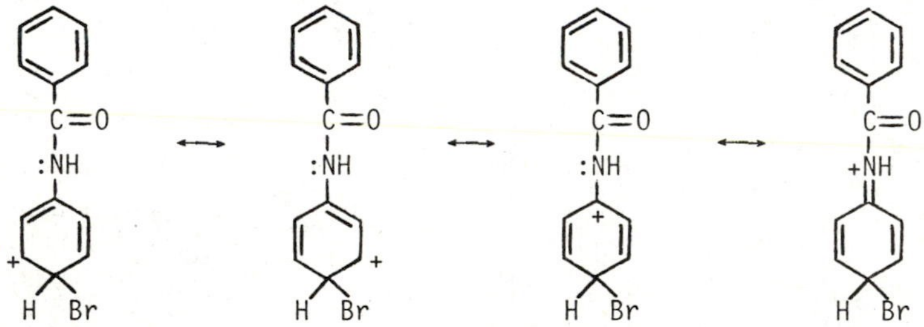

15.24 $\overset{\delta+}{CH_3}\cdots\overset{\delta-}{AlCl_4}$ → → + HCl + $AlCl_3$

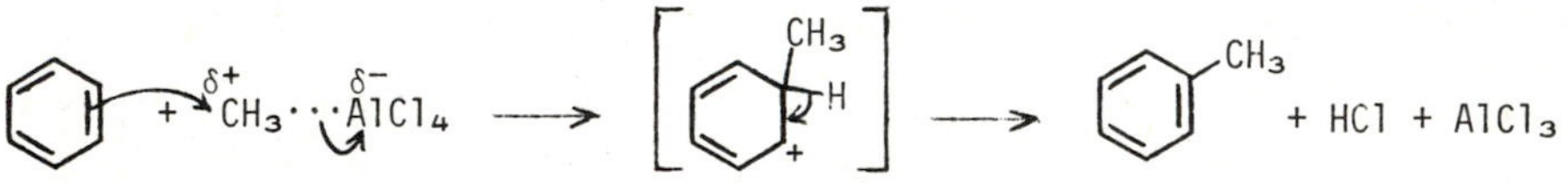

In this reaction a *carbocation* acts as the electrophile.

15.25 $(CH_3)_3COH$ $\xrightarrow{H^+}$ $(CH_3)_3\overset{+}{C}$ + H_2O → → $(CH_3)_3C$ ... + H^+

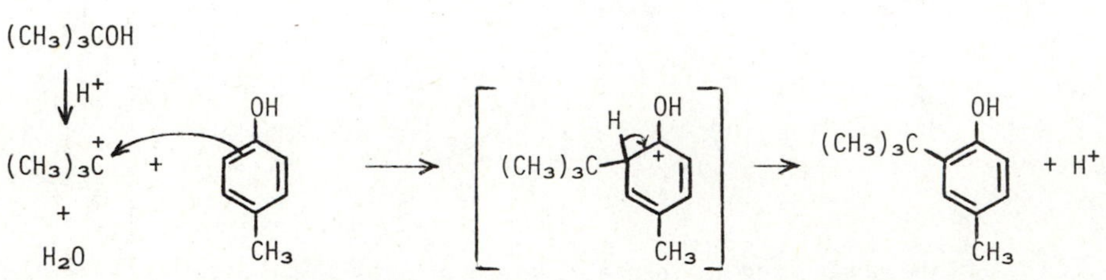

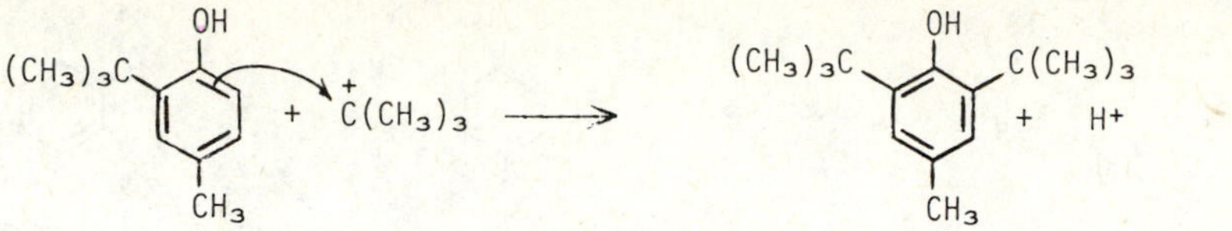

$(CH_3)_3C$–(4-methylphenol ring, OH) + $\overset{+}{C}(CH_3)_3$ ⟶ BHT ($(CH_3)_3C$, OH, $C(CH_3)_3$, CH_3) + H^+

BHT

Reaction of a tertiary alcohol with strong acid yields a carbocation via E_1 elimination (Chapter 10). The carbocation can react as an electrophile. Remember that the more powerful activating group determines orientation.

15.26 a) $\xrightarrow[FeCl_3]{Cl_2}$ 2-chloro-5-nitrophenol + 4-chloro-3-nitrophenol

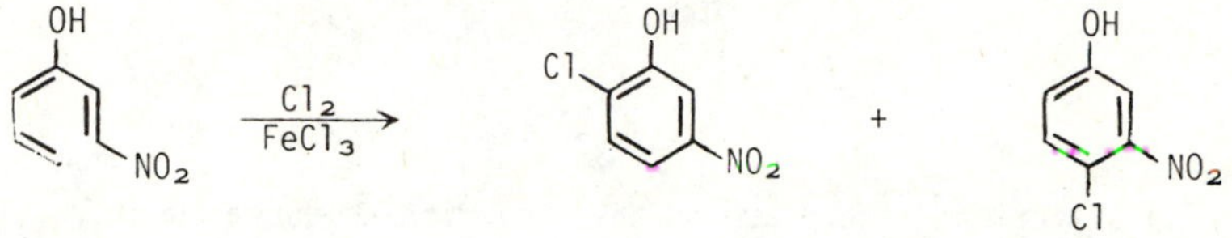

b) $\xrightarrow[FeCl_3]{Cl_2}$ 1-chloro-2,3-dimethylbenzene + 4-chloro-1,2-dimethylbenzene

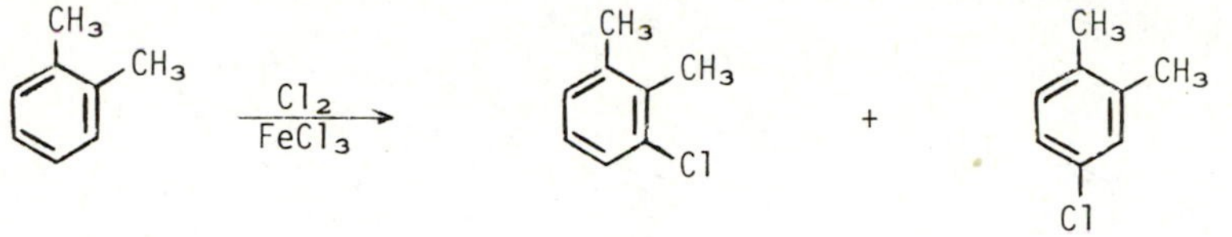

c) $\xrightarrow[CF_3COOH]{Cl_2O}$ 2-chloro-4-nitrobenzoic acid + 3-chloro-4-nitrobenzoic acid

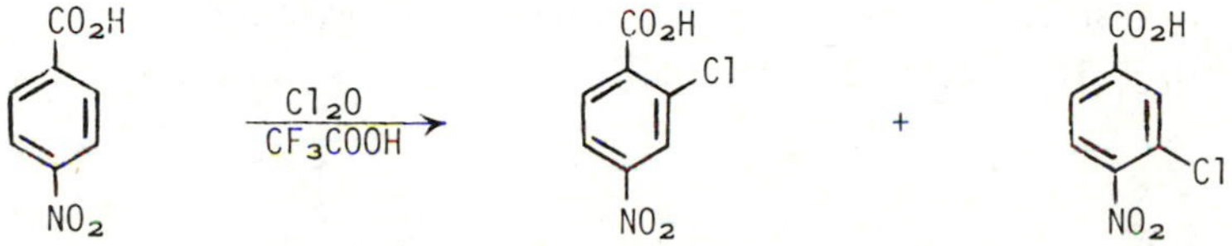

d) $\xrightarrow[FeCl_3]{Cl_2}$ 1-chloro-2,4-dimethylbenzene

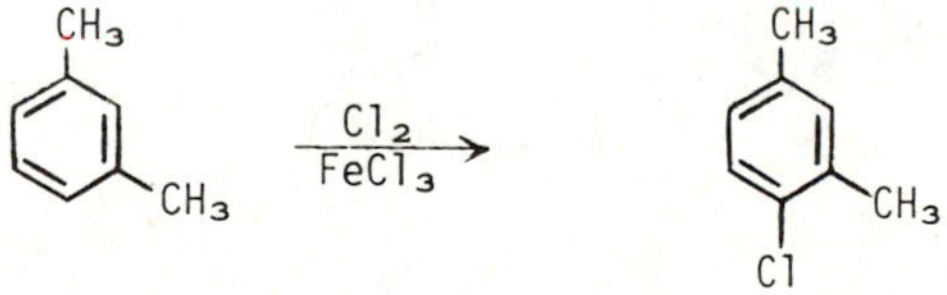

e) $\xrightarrow{Cl_2}$ 2,4-dibromo-6-chlorophenol

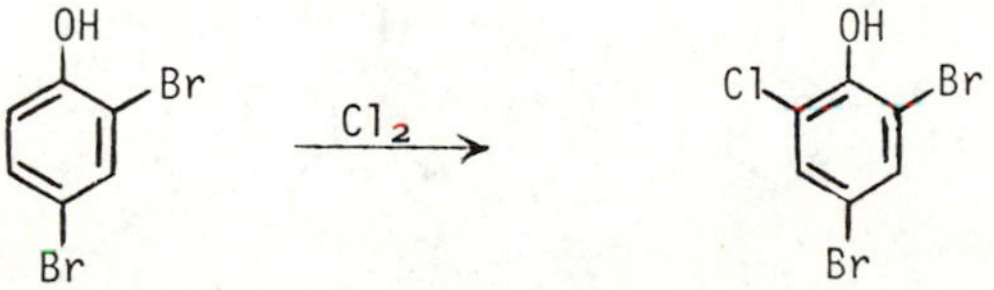

f) $\xrightarrow[FeCl_3]{Cl_2}$ 5-chloro-2-hydroxybenzoic acid + 3-chloro-2-hydroxybenzoic acid

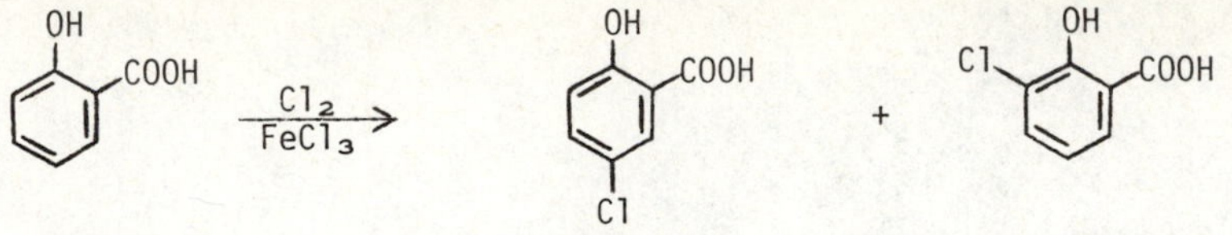

g) $\xrightarrow[FeCl_3]{Cl_2}$ 4-amino-3-chlorobenzenesulfonic acid

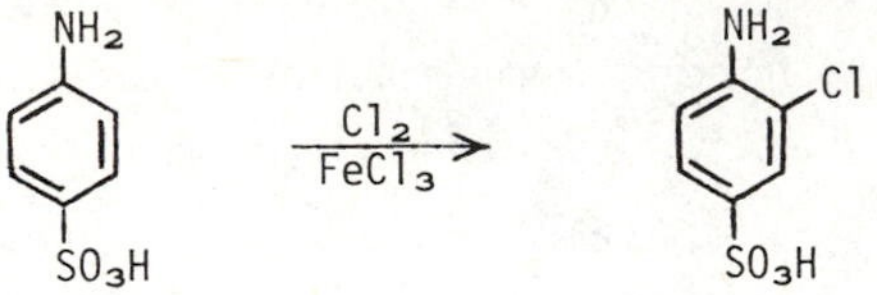

15.27 All substitutions take place on the ring with no substituents. (The other rings contain deactivating groups.)

a) $\xrightarrow{E^+}$ + H^+

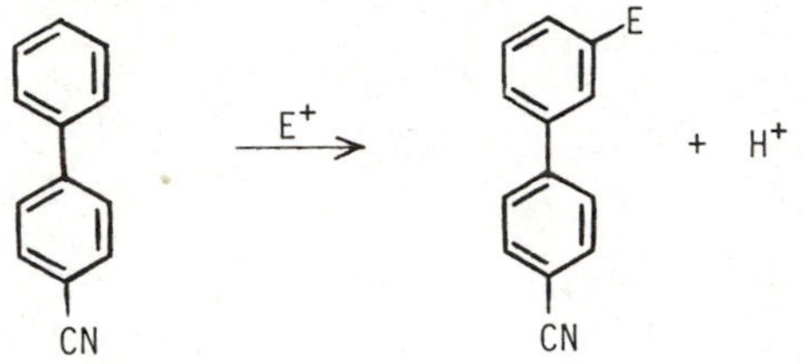

Because the -CN group can interact via resonance with the unsubstituted ring, it directs substitution to the *meta* positions of that ring.

b) $\xrightarrow{E^+}$ + + H^+

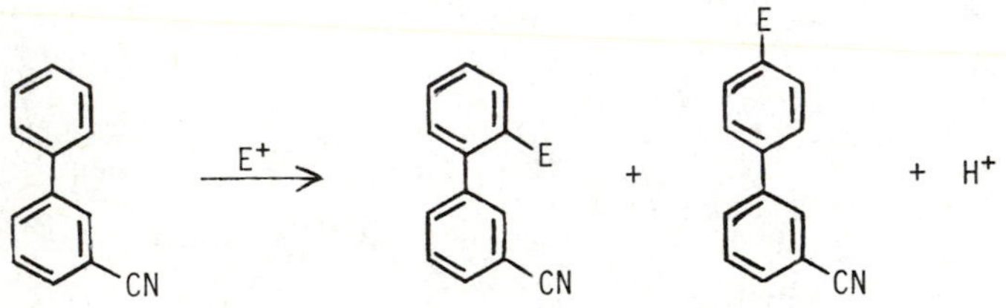

c) $\xrightarrow{E^+}$ + + H^+

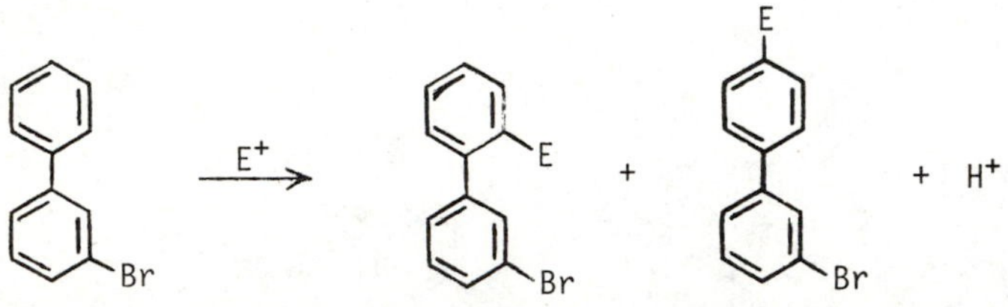

b), c) The -CN and -Br groups cannot interact via resonance with the unsubstituted ring. They also are too far from the unsubstituted ring to have an inductive effect. Electrophilic attack is directed by the substituted phenyl ring and occurs in the *ortho* and *para* positions of the unsubstituted ring.

15.28 When synthesizing substituted aromatic rings, it is necessary to introduce substituents in the proper order. A group that is introduced in the wrong order will not have the proper directing effect.

a) $C_6H_6 \xrightarrow[HOSO_2F]{H_2O_2}$ phenol (OH) $\xrightarrow{Cl_2}$ p-chlorophenol (OH, Cl) + [o-chlorophenol (OH, Cl)]

p-chlorophenol

b) $C_6H_6 \xrightarrow[H_2SO_4]{HNO_3}$ nitrobenzene (NO_2) $\xrightarrow[FeBr_3]{Br_2}$ m-bromonitrobenzene (NO_2, Br)

m-bromonitrobenzene

c) m-bromonitrobenzene (NO_2, Br) $\xrightarrow[2.\ ^{-}OH]{1.\ SnCl_2,\ H_3O^+}$ m-bromoaniline (NH_2, Br)

(from part b) *m*-bromoaniline

d) $C_6H_6 \xrightarrow[FeBr_3]{Br_2}$ bromobenzene (Br) $\xrightarrow[H_2SO_4]{HNO_3}$ p-bromonitrobenzene (Br, NO_2) $\xrightarrow[2.\ ^{-}OH]{1.\ SnCl_2,\ H_3O^+}$ p-bromoaniline (Br, NH_2)

p-bromoaniline

e) $C_6H_6 \xrightarrow[FeBr_3]{Br_2}$ bromobenzene (Br) $\xrightarrow[H_2SO_4]{SO_3}$ o-bromobenzenesulfonic acid (Br, SO_3H) + [p-bromobenzenesulfonic acid (Br, SO_3H)]

o-bromobenzene-sulfonic acid

f) $C_6H_6 \xrightarrow[H_2SO_4]{SO_3}$ benzenesulfonic acid (SO_3H) $\xrightarrow[CF_3COOH]{Cl_2O}$ m-chlorobenzenesulfonic acid (SO_3H, Cl)

m-chlorobenzene-sulfonic acid

15.29 a) toluene (CH_3) $\xrightarrow[H_2SO_4]{HNO_3}$ p-nitrotoluene (CH_3, NO_2) $\xrightarrow[FeBr_3]{Br_2}$ 2-bromo-4-nitrotoluene (CH_3, Br, NO_2)

2-bromo-4-nitrotoluene

b) benzene $\xrightarrow[H_2SO_4]{HNO_3}$ nitrobenzene (NO_2) $\xrightarrow[H_2SO_4]{HNO_3}$ 1,3-dinitrobenzene $\xrightarrow[H_2SO_4]{HNO_3}$ (O_2N, NO_2, NO_2)

1,3,5-trinitrobenzene

c) toluene (CH_3) $\xrightarrow[H_2SO_4]{HNO_3}$ (CH_3, NO_2) $\xrightarrow[2.\ ^-OH]{1.\ SnCl_2,\ H_3O^+}$ (CH_3, NH_2) $\xrightarrow{Cl_2}$ (CH_3, Cl, NH_2)

2-chloro-4-methylaniline

$-NH_2$ is a stronger director than $-CH_3$.

d) benzene $\xrightarrow[H_2SO_4]{HNO_3}$ (NO_2) $\xrightarrow[2.\ ^-OH]{1.\ SnCl_2,\ H_3O^+}$ (NH_2) $\xrightarrow{3\ Br_2}$ (NH_2, Br, Br, Br)

2,4,6-tribromoaniline

No catalyst is needed in parts c and d because aniline is highly activated toward substitution.

e) toluene (CH_3) $\xrightarrow[HOSO_2F]{H_2O_2}$ (CH_3, OH) $\xrightarrow{Cl_2}$ (CH_3, Cl, OH)

2-chloro-4-methylphenol

The -OH group is a stronger director than $-CH_3$.

15.30

(benzenediazonium-2-carboxylate: O, C, O^-, $\overset{+}{N}{\equiv}N$) $\xrightarrow{\Delta}$ [benzyne] $+ CO_2 + N_2$ $\xrightarrow{\text{furan (O)}}$ [transition state, O] $\rightarrow$ (product, O)

benzyne

A Diels-Alder reaction

15.31

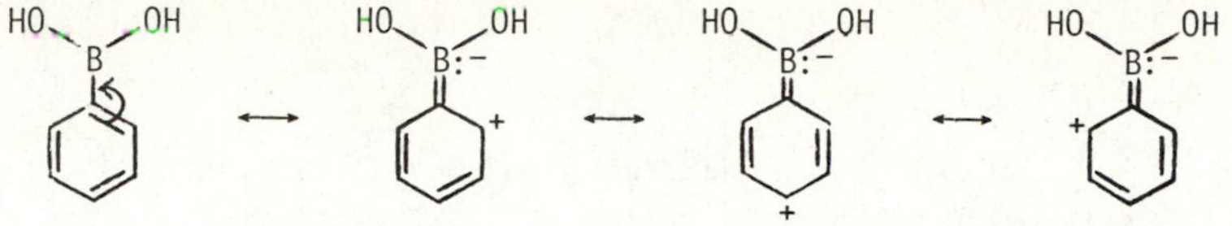

The trivalent boron in phenylboronic acid is a Lewis acid (electron pair acceptor). It is possible to write resonance forms for phenylboronic acid in which an electron pair from the phenyl ring is delocalized onto boron. In these resonance forms, the *ortho* and *para* positions of phenylboronic acid are the most electron-deficient; substitution occurs primarily at the *meta* position.

15.32 Resonance forms for the intermediate from attack at C-1:

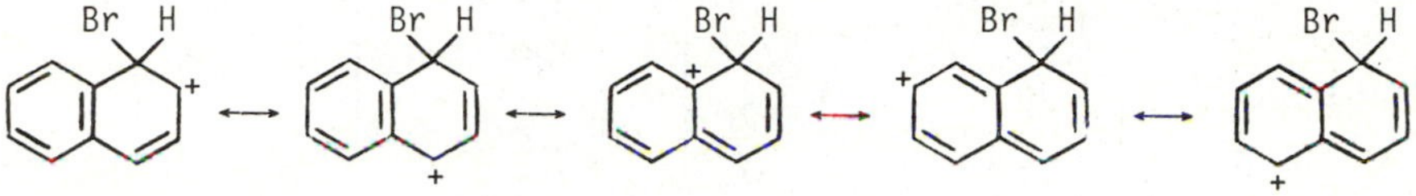

Resonance forms for the intermediate from attack at C2:

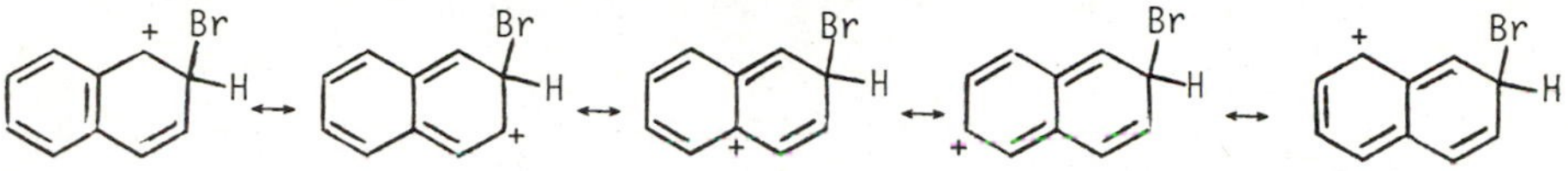

There are five resonance forms for attack at C-1 and five for attack at C2. Look carefully at the forms, however. In the first two resonance structures for C-1 attack, the second ring is still fully aromatic. The positive charge has been delocalized into the second ring in the other three forms, destroying the ring's aromaticity. For C2 attack, only the first resonance structure has a fully aromatic second ring. Since stabilization is lost when aromaticity is disrupted, the intermediate from C2 attack is less stable than the intermediate from C-1 attack, and C-1 attack is favored.

15.33

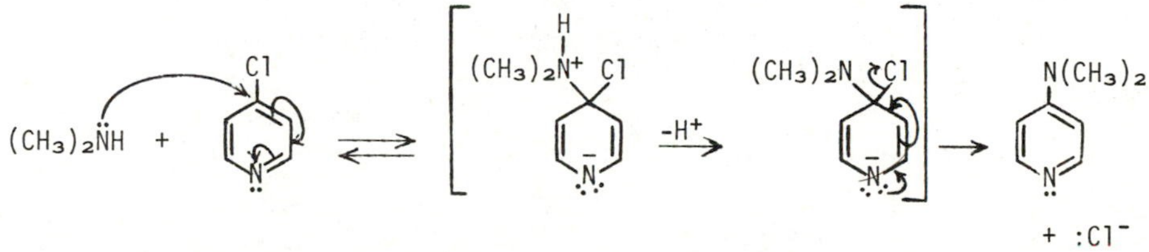

$$(CH_3)_2\ddot{N}H + \text{4-chloropyridine} \rightleftharpoons \left[(CH_3)_2\overset{H}{N^+}\ \text{Cl adduct} \xrightarrow{-H^+} (CH_3)_2N\ \text{Cl adduct} \right] \rightarrow \text{4-}N(CH_3)_2\text{pyridine} + :Cl^-$$

This reaction is an example of *nucleophilic aromatic substitution*. Dimethylamine is a nucleophile, and the nitrogen atom of the pyridine ring acts as an electron-withdrawing group that can stabilize the negatively-charged intermediate.

15.34

$$\text{(p-bromotoluene)} \xrightarrow{:\bar{N}H_2} [\text{benzyne}] + :Br^- \xrightarrow{:\bar{N}H_2} [\text{anions}] \xrightarrow{NH_3} \text{p-methylaniline} + \text{m-methylaniline}$$

The reaction of an aryl halide with potassium amide proceeds through a *benzyne* intermediate. The amide can add to either end of the triple bond to produce the two methylanilines observed.

15.35

$CH_3O:^-$ + (Cl, $COCH_3$, CO_2CH_3 substituted ring) ⟶ [CH_3O, Cl, $:\ddot{O}:^-$, $COCH_3$, CO_2CH_3] ⟶ (CH_3O, $COCH_3$, CO_2CH_3 substituted ring) + $:Cl^-$

This reaction is a nucleophilic aromatic substitution reaction. The *ortho* $-CO_2CH_3$ group can stabilize the negatively charged intermediate.

15.36

$:\ddot{O}H$ (phenol) $\xrightarrow{H^+}$ $:\overset{+}{O}H_2$ (protonated phenol) deactivated aromatic ring

In the presence of very strong acid, the -OH group of phenol can be protonated. When protonated, phenol can no longer activate an aromatic ring toward electrophilic substitution, and reaction does not occur.

15.37 Since the $-NH_2$ group of aniline is strongly activating and *ortho-para* directing, bromination produces 2,4,6-tribromoaniline. In nitration of aniline with HNO_3/H_2SO_4, the $-NH_2$ group is protonated by the mixture of strong acids to produce $-\overset{+}{N}H_3$. This substituent is deactivating and *meta* directing; the product of nitration of aniline is *m*-nitroaniline.

15.38

anthracene + benzyne $\xrightarrow{\Delta}$ [transition state] ⟶ Triptycene

The reaction between benzyne and anthracene is a Diels-Alder reaction.

What you should know:

After doing these problems, you should be able to:

1. Formulate the mechanisms of electrophilic aromatic substitution, nucleophilc aromatic substitution and benzyne-type elimination-addition reactions;
2. Predict reactivity and orientation of electrophilic aromatic substitution reactions by using inductive and resonance arguments;
3. Predict the position of electrophilic aromatic substitution of polysubstituted aromatic compounds.

CHAPTER 16 ARENES. SYNTHESIS AND REACTIONS OF ALKYLBENZENES

16.1 a) CH_3-benzene $\xrightarrow[AlCl_3]{CH_3CH_2Cl}$ *o*-ethyltoluene + *p*-ethyltoluene

o-ethyltoluene *p*-ethyltoluene

b) OH-benzene $\xrightarrow[AlCl_3]{CH_3CH_2Cl}$ *o*-ethylphenol + *p*-ethylphenol

o-ethylphenol *p*-ethylphenol

c) m-xylene $\xrightarrow[AlCl_3]{CH_3CH_2Cl}$ 1-ethyl-2,4-dimethylbenzene

1-ethyl-2,4-dimethylbenzene

d) Cl-benzene $\xrightarrow[AlCl_3]{CH_3CH_2Cl}$ *o*-chloroethylbenzene + *p*-chloroethylbenzene

o-chloroethylbenzene *p*-chloroethylbenzene

16.2

$CH_3\overset{CH_3}{C}=CH_2$ ⇅ H^+ — A carbocation is generated by protonation of 2-methyl-2-butene.

$CH_3\overset{CH_3}{\overset{+}{C}}CH_3$ + benzene ↓ — The pi electrons of the aromatic ring attack the carbocation.

(arenium ion with H and $C(CH_3)_3$)

↓ $-H^+$ — Loss of a proton yields *tert*-butylbenzene.

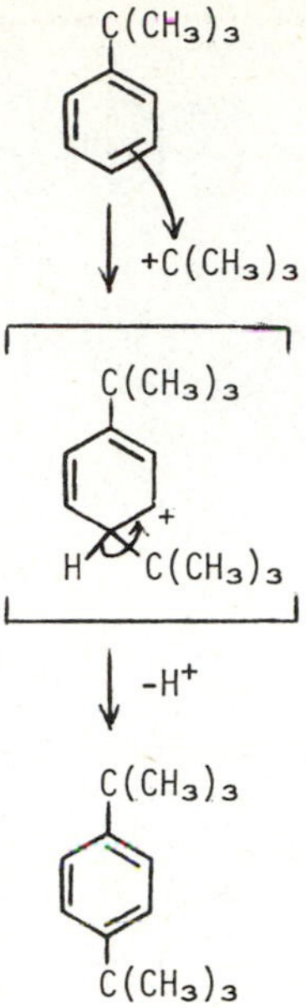

Tert-butylbenzene is activated toward further substitution.

The second substitution does not occur *ortho* to the first *tert*-butyl group for steric reasons.

16.3 most reactive ⟶ least reactive

phenol > toluene > *p*-bromotoluene > bromobenzene

Aniline and nitrobenzene do not undergo Friedel-Crafts alkylation.

16.4

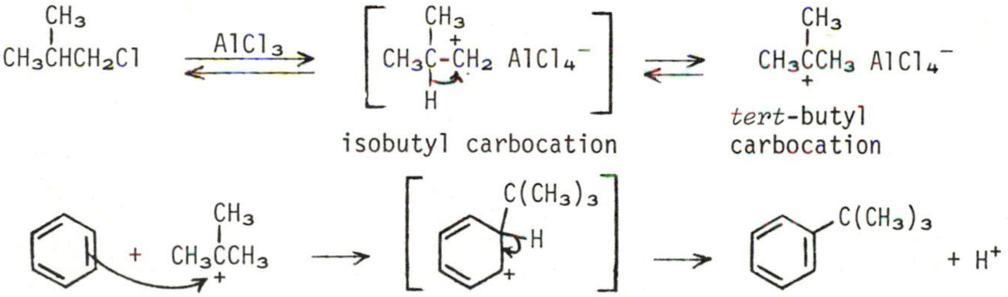

isobutyl carbocation

tert-butyl carbocation

The isobutyl carbocation is initially formed when 1-chloro-2-methylpropane and $AlCl_3$ react. This carbocation rearranges, via a hydride shift, to the more stable *tert*-butyl carbocation, which can then alkylate benzene to form *tert*-butylbenzene.

16.5

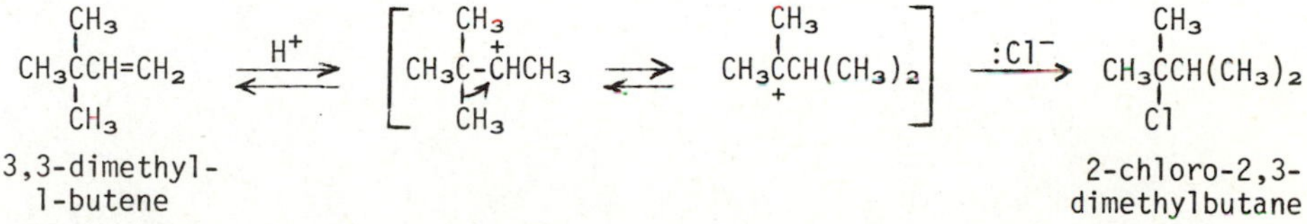

3,3-dimethyl-1-butene

2-chloro-2,3-dimethylbutane

This reaction illustrates the principle of carbocation rearrangements. Here, a more stable carbocation is produced by an alkyl shift.

16.6

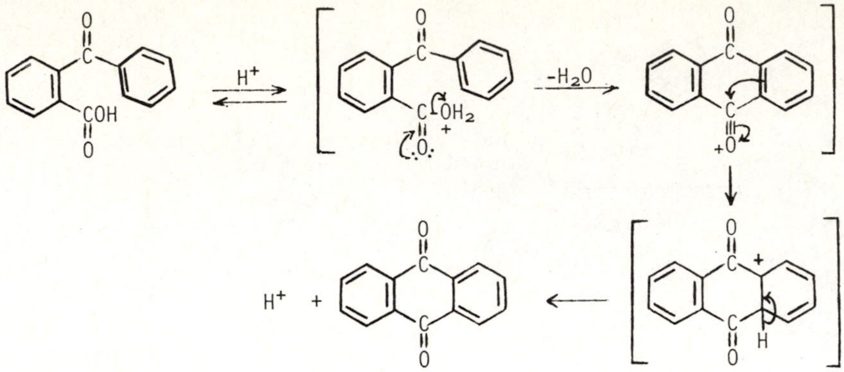

Concentrated H_2SO_4 protonates the carboxylic acid. Departure of H_2O produces an acylium ion, which acylates the aromatic ring.

16.7 A carbonyl group adjacent to an aromatic ring is deactivating. Once an aromatic ring has been acylated, it is much less reactive to further substitution.

16.8

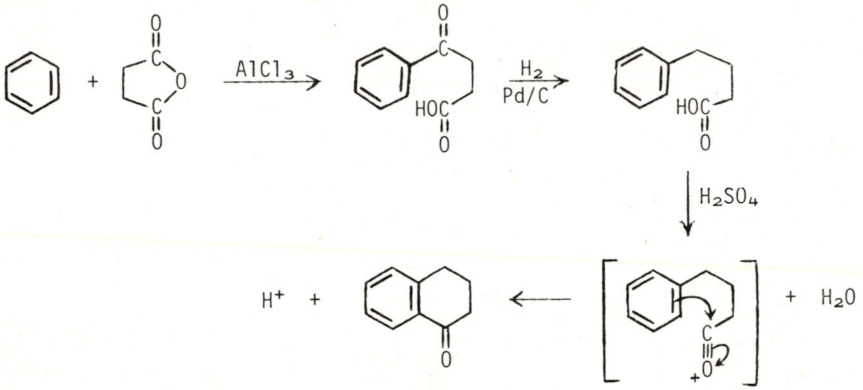

16.9

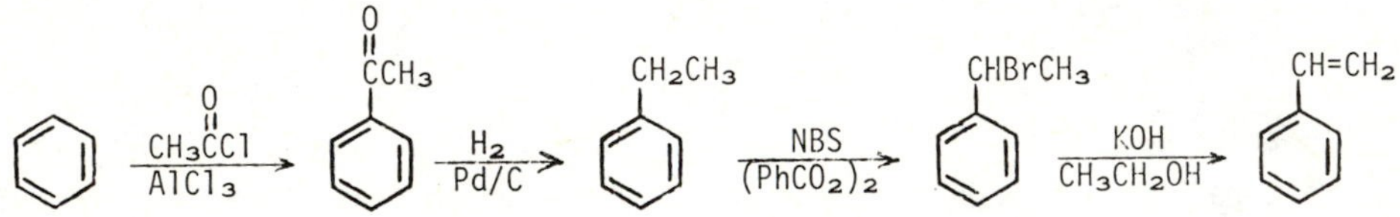

Friedel-Crafts *alkylation* can also be used to produce ethylbenzene, but polyethylbenzenes will also be formed.

16.10

bond	CH_3CH_2-H	(benzene)-CH_2-H	H_2C=$CHCH_2$-H
bond dissociation energy	98 kcal/mol	85 kcal/mol	87 kcal/mol

Bond dissociation energies measure the amount of energy that must be supplied to cleave a bond into two radical fragments. A radical is thus higher in energy and less stable than the compound it came from. Since the C-H bond dissociation energy is 98 kcal/mol for ethane and 85 kcal/mol for a methyl group C-H bond of toluene, less energy is required to form a benzyl radical than to form an ethyl radical. A benzyl radical is thus more stable than a primary alkyl radical by 13 kcal/mol. The bond dissociation energy of an allyl C-H bond is 87 kcal/mol, indicating that an allyl radical is nearly as stable as a benzyl radical.

16.11 a) (tetralin) $\xrightarrow[H_2O]{KMnO_4}$ (benzene-1,2-dicarboxylic acid: COOH, COOH)

b) (1-$CH(CH_3)_2$-3-NO_2-benzene) $\xrightarrow[H_2O]{KMnO_4}$ (3-NO_2-benzoic acid: COOH, NO_2)

c) (H_3CO-C$_6$H$_4$-C(Et)=C(Et)-C$_6$H$_4$-OCH_3) $\xrightarrow[H_2O]{KMnO_4}$ 2 H_3CO-C$_6$H$_4$-COOH

16.12 a) (benzene) $\xrightarrow[H_2SO_4]{HNO_3}$ (nitrobenzene, NO_2) $\xrightarrow[CF_3COOH]{Cl_2O}$ (3-chloronitrobenzene, NO_2, Cl)

m-chloronitrobenzene

b) (benzene) $\xrightarrow[AlCl_3]{CH_3\overset{O}{\overset{\|}{C}}Cl}$ (acetophenone, $\overset{O}{\overset{\|}{C}}CH_3$) $\xrightarrow[CF_3COOH]{Cl_2O}$ (3-chloroacetophenone, $\overset{O}{\overset{\|}{C}}CH_3$, Cl) $\xrightarrow[Pd/C]{H_2}$ (3-chloroethylbenzene, CH_2CH_3, Cl)

m-chloroethyl-
benzene

c) Two routes are possible:

$$\text{Benzene} \xrightarrow{CH_3CH_2\overset{O}{\overset{\|}{C}}Cl,\ AlCl_3} C_6H_5\overset{O}{\overset{\|}{C}}CH_2CH_3 \xrightarrow{H_2,\ Pd/C} C_6H_5CH_2CH_2CH_3 \xrightarrow{Cl_2,\ FeCl_3} p\text{-}ClC_6H_4CH_2CH_2CH_3$$

$$\text{Benzene} \xrightarrow{Cl_2,\ FeCl_3} C_6H_5Cl \xrightarrow{CH_3CH_2\overset{O}{\overset{\|}{C}}Cl,\ AlCl_3} p\text{-}ClC_6H_4\overset{O}{\overset{\|}{C}}CH_2CH_3 \xrightarrow{H_2,\ Pd/C} p\text{-}ClC_6H_4CH_2CH_2CH_3$$

p-chloro-
propylbenzene

16.13

$$\text{Benzene} \xrightarrow{CH_3Cl,\ AlCl_3} C_6H_5CH_3 \xrightarrow{H_2O_2,\ FSO_3H} o\text{-}CH_3C_6H_4OH \xrightarrow{CH_3\overset{O}{\overset{\|}{C}}O\overset{O}{\overset{\|}{C}}CH_3} o\text{-}CH_3C_6H_4O\overset{O}{\overset{\|}{C}}CH_3 \xrightarrow{KMnO_4} o\text{-}HOOCC_6H_4O\overset{O}{\overset{\|}{C}}CH_3$$

Aspirin

16.14 a) Friedel-Crafts acylation, like Friedel-Crafts alkylation, does not occur at an aromatic ring carrying a meta-directing group. We will learn in a later chapter how to synthesize this compound by another route.

b) There are two problems with this synthesis as it is written:

1. Rearrangement often occurs during Friedel-Crafts alkylations using primary halides.
2. Even if *p*-chloropropylbenzene could be synthesized, introduction of the second -Cl group would occur *ortho* to the alkyl group.

A possible route to this compound:

$$C_6H_5Cl \xrightarrow{CH_3CH_2\overset{O}{\overset{\|}{C}}Cl,\ AlCl_3} p\text{-}ClC_6H_4\overset{O}{\overset{\|}{C}}CH_2CH_3 \xrightarrow{Cl_2O,\ CF_3COOH} 3,4\text{-}Cl_2C_6H_3\overset{O}{\overset{\|}{C}}CH_2CH_3 \xrightarrow{H_2,\ Pd/C} 3,4\text{-}Cl_2C_6H_3CH_2CH_2CH_3$$

16.15 a) Bromobenzene $\xrightarrow[AlCl_3]{CH_3Cl}$ o-bromotoluene + p-bromotoluene

o-bromotoluene *p*-bromotoluene

b) 3-Bromophenol $\xrightarrow[AlCl_3]{CH_3Cl}$ 5-bromo-2-methylphenol + 3-bromo-4-methylphenol

5-bromo-2-methyl-phenol 3-bromo-4-methyl-phenol

c) 4-Chloroaniline $\xrightarrow[AlCl_3]{CH_3Cl}$ No reaction. $AlCl_3$ combines with $-\ddot{N}H_2$ to form a complex that deactivates the ring toward Friedel-Crafts alkylation.

d) 2,4-Dichloronitrobenzene $\xrightarrow[AlCl_3]{CH_3Cl}$ No reaction. The ring is deactivated.

e) 2,4-Dichlorophenol $\xrightarrow[AlCl_3]{CH_3Cl}$ 2,4-dichloro-6-methylphenol

2,4-dichloro-6-methylphenol

f) Benzoic acid (COOH) $\xrightarrow[AlCl_3]{CH_3Cl}$ No reaction.

g) p-Toluenesulfonic acid (SO_3H, CH_3) $\xrightarrow[AlCl_3]{CH_3Cl}$ No reaction. The ring is deactivated.

h) 2,5-Dibromotoluene $\xrightarrow[AlCl_3]{CH_3Cl}$ 1,4-dibromo-2,5-dimethylbenzene

1,4-dibromo-2,5-dimethylbenzene

i)

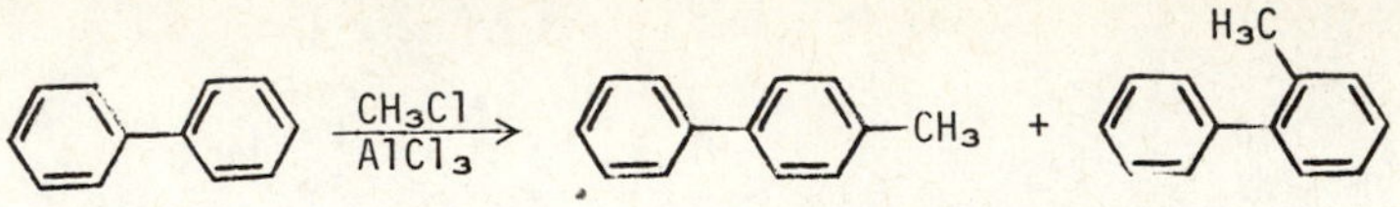

$\xrightarrow[AlCl_3]{CH_3Cl}$ ... $-CH_3$ + H_3C ...

+ polyalkylated biphenyls

j)

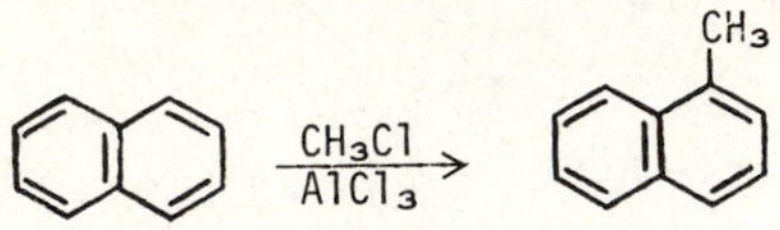

$\xrightarrow[AlCl_3]{CH_3Cl}$ CH_3 ... + polysubstituted products

Naphthalene is more reactive than benzene toward Friedel-Crafts alkylation, and reaction seldom stops at the mono-alkylated stage.

16.16 a)

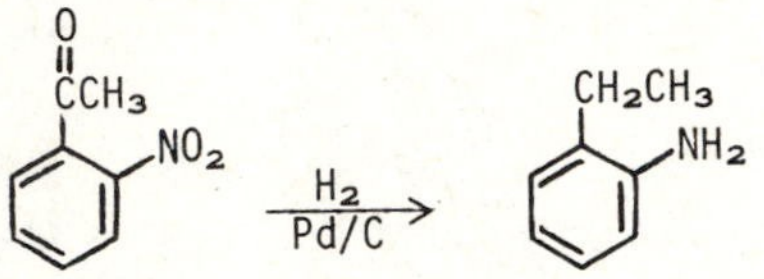

CCH_3 (O), NO_2 $\xrightarrow[Pd/C]{H_2}$ CH_2CH_3, NH_2

b)

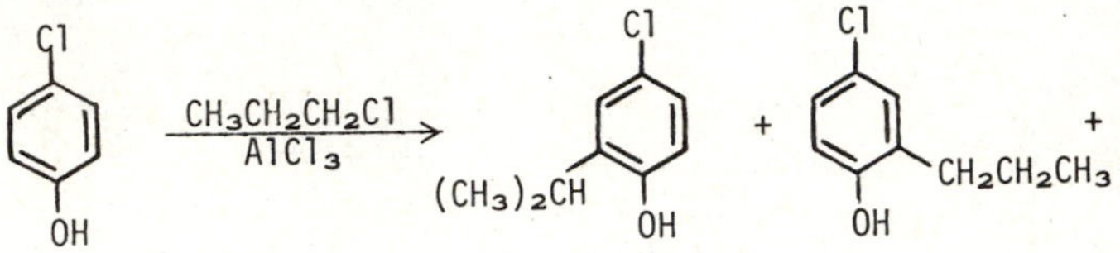

Cl, OH $\xrightarrow[AlCl_3]{CH_3CH_2CH_2Cl}$ Cl, $(CH_3)_2CH$, OH + Cl, $CH_2CH_2CH_3$, OH + disubstitution products

major product

c)

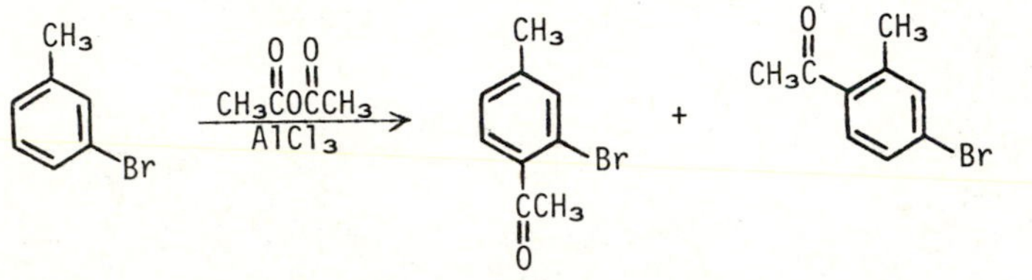

CH_3, Br $\xrightarrow[AlCl_3]{CH_3COCCH_3 \text{ (O O)}}$ CH_3, Br, CCH_3 (O) + CH_3C (O), CH_3, Br

major product

Friedel-Crafts acylation of alkylbenzenes usually occurs *para* to the alkyl substituent.

d)

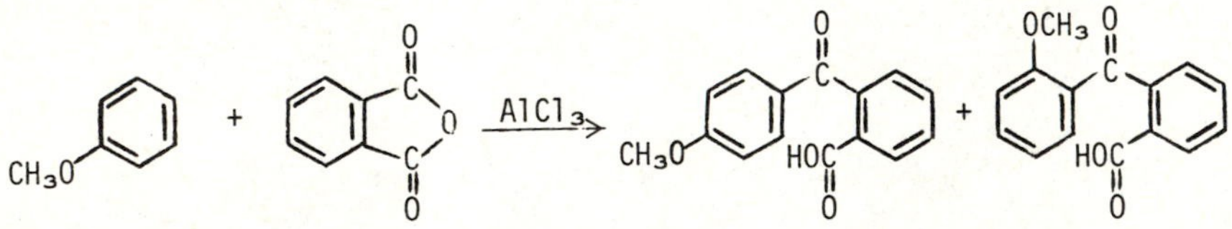

CH_3O + (phthalic anhydride) $\xrightarrow{AlCl_3}$ CH_3O, C (O), HOC (O) + OCH_3, C (O), HOC (O)

e)

products of d) $\xrightarrow{H_2SO_4}$ CH_3O, C (O), C (O) + OCH_3, C (O), C (O)

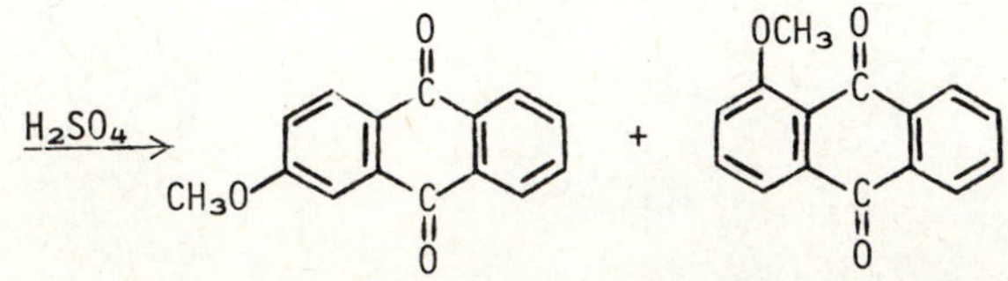

f) No reaction. Since this compound has no benzylic protons, it can't be oxidized.

g) No reaction. This aromatic compound has no benzylic protons to react with NBS.

16.17 $CHCl_3 + AlCl_3 \rightleftarrows \overset{+}{C}HCl_2\overset{-}{Al}Cl_4$

$C_6H_6 + \overset{+}{C}HCl_2\overset{-}{Al}Cl_4 \longrightarrow [\text{arenium ion}] \longrightarrow C_6H_5CHCl_2 + HCl + AlCl_3$

(dichloromethyl)-
benzene

(Dichloromethyl)benzene can react with two additional equivalents of benzene to produce triphenylmethane.

$C_6H_5CHCl_2 + 2\ C_6H_6 \xrightarrow{AlCl_3} (C_6H_5)_3CH$

triphenylmethane

16.18 Some of these compounds can be synthesized by more than one route.

a) $C_6H_6 \xrightarrow[AlCl_3]{CH_3Cl} C_6H_5CH_3 \xrightarrow[H_2SO_4]{HNO_3} o\text{-}CH_3C_6H_4NO_2 \xrightarrow[2.\ ^-OH]{1.\ SnCl_2,\ H_3O^+} o\text{-}CH_3C_6H_4NH_2$

o-methyl-
aniline

b) $C_6H_6 \xrightarrow[FSO_3H]{H_2O_2} C_6H_5OH \xrightarrow{3\left(\frac{HNO_3}{H_2SO_4}\right)}$ (OH, O_2N, NO_2, NO_2)

2,4,6-trinitrophenol

c) $C_6H_6 \xrightarrow[AlCl_3]{CH_3Cl} C_6H_5CH_3 \xrightarrow{3\left(\frac{HNO_3}{H_2SO_4}\right)}$ (CH_3, O_2N, NO_2, NO_2) $\xrightarrow{KMnO_4}$ (COOH, O_2N, NO_2, NO_2)

2,4,6-trinitrobenzoic acid

In b) and c) nitration is performed three times; the second and third nitrations require more severe conditions than the first.

d) Benzene $\xrightarrow[H_2SO_4]{HNO_3}$ nitrobenzene (NO_2) $\xrightarrow[FeBr_3]{2\ Br_2}$ 1,3-dibromo-5-nitrobenzene (NO_2, Br, Br) $\xrightarrow[2.\ ^-OH]{1.\ SnCl_2,\ H_3O^+}$ (NH_2, Br, Br)

3,5-dibromo-aniline

e) Benzene $\xrightarrow[AlCl_3]{CH_3Cl}$ toluene (CH_3) $\xrightarrow[H^+]{(CH_3)_3COH}$ (CH_3, $C(CH_3)_3$) $\xrightarrow{KMnO_4}$ (COOH, $C(CH_3)_3$)

p-tert-butyl-benzoic acid

f) Benzene $\xrightarrow[AlCl_3]{CH_3CH_2CH_2\overset{O}{\overset{\|}{C}}Cl}$ ($\overset{O}{\overset{\|}{C}}CH_2CH_2CH_3$) $\xrightarrow[H_2SO_4]{HNO_3}$ ($\overset{O}{\overset{\|}{C}}CH_2CH_2CH_3$, NO_2) $\xrightarrow[Pd/C]{H_2}$ ($CH_2CH_2CH_2CH_3$, NH_2)

m-butylaniline

16.19

Styrene ($C_6H_5CH{=}CH_2$) $\overset{H^+}{\rightleftharpoons}$ [$C_6H_5\overset{+}{C}HCH_3$]

Styrene is protonated to form the more stable carbocation.

$C_6H_5\overset{+}{C}HCH_3$ + $CH_2{=}CHC_6H_5$ $\longrightarrow$ $C_6H_5CH(CH_3)CH_2\overset{+}{C}HC_6H_5$

The carbocation is attacked by the double bond pi electrons of a second styrene to produce another stable carbocation...

...which can undergo electrophilic substitution on the ring of the first styrene to yield the observed product.

16.20

$$BF_3 + F\text{-}\overset{O}{\overset{\|}{C}}\text{-}H \rightleftharpoons \left[H\text{-}\overset{+}{C}{=}O \longleftrightarrow H\text{-}C{\equiv}\overset{+}{O} \;\; BF_4^- \longrightarrow \right] \longrightarrow C_6H_5CHO + H^+$$

16.21 a) Chlorination of toluene occurs at the *ortho* and *para* positions. To synthesize the given product, first oxidize toluene to benzoic acid and then chlorinate.

b) *p*-Nitrochlorobenzene is inert to Friedel-Crafts alkylation because the nitrochlorobenzene ring is deactivated.

c) The first two steps in the sequence are correct, but H_2/Pd reduces the nitro group as well as the ketone.

d) Nitration and reduction proceed as written, but Friedel-Crafts alkylation fails with amino-substituted aromatic rings. (Also, if alkylation were to succeed, it would yield rearranged product.)

16.22 The initial reaction of 2-chlorobutane with $AlCl_3$ produces an ion pair $[CH_3\overset{+}{C}HCH_2CH_3\overset{-}{Al}Cl_4]$, which dissociates to form the *sec*-butyl carbocation and $AlCl_4^-$. The planar, sp^2 hybridized carbocation is achiral and its reaction with benzene yields racemic product.

16.23

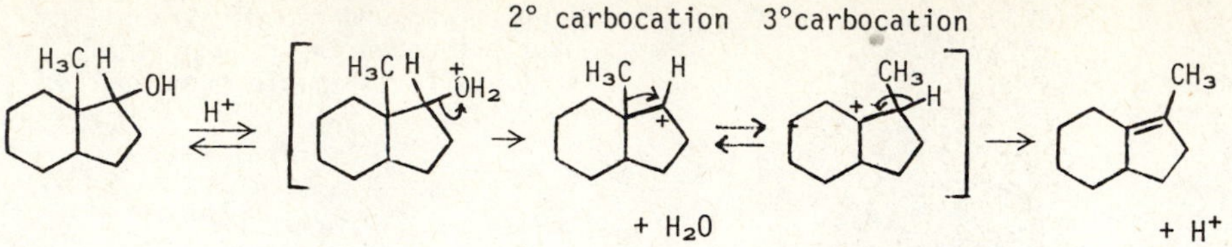

2° carbocation 3°carbocation

$+ H_2O$ $+ H^+$

This reaction is an example of an alkyl shift, which produces a more stable carbocation and a more stable (more substituted double bond) product.

16.24 a) Benzene $\xrightarrow[FeCl_3]{Cl_2}$ chlorobenzene $\xrightarrow[AlCl_3]{CH_3CH_2Cl}$ CH_2CH_3 / Cl $\xrightarrow[(PhCO_2)_2]{NBS}$ $CHBrCH_3$ / Cl $\xrightarrow[CH_3CH_2OH]{KOH}$ $HC{=}CH_2$ / Cl

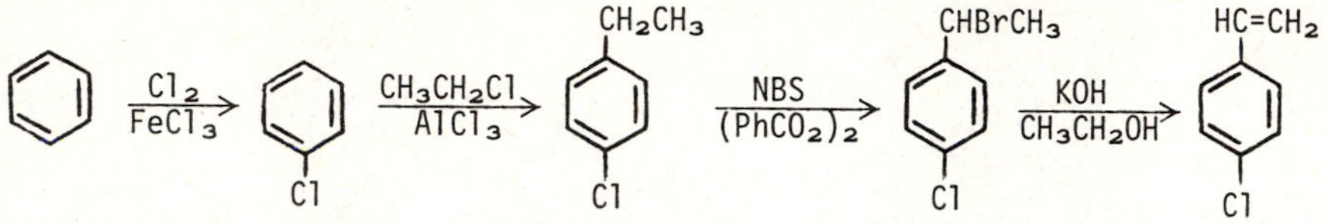

b) Benzene $\xrightarrow[AlCl_3]{CH_3Cl}$ CH_3 $\xrightarrow[AlCl_3]{CH_3Cl}$ CH_3 / CH_3 $\xrightarrow[(PhCO_2)_2]{2\ NBS}$ CH_2Br / CH_2Br $\xrightarrow{NaOH}$ CH_2OH / CH_2OH

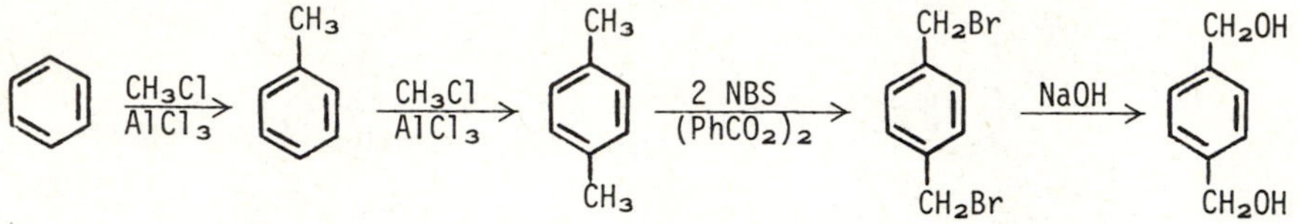

c) Benzene $\xrightarrow[AlCl_3]{CH_3CH_2Cl}$ CH_2CH_3 $\xrightarrow[(PhCO_2)_2]{NBS}$ $CHBrCH_3$ $\xrightarrow[CH_3CH_2OH]{KOH}$ $HC{=}CH_2$

1. BH_3
2. $H_2O_2/^-OH$

CH_2CH_2OH

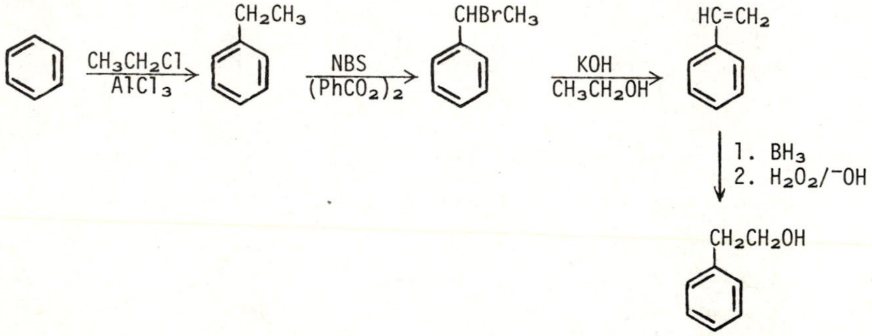

d) Benzene $\xrightarrow[AlCl_3]{CH_3CHClCH_3}$ CH_3 / H_3CCH $\xrightarrow[HOSO_2F]{H_2O_2}$ CH_3 / H_3CCH / OH $\xrightarrow[(PhCO_2)_2]{NBS}$ CH_3 / H_3CCBr / OH

KOH
CH_3CH_2OH

$H_3CC{=}CH_2$ / OH $\xrightarrow[AlCl_3]{CH_3Cl}$ $H_3CC{=}CH_2$ / CH_3 / OH $\xrightarrow{HBr}$ CH_3 / H_3CCBr / CH_3 / OH

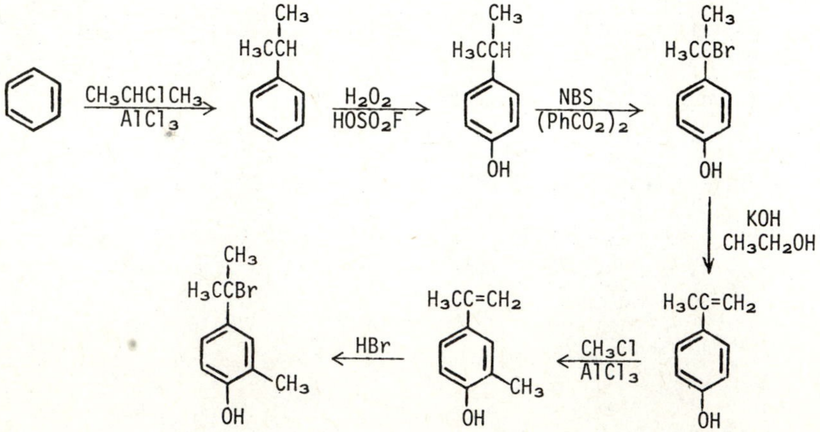

16.25

$$\text{C}_6\text{H}_6 \xrightarrow[\text{AlCl}_3]{(\text{CH}_3)_2\text{CHCl}} \text{C}_6\text{H}_5\text{-CH(CH}_3)_2 \xrightarrow[(\text{PhCO}_2)_2]{\text{NBS}} \text{C}_6\text{H}_5\text{-CBr(CH}_3)_2 \xrightarrow[\text{AlCl}_3]{\text{phenol}} \text{C}_6\text{H}_5\text{-C(CH}_3)_2\text{-C}_6\text{H}_4\text{-OH} \xrightarrow[\text{H}_2\text{SO}_4]{2\ (\text{CH}_3)_3\text{COH}} \text{MON-0585}$$

MON-0585

16.26

$$\text{C}_6\text{H}_6 \xrightarrow[\text{AlCl}_3]{\text{CH}_3\text{CH}_2\text{CH}_2\text{CH}_2\overset{\text{O}}{\overset{\|}{\text{C}}}\text{Cl}} \text{C}_6\text{H}_5\text{COCH}_2\text{CH}_2\text{CH}_2\text{CH}_3 \xrightarrow[\text{Pd/C}]{\text{H}_2} \text{C}_6\text{H}_5(\text{CH}_2)_4\text{CH}_3 \xrightarrow[(\text{PhCO}_2)_2]{\text{NBS}} \text{C}_6\text{H}_5\text{CHBr(CH}_2)_3\text{CH}_3 \xrightarrow{\text{NaOH}} \text{C}_6\text{H}_5\text{CH(OH)(CH}_2)_3\text{CH}_3$$

Fenipentol

This is an awkward route to a compound that we will learn to synthesize in two steps in Chapter 22.

16.27

$$2\ \text{C}_6\text{H}_6 \xrightarrow[\text{AlCl}_3]{(\text{CH}_3)_2\text{CCl}_2} \text{C}_6\text{H}_5\text{-C(CH}_3)_2\text{-C}_6\text{H}_5 \xrightarrow[\text{HOSO}_2\text{F}]{\text{H}_2\text{O}_2} \text{HO-C}_6\text{H}_4\text{-C(CH}_3)_2\text{-C}_6\text{H}_4\text{-OH}$$

Bisphenol A

16.28 1)

$$\text{H}_2\text{C=O} + \text{H}^+ \rightleftharpoons \left[\text{H}_2\text{C=}\overset{+}{\text{O}}\text{H} \longleftrightarrow \text{H}_2\overset{+}{\text{C}}\text{-OH}\right]$$

Formaldehyde is protonated to form a carbocation.

2)

$$\text{2,4,5-trichlorophenol} + \text{H}_2\text{C=}\overset{+}{\text{O}}\text{H} \longrightarrow [\text{arenium ion with CH}_2\text{-OH and H at C-6}] \longrightarrow \text{2,4,5-trichloro-6-(CH}_2\text{-OH)phenol} + \text{H}^+$$

The formaldehyde cation acts as the electrophile in a substitution reaction at the 6 position of 2,4,5-trichlorophenol.

3) [reaction scheme: 2,4,5-trichloro-6-(hydroxymethyl)phenol, H^+ ⇌ [protonated CH_2–$\overset{+}{O}H_2$ → benzylic $\overset{+}{C}H_2$ cation + H_2O]]

The product from step 2 is protonated by strong acid to produce a carbocation.

4) [reaction scheme: carbocation + 2,4,5-trichlorophenol → [arenium ion intermediate] → hexachlorophene + H^+]

hexachlorophene

This carbocation is attacked by a second molecule of 2,4,5-trichlorophenol to produce hexachlorophene.

16.29 a) [reaction scheme: 2,2,5,5-tetramethyltetrahydrofuran, H^+ ⇌ [protonated ether (+OH) ⇌ tertiary carbocation with OH]]

b) [reaction scheme: benzene + carbocation (OH) → [arenium ion intermediate] → aryl alcohol product + H^+]

c) [reaction scheme: aryl alcohol, H^+ ⇌ [protonated alcohol ($\overset{+}{O}H_2$) → carbocation + H_2O → cyclized arenium ion] → 1,1,4,4-tetramethyltetralin + H^+]

16.30 1) [reaction scheme: $H_2C{=}O$ + HCl, $ZnCl_2$ ⇌ [$H_2C{=}\overset{+}{O}H$ ⟷ $H_2\overset{+}{C}$–OH] $ZnCl_3^-$]

2) [reaction scheme: benzene + $H_2C{=}\overset{+}{O}H$ $ZnCl_3^-$ → [arenium ion with CH_2-OH] → $C_6H_5CH_2$-OH + HCl + $ZnCl_2$]

3) [reaction scheme: $C_6H_5CH_2$-OH, H^+ ⇌ [$C_6H_5CH_2$–$\overset{+}{O}H_2$ + :Cl^-] ⇌ $C_6H_5CH_2Cl$ + H_2O]

The function of zinc chloride is not completely understood.

16.31 1) $:\bar{C}\equiv\overset{+}{O} \underset{AlCl_3}{\overset{HCl}{\rightleftharpoons}} [\ H\text{-}C\equiv\overset{+}{O} \longleftrightarrow H\text{-}\overset{+}{C}=O\ ]\ \overset{-}{AlCl_4}$

2) H_3C-C$_6$H$_4$ + $H\text{-}C\equiv\overset{+}{O}\ \overset{-}{AlCl_4}$ ⟶ [intermediate] ⟶ H_3C-C$_6$H$_4$-CHO + HCl + $AlCl_3$

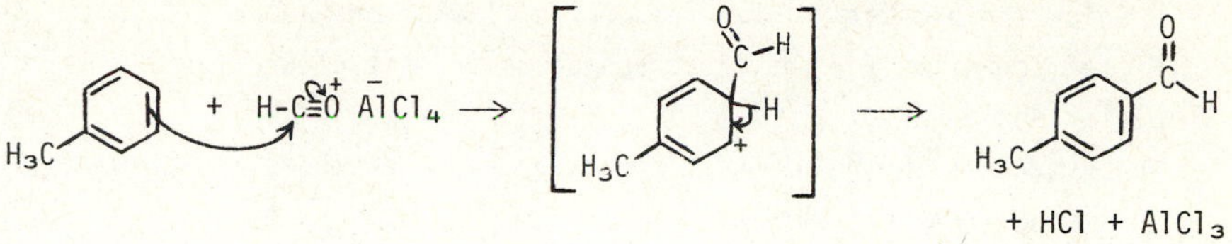

16.32 $\xrightarrow{H_3PO_4}$ $\xrightarrow{-H_2O}$

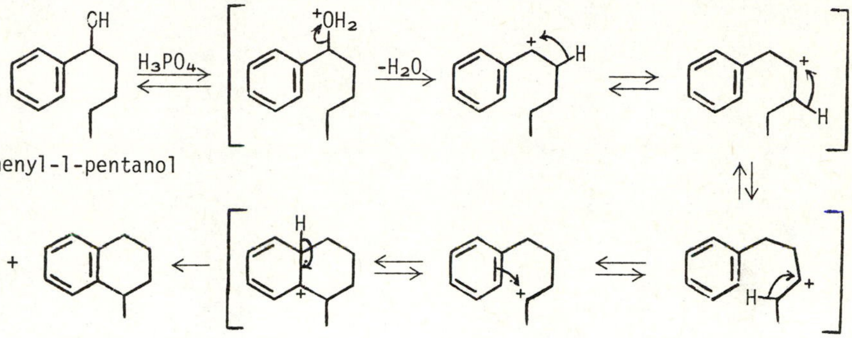

1-phenyl-1-pentanol

H^+ +

α-methyltetralin

Protonation of 1-phenyl-1-pentanol is followed by loss of water. The resulting carbocation undergoes a series of hydride shifts. All of the hydride shifts are reversible, and the carbocations are in equilibrium. Internal alkylation of the benzene ring, however, is irreversible. The pi electrons of the aromatic ring can attack one of these carbocations to yield the cyclic product, α-methyltetralin.

16.33 $\xrightarrow{AlCl_3}$ + $AlCl_4^-$

H^+ +

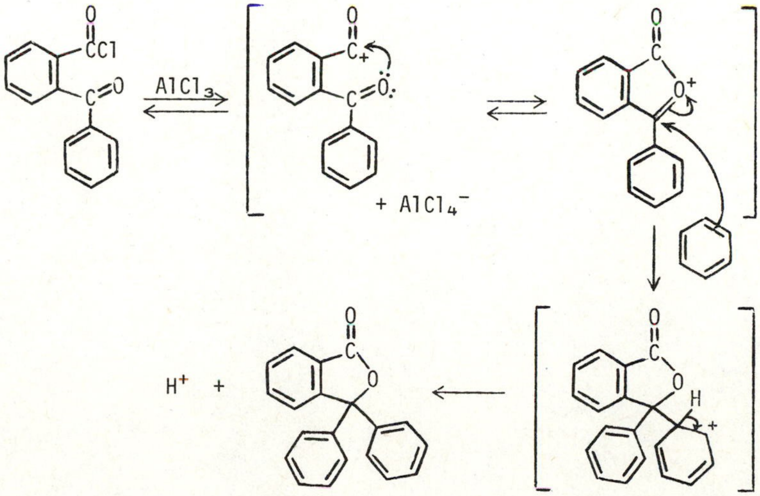

What you should know:

After doing these problems you should be able to:

1. Predict the products of Friedel-Crafts alkylations and acylations;
2. Write mechanisms for reactions analogous to Friedel-Crafts alkylations and acylations;
3. Understand the principle of carbocation rearrangements and draw the structures resulting from rearrangement;
4. Logically synthesize substituted arenes.

CHAPTER 17 ORGANIC REACTIONS. A BRIEF REVIEW

17.1 a) $:I^-$ + CH_3Cl ⟶ CH_3I + $:Cl^-$

nucleophile electrophile

b) benzene + Br_2 $\xrightarrow{AlBr_3}$ bromobenzene + HBr

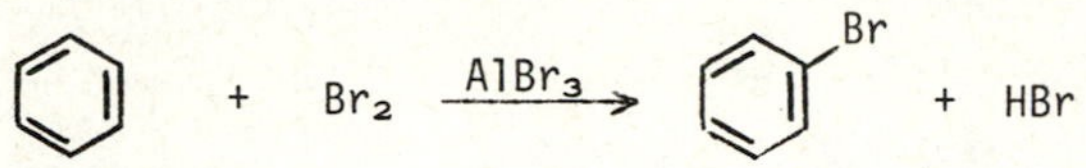

nucleophile

In this reaction, Br^+, formed by the action of $AlBr_3$ on Br_2, is the electrophile.

c) cyclohexene + $CH_3\overset{O}{\overset{\|}{C}}OOH$ ⟶ cyclohexene oxide + CH_3COOH

nucleophile electrophile

d) This reaction proceeds in two steps; we will identify the nucleophile and electrophile for each step:

Step 1 cyclohexene oxide ($\ddot{O}$:) + H^+ ⟶ protonated epoxide ($\overset{+}{O}H$)

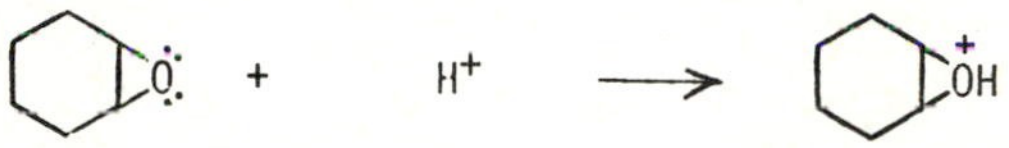

nucleophile electrophile

Step 2 protonated epoxide ($\overset{+}{O}H$) + $:Br^-$ ⟶ 2-bromocyclohexanol (Br, OH)

electrophile nucleophile

e) $CH_3C{\equiv}\bar{C}:$ + CO_2 ⟶ $CH_3C{\equiv}CCO_2^-$

nucleophile electrophile

f) $CH_3\bar{S}:$ + $CH_3CH_2\overset{O}{\overset{\|}{C}}OCH_3$ ⟶ CH_3SCH_3 + $CH_3CH_2\overset{O}{\overset{\|}{C}}O:^-$

nucleophile electrophile

g) $:N_3^-$ + CH_3CH_2Br ⟶ $CH_3CH_2N_3$ + $:Br^-$

nucleophile electrophile

h) 3 $CH_3CH{=}CH_2$ + BH_3 ⟶ $(CH_3CH_2CH_2)_3B$

nucleophile electrophile

i) 1-fluoro-2,4-dinitrobenzene (F, NO_2, NO_2) + $CH_3O:^-$ ⟶ 2,4-dinitroanisole (OCH_3, NO_2, NO_2) + $:F^-$

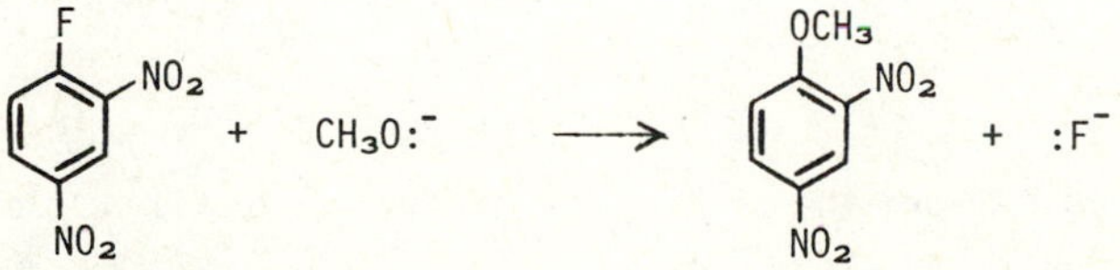

electrophile nucleophile

17.2 In most cases, cations (electron-deficient species) behave as electrophiles; anions (electron-rich species) behave as nucleophiles. The behavior of a neutral compound depends on the functional groups present in the compound and on the reaction under study.

a) CH_3Br can behave as an electrophile or as a nucleophile, depending on the reaction.

$$\underset{\text{electrophile}}{CH_3Br} + \underset{\text{nucleophile}}{:I^-} \longrightarrow CH_3I + :Br^-$$

$$\underset{\text{nucleophile}}{CH_3Br} + \underset{\text{electrophile}}{AlBr_3} \longrightarrow CH_3^+ + AlBr_4^-$$

b) Because of the polarity of the carbon-magnesium bond, $\overset{\delta-}{CH_3}\overset{\delta+}{Mg}Br$ behaves as a nucleophile.

c) The carbonyl carbon of acetone, $CH_3\overset{O^{\delta-}}{\overset{\|}{\underset{\delta+}{C}}}CH_3$, is electrophilic, and oxygen is nucleophilic.

d) The electron-rich benzene ring behaves as a nucleophile.

e) $CH_3\ddot{N}H_2$ acts as a nucleophile.

f) Aniline behaves as a nucleophile. Both the aromatic ring and the $-\ddot{N}H_2$ group are electron-rich.

g) The anion $:\bar{C}N$ behaves as a nucleophile.

h) Propyne can act as either a nucleophile or an electrophile. In cases where the electron-rich triple bond is reacting, propyne behaves as a nucleophile.

$$\underset{\text{nucleophile}}{CH_3C{\equiv}CH} + \underset{\text{electrophile}}{Br_2} \longrightarrow H_3CCBr{=}CHBr$$

When reaction occurs at the terminal hydrogen atom, $CH_3C{\equiv}CH$ acts as an electrophile.

$$\underset{\text{electrophile}}{CH_3C{\equiv}CH} + \underset{\text{nucleophile}}{:\bar{N}H_2} \longrightarrow CH_3C{\equiv}\bar{C}: + NH_3$$

i) $CH_3C{\equiv}\bar{C}:$ behaves as a nucleophile.

j) In reactions involving CO_2, attack occurs at the electron-deficient carbon atom, $\overset{\delta-}{O}{=}\overset{\delta+}{C}{=}\overset{\delta-}{O}$. CO_2, therefore, behaves as an electrophile.

k) $C_6H_5\overset{\delta+}{C}{\equiv}\overset{\delta-}{N}:$

The aromatic ring and the nitrogen behave as nucleophiles; the nitrile carbon behaves as an electrophile.

l) $(CH_3)_3C^+$ behaves as an electrophile.

17.3 The anion formed by attack of base on the ketone is stabilized by resonance with the neighboring carbonyl group.

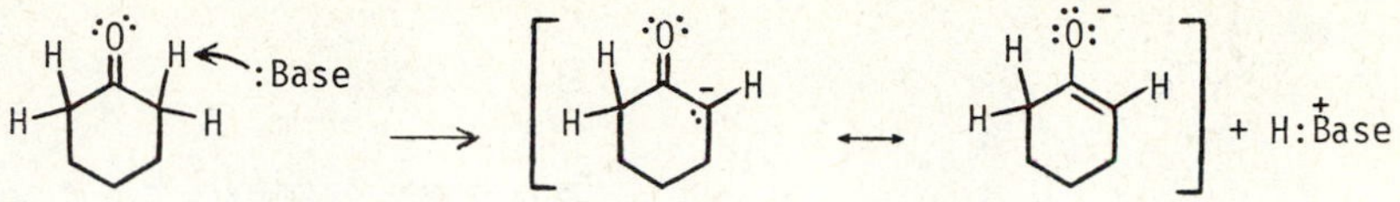

The enolate anion can act as a nucleophile to displace bromide ion from bromomethane is an S_N2 reaction.

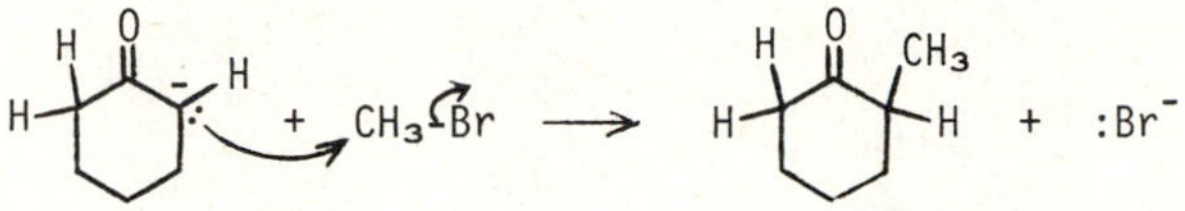

17.4 Initiation: Rad• + $H_2C{=}CH_2$ ⟶ Rad-$CH_2\dot{C}H_2$

Propagation: Rad-$CH_2\dot{C}H_2$ + $H_2C{=}CH_2$ ⟶ Rad-$CH_2CH_2CH_2\dot{C}H_2$ etc.

This reaction is a radical addition reaction. Propagation steps will continue until a termination step occurs.

17.5

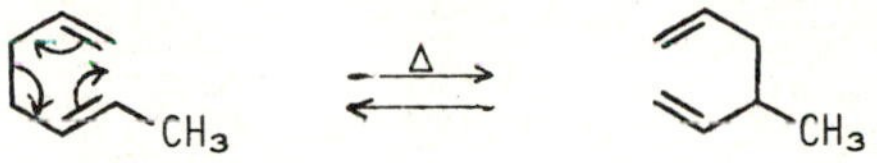

The Cope rearrangement is a pericyclic reaction. The reorganization of bonding electrons is indicated by arrows signifying the concerted movement of electron pairs.

17.6

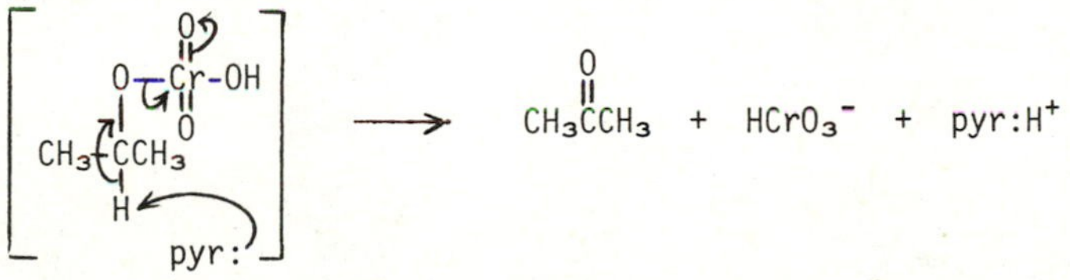

$CH_3\overset{O}{\overset{\|}{C}}CH_3$ + $HCrO_3^-$ + pyr:H^+

This is an example of an E_2 elimination. The base, pyridine, removes a proton; the chromium (IV) anion is eliminated simultaneously.

What you should know:

After doing these problems you should be able to:

1. Identify reactions as proceeding by polar, radical or pericyclic mechanisms;
2. Identify electrophiles and nucleophiles;
3. Formulate reaction mechanisms by using your knowledge of the types of reactions we have studied.

CHAPTER 18 ALICYCLIC MOLECULES

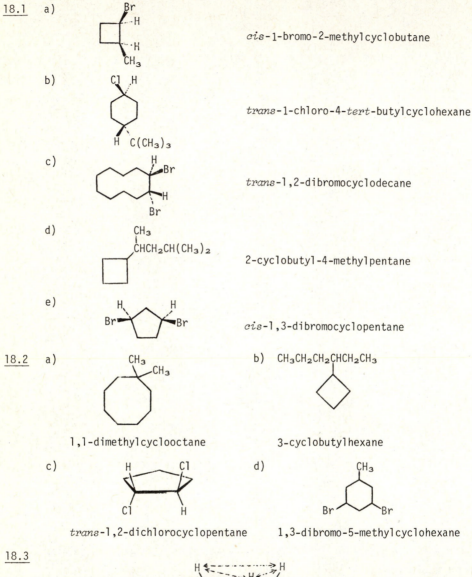

18.1 a) *cis*-1-bromo-2-methylcyclobutane

b) *trans*-1-chloro-4-*tert*-butylcyclohexane

c) *trans*-1,2-dibromocyclodecane

d) CH_3
$CHCH_2CH(CH_3)_2$

2-cyclobutyl-4-methylpentane

e) *cis*-1,3-dibromocyclopentane

18.2 a) 1,1-dimethylcyclooctane

b) $CH_3CH_2CH_2CHCH_2CH_3$

3-cyclobutylhexane

c) *trans*-1,2-dichlorocyclopentane

d) 1,3-dibromo-5-methylcyclohexane

18.3

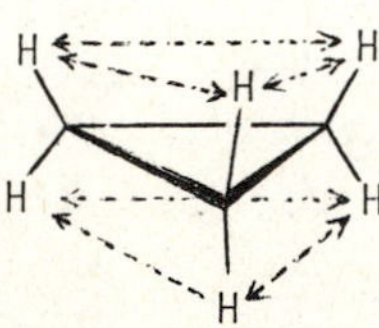

All hydrogen atoms on the same side of the cyclopropane ring are eclipsed by each other. If we draw each hydrogen-hydrogen interaction, we count six eclipsing interactions. Since each of these interactions "costs"

1.0 kcal/mol, all six cost 6.0 kcal/mol. (6.0 kcal/mol ÷ 27.6 kcal/mol) = 0.22; thus, 22% of the total strain energy of cyclopropane is due to eclipsing strain.

18.4

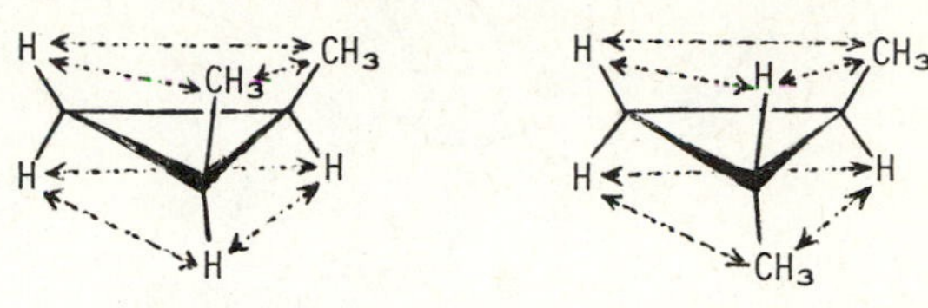

interaction	energy cost (kcal/mol)	*cis* isomer # of interactions	*cis* isomer total energy cost (kcal/mol)	*trans* isomer # of interactions	*trans* isomer total energy cost (kcal/mol)
H-H	1.0	3	3.0	2	2.0
H-CH_3	1.4	2	2.8	4	5.6
CH_3-CH_3	2.5	1	2.5	0	0
			8.3		7.6

The added energy "cost" of eclipsing interactions causes *cis*-1,2-dimethylcyclopropane to be of higher energy, and to be less stable than the *trans* isomer. Since the *cis* isomer is of higher energy, its heat of combustion is also greater.

18.5 Cyclopropane would be a more efficient fuel because its heat of combustion per $-CH_2-$ group is greater than that of propane.

18.6

$Cl_3C-C(=O)-\ddot{O}:^-$ (1) → $[Cl_3C:^-]$ + CO_2 (2) → $Cl_2C:$ + $:Cl^-$

In step 1 carbon dioxide is lost from sodium trichloroacetate. The three electron-withdrawing chlorine substituents stabilize the negative charge. In step 2, elimination of chloride ion produces dichlorocarbene. The mechanism is similar to α-elimination of HCl from chloroform.

18.7

cyclohexene + $:CHCH_3$ → A (H_3C, H; H, H) + B (H, CH_3; H, H)

Focus on the stereochemistry of the three-membered ring. Simmons-Smith reaction of 1,1-diiodoethane with the double bond occurs with *syn* stereochemistry and can produce two isomers. In one of these isomers (A), the methyl group is on the same side of the three-membered ring as the cyclohexane ring carbons. In B, the methyl group is on the side of the three-membered ring opposite to the cyclohexane ring carbons.

18.8

cis *trans*

Two types of interaction are present in *cis*-1,2-dimethylcyclobutane. One interaction occurs between the two methyl groups, which are almost eclipsed. The other is an across-the-ring interaction between methyl group at position 1 of the ring and a hydrogen at position 3. Because neither of these interactions are present in the *trans* isomer, it is more stable than the *cis* isomer.

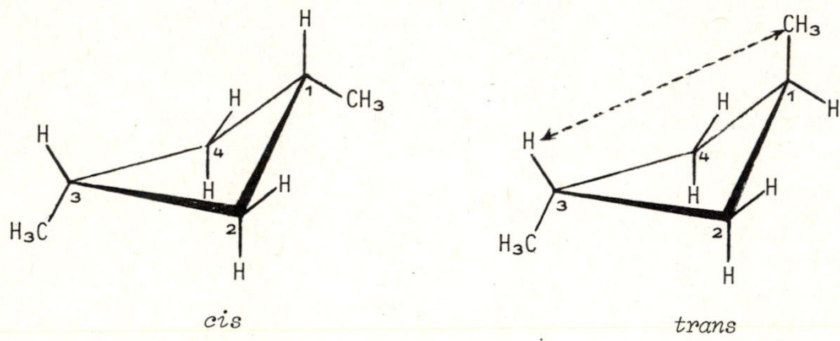

cis *trans*

In *trans*-1,3-dimethylcyclobutane an across-the-ring interaction occurs between the methyl group at position 1 of the ring and a hydrogen at position 3. Because no interactions are present in the *cis* isomer, it is more stable than the *trans* isomer.

18.9

If cyclopentane were planar, it would have ten hydrogen-hydrogen interactions with a total energy cost of 10.0 kcal/mol. The measured total strain energy of 6.5 kcal/mol indicates that 3.5 kcal/mol of eclipsing strain in cyclopentane has been relieved by puckering.

18.10 Be aware of the distinction between axial-equatorial and *cis-trans*. Axial substituents are parallel to the axis of the ring; equatorial substituents lie around the "equator" of the ring. *Cis* substituents are on the same side of the ring; *trans* substituents are on opposite sides of the ring.

a) 1,3-*trans*

ring-flip

axial, equatorial — equatorial, axial

b) 1,4-*cis*

ring-flip

axial, equatorial — equatorial, axial

c) 1,3-*cis*

ring-flip

axial, axial — equatorial, equatorial

d) 1,5-*trans* is the same as 1,3-*trans*.

e) 1,5-*cis* is the same as 1,3-*cis*.

f) 1,6-*trans*

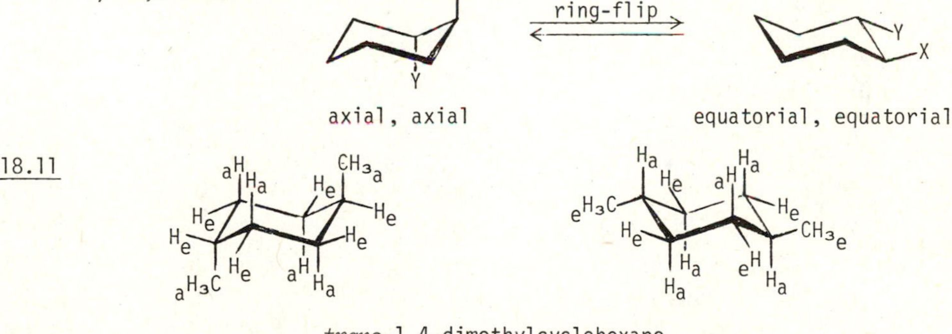

18.12

axial *tert*-butylcyclohexane — axial isopropylcyclohexane

The most stable conformation of axial *tert*-butylcyclohexane is pictured. One methyl group is positioned above the ring and competes for space with two axial ring protons. In the other axial alkylcyclohexanes, this methyl group is replaced by hydrogen, which has a much smaller space requirement. The steric strain caused by an axial *tert*-butyl group is therefore higher than the strain caused by axial methyl, ethyl or isopropyl groups.

18.13 The energy difference between an axial and an equatorial cyano group is very small because there are no 1-3 diaxial interactions for a cyano group.

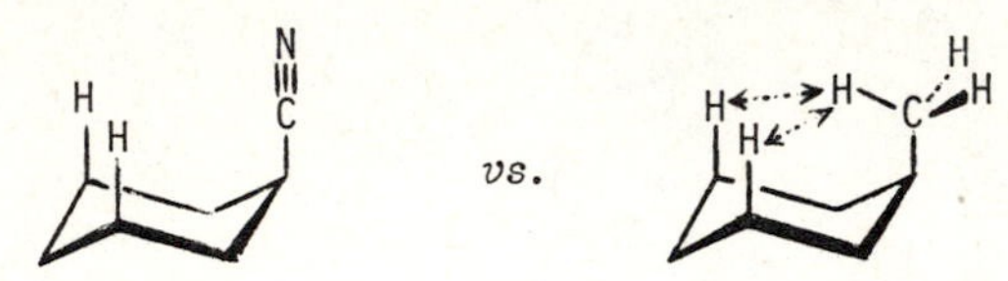

cyclohexanecarbonitrile methylcyclohexane

18.14 a)

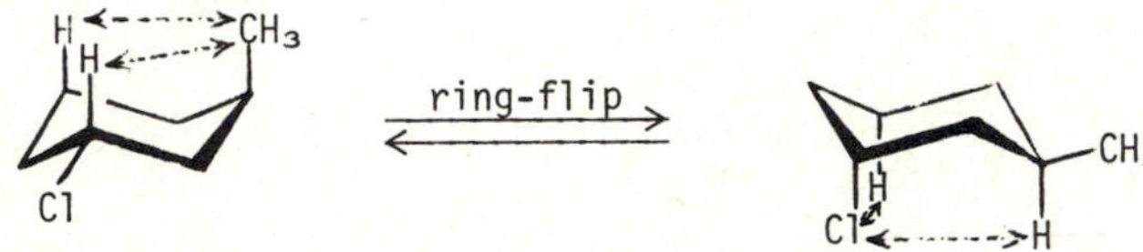

trans-1-chloro-3-methylcyclohexane

$2(H\text{-}CH_3) = 1.8$ kcal/mol $2(H\text{-}Cl) = 0.5$ kcal/mol

The second conformation is more stable than the first.

b)

cis-1-ethyl-2-methylcyclohexane

$2(H\text{-}CH_2CH_3) = 1.9$ kcal/mol $2(H\text{-}CH_3) = 1.8$ kcal/mol

The second conformation is slightly more stable than the first.

c)

cis-1-bromo-4-ethylcyclohexane

$2(H\text{-}CH_2CH_3) = 1.9$ kcal/mol $2(H\text{-}Br) = 0.5$ kcal/mol

The second conformation is more stable than the first.

d)

trans-4-methylcyclohexanol

$2(H\text{-}CH_3) = 1.8$ kcal/mol
$2(H\text{-}OH) = 1.0$ kcal/mol
Total = 2.8 kcal/mol

No diaxial interactions.

The second conformation is more stable than the first.

18.15

There are four stereoisomers of 1-chloro-3,5-dimethylcyclohexane. It is possible for; (1) both methyl groups to be *cis* to chlorine; (2) both methyl groups to be *trans* to chlorine; (3) one methyl group to be *cis* and the other to be *trans* to chlorine (two isomers). The geometric isomer in which all three groups are *cis* is the most stable because the three groups are equatorial.

18.16

In the chair conformation of *trans*-1,3-di-*tert*-butylcyclohexane, one *tert*-butyl group is axial and one is equatorial. The 1,3-diaxial interactions of the axial *tert*-butyl group make the chair form of *trans*-1,3-di-*tert*-butylcyclohexane 5.4 kcal/mol less stable than a cyclohexane with no axial substituents. Since a twist boat conformation is 5.5 kcal/mol less stable than a chair, it is almost as likely that the compound will assume a twist-boat conformation as a chair conformation. The twist-boat removes the 1,3-diaxial interaction present in the chair form of *trans*-1,3-di-*tert*-butylcyclohexane.

18.17

trans *cis*

The most stable conformations on the two isomers are pictured above; the *tert*-butyl group is always equatorial in the most stable conformation. The *cis* isomer should react faster under E_2 conditions because -Br and -H are in the *anti*-periplanar arrangement that favors E_2 elimination.

18.18

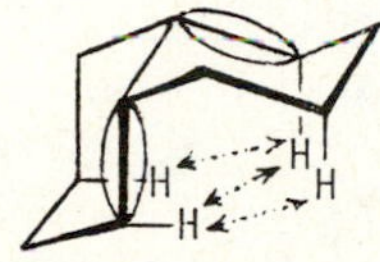

Trans-decalin is more stable than *cis*-decalin. Three 1,3-diaxial interactions cause *cis*-decalin to be of higher energy than *trans*-decalin. You may be able to visualize these interactions by thinking of the circled parts of *cis*-decalin as similar to axial methyl groups. The *gauche* interactions that occur with axial methyl groups also occur in *cis*-decalin.

18.19 a) 3-methylcyclohexene
b) *cis*-4-cyclopentene-1,3-diol
c) *cis*-1-iodo-2-methylcyclohexane
d) 1-cyclobutyl-3-methylpentane
e) cyclohexylcyclohexane
f) 5-ethyl-1,3-cycloheptadiene
g) 6,6-dimethylcyclononene
h) 1,2-dicyclopropylethane

18.20 a) CH_3
b) $CH_3CH_2CH_2C{=}CH_2$
c) Br H H Br
d) OH HO
e) 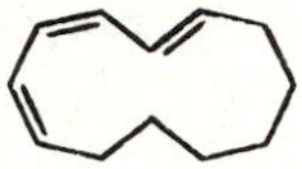
f) H_3C $CH(CH_3)_2$
g) $(\bigcirc)_3$CH

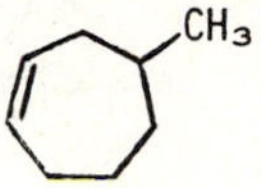

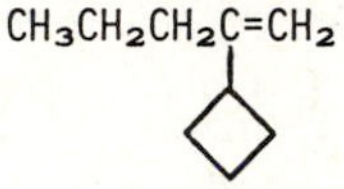

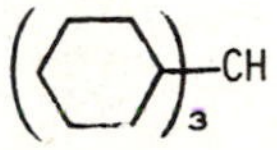

18.21 a) H_3C — 1. OsO_4 2. $NaHSO_3$, H_2O → OH OH H_3C H + enantiomer

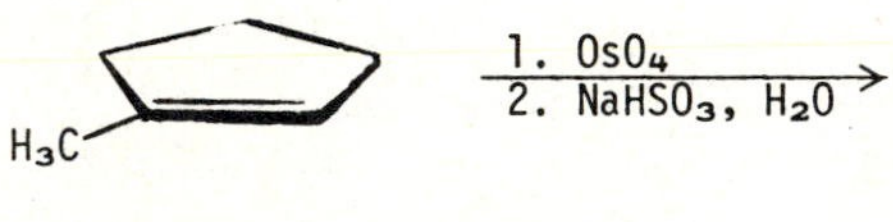

b) H_3C — 1. BH_3 2. H_2O_2, OH^- → H OH H_3C H + enantiomer

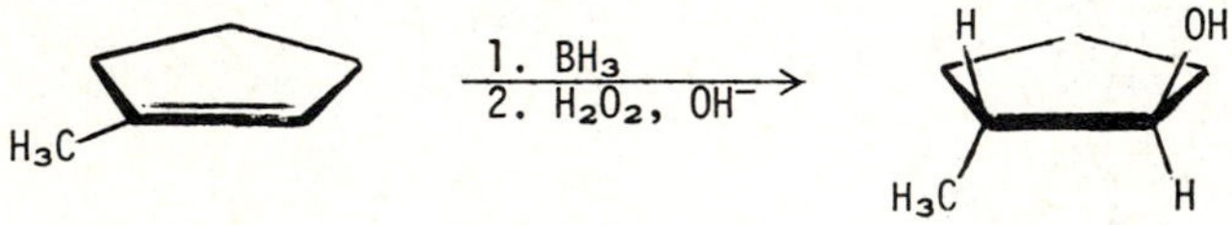

c) H_3C — Cl_2 / CCl_4 → Cl H H_3C Cl + enantiomer

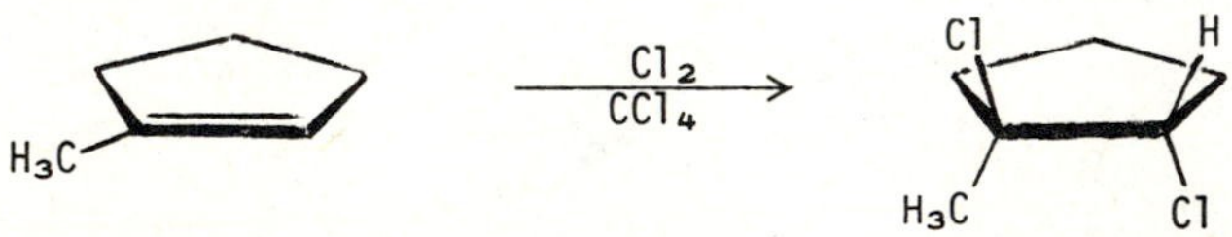

d) H_3C — HBr → Br H H_3C H

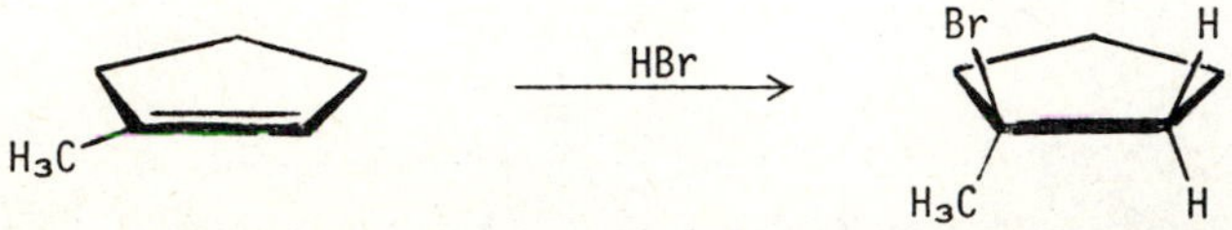

e) H_3C — DBr → Br H H_3C D + Br D H_3C H

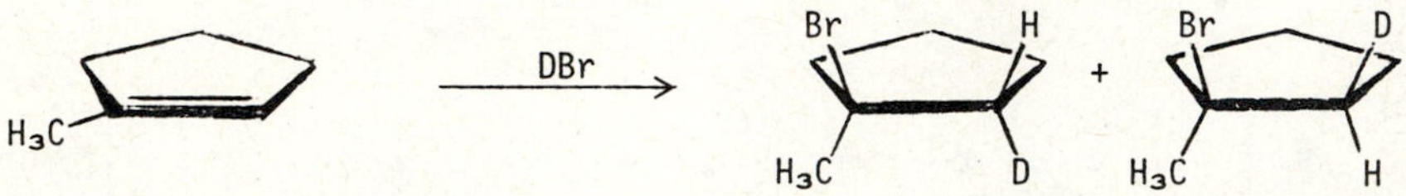

f)

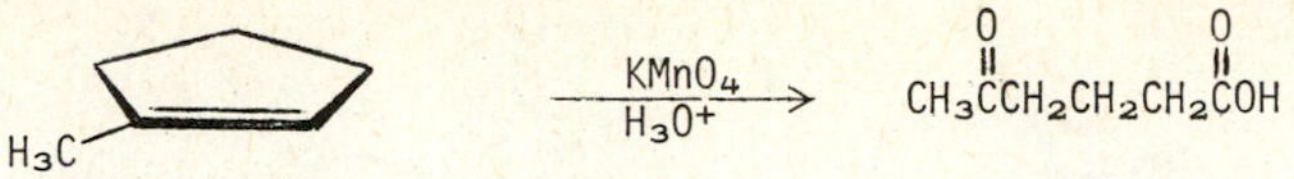

$\xrightarrow[H_3O^+]{KMnO_4}$ $CH_3\overset{O}{\overset{\|}{C}}CH_2CH_2CH_2\overset{O}{\overset{\|}{C}}OH$

g)

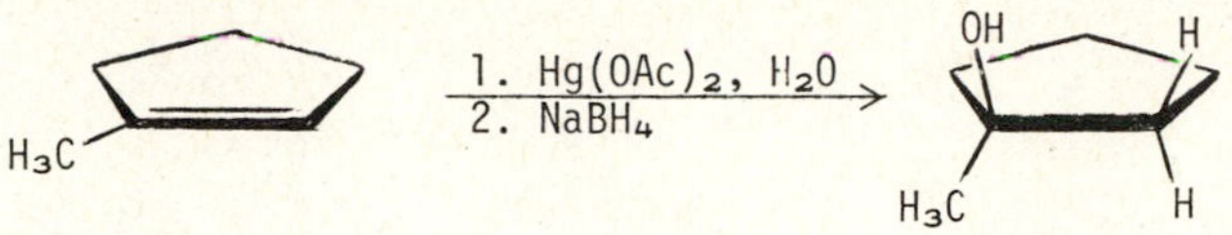

$\xrightarrow[2.\ NaBH_4]{1.\ Hg(OAc)_2,\ H_2O}$

h)

$\xrightarrow[Zn/Cu]{CH_2I_2}$ + enantiomer

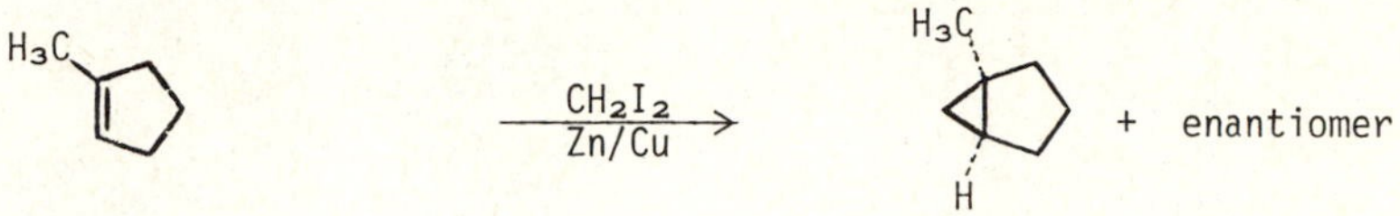

i)

$\xrightarrow[K^+\ ^-OtBu]{CHCl_3}$ + enantiomer

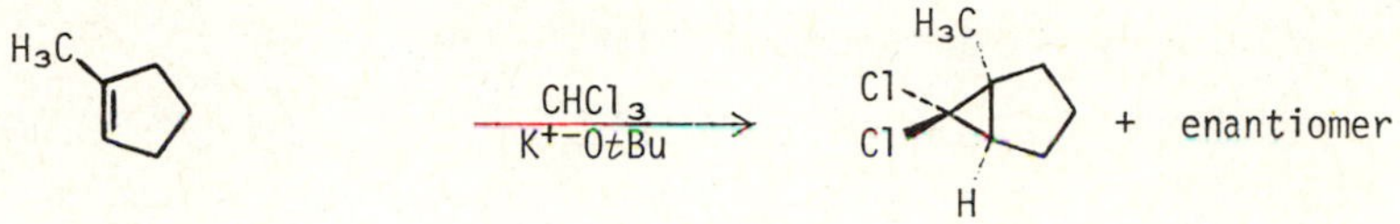

18.22 a) *Angle strain* is the energy cost caused by the distortion of the bond angles of a compound from their normal values.

b) *Steric strain* is the repulsive interaction that occurs when the spatial requirements of two atoms overlap.

c) The *heat of combustion* of an organic compound is the heat liberated when the compound reacts with oxygen to form CO_2 and H_2O. The heat of combustion for two compounds having the same number of carbon atoms is greater for a strained compound, indicating that the strained compound is of higher energy.

d) A substituent in an *axial position* of a cyclohexane ring is parallel to an "axis" that lies perpendicular to the plane of the ring.

e) A substituent in an *equatorial position* lies more or less in the plane of the ring.

18.23

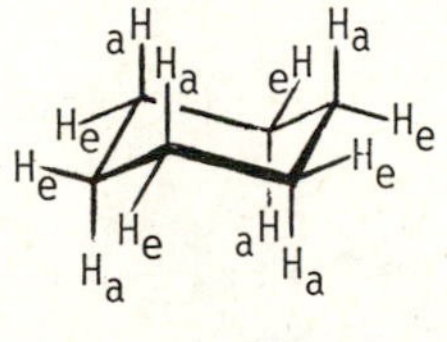

18.24 Since the methyl group of *N*-methylpiperidine prefers an equatorial conformation, the steric requirements of a methyl group must be greater than those of an electron pair.

18.25

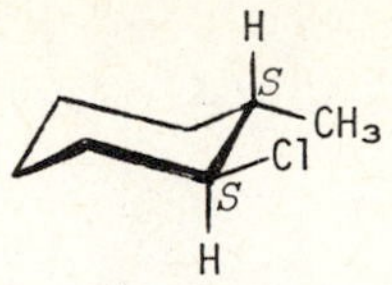

A 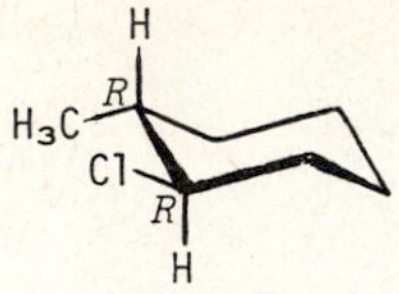   B

trans-1-chloro-2-methylcyclohexane

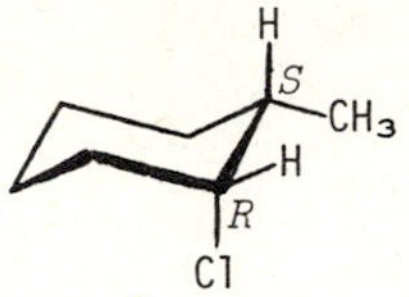

C 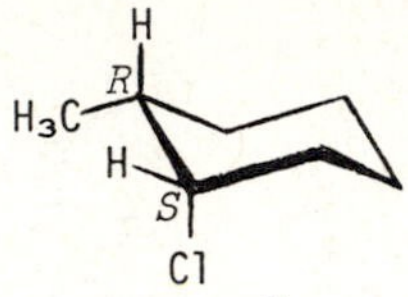   D

cis-1-chloro-2-methylcyclohexane

Stereoisomer	Enantiomeric with	Diastereomeric with
A	B	C,D
B	A	C,D
C	D	A,B
D	C	A,B

18.26 Draw the planar conformation of the *cis* and *trans* isomers.

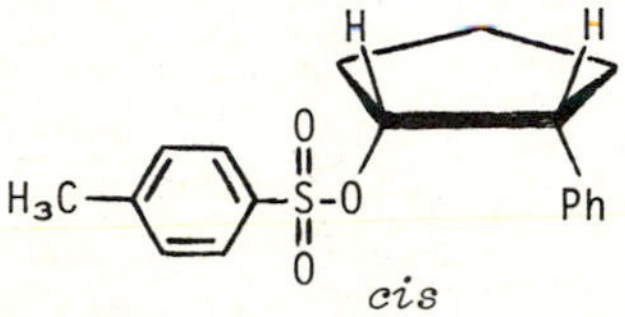

cis 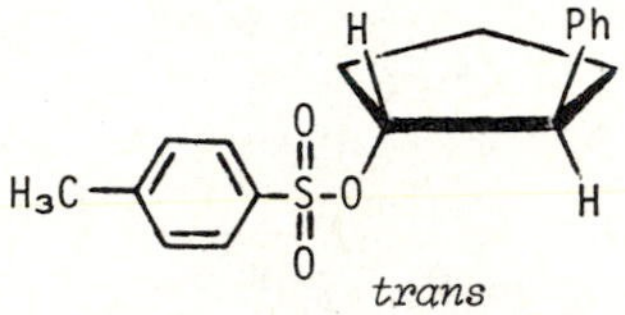   *trans*

The most favored elimination product has a double bond conjugated with the aromatic ring. The *cis* isomer can form this product by the usual *anti*-periplanar E_2 mechanism. In the case of the *trans* isomer, the two groups to be eliminated are on the same side of the ring. Elimination to yield the observed product occurs by a less favored *syn*-periplanar process, and the rate of reaction for the *trans* isomer is slower.

18.27

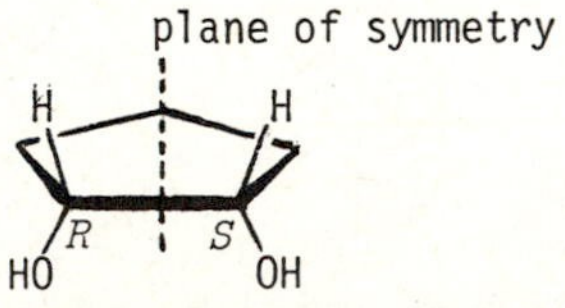

plane of symmetry

cis 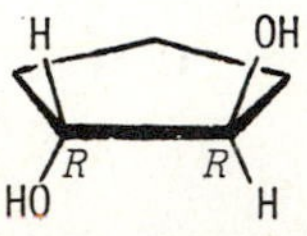 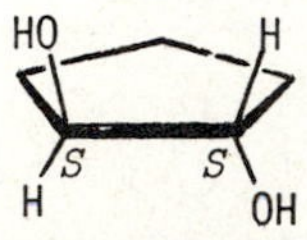   *trans*

Cis-1,2-cyclopentanediol has two chiral centers. A plane of symmetry separates the structure into two halves that are mirror images of each other. Consequently, *cis*-1,2-cyclopentanediol is a *meso* compound.

Trans-1,2-cyclopentanediol also has two chiral centers. The *trans* compound, however, is chiral and exists as a pair of non-interconvertible enantiomers.

18.28

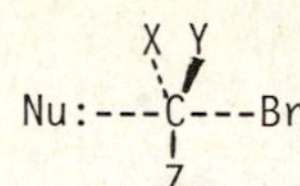

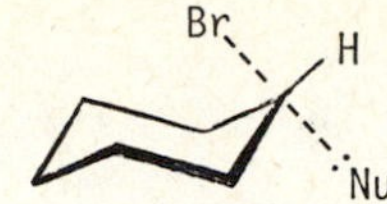

In the transition state for an S_N2 reaction, the attacking nucleophile and the leaving group are 180° apart. The remaining three groups bonded to the reacting carbon atom lie in a plane and are 120° apart. All atoms of bromocyclohexane can assume the appropriate geometry for S_N2 reaction to occur. For cyclopropane it is very difficult for the three groups bonded to the reacting carbon to lie in a plane and to be 120° apart because of the added strain to the cyclopropane ring. S_N2 reaction of bromocyclopropane thus occurs much more slowly than S_N2 reaction of bromocyclohexane.

18.29

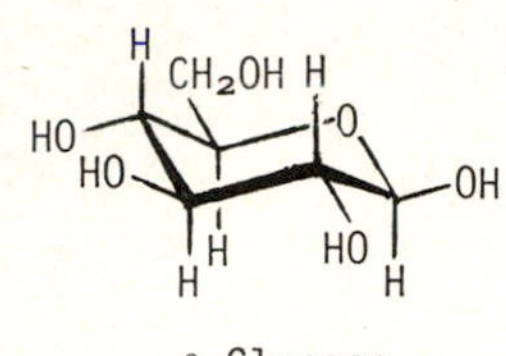

β-Glucose

18.30 To solve this problem; (1) Find the energy cost of a 1-3 diaxial interaction by using Table 18.2. (2) Convert this energy difference into a percent by using Table 3.7.

a)

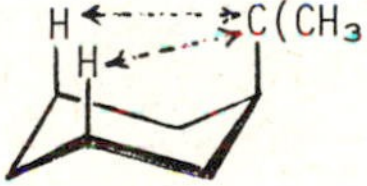

$2(H\text{-}C(CH_3)_3) = 5.4$ kcal/mol

$$\frac{\%\ \text{equatorial}}{\%\ \text{axial}} = \frac{>99.9}{<0.1}$$

< 0.1% axial

b)

2(H-Br) = 0.5 kcal/mol

$$\frac{\%\ \text{equatorial}}{\%\ \text{axial}} = \frac{70}{30}$$

30% axial

c)

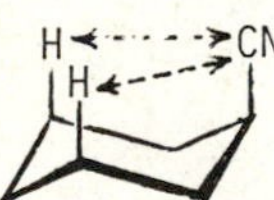

2(H-CN) = 0.2 kcal/mol

$$\frac{\%\ \text{equatorial}}{\%\ \text{axial}} = \frac{58}{42}$$

42% axial

d)

2(H-OH) = 1.0 kcal/mol

$$\frac{\%\ \text{equatorial}}{\%\ \text{axial}} = \frac{85}{15}$$

15% axial

18.31

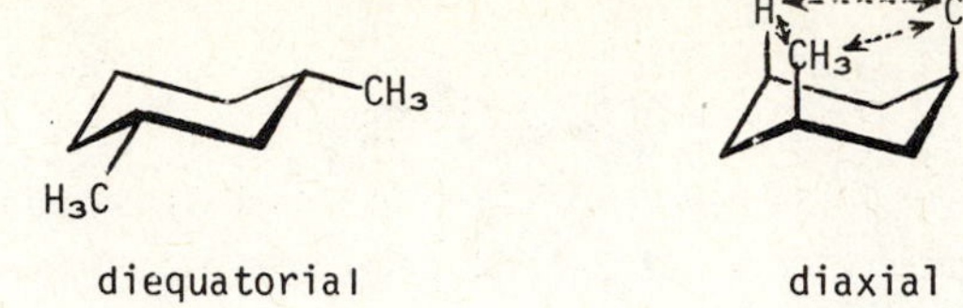

diequatorial   diaxial

Diaxial *cis*-1,3-dimethylcyclohexane contains three 1,3-diaxial interactions -- two $H\text{-}CH_3$ interactions (0.9 kcal/mol each) and one $CH_3\text{-}CH_3$ interaction. If the diaxial conformation is 5.4 kcal/mol less stable than the diequatorial, 5.4 - 2(0.9) = 3.6 kcal/mol of this strain energy must be due to the $CH_3\text{-}CH_3$ interaction. It is this interaction that is responsible for the large energy difference between the two conformations.

18.32

A   B

2 $H\text{-}CH_3$ interactions = 1.8 kcal/mol

2 $H\text{-}CH_3$ interaction	= 1.8 kcal/mol
1 $CH_3\text{-}CH_3$ interaction	= 3.6 kcal/mol
	5.4 kcal/mol

Conformation A is favored because it is 3.6 kcal/mol lower in energy than conformation B.

18.33

$(CH_3)_3C$ $\xrightarrow{Br_2}$ $(CH_3)_3C$

4-*tert*-butylcyclohexene

Since the *tert*-butyl group remains equatorial under most circumstances, a ring-flip to the diequatorial dibromo conformer does not occur.

18.34 Note: In working with decalins, it is essential to use models. Many structural features of decalins that are obvious with models are not easily visualized with drawings.

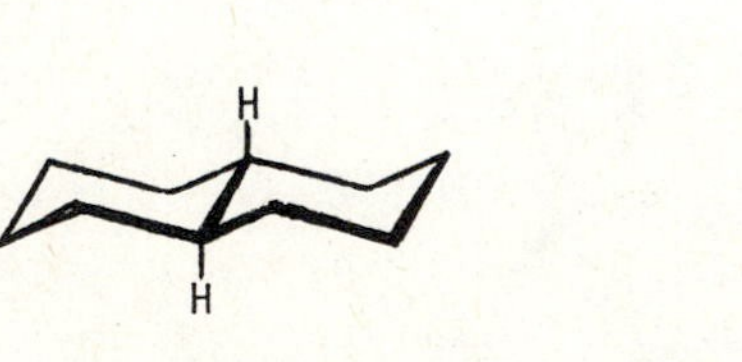

trans-decalin

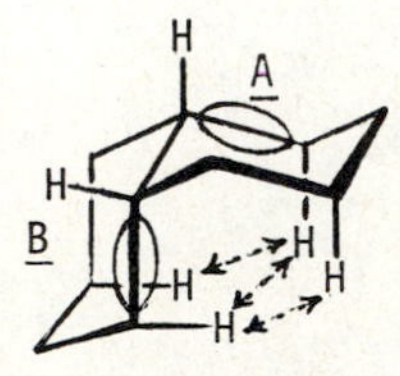

cis-decalin

No 1,3-diaxial interactions are present in *trans*-decalin.

At the ring junction of *cis*-decalin one ring acts as an axial substituent of the other (see circled bonds). The circled part of ring B has two 1,3-diaxial interactions with ring A (indicated by arrows). Similarly, the circled part of ring A has two 1,3-diaxial interactions with ring B; one of these interactions is the same as an interaction of part of the B ring with ring A. These three 1,3-diaxial interactions have a total energy cost of 2.7 kcal/mol. *Cis*-decalin is therefore less stable than *trans*-decalin by 2.7 kcal/mol.

18.35

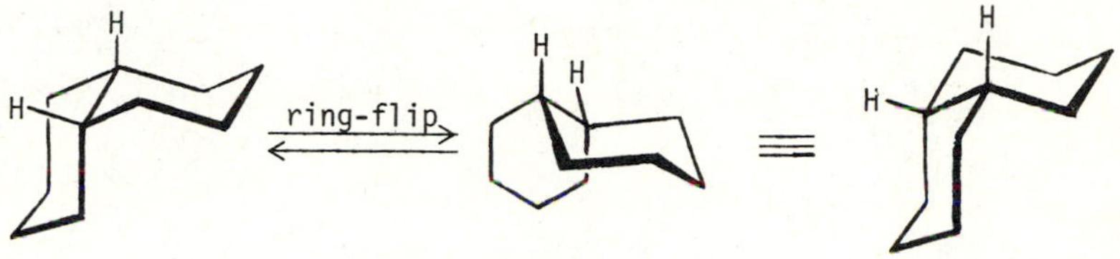

H H ring-flip H H ≡ H H

A ring-flip converts an axial substituent into an equatorial substituent and *vice versa*. At the ring junction of *trans*-decalin, each ring is a *trans-trans* diequatorial substituent of the other. If a ring-flip were to occur, the two rings would become axial substituents of each other. You can see with models that a diaxial ring junction is impossibly strained. Consequently, *trans*-decalin does not ring-flip.

The rings of *cis*-decalin are joined by an axial bond and an equatorial bond. After a ring-flip, the rings are still linked by an equatorial and and axial bond. No additional strain or interaction is introduced by a ring-flip of *cis*-decalin.

18.36

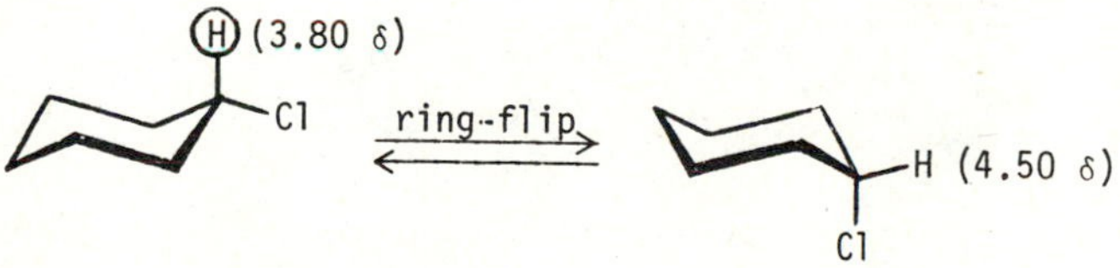

H (3.80 δ) Cl ring-flip H (4.50 δ) Cl

A B

At -150°C, the rate of interconversion of the equatorial and axial forms of chlorocyclohexane is relatively slow. Distinct absorptions for the equatorial proton (4.50 δ) and for the axial proton (3.80 δ) are observed. At room temperature, the rate of interconversion is faster than the NMR time scale (faster than 10^3 times per second). The absorptions from the equatorial proton and from the axial proton coalesce to one weighted average absorption at 3.95 δ. (This absortion is closer to 3.80 δ than to 4.50 δ because conformer A is more abundant than conformer B.)

18.37

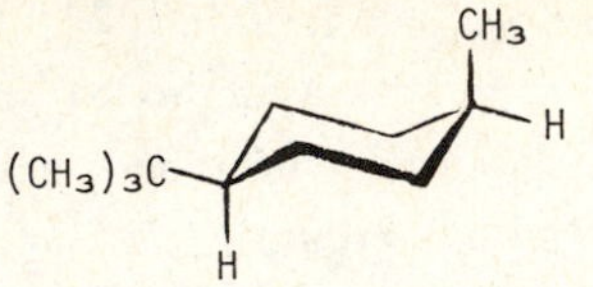

(CH$_3$)$_3$C, H, CH$_3$, H

cis

no chiral centers

(CH$_3$)$_3$C, H, H, CH$_3$

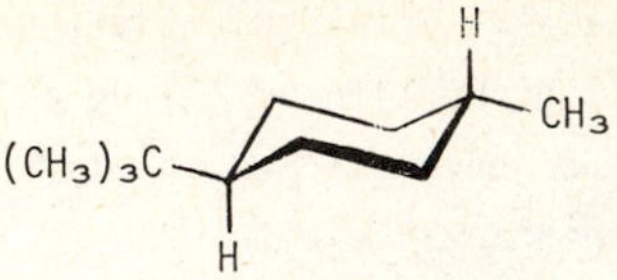

trans

no chiral centers

Both *cis*- and *trans*-1-methyl-4-*tert*-butylcyclohexane are achiral.

18.38 *cis* isomers:

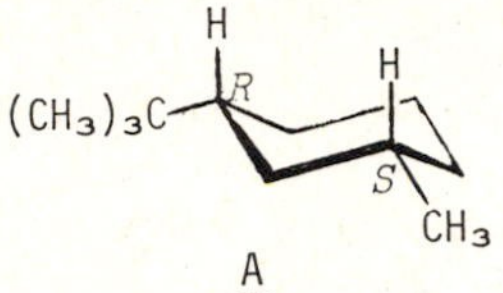

(CH$_3$)$_3$C, H, *R*, H, *S*, CH$_3$

A

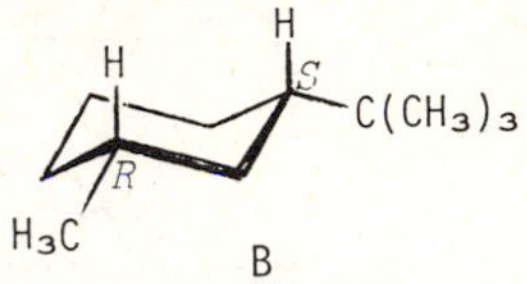

H, H, *S*, C(CH$_3$)$_3$, *R*, H$_3$C

B

trans isomers:

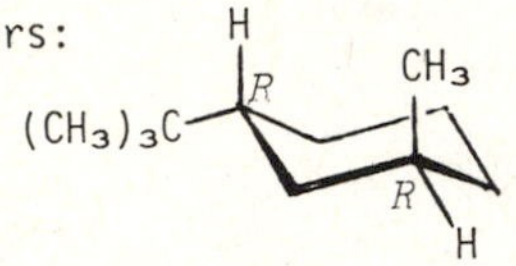

(CH$_3$)$_3$C, H, *R*, CH$_3$, *R*, H

C

CH$_3$, H, *S*, C(CH$_3$)$_3$, *S*, H

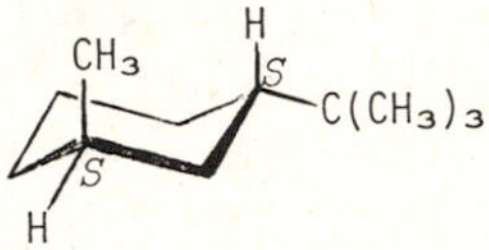

D

Each isomer has two chiral centers.

Isomer	Enantiomeric with	Diasteromeric with
A	B	C,D
B	A	C,D
C	D	A,B
D	C	A,B

18.39

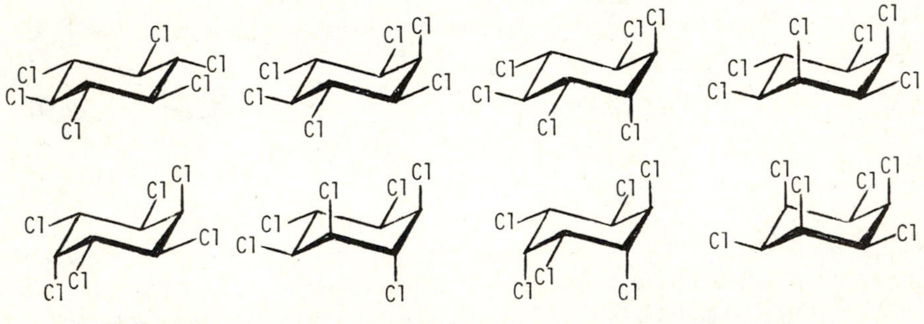

E_2 elimination requires periplanar geometry for the two groups being eliminated. In all but one of the isomers pictured above, at least one pair of neighboring -H and -Cl atoms is in the *anti* arrangement that is favored for E_2 elimination. The first isomer is the exception; since neighboring hydrogen and chlorine can assume neither *anti*-periplanar geometry nor the less favored *syn*-periplanar geometry, E_2 elimination does not occur.

18.40 Draw the four possible isomers of 4-*tert*-butylcyclohexane-1,3-diol. Make models of these isomers also.

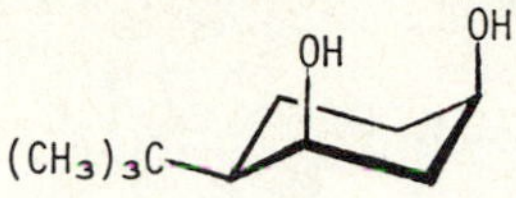

Only when the two hydroxyl groups are *cis*-diaxial (structure 1) can the acetal ring form. In any other conformation, the oxygen atoms are too far apart to be incorporated into a six-membered ring.

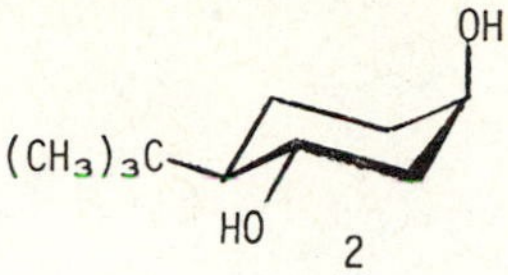

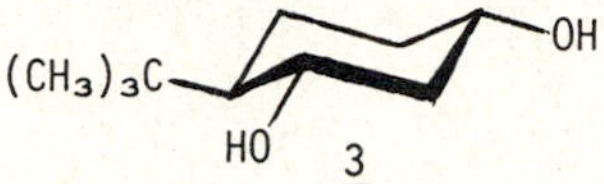

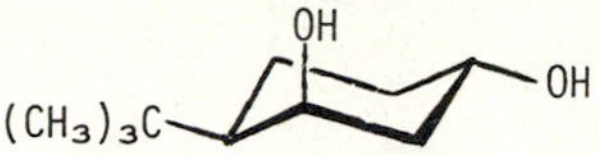

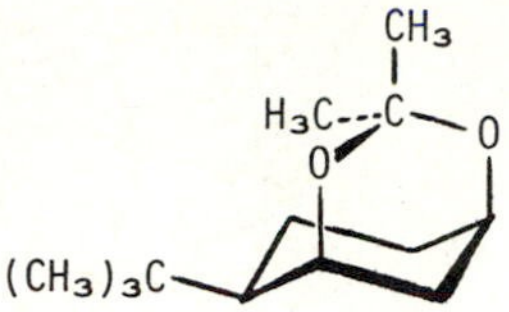

What you should know:

After doing these problems you should be able to:

1. Name cycloalkanes and draw the structures of cycloalkanes, given their names;
2. Use the concept of heat of combustion to predict the stablility of cyclic compounds;
3. Understand the geometry, properties and methods of preparation of three-membered rings;
4. Understand the geometry of four- and five-membered rings;
5. Know how to draw and label substituted cyclohexanes;
6. Predict the preferred conformation of substituted cyclohexanes by estimating the steric interactions present;
7. Use conformational analysis to predict reactivity of substituted cyclohexanes.

CHAPTER 19 ETHERS, EPOXIDES, AND SULFIDES

19.1 a) $(CH_3)_2CHOCH(CH_3)_2$

2-isopropoxypropane *or* diisopropyl ether

b) 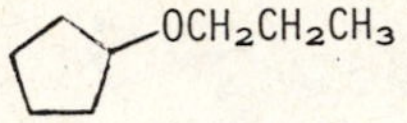 cyclopentyl–$OCH_2CH_2CH_3$

propoxycyclopentane *or* cyclopentyl propyl ether

c) 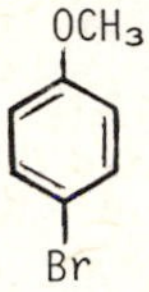 OCH_3 / Br (benzene ring, para)

p-bromoanisole *or* 4-bromo-1-methoxybenzene

d) 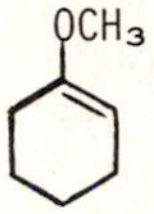 OCH_3 (cyclohexene ring)

1-methoxycyclohexene

e) $(CH_3)_2CHCH_2OCH_2CH_3$

1-ethoxy-2-methylpropane *or* ethyl isobutyl ether

f) $H_2C{=}CHCH_2OCH{=}CH_2$

allyl vinyl ether

19.2 The first step of the dehydration procedure is protonation of an alcohol; water is then displaced by another molecule of alcohol to form an ether. If two different alcohols are present either one can be protonated and either one can displace water, yielding a mixture of products.

If this procedure were used with ethanol and 1-propanol, the products would be diethyl ether, ethyl propyl ether, and dipropyl ether. If there were equimolar amounts of the alcohols, and if they were of equal reactivity, the product ratio would be diethyl ether : ethyl propyl ether : dipropyl ether = 1:2:1.

19.3 a) 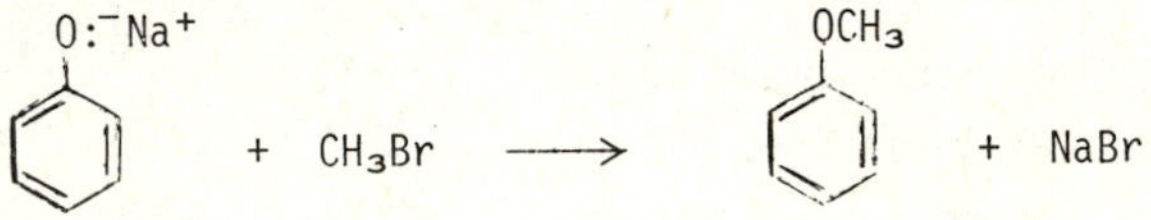 $C_6H_5O{:}^-\ Na^+ + CH_3Br \longrightarrow C_6H_5OCH_3 + NaBr$

anisole

b) 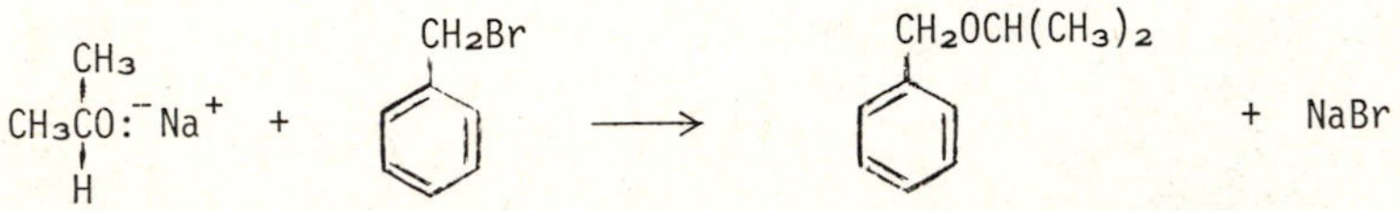 $CH_3\overset{CH_3}{\underset{H}{C}}O{:}^-\ Na^+ + C_6H_5CH_2Br \longrightarrow C_6H_5CH_2OCH(CH_3)_2 + NaBr$

benzyl isopropyl ether

c) $CH_3\overset{CH_3}{\underset{CH_3}{C}}CH_2O{:}^-\ Na^+ + CH_3CH_2Br \longrightarrow CH_3\overset{CH_3}{\underset{CH_3}{C}}CH_2OCH_2CH_3 + NaBr$

ethyl 2,2-dimethylpropyl ether

19.4 The compounds most reactive in the Williamson ether synthesis are also most reactive in any S_N2 process (review Chapter 10 if necessary).

most reactive ⟶ least reactive

a) $C_6H_5CH_2Br$ > CH_3CH_2Br > $CH_3CHBrCH_3$

primary benzylic > primary > secondary

b) $H_2C{=}CHCH_2Br$ > CH_3CH_2Br > CH_3CH_2Cl

primary allylic > primary, good leaving group > primary, poorer leaving group

19.5

The reaction mechanism of alkoxymercuration/demercuration of an alkene is analogous to other electrophilic additions we have studied. First, the cyclopentane pi electrons attack Hg^{2+} with formation of a cyclic mercurinium ion. Next, the nucleophilic alcohol displaces mercury. Markovnikov addition occurs because the carbon bearing the methyl group is better able to stabilize the partial positive charge arising from cleavage of the carbon-mercury bond. The ethoxyl and the mercuric groups are *trans* to each other. Finally, removal of mercury by $NaBH_4$ (the mechanism is not understood) results in the formation of 1-ethoxy-1-methylcyclopentane.

19.6 a) Either method of synthesis is appropriate.

1) Williamson

cyclohexanol $\xrightarrow{NaH}$ cyclohexoxide ($:\ddot{O}:^- Na^+$) + H_2 $\xrightarrow{BrCH_2CH_2CH_2CH_3}$ $OCH_2CH_2CH_2CH_3$ (cyclohexyl) + NaBr

butyl cyclohexyl ether

2) alkoxymercuration

cyclohexene $\xrightarrow[\text{2. } NaBH_4]{\text{1. } Hg(OCOCF_3)_2,\ HOCH_2CH_2CH_2CH_3}$ $OCH_2CH_2CH_2CH_3$ (cyclohexyl)

b) The Williamson synthesis is better.

$$C_6H_5OH \xrightarrow{NaH} C_6H_5\ddot{O}:^- Na^+ + H_2 \xrightarrow{BrCH_2CH_3} C_6H_5OCH_2CH_3 + NaBr$$

ethyl phenyl ether

c) Because both parts of the ether are somewhat hindered, use alkoxymercuration.

$$CH_3C(CH_3)=CH_2 \xrightarrow[2.\ NaBH_4]{1.\ Hg(OCOCF_3)_2,\ CH_3CH(OH)CH_2CH_3} (CH_3)_3C\text{-}O\text{-}CH(CH_3)CH_2CH_3$$

sec-butyl *tert*-butyl ether

d) The Williamson synthesis is better.

$$BrCH_2CH_2CH_2CH_2OH \xrightarrow{NaH} BrCH_2CH_2CH_2CH_2\ddot{O}:^- Na^+ + H_2 \longrightarrow \text{tetrahydrofuran} + NaBr$$

tetrahydrofuran

19.7 Let R be any other alkyl function.

$$(CH_3)_3C\text{-}OR \underset{}{\overset{CF_3COOH}{\rightleftharpoons}} \left[(CH_3)_3C\text{-}\overset{+}{O}(H)R + CF_3COO^- \rightleftharpoons HOR + CH_3\overset{+}{C}(CH_3)\text{-}CH_3 \right] \rightleftharpoons (CH_3)_2C=CH_2 + H^+$$

The first step of trifluoroacetic ether cleavage is protonation of the ether oxygen. The protonated intermediate collapses to form an alcohol and a tertiary carbocation. The carbocation loses a proton to form an alkene, isobutylene. This is an example of an E_1 elimination.

19.8

$$R\text{-}\ddot{O}\text{-}R' \overset{H^+}{\rightleftharpoons} \left[R\text{-}\overset{+}{\ddot{O}}(H)\text{-}R' \quad :X^- \right] \longrightarrow RX + R'OH$$

The first step in halogen acid cleavage of ethers involves protonation of oxygen. Halide then effects nucleophilic displacement to form an alcohol and an organic halide. The better the nucleophile, the more effective the displacement. Since $:I^-$ and $:Br^-$ are more nucleophilic than $:Cl^-$, ether cleavage proceeds more smoothly with HI or HBr than with HCl.

19.9

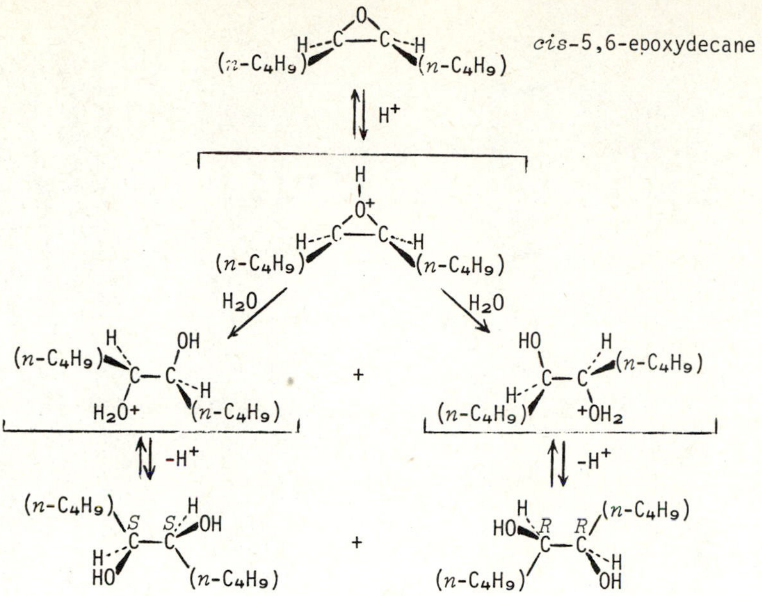

cis-5,6-epoxydecane

The product of acid hydrolysis of *cis*-5,6-epoxydecane is a racemic mixture.

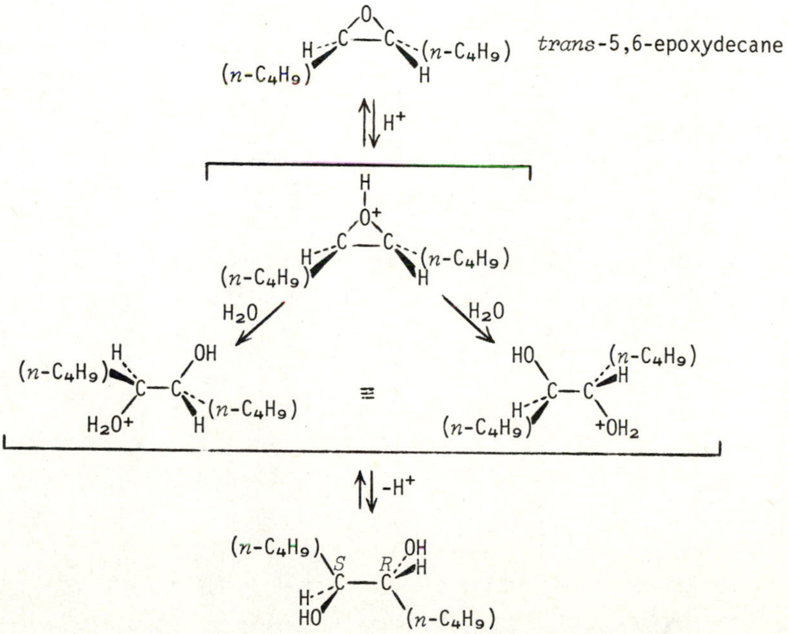

trans-5,6-epoxydecane

The product of acid hydrolysis of *trans*-5,6-epoxydecane is a *meso* compound. Acid hydrolysis of the enantiomer of *trans*-5,6-epoxydecane yields the same *meso* compound.

19.10

(CH$_3$)$_3$C ... H, H, O $\xrightarrow{H_3O^+}$ (CH$_3$)$_3$C ... OH, H, H, OH

cis-3-*tert*-butyl-1,2-epoxycyclohexane

(CH$_3$)$_3$C ... O, H, H $\xrightarrow{H_3O^+}$ (CH$_3$)$_3$C ... OH, H, H, OH

trans-3-*tert*-butyl-1,2-epoxycyclohexane

Both epoxycyclohexanes yield the same product on acid-catalyzed cleavage of the epoxide ring. The hydroxyl groups in the product have a *trans*-diaxial relationship.

19.11 a)

CH$_3$, O $\xrightarrow{HBr}$ Br, CH$_3$, OH, H

Attack of the nucleophile under acid conditions occurs at the more substituted carbon atom.

b)

$CH_3CH_2\overset{O}{CH-CH_2}$ $\xrightarrow[NH_3]{^-:NH_2}$ $CH_3CH_2\overset{OH}{C}HCH_2NH_2$

Attack of the basic nucleophile occurs at the less substituted carbon atom.

c)

$CH_3CH_2\overset{O}{CH-CH_2}$ $\xrightarrow{H_3{}^{18}O^+}$ $CH_3CH_2\overset{^{18}OH}{C}HCH_2OH$

Attack of the nucleophile under acid conditions occurs at the more substituted carbon atom.

19.12

15-crown-5   12-crown-4

The ion-to-oxygen distance in 15-crown-5 is about 40% longer than the ion-to-oxygen distance in 12-crown-4.

19.13

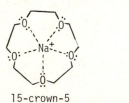

1.5 δ, O

CH$_3$CH$_2$CH-C-H, H

1.0 δ   ~2.9 δ   2.5 δ and 2.7 δ

1,2-epoxybutane

19.14 a) $CH_3CH_2SCH_3$
ethyl methyl sulfide

b) $(CH_3)_3CSCH_2CH_3$
ethyl *tert*-butyl sulfide

c) 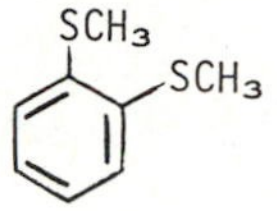 SCH_3, SCH_3
o-(dimethylthio)benzene

d) 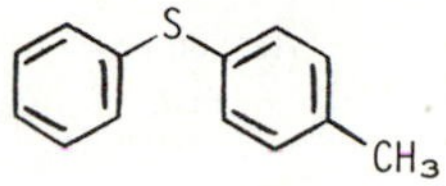 S, CH_3
phenyl *p*-tolyl sulfide *or*
p-(phenylthio)toluene

19.15 O^- $CH_3\overset{+}{S}CH_3$
dimethylsulfoxide

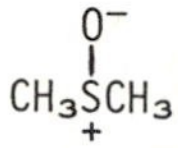

CH_3SCH_3
dimethyl sulfide

The boiling point of dimethylsulfoxide is high because it is a dipolar compound. Dimethylsulfoxide is miscible with water because it can hydrogen-bond with water.

19.16 a) 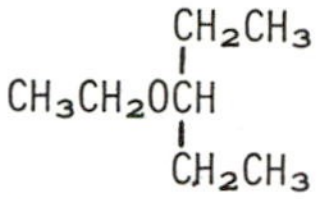 CH_2CH_3 / CH_3CH_2OCH / CH_2CH_3

b) 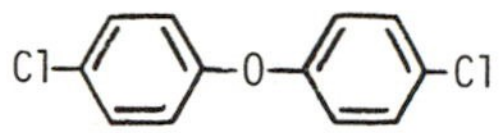 Cl–O–Cl

c) 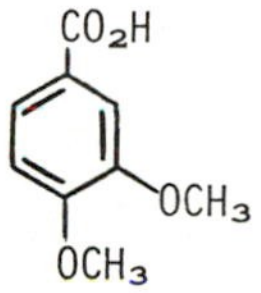 CO_2H, OCH_3, OCH_3

d) 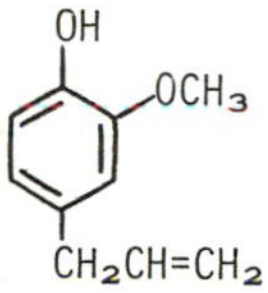 OH, OCH_3, $CH_2CH{=}CH_2$

e) 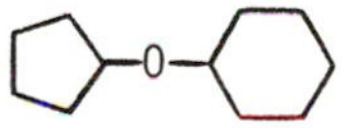 O

19.17 a) *o*-dimethoxybenzene

b) cyclopropyl isopropyl ether *or*
isopropoxycyclopropane

c) 2-methyltetrahydrofuran

d) cyclohexyl cyclopropyl sulfide

e) 2,2-dimethoxypropane

f) 1,2-epoxycyclopentane

g) *p*-nitroethoxybenzene *or*
ethyl *p*-nitrophenyl ether

h) 1,1-(dimethylthio)cyclohexane

19.18 CH_2CH_2, CH_3, CH_2, C, CH_3, HO, OH $\underset{}{\overset{H^+}{\rightleftharpoons}}$ [CH_2CH_2, CH_3, CH_2, C, CH_3, HO, $+OH_2$ $\overset{-H_2O}{\rightleftharpoons}$ CH_2CH_2, CH_3, CH_2, +C, CH_3, HO: $\rightarrow$ CH_2CH_2, CH_3, CH_2, C, CH_3, O, H^+] $\overset{-H^+}{\rightleftharpoons}$ O, CH_3, CH_3

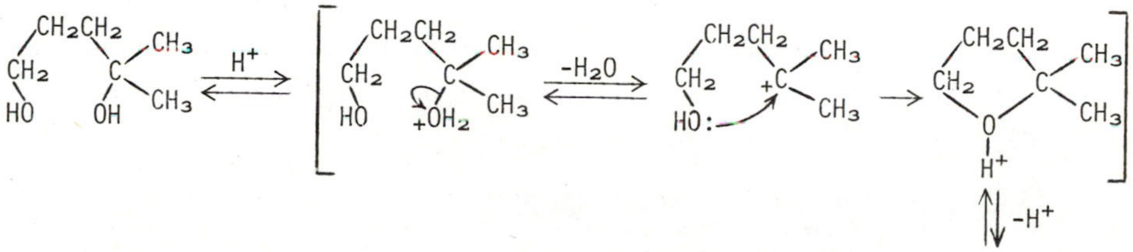

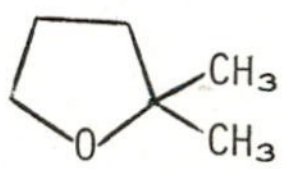

The reaction involves: (1) protonation of the tertiary hydroxyl group; (2) departure of water to form a tertiary carbocation; (3) nucleophilic attack on the cation by the other hydroxyl group.

19.19 a) cyclohexyl–OCH_2CH_3 $\xrightarrow[H_2O]{HI}$ cyclohexyl–OH + CH_3CH_2I

b) C_6H_5–$OC(CH_3)_3$ $\xrightarrow{CF_3COOH}$ C_6H_5–OH + $(CH_3)_2C=CH_2$

c) $H_2C=CH$–OCH_2CH_3 $\xrightarrow[H_2O]{HI}$ CH_3CH_2I + [$H_2C=CH$–OH] $\longrightarrow$ CH_3CHO

d) $(CH_3)_3CCH_2OCH_2CH_3$ $\xrightarrow[H_2O]{HI}$ $(CH_3)_3CCH_2OH$ + CH_3CH_2I

e) C_6H_5–CH_2OCH_3 $\xrightarrow[2.\ H_2O]{1.\ BCl_3}$ C_6H_5–CH_2OH + CH_3Cl

Remember that the halogen ends up on the less substituted ether fragment.

19.20 $\frac{1.06 \text{ g vanillin}}{152 \text{ g/mol}} = 6.97 \times 10^{-3}$ moles vanillin

$\frac{1.60 \text{ g AgI}}{234.8 \text{ g/mol}} = 6.81 \times 10^{-3}$ moles AgI

6.81×10^{-3} moles AgI $\rightarrow$ 6.81×10^{-3} moles $:I^-$ $\rightarrow$ 6.81×10^{-3} moles CH_3I $\rightarrow$ 6.81×10^{-3} moles $-OCH_3$

Thus, 6.97×10^{-3} moles of vanillin contains 6.81×10^{-3} moles of methoxyl groups. Since the ratio of moles vanillin to moles methoxyl is approximately 1:1, each vanillin contains one methoxyl group.

[Structure: benzene ring with CHO, OCH_3, OH]

vanillin

19.21 a) C_6H_5–OH $\xrightarrow{NaH}$ C_6H_5–$O:^- Na^+$ + H_2 $\xrightarrow{CH_3CH_2Br}$ C_6H_5–OCH_2CH_3 + NaBr

b) $CH_3CH=CH_2$ $\xrightarrow[2.\ NaBH_4]{1.\ C_6H_5OH,\ Hg(OCOCF_3)_2}$ C_6H_5–$OCH(CH_3)_2$

c) trans-$H_3C(H)C=C(H)CH_3$ $\xrightarrow[CHCl_3]{m\text{-}ClC_6H_4CO_3H}$ trans-2,3-dimethyloxirane (H_3C, H on each carbon of the epoxide ring)

d)

$$\text{cyclopentene} \xrightarrow[\text{2. } NaBH_4]{\text{1. } (CH_3)_3COH,\ Hg(OCOCF_3)_2} (CH_3)_3CO\text{-cyclopentyl}$$

e)

$$\text{cyclohexene} \xrightarrow[CHCl_3]{m\text{-}ClC_6H_4CO_3H} \text{cyclohexene oxide} \xrightarrow[\text{2. } H_2O]{\text{1. } LiAlD_4} \text{trans-2-deuteriocyclohexanol} \xrightarrow{NaH} \text{alkoxide } (O^-Na^+) + H_2$$

$$\xrightarrow{CH_3I} \text{trans-1-methoxy-2-deuteriocyclohexane} + NaI$$

f)

$$\text{cyclohexene} \xrightarrow[\text{2. } H_2O_2,\ OH^-]{\text{1. } BD_3} \text{cis-2-deuteriocyclohexanol} \xrightarrow{NaH} \text{alkoxide } (O^-Na^+) + H_2 \xrightarrow{CH_3I} \text{cis-1-methoxy-2-deuteriocyclohexane} + NaI$$

19.22

$$CH_3CH_2\text{-}\overset{+}{O}(CH_2CH_3)_2 + \text{cyclohexanol (HO:)} \longrightarrow \left[\begin{array}{c} CH_3CH_2 \\ | \\ H\text{-}O:^+ \\ \text{cyclohexyl} \end{array} + O(CH_2CH_3)_2 \right] \rightleftharpoons \text{cyclohexyl-}OCH_2CH_3 + H^+$$

Trialkyloxonium salts are more reactive alkylating agents than alkyl iodides because a neutral ether is a better leaving group than an iodide.

19.23

$$R\text{-}\ddot{O}\text{-}R' \underset{}{\overset{BCl_3}{\rightleftharpoons}} \left[R\text{-}\overset{^-BCl_3}{\underset{+}{O}}\text{-}R' \rightleftharpoons R\text{-}\overset{BCl_2}{\underset{+}{O}}\text{-}R' \;\; :Cl^- \right] \longrightarrow ROBCl_2 + R'Cl$$

$$\downarrow 3\ H_2O$$

$$ROH + B(OH)_3 + 2\ HCl$$

19.24 a)

$$\text{cyclohexene} \xrightarrow[\text{2. } NaBH_4]{\text{1. } (CH_3)_2CHOH,\ Hg(OCOCF_3)_2} \text{cyclohexyl-}OCH(CH_3)_2$$

b)

$$\text{4-methyl-1-methoxycyclohexane } (H_3C, OCH_3) \xrightarrow{HI} \text{4-methylcyclohexanol } (H_3C, OH) + CH_3I \xrightarrow{PBr_3} \text{1-bromo-4-methylcyclohexane } (H_3C, Br)$$

c) (4-tert-butylcyclohexene) $\xrightarrow[\text{CHCl}_3]{m\text{-ClC}_6\text{H}_4\text{CO}_3\text{H}}$ (epoxide, $C(CH_3)_3$) + (epoxide, $C(CH_3)_3$) $\xrightarrow[\text{H}_3\text{O}^+]{^-\text{:OH, H}_2\text{O} \text{ or}}$ (trans-diol: H, OH, H, -OH, $C(CH_3)_3$)

d) $CH_3CH_2CH_2CH_2C{\equiv}CH \xrightarrow[\text{Lindlar}]{H_2} CH_3CH_2CH_2CH_2CH{=}CH_2$

$\downarrow$ 1. BH_3 2. H_2O_2, OH^-

$CH_3CH_2CH_2CH_2CH_2CH_2O{:}^-\ Na^+ \xleftarrow{\text{NaH}} CH_3CH_2CH_2CH_2CH_2CH_2OH$

$\downarrow$ CH_3I

$CH_3CH_2CH_2CH_2CH_2CH_2OCH_3$ + NaI

e) $CH_3CH_2CH_2CH_2C{\equiv}CH \xrightarrow[\text{Lindlar}]{H_2} CH_3CH_2CH_2CH_2CH{=}CH_2$

$\downarrow$ 1. $Hg(OCOCF_3)_2$, CH_3OH 2. $NaBH_4$

$CH_3CH_2CH_2CH_2CH(OCH_3)CH_3$

19.25 a) (tetrahydrofuran) $\xrightarrow{\text{HI}}$ $HOCH_2CH_2CH_2CH_2I$

b) (1,4-dioxane) $\xrightarrow{\text{2 HI}}$ 2 $HOCH_2CH_2I$

19.26 (anisole, $C_6H_5O{-}CH_3$) + $:I^-$ $\longrightarrow$ $C_6H_5\ddot{O}{:}^-$ + CH_3I

This reaction is an S_N2 displacement and can't occur at an aryl carbon.

19.27 $CH_3C(CH_3){=}CH_2 \overset{H^+}{\rightleftharpoons}$ [$CH_3\overset{+}{C}(CH_3)CH_3$ + $H{-}\ddot{O}{-}R$ $\rightleftharpoons$ $CH_3C(CH_3)(\overset{+}{O}(H){-}R)CH_3$] $\rightleftharpoons$ $CH_3C(CH_3)(OR)CH_3$ + H^+

Notice that this reaction is the reverse of acid-catalyzed cleavage of a tertiary ether.

19.28 (catechol: OH, OH) $\xrightarrow{\text{2 NaH}}$ (disodium salt: $O{:}^-\ Na^+$, $O{:}^-\ Na^+$) + 2 H_2 $\xrightarrow{CH_2Br_2}$ (methylenedioxybenzene: O–CH_2–O) + 2 NaBr $\xrightarrow[\text{AlCl}_3]{H_2C{=}CHCH_2Cl}$ (O–CH_2–O, $CH_2CH{=}CH_2$)

safrole

19.29 $H_2C{=}CHOCH_2CH_3 \underset{}{\overset{H^+}{\rightleftharpoons}} [\, CH_3\overset{+}{C}HOCH_2CH_3 \longleftrightarrow CH_3CH{=}\overset{+}{O}CH_2CH_3 \,]$

$\updownarrow HOCH_2CH_3$

$H^+ + CH_3CH(OCH_2CH_3)_2 \rightleftharpoons [\, CH_3CH(\overset{+}{H}OCH_2CH_3)OCH_2CH_3 \,]$

Addition of ethanol occurs with the observed regiochemistry because the initial carbocation is stabilized by resonance involving the ether oxygen.

19.30 Cyclohexanone $\overset{H^+}{\rightleftharpoons}$ [protonated cyclohexanone $\overset{HOCH_2CH_2OH}{\rightleftharpoons}$ HO $H\overset{+}{O}CH_2CH_2OH$ adduct] $\overset{-H^+}{\rightleftharpoons}$ HO: :OCH$_2$CH$_2$OH (cyclohexane)

hemiacetal

In this sequence of reversible steps we have formed a hemiacetal. The steps are; (1) protonation of the ketone oxygen, (2) attack of a nucleophile on an electron-deficient carbon, and (3) loss of a proton. A similar sequence of steps is repeated in formation of the acetal.

HOCH$_2$CH$_2$O: :OH (cyclohexane) $\overset{H^+}{\rightleftharpoons}$ [HOCH$_2$CH$_2$O: :$\overset{+}{O}H_2$ $\overset{-H_2O}{\rightleftharpoons}$ H$_2$C-CH$_2$ +:O :OH $\rightleftharpoons$ H$_2$C-CH$_2$:O: :$\overset{+}{O}$H] $\overset{-H^+}{\rightleftharpoons}$ H$_2$C-CH$_2$ O O (cyclohexane)

acetal

19.31 The same reactions that were used to form the acetal in the previous problem apply here.

HO: H C=O: $\overset{H^+}{\rightleftharpoons}$ [HO: H C=$\overset{+}{O}$H $\rightleftharpoons$ $\overset{+}{O}$(H) ring, OH, H] $\overset{-H^+}{\rightleftharpoons}$ O ring, OH, H

hemiacetal

$\updownarrow H^+$

2-methoxy-tetrahydropyran (O ring, H, OCH$_3$) $\overset{-H^+}{\rightleftharpoons}$ [O ring, H, $\overset{+}{H}OCH_3$ $\rightleftharpoons$ $\overset{+}{O}$ ring, H, HOCH$_3$ $\overset{-H_2O}{\rightleftharpoons}$ O ring, $\overset{+}{O}H_2$, H]

19.32 anethole

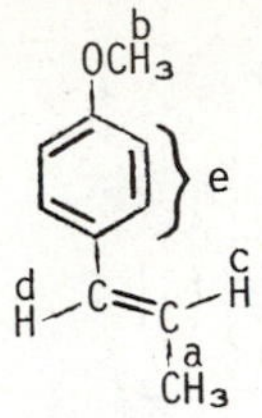

Proton(s)	δ	multiplicity	split by	J
a.	1.83	doublet	c	J_{ac} = 8
b.	3.75	singlet		
c.	6.08	quartet	a	J_{ca} = 8
d.	6.28	singlet		
e.	6.80, 7.23	multiplet		

J for coupling between protons c and d is very small.

19.33

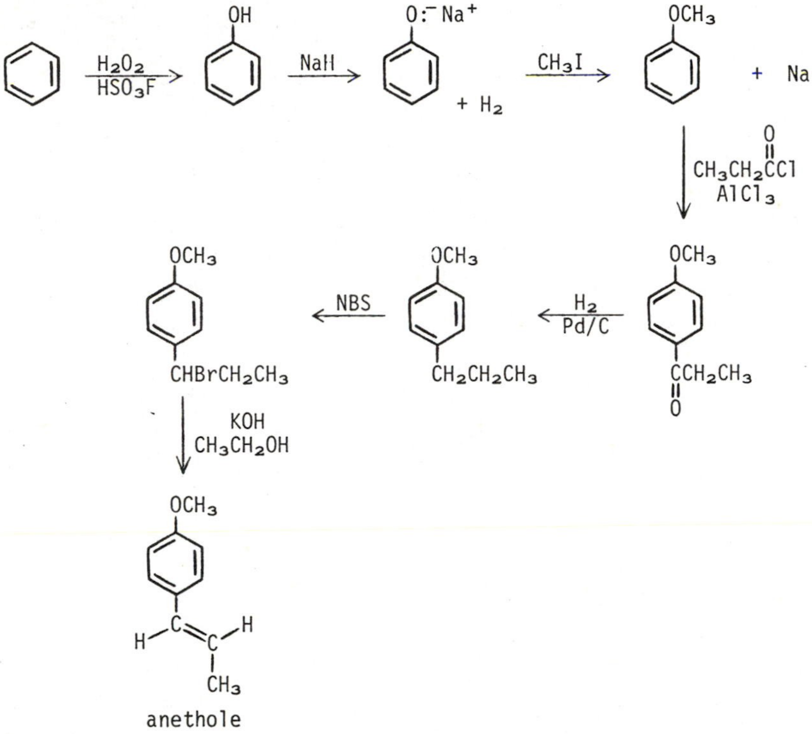

Benzene $\xrightarrow[HSO_3F]{H_2O_2}$ phenol (OH) $\xrightarrow{NaH}$ O^-Na^+ + H_2 $\xrightarrow{CH_3I}$ OCH_3 (anisole) + NaI

Anisole $\xrightarrow[AlCl_3]{CH_3CH_2CCl (O)}$ OCH_3-C₆H₄-CCH_2CH_3 (=O) $\xrightarrow[Pd/C]{H_2}$ OCH_3-C₆H₄-$CH_2CH_2CH_3$ $\xrightarrow{NBS}$ OCH_3-C₆H₄-$CHBrCH_2CH_3$ $\xrightarrow[CH_3CH_2OH]{KOH}$ OCH_3-C₆H₄-CH=CH-CH_3

anethole

19.34 M^+ = 116 corresponds to a sulfide of molecular formula $C_6H_{12}S$, indicating one degree of unsaturation.

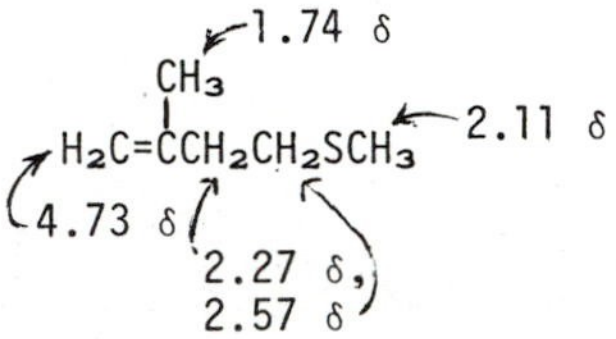

$H_2C=C(CH_3)CH_2CH_2SCH_3$

CH_3: 1.74 δ; $H_2C=$: 4.73 δ; CH_2CH_2: 2.27 δ, 2.57 δ; SCH_3: 2.11 δ

19.35 A molecular model of bornene shows that approach to the upper face of the double bond is hindered by a methyl group. Reaction with RCO_3H occurs at the lower face of the double bond to produce epoxide A.

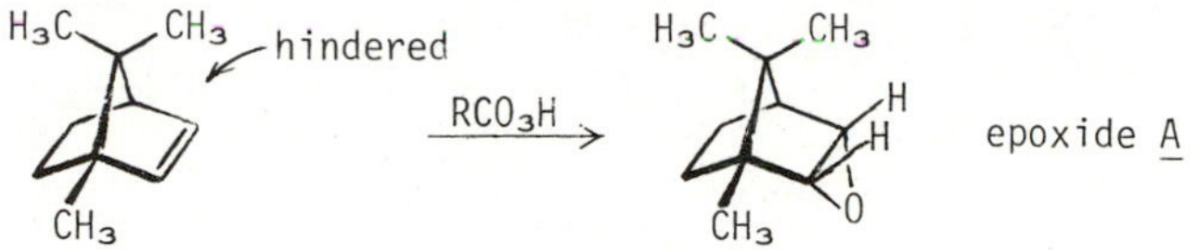

H_3C CH_3 hindered
RCO_3H
CH_3
epoxide A

In the reaction of Br_2 and H_2O with bornene, the intermediate bromonium ion also forms at the lower face. Reaction with water yields a bromohydrin which, when treated with base, forms epoxide B.

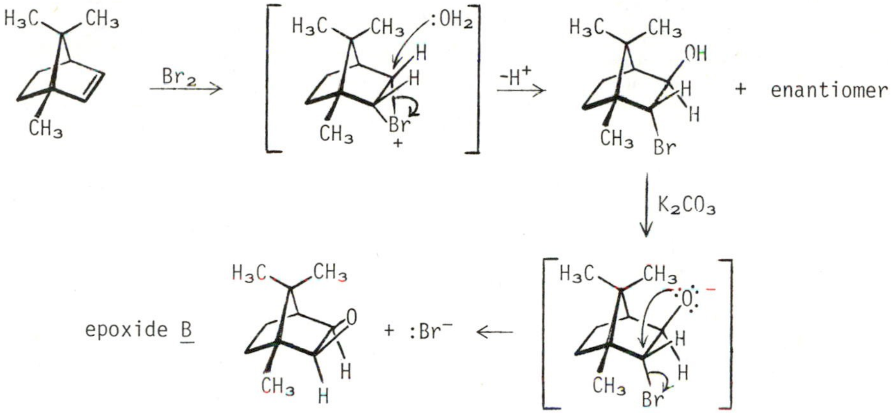

Br_2 → [bromonium ion, $:OH_2$] $\xrightarrow{-H^+}$ bromohydrin + enantiomer

K_2CO_3

epoxide B + $:Br^-$ ←

19.36 Disparlure, $C_{19}H_{38}O$, contains one degree of unsaturation, which the ^1H NMR absorption at 2.8 δ identifies as an epoxide ring.

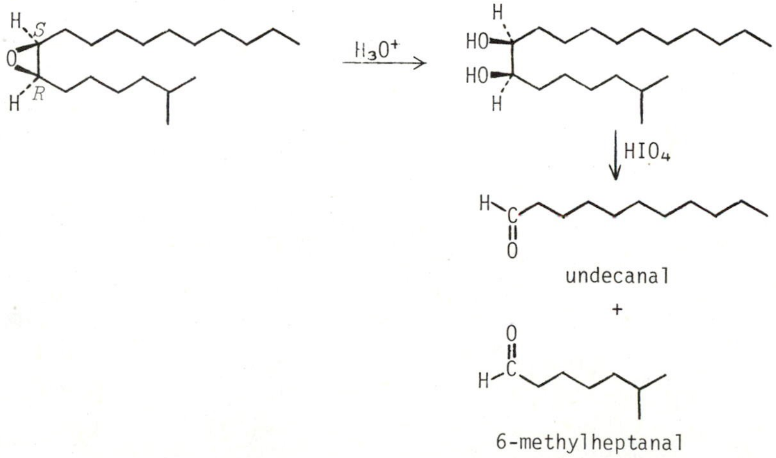

H_3O^+

HIO_4

undecanal

+

6-methylheptanal

19.37

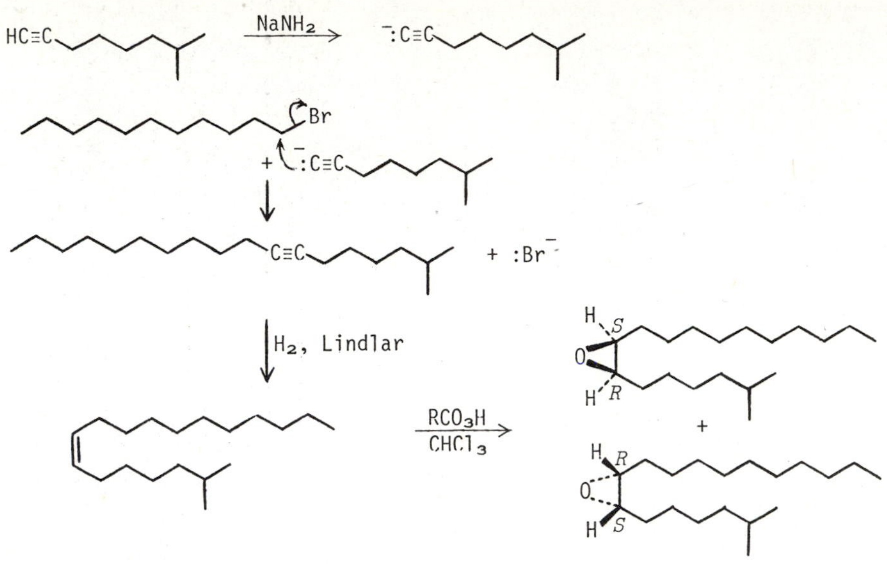

HC≡C– $\xrightarrow{NaNH_2}$ $^{-}$:C≡C–

Br

+ $^{-}$:C≡C–

C≡C– + :Br^{-}

H_2, Lindlar

$\xrightarrow[CHCl_3]{RCO_3H}$

H S O H R

+

H R O H S

What you should know:

After doing these problems, you should be able to:

1. Name ethers and sulfides, and draw the structures of ethers and sulfides corresponding to IUPAC names;
2. Prepare ethers using the reactions we have studied;
3. Formulate mechanisms for preparation and cleavage of ethers;
4. Predict the products of reactions involving ethers;
5. Formulate mechanisms and predict products of cleavage reactions of epoxides.

CHAPTER 20 ALCOHOLS AND THIOLS

20.1 a) 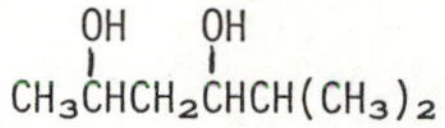 $CH_3CH(OH)CH_2CH(OH)CH(CH_3)_2$

5-methyl-2,4-hexanediol

b) 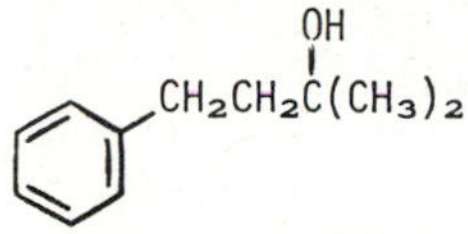 $C_6H_5CH_2CH_2C(OH)(CH_3)_2$

2-methyl-4-phenyl-2-butanol

c) 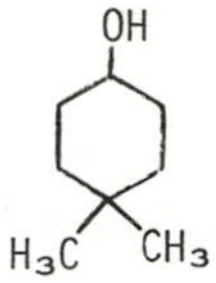 4,4-dimethylcyclohexanol

d) 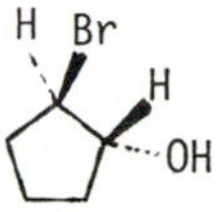 *trans*-2-bromocyclopentanol

e) 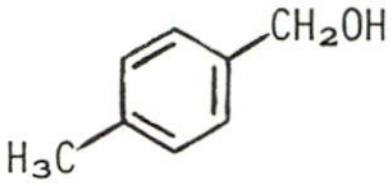 p-$CH_3C_6H_4CH_2OH$

p-methylbenzyl alcohol

20.2 a) 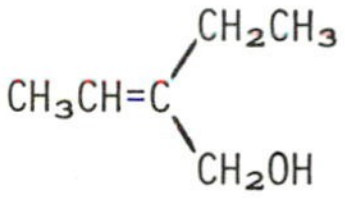 $CH_3CH{=}C(CH_2CH_3)CH_2OH$

2-ethyl-2-buten-1-ol

b)  3-cyclohexen-1-ol

c) 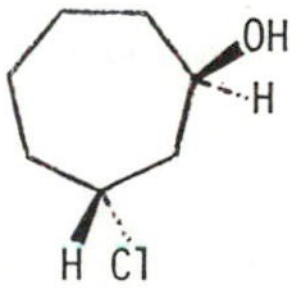 *trans*-3-chlorocycloheptanol

d) 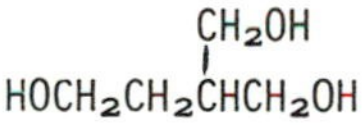 $HOCH_2CH_2CH(CH_2OH)CH_2OH$

2-(hydroxymethyl)-1,4-butanediol

20.3 In general, the boiling points of a series of isomers decrease with branching. The more nearly spherical a compound becomes, the less surface area it has relative to a straight chain compound of the same molecular weight and functional group type. A smaller surface area allows fewer van der Waals interactions, the weak forces that attract covalent molecules to each other. In addition, branching makes it more difficult for hydroxyl groups to approach each other from all directions to hydrogen bond.

A given volume of 2-methyl-2-propanol contains fewer hydrogen bonds than the same volume of 1-butanol. Since it take less energy to break the hydrogen bonds of 2-methyl-2-propanol, it is lower boiling than 1-butanol or 2-butanol.

20.4

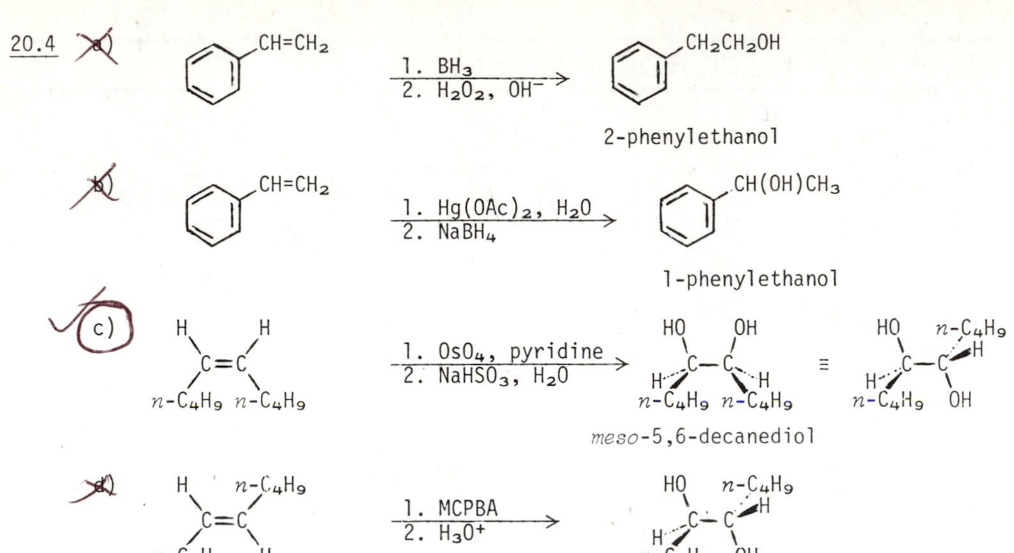

20.5 lowest oxidation state ⟶ highest oxidation state

a) cyclohexane < cyclohexanone , 1-methoxycyclohexene (OCH_3) < benzene

b) $CH_3CH_2NH_2 < H_2NCH_2CH_2NH_2 < CH_3CN$

20.6 a) cyclohexene $\xrightarrow{\text{NBS}}$ 3-bromocyclohexene (Br)

This reaction is an oxidation because it is an addition of bromine and a removal of hydrogen.

b) 3-bromocyclohexene (Br) $\xrightarrow{\text{Base}}$ 1,3-cyclohexadiene

This reaction is neither an oxidation nor a reduction.

c) benzene $\xrightarrow[\text{FeCl}_3]{\text{Cl}_2}$ chlorobenzene (Cl)

Because chlorine is added, this reaction is an oxidation.

d) $CH_3CH_2CH_2CH_2Br \xrightarrow{Mg} CH_3CH_2CH_2CH_2MgBr \xrightarrow{H^+} CH_3CH_2CH_2CH_3$

This reaction is a reduction; bromine is removed and hydrogen is added.

e) Benzene $\xrightarrow[AlCl_3]{CH_3Cl}$ toluene (benzene ring with CH_3)

This reaction is neither an oxidation nor a reduction.

20.7 a) $CH_3\overset{O}{\overset{\|}{C}}CH_2CH_2\overset{O}{\overset{\|}{C}}OCH_3 \xrightarrow[2.\ H_3O^+]{1.\ NaBH_4} CH_3\overset{OH}{\overset{|}{C}}HCH_2CH_2\overset{O}{\overset{\|}{C}}OCH_3$

$NaBH_4$ reduces aldehydes and ketones without disturbing other functional groups.

b) $CH_3\overset{O}{\overset{\|}{C}}CH_2CH_2\overset{O}{\overset{\|}{C}}OCH_3 \xrightarrow[2.\ H_3O^+]{1.\ LiAlH_4} CH_3\overset{OH}{\overset{|}{C}}HCH_2CH_2CH_2OH$

$LiAlH_4$ reduces both ketones and esters.

c) 2-methyl-5-(isopropenyl)cyclohex-2-enone (H_3C, CH_2, CH_3, O) $\xrightarrow[2.\ H_3O^+]{1.\ LiAlH_4}$ corresponding cyclohexenol (OH, CH_3, H_3C, CH_2)

$LiAlH_4$ reduces carbonyl functional groups without reducing double bonds.

d) 2-methyl-5-(isopropenyl)cyclohex-2-enone (H_3C, CH_2, CH_3, O) $\xrightarrow[Pd/C]{H_2}$ 2-methyl-5-isopropylcyclohexanone (O, CH_3, H_3C, CH_3)

H_2, Pd/C reduces double bonds without reducing carbonyl functional group.

20.8 a) $C_6H_5\overset{O}{\overset{\|}{C}}H$ *or* $C_6H_5\overset{O}{\overset{\|}{C}}OH$ *or* $C_6H_5\overset{O}{\overset{\|}{C}}OR \xrightarrow[2.\ H_3O^+]{1.\ LiAlH_4} C_6H_5CH_2OH$

b) $C_6H_5\overset{O}{\overset{\|}{C}}CH_3 \xrightarrow[2.\ H_3O^+]{1.\ LiAlH_4} C_6H_5\overset{OH}{\overset{|}{C}}HCH_3$

c) Cyclohexanone (=O) $\xrightarrow[2.\ H_3O^+]{1.\ LiAlH_4}$ cyclohexanol (OH, H)

d) $(CH_3)_2CHCHO$ *or* $(CH_3)_2CHCOOH$ *or* $(CH_3)_2CHCOOR \xrightarrow[2.\ H_3O^+]{1.\ LiAlH_4} (CH_3)_2CHCH_2OH$

20.9 a) 2-Methyl-2-propanol is a tertiary alcohol. To synthesize a tertiary alcohol, start with a ketone.

$$CH_3\overset{O}{\overset{\|}{C}}CH_3 \xrightarrow[2.\ H_3O^+]{1.\ CH_3MgBr} CH_3\underset{CH_3}{\underset{|}{\overset{OH}{\overset{|}{C}}}}CH_3$$

When two or more alkyl groups bonded to the carbon bearing the -OH group are the same, an alcohol can be synthesized from an ester and a Grignard reagent.

$$CH_3\overset{O}{\overset{\|}{C}}OR \xrightarrow[2.\ H_3O^+]{1.\ 2\ CH_3MgBr} CH_3\underset{CH_3}{\underset{|}{\overset{OH}{\overset{|}{C}}}}CH_3$$

2-methyl-2-propanol

b) Since 1-methylcyclohexanol is a tertiary alcohol, start with a ketone.

$$\text{cyclohexanone} \xrightarrow[2.\ H_3O^+]{1.\ CH_3MgBr} \text{1-methylcyclohexanol}$$

1-methylcyclohexanol

c) 3-Methyl-3-pentanol is a tertiary alcohol. Either a ketone or an ester can be used as a starting material.

$$CH_3CH_2\overset{O}{\overset{\|}{C}}CH_2CH_3 \xrightarrow[2.\ H_3O^+]{1.\ CH_3MgBr}$$

or

$$CH_3CH_2\overset{O}{\overset{\|}{C}}CH_3 \xrightarrow[2.\ H_3O^+]{1.\ CH_3CH_2MgBr}$$

or

$$CH_3\overset{O}{\overset{\|}{C}}OR \xrightarrow[2.\ H_3O^+]{1.\ 2\ CH_3CH_2MgBr}$$

$$CH_3CH_2\underset{CH_3}{\underset{|}{\overset{OH}{\overset{|}{C}}}}CH_2CH_3$$

3-methyl-3-pentanol

d) Three possible combinations of ketone plus Grignard reagent can be used to synthesize this tertiary alcohol.

$$CH_3CH_2\overset{O}{\overset{\|}{C}}CH_3 \xrightarrow[2.\ H_3O^+]{1.\ C_6H_5MgBr}$$

or

$$CH_3CH_2\overset{O}{\overset{\|}{C}}-C_6H_5 \xrightarrow[2.\ H_3O^+]{1.\ CH_3MgBr}$$

or

$$CH_3\overset{O}{\overset{\|}{C}}-C_6H_5 \xrightarrow[2.\ H_3O^+]{1.\ CH_3CH_2MgBr}$$

$$CH_3CH_2\underset{C_6H_5}{\underset{|}{\overset{OH}{\overset{|}{C}}}}CH_3$$

2-phenyl-2-butanol

e) *Formaldehyde* must be used to synthesize this primary alcohol.

$$CH_2O \xrightarrow[2.\ H_3O^+]{1.\ C_6H_5MgBr} C_6H_5CH_2OH$$

benzyl alcohol

20.10 The first equivalent of CH_3MgBr reacts with the hydroxyl hydrogen of 4-hydroxycyclohexanone.

$$\text{4-hydroxycyclohexanone (OH)} \xrightarrow{CH_3MgBr} \text{4-(OMgBr)cyclohexanone} + CH_4$$

The second equivalent of CH_3MgBr adds to the ketone in the expected manner to yield 1-methyl-1,4-cyclohexanediol.

$$\text{4-(OMgBr)cyclohexanone} \xrightarrow[2.\ H_3O^+]{1.\ CH_3MgBr} \text{1-methyl-1,4-cyclohexanediol (HO, CH}_3\text{; OH)}$$

20.11

$$CH_3CH_2OH \xrightarrow{CrO_3} \left[H_3C(H)(H)C\text{–}O\text{–}CrO_3H \quad \text{H} \leftarrow :Base \right] \xrightarrow{E_2} CH_3\overset{O}{\overset{\|}{C}}H + HCrO_3^-$$

The mechanism of oxidation of ethanol to acetaldehyde is like the mechanism pictured in the text. The base in this instance is water. Reversible nucleophilic addition of ethanol to the intermediate acetaldehyde yields a *hemiacetal*. Oxidation of this hemiacetal produces an ester.

$$CH_3\overset{O}{\overset{\|}{C}}H \underset{}{\overset{H^+}{\rightleftharpoons}} \left[CH_3\overset{^+OH}{\overset{\|}{C}}H \underset{}{\overset{H\ddot{O}CH_2CH_3}{\rightleftharpoons}} CH_3\overset{OH}{\overset{|}{C}}H\text{–}\overset{+}{H}OCH_2CH_3 \right] \rightleftharpoons CH_3\overset{OH}{\overset{|}{C}}H\text{–}OCH_2CH_3 + H^+$$

hemiacetal

$$CH_3CH_2O\text{–}\overset{OH}{\overset{|}{C}}H(CH_3) \xrightarrow{CrO_3} \left[CH_3CH_2O\text{–}C(H_3C)(H)\text{–}O\text{–}CrO_3H \quad \text{H} \leftarrow :Base \right] \xrightarrow{E_2} CH_3\overset{O}{\overset{\|}{C}}OCH_2CH_3 + HCrO_3^-$$

20.12

$$\text{dihydropyran} \underset{}{\overset{H^+}{\rightleftharpoons}} \left[\text{carbocation} \longleftrightarrow \text{oxonium ion} \overset{H\ddot{O}R}{\rightleftharpoons} \text{2-(}H\overset{+}{O}R\text{)tetrahydropyran} \right] \rightleftharpoons \text{2-(}\ddot{O}R\text{)tetrahydropyran} + H^+$$

Protonation of dihydropyran occurs on the non-oxygen-bearing carbon to give a resonance-stabilized carbocation intermediate.

20.13

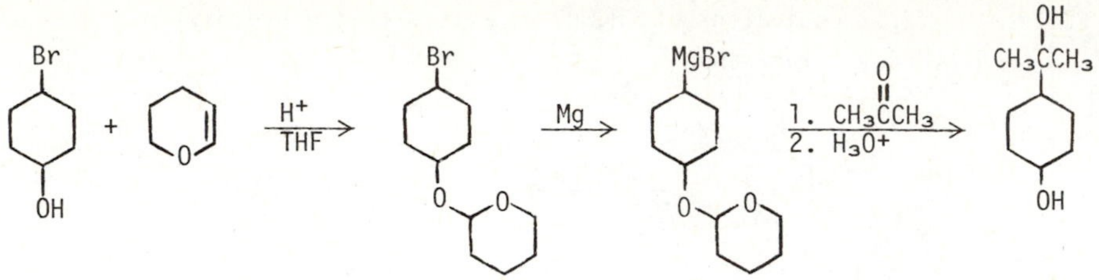

[Reaction scheme: 4-bromocyclohexanol + dihydropyran, H^+/THF → THP ether of 4-bromocyclohexanol; Mg → Grignard (MgBr); 1. CH_3COCH_3, 2. H_3O^+ → 4-(1-hydroxy-1-methylethyl)cyclohexanol ($CH_3C(OH)CH_3$ / OH)]

20.14 The infrared spectra of cholesterol and 5-cholesten-3-one each exhibit a unique absorption that makes it easy to distinguish between them. Cholesterol shows an -OH stretch at 3300-3600 cm^{-1}; 5-cholesten-3-one shows a C=O stretch at 1710 cm^{-1}. In the Jones oxidation of cholesterol to 5-cholesten-3-one the -OH band will disappear and will be replaced by a C=O band. When oxidation is complete no -OH absorption should be visible.

20.15 a) No protons are adjacent to the hydroxyl group of 2-methyl-2-propanol; the -OH signal is *unsplit*.

b) Since one proton is adjacent to the hydroxyl proton of cyclohexanol, the -OH absorption is a *doublet*.

c) The -OH absorption of ethanol appears as a *triplet* because of splitting by two adjacent protons.

d) The -OH absorption is split into a *doublet*.

e) The signal for the hydroxyl proton of cholesterol is split into a *doublet*.

f) The -OH absorption of 1-methylcyclohexanol is *unsplit*.

20.16 a) $CH_3CH_2CH(SH)CH_3$
2-butanethiol

b) [benzene ring with SH at para positions]
p-dimercaptobenzene *or* *p*-benzenedithiol

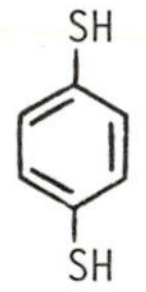

c) $(CH_3)_3CCH_2CH(SH)CH_2CH(CH_3)_2$
2,2,6-trimethyl-4-heptanethiol

d) [cyclopentene ring with SH]
2-cyclopentene-1-thiol

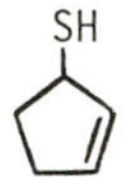

20.17 a) $CH_3CH{=}CHCO_2CH_3$ (methyl 2-butenoate) $\xrightarrow{1.\ LiAlH_4;\ 2.\ H_3O^+}$ $CH_3CH{=}CHCH_2OH$ $\xrightarrow{PBr_3}$ $CH_3CH{=}CHCH_2Br$ $\xrightarrow{1.\ (H_2N)_2C{=}S;\ 2.\ OH^-,\ H_2O}$ $CH_3CH{=}CHCH_2SH$
2-butene-1-thiol

b) $H_2C{=}CHCH{=}CH_2 \xrightarrow[\Delta]{HBr} CH_3CH{=}CHCH_2Br \xrightarrow[2.\ OH^-,\ H_2O]{1.\ (H_2N)_2C{=}S} CH_3CH{=}CHCH_2SH$

2-butene-1-thiol

20.18 a) 2-methyl-1,4-butanediol b) 3-ethyl-2-hexanol
c) *cis*-1,3-cyclobutanediol d) *cis*-2-methyl-4-cyclohepten-1-ol
e) *cis*-3-phenylcyclopentanol f) 2,3-dimethyl-3-hexanethiol

20.19 None of these alcohols have multiple bonds or rings.

$CH_3CH_2CH_2CH_2CH_2OH$ — 1-pentanol

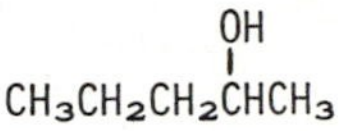

$CH_3CH_2CH_2CH(OH)CH_3$ — 2-pentanol

$CH_3CH_2CH(OH)CH_2CH_3$ — 3-pentanol

$CH_3CH_2CH(CH_3)CH_2OH$ — 2-methyl-1-butanol

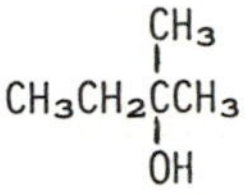

$CH_3CH_2C(CH_3)(OH)CH_3$ — 2-methyl-2-butanol

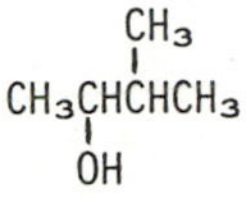

$CH_3CH(OH)CH(CH_3)CH_3$ — 3-methyl-2-butanol

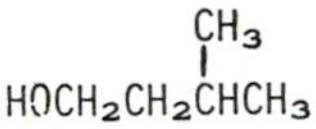

$HOCH_2CH_2CH(CH_3)CH_3$ — 3-methyl-1-butanol

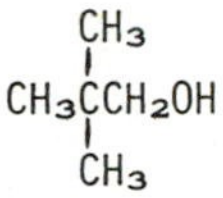

$CH_3C(CH_3)_2CH_2OH$ — 2,2-dimethyl-1-propanol

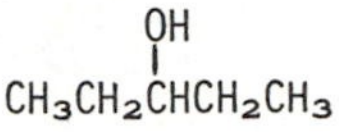

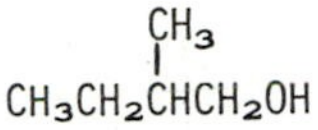

20.20 Primary alcohols react with Jones' reagent to form carboxylic acids, secondary alcohols yield ketones, and tertiary alcohols are unreactive to oxidation. Of the eight alcohols in the previous problem, only 2-methyl-2-butanol is unreactive to Jones oxidation.

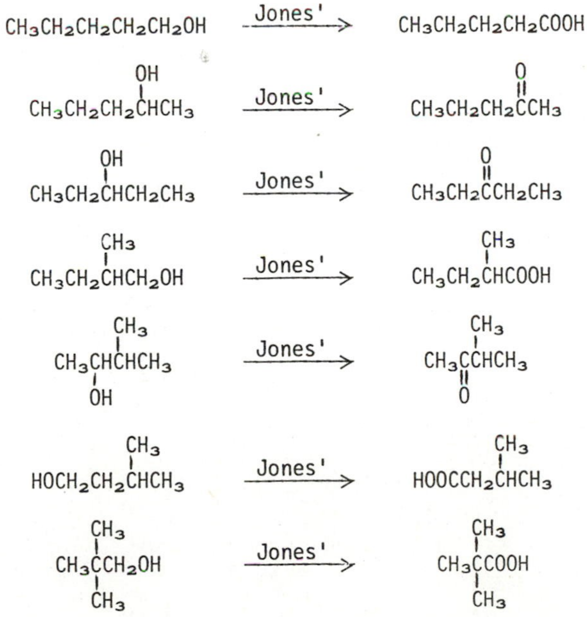

$$CH_3CH_2CH_2CH_2CH_2OH \xrightarrow{Jones'} CH_3CH_2CH_2CH_2COOH$$

$$CH_3CH_2CH_2CH(OH)CH_3 \xrightarrow{Jones'} CH_3CH_2CH_2C(=O)CH_3$$

$$CH_3CH_2CH(OH)CH_2CH_3 \xrightarrow{Jones'} CH_3CH_2C(=O)CH_2CH_3$$

$$CH_3CH_2CH(CH_3)CH_2OH \xrightarrow{Jones'} CH_3CH_2CH(CH_3)COOH$$

$$CH_3CH(OH)CH(CH_3)CH_3 \xrightarrow{Jones'} CH_3C(=O)CH(CH_3)CH_3$$

$$HOCH_2CH_2CH(CH_3)CH_3 \xrightarrow{Jones'} HOOCCH_2CH(CH_3)CH_3$$

$$CH_3C(CH_3)_2CH_2OH \xrightarrow{Jones'} CH_3C(CH_3)_2COOH$$

20.21 a) Benzene ring–CH_2CH_2OH $\xrightarrow[\text{pyridine}]{POCl_3}$ Benzene ring–$CH=CH_2$

2-phenylethanol → styrene

b) Benzene ring–CH_2CH_2OH $\xrightarrow[CH_2Cl_2]{PCC*}$ Benzene ring–CH_2CHO

phenylacetaldehyde

* $C_5H_6NCrO_3Cl$ (pyridinium chlorochromate)

c) Benzene ring–CH_2CH_2OH $\xrightarrow[H_2O,\ H_2SO_4]{CrO_3}$ Benzene ring–CH_2CO_2H

phenylacetic acid

d) Benzene ring–CH_2CH_2OH $\xrightarrow[H_2O]{KMnO_4}$ Benzene ring–COOH

benzoic acid

e) Benzene ring–$CH=CH_2$ $\xrightarrow[Pd/C]{H_2}$ Benzene ring–CH_2CH_3

(from a) → ethylbenzene

f) Benzene ring–$CH=CH_2$ $\xrightarrow[2.\ Zn,\ H_3O^+]{1.\ O_3}$ Benzene ring–CHO

(from a) → benzaldehyde

g) Benzene ring–$CH=CH_2$ $\xrightarrow[2.\ NaBH_4]{1.\ Hg(OAc)_2,\ H_2O}$ Benzene ring–$CH(OH)CH_3$

(from a) → 1-phenylethanol

h) Benzene ring–CH_2CH_2OH $\xrightarrow{PBr_3}$ Benzene ring–CH_2CH_2Br

2-bromo-1-phenylethane

20.22 a) 4-$C(CH_3)_3$-cyclohexanol (OH) $\xrightarrow[H_2O,\ H_2SO_4]{CrO_3}$ 4-$C(CH_3)_3$-cyclohexanone (O)

b) cis-4-tert-butylcyclohexanol (OH, $C(CH_3)_3$) $\xrightarrow{PBr_3}$ 1-bromo-4-tert-butylcyclohexane (Br, $C(CH_3)_3$)

c) 4-tert-butylcyclohexanol (OH, $C(CH_3)_3$) $\xrightarrow[\text{pyridine}]{POCl_3}$ 4-tert-butylcyclohexene ($C(CH_3)_3$)

d) $CH_3CH_2CH_2OH \xrightarrow{PCC} CH_3CH_2CHO$

e) $CH_3CH_2CH_2OH \xrightarrow[H_2O,\ H_2SO_4]{CrO_3} CH_3CH_2COOH$

f) $CH_3CH_2CH_2OH \xrightarrow[\text{pyridine}]{H_3C{-}C_6H_4{-}SO_2Cl} CH_3CH_2CH_2OSO_2{-}C_6H_4{-}CH_3$

g) $CH_3CH_2CH_2OH \xrightarrow[\text{pyridine}]{SOCl_2} CH_3CH_2CH_2Cl$

h) $CH_3CH_2CH_2OH \xrightarrow{NaH} CH_3CH_2CH_2O{:}^-\ Na^+ \ + \ H_2$

i) $CH_3CH_2CH_2OH \xrightarrow{PBr_3} CH_3CH_2CH_2Br \xrightarrow[\text{(from h)}]{CH_3CH_2CH_2O{:}^-\ Na^+} CH_3CH_2CH_2OCH_2CH_2CH_3$

20.23 In some of these problems, different combinations of Grignard reagent and carbonyl compound are possible. Remember that aqueous acid is added to the initial Grignard adduct to yield the alcohol.

a) $CH_3MgBr \ + \ CH_3CH_2CHO$

or

$CH_3CHO \ + \ CH_3CH_2MgBr$

$\longrightarrow$ $CH_3CH(OH)CH_2CH_3$

2-butanol

b) $C_6H_5{-}\overset{O}{\overset{\|}{C}}CH_3 \ + \ CH_3MgBr$

or

$C_6H_5{-}\overset{O}{\overset{\|}{C}}OR \ + \ 2\ CH_3MgBr$

or

$C_6H_5{-}MgBr \ + \ CH_3\overset{O}{\overset{\|}{C}}CH_3$

$\longrightarrow$ $C_6H_5{-}C(OH)(CH_3)_2$

2-phenyl-2-propanol

c) $H_2C{=}C(CH_3){-}MgBr + CH_2O \longrightarrow H_2C{=}C(CH_3)CH_2OH$

2-methyl-2-buten-1-ol

d) (benzophenone) $C_6H_5COC_6H_5$ + C_6H_5MgBr

or

C_6H_5COOR + 2 C_6H_5MgBr

$\longrightarrow (C_6H_5)_3C{-}OH$

triphenylmethanol

e) $CH_3MgBr + HCCH_2CH_2CH_2Br$ (with C=O) $\longrightarrow CH_3CH(OH)CH_2CH_2CH_2Br$

5-bromo-2-pentanol

20.24

(*S*)-2-octanol (Fischer projection: CH_3 top, H left, OH right, n-C_6H_{13} bottom) $\xrightarrow{SOCl_2,\ pyridine}$ (*R*)-2-chlorooctane (Cl left, H right)

(*S*)-2-octanol $\xrightarrow{p\text{-TosCl},\ pyridine}$ (*S*)-2-octyl-*p*-toluenesulfonate (H left, OTos right) $\xrightarrow{^{-}:OH}$ (*R*)-2-octanol (HO left, H right)

We saw in Problem 10.29 that treatment of a chiral secondary alcohol with hydroxide ion yields racemic product. To form optically pure (*R*)-2-octanol, the poor hydroxide leaving group must be converted to the very good tosyl leaving group. Reaction of (*S*)-2-octyl-*p*-toluenesulfonate with hydroxide ion proceeds with inversion (S_N2 mechanism) to give (*R*)-2-octanol.

20.25 This reaction is an intramolecular Williamson ether synthesis.

$ClCH_2CH_2CH_2CH_2\ddot{O}H \xrightarrow{NaH,\ ether} [ClCH_2CH_2CH_2CH_2\ddot{O}{:}^{-}\ Na^{+}] + H_2 \longrightarrow$ tetrahydrofuran (ring $CH_2CH_2CH_2CH_2O$) + NaCl

20.26 In these compounds we want to reduce some, but not all, of the functional groups present. To do this requires selection of the correct reducing agent.

a) $C_6H_5CH{=}CHCO_2H \xrightarrow[Rh/C]{H_2}$ cyclohexyl–$CH_2CH_2CO_2H$

b) $C_6H_5CH{=}CHCO_2H \xrightarrow[2.\ H_3O^+]{1.\ LiAlH_4} C_6H_5CH{=}CHCH_2OH$

c) $C_6H_5CH{=}CHCO_2H \xrightarrow[Pd/C]{H_2} C_6H_5CH_2CH_2CO_2H$

20.27 H_2C—CH_2 (ethylene oxide) + $\overset{\delta-}{:R}$-$\overset{\delta+}{MgX}$ ⟶ [R-CH_2CH_2-$\ddot{O}MgBr$] $\xrightarrow{H_3O^+}$ RCH_2CH_2OH

This reaction is a nucleophilic substitution.

20.28 The mechanism of Grignard addition to oxetane is the same as the mechanism in the previous problem. The reaction proceeds at a reduced rate because oxetane is less reactive than ethylene oxide. The four-membered ring oxetane is less strained, and therefore more stable, than the three-membered ethylene oxide ring.

20.29 [Mechanism scheme: alcohol + $POCl_3$ (pyr) ⟶ [dichlorophosphate ester + pyr:HCl ⟶ carbocation + $^{-}OP(O)Cl_2$] ⇌ [rearranged carbocation (methyl shift)] ⟶ alkene (CH_3) + H^+]

This is a carbocation rearrangement involving the shift of an alkyl group.

20.30

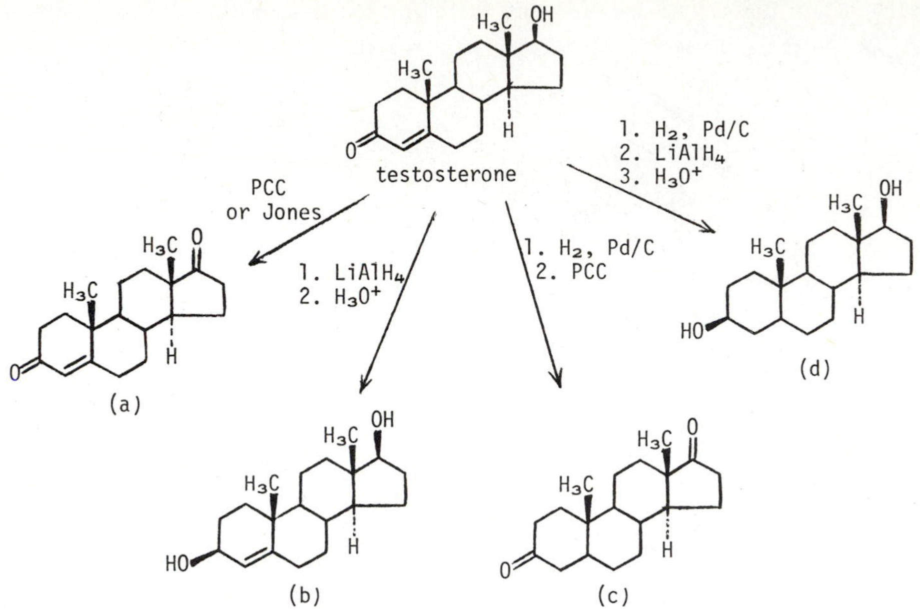

testosterone

PCC or Jones → (a)

1. $LiAlH_4$
2. H_3O^+ → (b)

1. H_2, Pd/C
2. PCC → (c)

1. H_2, Pd/C
2. $LiAlH_4$
3. H_3O^+ → (d)

20.31

$POCl_3$ pyridine (*syn*-elimination) ← trans-2-methylcyclopentanol → $POCl_3$ pyridine (*anti*-elimination)

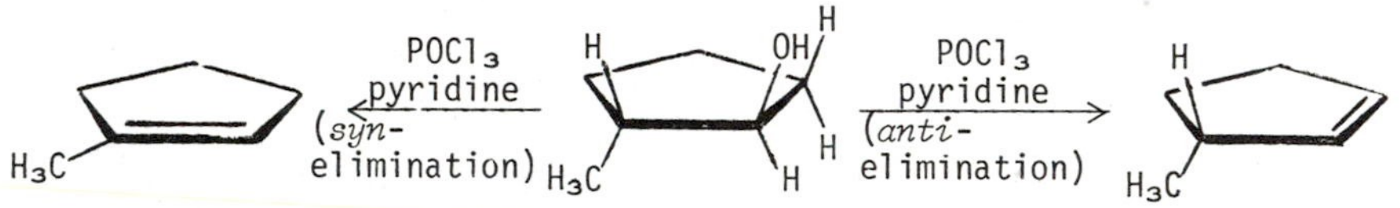

1-methylcyclopentene *trans*-2-methylcyclopentanol 3-methylcyclopentene

The more stable product of dehydration of *trans*-2-methylcyclopentanol is 1-methylcyclopentene, which can be formed only via *syn*-elimination. The product of *anti*-elimination is 3-methylcyclopentene. Since this product predominates, the requirement of *anti*-geometry must be more important than formation of the more stable product.

20.32

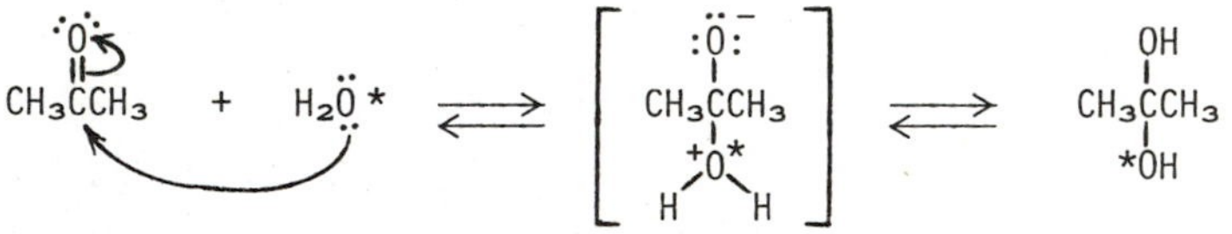

$$CH_3\overset{O}{\overset{\|}{C}}CH_3 + H_2O^* \rightleftharpoons \left[CH_3\overset{O^-}{C}(\overset{+}{O}^*H_2)CH_3 \right] \rightleftharpoons CH_3C(OH)(^*OH)CH_3$$

The mechanism of nucleophilic addition of water to a carbonyl carbon is similar to other mechanisms we have studied in this chapter. Since all of these steps are reversible, we can write the above mechanism in reverse to show how labeled oxygen is incorporated into acetone.

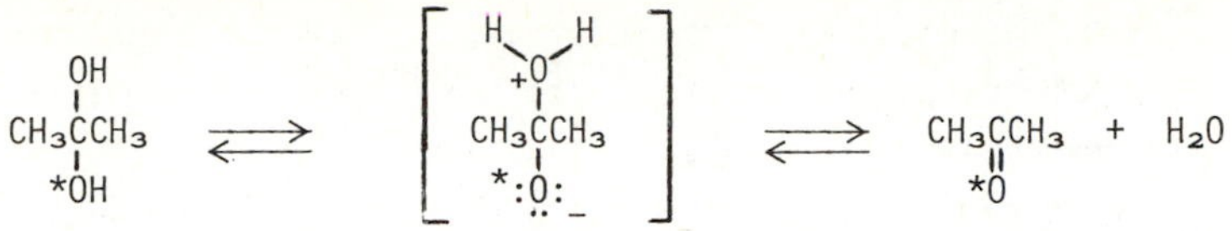

$$CH_3\underset{*OH}{\overset{OH}{C}}CH_3 \rightleftharpoons \left[CH_3\underset{*:\ddot{O}:^-}{\overset{H_2O^+}{C}}CH_3 \right] \rightleftharpoons CH_3\underset{*O}{\overset{}{C}}CH_3 + H_2O$$

This exchange is very slow in water but proceeds more rapidly when either acid or base is present.

20.33 We formulated a mechanism for this reaction in Problem 20.29. In this mechanism: (1) the poor -OH leaving group is converted to the much better dichlorophosphate leaving group; (2) departure of the leaving group produces a carbocation; (3) methyl migration from the adjacent carbon produces a more stable carbocation; and (4) elimination of H^+ gives the observed double bond.

When you make a model of epi-testosterone, you can see that the hydroxyl group and the adjacent methyl group have a *trans*-diaxial relationship to each other.

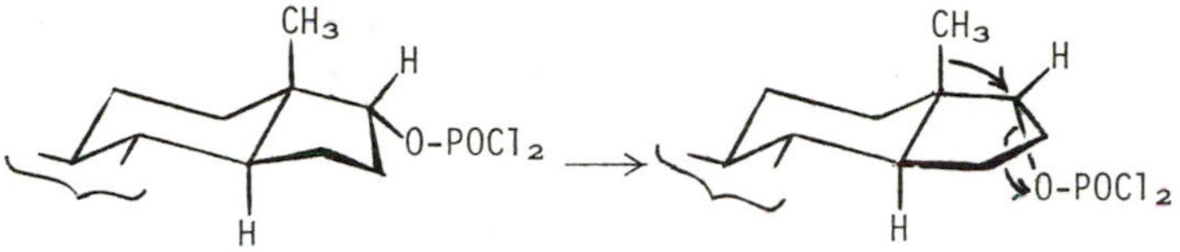

CH_3 H O-$POCl_2$ H → CH_3 H O-$POCl_2$ H

In this arrangement step (2), departure of -$OPOCl_2$, and step (3), methyl migration, can be concerted: the methyl group helps the -$OPOCl_2$ group to leave. Reaction of 17-epi-testosterone thus proceeds more rapidly than reaction of testosterone, which does not have such a favorable arrangement of migrating and leaving groups.

20.34

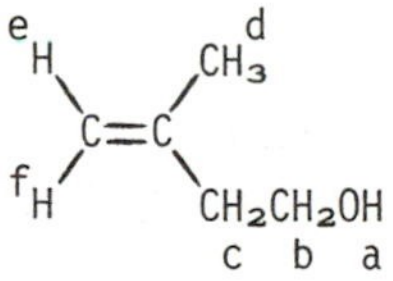

^eH, ^fH on $C{=}C$ with CH_3 (d) and CH_2CH_2OH (c b a) — 3-methyl-3-buten-1-ol

The peak absorbing at 1.75 δ (3H) is due to the d protons. This peak, which occurs in the allylic region of the spectrum, is unsplit.

The peak absorbing a 2.25 δ (3H) is due to protons c and a. The peak is a triplet because of splitting by the adjacent b protons. The absorption from proton a coincides with the center of the triplet and distorts the triplet from its usual 1:2:1 ratio of intensity.

The peak absorbing at 3.69 δ (2H) is due to the b protons. The adjacent electronegative oxygen causes the peak to be downfield, and the adjacent -CH_2- group splits the peak into a triplet.

The peak at 4.75 δ (2H) is due to protons e and f. These two protons are so similar that their absorptions overlap, and no splitting is observed.

20.35 1. $C_8H_{18}O_2$ has *no* double bonds or rings.

2. The IR band at 3350 cm^{-1} indicates the presence of a hydroxyl group.

3. The compound is probably symmetrical (simple NMR).

4. There is no splitting.

A structure that meets all of these criteria:

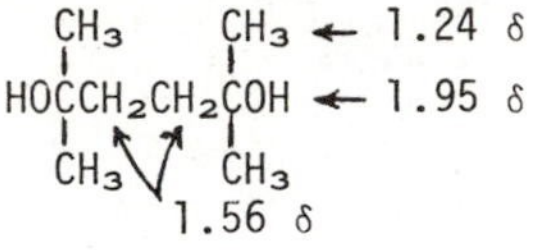

CH₃ CH₃ ← 1.24 δ
HOCCH₂CH₂COH ← 1.95 δ
CH₃ CH₃
1.56 δ

2,5-dimethyl-2,5-hexanediol

20.36

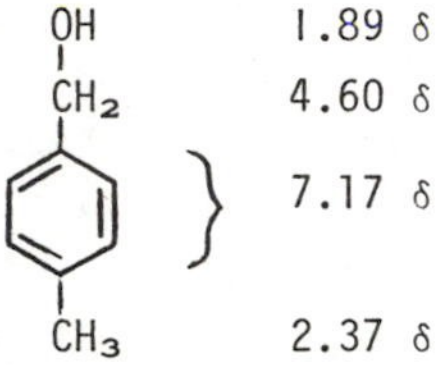

OH 1.89 δ
CH₂ 4.60 δ
(benzene ring) 7.17 δ
CH₃ 2.37 δ

p-methylbenzyl alcohol

20.37 Remember that the hydroboration-oxidation reaction proceeds with non-Markovnikov regiochemistry and *cis* stereochemistry; -H and -OH add to the same side of the double bond.

Make models and draw illustrations of the two possible products of hydroboration (only the first two rings). The product resulting from top face addition is:

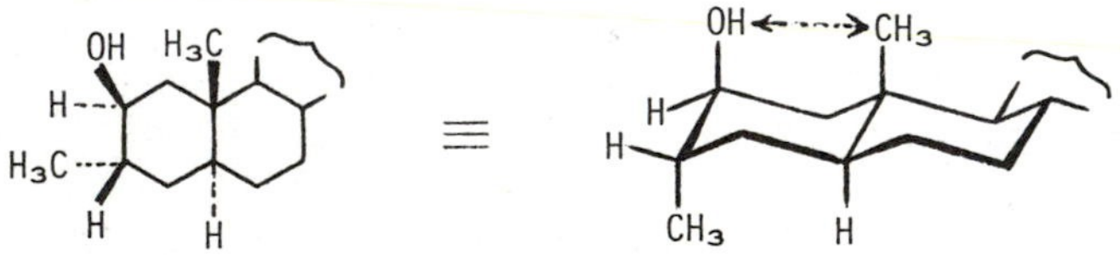

OH H₃C H H₃C H H ≡ OH ←--→ CH₃ H H CH₃ H

This product is disfavored because attack of the reagents is hindered by the presence of the axial methyl group on the same side of the ring. Addition from the bottom face is thus favored.

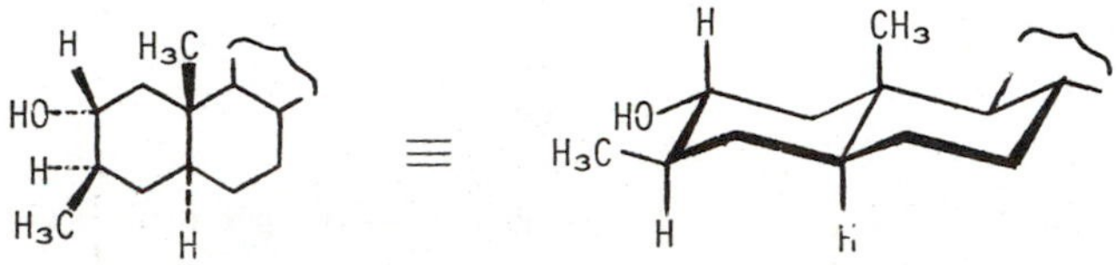

H H₃C HO H H₃C H ≡ H CH₃ HO H₃C H H

20.38

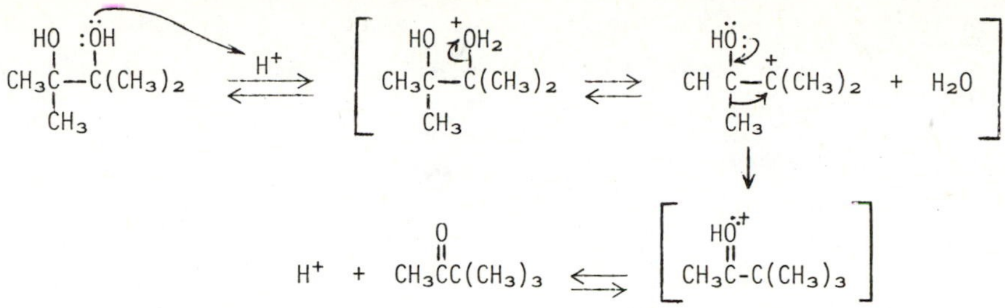

20.39 a) Compound A has one double bond or ring.

b) The infrared absorption at 3400 cm^{-1} indicates the presence of an alcohol. (The weak absorption at 1640 cm^{-1} is due to a C=C stretch.)

c) (1) The absorptions at 1.63 δ and 1.70 δ are due to unsplit methyl protons. Because the absorptions are shifted slightly downfield, the protons are adjacent to an unsaturated center.

(2) The broad singlet at 3.83 δ is due to an alcohol proton.

(3) The doublet at 4.15 δ is due to two protons attached to a carbon bearing an electronegative atom (oxygen, in this case).

(4) The proton absorbing at 5.70 δ is a vinylic proton.

d) 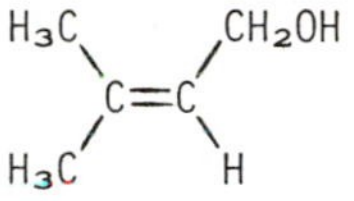

3-methyl-2-buten-1-ol

Compound A

20.40

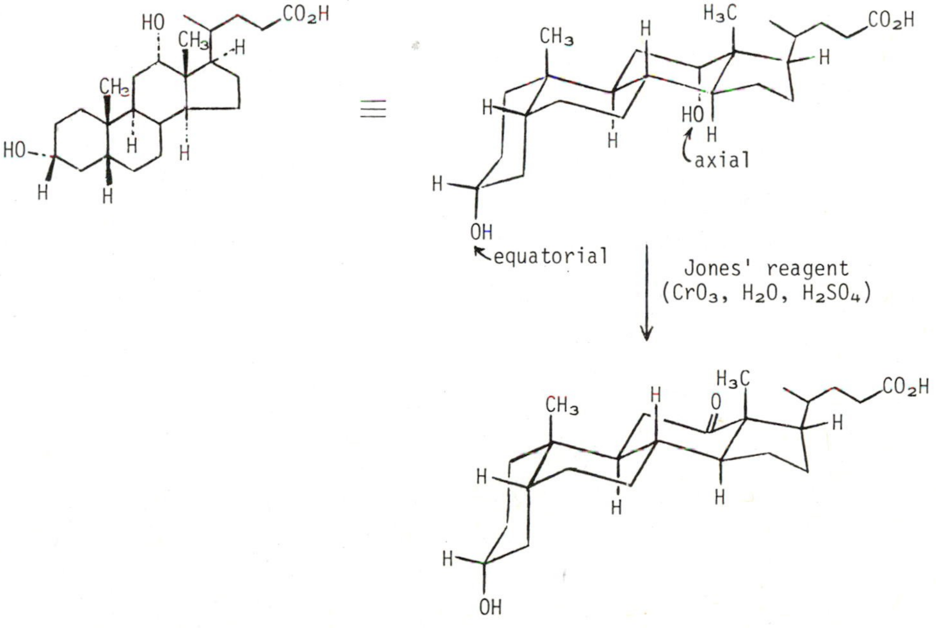

20.41

$$\text{cyclohexanone} \xrightarrow[\text{2. } H_3O^+]{\text{1. } LiAlH_4} \text{cyclohexanol (–OH)} \xrightarrow{PBr_3} \text{bromocyclohexane (–Br)} \xrightarrow{Mg} \text{cyclohexylmagnesium bromide (–MgBr)}$$

$$\xrightarrow[\text{2. } H_3O^+]{\text{1. cyclohexanone}} \text{1-cyclohexylcyclohexanol (OH)} \xrightarrow[\text{pyridine}]{POCl_3} \text{bicyclohexylidene}$$

bicyclohexylidene

20.42 $CH_3\text{-}S\text{-}S\text{-}CH_3$

What you should know:

After doing these problems, you should be able to:

1. Name and draw structures of alcohols and thiols;
2. Prepare alcohols from alkenes and from reaction of carbonyl compounds with Grignard reagents;
3. Formulate a mechanism for any nucleophilic addition reaction;
4. Predict the products of reactions involving alcohols.
5. Determine the structure of an alcohol from chemical and spectroscopic data.

CHAPTER 21 THE CHEMISTRY OF CARBONYL COMPOUNDS. AN OVERVIEW

21.1

$CH_3COCH_3 + {}^-:CN \rightleftharpoons [(H_3C)_2C(O^-)CN] \xrightarrow{H_2O} (H_3C)_2C(OH)CN + {}^-:OH$

21.2

cyclohexanone $\underset{}{\overset{H^+}{\rightleftharpoons}}$ [protonated cyclohexanone $\overset{:NH_2OH}{\rightleftharpoons}$ $HO\!-\!C\!-\!{}^+NH_2OH$ $\overset{-H^+}{\rightleftharpoons}$ $HO\!-\!C\!-\!NHOH$]

$\overset{H^+}{\rightleftharpoons}$ [$H_2\overset{+}{O}\!-\!C\!-\!NHOH$] $\rightleftharpoons$ [$H\!-\!\overset{+}{N}(OH)\!=\!$cyclohexylidene] + H_2O $\overset{-H^+}{\rightleftharpoons}$ $:N(OH)\!=\!$cyclohexylidene

21.3

cyclopentanone $\overset{H^+}{\rightleftharpoons}$ [protonated cyclopentanone $\overset{CH_3\ddot{N}HOH}{\rightleftharpoons}$ $HO\!-\!C\!-\!{}^+NH(CH_3)OH$ $\overset{-H^+}{\rightleftharpoons}$ $HO\!-\!C\!-\!N(CH_3)OH$]

$\overset{H^+}{\rightleftharpoons}$ [$H_2\overset{+}{O}\!-\!C\!-\!N(CH_3)OH$] $\rightleftharpoons$ [$H_3C\!-\!\overset{+}{N}(OH)\!=\!$cyclopentylidene] + H_2O $\overset{-H^+}{\rightleftharpoons}$ $H_3C\!-\!\overset{+}{N}(O^-)\!=\!$cyclopentylidene

This mechanism is very similar to that in Problem 21.2. Here, the proton lost in the final step comes from oxygen.

21.4 a)

$C_6H_5COOCH_3 + \overset{-}{C}H_3:\overset{+}{M}gBr \longrightarrow [C_6H_5C(O^-\ \overset{+}{M}gBr)(CH_3)OCH_3] \xrightarrow{H_3O^+} C_6H_5COCH_3 + CH_3OH$

b) [reaction scheme: acetophenone + $\bar{C}H_3:\overset{+}{M}gBr$ → [alkoxide–MgBr intermediate] $\xrightarrow{H_3O^+}$ 2-phenyl-2-propanol]

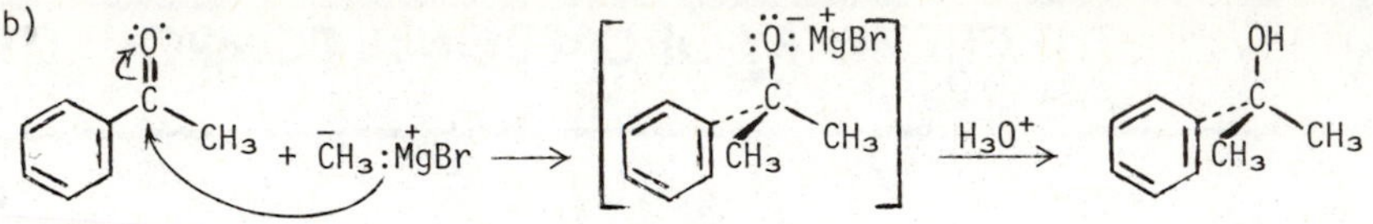

Step a) is a nucleophilic acyl substitution of methyl for methoxide, and step b) is a nucleophilic addition of Grignard reagent to form an alcohol.

21.5 [reaction scheme: methyl benzoate + $^-$:OH ⇌ [tetrahedral intermediate] ⇌ benzoic acid + $^-$:OCH$_3$ ⇌ benzoate anion + CH_3OH]

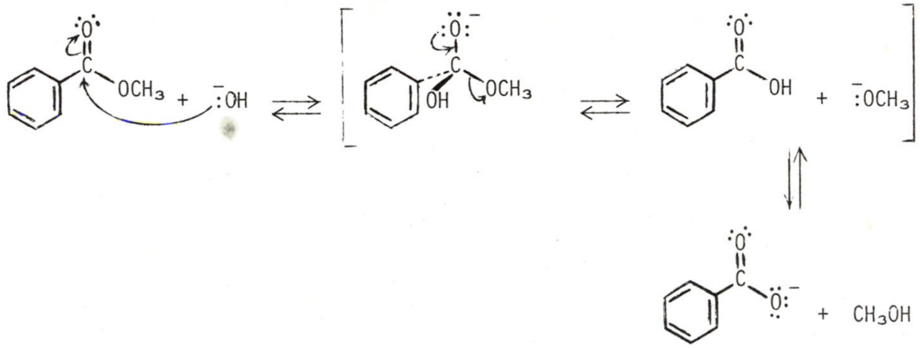

21.6 [reaction scheme: cyclohexanone + $^-$:OD ⇌ [enolate resonance forms] + HOD $\overset{D_2O}{\rightleftharpoons}$ 2-deuteriocyclohexanone + $^-$:OD]

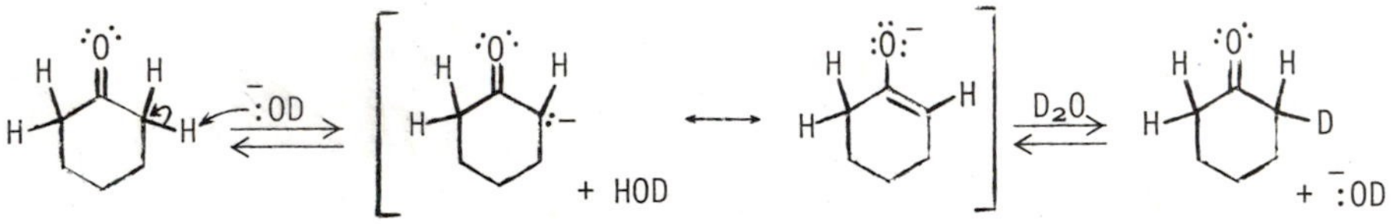

enolate

Treatment of cyclohexanone with deuteroxide forms cyclohexanone enolate. Reaction of the enolate with D_2O gives the monodeuterated product. Repetition of this sequence leads to replacement of all four alpha hydrogen atoms by deuterium atoms.

21.7 a) carboxylic acid, ester

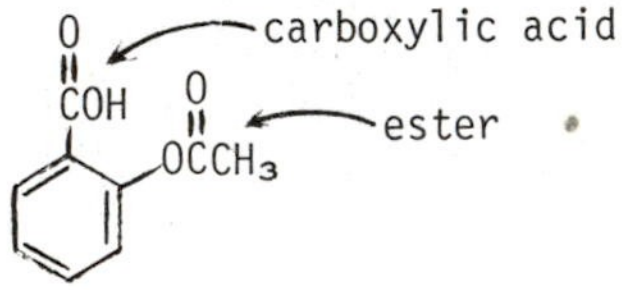

aspirin

b) ketone, ester

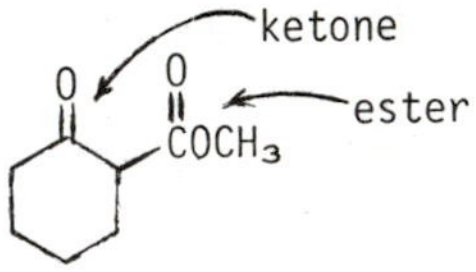

c) 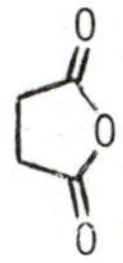 acid anhydride

d) lactam

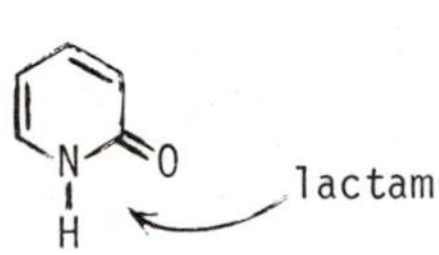

e) cocaine — esters

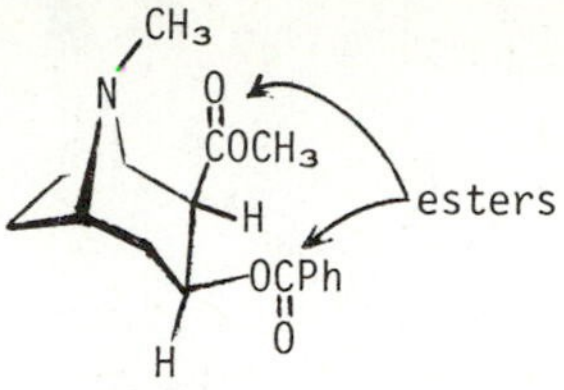

f) retinal — aldehyde

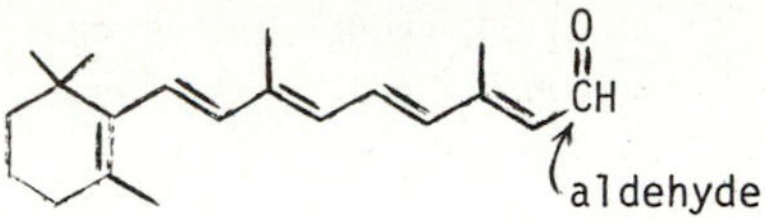

g) ketone, lactone

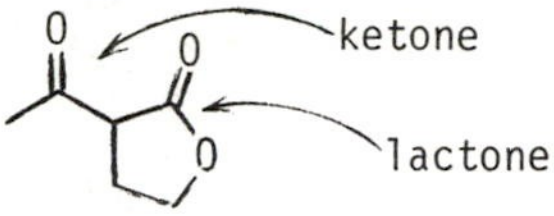

h) 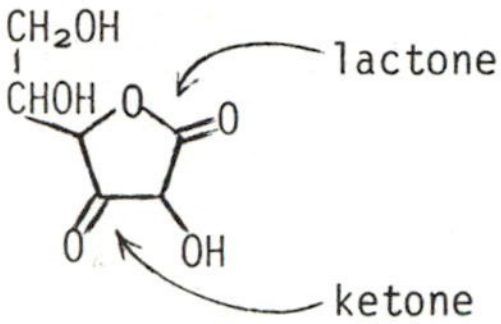 ascorbic acid — lactone, ketone

i) 6-aminopenicillanic acid — carboxylic acid, lactam

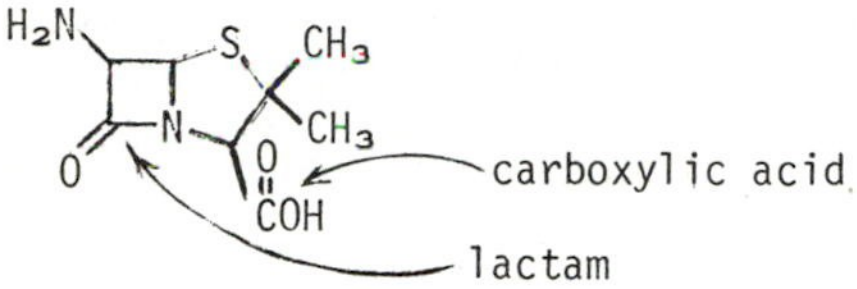

21.8

enolization — nucleophilic addition — protonation

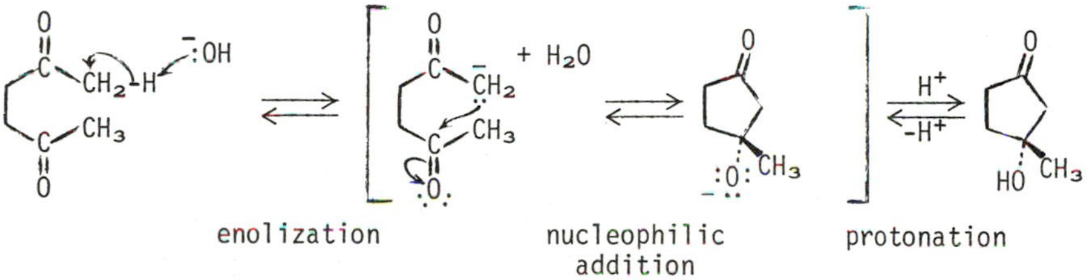

This sequence of reactions is known as the *aldol* condensation.

21.9

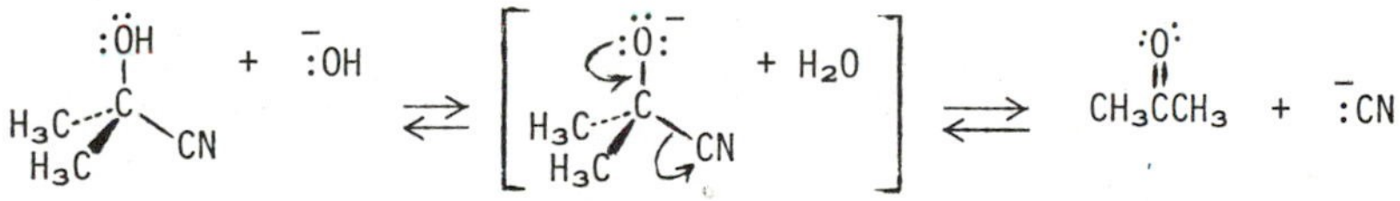

$:\ddot{O}H$ + $^{-}:OH$ ⇌ [... + H_2O] ⇌ $CH_3\overset{O}{\overset{\|}{C}}CH_3$ + $^{-}:CN$

The above steps are the reverse of Problem 21.1.

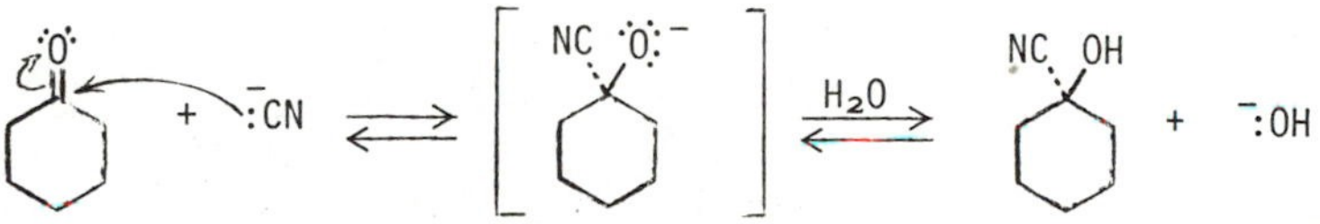

cyclohexanone + $^{-}:CN$ ⇌ [NC O^{-}] $\xrightarrow{H_2O}$ NC OH + $^{-}:OH$

This step is a nucleophilic addition of cyanide.

21.10 Even though the product looks unusual, this reaction is made up of steps with which you are familiar.

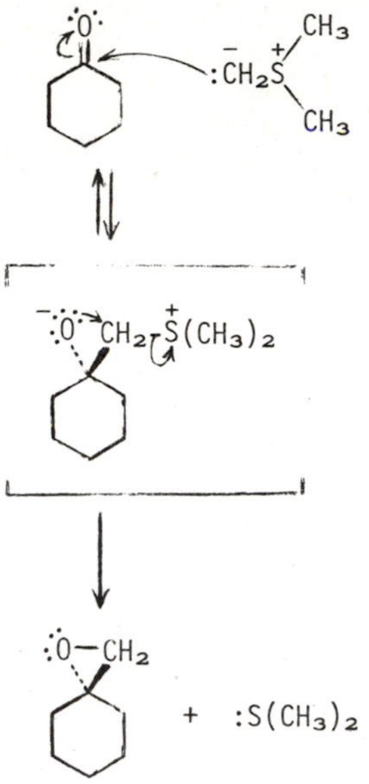

Nucleophilic addition of the ylide to a carbonyl group...

...is followed by S_N2 displacement of dimethyl sulfide by oxygen.

$:CH_2S(CH_3)_2$ $:S(CH_3)_2$

21.11 Ketones are relatively acidic because the proton alpha to the carbonyl can dissociate; the resulting *enolate* anion is stabilized by delocalization of the negative charge onto oxygen. Pictured below are the enolates of the three ketones.

acetone:

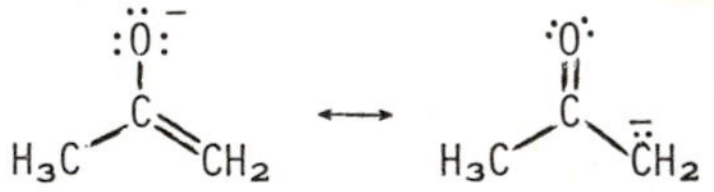

2,4-pentanedione:

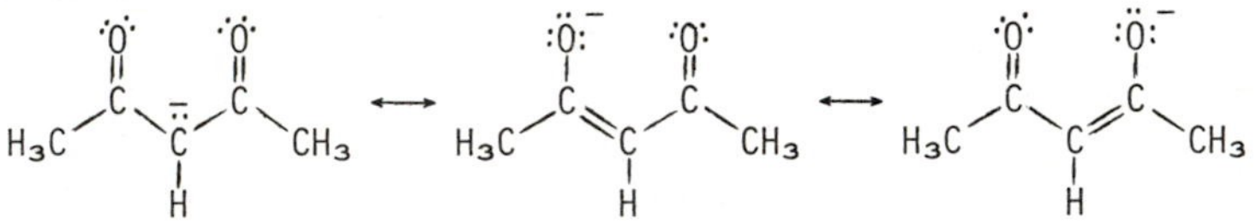

2,5-hexanedione:

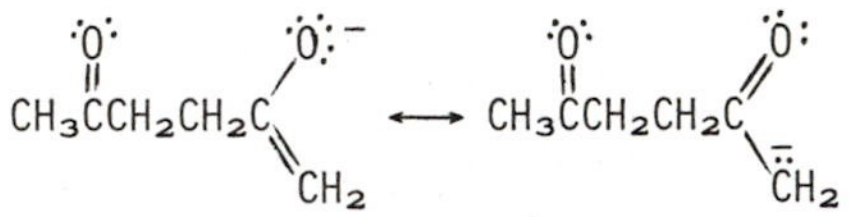

or

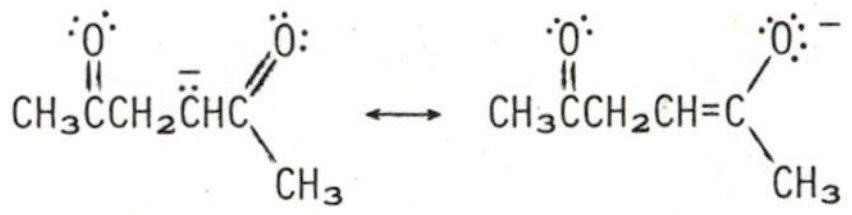

There are two resonance forms for the enolate of acetone and for each of the enolates of 2,5-hexanedione. For the enolate of 2,4-pentanedione, however, three resonance forms are possible. The negative charge resulting from dissociation of a proton alpha to both carbonyls can be stabilized by delocalization onto either of two oxygen atoms. Because this enolate is more stable, dissociation of a proton from 2,4-pentanedione is favored, and acidity is greater.

21.12 Most alcohols do not undergo elimination reactions when treated with hydroxide ion because no protons (other than the hydroxyl proton) are acidic enough to be removed by $^-$:OH. The alpha protons of β-hydroxyketones, however, are relatively acidic, and treatment with hydroxide forms an enolate anion.

$$(CH_3)_2C(OH)CH_2C(=O)CH_3 \underset{}{\overset{^-:OH}{\rightleftharpoons}} \left[(CH_3)_2C(OH)-\bar{C}HC(=O)CH_3 \right] + H_2O \rightleftharpoons (CH_3)_2C{=}CHC(=O)CH_3 + {}^-:OH$$

enolate enone

Elimination of hydroxide forms the conjugated keto-alkene.

What you should know:

After doing these problems, you should be able to:

1. Recognize the types of carbonyl functional groups;
2. Identify the reaction mechanism of a carbonyl-containing compound;
3. Write mechanisms for reactions involving the carbonyl functional group.

CHAPTER 22 ALDEHYDES AND KETONES. NUCLEOPHILIC ADDITION REACTIONS

22.1

a) $CH_3CH_2C(=O)CH(CH_3)_2$

2-methyl-3-pentanone

b) 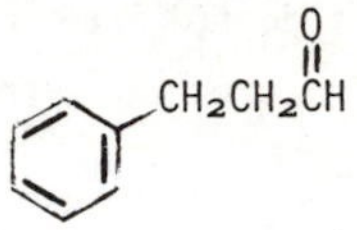 $C_6H_5CH_2CH_2CHO$

3-phenylpropanal

c) $CH_3C(=O)CH_2CH_2CH_2C(=O)CH_2CH_3$

2,6-octanedione

d) 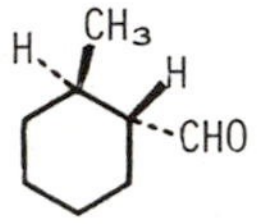 *trans*-2-methylcyclohexanecarbaldehyde

e) $HC(=O)CH_2CH_2CH_2CH(=O)$

pentanedial

f) 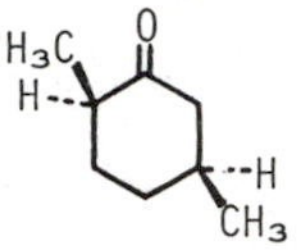 *cis*-2,5-dimethylcyclohexanone

g) $CH_3CH_2CH(CH_3)CH(CH_2CH_2CH_3)C(=O)CH_3$

4-methyl-3-propyl-2-hexanone

h) $CH_3CH{=}CHCH_2CH_2CHO$

4-hexenal

22.2

a) $H_2C{=}C(CH_3)CH_2CHO$

3-methyl-3-butenal

b) $CH_3CH(Cl)CH_2C(=O)CH_3$

4-chloro-2-pentanone

c) 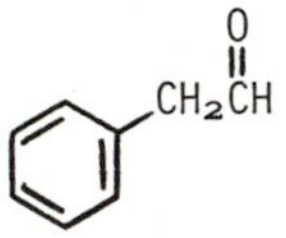 $C_6H_5CH_2CHO$

phenylacetaldehyde

d) 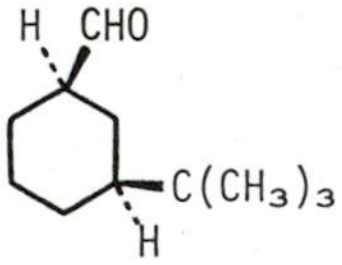 *cis*-3-*tert*-butylcyclohexanecarbaldehyde

e) $CH_3CH_2CH_2C(=O)CH(CH_3)CH_2OH$

1-hydroxy-2-methyl-3-hexanone

f) $CH_3CH_2CH(CH_3)CH_2CH_2CH(CHClCH_3)CHO$

2-(1-chloroethyl)-5-methylheptanal

22.3 a)

$$CH_3CH_2CH_2CH_2CH{=}CH_2 \xrightarrow[2.\ H_2O_2,\ OH^-]{1.\ BH_3} CH_3CH_2CH_2CH_2CH_2CH_2OH \xrightarrow[CH_2Cl_2]{PCC} CH_3CH_2CH_2CH_2CH_2\overset{O}{\overset{\|}{C}}H$$

b)

$$CH_3CH_2CH_2CH_2CH{=}CH_2 \xrightarrow[2.\ NaBH_4]{1.\ Hg(OAc)_2,\ H_2O} CH_3CH_2CH_2CH_2\overset{OH}{\overset{|}{C}}HCH_3 \xrightarrow[CH_2Cl_2]{PCC} CH_3CH_2CH_2CH_2\overset{O}{\overset{\|}{C}}CH_3$$

c)

$$CH_3CH_2CH_2CH_2CH{=}CH_2 \xrightarrow[2.\ Zn,\ H_3O^+]{1.\ O_3} CH_3CH_2CH_2CH_2\overset{O}{\overset{\|}{C}}H + CH_2O$$

d)

Benzene $\xrightarrow[AlCl_3]{CH_3\overset{O}{\overset{\|}{C}}Cl}$ $C_6H_5\overset{O}{\overset{\|}{C}}CH_3$

e)

Benzene $\xrightarrow[AlCl_3]{CH_3Cl}$ $C_6H_5CH_3$ $\xrightarrow{NBS}$ $C_6H_5CH_2Br$ $\xrightarrow{KOH}$ $C_6H_5CH_2OH$ $\xrightarrow[CH_2Cl_2]{PCC}$ $C_6H_5\overset{O}{\overset{\|}{C}}H$

f)

C_6H_5Br $\xrightarrow{Mg}$ C_6H_5MgBr $\xrightarrow[2.\ H_3O^+]{1.\ CH_2O}$ $C_6H_5CH_2OH$ $\xrightarrow[CH_2Cl_2]{PCC}$ $C_6H_5\overset{O}{\overset{\|}{C}}H$

22.4 An aromatic aldehyde is less reactive than an aliphatic aldehyde toward nucleophilic addition. The partial positive charge of the carbonyl carbon can be delocalized into the aromatic ring.

$C_6H_5-\overset{\delta-}{O}{=}\overset{\delta+}{C}-H$ ⟷ $C_6H_5-\overset{+}{C}H-O^-$ ⟷ ⟷ ⟷

The carbonyl carbon is thus less electron-poor and less reactive toward nucleophiles. In addition, the resonance stabilization of an aromatic aldehyde is lost when a nucleophile adds to the aldehyde functional group.

22.5

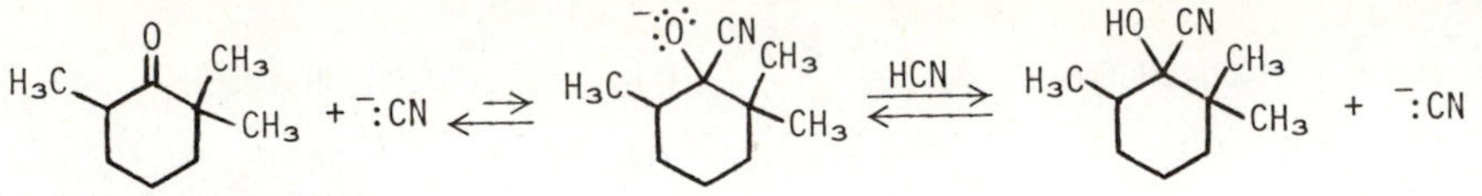

2,2,6-trimethyl-
cyclohexanone

Cyanohydrin formation is an equilibrium process in which the cyanide addition step is sterically sensitive. Addition of cyanide to 2,2,6-trimethylcyclohexanone is hindered by the three methyl groups, and the equilibrium lies far to the side of the unreacted ketone.

22.6

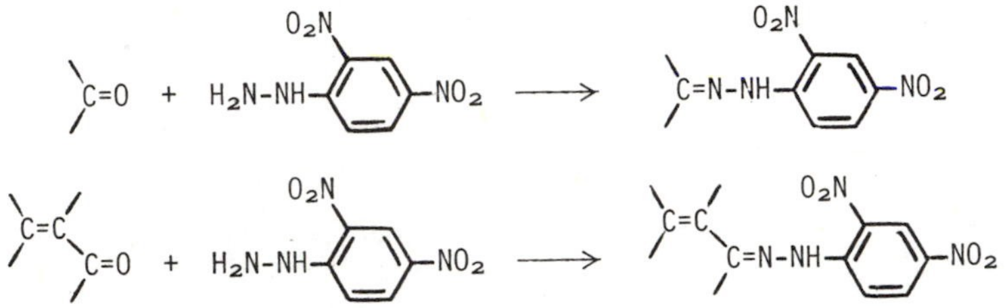

Both 2,4-DNPs have extended systems of conjugated double bonds, and both absorb light in the visible range of the electromagnetic spectrum. The 2,4-DNP of an α,β-unsaturated ketone has a longer system of conjugated double bonds, and absorption of light occurs at a lower energy (longer wavelength). For the 2,4-DNP of a saturated ketone, absorption occurs in the violet region of the visible spectrum (400-450 nm); the observed color is yellow. For the 2,4-DNP of an α,β-unsaturated ketone, the wavelength of absorption is shifted toward the blue-green region (450-500 nm); the observed color is red-orange.

22.7

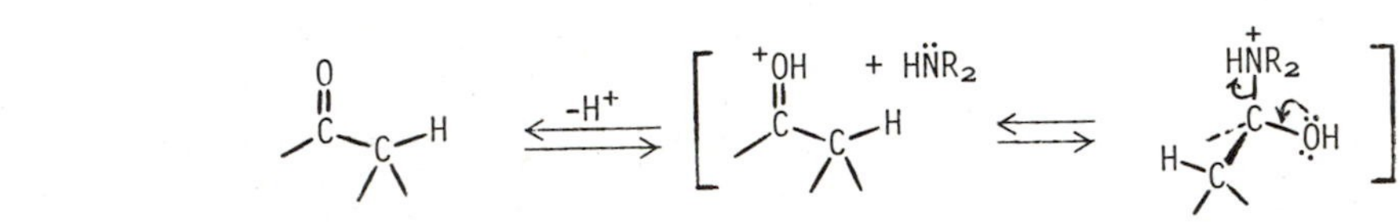

carbinolamine

Water is added to the enamine double bond. The nitrogen is then protonated and an amine is eliminated.

22.8 a)

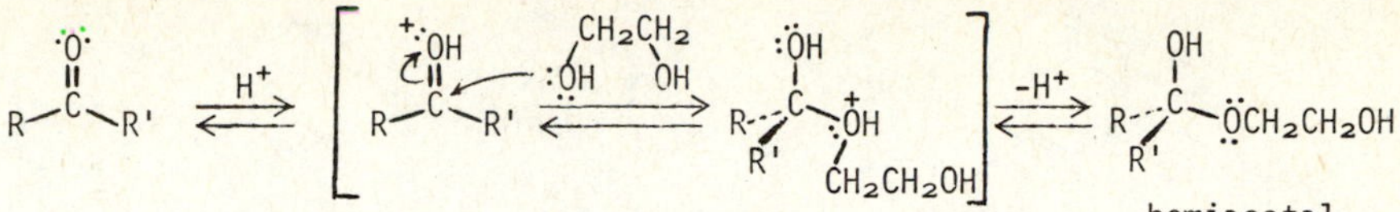

Formation of the hemiacetal is the first step.

hemiacetal

b)

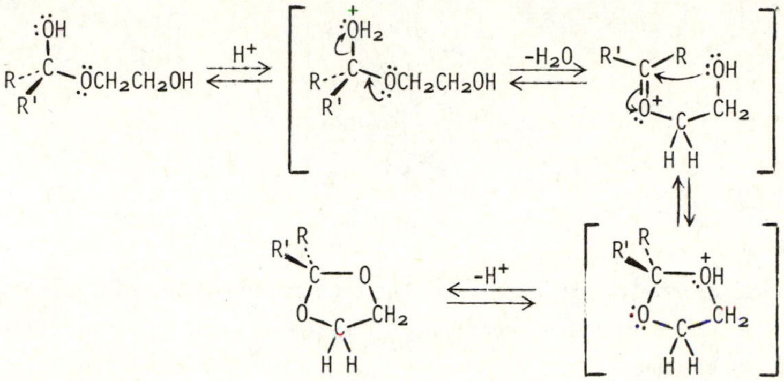

Protonation of the hemiacetal hydroxyl group is followed by loss of water. Attack by the second hydroxyl group of ethylene glycol forms the cyclic acetal ring.

22.9 In general, ketones are less reactive than aldehydes for both steric (excess crowding) and electronic reasons. If the keto-aldehyde in this problem were reduced with *one* equivalent of $NaBH_4$, the aldehyde functional group would be reduced in preference to the ketone.

For the same reason, reaction of the keto-aldehyde with *one* equivalent of ethylene glycol selectively forms the acetal of the aldehyde functional group. The ketone can then be reduced with $NaBH_4$ and the acetal protecting group cleaved.

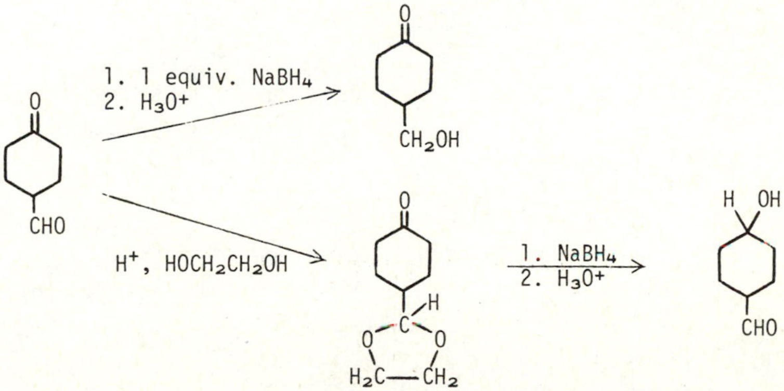

1. 1 equiv. $NaBH_4$
2. H_3O^+

H^+, $HOCH_2CH_2OH$

1. $NaBH_4$
2. H_3O^+

22.10 a) cyclohexanone $\xrightarrow[\text{2. } H_3O^+]{\text{1. } NaBH_4}$ cyclohexanol (H, OH) $\xrightarrow[\text{pyridine}]{POCl_3}$ cyclohexene $\xrightarrow[\text{Pd/C}]{H_2}$ cyclohexane

b) cyclohexanone $\xrightarrow[\text{KOH, } H_2O]{H_2NNH_2}$ cyclohexane (H, H) Wolff-Kishner reaction

c) cyclohexanone $\xrightarrow[H^+]{HSCH_2CH_2SH}$ dithioacetal ($H_2C{-}CH_2$, S, S) $\xrightarrow[\text{ethanol}]{\text{Raney Ni}}$ cyclohexane (H, H)

22.11 Because aldehydes are more reactive than ketones, the dithioacetal of the aldehyde is formed preferentially. Treatment with Raney nickel produces the desired ketone.

4-formylcyclohexanone (H–C=O) $\xrightarrow[H^+]{HSCH_2CH_2SH}$ 4-(1,3-dithiolan-2-yl)cyclohexanone (CH, S, S, $H_2C{-}CH_2$) $\xrightarrow[\text{ethanol}]{\text{Raney Ni}}$ 4-methylcyclohexanone (CH_3)

22.12 a)

cyclohexanone + $(C_6H_5)_3\overset{+}{P}\text{-}\overset{-}{C}HCH_3$ ⟶ ethylidenecyclohexane (=C(H)–CH_3) + $(C_6H_5)_3P{=}O$

(cyclohexylidene ylide ($\overset{-}{C}$–$\overset{+}{P}(C_6H_5)_3$) + CH_3CHO ⟶ ethylidenecyclohexane (=C(H)–CH_3) + $(C_6H_5)_3P{=}O$)

The second scheme is less satisfactory. The secondary halide used to form the ylide reacts slowly in the S_N2 displacement by triphenylphosphine.

b) $CH_3\overset{O}{\overset{\|}{C}}CH_3$ + $(C_6H_5)_3\overset{+}{P}\text{-}\overset{-}{C}HCH_2CH_2CH_3$ ⟶ $(CH_3)_2C{=}CHCH_2CH_2CH_3$ + $(C_6H_5)_3P{=}O$

2-methyl-2-hexene

c) 2-cyclohexenone + $(C_6H_5)_3\overset{+}{P}\text{-}\overset{-}{C}H_2$ ⟶ 3-methylenecyclohexene (=CH_2) + $(C_6H_5)_3P{=}O$

d)

$$C_6H_5\bar{\ddot{C}}H-\overset{+}{P}(C_6H_5)_3 + CH_3\overset{O}{\overset{\|}{C}}CH_3 \longrightarrow C_6H_5CH{=}C(CH_3)_2 + (C_6H_5)_3P{=}O$$

e)

$$C_6H_5\overset{O}{\overset{\|}{C}}-H + (C_6H_5)_3\overset{+}{P}-\bar{\ddot{C}}H(C_6H_5) \longrightarrow \textit{cis-}C_6H_5CH{=}CHC_6H_5 + \textit{trans-}C_6H_5CH{=}CHC_6H_5 + (C_6H_5)_3P{=}O$$

22.13 Suppose that tri*methyl*phosphine were to react with an alkyl halide.

$$(CH_3)_3P: \ + \ RCH_2CH_2X \longrightarrow (CH_3)_3\overset{+}{P}CH_2CH_2R \ X^-$$

phosphonium salt

Treatment of the phosphonium salt with strong base would yield two different ylides.

$$(CH_3)_3\overset{+}{P}CH_2CH_2R \ X^- \xrightarrow[\text{THF}]{\text{BuLi}} (CH_3)_3\overset{+}{P}-\bar{\ddot{C}}HCH_2R \quad \textit{and} \quad (CH_3)_2\overset{+}{P}(CH_2CH_2R)(\bar{:}CH_2)$$

Reaction of the ylides with a carbonyl compound would produce two different alkenes.

$$\gt C{=}O \ + \ (CH_3)_3\overset{+}{P}-\bar{\ddot{C}}HCH_2R \ \textit{or} \ (CH_3)_2\overset{+}{P}(CH_2CH_2R)(\bar{:}CH_2) \longrightarrow \gt C{=}CHCH_2R \ \textit{or} \ \gt C{=}CH_2$$

Another function of the phenyl groups of triphenylphosphine is stabilization of the ylide by delocalization of the positive charge on phosphorus.

22.14

$$2 \ (\beta\text{-ionylideneacetaldehyde}) \ + \ (C_6H_5)_3\overset{+}{P}-\bar{\ddot{C}}H-CH{=}C(CH_3)CH{=}CHCH{=}C(CH_3)-\bar{\ddot{C}}H-\overset{+}{P}(C_6H_5)_3 \longrightarrow \beta\text{-carotene}$$

β-ionylideneacetaldehyde

β-carotene

22.15

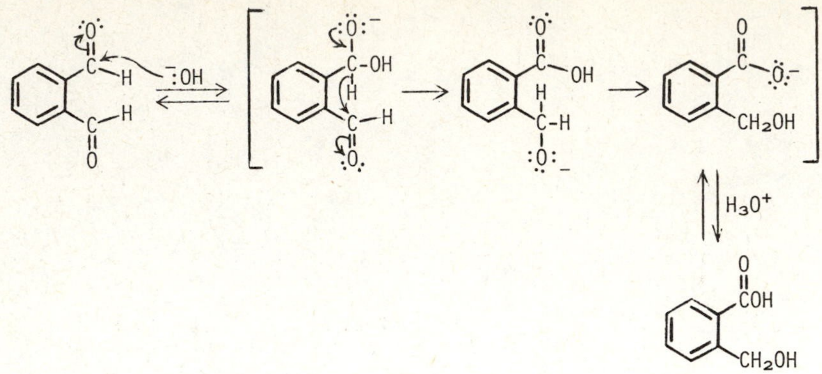

$\xrightarrow{H_3O^+}$

This is an internal Cannizzaro reaction.

22.16 a) $H_2C{=}CHCOCH_3 \xrightarrow[2.\ H_3O^+]{1.\ Li(CH_3CH_2CH_2)_2Cu} CH_3CH_2CH_2CH_2CH_2COCH_3$

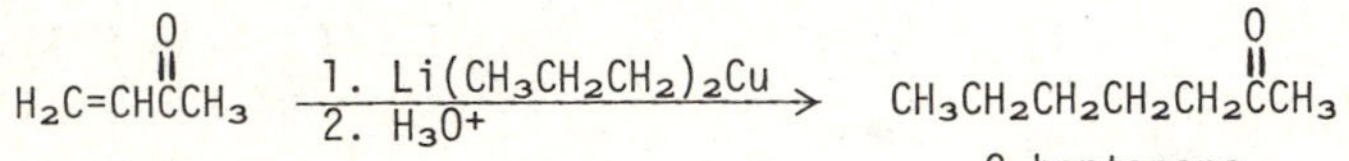

2-heptanone

b) $\xrightarrow[2.\ H_3O^+]{1.\ Li(CH_3)_2Cu}$

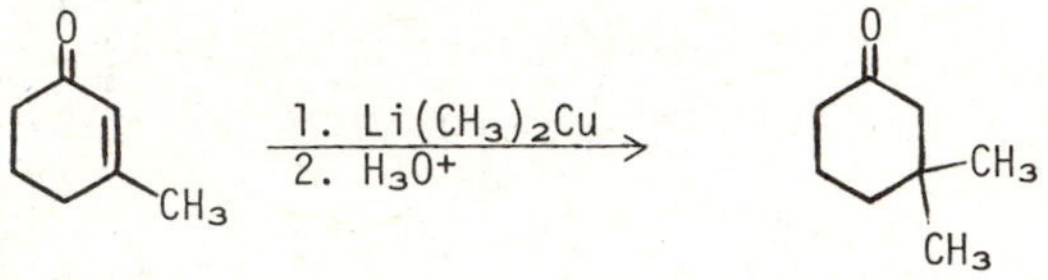

3,3-dimethylcyclohexanone

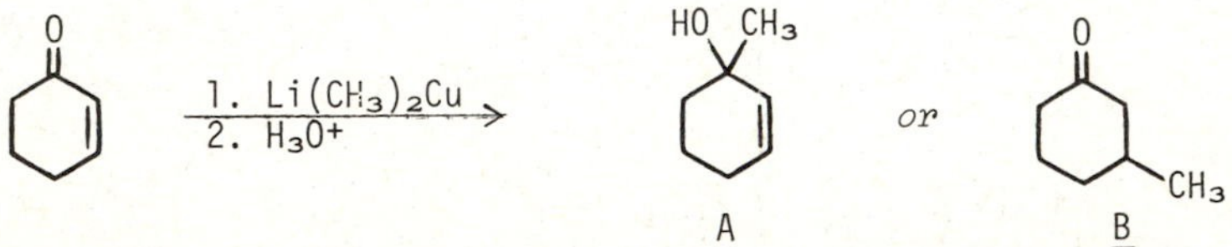

c) $\xrightarrow[2.\ H_3O^+]{1.\ Li(CH_3CH_2)_2Cu}$

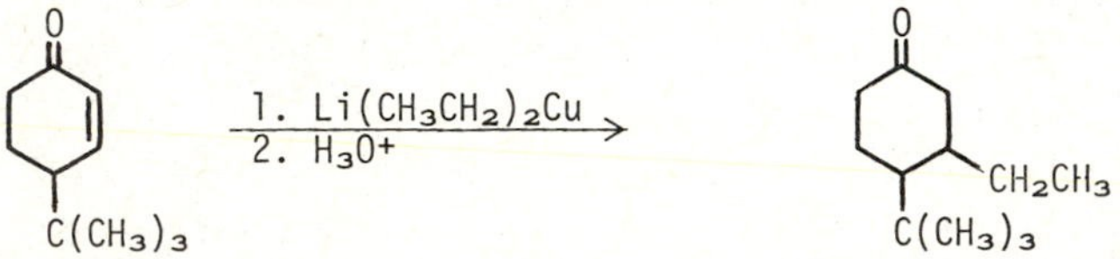

4-*tert*-butyl-3-ethylcyclohexanone

d) $\xrightarrow[2.\ H_3O^+]{1.\ Li(CH_2{=}CH)_2Cu}$

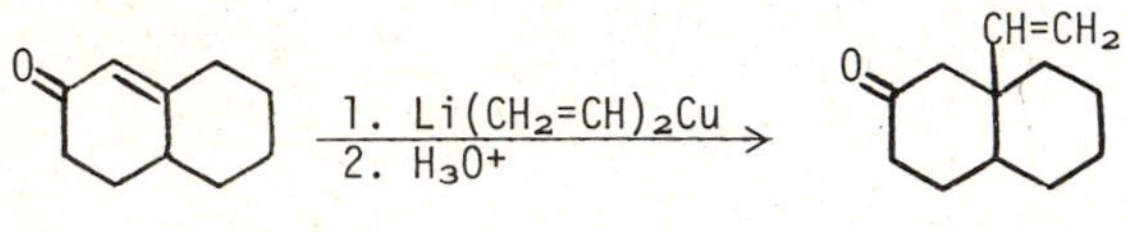

22.17 $\xrightarrow[2.\ H_3O^+]{1.\ Li(CH_3)_2Cu}$ A *or* B

2-cyclohexenone | 1-methyl-2-cyclohexen-1-ol (A) | 3-methylcyclohexanone (B)

2-Cyclohexenone is a cyclic α,β-unsaturated ketone whose carbonyl IR absorption occurs at 1690 cm^{-1}. If direct addition product A is formed, the carbonyl absorption will vanish and a hydroxyl absorption will appear at 3300 cm^{-1}. If conjugate addition produces B, the carbonyl absorption will shift to 1715 cm^{-1}, where 6-membered-ring saturated ketones absorb.

22.18 a) $H_2C{=}CHCH_2COCH_3$ absorbs at 1715 cm^{-1}. (4-penten-2-one is not an α,β-unsaturated ketone.)

b) $CH_3CH{=}CHCOCH_3$ is an α,β-unsaturated ketone and absorbs at 1685 cm^{-1}.

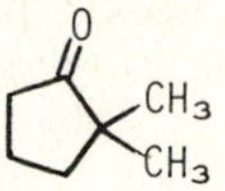

O
CH3
CH3

c

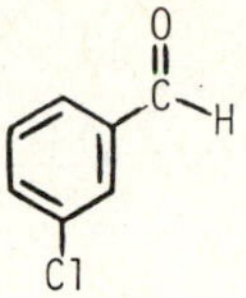

O
C
H
Cl

d

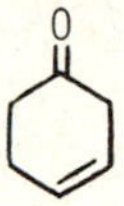

O

e

c) 2,2-Dimethylcyclopentanone (c) absorbs at 1750 cm^{-1} because it is a five-membered ring ketone.

d) *m*-Chlorobenzaldehyde shows a singlet absorption at 1705 cm^{-1} and a doublet at 2720 cm^{-1} and 2820 cm^{-1}. The aromatic ring absorbs at 1560 cm^{-1}.

e) 3-Cyclohexenone (e) absorbs at 1715 cm^{-1}.

22.19

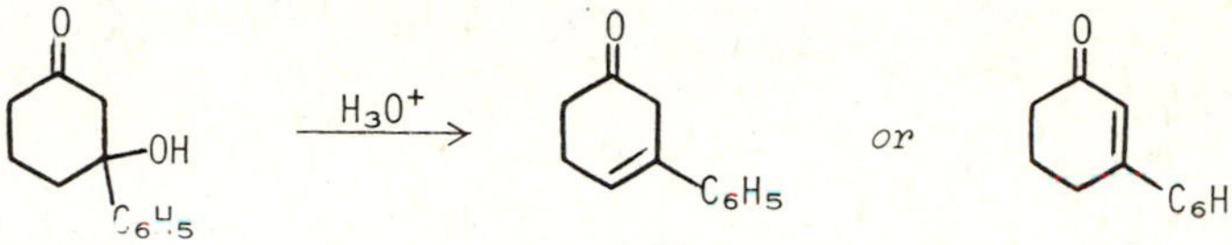

O
OH
C6H5
H3O+
O
C6H5
or
O
C6H5

A B

Compound A is a cyclic, non-conjugated keto-alkene whose infrared absorption should occur at 1715 cm^{-1}. Compound B is a conjugated, cyclic keto-alkene; additional conjugation with the phenyl ring should lower its IR absorption below 1685 cm^{-1}. Because the actual IR absorption occurs at 1670 cm^{-1}, B is the correct structure.

22.20 a)

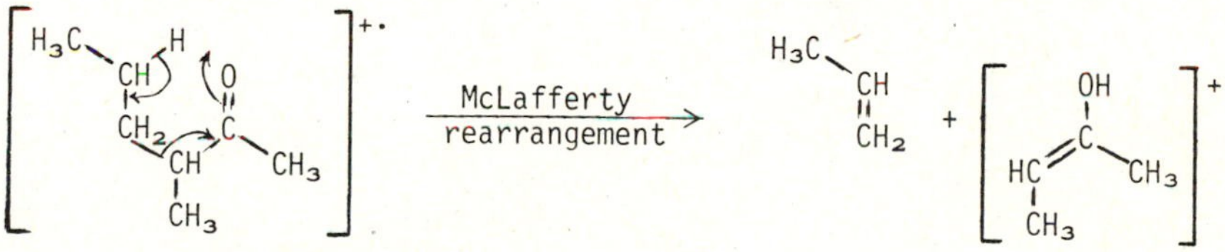

H3C
H
CH
O
CH2
C
CH
CH3
CH3
+·
McLafferty
rearrangement
H3C
CH
CH2
+
OH
C
HC
CH3
CH3
+·

m/*z* = 72

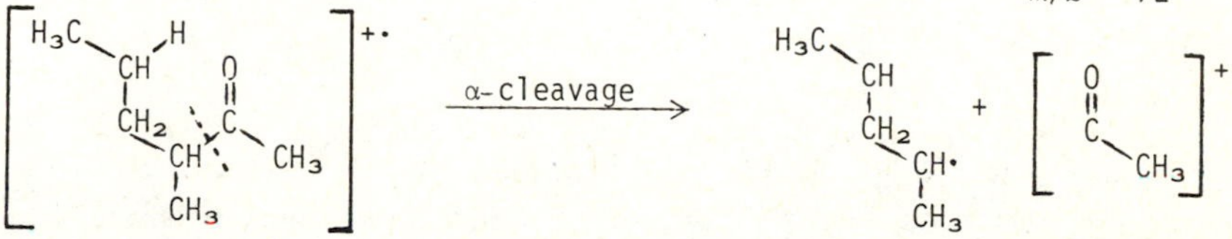

H3C
H
CH
O
CH2
C
CH
CH3
CH3
+·
α-cleavage
H3C
CH
CH2
CH·
CH3
+
O
C
CH3
+

m/*z* = 114

3-methyl-2-hexanone

m/*z* = 43

$[H_3C\text{–}CH(H)\text{–}CH(CH_3)\text{–}CH_2\text{–}C(=O)\text{–}CH_3]^{+\cdot}$ $\xrightarrow{\text{McLafferty rearrangement}}$ $H_3C\text{–}CH{=}CH\text{–}CH_3$ + $[H_2C{=}C(OH)\text{–}CH_3]^{+\cdot}$

m/z = 58

$[H_3C\text{–}CH(H)\text{–}CH(CH_3)\text{–}CH_2\text{–}C(=O)\text{–}CH_3]^{+\cdot}$ $\xrightarrow{\alpha\text{-cleavage}}$ $H_3C\text{–}CH_2\text{–}CH(CH_3)\text{–}\dot{C}H_2$ + $[O{=}C\text{–}CH_3]^{+}$

m/z = 114 *m/z* = 43

4-methyl-2-hexanone

Both isomers exhibit peaks at *m/z* = 43 due to α cleavage. The products of McLafferty rearrangement, however, occur at different values of *m/z*.

b) $[H_3C\text{–}CH(H)\text{–}CH_2\text{–}CH_2\text{–}C(=O)\text{–}CH_2CH_3]^{+\cdot}$ $\xrightarrow{\text{McLafferty rearrangement}}$ $H_3C\text{–}CH{=}CH_2$ + $[H_2C{=}C(OH)\text{–}CH_2CH_3]^{+\cdot}$

m/z = 72

$[H_3C\text{–}CH(H)\text{–}CH_2\text{–}CH_2\text{–}C(=O)\text{–}CH_2CH_3]^{+\cdot}$ $\xrightarrow{\alpha\text{-cleavage}}$ $H_3C\text{–}CH_2\text{–}CH_2\text{–}\dot{C}H_2$ + $[O{=}C\text{–}CH_2CH_3]^{+}$

m/z = 114 *m/z* = 57

3-heptanone

$[CH_2(H)\text{–}CH_2\text{–}CH_2\text{–}C(=O)\text{–}CH_2CH_2CH_3]^{+\cdot}$ $\xrightarrow{\text{McLafferty rearrangement}}$ $CH_2{=}CH_2$ + $[H_2C{=}C(OH)\text{–}CH_2CH_2CH_3]^{+\cdot}$

m/z = 86

$[CH_2(H)\text{–}CH_2\text{–}CH_2\text{–}C(=O)\text{–}CH_2CH_2CH_3]^{+\cdot}$ $\xrightarrow{\alpha\text{-cleavage}}$ $CH_3\text{–}CH_2\text{–}\dot{C}H_2$ + $[O{=}C\text{–}CH_2CH_2CH_3]^{+}$

m/z = 114 *m/z* = 71

4-heptanone

The isomers can be distinguished on the basis of both α-cleavage products (*m/z* = 57 *vs* *m/z* = 71) and McLafferty rearrangement products (*m/z* = 72 *vs* *m/z* = 86).

c) 3-methylpentanal:

$[H_3C\text{-}CH(H)\text{-}CH(CH_3)\text{-}CH_2\text{-}CHO]^{+\cdot}$ —McLafferty rearrangement→ $H_3C\text{-}CH{=}CH\text{-}CH_3$ + $[H_2C{=}C(OH)H]^{+\cdot}$

m/z = 43

$[H_3C\text{-}CH(H)\text{-}CH(CH_3)\text{-}CH_2\text{-}CHO]^{+\cdot}$ —α-cleavage→ $H_3C\text{-}CH_2\text{-}CH(CH_3)\text{-}\dot{C}H_2$ + $[O{=}C\text{-}H]^{+}$

m/z = 100 m/z = 29

3-methylpentanal

$[H_3C\text{-}CH(H)\text{-}CH_2\text{-}CH(CH_3)\text{-}CHO]^{+\cdot}$ —McLafferty rearrangement→ $H_3C\text{-}CH{=}CH_2$ + $[HC(CH_3){=}C(OH)H]^{+\cdot}$

m/z = 58

$[H_3C\text{-}CH(H)\text{-}CH_2\text{-}CH(CH_3)\text{-}CHO]^{+\cdot}$ —α-cleavage→ $H_3C\text{-}CH_2\text{-}CH_2\text{-}\dot{C}H\text{-}CH_3$ + $[O{=}C\text{-}H]^{+}$

m/z = 100 m/z = 29

2-methylpentanal

The fragments from McLafferty rearrangement, which occur at different values of m/z, serve to distinguish the two isomers.

22.21 a) CH_3COCH_2Br ($CH_3\overset{O}{\overset{\|}{C}}CH_2Br$)

b) 3,5-dinitrobenzaldehyde (CHO; O_2N, NO_2)

c) $CH_3CH_2CH_2CH_2\overset{O}{\overset{\|}{C}}CH(CH_3)_2$

d) 3,5-dimethylcyclohexanone (H_3C, CH_3)

e) $(CH_3)_3C\overset{O}{\overset{\|}{C}}C(CH_3)_3$

f) $CH_3\overset{CH_3}{\overset{|}{C}}{=}CCH\overset{O}{\overset{\|}{C}}CH_3$

g) $H\overset{O}{\overset{\|}{C}}CH_2CH_2\overset{O}{\overset{\|}{C}}H$

h) $C_6H_5\text{-}CH{=}CH\overset{O}{\overset{\|}{C}}H$

i) 6,6-dimethyl-2,4-cyclohexadienone (H_3C, H_3C, O)

j) $O_2N\text{-}C_6H_4\text{-}\overset{O}{\overset{\|}{C}}CH_3$

k) $O{=}C(H)\text{-}C(H)(H_3C)(OH)$

l) tetraphenylcyclopentadienone: (C_6H_5), (C_6H_5), (C_5H_5), (C_6H_5)

m) Fischer projection: CHO (O=C–H) at top; HO–C–H; H–C–OH; CH_2OH at bottom

22.22

$CH_3CH_2CH_2CH_2\overset{O}{\overset{\|}{C}}H$ pentanal

$CH_3CH_2\underset{CH_3}{\underset{|}{C}}H\overset{O}{\overset{\|}{C}}H$ 2-methylbutanal

$CH_3\overset{CH_3}{\overset{|}{C}}HCH_2\overset{O}{\overset{\|}{C}}H$ 3-methylbutanal

$(CH_3)_3C\overset{O}{\overset{\|}{C}}H$ 2,2-dimethylpropanal

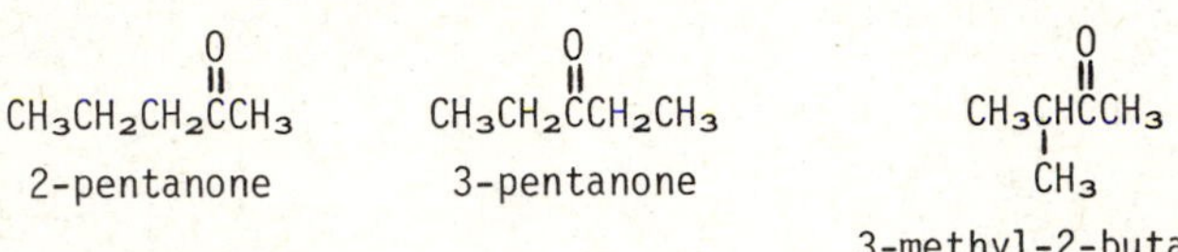

22.23 a) 3-methyl-3-cyclohexenone

b) (2*R*)-2,3-dihydroxypropanal (D-glyceraldehyde)

c) 5-isopropyl-2-methyl-2-cyclohexenone

d) 2-methyl-3-pentanone

e) 3-hydroxybutanal

f) *p*-benzenedicarbaldehyde

22.24 a) The α,β-unsaturated ketone C_6H_8O contains one ring, according to multiple bond and/or ring analysis. Possible structures:

O ; O, CH_3 ; O, CH_3 ; O, H_3C ; H_3C, O

b) $CH_3\overset{O}{\overset{\|}{C}}-\overset{O}{\overset{\|}{C}}CH_3$ and many other structures.

c) $\overset{O}{\overset{\|}{C}}CH_3$, CH_3 ; $\overset{O}{\overset{\|}{C}}CH_3$, CH_3 ; $\overset{O}{\overset{\|}{C}}CH_3$, H_3C ; $\overset{O}{\overset{\|}{C}}CH_2CH_3$

22.25 a)

$C_6H_5CH_2\overset{O}{\overset{\|}{C}}H \xrightarrow[2.\ H_3O^+]{1.\ NaBH_4} C_6H_5CH_2CH_2OH$

phenylacetaldehyde

b)

$C_6H_5CH_2\overset{O}{\overset{\|}{C}}H \xrightarrow[NH_4OH]{Ag_2O} C_6H_5CH_2\overset{O}{\overset{\|}{C}}OH$

c) $C_6H_5CH_2CHO \xrightarrow[H^+]{:NH_2OH} C_6H_5CH_2CH{=}N{-}OH$

d) $C_6H_5CH_2CHO \xrightarrow[2.\ H_3O^+]{1.\ CH_3MgBr} C_6H_5CH_2CH(OH)CH_3$

e) $C_6H_5CH_2CHO \xrightarrow[H^+]{CH_3OH} C_6H_5CH_2CH(OCH_3)OCH_3$

f) $C_6H_5CH_2CHO \xrightarrow[KOH]{H_2NNH_2} C_6H_5CH_2CH_3$

g) $C_6H_5CH_2CHO \xrightarrow{(C_6H_5)_3\overset{+}{P}{-}\overset{-}{\ddot{C}}H_2} C_6H_5CH_2CH{=}CH_2$

h) $C_6H_5CH_2CHO \xrightarrow[KCN]{HCN} C_6H_5CH_2CH(CN)OH$

i) $C_6H_5CH_2CHO \xrightarrow[2.\ H_3O^+]{1.\ \overset{+}{Na}:\overset{-}{C}{\equiv}CH} C_6H_5CH_2CH(OH)C{\equiv}CH$

22.26 a) $C_6H_5CCH_3$ (O over C) $\xrightarrow[2.\ H_3O^+]{1.\ NaBH_4} C_6H_5CH(OH)CH_3$

acetophenone

b) $C_6H_5COCH_3 \xrightarrow[NH_4OH]{Ag_2O}$ no reaction

c) $C_6H_5COCH_3 \xrightarrow[H^+]{:NH_2OH} C_6H_5C(=\ddot{N}{-}OH)CH_3$

d) $C_6H_5COCH_3 \xrightarrow[2.\ H_3O^+]{1.\ CH_3MgBr} C_6H_5C(OH)(CH_3)_2$

e) $C_6H_5COCH_3 \xrightarrow[H^+]{CH_3OH} C_6H_5C(OCH_3)_2CH_3$

f) $C_6H_5COCH_3 \xrightarrow[KOH]{H_2NNH_2} C_6H_5CH_2CH_3$

g) $C_6H_5COCH_3 \xrightarrow{(C_6H_5)_3\overset{+}{P}-\ddot{\overline{C}}H_2} C_6H_5C(=CH_2)CH_3$

h) $C_6H_5COCH_3 \xrightarrow[KCN]{HCN} C_6H_5C(OH)(CN)CH_3$

i) $C_6H_5COCH_3 \xrightarrow[2.\ H_3O^+]{1.\ \overset{+}{Na}:\overline{C}\equiv CH} C_6H_5C(OH)(CH_3)-C\equiv CH$

22.27 a) cyclohexenone $\xrightarrow[KOH]{H_2NNH_2}$ cyclohexene

b) cyclohexenone $\xrightarrow[2.\ H_3O^+]{1.\ Li(C_6H_5)_2Cu}$ 3-C_6H_5-cyclohexanone

c) cyclohexenone $\xrightarrow[2.\ H_3O^+]{1.\ (C_2H_5)_2AlCN}$ 3-CN-cyclohexanone $\xrightarrow{H_3O^+}$ 3-COOH-cyclohexanone

d) cyclohexenone $\xrightarrow[2.\ H_3O^+]{1.\ Li(CH_3)_2Cu}$ 3-CH_3-cyclohexanone $\xrightarrow[KOH]{H_2NNH_2}$ methylcyclohexane (CH_3)

or

cyclohexenone $\xrightarrow{(C_6H_5)_3\overset{+}{P}-\ddot{\overline{C}}H_2}$ (CH_2=) methylenecyclohexene $\xrightarrow[Pd/C]{H_2}$ methylcyclohexane (CH_3)

22.28 Glucose exists mainly as the cyclic hemiacetal, formed by the addition of the hydroxyl group of carbon atom 5 to the aldehyde functional group at carbon 1. A small amount of the open-chain aldehyde is in equilibrium with the cyclic hemiacetal. The open-chain aldehyde gives a positive Tollens' test, like any other aldehyde.

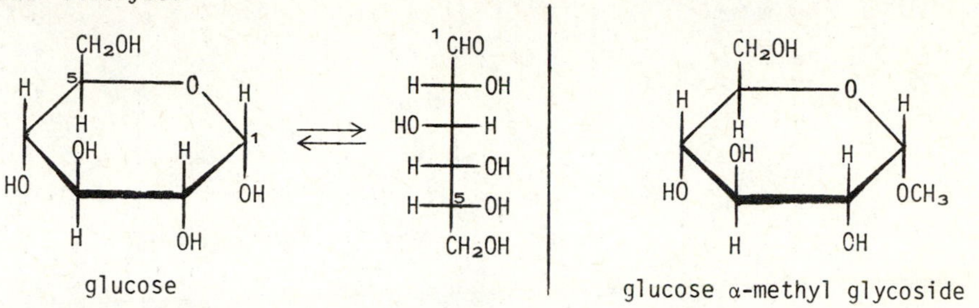

glucose

glucose α-methyl glycoside

Glucose α-methyl glycoside, an acetal, is not in equilibrium with an open-chain aldehyde and does not react with Tollens reagent.

22.29 a) Nucleophilic addition of hydrazine to one of the carbonyl groups, followed by elimination of water, produces a hydrazone.

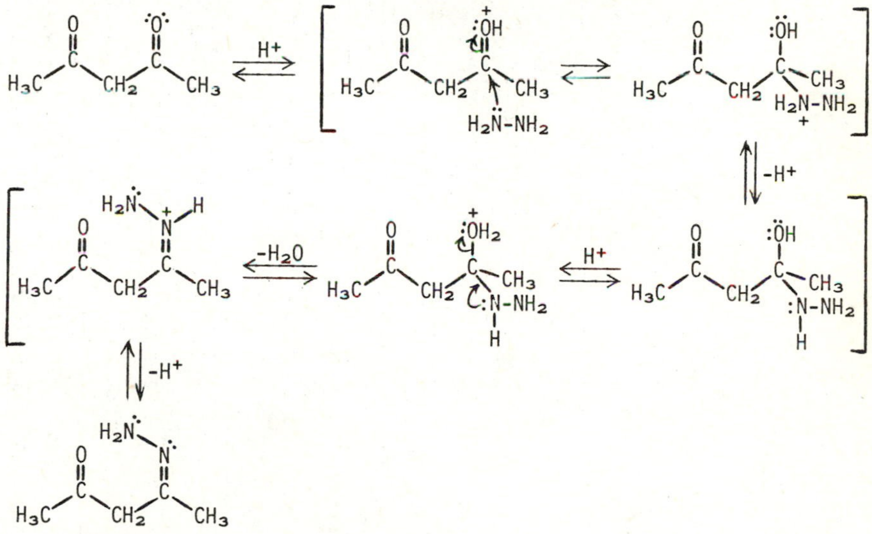

hydrazone

b) In a similar manner, the *other* nitrogen of hydrazine can add to the *other* carbonyl oxygen of 2,4-pentanedione to form the pyrazole.

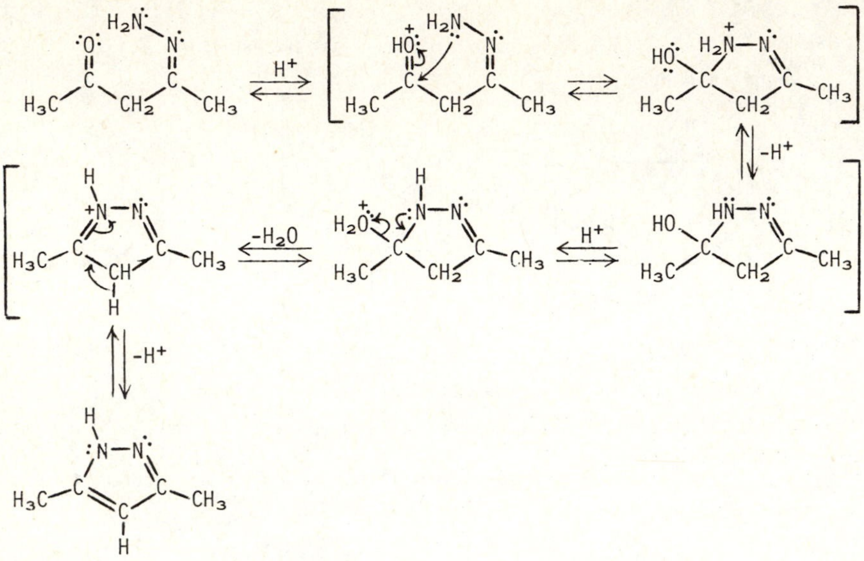

H^+ $-H^+$ $-H_2O$ H^+ $-H^+$

22.30 The same sequence of steps used in the previous problem leads to the formation of 3,5-dimethylisoxazole when hydroxylamine is the reagent.

a)

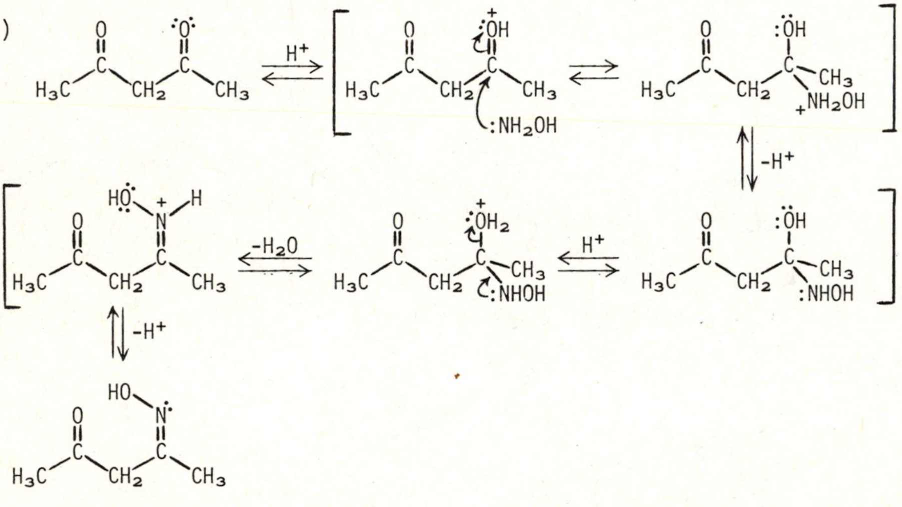

H^+ $:NH_2OH$ $-H^+$ H^+ $-H_2O$ $-H^+$

b) [reaction mechanism scheme: 2,4-pentanedione monoxime ⇌ (H⁺) protonated intermediates ⇌ cyclization ⇌ (−H⁺) ⇌ (H⁺) ⇌ (−H₂O) ⇌ (−H⁺) 3,5-dimethylisoxazole]

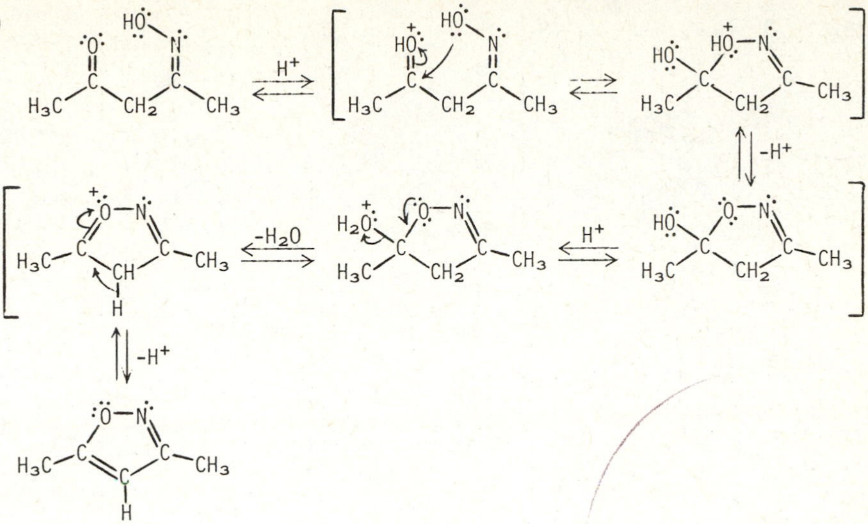

22.31 Remember:

RCH_2-X	+	$(C_6H_5)_3P$:	⟶	$(C_6H_5)_3\overset{+}{P}CH_2R\ X^-$
alkyl halide		triphenyl phosphine		phosphonium salt
$(C_6H_5)_3\overset{+}{P}CH_2R\ X^-$	+	$CH_3CH_2CH_2\overset{=}{C}H_2Li^+$	⟶	$(C_6H_5)_3\overset{+}{P}$-$\overset{..}{C}HR$
phosphonium salt		butyllithium		ylide
$(C_6H_5)_3\overset{+}{P}$-$\overset{..}{C}HR$	+	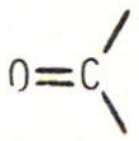 O=C<	⟶	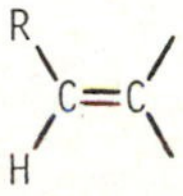 R(H)C=C<
ylide		carbonyl		alkene

	alkyl halide	carbonyl	product
a	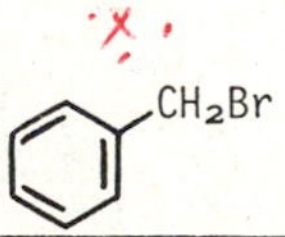 $C_6H_5CH_2Br$	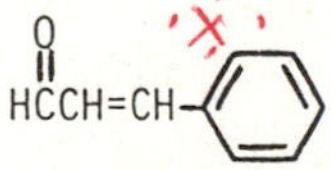 $HCCH=CH-C_6H_5$ (with C=O)	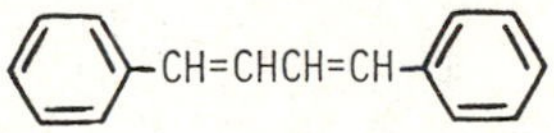 C_6H_5-CH=CHCH=CH-C_6H_5
b	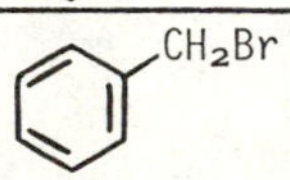 $C_6H_5CH_2Br$	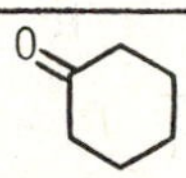 cyclohexanone	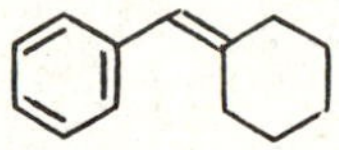 benzylidenecyclohexane
c	CH_3Br	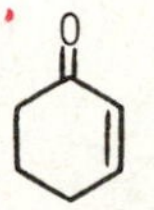 2-cyclohexen-1-one	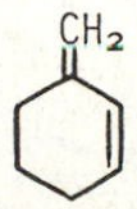 3-methylenecyclohexene (CH_2)
d	CH_3Br	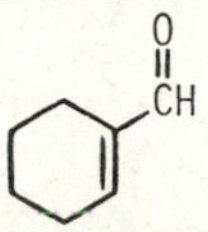 1-cyclohexenecarbaldehyde (CH=O)	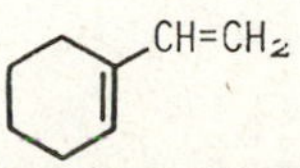 1-cyclohexenyl-CH=CH_2

22.32 a) $(C_6H_5)_3P: + Br\text{-}CH_2OCH_3 \longrightarrow (C_6H_5)_3\overset{+}{P}CH_2OCH_3\ Br^- \xrightarrow{BuLi} (C_6H_5)_3\overset{+}{P}\text{-}\overset{-}{C}HOCH_3 + LiBr$

b) [mechanism scheme: acid-catalyzed hydrolysis of the enol ether to cyclohexanecarbaldehyde; H^+, $:OH_2$, $-H^+$, H^+, $-H^+$; + CH_3OH]

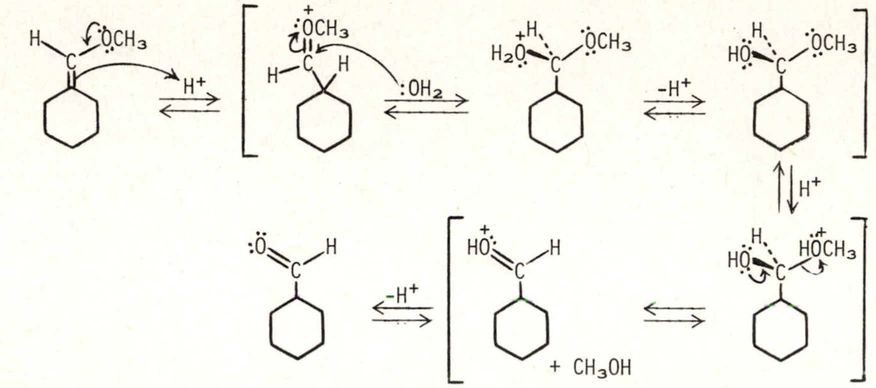

22.33 a) [mechanism scheme: HÖ–H_2C–H_2C–CH_2–CH=O, H^+, $-H^+$ → cyclic hemiacetal]

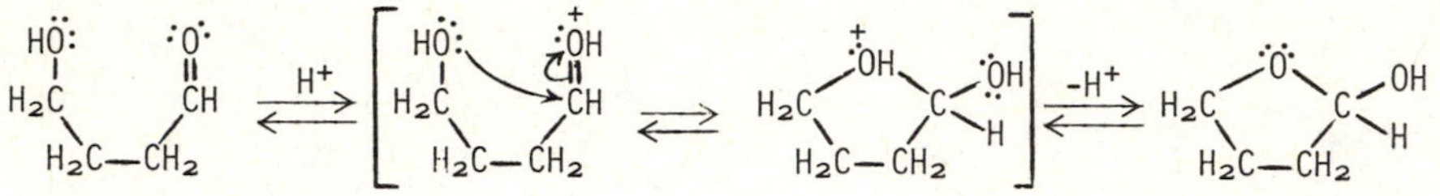

hemiacetal

b) [mechanism scheme: H^+, $-H_2O$, $HÖCH_3$, $-H^+$ → 2-methoxy cyclic acetal]

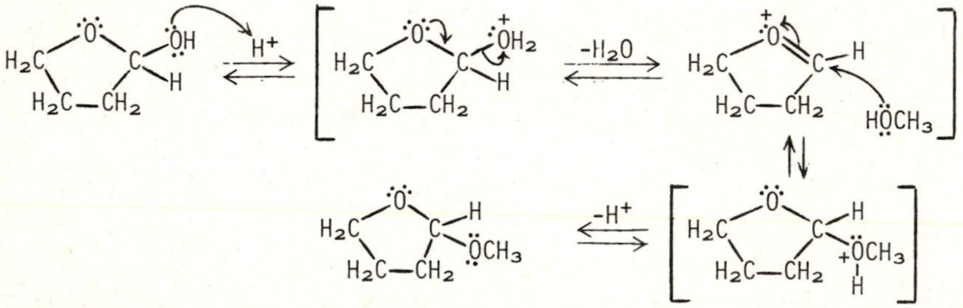

2-Methoxytetrahydropyran is a cyclic acetal. The hydroxyl oxygen of 4-hydroxybutanal reacts with the aldehyde to form the cyclic ether linkage.

22.34 [mechanism scheme: H_2O, $+ {}^-:OH$, $+ H_3O^+$, rotation, $:OH_2$, $+ H_3O^+$, $+ {}^-:OH$]

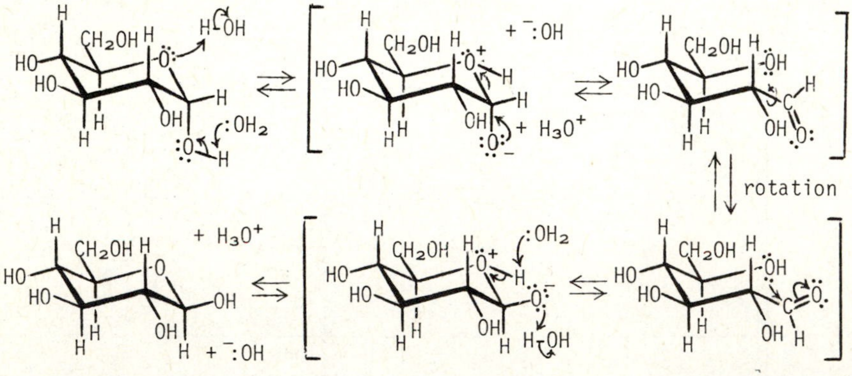

In this series of equilibrium steps, the hemiacetal ring of α-glucose opens to yield the free aldehyde. Rotation of the aldehyde group is followed by formation of the cyclic hemiacetal of β-glucose. The reaction is catalyzed by both acid and base.

22.35 a) $\ce{>C=O ->[H_2NNH_2][KOH, H_2O] >CH_2}$

Advantage; reaction is one-step.

Disadvantage; cannot be used when base-sensitive functional groups are present.

b) $\ce{>C=O ->[HSCH_2CH_2SH][H^+]}$ (dithiolane: H_2C-CH_2, S, S, C) $\ce{->[Raney][Ni] >CH_2}$

Advantage; reduction is only two steps.

Disadvantage; cannot be used when acid-sensitive functional groups are present.

c) $\ce{>C=O ->[1. NaBH_4][2. H_3O^+] >CH(OH) ->[POCl_3][pyridine] >C=C ->[H_2][Pd/C] >CH_2}$

d) $\ce{>C=O ->[1. NaBH_4][2. H_3O^+] >CH(OH) ->[PBr_3] >CH(Br) ->[Li(CH_3CH_2)_3BH] >CH_2}$

e) $\ce{>C=O ->[1. NaBH_4][2. H_3O^+] >CH(OH) ->[PBr_3] >CH(Br) ->[Mg] >CH(MgBr) ->[H_2O] >CH_2}$

Disadvantage; these three methods require several steps.

22.36 a) $\ce{->[1. Li(CH_3)_2Cu][2. H_3O^+]}$

b) $\ce{->[1. LiAlH_4][2. H_3O^+]}$ OH

c) $\ce{->[1. (C_2H_5)_2AlCN][2. H_3O^+]}$ CN

d) CH_3NH_2 →

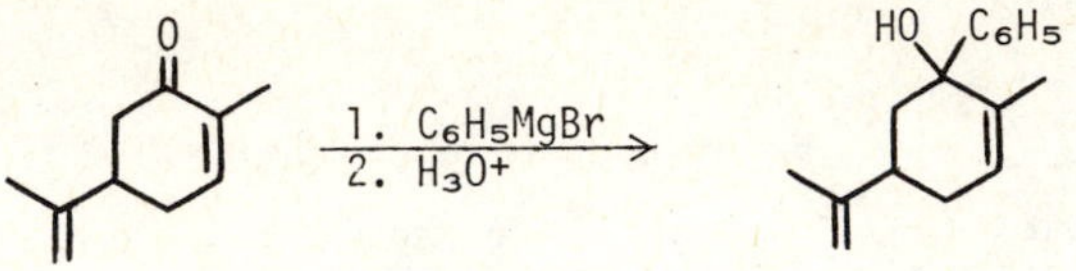

NHCH3

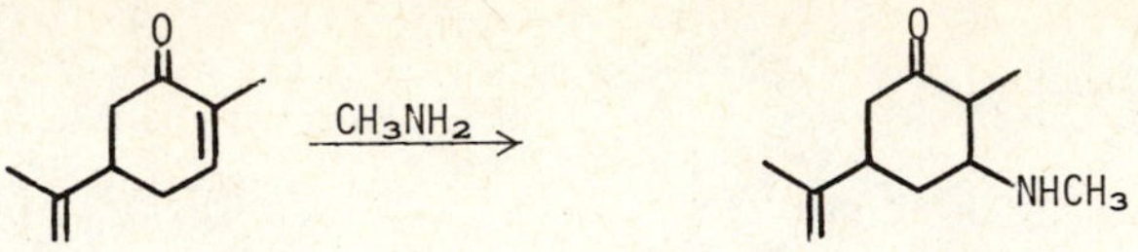

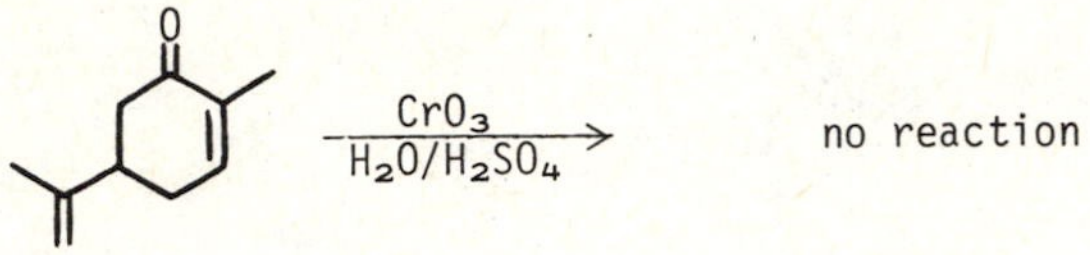

e) 1. C_6H_5MgBr 2. H_3O^+ →

HO C6H5

f) H_2/Pd →

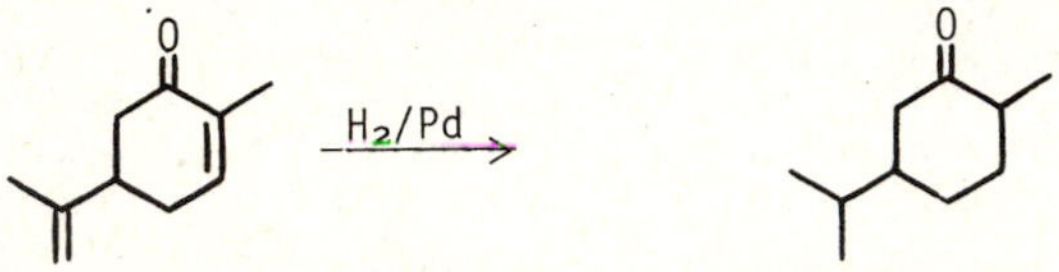

g) CrO_3 / H_2O/H_2SO_4 → no reaction

h) $(C_6H_5)_3\overset{+}{P}\overset{-}{\ddot{C}}HCH_3$ →

H CH3 C + H3C H C

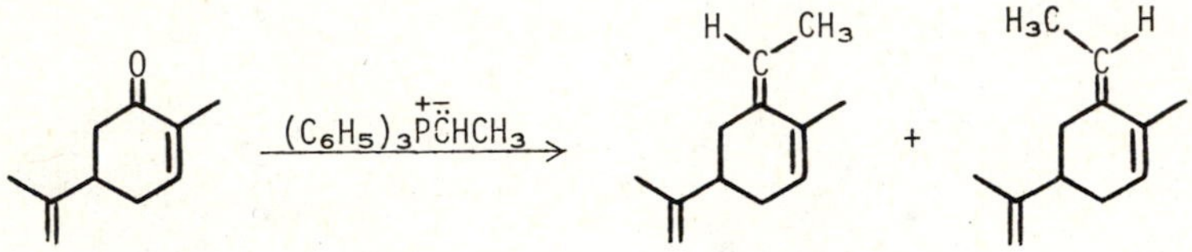

i) $HSCH_2CH_2SH$ / Raney Ni →

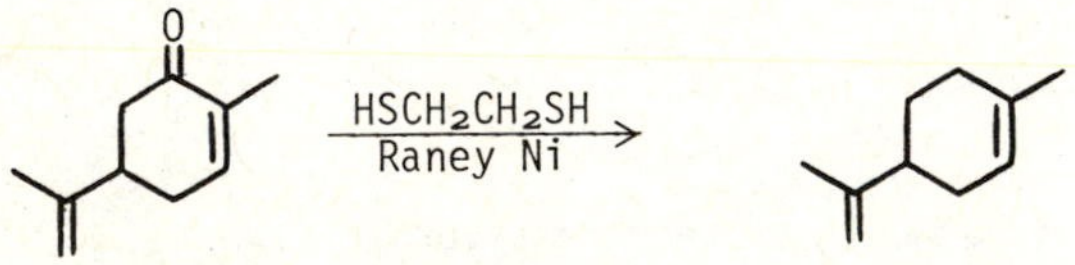

j) $HOCH_2CH_2OH$ / H^+ →

H2C—CH2 O O

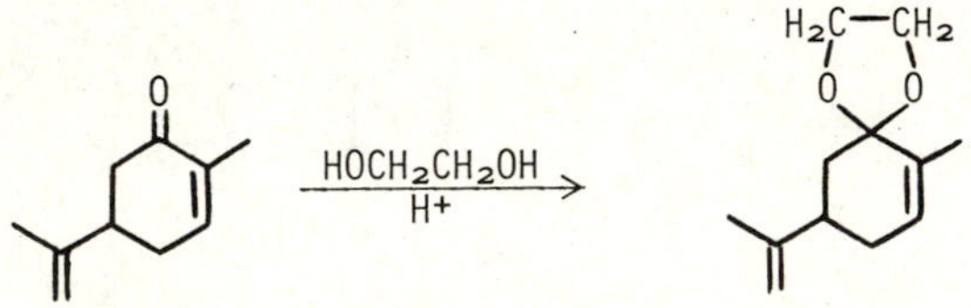

22.37

$(CH_3)_3\overset{O}{\overset{\|}{C}}H$

2,2-dimethylpropanal

compound A

22.38

$CH_3\underset{CH_3}{\underset{|}{C}}H\overset{O}{\overset{\|}{C}}CH_3$

3-methyl-2-butanone

compound B

22.39 Absorption: Due to:

	Absorption:	Due to:
a)	1750 cm^{-1}	5-membered ring ketone
	1685 cm^{-1}	α,β-unsaturated ketone
b)	1710 cm^{-1}	5-membered ring *and* α,β-unsaturated ketone
c)	1750 cm^{-1}	5-membered ring ketone
d)	1705 cm^{-1}; 2720 cm^{-1}; 2820 cm^{-1}	aromatic aldehyde
	1715 cm^{-1}	aliphatic ketone

Compounds in parts b-d also show aromatic ring IR absorptions in the range 1450 cm^{-1} - 1600 cm^{-1}.

22.40 1) Aluminum, a Lewis Acid, complexes with the carbonyl oxygen.

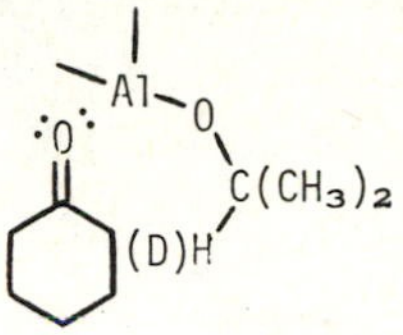

Al O C(CH_3)$_2$ (D)H

2) Complexation with aluminum makes the carbonyl functional group electrophilic and facilitates hydride transfer from isopropoxide.

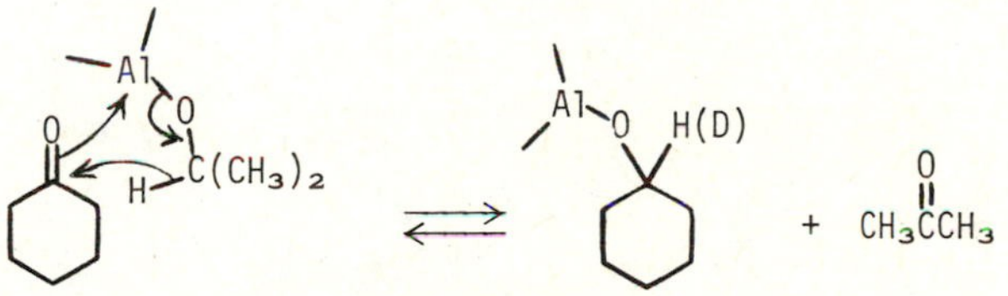

Al O C(CH_3)$_2$ H ⇌ Al O H(D) + CH_3CCH_3 (O)

3) Treatment of the reaction mixture with aqueous acid cleaves the aluminum-oxygen complex and produces cyclohexanol.

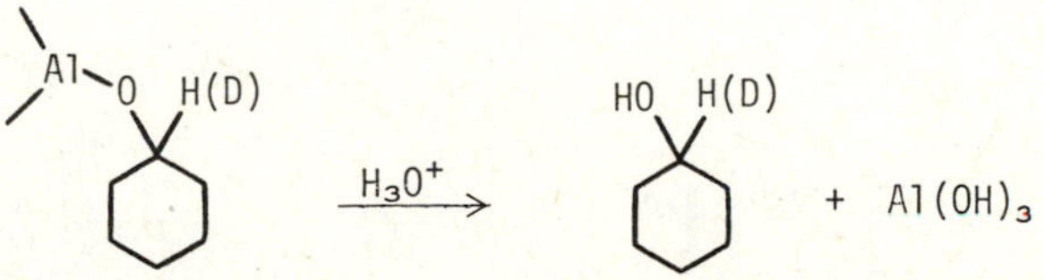

Al O H(D) $\xrightarrow{H_3O^+}$ HO H(D) + $Al(OH)_3$

Both the MPV reaction and the Cannizzaro reaction are hydride transfers in which a carbonyl group is reduced by a second oxygen-containing functional group, which is oxidized.

22.41 a) Basic silver ion does not oxidize secondary alcohols to ketones. Grignard addition to a conjugated ketone yield the 1,2 product, not the 1,4 product.

The correct scheme:

(bicyclic allylic alcohol, HO–) $\xrightarrow{PCC}$ (bicyclic enone, O=) $\xrightarrow[2.\ H_3O^+]{1.\ Li(CH_3)_2Cu}$ (bicyclic ketone with angular CH_3)

b) The Jones oxidation converts primary alcohols to carboxylic acids, not to aldehydes.

The correct scheme:

$C_6H_5CH{=}CHCH_2OH \xrightarrow{PCC} C_6H_5CH{=}CHCHO \xrightarrow[H^+]{HOCH_2CH_2OH} C_6H_5CH{=}CHCH$ (cyclic acetal, O–CH₂CH₂–O)

c) Treatment of a cyanohydrin with H_3O^+ produces a carboxylic acid, not an amine.

The correct scheme:

$CH_3\overset{O}{\overset{\|}{C}}CH_3 \xrightarrow{HCN} CH_3C(OH)(CN)CH_3 \xrightarrow[2.\ H_2O]{1.\ LiAlH_4} CH_3C(OH)(CH_2NH_2)CH_3$

d) Raney Nickel is used on thioacetals, not hydrazines, to convert a carbonyl carbon to a saturated carbon.

The correct scheme:

cyclohexanone $\xrightarrow[H^+]{HSCH_2CH_2SH}$ (spiro dithiolane, S S) $\xrightarrow{Raney\ Ni}$ cyclohexane (H, H)

or

cyclohexanone $\xrightarrow[KOH,\ H_2O]{H_2NNH_2}$ cyclohexane (H, H)

22.42

$CH_3\overset{O}{\overset{\|}{C}}CH_2CH_2\overset{O}{\overset{\|}{C}}OCH_3 \xrightarrow[H^+]{HOCH_2CH_2OH}$ $CH_3C(OCH_2CH_2O)CH_2CH_2\overset{O}{\overset{\|}{C}}OCH_3 \xrightarrow[2.\ H_3O^+]{1.\ LiAlH_4}$ $CH_3C(OCH_2CH_2O)CH_2CH_2CH_2OH$

$\xrightarrow[CH_2Cl_2]{PCC}$ (or directly from the ester: 1. DIBAH, 2. H_3O^+) $CH_3C(OCH_2CH_2O)CH_2CH_2\overset{O}{\overset{\|}{C}}H$

$\xrightarrow{(C_6H_5)_3\overset{+}{P}\overset{-}{\ddot{C}}(CH_3)_2}$ $CH_3C(OCH_2CH_2O)CH_2CH_2CH{=}C(CH_3)_2 \xrightarrow{H_3O^+}$ $CH_3\overset{O}{\overset{\|}{C}}CH_2CH_2CH{=}C(CH_3)_2$

6-methyl-5-hepten-2-one

22.43

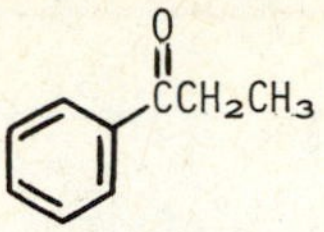

(structure: $C_6H_5C(=O)CH_2CH_3$)

propiophenone

22.44

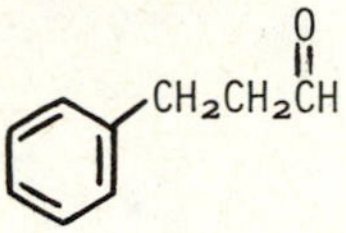

(structure: $C_6H_5CH_2CH_2CHO$)

3-phenylpropanal

22.45

(structure: bicyclo[3.3.0]octane-3,7-dione)

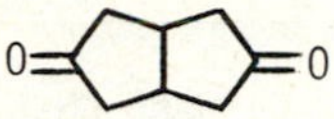

What you should know:

After doing these problems you should be able to:

1. Name aldehydes and ketones; draw structures of aldehydes and ketones, given names;
2. Prepare aldehydes and ketones;
3. Formulate mechanisms for nucleophilic addition reactions to aldehydes and ketones;
4. Choose the reagents necessary for transforming aldehydes and ketones;
5. Predict the products of reactions of aldehydes and ketones;
6. Use spectroscopic information to determine the structures of aldehydes and ketones.

CHAPTER 23 CARBOXYLIC ACIDS

23.1 a) $(CH_3)_2CHCH_2COOH$

3-methylbutanoic acid

b) $CH_3CHBrCH_2CH_2COOH$

4-bromopentanoic acid

c) H, COOH, H, CH_3 (cyclohexane ring)

trans-2-methylcyclohexane-carboxylic acid

d) $CH_3CH{=}CHCH{=}CHCOOH$

2,4-hexadienoic acid

e) 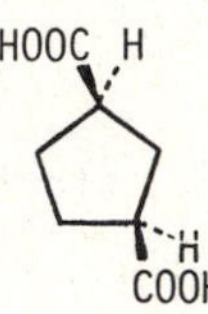

cis-1,3-cyclopentane-dicarboxylic acid

f) $CH_3CH_2CH(COOH)CH_2CH_2CH_3$

2-ethylpentanoic acid

g) $C_6H_5CH(CH_3)COOH$

2-phenylpropanoic acid

23.2 a) $CH_3CH_2CH_2CH(CH_3)CH(CH_3)COOH$

2,3-dimethylhexanoic acid

b) H, COOH, H, COOH (cyclobutane ring)

trans-1,2-cyclobutane-dicarboxylic acid

c) $CH_3CH_2CH_2CH_2CH_2CH{=}CHCH_2CH{=}CHCH_2CH_2CH_2CH_2CH_2CH_2CH_2COOH$ (both double bonds cis, H on the same side)

(9Z),(12Z)-octadecadienoic acid

d) 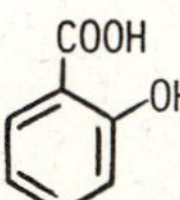

o-hydroxybenzoic acid

e) $(CH_3)_2CHCH_2CH_2COOH$

4-methylpentanoic acid

23.3 $\Delta G° = -RT \ln K_a = -2.303\ RT \log K_a$. Here R = 1.98 cal/mol·K; T = 298 K

$$\Delta G° = -2.303 \times 1.98\ \text{cal/mol·K} \times 298\ \text{K} \times \log K_a$$
$$= -1.36 \times 10^3\ \text{cal/mol} \times \log K_a$$
$$= -1.36\ \text{kcal/mol} \times \log K_a$$

for ethanol: $K_a = 10^{-16}$; $\log K_a = -16$

$\Delta G° = -1.36\ \text{kcal/mol} \times (-16) = \underline{+22\ \text{kcal/mol}}$

for acetic acid: $K_a \simeq 10^{-5}$; $\log K_a \simeq -5$

$\Delta G° = -1.36\ \text{kcal/mol} \times (-5) = \underline{+7\ \text{kcal/mol}}$

Dissociation of acetic acid is favored. Since $\Delta G°$ for acetic acid is a smaller number, less energy is required for dissociation of acetic acid than for dissociation of ethanol.

23.4 Naphthalene is insoluble in water; benzoic acid is only slightly soluble. The *salt* of benzoic acid is very soluble in water, however, and we can take advantage of this solubility in separating naphthalene from benzoic acid.

Dissolve the mixture in an organic solvent, and extract with a dilute aqueous solution of sodium hydroxide or sodium bicarbonate, which will neutralize benzoic acid. Naphthalene will remain in the organic layer, and all benzoic acid, now converted to the benzoate salt, will be in the aqueous layer. To recover benzoic acid, remove the aqueous layer, acidify it with dilute mineral acid, and extract with an organic solvent.

23.5

$$Cl_2CHCOH + H_2O \underset{}{\overset{K_a}{\rightleftarrows}} Cl_2CHCO^- + H_3O^+$$

(C=O shown above the carbonyl carbon)

$$K_a = \frac{[Cl_2CHCO_2^-][H_3O^+]}{[Cl_2CHCOOH]} = 5.5 \times 10^{-2}$$

	Initial molarity	Molarity after dissociation
$Cl_2CHC(=O)OH$	0.10	0.10 - y
$Cl_2CHC(=O)O^-$	0	y
H_3O^+	0	y

$$K_a = \frac{y \cdot y}{(0.10 - y)} = 5.5 \times 10^{-2}$$

Using the quadratic formula to solve for y, we find that y = 0.052.

$$\text{Percent dissociation} = \frac{0.052}{0.100} \times 100\% = 52\%$$

23.6 Weaker acid $\longrightarrow$ stronger acid

a) $CH_3CH_2COOH < BrCH_2COOH < FCH_2COOH$

Here, fluoride is the most electronegative group and can stabilize the carboxylate anion the best.

b) p-hydroxybenzoic acid (COOH, OH) < benzoic acid (COOH) < p-cyanobenzoic acid (COOH, CN)

The electron-withdrawing cyano group stabilizes the carboxylate anion. The hydroxyl group, which is a resonance electron donor, destabilizes the carboxylate anion.

c) $CH_3CH_2NH_2 < CH_3CH_2OH < CH_3CH_2COOH$

23.7 HOOC–COOH + H_2O $\rightleftharpoons$ H_3O^+ + HO(O=)C–C(=O)O$^-$ $\longleftrightarrow$ HO(O=)C–C(O$^-$)=O

pK_1 of oxalic acid is lower than that of a monocarboxylic acid because the carboxylate anion is stabilized both by resonance and by the inductive effect of the nearby second carboxylic acid group.

HO(O=)C–C(=O)O$^-$ $\longleftrightarrow$ HO(O=)C–C(O$^-$)=O + H_2O $\rightleftharpoons$ H_3O^+ + $^-$O(O=)C–C(=O)O$^-$ $\longleftrightarrow$ O=C(O$^-$)–C(O$^-$)=O

pK_2 of oxalic acid is higher than pK_1 for two reasons. (1) The first carboxylate group inductively destabilizes the negative charge resulting from dissociation of the second proton. (2) Electrostatic repulsion between the two adjacent negative charges destabilizes the dianion.

23.8 Inductive effects of functional groups are transmitted through bonds. As the length of the carbon chain increases, the effect of one functional group on another decreases. In this example, the influence of the second carboxyl group on the ionization of the first is barely felt by succinic and adipic acids.

23.9 *p*-(Trifluoromethyl)benzoic acid is a stronger acid than benzoic acid because the electron-withdrawing trifluoromethyl group stabilizes the carboxylate anion. A substituent that activates a reaction in which negative charge is developed, will *deactivate* a reaction in which positive charge is developed. Since a positively charged intermediate is formed in the Friedel-Crafts reaction, you would expect the trifluoromethyl substituent to be deactivating in this reaction.

23.10 Least acidic ⟶ Most acidic

a) p-methylbenzoic acid < benzoic acid < p-chlorobenzoic acid

(COOH / CH_3) < (COOH) < (COOH / Cl)

b) CH_3COOH < benzoic acid < p-nitrobenzoic acid

CH_3COOH < (COOH) < (COOH / NO_2)

23.11 Both of these methods are simple two-step transformations, and both methods can be used with compounds containing base-sensitive functional groups. Advantages of the cyanide hydrolysis method: (1) it can be used with acid-sensitive compounds; (2) it produces an optically active carboxylic acid from an optically active halide; (3) it can be used with compounds containing functional groups that interfere with Grignard reagent formation. The advantage of the Grignard method is that it can be used with tertiary halides and also with aryl and vinylic halides, which do not undergo S_N2 reactions.

23.12 a) Bromobenzene (Br) $\xrightarrow[\text{3. } H_3O^+]{\text{1. Mg, ether; 2. } CO_2\text{, ether}}$ benzoic acid (COOH)

b) $CH_3O\overset{O}{\overset{\|}{C}}CH_2CH_2CH_2Br \xrightarrow[\text{2. } H_3O^+]{\text{1. NaCN}} CH_3O\overset{O}{\overset{\|}{C}}CH_2CH_2CH_2COOH$

Grignard carboxylation can't be used because an ester group is present.

c) $(CH_3)_3CBr \xrightarrow[\text{3. } H_3O^+]{\text{1. Mg, ether; 2. } CO_2\text{, ether}} (CH_3)_3CCOOH$

d) $CH_3CH_2CH_2Br \xrightarrow[\text{3. } H_3O^+ \quad or \quad \text{2. } H_3O^+]{\text{1. Mg, ether; 2. } CO_2\text{, ether} \quad or \quad \text{1. NaCN}} CH_3CH_2CH_2COOH$

e) 1-Bromocyclohexene (Br) $\xrightarrow[\text{3. } H_3O^+]{\text{1. Mg, ether; 2. } CO_2\text{, ether}}$ 1-cyclohexenecarboxylic acid (COOH)

23.13 a) 2,5-dimethylhexanedioic acid b) 2,2-dimethylpropanoic acid

c) 3-propylhexanoic acid d) *p*-nitrobenzoic acid

e) 1-cyclodecenecarboxylic acid f) 4,5-dibromopentanoic acid

23.14 a) [cyclohexane-1,2-dicarboxylic acid structure: ring with H and COOH on each of two adjacent carbons]

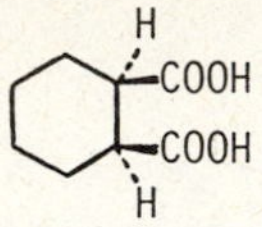

b) $HOOCCH_2CH_2CH_2CH_2CH_2COOH$

c) $CH_3C{\equiv}CCH{=}CHCOOH$

d) $CH_3CH_2CH_2CH_2CH(CH_2CH_3)CH_2CH(CH_2CH_2CH_3)COOH$

e) [benzene ring with COOH, COOH on adjacent carbons and Cl adjacent to the second COOH]

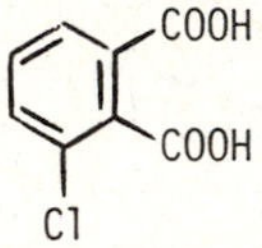

f) $(C_6H_5)_3CCOOH$

g) $CH_3CH_2CH_2CH(CH(CH_3)_2)CHBrCH_2CHBrCOOH$

23.15 Acetic acid molecules are strongly associated because of hydrogen bonding. Molecules of the ethyl ester are much more weakly associated, and less heat is required to overcome the attractive forces between molecules of the ethyl ester. Even though the ethyl ester has a greater molecular weight, it boils at a lower temperature than the acid.

23.16

$CH_3CH_2CH_2CH_2CH_2COOH$
hexanoic acid

$CH_3CH_2CH_2CH(CH_3)COOH$
2-methylpentanoic acid

$CH_3CH_2CH(CH_3)CH_2COOH$
3-methylpentanoic acid

$CH_3CH(CH_3)CH_2CH_2COOH$
4-methylpentanoic acid

$CH_3CH_2CH(COOH)CH_2CH_3$
2-ethylbutanoic acid

$CH_3CH_2C(CH_3)_2COOH$
2,2-dimethylbutanoic acid

$CH_3CH(CH_3)CH(CH_3)COOH$
2,3-dimethylbutanoic acid

$CH_3C(CH_3)_2CH_2COOH$
3,3-dimethylbutanoic acid

23.17 Least acidic ⟶ most acidic

a) CH_3COOH < $HCOOH$ < $HOOCCOOH$
acetic acid formic acid oxalic acid

b) *p*-bromobenzoic acid < *p*-nitrobenzoic acid < 2,4-dinitrobenzoic acid

c) $C_6H_5CH_2CH_2COOH$ < $C_6H_5CH_2COOH$ < $(C_6H_5)_2CHCOOH$

d) FCH_2CH_2COOH < ICH_2COOH < FCH_2COOH

23.18 Remember that the conjugate base of a weak acid is a strong base. In other words, the stronger the acid, the weaker the base derived from that acid.

Least basic ⟶ most basic

a) $CH_3C(=O){-}\ddot{O}{:}^- Na^+ < NH_3 < NaOH$

b) p-$O_2N{-}C_6H_4{-}C(=O){-}\ddot{O}{:}^- Na^+ < C_6H_5{-}C(=O){-}\ddot{O}{:}^- Na^+ < \overset{+}{Na}\,{:}\overset{-}{C}{\equiv}CH$

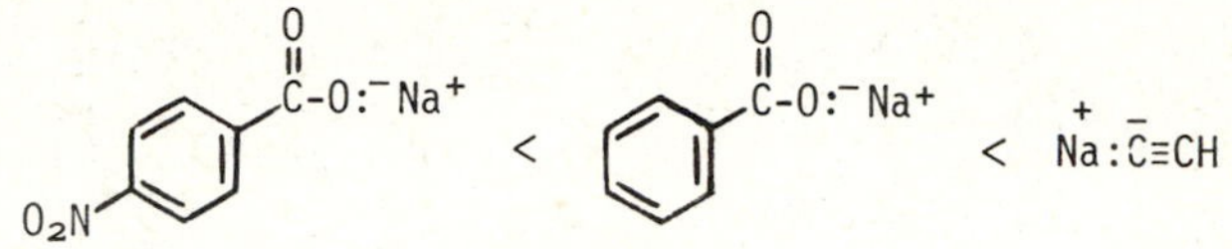

c) $HC(=O){-}\ddot{O}{:}^- Li^+ < H\ddot{O}{:}^- Li^+ < (CH_3)_3C\ddot{O}{:}^- Li^+$

23.19 Two factors are responsible for the difference in pK_2 values between these benzenedicarboxylic acids. (1) The mono-anion of phthalic acid is stabilized by hydrogen bonding of the remaining -COOH proton with the adjacent carboxylate group. This type of hydrogen bonding, which is not possible for terephthalic acid, stabilizes the phthalate monoanion. (2) The dianion resulting

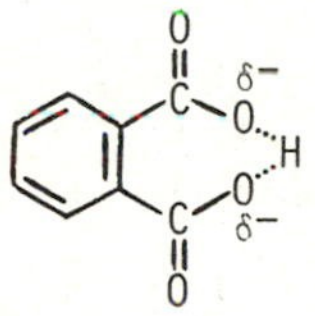

[structure: phthalate monoanion with the H shared between the two carboxylate oxygens ($O^{\delta-}\cdots H\cdots O^{\delta-}$)]

from the dissociation of the second proton of phthalic acid has two negative charges close to one another. The resulting electrostatic repulsion destabilizes the phthalate dianion. The energy difference between mono- and dianion is greater for phthalic acid, and pK_2 of phthalic acid is greater than pK_2 of terephthalic acid.

23.20 a) $CH_3CH_2CH_2COH$ (C=O) $\xrightarrow[2.\ H_3O^+]{1.\ BH_3}$ $CH_3CH_2CH_2CH_2OH$
1-butanol

b) $CH_3CH_2CH_2CH_2OH$ (from a) $\xrightarrow{PBr_3}$ $CH_3CH_2CH_2CH_2Br$
1-bromobutane

c) $CH_3CH_2CH_2CH_2Br$ (from b) $\xrightarrow{NaCN}$ $CH_3CH_2CH_2CH_2C{\equiv}N$ $\xrightarrow{H_3O^+}$ $CH_3CH_2CH_2CH_2COOH$
pentanoic acid

d) $CH_3CH_2CH_2CH_2Br$ (from b) $\xrightarrow[ethanol]{KOH}$ $CH_3CH_2CH{=}CH_2$
1-butene

e) $CH_3CH_2CH_2COH$ (C=O) $\xrightarrow[CCl_4]{HgO,\ Br_2}$ $CH_3CH_2CH_2Br$
1-bromopropane

23.21

p-Methylbenzoic acid (COOH, CH_3):

- 1. BH_3 2. H_3O^+ → (a) CH_2OH / CH_3 benzene
- NBS, CCl_4 → (b) COOH / CH_2Br benzene
- HgO, I_2, CCl_4 → (c) I / CH_3 benzene
- CH_3MgBr, ether → (d) $O{=}C{-}OMgBr$ / CH_3 benzene + CH_4

23.22 a) $CH_3CH_2Br \xrightarrow{Mg} CH_3CH_2MgBr \xrightarrow[2.\ H_3O^+]{1.\ ^{13}CO_2} CH_3CH_2{}^{13}COOH$

b) $CH_3Br \xrightarrow{Mg} CH_3MgBr \xrightarrow[2.\ H_3O^+]{1.\ ^{13}CO_2} CH_3{}^{13}\overset{O}{\overset{\|}{C}}OH \xrightarrow[2.\ H_3O^+]{1.\ BH_3} CH_3{}^{13}CH_2OH$

$\xrightarrow{PBr_3} CH_3{}^{13}CH_2Br \xrightarrow{Mg} CH_3{}^{13}CH_2MgBr \xrightarrow[2.\ H_3O^+]{1.\ CO_2} CH_3{}^{13}CH_2\overset{O}{\overset{\|}{C}}OH$

23.23 $(CH_3)_3CCH_2COOH$

3,3-dimethylbutanoic acid

23.24 a) Methylenecyclohexane (CH_2) $\xrightarrow[\text{peroxides}]{HBr}$ (bromomethyl)cyclohexane (H, CH_2Br) $\xrightarrow[2.\ H_3O^+]{1.\ ^-{:}CN}$ *or* $\xrightarrow{1.\ Mg\ 2.\ CO_2\ 3.\ H_3O^+}$ cyclohexylacetic acid (CH_2COOH)

b) Methylenecyclohexane (CH_2) $\xrightarrow{HBr}$ 1-bromo-1-methylcyclohexane (H_3C, Br) $\xrightarrow{Mg}$ (H_3C, MgBr) $\xrightarrow[2.\ H_3O^+]{1.\ CO_2}$ 1-methylcyclohexanecarboxylic acid (H_3C, COOH)

23.25 Either ^{13}C NMR or ^1H NMR can be used to distinguish among these three isomeric carboxylic acids.

Compound	Number of ^{13}C NMR signals	Number of ^1H NMR signals	Splitting of ^1H NMR signals
$CH_3(CH_2)_3COOH$	5	5	1 triplet, peak area 3, 1.0 δ 1 triplet, peak area 2, 2.4 δ 2 multiplets, peak area 4, 1.5 δ 1 singlet, peak area 1, 12.0 δ
$(CH_3)_2CHCH_2COOH$	4	4	1 doublet, peak area 6, 1.0 δ 1 doublet, peak area 2, 2.4 δ 1 multiplet, peak area 1, 1.6 δ 1 singlet, peak area 1, 12.0 δ
$(CH_3)_3CCOOH$	3	2	1 singlet, peak area 9, 1.3 δ 1 singlet, peak area 1, 12.1 δ

23.26 a) Grignard carboxylation cannot be used to prepare the carboxylic acid because of the presence of the acidic hydroxyl group. Use *nitrile hydrolysis*.

b) Either method produces the carboxylic acid in suitable yield. *Grignard carboxylation* is a better reaction for preparing a carboxylic acid from a secondary bromide. *Nitrile hydrolysis* produces a chiral carboxylic acid from a chiral bromide.

c) Neither method of acid synthesis yields the desired product. Any Grignard reagent formed will react with the carbonyl functional group present in the starting material. Reaction with cyanide occurs at the carbonyl functional group, producing a cyanohydrin, as well as at halogen. However, if the ketone is first protected by forming an acetal, *either method* can be used for producing a carboxylic acid.

d) Since the hydroxyl proton interferes with formation of the Grignard reagent, *nitrile hydrolysis* must be used to form the carboxylic acid.

23.27 a) BH_3 is a reducing agent, not an oxidizing agent. To obtain benzoic acid from toluene, use $KMnO_4$.

b) Use CO_2, instead of NaCN, to form the carboxylic acid, or eliminate Mg from this reaction scheme and form the acid by nitrile hydrolysis.

c) Reduction of a carboxylic acid with $LiAlH_4$ yields an alcohol, not an alkyl group. To obtain the desired product:

$$C_6H_5CH_2COOH \xrightarrow[2.\ H_3O^+]{1.\ BH_3} C_6H_5CH_2CH_2OH \xrightarrow{PBr_3} C_6H_5CH_2CH_2Br \xrightarrow{Li(CH_3CH_2)_3BH} C_6H_5CH_2CH_3$$

d) Acid hydrolysis of the nitrile will also dehydrate the tertiary alcohol. Use basic hydrolysis to form the carboxylic acid.

23.28

$$C_6H_5CH_3 \xrightarrow[H_2SO_4]{HNO_3} p\text{-}CH_3C_6H_4NO_2 \xrightarrow[H_2O]{KMnO_4} p\text{-}HOOCC_6H_4NO_2 \xrightarrow[2.\ ^-OH]{1.\ SnCl_2,\ H_3O^+} p\text{-}HOOCC_6H_4NH_2$$

PABA

Notice that the order of the reactions is very important. If toluene is oxidized first, the nitro group will be introduced in the *meta* position. If the nitro group is reduced first, oxidation to the carboxylic acid will reoxidize the $-NH_2$ group.

23.29 Before starting this type of problem, look at the starting material and identify the functional groups present. Lithocholic acid contains only alcohol and carboxylic acid functional groups. The given reagents can react with one, both, or neither functional group. Remember to keep track of stereochemistry.

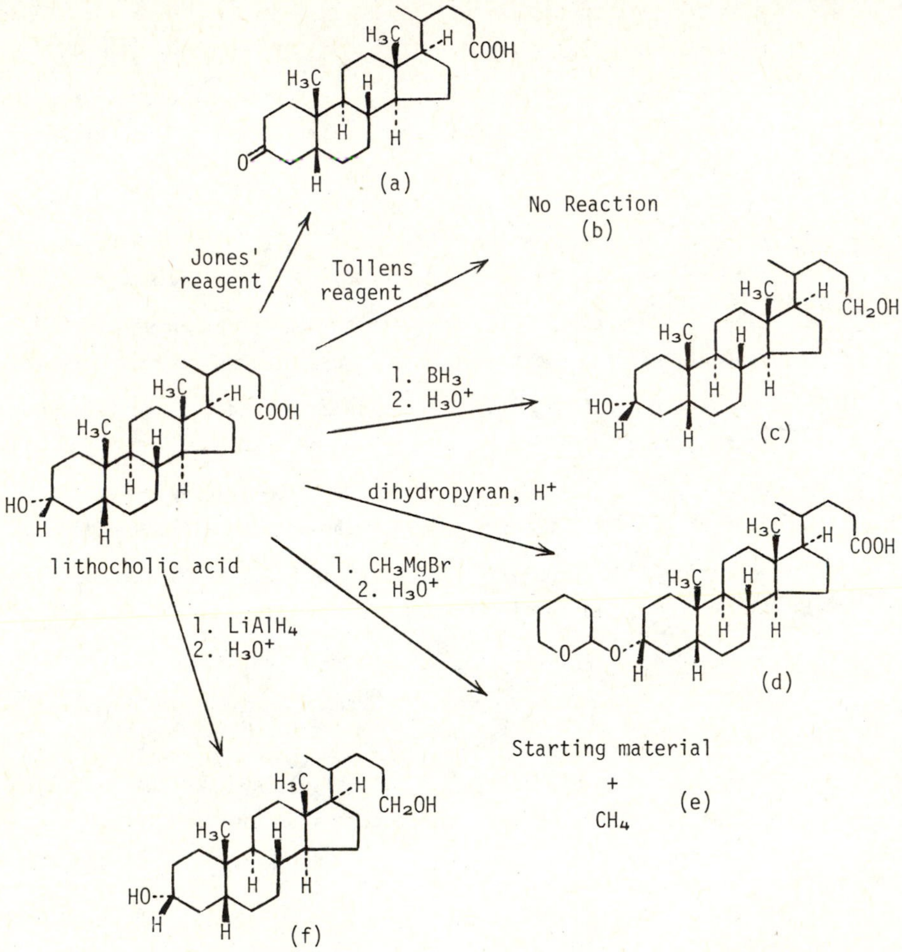

(a)

No Reaction
(b)

Jones' reagent

Tollens reagent

1. BH_3
2. H_3O^+

(c)

dihydropyran, H^+

lithocholic acid

1. CH_3MgBr
2. H_3O^+

(d)

1. $LiAlH_4$
2. H_3O^+

Starting material
+
CH_4
(e)

(f)

23.30 $\xrightarrow[H_2O]{KMnO_4}$ $\xrightarrow[Rh/C]{H_2}$ +

major product
Tranxemic acid

minor product

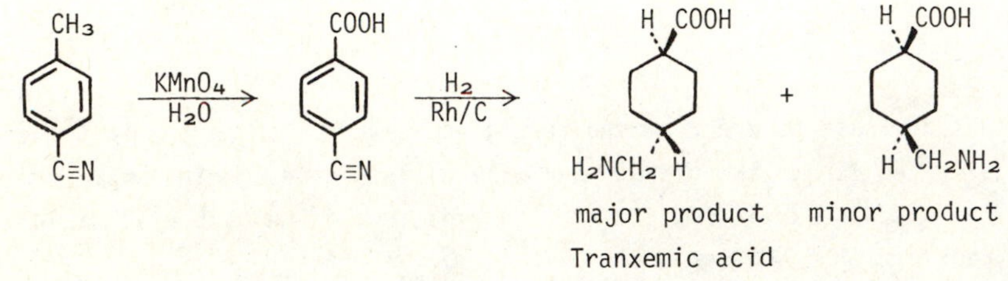

23.31 $\xrightarrow[FeBr_3]{Br_2}$ —Br + *ortho* isomer $\xrightarrow{Mg}$ —MgBr $\xrightarrow[2.\ H_3O^+]{1.\ CH_2O}$ —CH_2OH $\xrightarrow{PCC}$ —CHO $\xrightarrow[FeCl_3]{Cl_2}$ —CHO (Cl) $\xrightarrow{HCN,\ KCN}$ —CH(OH)C≡N (Cl) $\xrightarrow{H_3O^+}$ —CH(OH)COOH (Cl) $\xrightarrow[2.\ H_2O]{1.\ SOCl_2,\ pyr}$ —CH(Cl)COOH (Cl)

fenclorac

Other routes to this compound are possible. Notice that the aldehyde functional group and the cyclohexyl group both serve to direct the aromatic chlorine to the correct position of the ring. Also, reaction of the hydroxy-acid with $SOCl_2$ converts -OH to -Cl and -COOH to -COCl. Treatment with H_2O regenerates the carboxylic acid.

23.32 a), b) Use either ^1H NMR or ^{13}C NMR to distinguish between the isomers.

	Compound	# of ^{13}C Absorptions	# of ^1H Absorptions
a)	1,3-benzenedicarboxylic acid (COOH, COOH)	5	4
	1,4-benzenedicarboxylic acid (COOH, COOH)	3	2
b)	$HOOCCH_2CH_2COOH$	2	2
	$CH_3CH(COOH)_2$	3	3

c) Use ^1H NMR to distinguish between these two compounds. The carboxylic acid proton of the first compound absorbs near 12 δ; the aldehyde proton of the second compound absorbs near 10 δ and is split into a triplet.

d) The cyclic acid shows four absorptions in both its ^1H NMR and ^{13}C NMR spectra. The unsaturated acid show six absorptions in its ^{13}C NMR and five in its ^1H NMR spectrum; one of the ^1H NMR signals occurs in the vinyl (4.5 - 6.5 δ) region of the spectrum.

<u>23.33</u> β-Keto acids decarboxylate via a cyclic mechanism; the initial product is an enol. If the bicyclic β-keto acid of this problem were to decarboxylate, the initial enol would have a double bond at the bridgehead of the bicyclic ring system.

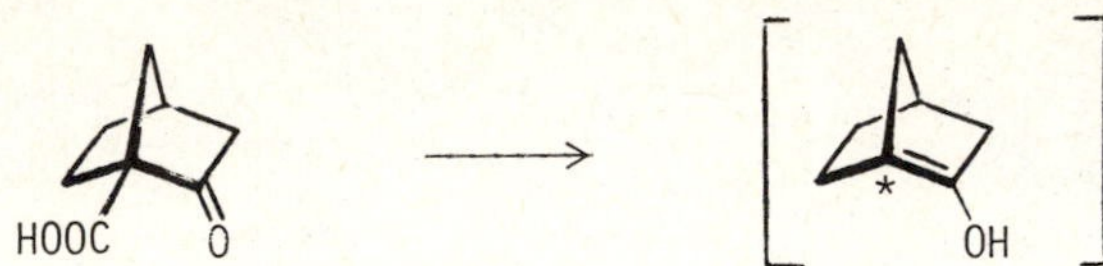

If you build a model of the enol, you will find that it is almost impossible to form the bridgehead double bond. (sp^2 Hybridization requires that all three bonds attached to the starred carbon lie in a plane.) Since the strain of a ring containing a bridgehead double bond is very large, any reaction producing such a bond is unlikely to occur.

<u>23.34</u>

$$CH_3CH_2OCH_2\overset{O}{\overset{\|}{C}}OH$$

<u>23.35</u> Both of these compounds contain four different kinds of protons (remember that the $H_2C=$ protons are non-equivalent). The carboxylic acid proton absorptions are easy to identify; the other three absorptions in each spectrum are more complex.

It is possible to assign the spectra correctly by studying the methyl group absorptions. The methyl group peak of crotonic acid is split into a doublet (J = 10) by the geminal ($CH_3CH=$) proton. (A small "second-order" splitting by the *trans* vinylic proton is also visible; on close inspection the doublet is seen to be a doublet of doublets.) The methyl group absorption of methacrylic acid appears to be a singlet, but second-order splitting by the $H_2C=$ protons shows it to be a quartet having very small J values. The first spectrum is that of crotonic acid and the second spectrum is that of methacrylic acid.

<u>What you should know</u>:

After completing these problems you should be able to:

1. Name carboxylic acids by several different systems; draw the structure of a carboxylic acid given the name;
2. Predict the effects of substituents on the acidity of carboxylic acids;
3. Synthesize carboxylic acids by several routes;
4. Recognize the types of reactions carboxylic acids undergo, and predict the products of these reactions;
5. Identify the carboxylic acid functional group spectroscopically.

CHAPTER 24 CARBOXYLIC ACID DERIVATIVES. NUCLEOPHILIC ACYL SUBSTITUTION REACTIONS

24.1 a) $(CH_3)_2CHCH_2CH_2C(=O)Cl$
4-methylpentanoyl chloride

b) Cyclohexyl–$CH_2C(=O)NH_2$
cyclohexylacetamide

c) $CH_3CH_2CH(CH_3)CN$
2-methylbutanenitrile

d) $C_6H_5C(=O)–O–C(=O)C_6H_5$
benzoic anhydride

e) Cyclopentyl–$C(=O)OCH(CH_3)_2$
isopropyl cyclopentanecarboxylate

f) Cyclopentyl–$OC(=O)CH(CH_3)_2$
cyclopentyl 2-methylpropanoate

g) $H_2C{=}CHCH_2CH_2C(=O)NH_2$
4-pentenamide

h) $CH_3CH_2CH(CN)CH_2CH_3$
2-ethylbutanenitrile

i) $(H_3C)_2C{=}C(CH_3)C(=O)Cl$
2,3-dimethyl-2-butenoyl chloride

j) $CF_3C(=O)OC(=O)CF_3$
trifluoroacetic anhydride

24.2 a) $CH_3CH_2CH{=}CHCN$ Correct name: 2-pentenenitrile
The nitrile carbon is at position 1.

b) $CH_3CH_2CH_2CONHCH_3$ Correct name: *N*-methylbutanamide
You must specify that the methyl group is bonded to nitrogen.

c) $(CH_3)_2CHCH_2CH(CH_3)COCl$ Correct name: 2,4-dimethylpentanoyl chloride
The prefix "di" must be put before "methyl".

d) 1-methylcyclohexane with CO_2CH_3 at C1 Correct name: methyl 1-methylcyclohexanecarboxylate

The methyl group on the cyclohexane ring is at position 1.

24.3 Most reactive ⟶ least reactive

a) CH_3CCl (C=O) > CH_3COCH_3 (C=O) > CH_3CNH_2 (C=O)

b) $CH_3COCH(CF_3)_2$ (C=O) > $CH_3COCH_2CCl_3$ (C=O) > CH_3COCH_3 (C=O)

The most reactive acyl derivatives contain strongly electron-withdrawing groups in the alkyl portion of the structure.

24.4

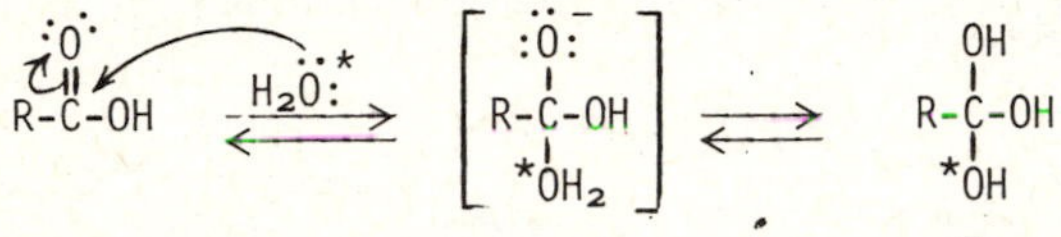

$$\text{R-C(=O)-OH} \underset{}{\overset{H_2\ddot{O}:^*}{\rightleftharpoons}} \left[\text{R-C(}:\ddot{O}:^-\text{)(OH)(}^*OH_2\text{)}\right] \rightleftharpoons \text{R-C(OH)(OH)(}^*OH\text{)}$$

T

The tetrahedral intermediate T can eliminate any one of the three -OH groups to reform either the original carboxylic acid or labeled carboxylic acid. Further reaction of water with mono-labeled carboxylic acid leads to the doubly labeled product.

24.5

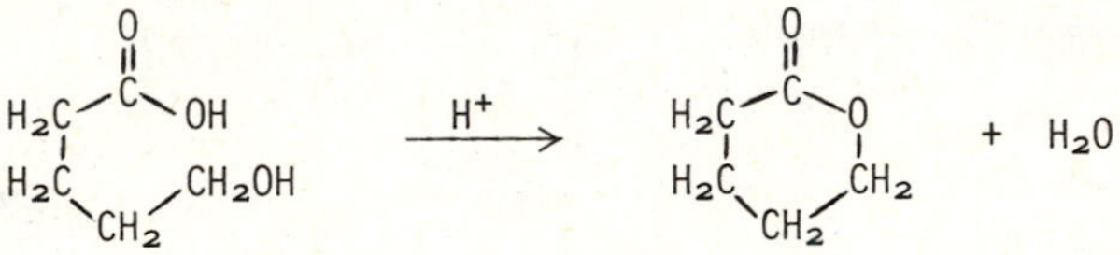

$$H_2C(-C(=O)-OH)-H_2C-CH_2-CH_2OH \xrightarrow{H^+} \text{cyclic } H_2C-C(=O)-O-CH_2-CH_2-H_2C + H_2O$$

a lactone

24.6 The carbonyl carbon of a carboxylate anion is relatively electron-rich and reacts poorly with nucleophiles.

24.7

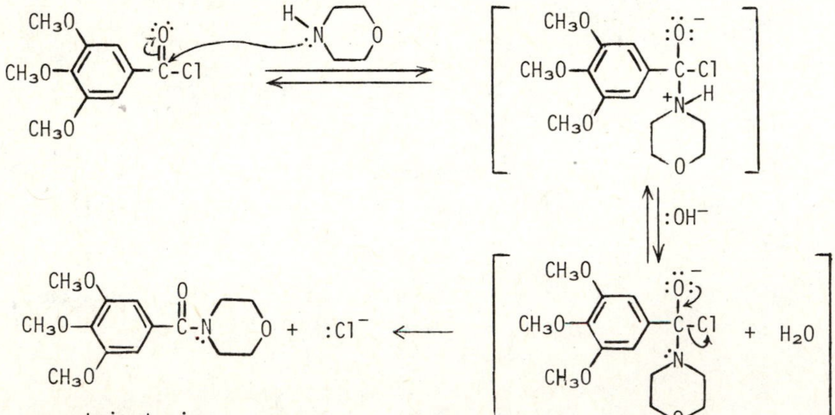

3,4,5-$(CH_3O)_3C_6H_2$-C(=O)-Cl + morpholine (H-N) ⇌ [3,4,5-$(CH_3O)_3C_6H_2$-C($:\ddot{O}:^-$)(Cl)-$^+$N(H)(morpholine)]

⇅ $:OH^-$

3,4,5-$(CH_3O)_3C_6H_2$-C(=O)-N(morpholine) + $:Cl^-$ ⟵ [3,4,5-$(CH_3O)_3C_6H_2$-C($:\ddot{O}:^-$)(Cl)-N(morpholine) + H_2O]

trimetozine

24.8 For primary and secondary amines:

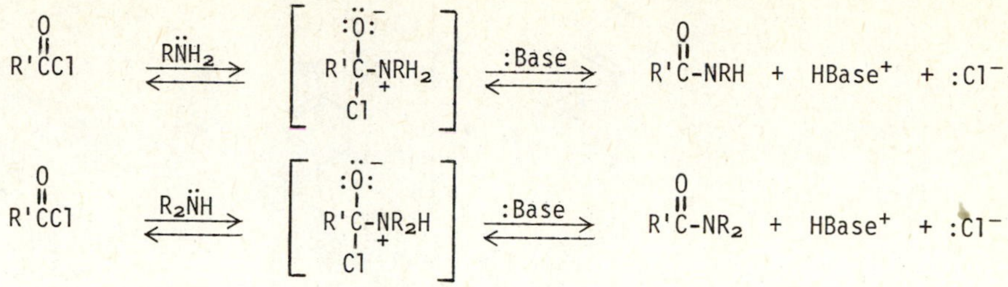

$$R'\overset{O}{\overset{\|}{C}}Cl \underset{\longleftarrow}{\xrightarrow{R\ddot{N}H_2}} \left[R'\overset{:\ddot{O}:^-}{\underset{Cl}{\overset{|}{\underset{|}{C}}}}-\overset{+}{N}RH_2 \right] \underset{\longleftarrow}{\xrightarrow{:Base}} R'\overset{O}{\overset{\|}{C}}-NRH + HBase^+ + :Cl^-$$

$$R'\overset{O}{\overset{\|}{C}}Cl \underset{\longleftarrow}{\xrightarrow{R_2\ddot{N}H}} \left[R'\overset{:\ddot{O}:^-}{\underset{Cl}{\overset{|}{\underset{|}{C}}}}-\overset{+}{N}R_2H \right] \underset{\longleftarrow}{\xrightarrow{:Base}} R'\overset{O}{\overset{\|}{C}}-NR_2 + HBase^+ + :Cl^-$$

For triethylamine:

$$R'\overset{O}{\overset{\|}{C}}Cl \underset{\longleftarrow}{\xrightarrow{(C_2H_5)_3N:}} \left[R'\overset{:\ddot{O}:^-}{\underset{Cl}{\overset{|}{\underset{|}{C}}}}-\overset{+}{N}(C_2H_5)_3 \right]$$

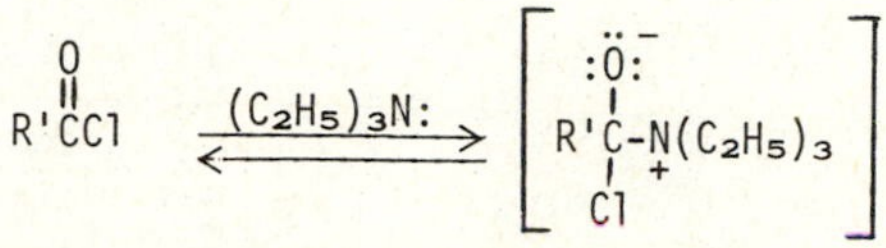

Triethylamine, like other amines, is a base whose lone pair of electrons can scavenge protons. Unlike primary and secondary amines, triethylamine does not form an amide; no nitrogen-bonded proton can be removed to yield neutral amide product. The tetrahedral adduct formed by nucleophilic addition of triethylamine to an acid chloride reverts to starting material instead of forming an amide.

24.9 In the slow addition of Grignard reagent to a solution of an acid chloride, each drop of Grignard reagent is surrounded by a large amount of acid chloride. The Grignard reagent is consumed immediately in converting the reactive acid chloride to less reactive ketone. No Grignard reagent remains to react with the ketone to form a tertiary alcohol. If the acid chloride were slowly added to a solution of Grignard reagent, however, excess reagent would convert the initially formed ketone into a tertiary alcohol.

24.10

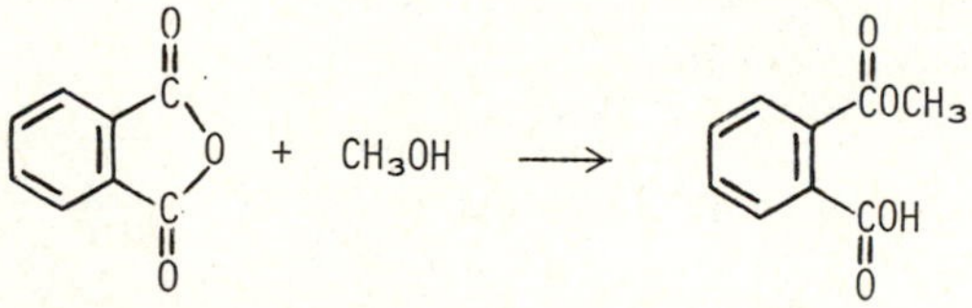

phthalic anhydride + CH_3OH $\longrightarrow$ (benzene ring bearing $\overset{O}{\overset{\|}{C}}OCH_3$ and $\underset{O}{\underset{\|}{C}}OH$ on adjacent carbons)

The second half of a cyclic anhydride becomes a carboxylic acid functional group.

24.11

$$CH_3\overset{O}{\overset{\|}{C}}-O-\overset{O}{\overset{\|}{C}}CH_3 + H_2N-C_6H_4-OCH_2CH_3 \rightleftarrows \left[CH_3\overset{O}{\overset{\|}{C}}-O-\overset{:\ddot{O}:^-}{\overset{|}{\underset{\underset{C_6H_4OCH_2CH_3}{|}}{\underset{+NH}{|}}{C}}}CH_3 \underset{\longleftarrow}{\overset{^-:OH}{\longrightarrow}} CH_3\overset{O}{\overset{\|}{C}}-O-\overset{:\ddot{O}:^-}{\overset{|}{\underset{\underset{C_6H_4OCH_2CH_3}{|}}{\underset{:NH}{|}}{C}}}CH_3 + H_2O \right]$$

$$\downarrow$$

$$CH_3\overset{O}{\overset{\|}{C}}O:^- + CH_3\overset{O}{\overset{\|}{C}}\ddot{N}H-C_6H_4-OCH_2CH_3$$

phenacetin

24.12 One equivalent of base must be added to the reaction mixture when an amine reacts with an anhydride. This base removes a proton from nitrogen after the formation of the initial tetrahedral intermediate. If base were not added, the amine starting material would serve as the base. Instead of going to completion, the reaction would stop when half of the starting amine has been converted to amide; the rest of the amine would be protonated and would no longer be nucleophilic.

24.13

$$CH_3\overset{O}{\overset{\|}{C}}-^{18}OH \xrightarrow[2.\ H_3O^+]{1.\ BH_3} CH_3CH_2-^{18}OH \xrightarrow[\text{pyridine}]{CH_3CH_2\overset{O}{\overset{\|}{C}}Cl} CH_3CH_2\overset{O}{\overset{\|}{C}}-^{18}OCH_2CH_3 + H_2O$$

ethyl propanoate

24.14 The products of acidic hydrolysis of an ester are a carboxylic acid and an alcohol. Under acidic conditions, these products can reform an ester by the Fischer esterification route.

The products of basic hydrolysis of an ester are an alcohol and a carboxylate anion, which is stable to nucleophilic attack.

24.15

$$\text{(γ-butyrolactone)} \xrightarrow[2.\ H_3O^+]{1.\ DIBAH} HOCH_2CH_2CH_2\overset{O}{\overset{\|}{C}}H$$

24.16

$$\text{C}_6\text{H}_{11}\text{Br} \xrightarrow{Mg} \text{C}_6\text{H}_{11}\text{MgBr} \xrightarrow[2.\ H_3O^+]{1.\ CO_2} \text{C}_6\text{H}_{11}\text{COOH} \xrightarrow[CHCl_3]{SOCl_2} \text{C}_6\text{H}_{11}\text{COCl}$$

$$\text{C}_6\text{H}_{11}\text{COCl} \xrightarrow{\text{excess } (CH_3)_2NH} \text{C}_6\text{H}_{11}\text{C(=O)-N(CH}_3)_2 \xrightarrow[2.\ H_2O]{1.\ LiAlH_4} \text{C}_6\text{H}_{11}\text{CH}_2\text{N(CH}_3)_2$$

24.17

$$R\text{-}C{\equiv}N: \underset{}{\overset{H^+}{\rightleftharpoons}} \left[R\text{-}\overset{+}{C}{\equiv}\ddot{N}H \overset{:OH_2}{\rightleftharpoons} R\text{-}\underset{}{\overset{^+OH_2}{C}}{=}\ddot{N}H \overset{-H^+}{\rightleftharpoons} R\text{-}\overset{OH}{C}{=}\ddot{N}H \rightleftharpoons R\text{-}\overset{O}{\overset{\|}{C}}\text{-}\ddot{N}H_2 \right]$$

amide

$\updownarrow H^+$

$$\left[R\text{-}\overset{OH}{\underset{OH}{C}}\text{-}\ddot{N}H_2 \overset{-H^+}{\rightleftharpoons} R\text{-}\overset{OH}{\underset{H_2\ddot{O}_+}{C}}\text{-}\ddot{N}H_2 \overset{H_2O:}{\rightleftharpoons} R\text{-}\overset{OH}{C}{=}\overset{+}{N}H_2 \right]$$

$\updownarrow H^+$

$$\left[R\text{-}\overset{:\ddot{O}H}{\underset{OH}{C}}\text{-}\overset{+}{N}H_3 \right] \rightleftharpoons R\text{-}\overset{^+\ddot{O}H}{\overset{\|}{C}}\text{-}OH + NH_3 \longrightarrow R\text{-}\overset{O}{\overset{\|}{C}}\text{-}OH + \overset{+}{N}H_4$$

An amide is an intermediate in the acidic hydrolysis of nitrile to a carboxylic acid.

24.18 Write the mechanism for addition of each equivalent of Grignard reagent to the nitrile. The nitrogen atom of the imine anion bears the negative charge.

$$R\text{-}C{\equiv}N: + \bar{R}':\overset{+}{Mg}X \longrightarrow R\text{-}\overset{\cdot\bar{N}\cdot\overset{+}{Mg}X}{\overset{\|}{C}}\text{-}R'$$

imine anion

A possible mechanism for addition of a second equivalent of Grignard reagent is:

$$R\text{-}\overset{:\bar{N}\cdot\overset{+}{Mg}X}{\overset{\|}{C}}\text{-}R' + \bar{R}'':\overset{+}{Mg}X \longrightarrow R\text{-}\overset{:\ddot{N}:^{2-}\ (MgX)_2^{2+}}{\underset{R''}{C}}\text{-}R'$$

A second addition of Grignard reagent is unlikely to occur for two reasons. (1) The imine carbon is relatively electron-rich and unreactive toward nucleophiles. (2) Nitrogen would have to bear two negative charges in the intermediate.

24.19

	Absorption	Functional group present
a)	1735 cm^{-1}	aliphatic ester *or* 6-membered ring lactone
b)	1810 cm^{-1}	aliphatic acid chloride
c)	2500-3300 cm^{-1} and 1710 cm^{-1}	carboxylic acid
d)	2250 cm^{-1}	aliphatic nitrile
e)	1715 cm^{-1}	aliphatic ketone

24.20 To solve this type of problem;

1. Use the IR absorption to determine the functional group(s) present.
2. Draw the functional group.
3. Use the remaining atoms to complete the structure.

a) 1. IR 2250 cm^{-1} corresponds to a nitrile.

2. $-C{\equiv}N$

3. CH_3CH_2CN is the structure of the compound.

b) 1. IR 1735 cm^{-1} corresponds to an aliphatic ester.

2. $-\overset{\overset{\displaystyle O}{\|}}{C}-O-$

3. The remaining five carbons and twelve hydrogens can be arranged in a number of ways to produce a satisfactory structure for this compound. For example:

$$CH_3CH_2CH_2\overset{\overset{\displaystyle O}{\|}}{C}OCH_2CH_3 \quad or \quad CH_3\overset{\overset{\displaystyle O}{\|}}{C}OCH_2CH_2CH_2CH_3$$

This compound cannot be a lactone; the structural formula is inconsistent with the presence of a ring.

c) $CH_3\overset{\overset{\displaystyle O}{\|}}{C}N(CH_3)_2$

d) $CH_3CH{=}CH\overset{\overset{\displaystyle O}{\|}}{C}Cl$ *or* $H_2C{=}C(CH_3)\overset{\overset{\displaystyle O}{\|}}{C}Cl$

24.21

a) *p*-methylbenzamide

b) 4-ethyl-2-hexenenitrile

c) dimethyl succinate *or* dimethyl butanedioate

d) isopropyl 3-phenylpropanoate

e) phenyl benzoate

f) *N*-methyl-3-bromobutanamide

g) 3,5-dibromobenzoyl chloride

h) 1-cyclopentenecarbonitrile

i) methyl 2,5-cyclohexadienecarboxylate

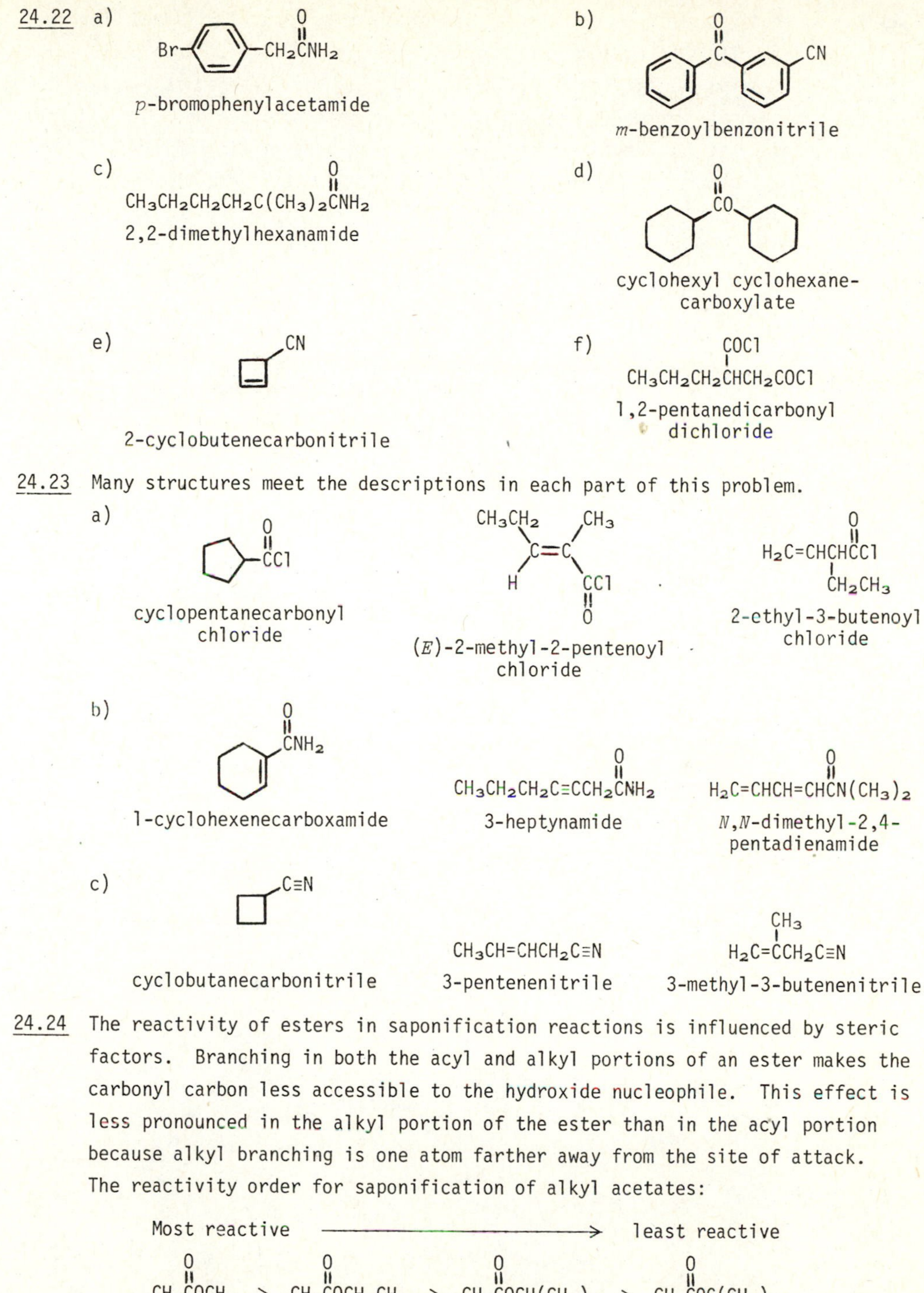

24.22 a) $Br-C_6H_4-CH_2CNH_2$ (C=O)

p-bromophenylacetamide

b) *m*-benzoylbenzonitrile

c) $CH_3CH_2CH_2CH_2C(CH_3)_2CNH_2$ (C=O)

2,2-dimethylhexanamide

d) cyclohexyl cyclohexane-carboxylate

e) 2-cyclobutenecarbonitrile

f) $CH_3CH_2CH_2CH(COCl)CH_2COCl$

1,2-pentanedicarbonyl dichloride

24.23 Many structures meet the descriptions in each part of this problem.

a) cyclopentanecarbonyl chloride

(*E*)-2-methyl-2-pentenoyl chloride

$H_2C{=}CHCH(CH_2CH_3)CCl$ (C=O)

2-ethyl-3-butenoyl chloride

b) 1-cyclohexenecarboxamide

$CH_3CH_2CH_2C{\equiv}CCH_2CNH_2$ (C=O)

3-heptynamide

$H_2C{=}CHCH{=}CHCN(CH_3)_2$ (C=O)

N,*N*-dimethyl-2,4-pentadienamide

c) cyclobutanecarbonitrile

$CH_3CH{=}CHCH_2C{\equiv}N$

3-pentenenitrile

$H_2C{=}C(CH_3)CH_2C{\equiv}N$

3-methyl-3-butenenitrile

24.24 The reactivity of esters in saponification reactions is influenced by steric factors. Branching in both the acyl and alkyl portions of an ester makes the carbonyl carbon less accessible to the hydroxide nucleophile. This effect is less pronounced in the alkyl portion of the ester than in the acyl portion because alkyl branching is one atom farther away from the site of attack. The reactivity order for saponification of alkyl acetates:

Most reactive ⟶ least reactive

$$CH_3\overset{O}{\overset{\|}{C}}OCH_3 > CH_3\overset{O}{\overset{\|}{C}}OCH_2CH_3 > CH_3\overset{O}{\overset{\|}{C}}OCH(CH_3)_2 > CH_3\overset{O}{\overset{\|}{C}}OC(CH_3)_3$$

24.25 A negatively charged tetrahedral intermediate is initially formed when the nucleophile $^{-}$:OH attacks the carbonyl carbon of an ester. An electron-withdrawing substituent can delocalize the negative charge, stabilizing the tetrahedral intermediate and increasing the rate of reaction. Contrast this effect with substituent effects in electrophilic aromatic substitution, in which positive charge developed in the intermediate is stabilized by electron-*donating* substituents. Substituents that are deactivating in electrophilic aromatic substitution are activating in ester hydrolysis, as the observed reactivity order shows. The substituents -CN and -CHO are electron-withdrawing; $-NH_2$ is strongly electron-donating.

Most reactive ⟶ least reactive

$X = -NO_2 > -CN > -CHO > -Br > -H > -CH_3 > -OCH_3 > -NH_2$

24.26

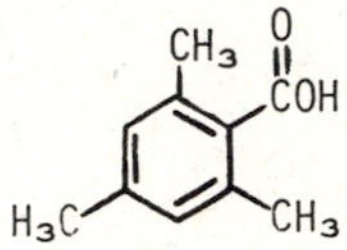
CH3
COH
H3C
CH3

mesitoic acid

Mesitoic acid has two methyl groups *ortho* to the carboxylic acid functional group. These bulky methyl groups block the approach of the alcohol and prevent esterification from occurring under Fischer esterification conditions. Two possible routes to the methyl ester:

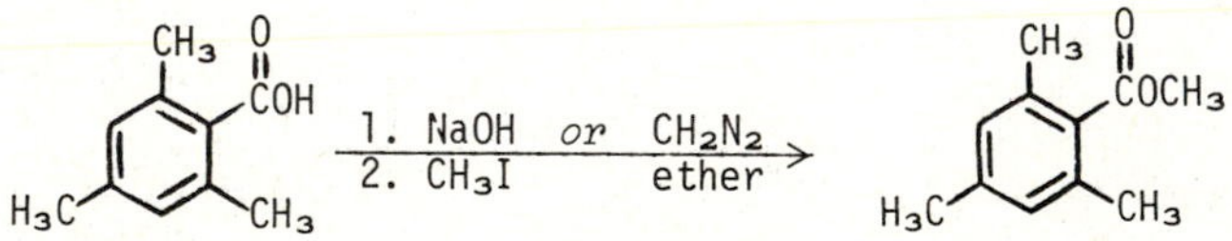
1. NaOH
2. CH3I
or
CH2N2
ether
COCH3

24.27 a)

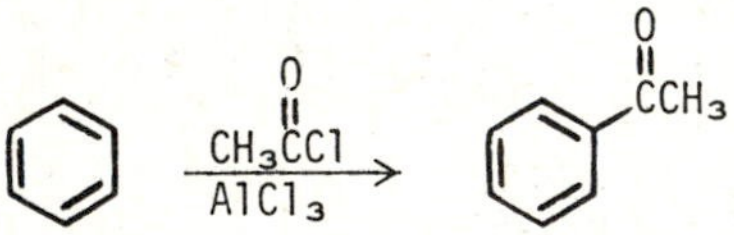
CH3CCl
AlCl3
CCH3

b)

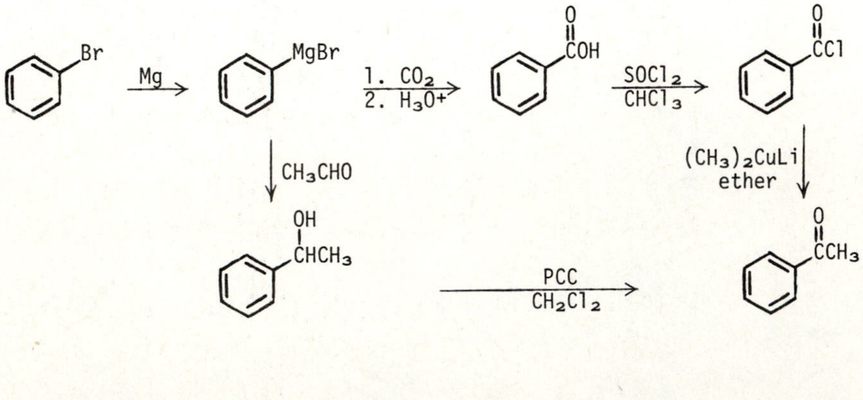
Br
Mg
MgBr
1. CO2
2. H3O+
COH
SOCl2
CHCl3
CCl
CH3CHO
(CH3)2CuLi
ether
OH
CHCH3
PCC
CH2Cl2
CCH3

c) Methyl benzoate $\xrightarrow[H_2O]{^-:OH}$ benzoic acid $\xrightarrow[CHCl_3]{SOCl_2}$ benzoyl chloride $\xrightarrow[ether]{(CH_3)_2CuLi}$ acetophenone

$C_6H_5COCH_3$ (ester: $C_6H_5C(=O)OCH_3$) → $C_6H_5C(=O)OH$ → $C_6H_5C(=O)Cl$ → $C_6H_5C(=O)CH_3$

Reaction of an ester with Grignard reagent produces a tertiary alcohol, not a ketone.

d) $C_6H_5C{\equiv}N \xrightarrow[2.\ H_3O^+]{1.\ CH_3MgBr} C_6H_5C(=O)CH_3$

e) $C_6H_5CH{=}CH_2 \xrightarrow[2.\ NaBH_4]{1.\ Hg(OAc)_2,\ H_2O} C_6H_5CH(OH)CH_3 \xrightarrow[CH_2Cl_2]{PCC} C_6H_5C(=O)CH_3$

24.28 Dimethyl carbonate is a diester. Use your knowledge of the Grignard reaction to work your way through this problem.

$CH_3O\text{-}C(=O)\text{-}OCH_3 + C_6H_5^-\ ^+MgBr \longrightarrow [CH_3O\text{-}C(O^-\ ^+MgBr)(C_6H_5)\text{-}OCH_3] \longrightarrow CH_3O\text{-}C(=O)\text{-}C_6H_5 + {}^-:OCH_3$

$CH_3O\text{-}C(=O)\text{-}C_6H_5 + C_6H_5^-\ ^+MgBr \longrightarrow [CH_3O\text{-}C(O^-\ ^+MgBr)(C_6H_5)_2] \longrightarrow (C_6H_5)_2C{=}O + {}^-:OCH_3$

$(C_6H_5)_2C{=}O + C_6H_5^-\ ^+MgBr \longrightarrow [(C_6H_5)_3C\text{-}O^-\ ^+MgBr] \xrightarrow{H_3O^+} (C_6H_5)_3C\text{-}OH$

triphenylmethanol

24.29 $R\text{-}C(=O)\text{-}CI_3 + :OH^- \rightleftarrows \left[R\text{-}C(O^-)(OH)\text{-}CI_3 \longrightarrow R\text{-}C(=O)\text{-}OH + {}^-:CI_3\right] \longrightarrow R\text{-}C(=O)\text{-}O:^- + HCI_3$

24.30 The nucleophile NH_3 replaces chlorine because chloride is a better leaving group than the methoxyl group.

$$CH_3O\overset{O}{\overset{\|}{C}}-Cl + :NH_3 \rightleftharpoons \left[CH_3O\overset{:\ddot{O}:^-}{\overset{|}{\underset{+NH_3}{\underset{|}{C}}}}-Cl \underset{}{\overset{:NH_3}{\rightleftharpoons}} CH_3O\overset{:\ddot{O}:^-}{\overset{|}{\underset{NH_2}{\underset{|}{C}}}}-Cl + NH_4^+ \right] \longrightarrow CH_3O\overset{O}{\overset{\|}{C}}-NH_2 + NH_4^+Cl^-$$

24.31 Acid chlorides generally react rapidly with nucleophiles. Both of these compounds, however, contain electron-donating substituents, which slow up the rate of reaction relative to an unsubstituted acid chloride. Because methoxyl groups are less strongly electron-donating than amino groups, CH_3OCOCl reacts faster with nucleophiles.

24.32

$$(CH_3)_3CO\overset{:\ddot{O}:}{\overset{\|}{C}}-Cl + :\bar{N}_3 \rightleftharpoons \left[(CH_3)_3CO\overset{:\ddot{O}:^-}{\overset{|}{\underset{N_3}{\underset{|}{C}}}}-Cl \right] \longrightarrow (CH_3)_3CO\overset{O}{\overset{\|}{C}}-N_3 + :Cl^-$$

24.33 a)

$$CH_3CH_2\overset{O}{\overset{\|}{C}}Cl \xrightarrow[\text{ether}]{(C_6H_5)_2CuLi} CH_3CH_2\overset{O}{\overset{\|}{C}}-C_6H_5$$

b)

$$CH_3CH_2\overset{O}{\overset{\|}{C}}Cl \xrightarrow[2.\ H_3O^+]{1.\ LiAlH_4} CH_3CH_2CH_2OH$$

c)

$$CH_3CH_2\overset{O}{\overset{\|}{C}}Cl \xrightarrow[2.\ H_3O^+]{1.\ 2\ CH_3MgBr} CH_3CH_2\overset{OH}{\overset{|}{\underset{CH_3}{\underset{|}{C}}}}CH_3$$

d)

$$CH_3CH_2\overset{O}{\overset{\|}{C}}Cl \xrightarrow{H_2/Pd/BaSO_4} CH_3CH_2\overset{O}{\overset{\|}{C}}H$$

e)

$$CH_3CH_2\overset{O}{\overset{\|}{C}}Cl \xrightarrow{H_3O^+} CH_3CH_2\overset{O}{\overset{\|}{C}}OH + HCl$$

f)

$$CH_3CH_2\overset{O}{\overset{\|}{C}}Cl + \text{cyclohexanol (OH)} \xrightarrow{\text{pyridine}} CH_3CH_2\overset{O}{\overset{\|}{C}}O-\text{cyclohexyl}$$

g)

$$CH_3CH_2\overset{O}{\overset{\|}{C}}Cl + \text{aniline } (NH_2) \xrightarrow{NaOH} CH_3CH_2\overset{O}{\overset{\|}{C}}-\underset{H}{\underset{|}{N}}-C_6H_5$$

h)

$$CH_3CH_2\overset{O}{\overset{\|}{C}}Cl + \text{cholesterol (HO, } H_3C, H_3C) \longrightarrow CH_3CH_2\overset{O}{\overset{\|}{C}}O-\text{cholesteryl (} H_3C, H_3C)$$

24.34 a) $CH_3CH_2\overset{O}{\overset{\|}{C}}OCH_3 \xrightarrow[\text{ether}]{(C_6H_5)_2CuLi}$ No reaction

b) $CH_3CH_2\overset{O}{\overset{\|}{C}}OCH_3 \xrightarrow[2.\ H_3O^+]{1.\ LiAlH_4} CH_3CH_2CH_2OH$

c) $CH_3CH_2\overset{O}{\overset{\|}{C}}OCH_3 \xrightarrow[2.\ H_3O^+]{1.\ 2\ CH_3MgBr} CH_3CH_2C(OH)(CH_3)CH_3$

d) $CH_3CH_2\overset{O}{\overset{\|}{C}}OCH_3 \xrightarrow{H_2/Pd/BaSO_4}$ No reaction

e) $CH_3CH_2\overset{O}{\overset{\|}{C}}OCH_3 \xrightarrow{H_3O^+} CH_3CH_2\overset{O}{\overset{\|}{C}}OH + CH_3OH$

f) $CH_3CH_2\overset{O}{\overset{\|}{C}}OCH_3$ + cyclohexanol (OH) $\xrightarrow{\text{pyridine}}$ No reaction

g) $CH_3CH_2\overset{O}{\overset{\|}{C}}OCH_3$ + aniline (NH_2) $\xrightarrow{NaOH}$ $CH_3CH_2\overset{O}{\overset{\|}{C}}-\underset{H}{N}-C_6H_5$

h) $CH_3CH_2\overset{O}{\overset{\|}{C}}OCH_3$ + cholesterol (HO, H_3C, H_3C) $\longrightarrow$ No reaction

24.35 a) $CF_3\overset{O}{\overset{\|}{C}}O-\overset{\ddot{O}:}{\overset{\|}{C}}CF_3$ + $R\overset{O}{\overset{\|}{C}}\ddot{O}H$ $\rightleftarrows$ $\left[CF_3\overset{O}{\overset{\|}{C}}O-\underset{R\overset{O}{\overset{\|}{C}}-\overset{+}{\ddot{O}}H}{\overset{:\ddot{O}:^-}{\overset{|}{C}}}CF_3 \underset{}{\overset{-H^+}{\rightleftarrows}} CF_3\overset{O}{\overset{\|}{C}}O-\underset{R\overset{O}{\overset{\|}{C}}O}{\overset{:\ddot{O}:^-}{\overset{|}{C}}}CF_3 \right] \longrightarrow R\overset{O}{\overset{\|}{C}}O\overset{O}{\overset{\|}{C}}CF_3 + CF_3\overset{O}{\overset{\|}{C}}O:^-$

b) The trifluoroacetate portion of the mixed anhydride is an especially good leaving group. The electron-withdrawing fluorine atoms, which stabilize the negative charge of the trifluoroacetate anion, make the mixed anhydride more reactive than other anhydrides.

c) Because trifluoroacetate is a better leaving group than other carboxylate anions, the reaction proceeds as indicated.

24.36 a) $CH_3CH_2CH_2CH_2C{\equiv}N \xrightarrow[2.\ H_2O]{1.\ LiAlH_4} CH_3CH_2CH_2CH_2CH_2NH_2$

b) $CH_3CH_2CH_2CH_2C{\equiv}N \xrightarrow{H_3O^+} CH_3CH_2CH_2CH_2\overset{O}{\overset{\|}{C}}OH \xrightarrow[CHCl_3]{SOCl_2} CH_3CH_2CH_2CH_2\overset{O}{\overset{\|}{C}}Cl$

$\downarrow$ 2 $HN(CH_3)_2$

$CH_3CH_2CH_2CH_2CH_2N(CH_3)_2 \xleftarrow[2.\ H_2O]{1.\ LiAlH_4} CH_3CH_2CH_2CH_2\overset{O}{\overset{\|}{C}}N(CH_3)_2$

c) $CH_3CH_2CH_2CH_2C{\equiv}N \xrightarrow[2.\ H_3O^+]{1.\ CH_3MgBr} CH_3CH_2CH_2CH_2\overset{O}{\overset{\|}{C}}CH_3$

$\downarrow$ 1. CH_3MgBr 2. H_3O^+

$CH_3CH_2CH_2CH_2C(OH)(CH_3)CH_3$

d) $CH_3CH_2CH_2CH_2\overset{O}{\overset{\|}{C}}CH_3 \xrightarrow[2.\ H_3O^+]{1.\ NaBH_4} CH_3CH_2CH_2CH_2C(OH)(H)CH_3$

(from part c)

e) $CH_3CH_2CH_2CH_2C{\equiv}N \xrightarrow[2.\ H_3O^+]{1.\ DIBAH} CH_3CH_2CH_2CH_2\overset{O}{\overset{\|}{C}}H$

24.37 First, convert cyclohexanol to bromocyclohexane by reaction with PBr_3. The first three transformations start with bromocyclohexane.

a) $C_6H_{11}Br \xrightarrow{NaCN} C_6H_{11}CN \xrightarrow[2.\ H_3O^+]{1.\ DIBAH} C_6H_{11}\overset{O}{\overset{\|}{C}}H$

b) $C_6H_{11}Br \xrightarrow{Mg} C_6H_{11}MgBr \xrightarrow[2.\ H_3O^+]{1.\ CO_2} C_6H_{11}\overset{O}{\overset{\|}{C}}OH \xrightarrow{CH_2N_2} C_6H_{11}\overset{O}{\overset{\|}{C}}OCH_3$

$\downarrow$ 1. DIBAH 2. H_3O^+

$C_6H_{11}\overset{O}{\overset{\|}{C}}H$

c) $C_6H_{11}Br \xrightarrow{Mg} C_6H_{11}MgBr \xrightarrow[2.\ H_3O^+]{1.\ CH_2O} C_6H_{11}CH_2OH \xrightarrow[CH_2Cl_2]{PCC} C_6H_{11}\overset{O}{\overset{\|}{C}}H$

d) cyclohexanol $\xrightarrow[CH_2Cl_2]{PCC}$ cyclohexanone $\xrightarrow{HCN}$ cyclohexanone cyanohydrin (OH, CN) $\xrightarrow[2.\ H_3O^+]{1.\ POCl_3}$ cyclohexene-1-carboxylic acid (COH with C=O)

$\downarrow$ H_2, Pd/C

cyclohexanecarbaldehyde $\xleftarrow{H_2/Pd/BaSO_4}$ cyclohexanecarbonyl chloride (CCl with C=O) $\xleftarrow[CHCl_3]{SOCl_2}$ cyclohexanecarboxylic acid (COH with C=O)

e) cyclohexanol $\xrightarrow[CH_2Cl_2]{PCC}$ cyclohexanone $\xrightarrow{(C_6H_5)_3\overset{+}{P}\overset{-}{C}H_2}$ methylenecyclohexane ($=CH_2$)

$\downarrow$ 1. BH_3 2. H_2O_2, ^-OH

cyclohexanecarbaldehyde (CH with C=O) $\xleftarrow[CH_2Cl_2]{PCC}$ cyclohexylmethanol (CH_2OH)

24.38 succinic anhydride $\underset{}{\overset{NH_4^+}{\rightleftharpoons}}$ [protonated anhydride (+:OH) attacked by $:NH_3$ $\rightleftharpoons$ H_2C–C(=O)–ÖH / H_2C–C(=O)–$\overset{+}{N}H_3$ $\rightleftharpoons$ H_2C–C(=$\overset{+}{O}$H)–ÖH / H_2C–C(=O)–$\ddot{N}H_2$]

$\upharpoonleft\downharpoonright$

succinimide (H_2C–C(=O)–NH–C(=O)–CH_2) $\underset{}{\overset{-H^+}{\rightleftharpoons}}$ [protonated imide ($\overset{+}{O}$H) $\rightleftharpoons$ HÖ: :$\overset{+}{O}H_2$ tetrahedral intermediate with :NH $\rightleftharpoons$ HÖ: :ÖH tetrahedral intermediate with $+NH_2$]

24.39 Start at the end of the sequence of reactions and work backwards. If necessary, cover up all pieces of information you are not using at the moment to keep them from distracting you.

a) Because the *keto acid* $C_9H_{13}NO_3$ loses CO_2 on heating, it must be a β-keto acid (if you've forgotten about β-keto acids, review them in section 23.9). Neglecting stereoisomerism, we can then draw the structure of the β-keto acid as:

[structure: bicyclic β-keto acid with CH_3N bridge, COH (C=O), H, and ketone O]

b) When ecgonine ($C_9H_{15}NO_3$) is treated with CrO_3, the keto acid $C_9H_{13}NO_3$ is produced. Since CrO_3 is an oxidizing agent (used for oxidizing alcohols to carbonyl compounds), we can determine that ecgonine has the following structure. Again, the stereochemistry is unspecified.

Ecgonine

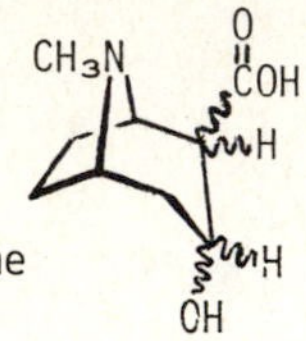

c) Ecgonine contains carboxylic acid and alcohol functional groups. The other products of hydroxide treatment of cocaine are a carboxylic acid (benzoic acid) and an alcohol (methanol). Cocaine thus contains two ester functional groups, which are saponified on reaction with hydroxide.

Cocaine

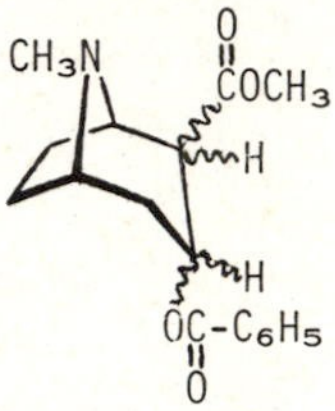

d) The complete reaction sequence:

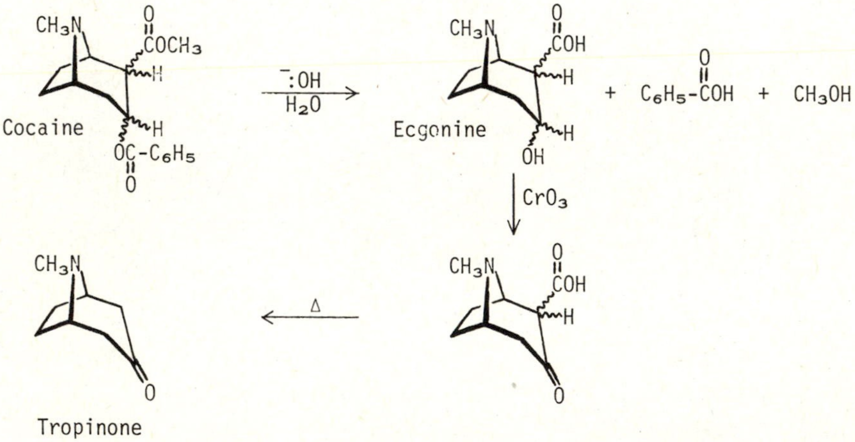

Cocaine $\xrightarrow[H_2O]{^-:OH}$ Ecgonine + $C_6H_5-\overset{O}{\overset{\|}{C}}OH$ + CH_3OH

Ecgonine $\xrightarrow{CrO_3}$ keto acid $\xrightarrow{\Delta}$ Tropinone

<u>24.40</u> This synthesis requires a *nucleophilic aromatic substitution* reaction, which we studied in section 15.13.

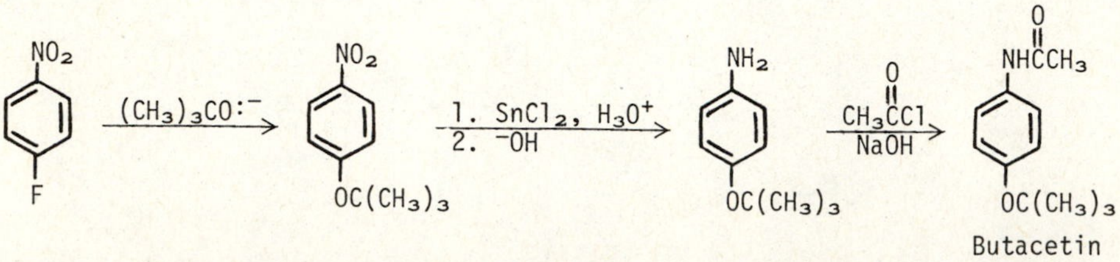

p-fluoronitrobenzene $\xrightarrow{(CH_3)_3CO:^-}$ NO_2–C_6H_4–$OC(CH_3)_3$ $\xrightarrow[2.\ ^-OH]{1.\ SnCl_2,\ H_3O^+}$ NH_2–C_6H_4–$OC(CH_3)_3$ $\xrightarrow[NaOH]{CH_3\overset{O}{\overset{\|}{C}}Cl}$ $CH_3\overset{O}{\overset{\|}{C}}NH$–$C_6H_4$–$OC(CH_3)_3$

Butacetin

24.41

$$O_2N\text{-}C_6H_3(OH)\text{-}COOH + SOCl_2 \longrightarrow O_2N\text{-}C_6H_3(OH)\text{-}COCl \xrightarrow{C_6H_5OH} O_2N\text{-}C_6H_3(OH)\text{-}CO_2C_6H_5$$

$$\xrightarrow{H_2,\ Pd/C} H_2N\text{-}C_6H_3(OH)\text{-}CO_2C_6H_5$$

Phenyl 4-aminosalicylate

24.42 a)

	$CH_3CH_2CONHCH_3$	$CH_3CON(CH_3)_2$
	N-methylpropanamide	*N*,*N*-dimethylacetamide
IR	1680 cm^{-1} (*N*-substituted amide)	1650 cm^{-1} (*N*,*N*-disubstituted amide)
^1H NMR	1 methyl group 1 ethyl group	3 methyl groups

b)

	$HOCH_2CH_2CH_2CH_2C{\equiv}N$	$C_4H_7CONH_2$ (cyclobutyl–CNH_2 with C=O)
	5-hydroxypentanenitrile	cyclobutanecarboxamide
IR	3300-3400 cm^{-1} (hydroxyl) 2250 cm^{-1} (nitrile)	1690 cm^{-1} (amide)

c)

	$ClCH_2CH_2CH_2COH$ (C=O)	$HOCH_2CH_2CH_2CCl$ (C=O)
	4-chlorobutanoic acid	4-hydroxybutanoyl chloride
IR	1710 cm^{-1} (carboxylic acid)	1810 cm^{-1} (acid chloride)

d)

	$CH_3CH_2COCH_2CH_3$ (C=O)	$CH_3COCH_2CH_2CH_3$ (C=O)
	ethyl propanoate	propyl acetate
^1H NMR	2 triplets 2 quartets	1 singlet 1 triplet 1 quartet 1 multiplet

24.43 H_3C–C_6H_4–Br (meta) $\xrightarrow{Mg}$ H_3C–C_6H_4–MgBr $\xrightarrow[2.\ H_3O^+]{1.\ CO_2}$ H_3C–C_6H_4–COOH

$\xrightarrow[CHCl_3]{SOCl_2}$ H_3C–C_6H_4–$\overset{O}{\overset{\|}{C}}Cl$

$\xrightarrow[NaOH]{HN(CH_2CH_3)_2}$ H_3C–C_6H_4–$\overset{O}{\overset{\|}{C}}N(CH_2CH_3)_2$

N,N-diethyl-*m*-toluamide

24.44 $Cl_2CH\overset{O}{\overset{\|}{C}}OH$ + 2 Cl–C_6H_4–O:$^-$ ⟶ $(Cl–C_6H_4–O)_2CH\overset{O}{\overset{\|}{C}}OH$

$\xrightarrow[CHCl_3]{SOCl_2}$ $(Cl–C_6H_4–O)_2CH\overset{O}{\overset{\|}{C}}Cl$

$\xrightarrow[pyridine]{HO–C_5H_9N–CH_3}$ $(Cl–C_6H_4–O)_2CH\overset{O}{\overset{\|}{C}}O$–$C_5H_9N$–$CH_3$

Lifibrate

This problem is made quite easy by the fact that you can start with any six-carbon compound. The reactions used to synthesize Lifibrate are:(1) two S_N2 displacements of chloride by a substituted phenoxide anion (2) formation of an acyl chloride (3) an esterification.

24.45 $R–\overset{:O:}{\overset{\|}{C}}–OCH_2CH_3$ $\overset{H^+}{\rightleftharpoons}$ $\left[R–\overset{\overset{+}{:OH}}{\overset{\|}{C}}–OCH_2CH_3 + H\ddot{O}CH_3 \rightleftharpoons R–\overset{:\ddot{O}H}{\overset{|}{C}}(–\ddot{O}CH_2CH_3)–\overset{+}{H\ddot{O}}CH_3 \rightleftharpoons R–\overset{:\ddot{O}H}{\overset{|}{C}}(–\ddot{O}CH_2CH_3)–:\ddot{O}CH_3 \right]$

$\rightleftharpoons$ $\left[R–\overset{H\ddot{O}:}{\overset{|}{C}}(–\overset{+}{\underset{H}{O}}CH_2CH_3)–:\ddot{O}CH_3 \rightleftharpoons R–\overset{\overset{+}{:OH}}{\overset{\|}{C}}–\ddot{O}CH_3 + HOCH_2CH_3 \right]$ $\overset{-H^+}{\rightleftharpoons}$ $R–\overset{O}{\overset{\|}{C}}–OCH_3$

In acidic methanol, the ethyl ester can react to yield a methyl ester. Conversion of the ethyl ester to the methyl ester occurs because of the large excess of methanol, the solvent.

23.46 $CH_3CHCl\overset{O}{\overset{\|}{C}}OCH_3$

23.47 $CH_3CH_2CH_2C{\equiv}N$

What you should know:

After doing these problems you should be able to:

1. Name and draw acid chlorides, anhydrides, esters, amides and nitriles;
2. Formulate nucleophilic acyl substitution mechanisms for reactions of carboxylic acid derivatives;
3. Understand the relative reactivity of carboxylic acid derivatives;
4. Predict the products of reactions of carboxylic acid derivatives;
5. Identify carboxylic acid derivatives by spectroscopic methods.

CHAPTER 25 CARBONYL α-SUBSTITUTION REACTIONS

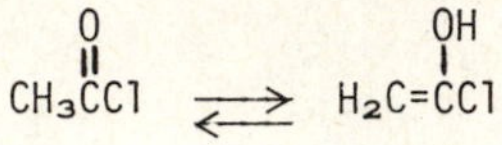

25.1 a) cyclopentanone ⇌ 1-cyclopentenol (*enol*)

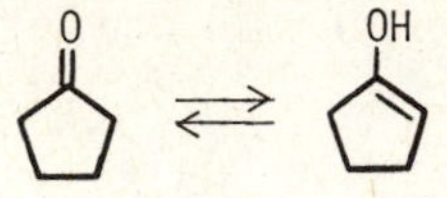

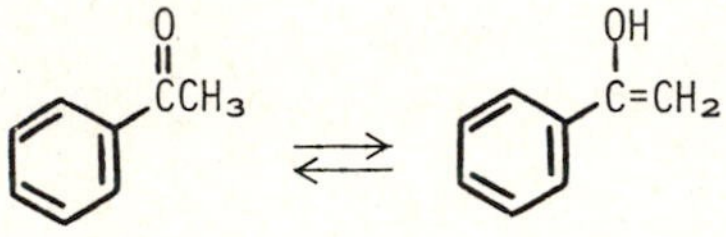

b) CH_3CCl (C=O) ⇌ $H_2C{=}CCl$ (OH) *enol*

c) $CH_3COCH_2CH_3$ (C=O) ⇌ $H_2C{=}COCH_2CH_3$ (OH) *enol*

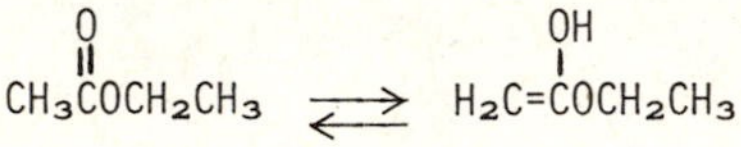

d) CH_3CH (C=O) ⇌ $H_2C{=}CH$ (OH) *enol*

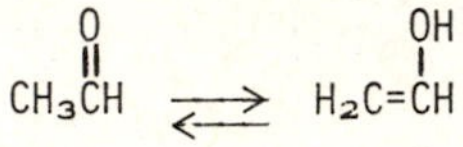

e) CH_3COH (C=O) ⇌ $H_2C{=}COH$ (OH) *enol*

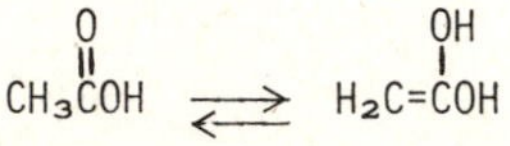

f) $C_6H_5CH_2CCH_3$ (C=O) ⇌ $C_6H_5CH{=}CCH_3$ (OH) *enol* or $C_6H_5CH_2C{=}CH_2$ (OH) *enol*

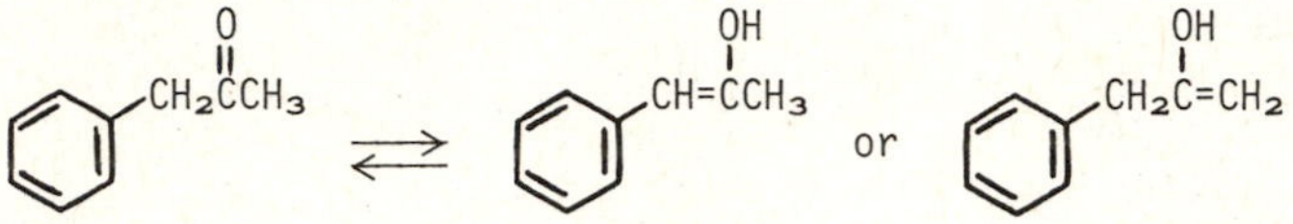

g) $C_6H_5CCH_3$ (C=O) ⇌ $C_6H_5C{=}CH_2$ (OH) *enol*

25.2 1,3-cyclohexanedione ⇌ 3-hydroxy-2-cyclohexenone ⇌ 3-hydroxy-2-cyclohexenone ⇌ 3-hydroxy-3-cyclohexenone ⇌ 3-hydroxy-3-cyclohexenone

equivalent; more stable

equivalent; less stable

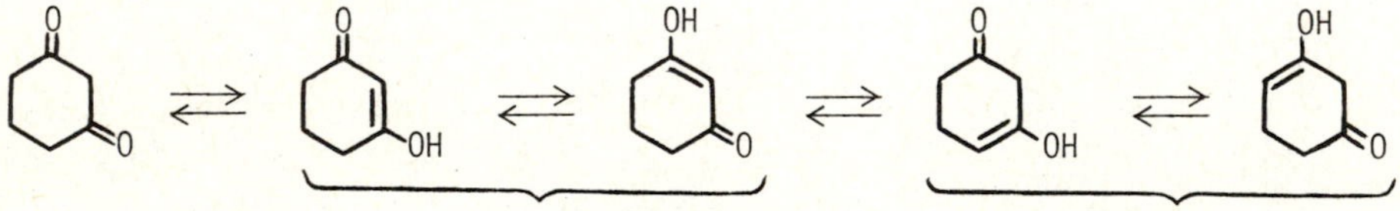

The first two mono-enols are more stable because the enol double bond is conjugated with the carbonyl group.

25.3 $H_3C{-}C({=}\ddot{O}{:}){-}CH_3 + D_3O^+ \rightleftharpoons [D_2\ddot{O}{:} \; H{-}H_2C{-}C({=}\overset{+}{O}D){-}CH_3] \rightleftharpoons H_2C{=}C(\ddot{O}D){-}CH_3 + D_2\overset{+}{O}H$

enol

$H_2C{=}C(\ddot{O}D){-}CH_3$ (*enol*) + $D_2\overset{+}{O}{-}D \rightleftharpoons [D_2\ddot{O}{:} + DH_2C{-}C({=}\overset{+}{O}D){-}CH_3] \rightleftharpoons DH_2C{-}C({=}\ddot{O}{:}){-}CH_3 + D_3O^+$

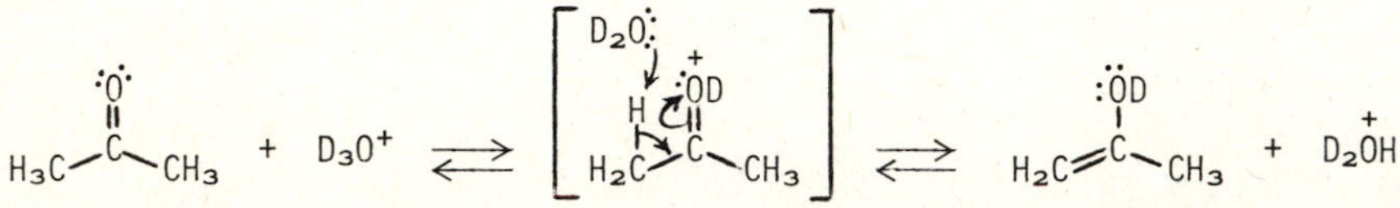

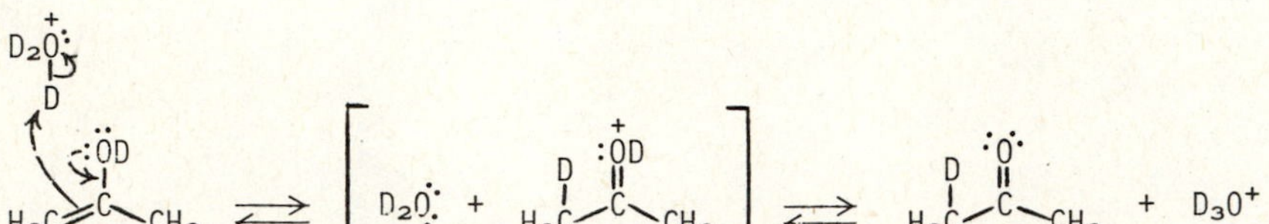

25.4 $CH_3\overset{*}{C}H(Ph)C(=O)CH_3 \underset{}{\overset{H_3O^+}{\rightleftharpoons}} [CH_3C(Ph)=C(OH)CH_3]$

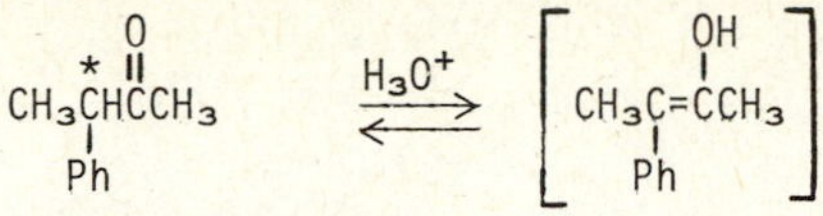

3-phenyl-2-butanone — *enol*

Loss of the proton at carbon 3 during enolization results in a loss of stereochemical configuration. Reattachment of a proton at carbon 3 can occur from either side of the sp^2 carbon, producing racemic 3-phenyl-2-butanone.

25.5 $CH_3CH_2C(CH_3)(Ph)C(=O)CH_3$ — $CH_3CH_2C(CH_3)(Ph)C(OH)=CH_2$

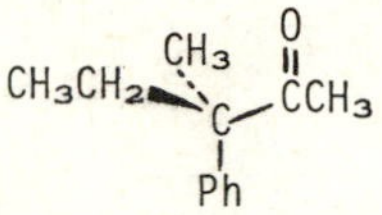

(*R*)-3-methyl-3-phenyl-2-pentanone 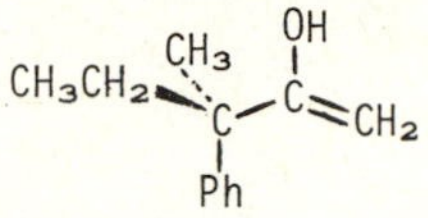 — *enol*

(*R*)-3-methyl-3-phenyl-2-pentanone can enolize only toward the methyl group since there is no proton at carbon 3. Because the chiral center is not involved in the enolization process, the ketone is not racemized by acid treatment.

25.6 Hell-Volhard-Zelinskii bromination involves formation of an intermediate acid bromide enol. As in problem 25.4, enolization results in loss of configuration at the chiral center. Bromination can occur from either side of the enol double bond, producing racemic 2-bromo-2-phenylpropanoic acid.

25.7 a) 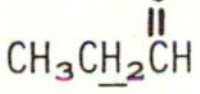 $CH_3\underline{CH_2}CHO$ b) 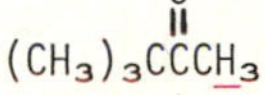 $(CH_3)_3CC(=O)\underline{CH_3}$

c) 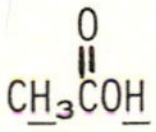 $\underline{CH_3}COO\underline{H}$ d) 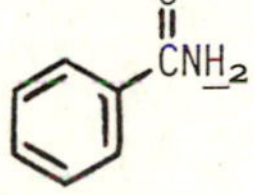 $C_6H_5CON\underline{H_2}$

e) $CH_3CH_2\underline{CH_2}CN$ f) 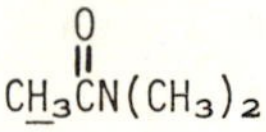 $\underline{CH_3}CON(CH_3)_2$

g) 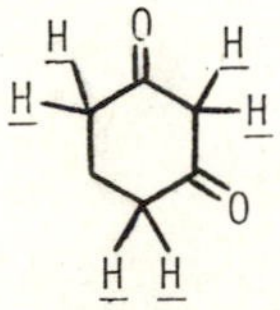 cyclohexane-1,3-dione with underlined H's on C2 (H, H), C4 (H, H), C6 (H, H)

25.8 2-methylcyclohexanone + $^-$:OH $\rightleftharpoons$ [enolate] $\rightleftharpoons$ racemic 2-methylcyclohexanone + $^-$:OH

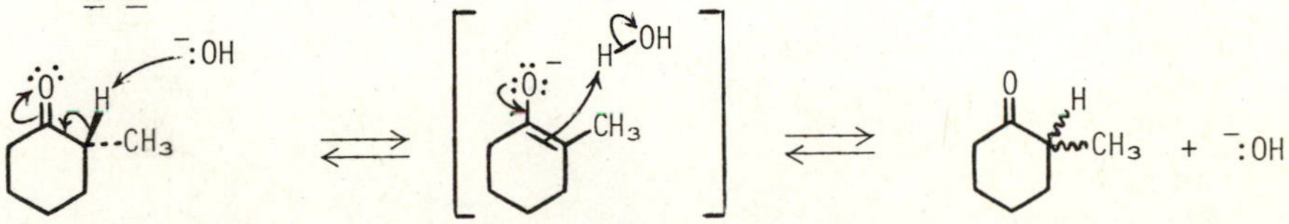

Carbon 2 loses its chirality in the step in which the enol double bond is formed. Subsequent steps occur with equal probability from either side of sp^2-hybridized carbon 2, resulting in racemic product. 3-Methylcyclohexanone is not racemized by base because its chiral center is not involved in the enolization process.

25.9 If both base-promoted chlorination and base-promoted bromination occur at the same rate, then the step involving halogen must come after the slow, or rate-limiting, step. Formation of the enolate is the rate-limiting step and is dependent on the concentrations of ketone and base.

$$\text{rate} = k[\text{base}][\text{ketone}]$$

25.10 Halogenation in acid medium:

$$\text{H–C(–)(–)–C(=O)–} + X_2 \xrightarrow{H^+} \text{X–C(–)(–)–C(=O)–} + H^+ + X:^-$$

Halogenation in basic medium:

$$\text{H–C(–)(–)–C(=O)–} + X_2 + B:^- \longrightarrow \text{X–C(–)(–)–C(=O)–} + BH + X:^-$$

Ketone halogenation in acid medium requires a small amount of acid to form the enol initially. However, hydrogen ions are generated as a byproduct of halogenation, and they carry on the catalysis of enol formation.

In basic medium, the function of the base is to form the enolate anion by removing a proton from the ketone. Each molecule of ketone requires one molecule of base. Because base is consumed in stoichiometric amounts, the reaction in basic medium is not base-catalyzed.

25.11

$$CH_3\overset{O}{\overset{\|}{C}}\text{-}CX_3 + :\bar{O}H \rightleftharpoons \left[CH_3\overset{:\ddot{O}:^-}{\overset{|}{\underset{:\ddot{O}H}{\underset{|}{C}}}}\text{-}CX_3\right] \longrightarrow CH_3\overset{O}{\overset{\|}{C}}\text{-}\ddot{O}H + :\bar{C}X_3 \longrightarrow CH_3\overset{O}{\overset{\|}{C}}\text{-}\ddot{O}:^- + HCX_3$$

This reaction is a nucleophilic acyl substitution.

25.12

Acid catalyzes the formation of enol.

$$RCH_2\overset{O}{\overset{\|}{C}}CH_3 + H^+ \rightleftharpoons RCH{=}\overset{:\ddot{O}H}{\overset{|}{C}}CH_3$$

The pi electrons from the enol double bond attack phenylselenenyl chloride.

$$RCH{=}\overset{OH}{\overset{|}{C}}CH_3 + C_6H_5\ddot{Se}\text{-}Cl \longrightarrow C_6H_5\text{-}Se\text{-}\overset{R}{\overset{|}{C}H}\text{-}\overset{\overset{+}{O}H}{\overset{\|}{C}}CH_3 + :Cl^-$$

Loss of a proton from oxygen yields the α-phenylseleno ketone. This reaction is similar to acid-catalyzed α-halogenation of ketones.

$$\rightleftharpoons C_6H_5\text{-}Se\text{-}\overset{R}{\overset{|}{C}H}\text{-}\overset{O}{\overset{\|}{C}}CH_3 + H^+$$

25.13 a)

$$CH_2(CO_2C_2H_5)_2 \xrightarrow[2.\ PhCH_2Br]{1.\ \overset{+}{Na}\overset{..}{\underset{..}{O}}CH_2CH_3} PhCH_2CH(CO_2C_2H_5)_2 + NaBr$$

$$\downarrow H_3O^+,\ \Delta$$

$$PhCH_2CH_2CO_2H + CO_2 + 2\ C_2H_5OH$$

3-phenylpropanoic acid

b)

$$CH_2(CO_2C_2H_5)_2 \xrightarrow[2.\ CH_3CH_2CH_2Br]{1.\ \overset{+}{Na}\overset{..}{\underset{..}{O}}CH_2CH_3} CH_3CH_2CH_2CH(CO_2C_2H_5)_2 + NaBr$$

$$\downarrow 1.\ \overset{+}{Na}\overset{..}{\underset{..}{O}}CH_2CH_3;\ 2.\ CH_3Br$$

$$CH_3CH_2CH_2CH(CH_3)CO_2H \xleftarrow{H_3O^+,\ \Delta} CH_3CH_2CH_2C(CH_3)(CO_2C_2H_5)_2 + NaBr$$

2-methylpentanoic acid

$$+ CO_2 + 2\ C_2H_5OH$$

c)

$$CH_2(CO_2C_2H_5)_2 \xrightarrow[2.\ (CH_3)_2CHCH_2Br]{1.\ \overset{+}{Na}\overset{..}{\underset{..}{O}}CH_2CH_3} (CH_3)_2CHCH_2CH(CO_2C_2H_5)_2 + NaBr$$

$$\downarrow H_3O^+,\ \Delta$$

$$(CH_3)_2CHCH_2CH_2CO_2H + CO_2 + 2\ C_2H_5OH$$

4-methylpentanoic acid

d)

$$CH_2(CO_2C_2H_5)_2 \xrightarrow[2.\ BrCH_2CH_2CH_2Br]{1.\ 2\ \overset{+}{Na}\overset{..}{\underset{..}{O}}CH_2CH_3} \text{(cyclobutane-1,1-)}(CO_2C_2H_5)_2 + NaBr$$

$$\downarrow H_3O^+,\ \Delta$$

$$\text{cyclobutyl}{-}CO_2CH_2CH_3 \xleftarrow[H^+]{CH_3CH_2OH} \text{cyclobutyl}{-}CO_2H + CO_2 + 2\ C_2H_5OH$$

ethyl cyclobutanecarboxylate

25.14 Since malonic ester has only two acidic hydrogen atoms, it can be alkylated only two times. Formation of trialkylated acetic acids is not possible.

25.15

$$CH_3\overset{O}{\overset{\|}{C}}CH_2\overset{O}{\overset{\|}{C}}OC_2H_5 \xrightarrow[2.\ BrCH_2CH_2CH_2CH_2Br]{1.\ 2\ \overset{+}{Na}\overset{..}{\underset{..}{O}}C_2H_5} \text{(cyclopentane-1,1-)}(\overset{O}{\overset{\|}{C}}CH_3)(\overset{O}{\overset{\|}{C}}OC_2H_5) + 2\ NaBr$$

$$\downarrow H_3O^+,\ \Delta$$

$$\text{cyclopentyl}{-}\overset{O}{\overset{\|}{C}}CH_3 + C_2H_5OH + CO_2$$

25.16 Compounds synthesized by the acetoacetic acid method (1) must be methyl ketones, (2) must not have a trisubstituted carbon next to the carbonyl functional group, and (3) must use primary, methyl, benzylic, or allylic halides or tosylates as alkylating agents.

a)

$$CH_3\overset{O}{\overset{\|}{C}}CH_2\overset{O}{\overset{\|}{C}}OC_2H_5 \xrightarrow[2.\ CH_3Br]{1.\ Na^+\ {}^-\ddot{O}CH_2CH_3} CH_3\overset{O}{\overset{\|}{C}}\underset{CH_3}{\underset{|}{C}H}\overset{O}{\overset{\|}{C}}OC_2H_5 + NaBr$$

$$\downarrow H_3O^+,\ \Delta$$

$$CH_3\overset{O}{\overset{\|}{C}}CH_2CH_3 + CO_2 + C_2H_5OH$$

2-butanone

b) Phenylacetone cannot be synthesized by the acetoacetic ester route because aryl halides cannot be used as alkylating agents.

c) The product of an acetoacetic ester synthesis must contain three carbon atoms from the original acetoacetic ester starting material. This is not possible for $PhCOCH_3$.

d) 3,3-Dimethyl-2-butanone can not be synthesized by the acetoacetic ester route because the carbon *alpha* to the carbonyl group has three alkyl substituents.

e)

$$CH_3\overset{O}{\overset{\|}{C}}CH_2\overset{O}{\overset{\|}{C}}OC_2H_5 \xrightarrow[2.\ PhCHBrCH_3]{1.\ Na^+\ {}^-\ddot{O}CH_2CH_3} CH_3\overset{O}{\overset{\|}{C}}\underset{\underset{Ph}{CHCH_3}}{\underset{|}{C}H}\overset{O}{\overset{\|}{C}}OCH_2CH_3 + NaBr$$

$$\downarrow H_3O^+,\ \Delta$$

$$CH_3\overset{O}{\overset{\|}{C}}CH_2\underset{Ph}{\underset{|}{C}H}CH_3 + CO_2 + C_2H_5OH$$

4-phenyl-2-pentanone

25.17 a)

$$PhCH_2\overset{O}{\overset{\|}{C}}CH_3 \xrightarrow[2.\ CH_3I]{1.\ LDA} PhCH(CH_3)\overset{O}{\overset{\|}{C}}CH_3$$

3-phenyl-2-butanone

Alkylation occurs at the carbon next to the phenyl group because the phenyl group can stabilize the enolate anion intermediate.

b) $CH_3CH_2CH_2CH_2C{\equiv}N \xrightarrow[\text{2. } CH_3CH_2I]{\text{1. LDA}} CH_3CH_2CH_2CH(CH_2CH_3)C{\equiv}N$

2-ethylpentanenitrile

c) cyclohexanone $\xrightarrow[\text{2. } H_2C{=}CHCH_2Br]{\text{1. LDA}}$ 2-($CH_2CH{=}CH_2$)cyclohexanone

2-allylcyclohexanone

d) cyclohexanone $\xrightarrow[\text{2. excess } CH_3I]{\text{1. excess } \overset{+}{Na}\,\overset{-}{\ddot{O}}CH_2CH_3}$ 2,6-(H_3C, H_3C, CH_3, CH_3)cyclohexanone

2,2,6,6-tetramethylcyclohexanone

25.18 A nitroso compound is analogous to a carbonyl compound. If there are protons *alpha* to the nitroso group, enolization similar to that observed for carbonyl compounds can occur, leading to formation of an oxime. Here, the enolization equilibrium favors the oxime. If no protons are adjacent to the nitroso group, enolization to the oxime cannot occur, and the nitroso compound is stable.

25.19 The enol form of 2,4-pentanedione is favored over the ketone form for two reasons. (1) Conjugation of the enol double bond with the second carbonyl group lowers the energy of the enol form relative to the ketone form. (2) Hydrogen bonding of the enol hydrogen to the other carbonyl group provides further stabilization. Neither of these types of enol stabilization is available to acetone.

$$CH_3\overset{O}{\overset{\|}{C}}CH_2\overset{O}{\overset{\|}{C}}CH_3 \rightleftarrows CH_3\overset{OH}{\overset{|}{C}}{=}CH\overset{O}{\overset{\|}{C}}CH_3$$

$H_3C{-}C(O{-}H\cdots O){=}CH{-}C{-}CH_3$ (intramolecular hydrogen bond: O—H····O)

25.20 a) $CH_3\overset{\ddot{O}}{\overset{\|}{C}}{-}\overset{-}{\ddot{C}}H{-}\overset{\ddot{O}}{\overset{\|}{C}}CH_3 \longleftrightarrow CH_3\overset{:\ddot{O}:^-}{\overset{|}{C}}{=}CH{-}\overset{\ddot{O}}{\overset{\|}{C}}CH_3 \longleftrightarrow CH_3\overset{\ddot{O}}{\overset{\|}{C}}{-}CH{=}\overset{:\ddot{O}:^-}{\overset{|}{C}}CH_3$

b) $:\overset{-}{C}H_2C{\equiv}N: \longleftrightarrow H_2C{=}C{=}\overset{-}{\ddot{N}}:$

c) $CH_3CH{=}CH{-}\overset{-}{\ddot{C}}H{-}\overset{\ddot{O}}{\overset{\|}{C}}CH_3 \longleftrightarrow CH_3CH{=}CH{-}CH{=}\overset{:\ddot{O}:^-}{\overset{|}{C}}CH_3 \longleftrightarrow CH_3\overset{-}{\ddot{C}}H{-}CH{=}CH{-}\overset{\ddot{O}}{\overset{\|}{C}}CH_3$

d) $:N{\equiv}C{-}\overset{-}{\ddot{C}}H{-}\overset{\ddot{O}}{\overset{\|}{C}}OCH_3 \longleftrightarrow :N{\equiv}C{-}CH{=}\overset{:\ddot{O}:^-}{\overset{|}{C}}OCH_3 \longleftrightarrow :\overset{-}{\ddot{N}}{=}C{=}CH{-}\overset{\ddot{O}}{\overset{\|}{C}}OCH_3$

e) [Resonance structures of the enolate anion of phenylacetone (1-phenyl-2-propanone): $C_6H_5\ddot{C}HCOCH_3 \longleftrightarrow C_6H_5\ddot{C}HCOCH_3 \longleftrightarrow C_6H_5CH{=}C(\ddot{O}^-)CH_3 \longleftrightarrow C_6H_5CH{=}C(\ddot{O}^-)CH_3$, then $\updownarrow$, with the negative charge delocalized to the ortho and para positions of the ring in the structures $\longleftrightarrow$ bonded to $CHCCH_3$ (C=O)]

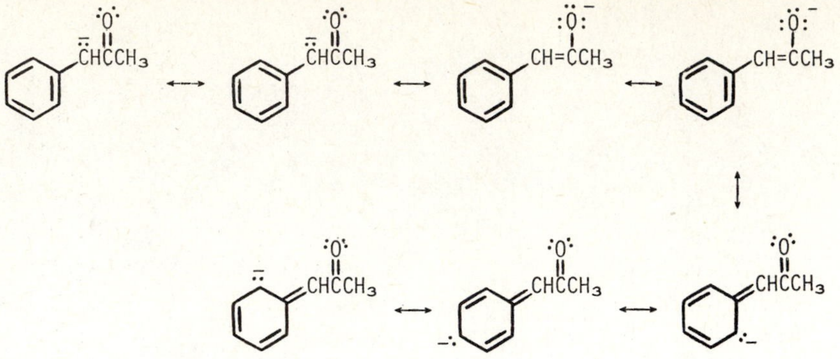

25.21 Acidic hydrogens are italicized.

a) $\mathit{H}OC\mathit{H}_2C(=O)C\mathit{H}_3$

b) $\mathit{H}OCH_2C\mathit{H}_2C(=O)C(CH_3)_3$

c) [cyclopentane-1,3-dione with all six ring hydrogens italicized] All hydrogens are somewhat acidic, although those between the carbonyls are the most acidic.

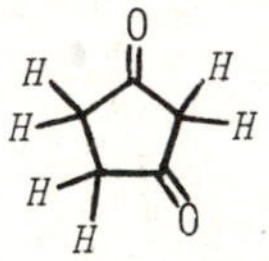

d) $C\mathit{H}_3CH{=}CHC(=O)H$

e) $(CH_3)_2NH$ No hydrogens are acidic, although the amine hydrogen is more acidic than the methyl hydrogens.

f) [structure of cortisone with acidic hydrogens italicized: $C\mathit{H}_2O\mathit{H}$, O*H*, ring hydrogens alpha to carbonyls] There are eleven acidic hydrogens in cortisone.

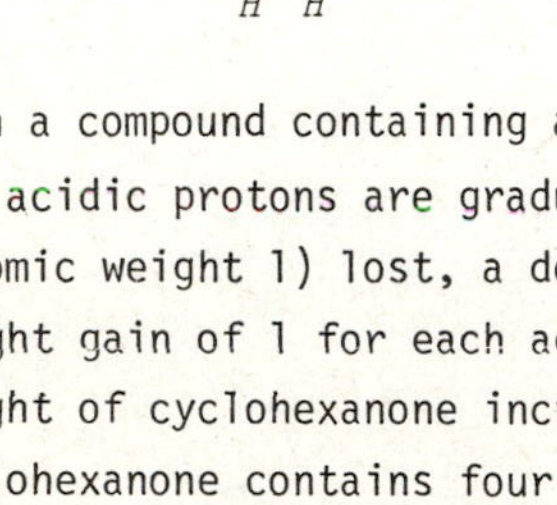

25.22 When a compound containing acidic hydrogen atoms is treated with NaOD in D_2O, all acidic protons are gradually replaced by deuterons. For each proton (atomic weight 1) lost, a deuteron (atomic weight 2) is added, for a molecular weight gain of 1 for each acidic hydrogen replaced. Since the molecular weight of cyclohexanone increases from 98 to 102 after $NaOD/D_2O$ treatment, cyclohexanone contains four acidic hydrogen atoms.

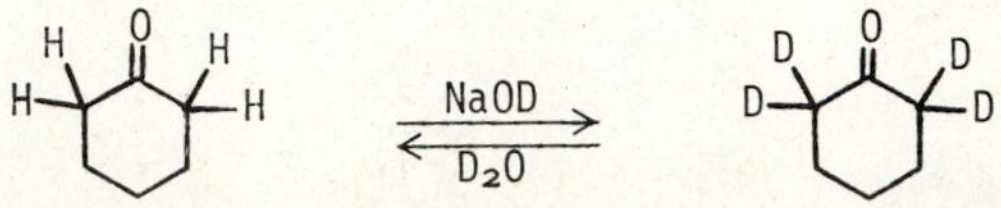

[cyclohexanone with four α-H] $\underset{D_2O}{\overset{NaOD}{\rightleftharpoons}}$ [2,2,6,6-tetradeuteriocyclohexanone]

25.23

2-methylcycloheptanone (structure: cycloheptanone ring with H, H on C-2... H and CH_3 on the alpha carbons)

3-methylcycloheptanone (structure: cycloheptanone ring with H, H on both alpha carbons, CH_3 on C-3)

2-Methylcycloheptanone contains three acidic hydrogen atoms; 3-methylcycloheptanone contains four. After treatment with NaOD in D_2O, the molecular weight of the 2-methyl isomer increases by 3 and that of the 3-methyl isomer increases by 4. A mass spectrum of the deuterated ketones can then distinguish between them.

25.24 Least acidic ⟶ most acidic

$$(CH_3CH_2)_2NH < CH_3\overset{O}{\overset{\|}{C}}CH_3 < CH_3CH_2OH < CH_3\overset{O}{\overset{\|}{C}}CH_2\overset{O}{\overset{\|}{C}}CH_3 < CH_3CH_2\overset{O}{\overset{\|}{C}}OH < Cl_3\overset{O}{\overset{\|}{C}}OH$$

25.25 a)

$$CH_2(CO_2C_2H_5)_2 \xrightarrow[2.\ CH_3CH_2CH_2Br]{1.\ \overset{+}{Na}\ddot{\overset{-}{O}}CH_2CH_3} CH_3CH_2CH_2CH(CO_2C_2H_5)_2 + NaBr$$

$$\downarrow H_3O^+,\ \Delta$$

$$\underset{\text{ethyl pentanoate}}{CH_3CH_2CH_2CH_2CO_2C_2H_5} \xleftarrow[H^+]{C_2H_5OH} CH_3CH_2CH_2CH_2CO_2H + CO_2 + 2\ C_2H_5OH$$

b) Synthesis of ethyl 3-methylbutanoate by a malonic ester route would occur in poor yield, since the halide needed for alkylation, CH_3CHXCH_3, undergoes elimination, instead of substitution, in the presence of $\ddot{\overset{-}{C}}H(CO_2Et)_2$. A possible route:

$$(CH_3)_2CHCH_2Br \xrightarrow{Mg} (CH_3)_2CHCH_2MgBr \xrightarrow[2.\ H_3O^+]{1.\ CO_2} (CH_3)_2CHCH_2COOH$$

$$\downarrow C_2H_5OH,\ H^+$$

$$\underset{\text{ethyl 3-methylbutanoate}}{(CH_3)_2CHCH_2CO_2C_2H_5}$$

c)

$$CH_2(CO_2C_2H_5)_2 \xrightarrow[2.\ CH_3CH_2Br]{1.\ \overset{+}{Na}\ddot{\overset{-}{O}}CH_2CH_3} CH_3CH_2CH(CO_2C_2H_5)_2 + NaBr$$

$$\searrow 1.\ \overset{+}{Na}\ddot{\overset{-}{O}}CH_2CH_3;\ 2.\ CH_3Br$$

$$CH_3CH_2\overset{CH_3}{\overset{|}{C}}HCOOH + 2\ C_2H_5OH + CO_2 \xleftarrow[\Delta]{H_3O^+} CH_3CH_2\overset{CH_3}{\overset{|}{C}}(CO_2C_2H_5)_2 + NaBr$$

$$\downarrow C_2H_5OH,\ H^+$$

$$\underset{\text{ethyl 2-methylbutanoate}}{CH_3CH_2\overset{CH_3}{\overset{|}{C}}HCO_2C_2H_5}$$

d) The malonic ester route can't be used to synthesize *alpha*-trisubstituted carboxylic acids. A possible route:

$$(CH_3)_3CBr \xrightarrow{Mg} (CH_3)_3CMgBr \xrightarrow[2.\ H_3O^+]{1.\ CO_2} (CH_3)_3CCOOH \xrightarrow[H^+]{C_2H_5OH} (CH_3)_3CCO_2C_2H_5$$

ethyl 2,2-dimethylpropanoate

25.26 a) [reaction scheme: 3-cyclohexenone + H_3O^+ ⇌ [protonated ketone, $:OH_2$ removes α-H ⇌ enol + H_3O^+]]

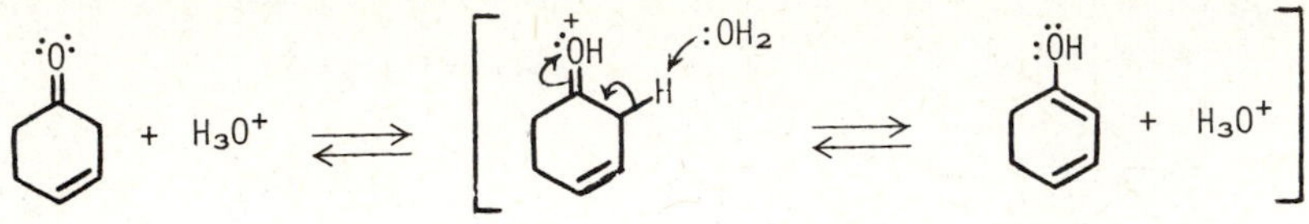

enol

b) [reaction scheme: [enol + $H-\overset{+}{O}H_2$ ⇌ protonated 2-cyclohexenone + H_2O] ⇌ 2-cyclohexenone + H_3O^+]

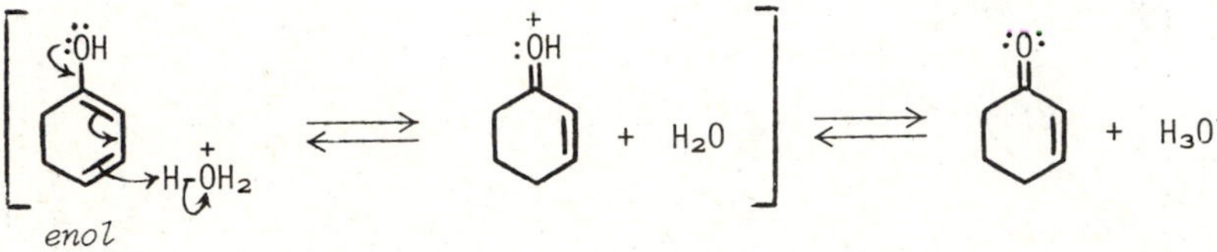

enol

25.27 [reaction scheme: 3-cyclohexenone + $^-:B$ ⇌ [resonance forms of enolate] + BH ⇌ 2-cyclohexenone + $B:^-$]

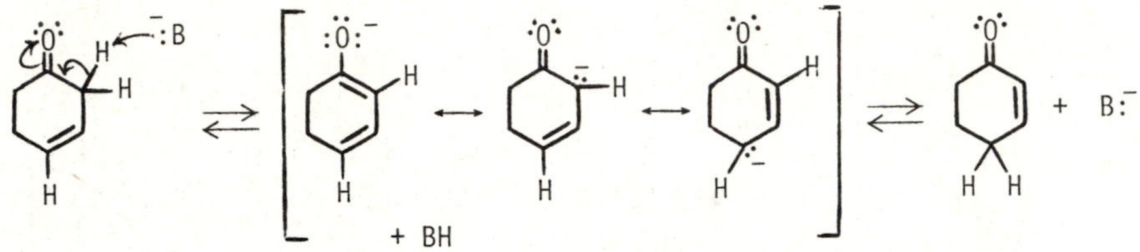

The enolate of 3-cyclohexenone can be reprotonated at three different positions. Reprotonation at the *gamma* position yields the α,β-unsaturated ketone.

25.28 [reaction scheme: 2-methyl-2-cyclopentenone $\underset{\text{abstraction of } \gamma \text{ proton}}{\overset{^-:OH}{\rightleftharpoons}}$ [resonance forms of dienolate] $\overset{H_2O}{\rightleftharpoons}$ 2-methyl-3-cyclopentenone $\underset{\text{abstraction of } \alpha \text{ proton}}{\overset{^-:OH}{\rightleftharpoons}}$ [resonance forms of enolate] $\overset{H_2O}{\rightleftharpoons}$ 5-methyl-2-cyclopentenone]

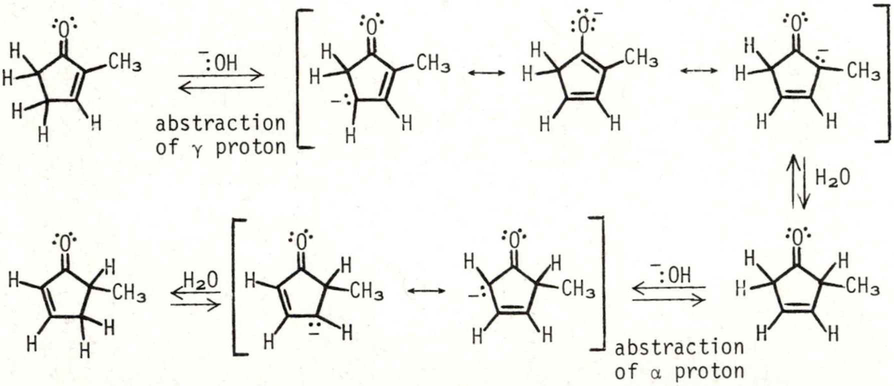

All protons in the five-membered ring can be exchanged by base treatment.

25.29 Use a *malonic ester synthesis* if the product you want is an α-substituted carboxylic acid or derivative.

Use an *acetoacetic acid synthesis* if the product you want is an α-substituted methyl ketone.

a) $CH_2(CO_2C_2H_5)_2 \xrightarrow[2.\ 2\ CH_3Br]{1.\ 2\ Na\overset{+}{}\ddot{\overset{-}{O}}CH_2CH_3} (CH_3)_2C(CO_2C_2H_5)_2 + 2\ NaBr$

b) $CH_2(CO_2C_2H_5)_2 \xrightarrow[\text{2. } BrCH_2CH_2CH_2CH_2Br]{\text{1. 2 } Na^+\,^-OCH_2CH_3}$ 1,1-bis($CO_2C_2H_5$)cyclopentane + 2 NaBr

$\downarrow H_3O^+, \Delta$

cyclopentyl-CH_2OH $\xleftarrow[\text{2. } H_3O^+]{\text{1. } BH_3}$ cyclopentyl-COOH + CO_2 + 2 C_2H_5OH

c) $CH_3\overset{O}{\overset{\|}{C}}CH_2\overset{O}{\overset{\|}{C}}OC_2H_5 \xrightarrow[\text{2. } BrCH_2CH_2CH_2CH_2Br]{\text{1. 2 } Na^+\,^-OCH_2CH_3}$ 1-($\overset{O}{\overset{\|}{C}}CH_3$)-1-($CO_2C_2H_5$)cyclopentane + 2 NaBr

$\downarrow H_3O^+, \Delta$

cyclopentyl-$\overset{OH}{\overset{|}{C}}HCH_3$ $\xleftarrow[\text{2. } H_3O^+]{\text{1. } NaBH_4}$ 1-($\overset{O}{\overset{\|}{C}}CH_3$)-1-H-cyclopentane + CO_2 + C_2H_5OH

d) $CH_2(CO_2C_2H_5)_2 \xrightarrow[\text{2. } BrCH_2CO_2C_2H_5]{\text{1. } Na^+\,^-OCH_2CH_3}$ $C_2H_5O_2CCH_2CH(CO_2C_2H_5)_2$ + NaBr

$\downarrow H_3O^+, \Delta$

$HOOCCH_2CH_2COOH$ + CO_2 + 3 C_2H_5OH

e) $CH_3\overset{O}{\overset{\|}{C}}CH_2\overset{O}{\overset{\|}{C}}OC_2H_5 \xrightarrow[\text{2. } H_2C=CHCH_2Br]{\text{1. } Na^+\,^-OCH_2CH_3}$ $H_2C=CHCH_2\underset{CO_2C_2H_5}{\underset{|}{C}}H\overset{O}{\overset{\|}{C}}CH_3$ + NaBr

$\downarrow H_3O^+, \Delta$

$H_2C=CHCH_2CH_2\overset{O}{\overset{\|}{C}}CH_3$ + CO_2 + C_2H_5OH

25.30 a) 1,1-cyclohexanedicarboxylic acid (COOH, COOH) $\xrightarrow{\Delta}$ cyclohexane-COOH + CO_2

b) $(CH_3)_2CHCO_2C_2H_5 \xrightarrow[\text{2. } C_6H_5SeBr]{\text{1. LDA}}$ $(CH_3)_2\overset{SeC_6H_5}{\overset{|}{C}}CO_2C_2H_5$ (A) $\xrightarrow{H_2O_2}$ $H_2C=\overset{CH_3}{\overset{|}{C}}CO_2C_2H_5$ (B)

c) 1,3-cyclopentanedione $\xrightarrow[\text{2. } CH_3I]{\text{1. } Na^+\,^-OC_2H_5}$ 2-methyl-1,3-cyclopentanedione (CH_3)

d) $CH_3CH_2CH_2COOH \xrightarrow[PBr_3]{Br_2}$ $CH_3CH_2CHBrCOBr$ (A) $\xrightarrow{H_2O}$ $CH_3CH_2CHBrCOOH$ (B)

e)

$$C_6H_5\overset{O}{\overset{\|}{C}}CH_3 \xrightarrow[I_2]{^-:OH,\ H_2O} C_6H_5\overset{O}{\overset{\|}{C}}O:^- + HCI_3$$

25.31 First, treat geraniol with PBr_3 to form geranyl bromide, $(CH_3)_2C{=}CHCH_2CH_2C(CH_3){=}CHCH_2Br$.

a) $$CH_3CO_2C_2H_5 \xrightarrow[\text{2. geranyl bromide}]{\text{1. LDA}} (CH_3)_2C{=}CHCH_2CH_2C(CH_3){=}CHCH_2CH_2CO_2C_2H_5$$

ethyl geranylacetate

Alternatively:

$$CH_2(CO_2C_2H_5)_2 \xrightarrow[\text{2. geranyl bromide}]{\text{1. } \overset{+}{Na}\overset{-}{\ddot{O}}C_2H_5} (CH_3)_2C{=}CHCH_2CH_2C(CH_3){=}CHCH_2CH(CO_2C_2H_5)_2$$

$$\downarrow H_3O^+,\ \Delta$$

$$(CH_3)_2C{=}CHCH_2CH_2C(CH_3){=}CHCH_2CH_2COOH$$

$$\downarrow C_2H_5OH,\ H^+$$

$$(CH_3)_2C{=}CHCH_2CH_2C(CH_3){=}CHCH_2CH_2CO_2C_2H_5$$

ethyl geranylacetate

b) $$CH_3\overset{O}{\overset{\|}{C}}CH_2CO_2C_2H_5 \xrightarrow[\text{2. geranyl bromide}]{\text{1. } \overset{+}{Na}\overset{-}{\ddot{O}}C_2H_5} (CH_3)_2C{=}CHCH_2CH_2C(CH_3){=}CHCH_2\underset{CO_2C_2H_5}{\underset{|}{C}H}\overset{O}{\overset{\|}{C}}CH_3$$

$$\downarrow H_3O^+,\ \Delta$$

$$(CH_3)_2C{=}CHCH_2CH_2C(CH_3){=}CHCH_2CH_2\overset{O}{\overset{\|}{C}}CH_3 + C_2H_5OH + CO_2$$

geranylacetone

25.32 a) cyclohexanone $\xrightarrow{Ph_3\overset{+}{P}\overset{-}{\ddot{C}}H_2}$ methylenecyclohexane (ring$=CH_2$)

b) product of a) $\xrightarrow[\text{peroxides}]{HBr}$ cyclohexyl–CH_2Br

c) cyclohexanone $\xrightarrow[\text{2. } C_6H_5CH_2Br]{\text{1. LDA}}$ 2-($CH_2C_6H_5$)cyclohexanone

d) $$CH_2(CO_2C_2H_5)_2 \xrightarrow[\text{2. product of b)}]{\text{1. } \overset{+}{Na}\overset{-}{\ddot{O}}CH_2CH_3} \text{cyclohexyl–}CH_2CH(CO_2C_2H_5)_2 + NaBr \xrightarrow[\Delta]{H_3O^+} \text{cyclohexyl–}CH_2CH_2COOH + 2\ C_2H_5OH + CO_2$$

e) cyclohexanone $\xrightarrow[H_3O^+]{NaCN}$ 1-hydroxycyclohexanecarbonitrile (HO, CN) $\xrightarrow[pyridine]{POCl_3}$ 1-cyclohexenecarbonitrile (CN) $\xrightarrow{H_3O^+}$ 1-cyclohexenecarboxylic acid (COOH)

f) Method 1: cyclohexanone $\xrightarrow[CH_3COOH]{Br_2}$ 2-bromocyclohexanone $\xrightarrow[\Delta]{pyridine}$ 2-cyclohexenone

Method 2: cyclohexanone $\xrightarrow[2.\ C_6H_5SeBr]{1.\ LDA}$ 2-(SeC_6H_5)cyclohexanone $\xrightarrow{H_2O_2}$ 2-cyclohexenone

g) product from d) $\xrightarrow[2.\ H_2O]{1.\ Br_2,\ PBr_3}$ $CH_2CHBrCOOH$ (on cyclohexane) $\xrightarrow[H^+]{CH_3OH}$ $CH_2CHBrCO_2CH_3$ (on cyclohexane) $\xrightarrow[\Delta]{C_5H_5N}$ $CH{=}CHCO_2CH_3$ (on cyclohexane)

product from d) $\xrightarrow[H^+]{CH_3OH}$ $CH_2CH_2CO_2CH_3$ (on cyclohexane) $\xrightarrow[2.\ C_6H_5SeBr]{1.\ LDA}$ $CH_2CH(SeC_6H_5)CO_2CH_3$ (on cyclohexane) $\xrightarrow{H_2O_2}$ $CH{=}CHCO_2CH_3$ (on cyclohexane)

h) product from e) $\xrightarrow[Pd/C]{H_2}$ cyclohexanecarboxylic acid (COOH)

25.33 Treatment of either the *cis* or *trans* isomer with base causes enolization *alpha* to the carbonyl group and results in loss of configuration at the α-position. Reattachment of the proton at carbon 2 produces either of the diastereomeric 4-*tert*-butyl-2-methylcyclohexanones. In both diastereomers the *tert*-butyl group of carbon 4 occupies the equatorial position for steric reasons. The methyl group of the *cis* isomer is also equatorial; the methyl group of the *trans* isomer is axial. The *trans* isomer is less stable because of interactions of the axial methyl group with the ring protons.

$(CH_3)_3C$–, O, CH_3, H, H, H (*cis*) $\underset{}{\overset{^{-}:OH}{\rightleftharpoons}}$ $(CH_3)_3C$–, O, H, H, H, CH_3 (*trans*)

cis *trans*

25.34 Protons *alpha* to a carbonyl group are acidic. We also know from Problem 25.28 that protons *gamma* to an enone carbonyl group are acidic. Thus for 2-methyl-2-cyclopentenone, protons at the starred positions are acidic.

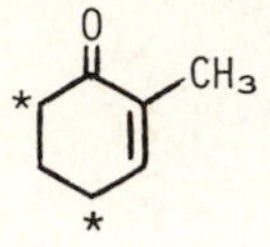

Isomerization of a 2-substituted 2-cyclohexenone to a 6-substituted 2-cyclohexenone requires removal of a proton from the 5-position of the 2-substituted isomer. Since protons in this position are not acidic, double bond isomerization does not occur.

25.35 a) Reaction with Br_2 at the *alpha* position occurs with aldehydes and ketones, not with esters.

b) Aryl halides can't be used in malonic ester syntheses because they do not undergo S_N2 displacement reactions.

c) The product of this reaction sequence, $H_2C{=}CHCH_2CH_2COCH_3$, is a methyl ketone, not a carboxylic acid.

d) (1) A strong base such as LDA must be used to form the nitrile anion; (2) H_2O_2, not pyridine, is used to remove the phenylselenyl group and to form the double bond.

e) Base-promoted bromination of a ketone yields a mixture of mono- to tetrabromo products. Reaction of the brominated ketones with pyridine and heat produces a mixture of compounds.

f) When $(CH_3)_2CHCOCH_3$ reacts with LDA, the less substituted enolate anion, $(CH_3)_2CHCO\ddot{C}H_2^-$, is more likely to form. Alkylation of this enolate with $C_6H_5CH_2Br$, yields $(CH_3)_2CHCOCH_2CH_2C_6H_5$ as the major product.

25.36

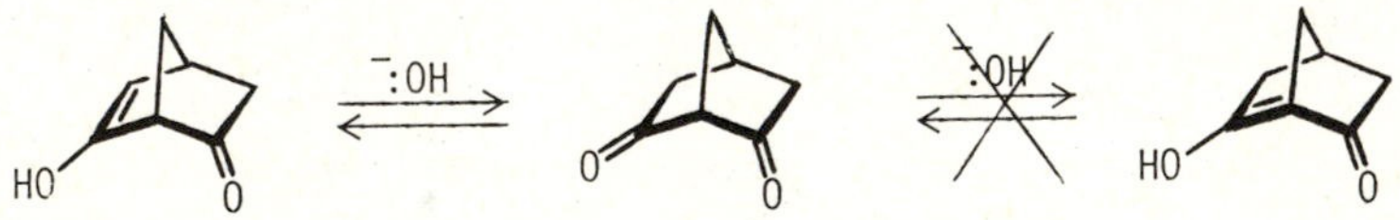

Enolization in the direction predicted for β-diketones would produce an enol with a double bond at the "bridgehead" of the fused ring system. Since ring systems with bridgehead double bonds are highly strained (see Problem 23.33), enolization occurs instead in the opposite direction. This diketone thus resembles a monoketone, rather than a diketone, in its pK_a and degree of enolization.

25.37 Both methylmagnesium bromide and *tert*-butylmagnesium bromide give the expected carbonyl addition products. The yield of the *tert*-butylmagnesium bromide addition product is very low, however, because of the difficulty of approach of the bulky *tert*-butyl Grignard reagent to the carbonyl carbon. More favorable is the acid-base reaction between the Grignard reagent and a carbonyl *alpha* proton.

$$\text{cyclohexanone} + BrMgC(CH_3)_3 \longrightarrow \text{1-(OMgBr)cyclohexene} + HC(CH_3)_3$$

When D_2O is added to the reaction mixture, the deuterated ketone is produced.

$$\text{1-(OMgBr)cyclohexene} \xrightarrow{D_3O^+} \text{2-deuteriocyclohexanone}$$

An additional process that competes with addition of hindered Grignard reagents is reduction of the carbonyl group.

$$\text{cyclohexanone} + BrMgC(CH_3)_2CH_2H \longrightarrow \text{cyclohexyl-OMgBr (H)} + \underset{CH_2}{\overset{C(CH_3)_2}{\|}} \xrightarrow{H_3O^+} \text{cyclohexanol (OH)} + \underset{CH_2}{\overset{C(CH_3)_2}{\|}}$$

25.38 Laurene and its isomer differ only in the configuration at the carbon *alpha* to the methylene group. This chiral center, which is *alpha* to a carbonyl group in the starting material is epimerized by the basic Wittig reagent to yield an equilibrium mixture of two diastereomeric ketones. One of these ketones reacts preferentially with the Wittig reagent to give only one epimeric product.

$$\text{laurene} + \ddot{C}H_2\overset{+}{P}Ph_3 \rightleftharpoons \left[\text{enolate} + CH_3\overset{+}{P}Ph_3\right]$$

$$\text{enolate} \rightleftharpoons \text{epimeric ketone} + \ddot{C}H_2\overset{+}{P}Ph_3 \longrightarrow \text{methylene product}$$

25.39

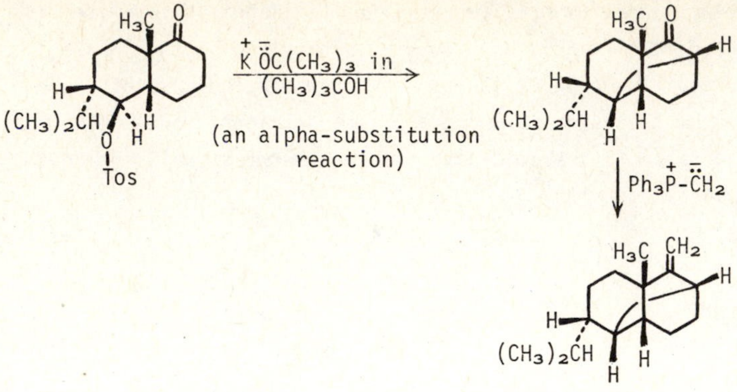

$\overset{+}{K}\ \overset{-}{O}C(CH_3)_3$ in $(CH_3)_3COH$

(an alpha-substitution reaction)

$Ph_3\overset{+}{P}-\overset{-}{C}H_2$

sativene

25.40

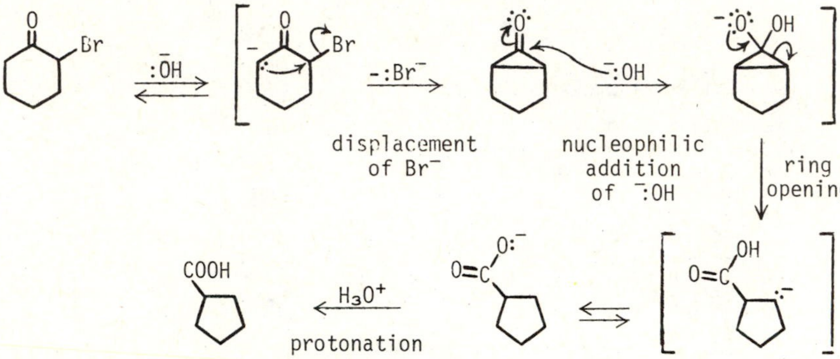

$:\overset{-}{O}H$

$-:Br^-$

displacement of Br^-

$^-:OH$

nucleophilic addition of $^-:OH$

ring opening

H_3O^+

protonation

25.41

$:\overset{-}{C}H_2-\overset{+}{N}\equiv N$

nucleophilic addition of CH_2N_2

carbocation rearrangement and loss of N_2

$+\ N_2$

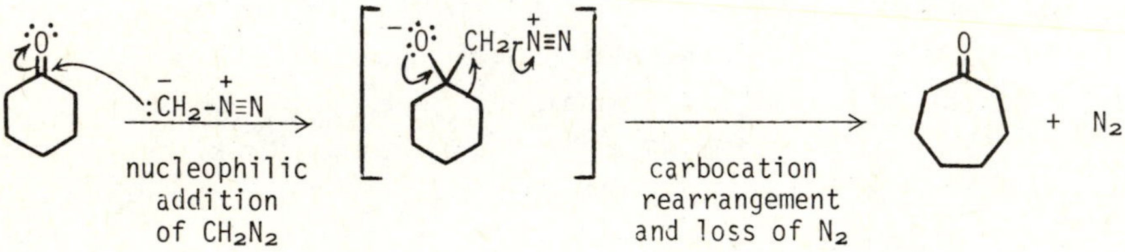

What you should know:

After doing these problems, you should be able to:

1. Locate the acidic protons of carbonyl compounds;
2. Write the mechanisms for acid- and base-catalyzed enolization of aldehydes and ketones;
3. Propose mechanisms for reactions involving enolates;
4. Synthesize compounds by the malonic ester and acetoacetic ester routes and by LDA alkylation.

CHAPTER 26 CARBONYL CONDENSATION REACTIONS

26.1 When you are first learning the aldol condensation, write out all the steps.

(1) Form the enolate of one molecule of the carbonyl compound.

$$CH_3CH_2CH(H)CHO + {}^-{:}OH \rightleftarrows CH_3CH_2\bar{C}HCHO + H_2O$$

(2) Have the enolate attack the electrophilic carbonyl of the other molecule.

$$CH_3CH_2CH_2CHO + CH_3CH_2\bar{C}HCHO \rightleftarrows CH_3CH_2CH_2C(H)(O^-)-CH(CH_2CH_3)CHO$$

(3) Protonate the anionic oxygen.

$$CH_3CH_2CH_2C(H)(O^-)-CH(CH_2CH_3)CHO + HOC_2H_5 \rightleftarrows CH_3CH_2CH_2C(H)(OH)-CH(CH_2CH_3)CHO + {}^-{:}OC_2H_5$$

Practice writing out these steps for the other aldol condensations.

b)

$$2\ CH_3CH_2COCH_3 \xrightleftharpoons{NaOH,\ C_2H_5OH} CH_3CH_2C(OH)(CH_3)-CH(CH_3)COCH_3 + CH_3CH_2C(OH)(CH_3)-CH_2COCH_2CH_3$$

c) 2 cyclopentanone $\xrightleftharpoons{NaOH,\ C_2H_5OH}$ 2-(1-hydroxycyclopentyl)cyclopentanone (HO, O)

d) $2\ C_6H_5COCH_3 \xrightleftharpoons{NaOH,\ C_2H_5OH} C_6H_5C(OH)(CH_3)-CH_2COC_6H_5$

e) $2\ C_6H_5CH_2CHO \xrightleftharpoons{NaOH,\ C_2H_5OH} C_6H_5CH_2C(H)(OH)-CH(C_6H_5)CHO$

26.2 The reactive nucleophile in the acid-catalyzed aldol condensation is the *enol* of the carbonyl compound. The electrophile is the protonated carbonyl compound.

(1) $CH_3CH{=}\ddot{O}: + H^+ \rightleftharpoons [\, H{-}CH_2CH{=}\overset{+}{O}H \rightleftharpoons CH_2{=}CH{-}OH + H^+ \,]$

(2) $[\, CH_3CH{=}\overset{+}{O}H + CH_2{=}CH{-}\ddot{O}H \rightleftharpoons CH_3C(H)(\ddot{O}H)CH_2CH{=}\overset{+}{O}H \,]$

electrophile nucleophile

(3) $[\, CH_3C(H)(OH)CH_2CH{=}\overset{+}{O}H \,] \rightleftharpoons CH_3C(H)(OH)CH_2CH{=}O + H^+$

26.3

$$CH_3C(OH)(CH_3){-}CH_2CCH_3 \; (C{=}O)$$

4-methyl-4-hydroxy-2-pentanone

The steps for the reverse aldol are the reverse of those described in Prob. 26.1.

(1) Deprotonate the alcohol oxygen.

$$CH_3C(\ddot{O}{-}H)(CH_3)CH_2\overset{O}{\overset{\|}{C}}CH_3 + {}^-{:}OH \rightleftharpoons CH_3C(:\ddot{O}:^-)(CH_3)CH_2\overset{O}{\overset{\|}{C}}CH_3 + H_2O$$

(2) Eliminate the enolate anion.

$$CH_3C(:\ddot{O}:^-)(CH_3){-}CH_2\overset{O}{\overset{\|}{C}}CH_3 \rightleftharpoons CH_3\overset{O}{\overset{\|}{C}}CH_3 + {}^-{:}CH_2\overset{O}{\overset{\|}{C}}CH_3$$

(3) Reprotonate the enolate anion.

$${}^-{:}CH_2\overset{O}{\overset{\|}{C}}CH_3 + C_2H_5OH \rightleftharpoons CH_3\overset{O}{\overset{\|}{C}}CH_3 + {}^-{:}OC_2H_5$$

Acetone is the product of this reaction.

26.4 a)

cyclopentanone $\xrightarrow{NaOH,\ C_2H_5OH}$ [2-(1-hydroxycyclopentyl)cyclopentanone (OH, O)] $\longrightarrow$ 2-cyclopentylidenecyclopentanone (O) $+ H_2O$

b)

2 [2-methylcyclohexanone] $\xrightleftharpoons{NaOH,\ C_2H_5OH}$ [aldol adduct] $\longrightarrow$ [2-(2-methylcyclohexylidene)-6-methylcyclohexanone] + H_2O

c)

$2\ C_6H_5\overset{O}{\overset{\|}{C}}CH_3 \xrightleftharpoons{NaOH,\ C_2H_5OH} \left[C_6H_5\overset{OH}{\underset{CH_3}{C}}-CH_2\overset{O}{\overset{\|}{C}}C_6H_5\right]$

$\downarrow$

$C_6H_5\overset{CH_3}{C}=CH\overset{O}{\overset{\|}{C}}C_6H_5 + H_2O$

d)

$2\ (CH_3)_2CHCH_2\overset{O}{\overset{\|}{C}}H \xrightleftharpoons{NaOH,\ C_2H_5OH} \left[(CH_3)_2CHCH_2\overset{OH}{\underset{H}{C}}-\underset{CH(CH_3)_2}{C}H\overset{O}{\overset{\|}{C}}H\right]$

$\downarrow$

$(CH_3)_2CHCH_2CH=\underset{CH(CH_3)_2}{C}\overset{O}{\overset{\|}{C}}H + H_2O$

e)

2 [3-methylcyclohexanone] $\xrightleftharpoons{NaOH,\ C_2H_5OH}$ [two aldol adducts] → [four condensation products] + H_2O

minor major

26.5 a)

$H_3C-\underset{H_3C}{\overset{HO}{C}}-\underset{CH_3}{\overset{CH_3}{C}}-\overset{O}{\overset{\|}{C}}H$

2,2,3-Trimethyl-3-hydroxybutanal is not an aldol self-condensation product. (Note that no aldol *self*-condensation can yield a product with an odd number of carbons.)

b) $CH_3CH_2CH_2C(OH)(CH_3)CHO$ 2-methyl-2-hydroxypentanal

This is not an aldol product. The hydroxyl group in an aldol product must be β, not α, to the carbonyl group.

c) $CH_3CH_2C(CH_2CH_3)=C(CH_3)C(=O)CH_2CH_3$ 5-ethyl-4-methyl-4-hepten-3-one

This product results from the aldol self-condensation of 3-pentanone, followed by dehydration.

26.6 a)

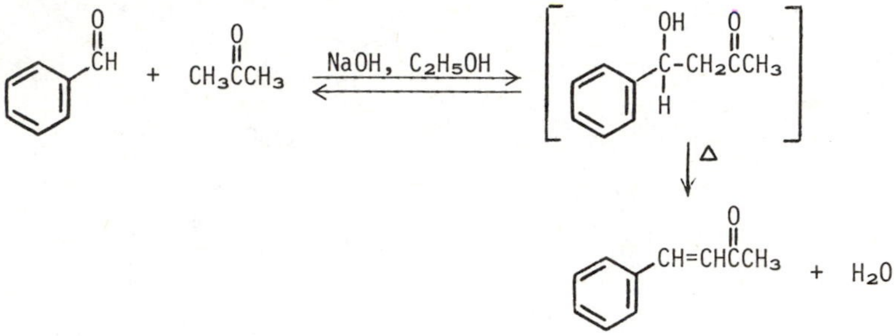

$$C_6H_5CHO + CH_3CCH_3(=O) \xrightleftharpoons{NaOH,\ C_2H_5OH} [C_6H_5CH(OH)CH_2CCH_3(=O)] \xrightarrow{\Delta} C_6H_5CH=CHCCH_3(=O) + H_2O$$

4-phenyl-3-buten-2-one

This mixed aldol will succeed because one of the components, benzaldehyde, has no α-hydrogen atoms.

b)

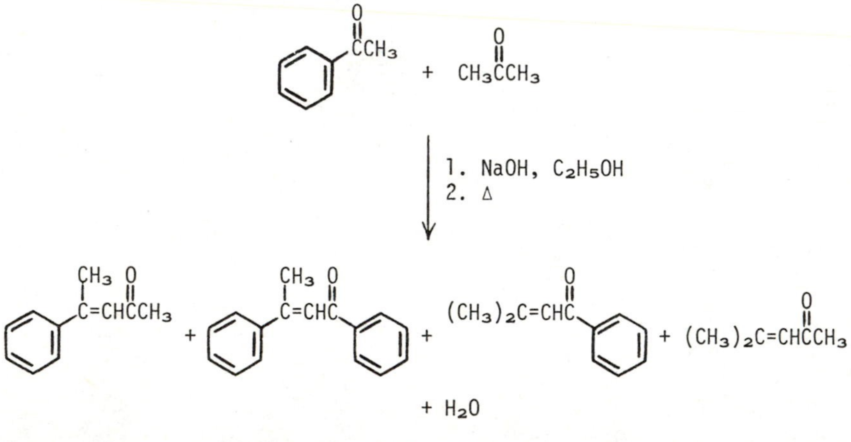

$$C_6H_5CCH_3(=O) + CH_3CCH_3(=O) \xrightarrow{1.\ NaOH,\ C_2H_5OH;\ 2.\ \Delta}$$

$$C_6H_5C(CH_3)=CHCCH_3(=O) + C_6H_5C(CH_3)=CHC(=O)C_6H_5 + (CH_3)_2C=CHC(=O)C_6H_5 + (CH_3)_2C=CHCCH_3(=O) + H_2O$$

Four products result from the aldol condensation of acetone and acetophenone.

c)

$CH_3CH_2\overset{O}{\overset{\|}{C}}H$

1. NaOH, C_2H_5OH
2. Δ

$CHCH_2CH_3$ + + $\overset{O}{\overset{\|}{HC}}$ CCH_3 + $CH_3CH_2CH{=}C(CH_3)\overset{O}{\overset{\|}{C}}H$ + H_2O

A mixture of products is formed because both carbonyl partners contain α-hydrogen atoms.

d) This mixed aldol reaction succeeds because the anion of 2,4-pentanedione is formed more readily than the anion of cyclohexanone.

+ $CH_2(\overset{O}{\overset{\|}{C}}CH_3)_2$ $\xrightleftharpoons{NaOH,\ C_2H_5OH}$ [HO $CH(\overset{O}{\overset{\|}{C}}CH_3)_2$] $\xrightarrow{\Delta}$ $C(\overset{O}{\overset{\|}{C}}CH_3)_2$ + H_2O

26.7

H_2O + [$CH_3\overset{O}{\overset{\|}{C}}\bar{C}H\overset{O}{\overset{\|}{C}}CH_3$] $\xleftrightharpoons{:\bar{O}H}$ $CH_3\overset{O}{\overset{\|}{C}}CH_2\overset{O}{\overset{\|}{C}}CH_3$ $\xrightleftharpoons{:\bar{O}H}$ [$\bar{C}H_2\overset{O}{\overset{\|}{C}}CH_2\overset{O}{\overset{\|}{C}}CH_3$] + H_2O

A — 2,4-pentanedione — B

H_2O + $H_2C{-}C{=}O$, $C{=}CH$, CH_3 ⟵ [$H_2C{-}C{=}O$, $HO{-}C{-}CH_2$, CH_3 $\xleftrightharpoons{C_2H_5OH}$ $H_2C{-}C{=}O$, $:\bar{O}{-}C{-}CH_2$, CH_3]

2,4-Pentanedione is in equilibrium with two enolates after treatment with base. Enolate B can undergo internal aldol condensation to form cyclic product. Formation of enolate A is more likely, however, because A is a stabilized carbanion.

26.8

$\xrightleftharpoons{NaOH,\ C_2H_5OH}$ [⇄ ≡]

C_2H_5OH

H_2O + ⟵ [OH]

26.9 $CH_2(CO_2C_2H_5)_2 + {}^{-}{:}OC_2H_5 \rightleftharpoons {}^{-}{:}CH(CO_2C_2H_5)_2 + HOC_2H_5$

$$C_6H_5CH{=}O + {}^{-}{:}CH(CO_2C_2H_5)_2 \rightleftharpoons \left[C_6H_5CH(O^-)CH(CO_2C_2H_5)_2\right]$$

$$\xrightleftharpoons{HOC_2H_5} \left[C_6H_5CH(OH)CH(CO_2C_2H_5)_2\right]$$

$$\xrightleftharpoons{{}^{-}{:}OC_2H_5} \left[C_6H_5CH(OH)\text{-}C(CO_2C_2H_5){=}C(O^-)OC_2H_5\right]$$

$$\left[C_6H_5CH(OH)\text{-}C(CO_2C_2H_5){=}C(O^-)OC_2H_5\right] \rightleftharpoons C_6H_5CH{=}C(CO_2C_2H_5)_2 + {}^{-}{:}OH$$

26.10 Here is an outline of the Perkin reaction.

a) Acetate ion abstracts an α-proton from acetic anhydride to form the enolate anion.

$$CH_3CO_2^- + H\text{-}CH_2COCOCH_3 \rightleftharpoons CH_3COOH + {}^{-}{:}CH_2C(O)OC(O)CH_3$$

b) The enolate anion attacks the electrophilic carbon of benzaldehyde.

$$C_6H_5CH{=}O + {}^{-}{:}CH_2C(O)OC(O)CH_3 \rightleftharpoons \left[C_6H_5CH(O^-)CH_2C(O)OC(O)CH_3\right]$$

c) Treatment of the adduct with water cleaves the anhydride. The β-hydroxy acid loses water to yield the unsaturated aromatic carboxylate anion.

$$\left[C_6H_5CH(O^-)CH_2C(O)OC(O)CH_3\right] + H_2O \rightleftharpoons \left[C_6H_5CH(OH)CH_2CO_2^- + CH_3COOH\right]$$

$$\downarrow$$

$$C_6H_5CH{=}CHCO_2^- + H_2O$$

26.11

$$\left[C_6H_5\overset{O}{\overset{\|}{C}}-\overset{-}{\ddot{C}}HCO_2C_2H_5 \right] \underset{}{\overset{^{-}:OH}{\rightleftharpoons}} C_6H_5\overset{:\ddot{O}:}{\overset{\|}{C}}-CH_2CO_2C_2H_5 \overset{^{-}:OH}{\rightleftharpoons} \left[C_6H_5\overset{:\ddot{O}:^{-}}{\overset{|}{C}}(OH)-CH_2CO_2C_2H_5 \right]$$

$$\rightleftharpoons$$

$$C_6H_5\overset{O}{\overset{\|}{C}}O:^{-} + CH_3CO_2C_2H_5 \longleftarrow \left[C_6H_5\overset{O}{\overset{\|}{C}}OH + ^{-}:CH_2CO_2C_2H_5 \right]$$

Hydroxide can react at two different sites of the β-keto ester. Abstraction of the acidic α-proton is more favorable but is reversible and does not lead to product. Addition of hydroxide ion to the carbonyl group, followed by irreversible elimination of ethyl acetate, accounts for the observed product.

26.12 Two different reactions are possible when ethyl acetoacetate reacts with ethoxide anion. One reaction involves attack of ethoxide ion on the carbonyl carbon, followed by elimination of the anion of ethyl acetate.

$$CH_3\overset{:\ddot{O}:}{\overset{\|}{C}}CH_2\overset{O}{\overset{\|}{C}}OC_2H_5 \;\; ^{-}:OC_2H_5 \rightleftharpoons \left[CH_3\overset{:\ddot{O}:^{-}}{\overset{|}{C}}(OC_2H_5)-CH_2\overset{O}{\overset{\|}{C}}OC_2H_5 \right] \rightleftharpoons CH_3\overset{O}{\overset{\|}{C}}OC_2H_5 + ^{-}:CH_2\overset{O}{\overset{\|}{C}}OC_2H_5$$

$$^{-}:CH_2\overset{O}{\overset{\|}{C}}OC_2H_5 + C_2H_5OH \rightleftharpoons CH_3\overset{O}{\overset{\|}{C}}OC_2H_5 + C_2H_5O:^{-}$$

This is a reverse Claisen reaction. More likely, however, is the acid-base reaction of ethoxide ion and an activated α-proton of ethyl acetoacetate.

$$CH_3\overset{O}{\overset{\|}{C}}CH_2\overset{O}{\overset{\|}{C}}OC_2H_5 + ^{-}:OC_2H_5 \rightleftharpoons CH_3\overset{O}{\overset{\|}{C}}\overset{-}{\ddot{C}}H\overset{O}{\overset{\|}{C}}OC_2H_5 + HOC_2H_5$$

The resonance-stabilized acetoacetate anion is no longer reactive toward nucleophiles, and no further reaction occurs at room temperature. Elevated temperatures are required to make the cleavage reaction proceed.

Ethyl α,α-dimethylacetoacetate has no activated α-protons. No acid-base reaction competes with the cleavage reaction, which proceeds readily.

$$CH_3\overset{:\ddot{O}:}{\overset{\|}{C}}C(CH_3)_2\overset{O}{\overset{\|}{C}}OC_2H_5 \;\; ^{-}:OC_2H_5 \rightleftharpoons \left[CH_3\overset{:\ddot{O}:^{-}}{\overset{|}{C}}(OC_2H_5)-C(CH_3)_2\overset{O}{\overset{\|}{C}}OC_2H_5 \right] \rightleftharpoons CH_3\overset{O}{\overset{\|}{C}}OC_2H_5 + (CH_3)_2\overset{-}{\ddot{C}}\overset{O}{\overset{\|}{C}}OC_2H_5$$

$$(CH_3)_2\overset{-}{\ddot{C}}\overset{O}{\overset{\|}{C}}OC_2H_5 + C_2H_5OH \rightleftharpoons (CH_3)_2CH\overset{O}{\overset{\|}{C}}OC_2H_5 + C_2H_5O:^{-}$$

26.13 Diethyl oxalate gives good yields in mixed Claisen reactions because there are no hydrogen atoms *alpha* to the carbonyl carbon.

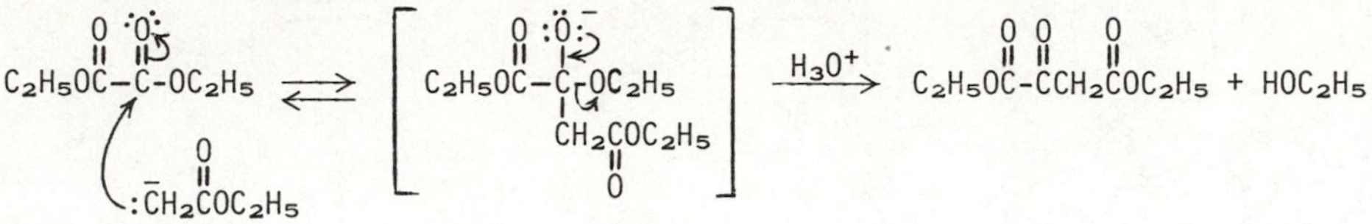

$C_2H_5OC(=O)-C(=O)-OC_2H_5 + {:}\bar{C}H_2COC_2H_5 \rightleftharpoons [C_2H_5OC(=O)-C(O^-)(OC_2H_5)-CH_2COOC_2H_5] \xrightarrow{H_3O^+} C_2H_5OC(=O)-C(=O)CH_2C(=O)OC_2H_5 + HOC_2H_5$

26.14

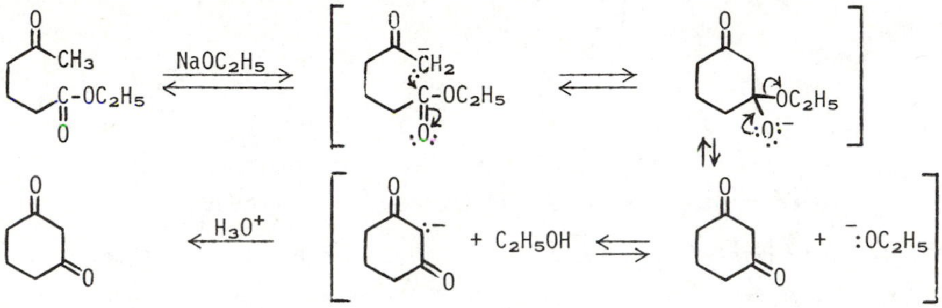

$\xrightarrow{NaOC_2H_5}$ ⇌ ⇅ $+ C_2H_5OH$ ⇌ $+ {^-}{:}OC_2H_5$ $\xleftarrow{H_3O^+}$

26.15

NaOCH$_3$, CH$_3$OH

C1-C6 bond formation

$\xrightarrow{H_3O^+}$ + C_2H_5OH

NaOCH$_3$, CH$_3$OH

C2-C7 bond formation

$\xrightarrow{H_3O^+}$ + C_2H_5OH

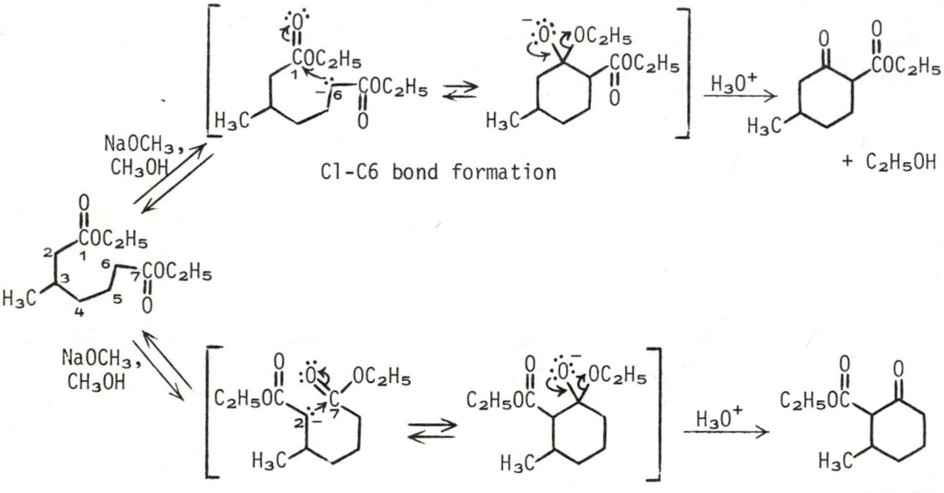

Unlike diethyl heptanedioate, diethyl 3-methylheptanedioate is unsymmetrical. Two different enolates can form, and each can cyclize to a different product.

26.16 Michael reactions occur between stabilized enolate anions and α,β-unsaturated enones or similar compounds. Learn to locate these components in possible Michael products. Usually, it is easier to recognize the enolate nucleophile; in a) the nucleophile is the ethyl acetoacetate anion.

$$CH_3\overset{O}{\overset{\|}{C}}\underset{\underset{CO_2CH_3}{|}}{C}H\text{-}CH_2CH_2\overset{O}{\overset{\|}{C}}C_6H_5 \quad \text{comes from} \quad CH_3\overset{O}{\overset{\|}{C}}\underset{\underset{CO_2CH_3}{|}}{C}H_2$$

The rest of the compound is the Michael acceptor. Draw a double bond in conjugation with the electron-withdrawing group in this part of the molecule.

$$CH_3\overset{O}{\overset{\|}{C}}\underset{\underset{CO_2CH_3}{|}}{C}H\text{-}CH_2CH_2\overset{O}{\overset{\|}{C}}C_6H_5 \quad \text{comes from} \quad CH_2{=}CH\overset{O}{\overset{\|}{C}}C_6H_5$$

$$\underset{\text{Michael donor}}{CH_3\overset{O}{\overset{\|}{C}}\underset{\underset{CO_2CH_3}{|}}{C}H_2} \quad + \quad \underset{\text{Michael acceptor}}{CH_2{=}CH\overset{O}{\overset{\|}{C}}C_6H_5} \xrightarrow[\text{2. } H_3O^+]{\text{1. } NaOC_2H_5} CH_3\overset{O}{\overset{\|}{C}}\underset{\underset{CO_2CH_3}{|}}{C}HCH_2CH_2\overset{O}{\overset{\|}{C}}C_6H_5$$

b) Since Michael products are often decarboxylated after the addition reaction, it is more difficult to recognize the original enolate anion.

$$\underset{\text{Michael donor}}{CH_3\overset{O}{\overset{\|}{C}}\underset{\underset{CO_2CH_3}{|}}{C}H_2} \quad + \quad \underset{\text{Michael acceptor}}{CH_2{=}CH\overset{O}{\overset{\|}{C}}CH_3} \xrightarrow[\text{2. } H_3O^+]{\text{1. } NaOC_2H_5} CH_3\overset{O}{\overset{\|}{C}}\underset{\underset{CO_2CH_3}{|}}{C}HCH_2CH_2\overset{O}{\overset{\|}{C}}CH_3$$

$$CH_3\overset{O}{\overset{\|}{C}}\underset{\underset{CO_2CH_3}{|}}{C}HCH_2CH_2\overset{O}{\overset{\|}{C}}CH_3 \xrightarrow[\Delta]{H_3O^+} CH_3\overset{O}{\overset{\|}{C}}CH_2CH_2CH_2\overset{O}{\overset{\|}{C}}CH_3$$

c) $$\underset{\text{Michael donor}}{(CH_3O_2C)_2CH_2} \quad + \quad \underset{\text{Michael acceptor}}{CH_2{=}CHCN} \xrightarrow[\text{2. } H_3O^+]{\text{1. } NaOC_2H_5} (CH_3O_2C)_2CHCH_2CH_2CN$$

d) $$\underset{\text{Michael donor}}{CH_3CH_2NO_2} \quad + \quad \underset{\text{Michael acceptor}}{CH_2{=}CH\overset{O}{\overset{\|}{C}}OCH_3} \xrightarrow[\text{2. } H_3O^+]{\text{1. } NaOC_2H_5} CH_3\overset{NO_2}{\overset{|}{C}}HCH_2CH_2\overset{O}{\overset{\|}{C}}OCH_3$$

e) $$\underset{\text{Michael donor}}{(CH_3O_2C)_2CH_2} \quad + \quad \underset{\text{Michael acceptor}}{CH_2{=}CHNO_2} \xrightarrow[\text{2. } H_3O^+]{\text{1. } NaOC_2H_5} (CH_3O_2C)_2CHCH_2CH_2NO_2$$

26.17

$$CH_3\overset{O}{\overset{\|}{C}}\bar{C}H_2 + CH_2{=}CHC{\equiv}N \xrightarrow[\text{Addition}]{\text{Michael}} CH_3\overset{O}{\overset{\|}{C}}CH_2CH_2\bar{C}HC{\equiv}N \rightleftarrows CH_3\overset{O}{\overset{\|}{C}}\bar{C}HCH_2CH_2C{\equiv}N \ (\underline{A})$$

$$\updownarrow$$

$$:\bar{C}H_2\overset{O}{\overset{\|}{C}}CH_2CH_2CH_2C{\equiv}N \ (\underline{B})$$

Michael reaction of a mono-ketone enolate yields products A and B, with reactivity comparable to that of the starting enolate. A and B can also add to the Michael acceptor. This side reaction lowers the yield of the desired product.

26.18

N, CH_2, CH_3

Crowding between the methyl group and the pyrrolidine ring disfavors this enamine.

N, CH_2, CH_3 —ring flip→ CH_3, N, CH_2

The crowding in this enamine can be relieved by a ring-flip, which puts the methyl group in an axial position. This enamine is the only one formed.

26.19 a)

cyclopentanone + pyrrolidine (N–H) → enamine + H_2O —$CH_2{=}CHCN$→ [iminium (N^+) with $CH_2\bar{C}HCN$] —H_3O^+→ pyrrolidine (N–H) + 2-(CH_2CH_2CN)cyclopentanone

b)

2-methylcyclohexanone (H_3C) + pyrrolidine (N–H) → H_3C-enamine —$CH_2{=}CH\overset{O}{\overset{\|}{C}}OCH_3$→ [$H_3C$, N^+, $CH_2\bar{C}H\overset{O}{\overset{\|}{C}}OCH_3$] —$H_3O^+$→ pyrrolidine (N–H) + H_3C-cyclohexanone-$CH_2CH_2\overset{O}{\overset{\|}{C}}OCH_3$

26.20 a)

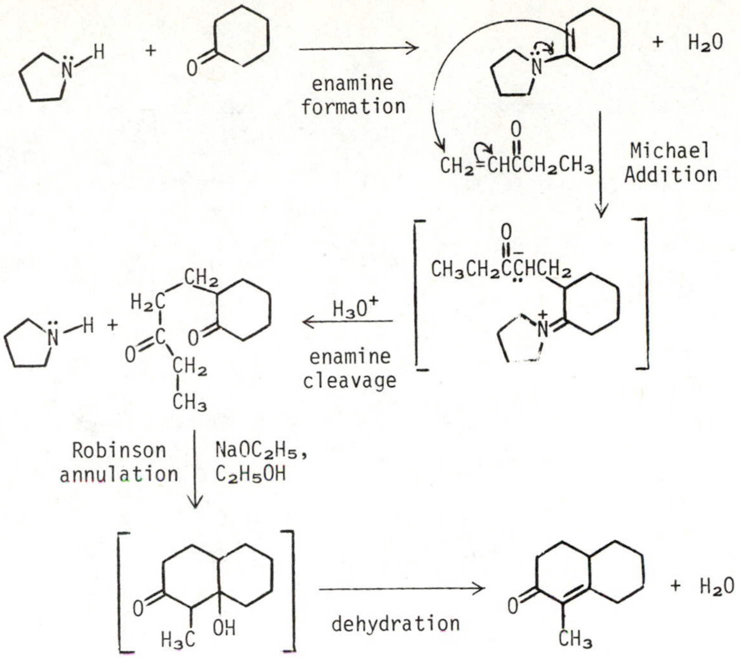

Pyrrolidine + cyclohexanone —enamine formation→ enamine + H_2O

$CH_2{=}CHCCH_2CH_3$ (O on C) —Michael Addition→

[$CH_3CH_2\overset{O}{C}\overset{-}{C}HCH_2$– iminium]

—H_3O^+, enamine cleavage→ pyrrolidine + 2-(3-oxopentyl)cyclohexanone (H_2C, CH_2, $C{=}O$, CH_2, CH_3)

Robinson annulation; $NaOC_2H_5$, C_2H_5OH

[H_3C, OH] —dehydration→ CH_3 + H_2O

b)

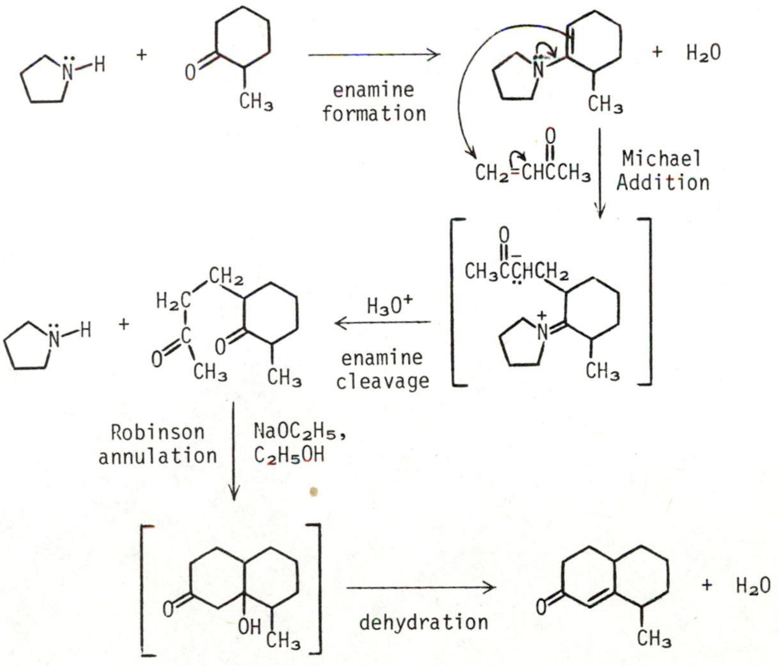

Pyrrolidine + 2-methylcyclohexanone (CH_3) —enamine formation→ enamine (CH_3) + H_2O

$CH_2{=}CHCCH_3$ (O on C) —Michael Addition→

[$CH_3\overset{O}{C}\overset{-}{C}HCH_2$– iminium, CH_3]

—H_3O^+, enamine cleavage→ pyrrolidine + (CH_2, H_2C, $C{=}O$, CH_3; O, CH_3)

Robinson annulation; $NaOC_2H_5$, C_2H_5OH

[OH, CH_3] —dehydration→ CH_3 + H_2O

26.21 a) $(CH_3)_3CCHO$ has no *alpha* protons and does not undergo aldol self-condensation.

b) 2 cyclobutanone $\xrightleftharpoons{NaOH,\ C_2H_5OH}$ [2-(1-hydroxycyclobutyl)cyclobutanone] $\longrightarrow$ 2-cyclobutylidenecyclobutanone + H_2O

c) Benzophenone does not undergo aldol self-condensation.

d) $O{=}C(CH_3)CH_2CH_2C(CH_3){=}O \xrightleftharpoons{NaOH,\ C_2H_5OH}$ [3-hydroxy-3-methylcyclopentanone] $\longrightarrow$ 3-methyl-2-cyclopentenone + H_2O

Non-cyclic aldol products can also form.

e) $2\ CH_3(CH_2)_8\overset{O}{\overset{\|}{C}}H \xrightleftharpoons{NaOH,\ C_2H_5OH} \left[CH_3(CH_2)_8\underset{H}{\overset{OH}{C}}{-}\underset{(CH_2)_7CH_3}{C}H\overset{O}{\overset{\|}{C}}H \right] \longrightarrow CH_3(CH_2)_7CH{=}\underset{CH_3(CH_2)_7}{C}\overset{O}{\overset{\|}{C}}H$

$+\ H_2O$

f) $C_6H_5CH{=}CHCHO$ does not undergo aldol reactions.

26.22

$$\text{H}_2\text{C}{-}\text{CH}_2{-}\text{CH}_2\overset{\text{O}}{\overset{\|}{\text{C}}}\text{H} \;(\text{H}_2\text{C}{-}\overset{\text{O}}{\overset{\|}{\text{C}}}\text{H}) \xrightleftharpoons{\text{NaOH, } C_2H_5OH} \left[\text{H}_2\text{C}{-}\text{CH}_2{-}\overset{-}{\text{C}}\text{H}\overset{\text{O}}{\overset{\|}{\text{C}}}\text{H},\ \text{H}_2\text{C}{-}\text{CH}{=}\ddot{\text{O}} \xrightleftharpoons{C_2H_5OH} \text{H}_2\text{C}{-}\text{CH}_2{-}\text{CH}\overset{\text{O}}{\overset{\|}{\text{C}}}\text{H},\ \text{H}_2\text{C}{-}\text{C(HO)(H)} \right]$$

$$\downarrow$$

$$\text{H}_2\text{C}{-}\text{CH}_2{-}\text{C}{-}\overset{\text{O}}{\overset{\|}{\text{C}}}\text{H},\ \text{H}_2\text{C}{-}\text{C(H)}{=} \quad + \quad H_2O$$

26.23 a)

$$\text{C}_6\text{H}_5\overset{\text{O}}{\overset{\|}{\text{C}}}\text{H} + \text{CH}_3\overset{\text{O}}{\overset{\|}{\text{C}}}\text{C}_6\text{H}_5 \xrightarrow[\text{2. } -H_2O]{\text{1. NaOH, } C_2H_5OH} \text{C}_6\text{H}_5\text{CH{=}CH}\overset{\text{O}}{\overset{\|}{\text{C}}}\text{C}_6\text{H}_5$$

b)

$$\text{H}_2\text{C}(\overset{\text{O}}{\overset{\|}{\text{C}}}\text{CH}_3){-}\text{H}_2\text{C}{-}\text{CH}_2{-}\text{C(H)}{=}\text{O} \xrightarrow[\text{2. } -H_2O]{\text{1. NaOH, } C_2H_5OH} \text{cyclohexenone}$$

c)

$$\text{C}_6\text{H}_4(\text{CH}_2\overset{\text{O}}{\overset{\|}{\text{C}}}\text{H})(\text{CH}_2\overset{\text{O}}{\overset{\|}{\text{C}}}\text{H}) \xrightarrow[\text{2. } -H_2O]{\text{1. NaOH, } C_2H_5OH} \text{(indene: CH}_2\text{, CH=C}{-}\overset{\text{O}}{\overset{\|}{\text{C}}}\text{H)}$$

d)

$$\text{C}_6\text{H}_5\text{CH}_2\overset{\text{O}}{\overset{\|}{\text{C}}}\text{CH}_2\text{C}_6\text{H}_5 + \text{C}_6\text{H}_5{-}\underset{\text{O}}{\underset{\|}{\text{C}}}{-}\underset{\text{O}}{\underset{\|}{\text{C}}}{-}\text{C}_6\text{H}_5 \xrightarrow[\text{2. } -2\ H_2O]{\text{1. NaOH, } C_2H_5OH} \text{(tetraphenylcyclopentadienone: O; } C_6H_5,\ C_6H_5,\ C_6H_5,\ C_6H_5)$$

26.24 a)

$$\text{cyclohexanone} + 2\ \text{C}_6\text{H}_5\overset{\text{O}}{\overset{\|}{\text{C}}}\text{H} \xrightarrow[\text{2. } -2\ H_2O]{\text{1. NaOH, } C_2H_5OH} \text{C}_6\text{H}_5\text{CH}{=}(\text{cyclohexanone, O}){=}\text{CHC}_6\text{H}_5$$

b)

$$\text{cyclohexanone} + \text{C}_2\text{H}_5\text{O}\overset{\text{O}}{\overset{\|}{\text{C}}}\text{OC}_2\text{H}_5 \xrightarrow[\text{2. } H_3O^+]{\text{1. NaOH, } C_2H_5OH} \text{(2-oxocyclohexyl)}\overset{\text{O}}{\overset{\|}{\text{C}}}\text{OC}_2\text{H}_5$$

c) 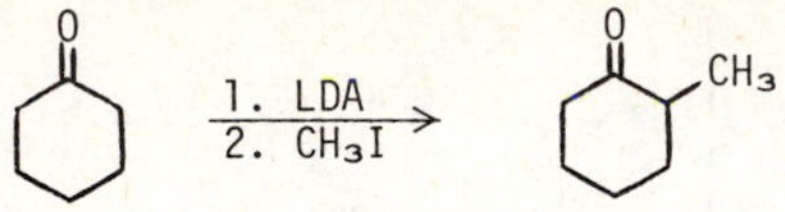 Cyclohexanone $\xrightarrow{\text{1. LDA; 2. } CH_3I}$ 2-methylcyclohexanone + dialkylated product

or

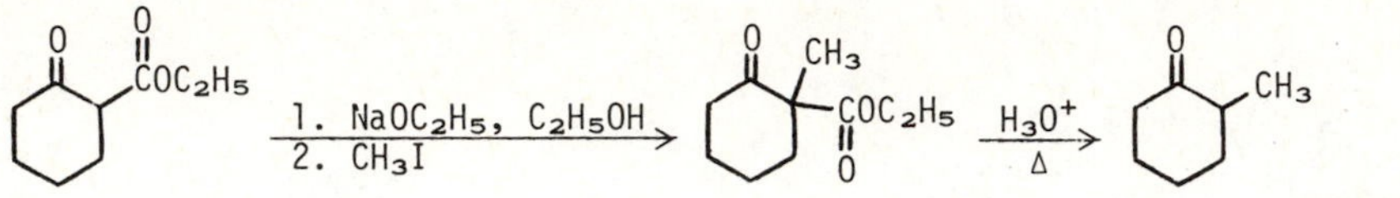

2-(ethoxycarbonyl)cyclohexanone (from part b) $\xrightarrow{\text{1. } NaOC_2H_5,\ C_2H_5OH;\ \text{2. } CH_3I}$ ethyl 1-methyl-2-oxocyclohexanecarboxylate $\xrightarrow[\Delta]{H_3O^+}$ 2-methylcyclohexanone + CO_2 + C_2H_5OH

from part b)

Alkylation of the β-ketoester is a better reaction.

d) 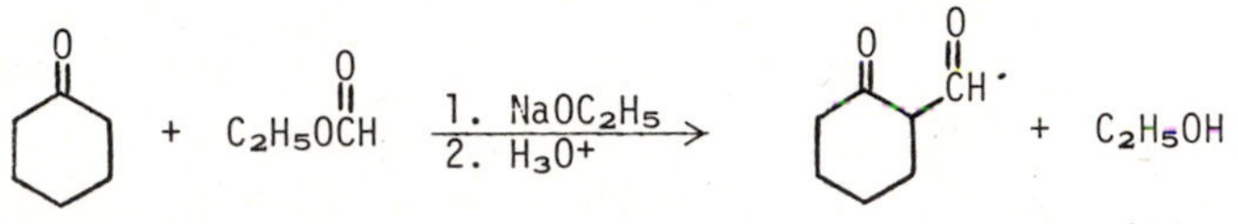 Cyclohexanone + C_2H_5OCH (ethyl formate, C=O) $\xrightarrow{\text{1. } NaOC_2H_5;\ \text{2. } H_3O^+}$ 2-formylcyclohexanone (CH·) + C_2H_5OH

e) 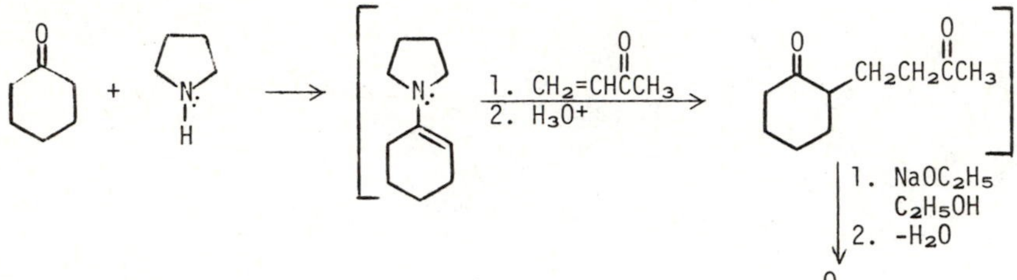 Cyclohexanone + pyrrolidine (N–H) ⟶ [enamine] $\xrightarrow{\text{1. } CH_2{=}CHCCH_3 \text{ (C=O)};\ \text{2. } H_3O^+}$ [2-($CH_2CH_2CCH_3$)cyclohexanone] $\xrightarrow{\text{1. } NaOC_2H_5,\ C_2H_5OH;\ \text{2. } -H_2O}$ octalone (bicyclic enone)

It is necessary to form the enamine in order to turn cyclohexanone into a good Michael donor.

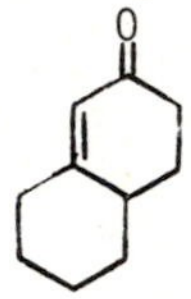

f) 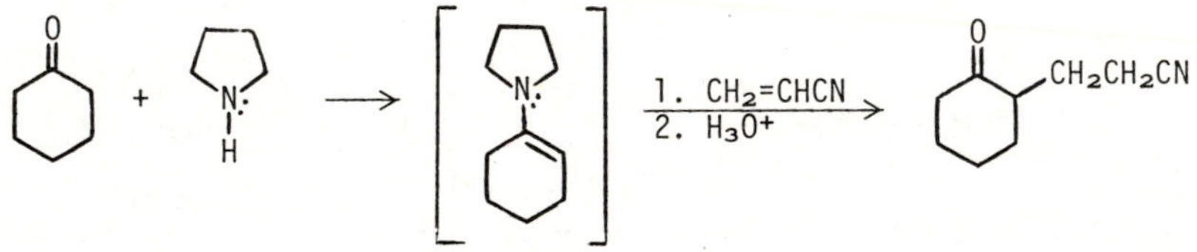 Cyclohexanone + pyrrolidine (N–H) ⟶ [enamine] $\xrightarrow{\text{1. } CH_2{=}CHCN;\ \text{2. } H_3O^+}$ 2-(CH_2CH_2CN)cyclohexanone

26.25

Bicyclic alkene $\xrightarrow{\text{1. } O_3;\ \text{2. Zn, } H_3O^+}$ cyclodecane-1,6-dione $\rightleftharpoons^{NaOC_2H_5,\ C_2H_5OH}$ [enolate anion]

$\uparrow\downarrow$ C_2H_5OH

H_2O + bicyclic enone ⟵ [bicyclic ketol, OH]

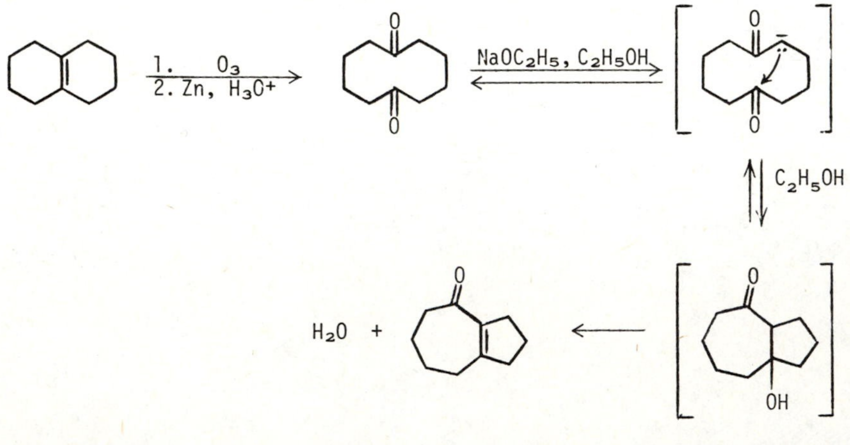

26.26

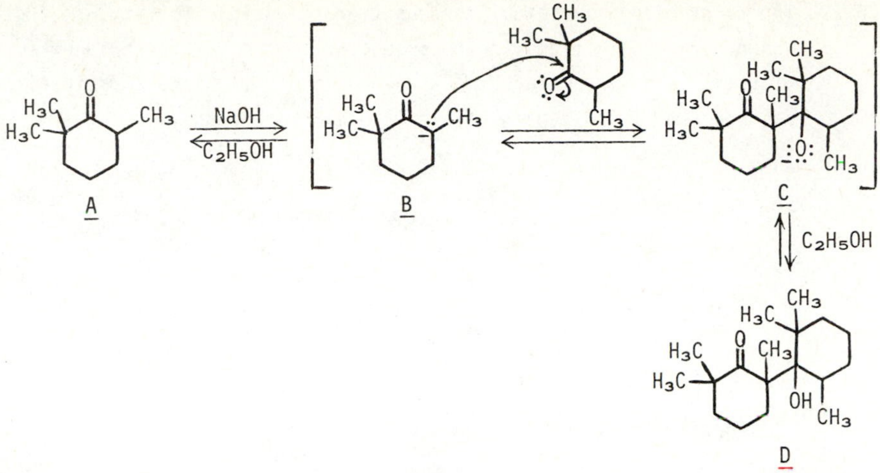

A ⇌ ($NaOH$ / C_2H_5OH) [B ⇌ C] ⇌ (C_2H_5OH) D

An aldol condensation involves a series of reversible equilibrium steps. Formation of product is favored by the dehydration of the β-hydroxy ketone to form a conjugated enone. Dehydration to form *conjugated* product can't occur with D. In addition, the B ⇌ C equilibrium favors B because of steric hindrance in the nucleophilic addition step.

26.27

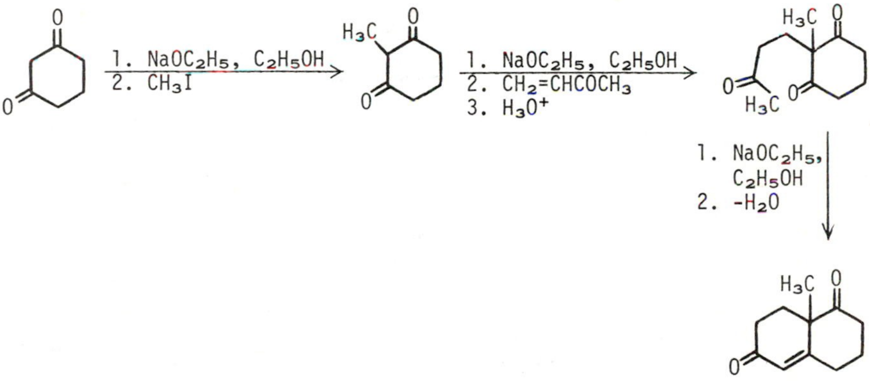

1,3-cyclohexanedione → (1. $NaOC_2H_5$, C_2H_5OH; 2. CH_3I) 2-methyl-1,3-cyclohexanedione → (1. $NaOC_2H_5$, C_2H_5OH; 2. CH_2=$CHCOCH_3$; 3. H_3O^+) 2-methyl-2-(3-oxobutyl)-1,3-cyclohexanedione → (1. $NaOC_2H_5$, C_2H_5OH; 2. $-H_2O$) bicyclic enedione

26.28

$$2\ CH_3\overset{O}{\overset{\|}{C}}H \xrightarrow[2.\ -H_2O]{1.\ NaOH,\ C_2H_5OH} CH_3CH{=}CH\overset{O}{\overset{\|}{C}}H \xrightarrow[2.\ H_3O^+]{1.\ LiAlH_4} CH_3CH{=}CHCH_2OH$$

$$CH_3CH{=}CHCH_2OH \xrightarrow{H_2,\ Pd/C} CH_3CH_2CH_2CH_2OH$$

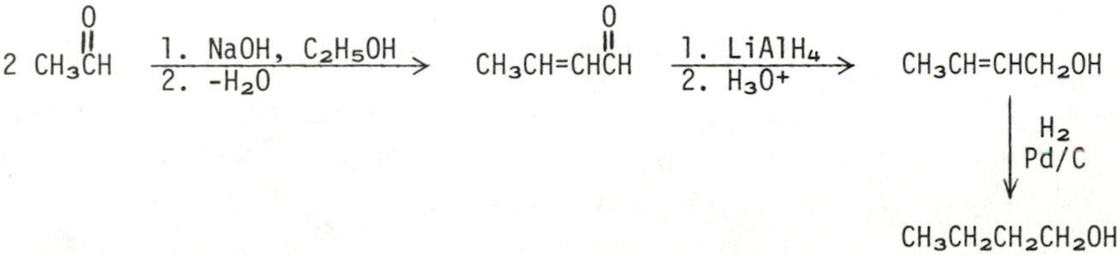

26.29 The first step of an aldol condensation is enolate formation. The ketone shown here does not enolize because double bonds at the bridgehead of small bicyclic ring systems are too strained to form. Since the bicyclic ketone does not enolize, it does not undergo aldol condensation.

26.30 a)

$$CH_3CO_2CH_3 + CH_3CH_2CO_2CH_3 \xrightarrow{NaOC_2H_5,\ C_2H_5OH}$$

$$CH_3\overset{O}{\overset{\|}{C}}CH_2CO_2CH_3 + CH_3CH_2\overset{O}{\overset{\|}{C}}\underset{CH_3}{\underset{|}{C}H}CO_2CH_3 + CH_3\overset{O}{\overset{\|}{C}}\underset{CH_3}{\underset{|}{C}H}CO_2CH_3 + CH_3CH_2\overset{O}{\overset{\|}{C}}CH_2CO_2CH_3$$

self-condensation products — mixed condensation products

Approximately equal amounts of each product will form.

b)

$$C_6H_5CO_2CH_3 + C_6H_5CH_2CO_2CH_3 \xrightarrow[C_2H_5OH]{NaOC_2H_5} C_6H_5CH_2\overset{O}{\overset{\|}{C}}\underset{C_6H_5}{\underset{|}{C}H}CO_2CH_3 + C_6H_5\overset{O}{\overset{\|}{C}}\underset{C_6H_5}{\underset{|}{C}H}CO_2CH_3$$

self-condensation product — mixed condensation product

The mixed condensation product predominates.

c)

$$CH_3O\overset{O}{\overset{\|}{C}}OCH_3 + \text{cyclohexanone} \xrightarrow[C_2H_5OH]{NaOC_2H_5} \text{2-(}\overset{O}{\overset{\|}{C}}OCH_3\text{)cyclohexanone}$$

This is the only *Claisen* monocondensation product.

d)

$$C_6H_5\overset{O}{\overset{\|}{C}}H + CH_3CO_2CH_3 \xrightarrow[C_2H_5OH]{NaOC_2H_5} CH_3\overset{O}{\overset{\|}{C}}CH_2CO_2CH_3 + C_6H_5CH{=}CHCO_2CH_3$$

self-condensation product — mixed condensation product

The mixed Claisen product is the major product.

26.31 a) Several other products are formed in addition to the one pictured. Self-condensation of acetaldehyde and acetone (less likely) can occur, and an additional mixed product is formed.

b) *Conjugate* addition occurs between a carbon nucleophile and an α,β-unsaturated carbonyl electrophile (the Michael reaction). Addition occurs at the double bond, not at the carbonyl group.

$$CH_3\overset{O}{\overset{\|}{C}}CH{=}CH_2 + CH_2(CO_2C_2H_5)_2 \xrightarrow{base} CH_3\overset{O}{\overset{\|}{C}}CH_2CH_2CH(CO_2C_2H_5)_2$$

c) There are two problems with this reaction. (1) Michael reactions occur in low yield with mono-ketones. Formation of the enamine, followed by the Michael reaction, gives a higher yield of product. (2) Addition can occur on either side of the ketone to give a mixture of products.

d) Internal aldol condensation of 2,6-heptanedione can produce a four-membered ring or a six-membered ring. The six-membered ring is more likely to form because it is less strained.

$$\left[\bar{C}H_2CCH_2CH_2CH_2CCH_3 \rightleftarrows \text{(cyclic alkoxide)} \underset{}{\overset{C_2H_5OH}{\rightleftarrows}} \text{(cyclic alcohol)} \right] \longrightarrow \text{3-methylcyclohexenone} + H_2O$$

e) Michael addition occurs at the most acidic position of the Michael donor.

$$\text{(1,3-cyclohexanedione enolate)} + CH_2{=}CHCO_2CH_3 \longrightarrow \text{2-}(CH_2CH_2CO_2CH_3)\text{-1,3-cyclohexanedione}$$

26.32 If cyclopentanone and base are mixed first, aldol self-condensation of cyclopentanone can occur before ethyl formate is added. If both carbonyl components are mixed together before adding base, the more favorable mixed Claisen condensation occurs with little competition from the aldol self-condensation reaction.

26.33 $ClCH_2CO_2C_2H_5 + {}^-:OC_2H_5 \rightleftarrows Cl\bar{C}HCO_2C_2H_5 + HOC_2H_5$

enolate formation

$$\text{cyclohexanone} + Cl\bar{C}HCO_2C_2H_5 \rightleftarrows \left[\text{alkoxide intermediate} \right] \longrightarrow \text{epoxide } (O{-}CHCO_2C_2H_5) + :Cl^-$$

carbonyl condensation

S_N2 displacement

26.34

$$RCOCH_2CO_2^- \xrightarrow{\Delta} RCO\bar{C}H_2 + O{=}C{=}O$$

decarboxylation

$$RCOCH_2CH(O^-)R' \rightleftarrows RCO\bar{C}H_2 + HCOR'$$

reverse aldol condensation

26.35 a) $H_2C(CO_2C_2H_5)-H_2C(CO_2C_2H_5)$ $\underset{}{\overset{NaOC_2H_5,\ C_2H_5OH}{\rightleftharpoons}}$ [$H\bar{C}:(CO_2C_2H_5)-H_2C(CO_2C_2H_5)$]

[$C_2H_5O_2C$, $H\bar{C}:$, H_2C, $C(=O)OC_2H_5$, CH_2, $\bar{:}CH$, $CO_2C_2H_5$, $C(=O)OC_2H_5$ $\rightleftharpoons$ $C_2H_5O_2C$, $\bar{C}$, $C=O$, CH_2, H_2C, $\bar{C}$, $CO_2C_2H_5$, $C=O$ + 2 C_2H_5OH]

H_3O^+, Δ

H_2C, $C=O$, CH_2, H_2C, CH_2, $C=O$ + 2 CO_2 + 2 C_2H_5OH

b) CO_2CH_3, CO_2CH_3 $\xrightarrow{1.\ LiAlH_4;\ 2.\ H_3O^+}$ CH_2OH, CH_2OH $\xrightarrow{2\ PBr_3}$ CH_2Br, CH_2Br

$CH_2(CO_2CH_3)_2$, 2 $NaOC_2H_5$

—COOH + CO_2 + 2 C_2H_5OH $\xleftarrow[\Delta]{H_3O^+}$ CH_2, $C(CO_2CH_3)_2$, CH_2 + 2 NaBr

CH_3OH, H^+

—CO_2CH_3

26.36 a) $CH_3C(=O)CH_2CO_2C_2H_5$ $\underset{}{\overset{NaOC_2H_5,\ C_2H_5OH}{\rightleftharpoons}}$ [$CH_3C(=O)\bar{C}H(CO_2C_2H_5)$ + $HC(=O)H$ $\rightleftharpoons$ $HO-CH_2-\bar{C}(CCH_3=O)(CO_2C_2H_5)$]

enolate formation

aldol condensation

$H_2C=C(CCH_3=O)(CO_2C_2H_5)$ + $^-:OH$

b) $C_2H_5O_2C$–$\bar{C}H$–C(=O)–CH_3 + H_2C=C($CO_2C_2H_5$)–C(=O)–CH_3 ⇌ (Michael Addition) ⇌ ⇌

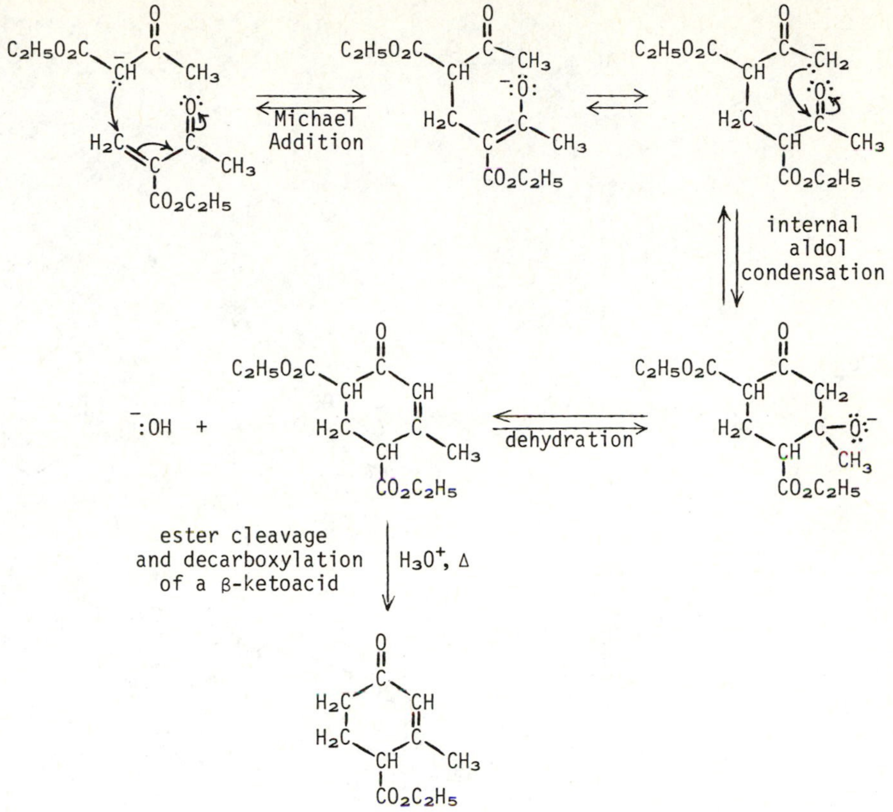

(internal aldol condensation)

$^-$:OH + [cyclohexenone with $C_2H_5O_2C$, CH_3, $CO_2C_2H_5$] ⇌ (dehydration)

ester cleavage and decarboxylation of a β-ketoacid

H_3O^+, Δ

26.37

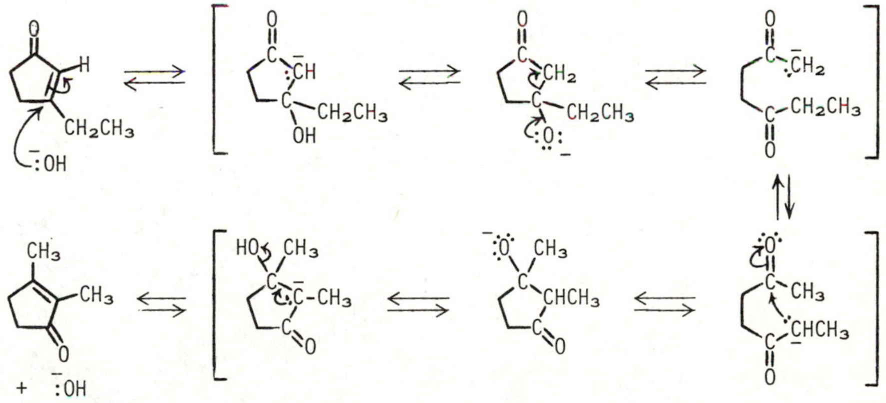

+ $^-$:OH

26.38

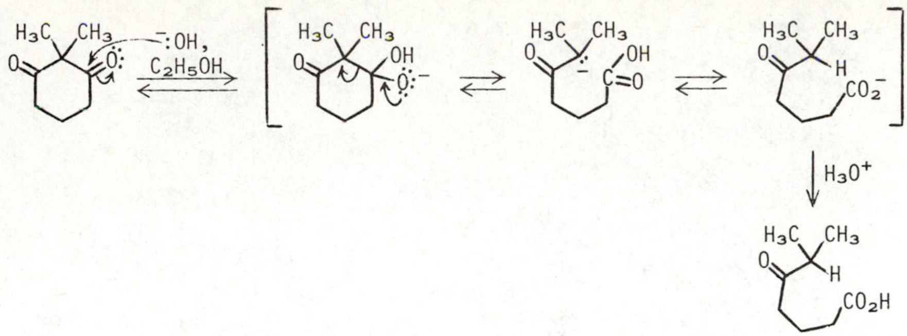

$^{-}$:OH, C_2H_5OH

H_3O^+

26.39

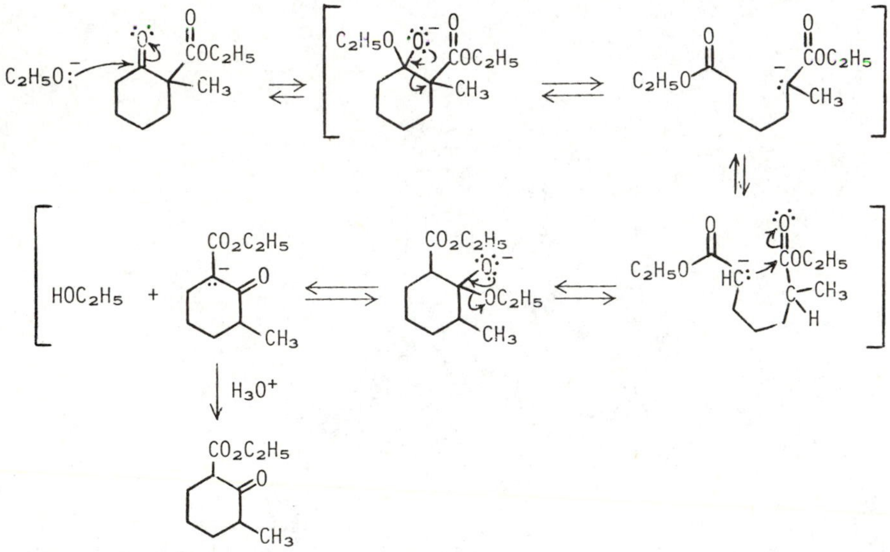

H_3O^+

26.40

$^{-}$:OC_2H_5, C_2H_5OH

H_3O^+

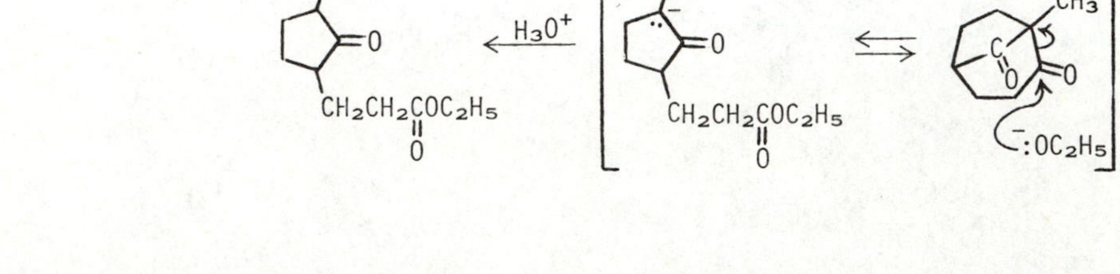

26.41 $CH_2(CO_2C_2H_5)_2 + {}^{-}:OC_2H_5 \rightleftharpoons :\bar{C}H(CO_2C_2H_5)_2 + HOC_2H_5$

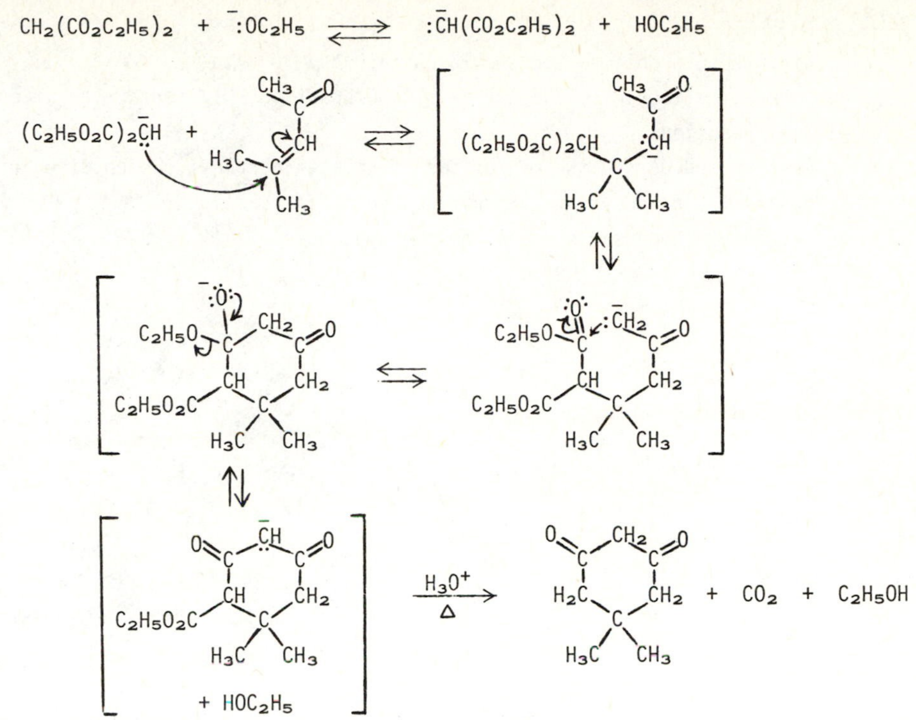

$(C_2H_5O_2C)_2\bar{C}H$ +

CH3 C=O, CH, H3C–C, CH3

⇄ [$(C_2H_5O_2C)_2CH$, CH3 C=O, :CH⁻, C, H3C CH3]

⇅

[:O:⁻, C2H5O–C, CH2, C=O, CH, CH2, C2H5O2C, C, H3C CH3] ⇄ [:O:, C2H5O–C, :CH2⁻, C=O, CH, CH2, C2H5O2C, C, H3C CH3]

⇅

[O=C, ⁻CH, C=O, CH, CH2, C2H5O2C, C, H3C CH3 + HOC_2H_5] $\xrightarrow[\Delta]{H_3O^+}$ O=C, CH2, C=O, H2C, CH2, C, H3C CH3 + CO_2 + C_2H_5OH

26.42 OCH3, O, H, H, CH3O, O, Cl $\underset{HOC(CH_3)_3}{\overset{\overset{+}{K}\,\ddot{O}C(CH_3)_3}{\rightleftharpoons}}$ [OCH3, O, H, CH3O, O⁻, Cl]

1. $CH_3OC{\equiv}C\overset{O}{\overset{\|}{C}}CH{=}CHCH_3$
2. H_3O^+

OCH3, O, OCH3, C=CH, H, C=O, CH3O, O, Cl, CH3CH=CH

⇅ $(CH_3)_3C\ddot{O}\overset{+}{K}$

[OCH3, O, OCH3, C=CH, C=O, CH3O, O⁻, Cl, CH3CH=CH] $\xrightarrow{H_3O^+}$ OCH3, O, OCH3, CH3, =O, CH3O, O, H, Cl

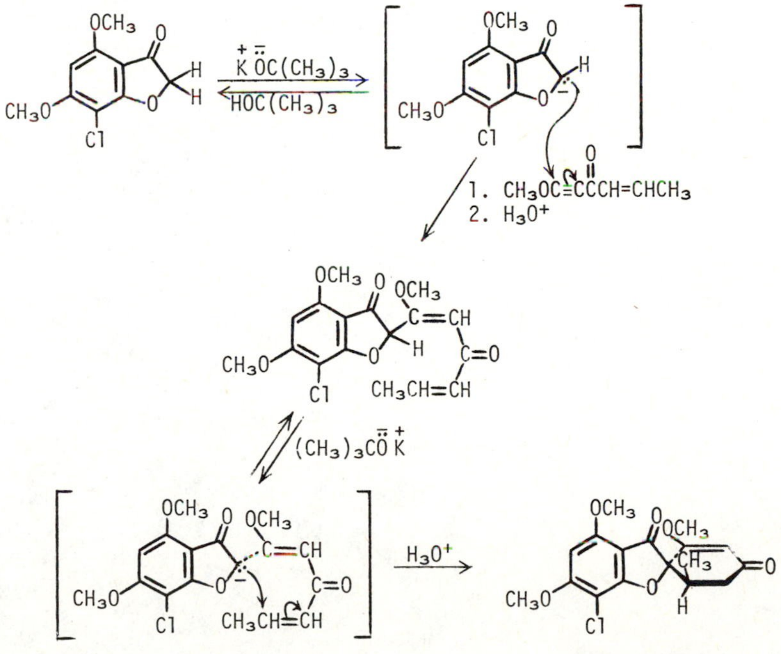

griseofulvin

Two Michael reactions are involved in the key step that forms the *spiro* ring.

What you should know:

After doing these problems, you should be able to:

1. Formulate the mechanisms of the aldol condensation, Claisen condensation, and Michael reaction;
2. Understand and derive the mechanisms of related carbonyl condensation reactions;
3. Recognize the products of carbonyl condensation reactions;
4. Use carbonyl condensation reactions in synthesis.

CHAPTER 27 CARBOHYDRATES

27.1

a)
```
 CHO
 |
HOCH
 |
HCOH
 |
 CH2OH
```
threose
an *aldotetrose*

b)
```
 CH2OH
 |
 C=O
 |
HCOH
 |
HCOH
 |
 CH2OH
```
ribulose
a *ketopentose*

c)
```
 CH2OH
 |
 C=O
 |
HOCH
 |
HOCH
 |
HCOH
 |
 CH2OH
```
tagatose
a *ketohexose*

d)
```
 CHO
 |
 CH2
 |
HCOH
 |
HCOH
 |
 CH2OH
```
2-deoxyribose
an *aldopentose*

27.2

a)
```
      CHO
HO ---S--- H
HO ---S--- H
      CH2OH
```
L-erythrose

b)
```
      CHO
H  ---R--- OH
HO ---S--- H
H  ---R--- OH
      CH2OH
```
D-xylose

c)
```
      CH2OH
      C=O
HO ---S--- H
H  ---R--- OH
      CH2OH
```
D-xylulose

27.3

```
      CHO
H  ---R--- OH
HO ---S--- H
HO ---S--- H
      CH2OH
```
L-(+)-arabinose

27.4

a)
```
      CHO
HO ------- H
H  ------- OH
HO ------- H
      CH2OH
```
L-xylose

b)
```
      CHO
HO ------- H
H  ------- OH
H  ------- OH
HO ------- H
      CH2OH
```
L-galactose

c)
```
      CHO
HO ------- H
H  ------- OH
HO ------- H
HO ------- H
      CH2OH
```
L-glucose

27.5 Thirty-two aldoheptoses are possible (sixteen D and sixteen L).

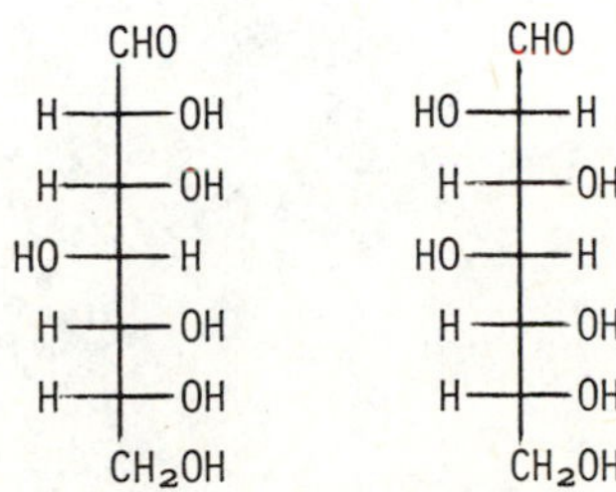

27.6

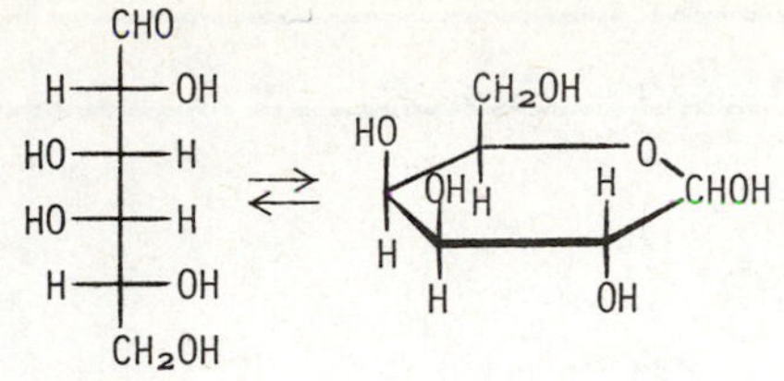

D-galactose

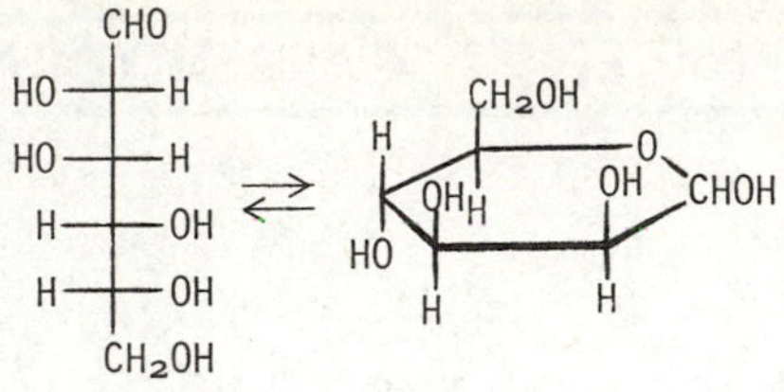

D-mannose

27.7

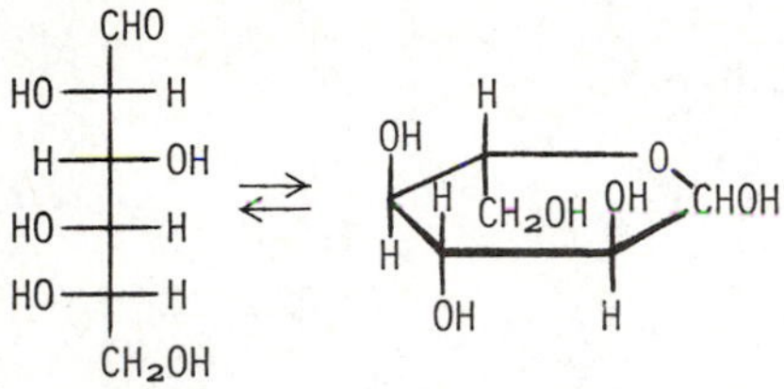

L-glucose

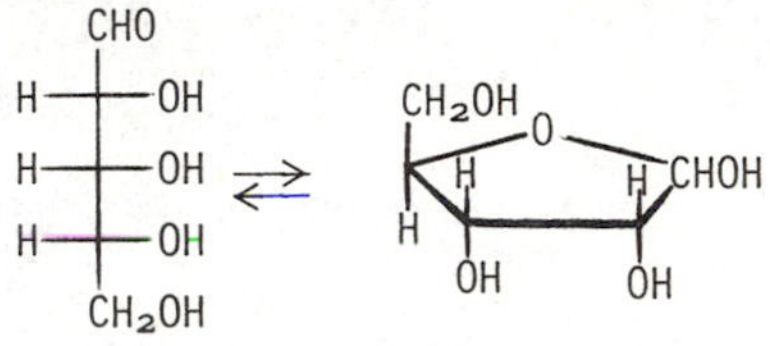

D-ribose

27.8

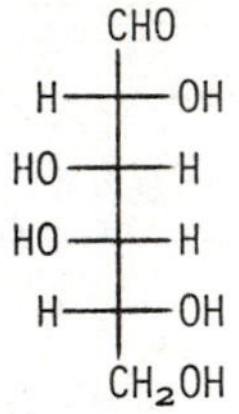

D-galactose

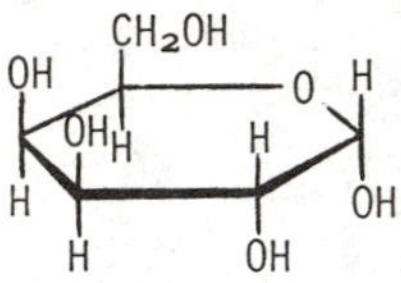

α-D-galactose

$[\alpha]_D$ = +150.7°

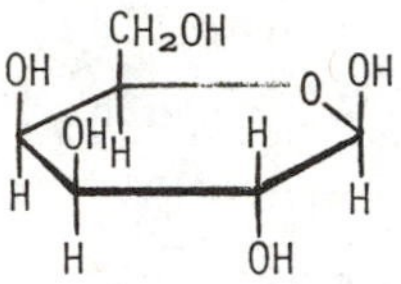

β-D-galactose

$[\alpha]_D$ = +52.8°

Let x be the fraction of D-galactose present as the α anomer and y be the fraction of D-galactose present as the β anomer.

$$150.7°x + 52.8°y = 80.2° \qquad x + y = 1;\ y = 1 - x$$

$$150.7°x + 52.8°(1 - x) = 80.2°$$

$$97.9°x = 27.4°$$

$$x = 0.280$$

$$y = 0.720$$

28.0% of D-galactose is present as the α anomer, and 72.0% is present as the β anomer.

27.9

β-D-galactopyranose

β-D-mannopyranose

β-D-Galactopyranose and β-D-mannopyranose each have one hydroxyl group in the axial position. Galactose and mannose are of equal stability.

27.10

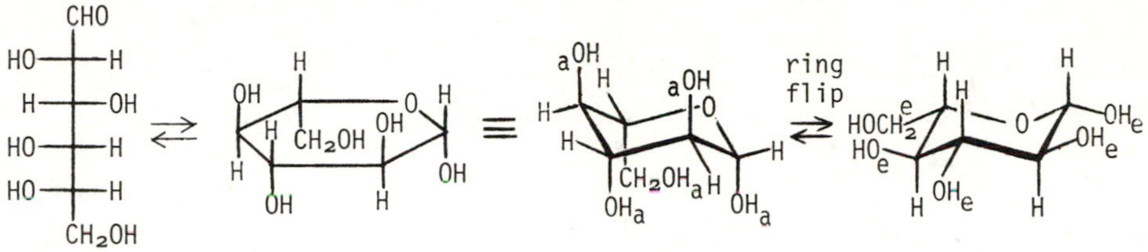

L-glucose β-L-glucopyranose

All substitutents are equatorial in the more stable conformation of β-L-glucopyranose.

27.11

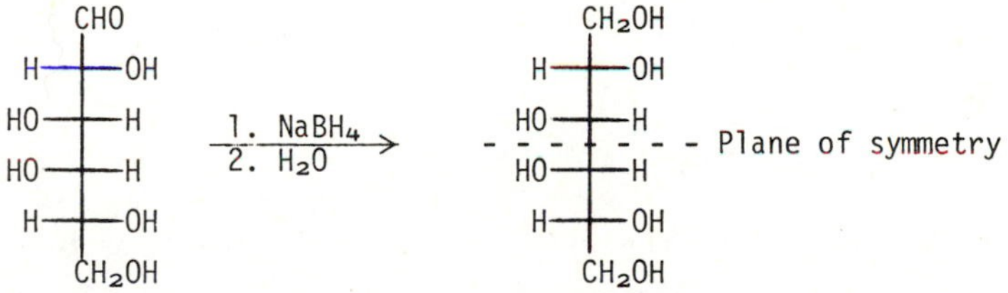

D-galactose galactitol

Reaction of D-galactose with $NaBH_4$ yields an alditol that has a plane of symmetry. Galactitol is a *meso* compound.

27.12

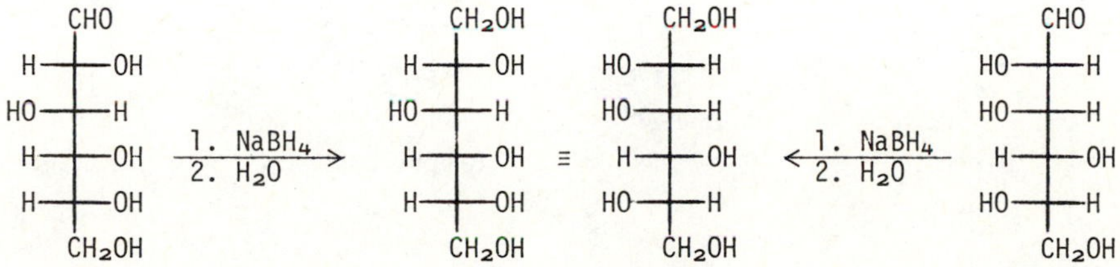

D-glucose D-glucitol L-gulitol L-gulose

Reaction of an aldose with $NaBH_4$ produces a polyol (alditol). Because an alditol has the same functional group at both ends, the number of stereoisomers of an n-carbon alditol is one-half the number of stereoisomers of the parent aldose, and two different aldoses can yield the same alditol. Here, L-gulose and D-glucose form the same alditol (rotate the Fischer projection of L-gulitol 180° to see the identity).

27.13 D-Allose and D-galactose yield *meso* aldaric acids. All other D-hexoses produce optically active aldaric acids on oxidation.

27.14

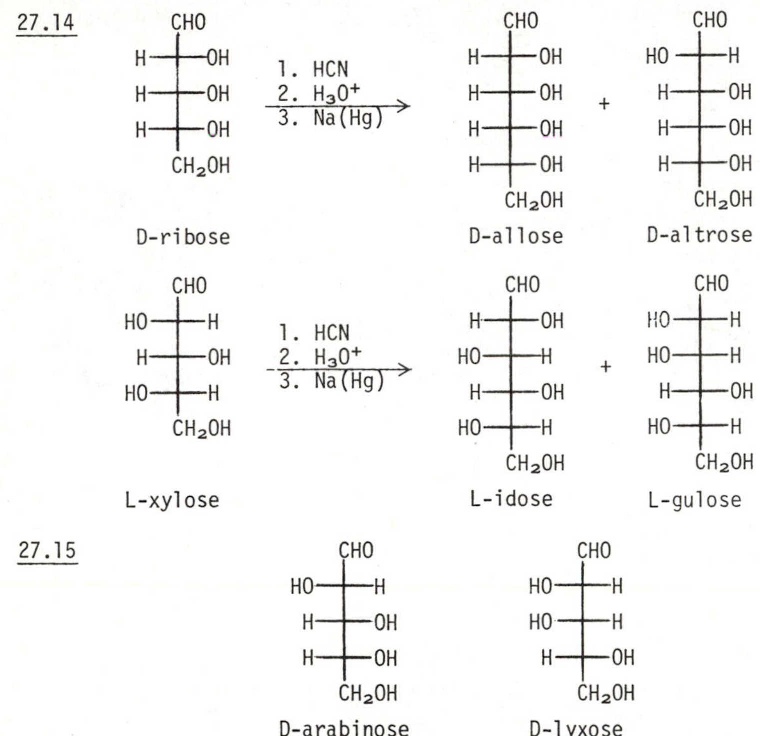

27.15

D-Glucose has the same configuration at C3, C4, and C5 as D-arabinose. D-Lyxose is the only other aldopentose that yields an optically active aldaric acid on nitric acid oxidation.

27.16 a) CHO; H–C–OH; H–C–OH; H–C–OH; CH_2OH — D-ribose

b) CHO; H–C–OH; HO–C–H; H–C–OH; CH_2OH — D-xylose

c) CHO; H–C–OH; H–C–OH; CH_2OH — D-erythrose

d) CHO; HO–C–H; H–C–OH; CH_2OH — D-threose

27.17

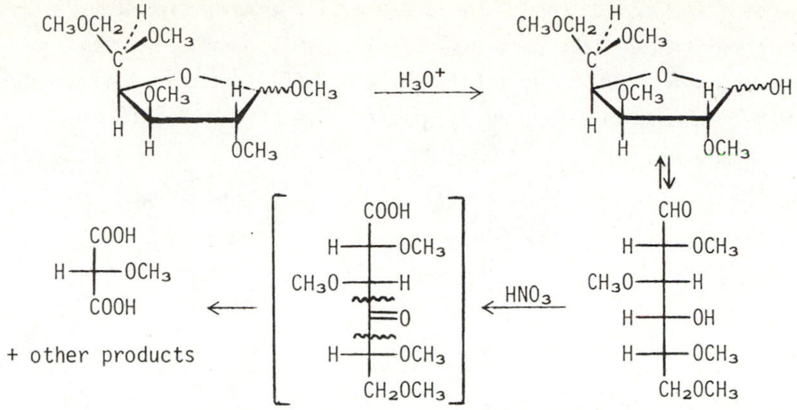

If glucose were a furanose, the hemiacetal linkage would occur between the aldehyde and the hydroxyl at C4. Treatment of the acetal with H_3O^+, followed by nitric acid oxidation of the 4-hydroxy tetramethyl ether, yields methoxymalonic acid and other products.

27.18 Degradation of D-glucose does not affect the stereochemistry at C2 and C3. The product of degradation, (2*R*,3*R*)-2,3-dimethoxytartaric acid, is optically active.

27.19

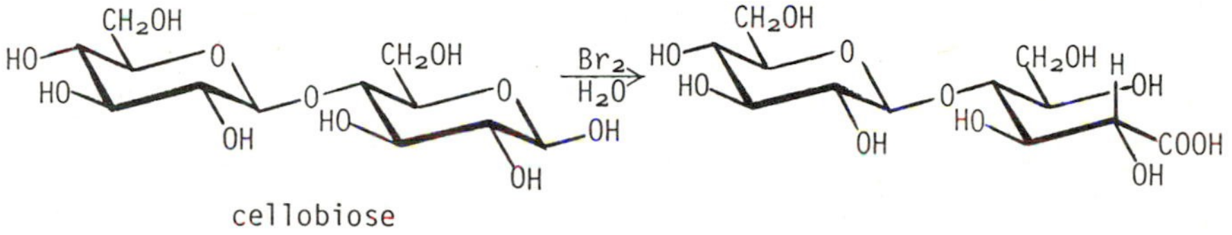

cellobiose

27.20

cellobiose

$CH_3OCH(COOH)_2$ + other products

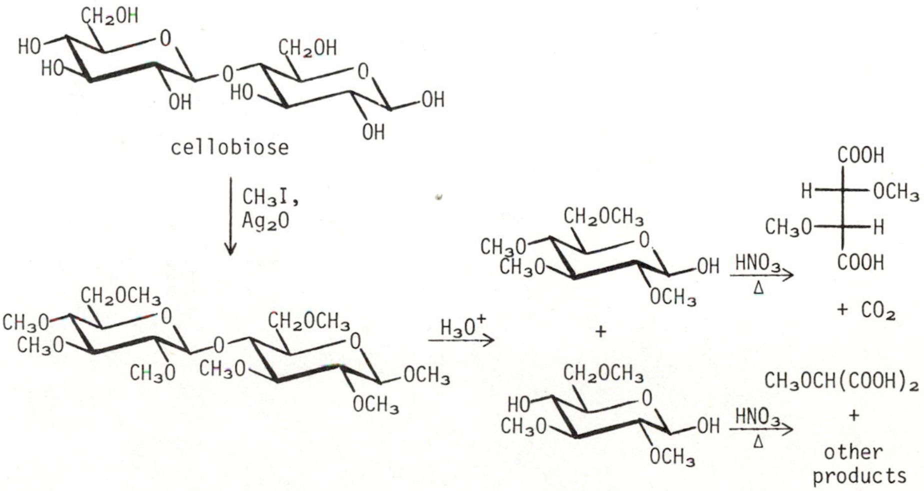

Treatment of cellobiose with methyl iodide methylates all hydroxyl groups. Mild acid hydrolysis cleaves the glycosidic glucose-glucose bond, as well as the methyl glycoside bond. The tri-*O*-methyl glucopyranose thus comes from the glucopyranose that contains the hemiacetal group. The two *O*-methylated glucopyranoses can be subjected to nitric acid oxidation, and the fragments can be analyzed to determine the position of the glycosidic bond.

27.21

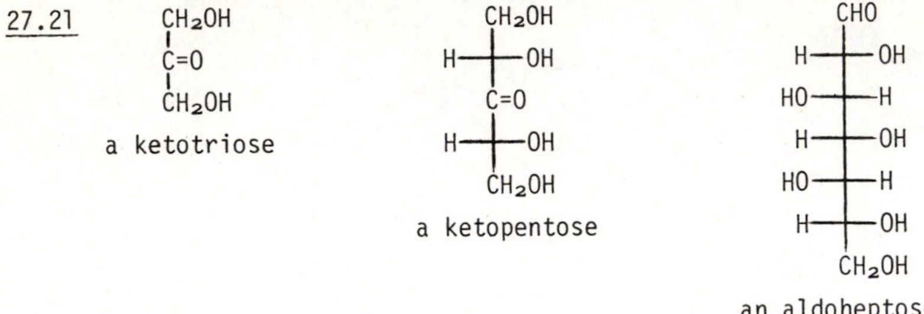

a ketotriose a ketopentose an aldoheptose

27.22 and 27.23

L-ascorbic acid

27.24

β-D-talopyranose

a) 1. $NaBH_4$ 2. H_2O

b) HNO_3

c) Br_2, H_2O

d) CH_3CH_2OH, H^+

e) CH_3I, Ag_2O

f) $(CH_3CO)_2O$ pyridine

27.25 - 27.27 Four D-2-ketohexoses are possible.

D-psicose D-fructose D-sorbose D-tagatose

D-psicose → 1. $NaBH_4$ 2. H_2O → D-allitol + D-altritol

D-sorbose → 1. $NaBH_4$ 2. H_2O → D-gulitol + D-iditol

27.28

HNO_3

Na(Hg)

D-glucose

Na(Hg)

rotate 180° ≡

L-gulose

27.29

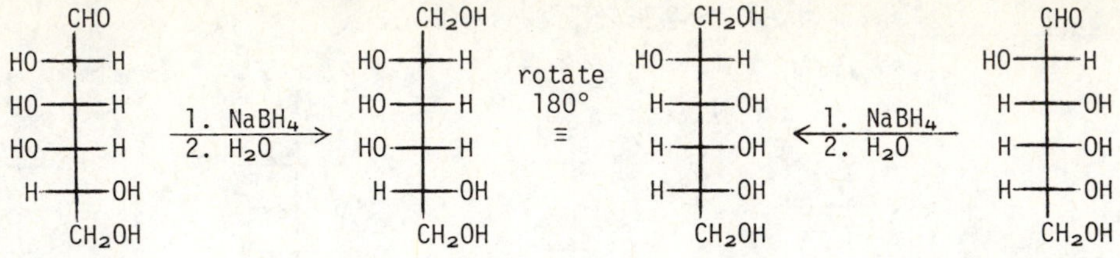

D-talose: CHO; HO—H; HO—H; HO—H; H—OH; CH_2OH

$\xrightarrow{\text{1. NaBH}_4 \;\; \text{2. H}_2\text{O}}$

CH_2OH; HO—H; HO—H; HO—H; H—OH; CH_2OH

rotate 180° ≡

CH_2OH; HO—H; H—OH; H—OH; H—OH; CH_2OH

$\xleftarrow{\text{1. NaBH}_4 \;\; \text{2. H}_2\text{O}}$

D-altrose: CHO; HO—H; H—OH; H—OH; H—OH; CH_2OH

27.30

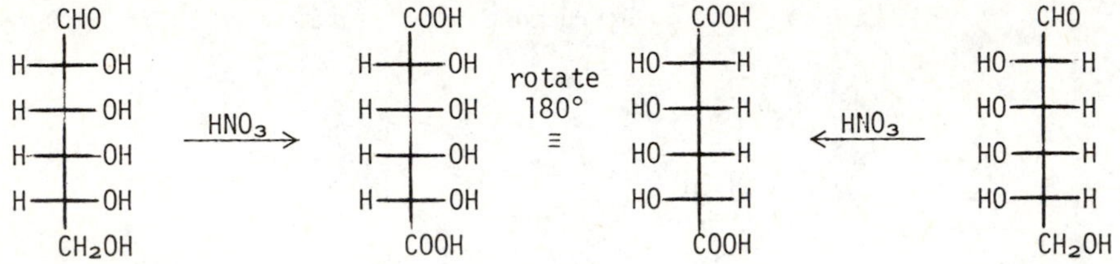

D-allose: CHO; H—OH; H—OH; H—OH; H—OH; CH_2OH

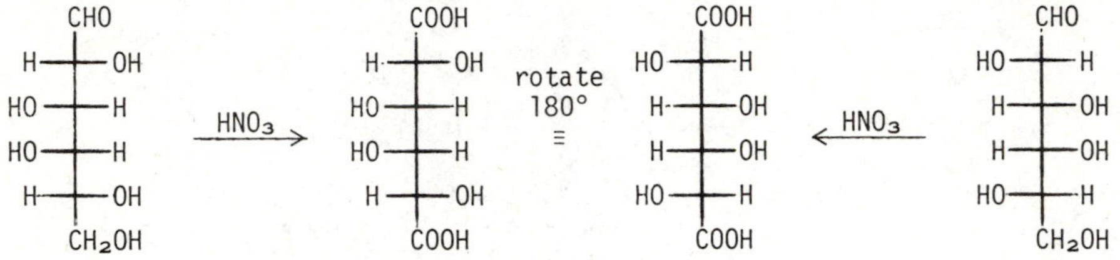

$\xrightarrow{HNO_3}$

COOH; H—OH; H—OH; H—OH; H—OH; COOH

rotate 180° ≡

COOH; HO—H; HO—H; HO—H; HO—H; COOH

$\xleftarrow{HNO_3}$

L-allose: CHO; HO—H; HO—H; HO—H; HO—H; CH_2OH

D-galactose: CHO; H—OH; HO—H; HO—H; H—OH; CH_2OH

$\xrightarrow{HNO_3}$

COOH; H—OH; HO—H; HO—H; H—OH; COOH

rotate 180° ≡

COOH; HO—H; H—OH; H—OH; HO—H; COOH

$\xleftarrow{HNO_3}$

L-galactose: CHO; HO—H; H—OH; H—OH; HO—H; CH_2OH

27.31

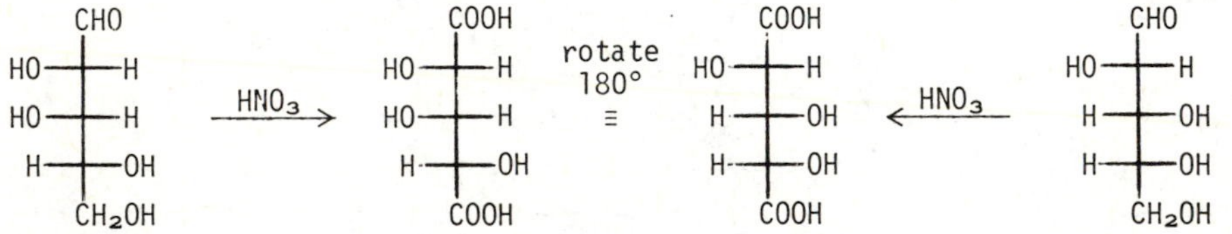

D-lyxose: CHO; HO—H; HO—H; H—OH; CH_2OH

$\xrightarrow{HNO_3}$

COOH; HO—H; HO—H; H—OH; COOH

rotate 180° ≡

COOH; HO—H; H—OH; H—OH; COOH

$\xleftarrow{HNO_3}$

D-arabinose: CHO; HO—H; H—OH; H—OH; CH_2OH

27.32

CH_2OH, HO, HO, O, OH, O, CH_2, HO, HO, O, OH, OH

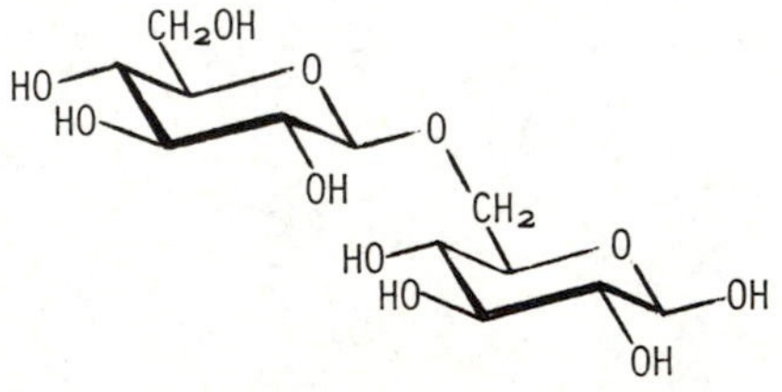

gentiobiose

6-*O*-(β-D-glucopyranosyl)-β-D-glucopyranose

27.33

amygdalin

27.34 There are three possible structures for trehalose. The two glucopyranose rings can be connected (α,α), (β,β), or (α,β). Since trehalose is not cleaved by β-glycosidase, it must be α,α.

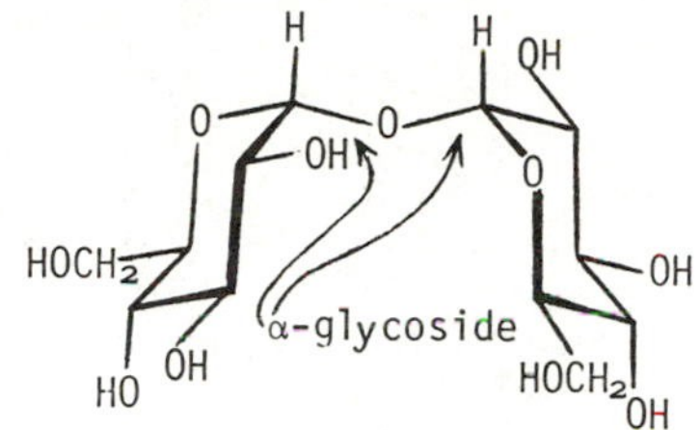

trehalose

1-*O*-(α-D-glucopyranosyl)-α-D-glucopyranose

27.35

β-glycoside

β-glycoside

neotrehalose

1-*O*-(β-D-glucopyranosyl)-β-D-glucopyranose

α-glycoside

β-glycoside

isotrehalose

1-*O*-(α-D-glucopyranosyl)-β-D-glucopyranose

27.36

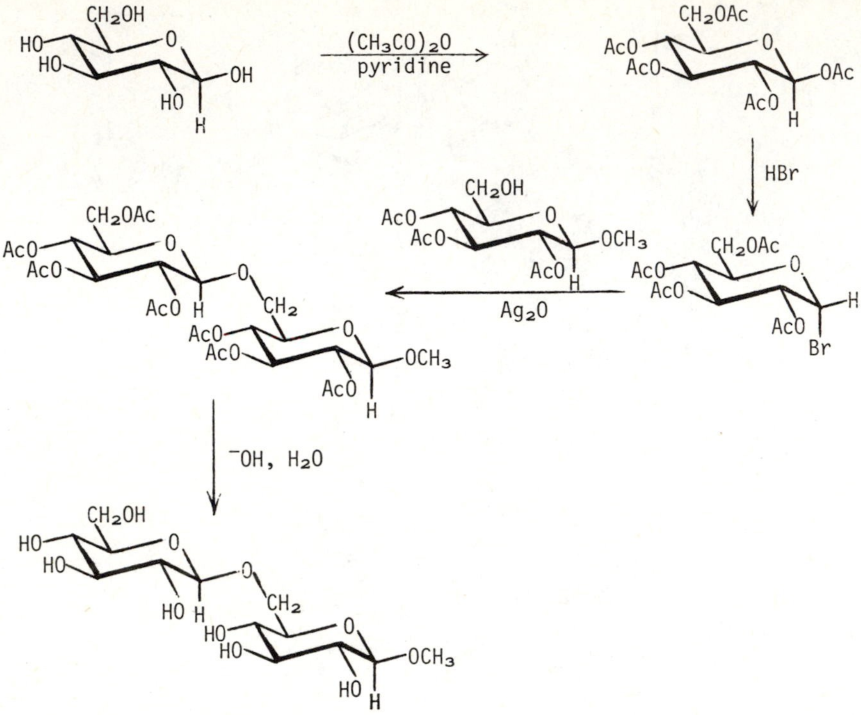

(CH$_3$CO)$_2$O, pyridine → HBr → Ag$_2$O → $^-$OH, H$_2$O

gentiobiose methyl glycoside

The key step in this synthesis is the Koenigs-Knorr reaction (Sec. 27.8). Treatment of penta-*O*-acetyl-β-D-glucopyranose with HBr converts the anomeric β-acetate into an α-glycosyl bromide, which reacts with methyl-β-D-2,3,4-tri-*O*-acetylglucopyranoside in the presence of Ag_2O. Reaction occurs at the 6-hydroxyl position of the methyl pyranoside because all other positions are blocked. Cleavage of the acetyl groups does not affect the glycosidic bonds because they are stable to base.

27.37

Glucopyranose is in equilibrium with glucofuranose.

Reaction with two equivalents of acetone occurs *via* the mechanism we learned for acetal formation (Sec. 22.11).

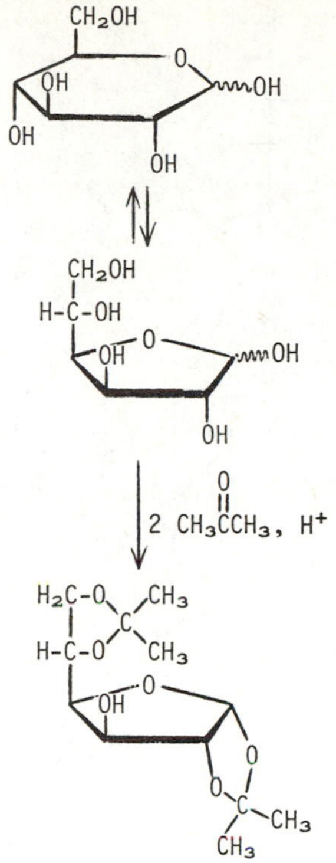

$2\ CH_3\overset{O}{\overset{\|}{C}}CH_3,\ H^+$

A five-membered acetal ring forms much more readily when the hydroxyl groups that are to be part of the acetal ring are *cis* to one another. In glucopyranose the C3 hydroxyl is *trans* to the C2 hydroxyl, and acetal formation occurs between acetone and the C-1 and C2 hydroxyls. Since the C-1 hydroxyl group is part of the acetone acetal, the furanose is no longer in equilibrium with the free aldehyde, and the diacetone derivative is not a reducing sugar.

27.38

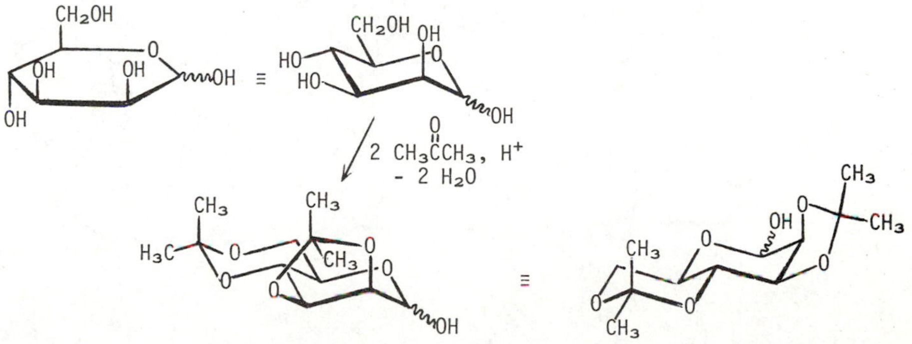

$2\ CH_3\overset{O}{\overset{\|}{C}}CH_3,\ H^+$
$-\ 2\ H_2O$

2,3:4,6-diacetone mannopyranoside

Acetone forms an acetal with the hydroxyl groups at C2 and C3 of D-mannopyranoside because the hydroxyl groups at these positions are *cis* to one another. The pyranoside ring is still a hemiacetal that is in equilibrium with free aldehyde, which is reducing toward Tollens reagent.

27.39

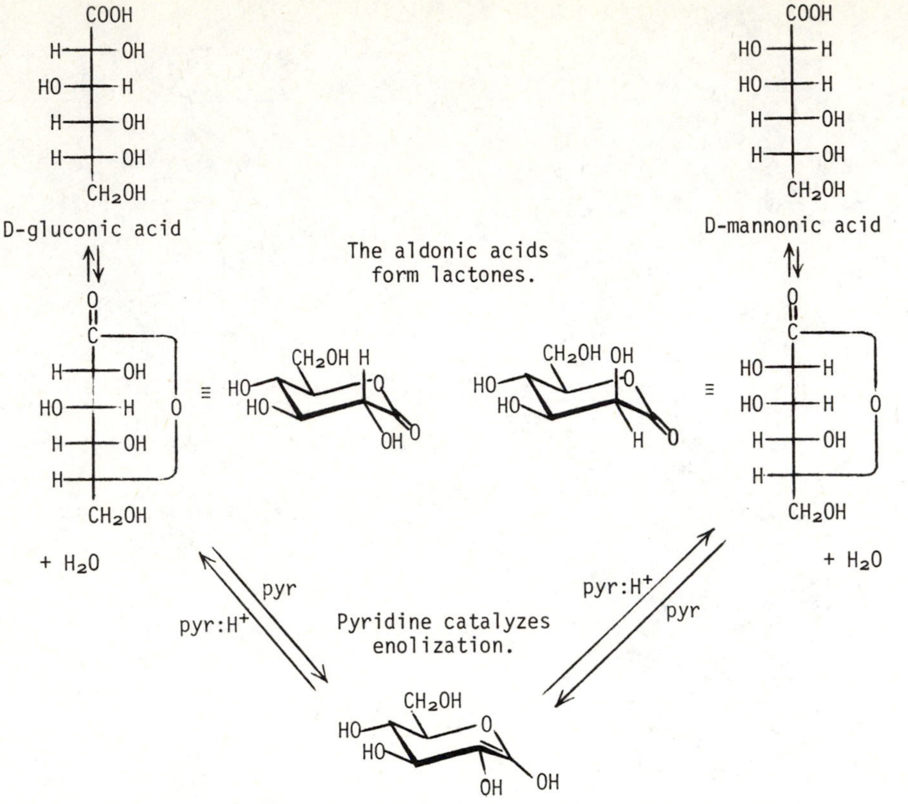

COOH
H—OH
HO—H
H—OH
H—OH
CH_2OH
D-gluconic acid

COOH
HO—H
HO—H
H—OH
H—OH
CH_2OH
D-mannonic acid

The aldonic acids form lactones.

$+ H_2O$ $+ H_2O$

pyr pyr:H^+

Pyridine catalyzes enolization.

pyr:H^+ pyr

Epimerization at C2 occurs because the enediol can be reprotonated on either side of the double bond.

27.40

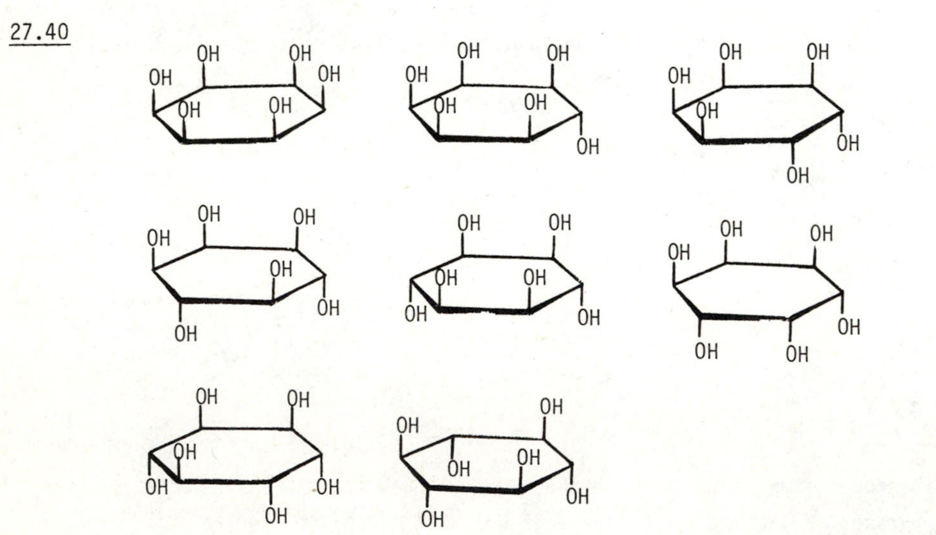

27.41

```
        CHO                        COOH
    H ──┼── OH                 H ──┼── OH
    H ──┼── OH     HNO3 →      H ──┼── OH
    H ──┼── OH                 H ──┼── OH
       CH2OH                       COOH
         A                          B

         │ 1. HCN
         │ 2. H3O+
         ↓ 3. Na(Hg)

      COOH                  CHO            CHO                   COOH
 HO ──┼── H           HO ──┼── H       H ──┼── OH            H ──┼── OH
  H ──┼── OH  ← HNO3   H ──┼── OH   +  H ──┼── OH   HNO3 →   H ──┼── OH
  H ──┼── OH           H ──┼── OH      H ──┼── OH            H ──┼── OH
  H ──┼── OH           H ──┼── OH      H ──┼── OH            H ──┼── OH
      COOH                CH2OH          CH2OH                 COOH
       E                    C              D                     F
```

What you should know:

After doing these problems, you should be able to:

1. Classify carbohydrates as aldoses, ketoses, D and L sugars, monosaccharides or polysaccharides;
2. Draw monosaccharides as Fischer and Haworth projections and as chair conformations;
3. Predict the products of reactions of monosaccharides;
4. Understand the logic of the Fischer proof of the structure of glucose;
5. Deduce the structure of an unknown sugar.

CHAPTER 28 ALIPHATIC AMINES

28.1 a) (cyclohexyl–NH_2)
primary amine

b) (cyclohexyl–$N(CH_3)$–CH_3)
tertiary amine

c) (phenyl–$CH_2\overset{+}{N}(CH_3)_3$ I^-)
quaternary ammonium salt

d) $[(CH_3)_2CH]_2NH$
secondary amine

e) (pyrrolidine, N–H)
secondary amine

28.2 a) $CH_3N(H)CH_2CH_3$
N-methylethylamine

b) (cyclohexyl$)_3$N
tricyclohexylamine

c) $CH_3N(C_6H_{11})CH_2CH_2CH_3$
N-methyl-*N*-propylcyclohexylamine

d) (pyrrolidine, N–CH_3)
N-methylpyrrolidine

e) $[(CH_3)_2CH]_2NH$
diisopropylamine

f) $H_2NCH_2CH_2CH(CH_3)NH_2$
1,3-butanediamine

g) (cyclopentane with cis H, NH_2, NH_2, H)
cis-1,2-cyclopentanediamine

28.3 a) $(CH_3CH_2)_3N$
triethylamine

b) $(H_2C{=}CHCH_2)_3N$
triallylamine

c) (phenyl–$NHCH_3$)
N-methylaniline

d) (cyclopentyl–$N(CH_3)CH_2CH_3$)
N-ethyl-*N*-methylcyclopentylamine

e) (cyclohexyl–NH–$CH(CH_3)CH_3$)
N-isopropylcyclohexylamine

f) (pyrrole, N–CH_2CH_3)
N-ethylpyrrole

28.4 Pyramidal inversion of nitrogen requires the hybridization at nitrogen to change momentarily from sp^3 to sp^2. The 120° bond angle required for substituents bonded to sp^2-hybridized atoms results in a high degree of strain for the three-membered aziridine ring. The energy barrier for inversion is relatively high, and the rate of inversion of (+)-1-chloro-2,2-diphenylaziridine is slow.

28.5

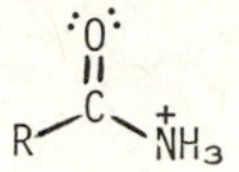

$R-C(=\ddot{O}:)-\overset{+}{N}H_3$	$R-C(=\overset{+}{\ddot{O}}-H)-\ddot{N}H_2 \longleftrightarrow R-C(-\ddot{O}-H)=\overset{+}{N}H_2$
N-protonation	*O*-protonation
(no resonance stabilization)	(resonance stabilization)

Protonation occurs at oxygen because an *O*-protonated amide can be stabilized by resonance.

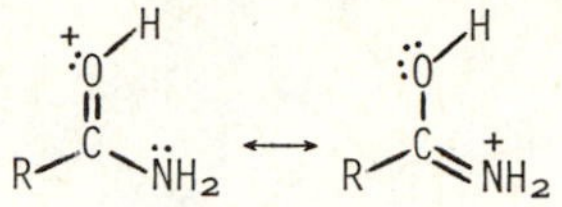

28.6

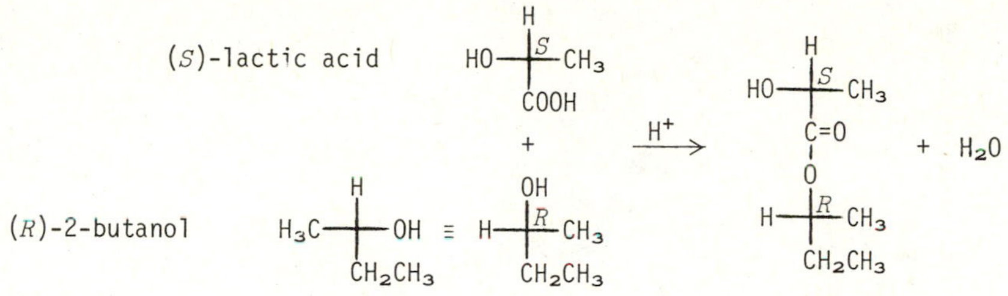

(*S*)-lactic acid: HO–C(H)(*S*)(CH$_3$)–COOH

\+

(*R*)-2-butanol: H_3C–C(H)(OH)–CH_2CH_3 ≡ H–C(OH)(*R*)(CH$_3$)–CH_2CH_3

$\xrightarrow{H^+}$ HO–C(H)(*S*)(CH$_3$)–C(=O)–O–C(H)(*R*)(CH$_3$)–CH_2CH_3 + H_2O

The stereochemistry at the two chiral centers in the product is the same as the stereochemistry of the reactants because no bonds were formed or broken at the chiral centers.

28.7

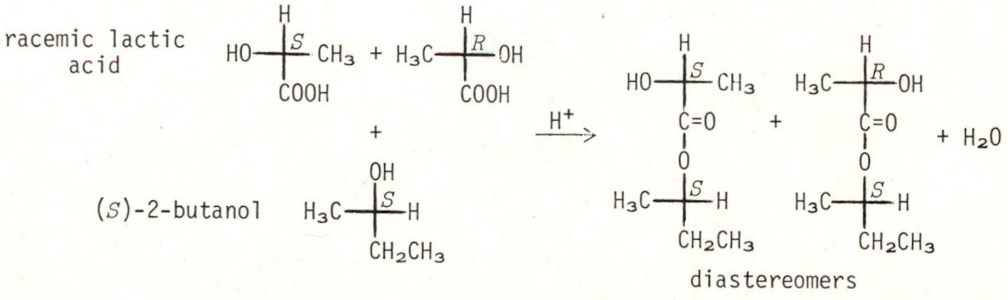

racemic lactic acid: HO–C(H)(*S*)(CH$_3$)–COOH + H_3C–C(H)(*R*)(OH)–COOH

\+

(*S*)-2-butanol: H_3C–C(OH)(*S*)(H)–CH_2CH_3

$\xrightarrow{H^+}$ HO–C(H)(*S*)(CH$_3$)–C(=O)–O–C(*S*)(H)(H_3C)–CH_2CH_3 + H_3C–C(H)(*R*)(OH)–C(=O)–O–C(*S*)(H)(H_3C)–CH_2CH_3 + H_2O

diastereomers

The above reaction can be used to resolve racemic lactic acid. The diastereomeric esters can be separated on the basis of their different physical or chemical behavior. For example, they may have different solubilities, melting or boiling points, or retention times on a chromatography column. After separation, the esters can be hydrolyzed to regenerate (*R*)- or (*S*)-lactic acid and (*S*)-2-butanol. Lactic acid can be separated from 2-butanol by extraction with aqueous base, reacidification, and extraction of the acidified mixture with an organic solvent.

28.8

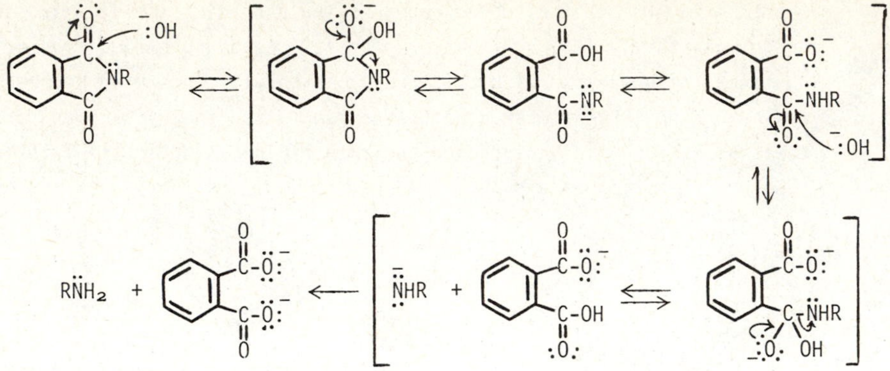

28.9

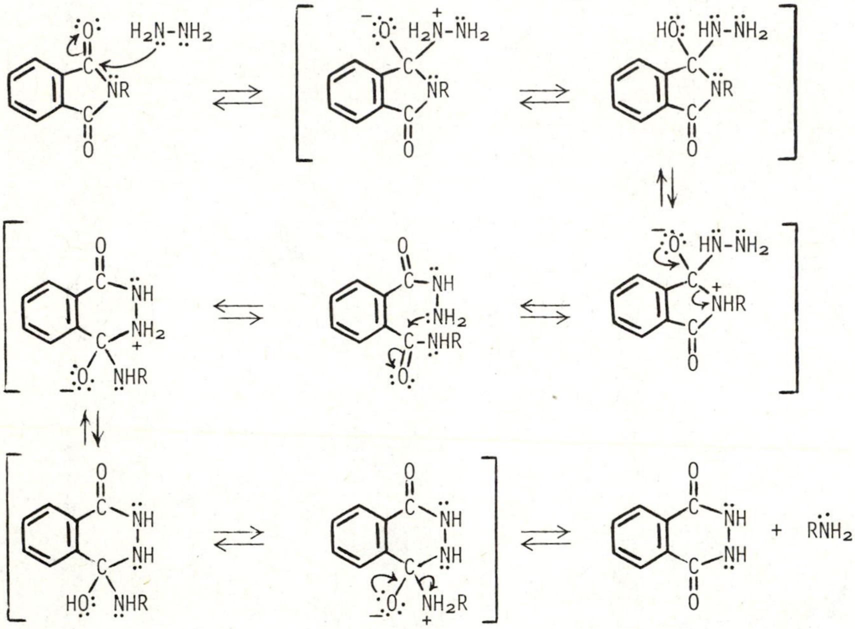

28.10

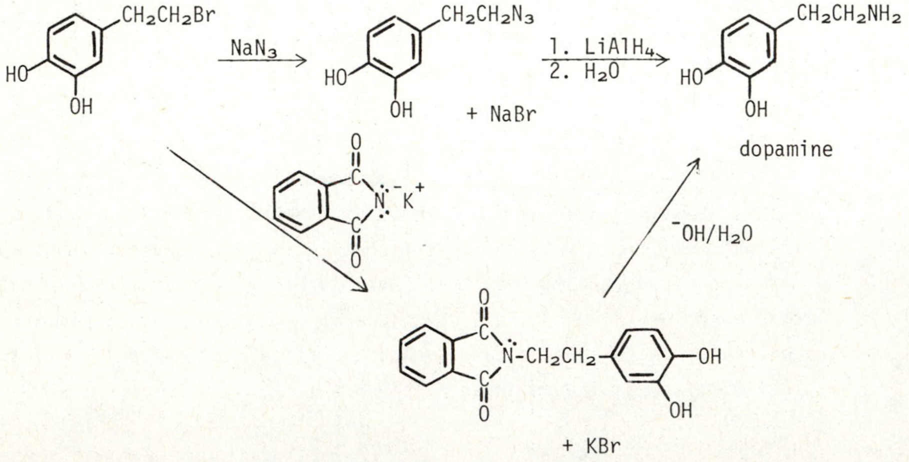

NaN$_3$

+ NaBr

1. LiAlH$_4$
2. H$_2$O

dopamine

$^-$OH/H$_2$O

+ KBr

28.11

$$\text{cyclohexanone} + :NH(CH_3)_2 \rightleftharpoons \left[\underset{\text{tetrahedral intermediate}}{^{-}:\ddot{O}:\ \overset{+}{N}H(CH_3)_2} \rightleftharpoons \underset{\text{carbinolamine}}{H\ddot{O}\ N(CH_3)_2} \right]$$

$$\updownarrow$$

$$\underset{}{H_3C\ \ CH_3,\ N,\ H} \xleftarrow{NaBH_3CN} \left[\underset{\text{enamine}}{H_3C\ \ddot{N}\ CH_3} + H_2O \leftrightarrows \underset{\text{iminium ion}}{H_3C\ \overset{+}{N}\ CH_3} + {}^{-}:OH \right]$$

28.12

$$C_6H_5\text{-}\overset{OH}{\overset{|}{C}H}\text{-}\overset{O}{\overset{\|}{C}}CH_3 + CH_3NH_2 \xrightarrow[CH_3OH]{NaBH_3CN} C_6H_5\text{-}\overset{OH}{\overset{|}{C}H}\text{-}\overset{NHCH_3}{\overset{|}{C}H}CH_3$$

ephedrine

28.13

$$R\text{-}\overset{O}{\overset{\|}{C}}\text{-}\ddot{N}: \qquad H_2C: \qquad H_2\overset{+}{C}\text{—}H$$

nitrene carbene carbocation

A nitrene, like a carbene and a carbocation, has six electrons in its valence shell. Both a nitrene and a carbene have nonbonded electron pairs and are uncharged; a carbocation has no nonbonded electron pairs and is positively charged.

28.14

$$\underset{\text{nitrene}}{R:\overset{\cdot\cdot O\cdot\cdot}{\overset{\cdot\cdot}{C}}:\ddot{N}} \longrightarrow \underset{\text{isocyanate}}{\ddot{O}::C::\ddot{N}:R}$$

All atoms of the isocyanate have complete octets. The R group shares its electron pair with nitrogen, and nitrogen shares an electron pair with the adjacent carbon.

$$\ddot{O}=C=\ddot{N}\text{-}R + H_2\ddot{O}: \rightleftharpoons \left[H_2\overset{+}{\ddot{O}}\text{-}\overset{O}{\overset{\|}{C}}\text{-}\overset{-}{\ddot{N}}R \rightleftharpoons H\ddot{O}\text{-}\overset{O}{\overset{\|}{C}}\text{-}\ddot{N}HR \rightleftharpoons {}^{-}:\ddot{O}\text{-}\overset{O}{\overset{\|}{C}}\text{-}\overset{+}{N}H_2R \right] \longrightarrow CO_2 + R\ddot{N}H_2$$

28.15

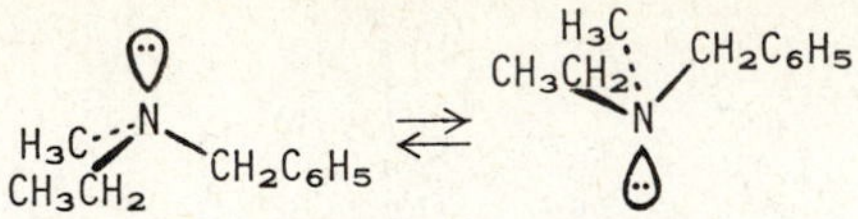

H_3C, CH_3CH_2, N (lone pair), $CH_2C_6H_5$ ⇌ H_3C, CH_3CH_2, N (lone pair), $CH_2C_6H_5$

tertiary amine

$CH_2C_6H_5$, N^+I^-, H_3C, CH_3CH_2, $CH_2CH{=}CH_2$ — $CH_2C_6H_5$, N^+I^-, CH_3, $H_2C{=}CHCH_2$, CH_2CH_3

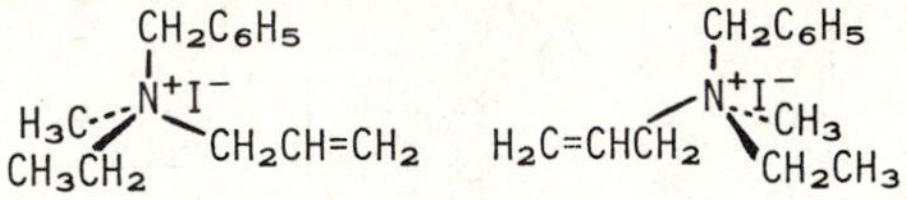

quaternary ammonium salt

Tertiary and quaternary amines such as those pictured above are chiral. Most tertiary amines undergo lone-pair inversion, which interconverts the enantiomers. Quaternary amine interconversion, which would require breaking and reforming bonds, does not occur, and each enantiomer is stable and resolvable.

28.16

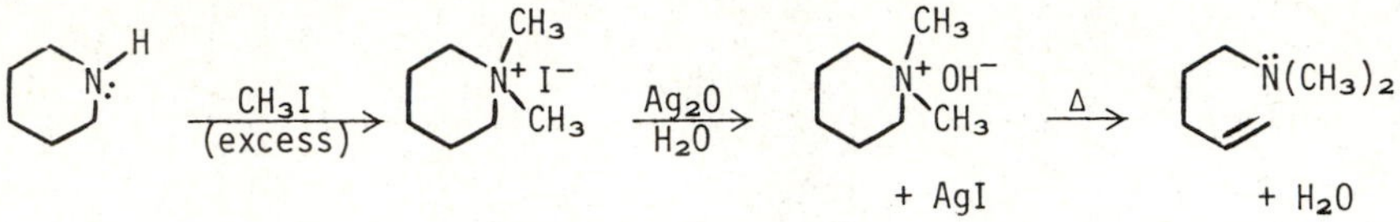

piperidine (N–H) $\xrightarrow[\text{(excess)}]{CH_3I}$ N,N-dimethylpiperidinium I^- $\xrightarrow[H_2O]{Ag_2O}$ N,N-dimethylpiperidinium OH^- + AgI $\xrightarrow{\Delta}$ $\ddot{N}(CH_3)_2$ alkene + H_2O

28.17

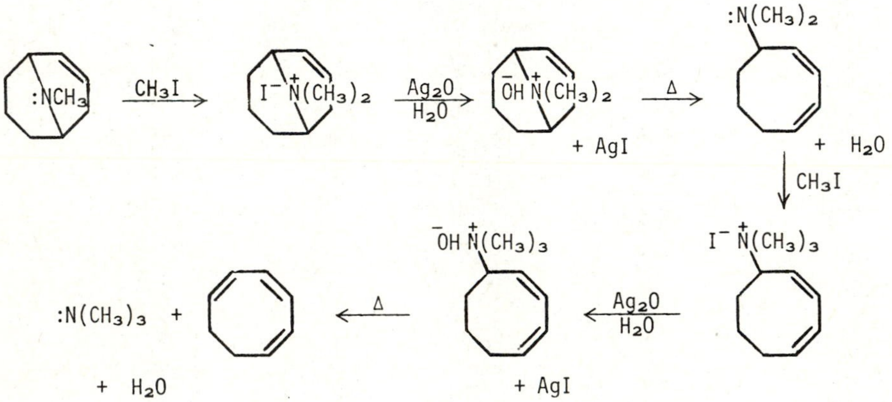

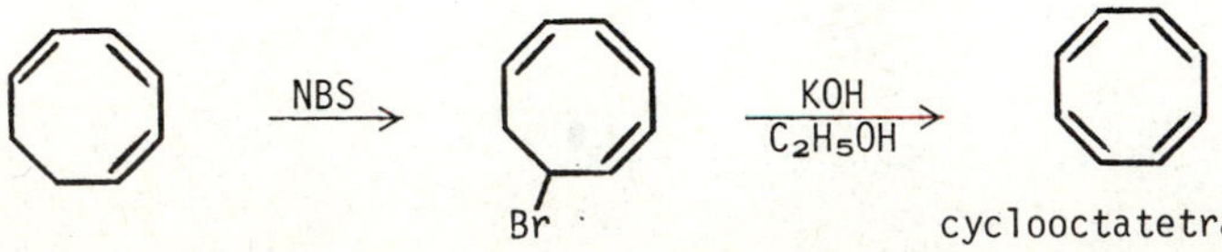

:NCH_3 bicyclic amine $\xrightarrow{CH_3I}$ $I^-\ \overset{+}{N}(CH_3)_2$ $\xrightarrow[H_2O]{Ag_2O}$ $^-OH\ \overset{+}{N}(CH_3)_2$ + AgI $\xrightarrow{\Delta}$:$N(CH_3)_2$ cyclooctadiene + H_2O

$\downarrow CH_3I$

$I^-\ \overset{+}{N}(CH_3)_3$ $\xrightarrow[H_2O]{Ag_2O}$ $^-OH\ \overset{+}{N}(CH_3)_3$ + AgI $\xrightarrow{\Delta}$:$N(CH_3)_3$ + cyclooctatriene + H_2O

cyclooctatriene $\xrightarrow{NBS}$ Br-substituted cyclooctatriene $\xrightarrow[C_2H_5OH]{KOH}$ cyclooctatetraene

28.18

CH_3O (1) (2) (4) N CH_3 H (3)

dextromethorphan

(1) aromatic ring
(2) quaternary carbon
(3) two carbons
(4) tertiary amine

28.19

$NHCH_3$
CH_2CHCH_3

methamphetamine

Mass spectrum: The odd-numbered molecular ion indicates an odd number of nitrogen atoms.

Infrared: The weak band at 3350 cm^{-1} shows that methamphetamine is a secondary amine.

^1H *NMR:*	Absorption	Due to
	7.15 δ (5H)	monosubstituted benzene ring
	2.65 δ (3H)	two benzylic protons one proton adjacent to nitrogen
	2.35 δ (3H)	*N*-methyl group
	0.8 δ (1H)	*N*-H proton (disappears when D_2O is added)
	1.00 δ (3H)	methyl group split by one adjacent proton

28.20 a)

tertiary amide
$(C_2H_5)_2N$-C(=O)
CH_3
tertiary amine
secondary amine
N-H

Lysergic acid diethylamide

b)

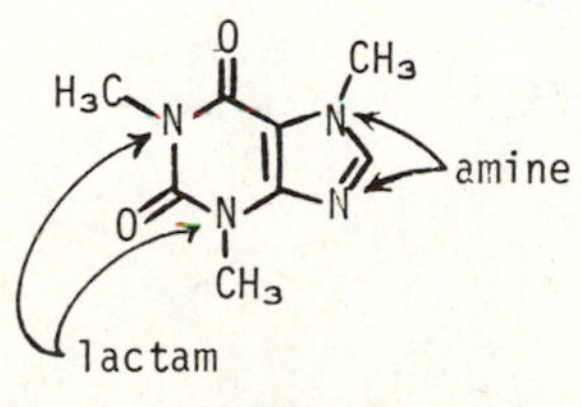

Caffeine

All nitrogen atoms are tertiary.

28.21

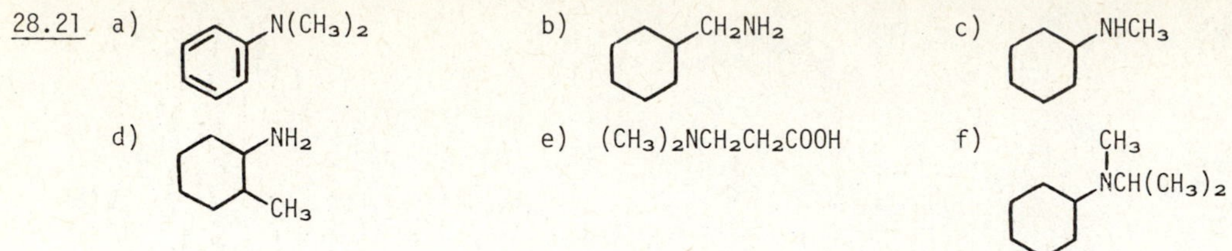

28.22 a) 2,4-dibromoaniline b) (2-cyclopentyl)ethylamine

c) *N*-ethylcyclopentylamine d) *N*,*N*-dimethylcyclopentylamine

e) *N*-propylpyrrolidine f) 4-aminobutanenitrile

28.23 H_3C CH_3 H_3C CH_3 N N H H H H N N H_3C CH_3 H_3C CH_3 $(CH_3)_3N:$

dimethylamine trimethylamine

Even though dimethylamine has a lower molecular weight than trimethylamine, it boils at a higher temperature. Liquid dimethylamine forms hydrogen bonds that must be broken in the boiling process. Since extra energy must be added to break these hydrogen bonds, dimethylamine has a higher boiling point than trimethylamine, which does not form hydrogen bonds.

28.24 a) $CH_3CH_2CH_2CH_2OH \xrightarrow{PBr_3} CH_3CH_2CH_2CH_2Br \xrightarrow{NaN_3} CH_3CH_2CH_2CH_2N_3$

$\downarrow$ 1. $LiAlH_4$ 2. H_2O

$CH_3CH_2CH_2CH_2NH_2$

butylamine

b) $CH_3CH_2CH_2CH_2OH \xrightarrow{Jones'} CH_3CH_2CH_2\overset{O}{\overset{\|}{C}}OH \xrightarrow[CHCl_3]{SOCl_2} CH_3CH_2CH_2\overset{O}{\overset{\|}{C}}\text{-}Cl$

$\downarrow$ $CH_3CH_2CH_2CH_2NH_2$ [from a)] NaOH

$(CH_3CH_2CH_2CH_2)_2NH \xleftarrow[2.\ H_2O]{1.\ LiAlH_4} CH_3CH_2CH_2\overset{O}{\overset{\|}{C}}\text{-}NHCH_2CH_2CH_2CH_3$

dibutylamine

c)

$$CH_3CH_2CH_2CH_2OH \xrightarrow{\text{Jones'}} CH_3CH_2CH_2\overset{O}{\overset{\|}{C}}OH \xrightarrow{SOCl_2} CH_3CH_2CH_2\overset{O}{\overset{\|}{C}}\text{-}Cl$$

$$\downarrow 2\ NH_3$$

$$\underset{\text{propylamine}}{CH_3CH_2CH_2NH_2} \xleftarrow[H_2O,\ \Delta]{Br_2,\ {}^-OH} CH_3CH_2CH_2\overset{O}{\overset{\|}{C}}\text{-}NH_2$$

or $$CH_3CH_2CH_2\overset{O}{\overset{\|}{C}}\text{-}Cl \xrightarrow{NaN_3} CH_3CH_2CH_2\overset{O}{\overset{\|}{C}}N_3 \xrightarrow[2.\ H_2O]{1.\ \Delta} CH_3CH_2CH_2NH_2$$

d)

$$CH_3CH_2CH_2CH_2OH \xrightarrow{PBr_3} CH_3CH_2CH_2CH_2Br \xrightarrow{NaCN} CH_3CH_2CH_2CH_2CN$$

$$\downarrow 1.\ LiAlH_4;\ 2.\ H_2O$$

$$\underset{\text{pentylamine}}{CH_3CH_2CH_2CH_2CH_2NH_2}$$

e)

$$CH_3CH_2CH_2CH_2OH \xrightarrow{PCC} CH_3CH_2CH_2\overset{O}{\overset{\|}{C}}H \xrightarrow[NaBH_3CN]{(CH_3)_2NH} \underset{N,N\text{-dimethylbutylamine}}{CH_3CH_2CH_2CH_2N(CH_3)_2}$$

f)

$$\underset{\text{[from c)]}}{CH_3CH_2CH_2NH_2} \xrightarrow[CH_3I]{\text{excess}} CH_3CH_2CH_2\overset{+}{N}(CH_3)_3\ I^- \xrightarrow[H_2O]{Ag_2O} CH_3CH_2CH_2\overset{+}{N}(CH_3)_3\ {}^-OH$$

$$\downarrow \Delta$$

$$\underset{\text{propene}}{CH_3CH{=}CH_2}$$

28.25 a)

$$CH_3CH_2CH_2CH_2\overset{O}{\overset{\|}{C}}OH \xrightarrow{SOCl_2} CH_3CH_2CH_2CH_2\overset{O}{\overset{\|}{C}}\text{-}Cl \xrightarrow{2\ NH_3} \underset{\text{pentamide}}{CH_3CH_2CH_2CH_2\overset{O}{\overset{\|}{C}}NH_2}$$

b)

$$\underset{\text{[from a)]}}{CH_3CH_2CH_2CH_2\overset{O}{\overset{\|}{C}}NH_2} \xrightarrow[H_2O,\ \Delta]{Br_2,\ {}^-OH} \underset{\text{butylamine}}{CH_3CH_2CH_2CH_2NH_2}$$

c)

$$\underset{\text{[from a)]}}{CH_3CH_2CH_2CH_2\overset{O}{\overset{\|}{C}}NH_2} \xrightarrow[2.\ H_2O]{1.\ LiAlH_4} \underset{\text{pentylamine}}{CH_3CH_2CH_2CH_2CH_2NH_2}$$

d)

$$CH_3CH_2CH_2CH_2\overset{O}{\overset{\|}{C}}OH \xrightarrow[2.\ H_2O]{1.\ Br_2,\ PBr_3} \underset{\text{2-bromopentanoic acid}}{CH_3CH_2CH_2CH_2\overset{Br}{\overset{|}{C}}H\overset{O}{\overset{\|}{C}}OH}$$

e) $CH_3CH_2CH_2CH_2\overset{O}{\overset{\|}{C}}OH \xrightarrow[2.\ H_3O^+]{1.\ LiAlH_4} CH_3CH_2CH_2CH_2CH_2OH \xrightarrow{PBr_3} CH_3CH_2CH_2CH_2CH_2Br$

$\downarrow NaCN$

$CH_3CH_2CH_2CH_2CH_2CN$
hexanenitrile

f) $CH_3CH_2CH_2CH_2CH_2CN \xrightarrow[2.\ H_2O]{1.\ LiAlH_4} CH_3CH_2CH_2CH_2CH_2NH_2$
[from e)] hexylamine

28.26 $CH_3\overset{O}{\overset{\|}{C}}CH_2Br + :NH_3 \rightleftharpoons \left[CH_3\overset{O}{\overset{\|}{C}}CH_2\overset{+}{N}H_3\ Br^- \right] \underset{}{\overset{NH_3}{\rightleftharpoons}} CH_3\overset{O}{\overset{\|}{C}}CH_2NH_2 + \overset{+}{N}H_4\ Br^-$

H₂C–C(=O)–CH₃ / H₂N· ·NH₂ / H₃C–C(=O)–CH ⇌ [cyclic intermediate with $^-$:O:, CH₃, $\overset{+}{N}H_2$, $H_2\overset{+}{N}$, CH₂, H₃C, :O:$^-$ ⇌ cyclic intermediate with HO, CH₃, NH, HN, CH₂, H₃C, OH] ⇌ dihydropyrazine (H₂C, C–CH₃, =N·, ·N=, C–CH₃, CH₂)

$+\ 2\ H_2O$

28.27 a) $CH_3CH_2CH_2CH_2\overset{O}{\overset{\|}{C}}NH_2 \xrightarrow[2.\ H_2O]{1.\ LiAlH_4} CH_3CH_2CH_2CH_2CH_2NH_2$

b) $CH_3CH_2CH_2CH_2CN \xrightarrow[2.\ H_2O]{1.\ LiAlH_4} CH_3CH_2CH_2CH_2CH_2NH_2$

c) $CH_3CH_2CH{=}CH_2 \xrightarrow[peroxides]{HBr} CH_3CH_2CH_2CH_2Br \xrightarrow{NaCN} CH_3CH_2CH_2CH_2CN$

$\downarrow$ 1. $LiAlH_4$ 2. H_2O

$CH_3CH_2CH_2CH_2CH_2NH_2$

d) $CH_3CH_2CH_2CH_2CH_2\overset{O}{\overset{\|}{C}}NH_2 \xrightarrow[H_2O,\ \Delta]{Br_2,\ ^-OH} CH_3CH_2CH_2CH_2CH_2NH_2$

e) $CH_3CH_2CH_2CH_2OH \xrightarrow{PBr_3} CH_3CH_2CH_2CH_2Br \xrightarrow{NaCN} CH_3CH_2CH_2CH_2CN$

$\downarrow$ 1. $LiAlH_4$ 2. H_2O

$CH_3CH_2CH_2CH_2CH_2NH_2$

f) $CH_3CH_2CH_2CH_2CH{=}CHCH_2CH_2CH_2CH_3 \xrightarrow[2.\ Zn,\ H_3O^+]{1.\ O_3} 2\ CH_3CH_2CH_2CH_2\overset{O}{\overset{\|}{C}}H$

$CH_3CH_2CH_2CH_2\overset{O}{\overset{\|}{C}}H \xrightarrow[NaBH_3CN]{NH_3} CH_3CH_2CH_2CH_2CH_2NH_2$

g) $CH_3CH_2CH_2CH_2COOH \xrightarrow{SOCl_2} CH_3CH_2CH_2CH_2COCl \xrightarrow{2\ NH_3} CH_3CH_2CH_2CH_2CONH_2$

$\xrightarrow{1.\ LiAlH_4;\ 2.\ H_2O}$ $CH_3CH_2CH_2CH_2CH_2NH_2$

28.28

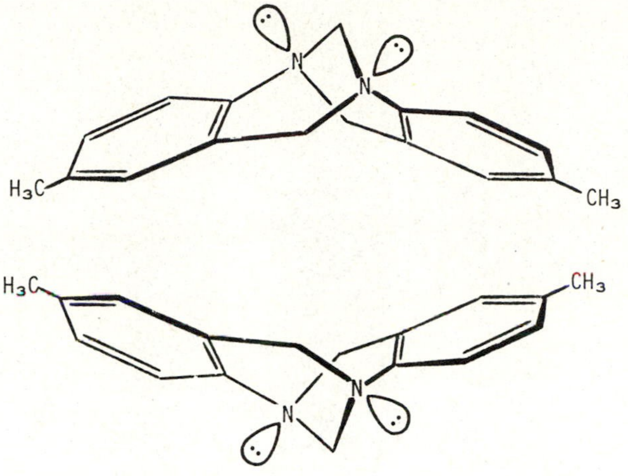

H_3C CH_3 N N

Tröger's base

H_3C CH_3 N N

The enantiomers of Tröger's base do not interconvert. Because of the rigid ring system, the substituents bonded to nitrogen can't be forced into the planar *sp*² geometry necessary for inversion at nitrogen to occur. Since inversion is not possible, the enantiomers are resolvable.

28.29 a) $\xrightarrow[CH_3I]{excess}$ N^+ I^- H_3C CH_3 A $\xrightarrow[H_2O]{Ag_2O}$ N^+ ^-OH H_3C CH_3 B

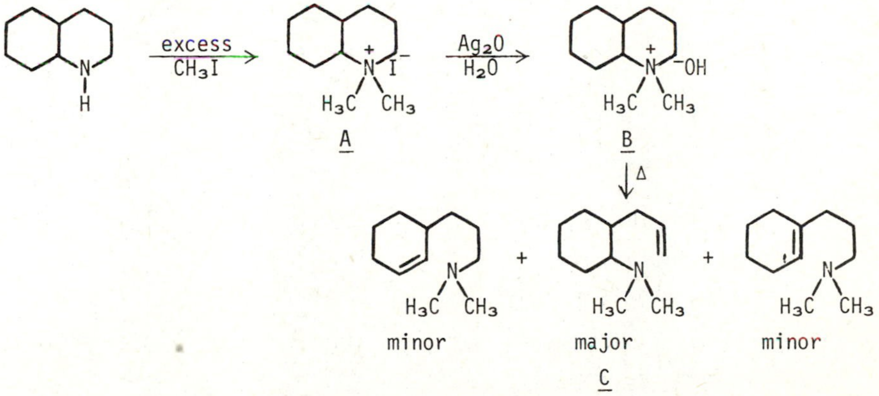

$\downarrow \Delta$

N H_3C CH_3 minor + N H_3C CH_3 major C + N H_3C CH_3 minor

b) CCl (O) $\xrightarrow{NaN_3}$ CN_3 (O) A $\xrightarrow{\Delta}$ N=C=O + N_2 B $\xrightarrow{H_2O}$ NH_2 + CO_2 C

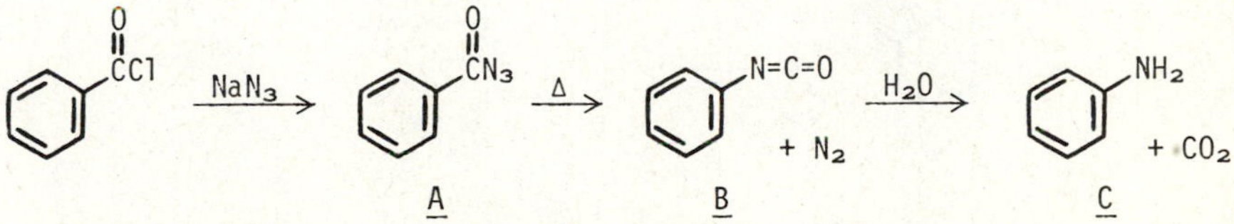

c) Phthalimide $\xrightarrow{KOH}$ potassium phthalimide (A) + H_2O $\xrightarrow{C_6H_5CH_2Br}$ N-benzylphthalimide, $N\text{-}CH_2C_6H_5$ (B) + KBr

B $\xrightarrow{KOH,\ H_2O}$ dipotassium phthalate (CO^-K^+, CO^-K^+) + $C_6H_5CH_2NH_2$ (C)

d) $BrCH_2CH_2CH_2CH_2Br$ $\xrightarrow{1\ equiv.\ CH_3NH_2}$

$CH_3NHCH_2CH_2CH_2CH_2NHCH_3$ (minor) + N-methylpyrrolidine (H_2C-CH_2, H_2C, CH_2, N–CH_3) (major) + $BrCH_2CH_2CH_2CH_2NHCH_3$ (minor)

28.30 Phthalic anhydride + $:NH_3$ ⇌ [tetrahedral intermediate ($^-:\ddot{O}:$, $\overset{+}{N}H_3$) ⇌ amide with $C\text{-}NH_2$ and $C\text{-}OH$, $:\ddot{O}:$ ⇌ cyclic intermediate ($\overset{+}{N}H_2$, OH, $^-:\ddot{O}:$)] ⇌ phthalimide (NH) + H_2O

28.31 a) The Hofmann rearrangement of amides yields an amine containing one less carbon atom than the starting amide. Here, the product of Hofmann rearrangement is $CH_3CH_2NH_2$, not $CH_3CH_2CH_2NH_2$.

b) No reaction occurs between a tertiary amine and a carbonyl group. To obtain the given product, use $(CH_3)_2NH$.

Cyclohexanone $\xrightarrow{(CH_3)_2\ddot{N}H}$ [cyclohexylidene $\overset{+}{N}(CH_3)_2$] $\xrightarrow{NaBH_3CN}$ cyclohexyl-$\ddot{N}(CH_3)_2$

c) Both elimination and substitution products are obtained when the tertiary bromide $(CH_3)_3CBr$ reacts with ammonia.

$$(CH_3)_3CBr \xrightarrow{:NH_3} \underset{\text{major}}{(CH_3)_2C{=}CH_2} + \underset{\text{minor}}{(CH_3)_3C\ddot{N}H_2}$$

d) The isocyanate intermediate in the Hofmann rearrangement results from treatment of an amide with Br_2 and ^-OH. Heat is used to decarboxylate the carbamic acid intermediate.

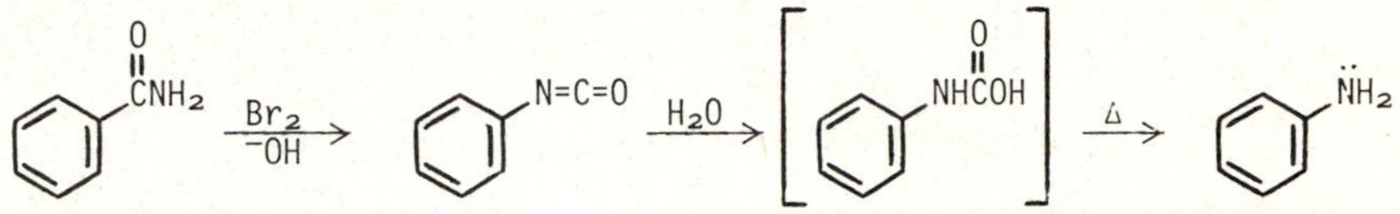

$$C_6H_5CONH_2 \xrightarrow[^-OH]{Br_2} C_6H_5N{=}C{=}O \xrightarrow{H_2O} [C_6H_5NHCOOH] \xrightarrow{\Delta} C_6H_5\ddot{N}H_2$$

e) The amine in this problem is being subjected to the conditions of Hofmann elimination. The major alkene product, $CH_3CH_2CH_2CH{=}CH_2$, contains the *less* substituted double bond. The product shown is the minor product.

28.32 The base-insoluble Hinsberg product indicates that coniine is a secondary amine.

The structural formula indicates that coniine has one double bond or ring; the Hofmann elimination product shows that the nitrogen atom is part of a ring.

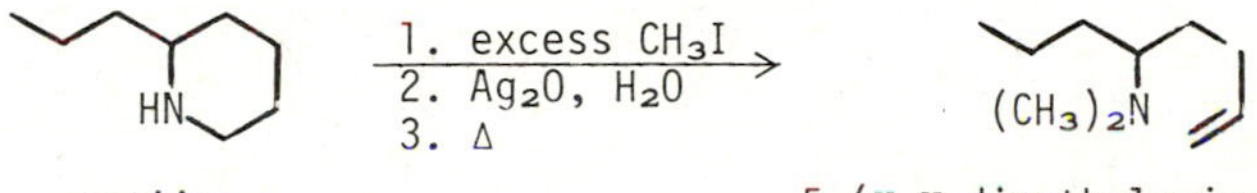

coniine $\xrightarrow{\text{1. excess } CH_3I;\ \text{2. } Ag_2O,\ H_2O;\ \text{3. } \Delta}$ $(CH_3)_2N$ ...

coniine

5-(*N*,*N*-dimethylamino)-1-octene

28.33 atropine $\xrightarrow{^-OH,\ H_2O}$ tropine + $C_6H_5CH(CH_2OH)COOH$ tropic acid

$OC{-}CHC_6H_5$ O CH_2OH

atropine

tropine

tropine $\xrightarrow{H_2SO_4}$ tropidene

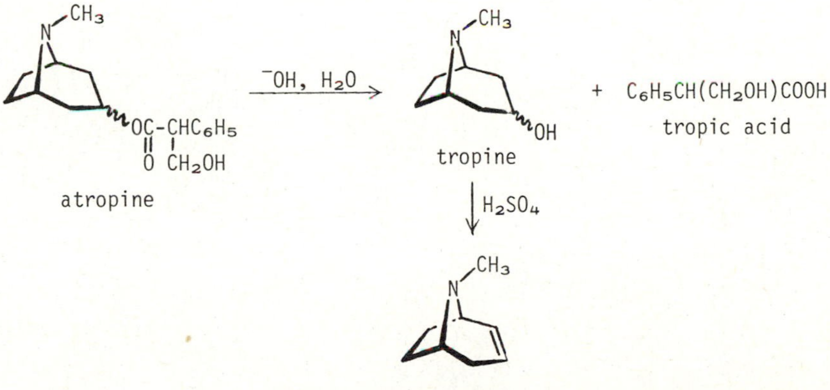

tropidene

28.34 $\xrightarrow{\text{1. } CH_3I;\ \text{2. } Ag_2O,\ H_2O;\ \text{3. } \Delta}$ $N(CH_3)_2$ $\xrightarrow{\text{1. } CH_3I;\ \text{2. } Ag_2O,\ H_2O;\ \text{3. } \Delta}$ + $(CH_3)_3N:$

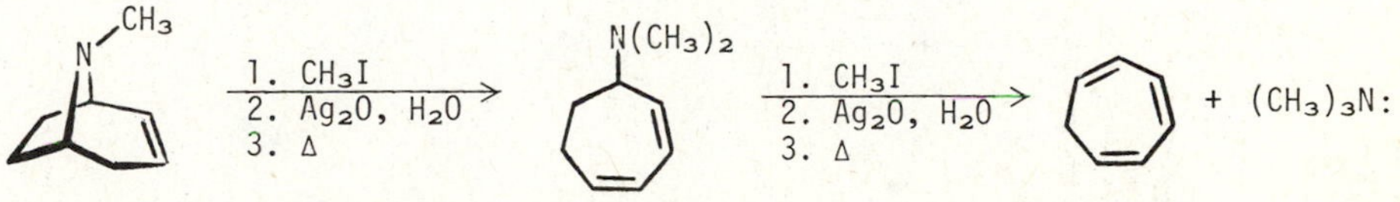

28.35

Benzaldehyde + $:NH_2CH_3$ ⇌ [C_6H_5–C(O$^-$)(H)–$\overset{+}{N}H_2CH_3$ ⇌ C_6H_5–C(OH)(H)–NHCH$_3$ ⇌ C_6H_5CH=NCH$_3$ + H_2O]

$\xrightarrow{NaBH_3CN,\ CH_3OH}$ $C_6H_5CH_2\ddot{N}HCH_3$

Benzaldehyde + $CH_3\ddot{N}HCH_2C_6H_5$ ⇌ [C_6H_5–C(O$^-$)(H)–$\overset{+}{N}H$(CH$_3$)CH$_2$C$_6$H$_5$] ⇌ [C_6H_5–C(OH)(H)–N(CH$_3$)CH$_2$C$_6$H$_5$]

$\xrightarrow{-OH^-}$ [$C_6H_5CH=\overset{+}{N}(CH_3)CH_2C_6H_5$]

$\xrightarrow{NaBH_3CN,\ CH_3OH}$ $C_6H_5CH_2\ddot{N}(CH_3)CH_2C_6H_5$

28.36

$CH_3\overset{O}{\overset{\|}{C}}CH_2\overset{O}{\overset{\|}{C}}OCH_3$ $\xrightarrow[\text{3. } H_3O^+,\ \Delta]{\text{1. } NaOC_2H_5 \quad \text{2. cyclopentyl bromide (Br)}}$ cyclopentyl–$CH_2\overset{O}{\overset{\|}{C}}CH_3$ $\xrightarrow[NaBH_3CN]{NH_2CH_3}$ cyclopentyl–$CH_2\underset{}{C}H(HNCH_3)CH_3$

cyclopentamine

28.37

$C_6H_5CH_2COCH_2CH_3$ $\xrightarrow[NaBH_3CN]{\text{pyrrolidine (N-H)}}$ $C_6H_5CH_2CH(N\text{-pyrrolidinyl})CH_2CH_3$

prolitane

28.38 a)

$CH_3(CH_2)_3NH-C_6H_4-COCH_2CH_2N(CH_3)_2$ (tetracaine) $\xleftarrow[NaBH_3CN]{CH_3CH_2CH_2CHO}$ $H_2N-C_6H_4-COCH_2CH_2N(CH_3)_2$

tetracaine

1. $SnCl_2$, H_3O^+
2. ^-OH, H_2O

b)

$O_2N-C_6H_4-CCl$ (O) $\xrightarrow[pyridine]{HOCH_2CH_2N(CH_3)_2}$ $O_2N-C_6H_4-COCH_2CH_2N(CH_3)_2$

$SOCl_2$

$O_2N-C_6H_4-COH$ (O)

$KMnO_4$, H_2O

c)

$O_2N-C_6H_4-CH_3 \xleftarrow[H_2SO_4]{HNO_3} C_6H_5-CH_3 \xleftarrow[AlCl_3]{CH_3Cl} C_6H_6$

28.39

O CN $\xrightarrow{H_2/Pt}$ [O NH₂ ⇌ HO N H] → [N + H_2O] $\xrightarrow{H_2/Pt}$ N H

28.40

$CH_3CH_2CH_2C(=O)CH_2CO_2C_2H_5$ + $CH_2=CHCN$ $\xrightarrow[2.\ H_3O^+]{1.\ NaOC_2H_5,\ C_2H_5OH}$ $CH_3CH_2CH_2C(=O)CH(CO_2C_2H_5)CH_2CH_2CN$

H_3O^+, Δ

$CH_3CH_2CH_2C(=O)CH_2CH_2CH_2CN$ + CO_2 + C_2H_5OH

$\xrightarrow{H_2/Pt}$ [$CH_3CH_2CH_2C(=O)CH_2CH_2CH_2CH_2-NH_2$] → $CH_3CH_2CH_2$ HN (piperidine ring)

coniine

What you should know:

After doing these problems, you should be able to:

1. Name amines and draw their structures;
2. Understand the geometry and basicity of amines;
3. Synthesize amines by several routes;
4. Formulate the mechanisms of reactions involving amines, including the Gabriel amine synthesis, reductive amination, the Hofmann rearrangement of amides, and the Hofmann elimination;
5. Deduce the structure of an amine by spectroscopic techniques and by examination of the products of Hofmann elimination.

29.1

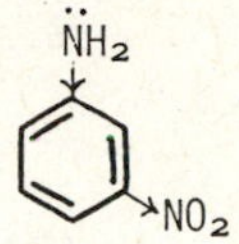

$\ddot{N}H_2$ / NO_2 (m-nitroaniline structure)

The inductive effect of the electron-withdrawing nitro group makes the amine nitrogen of *m*-nitroaniline less electron-rich and less basic than aniline.

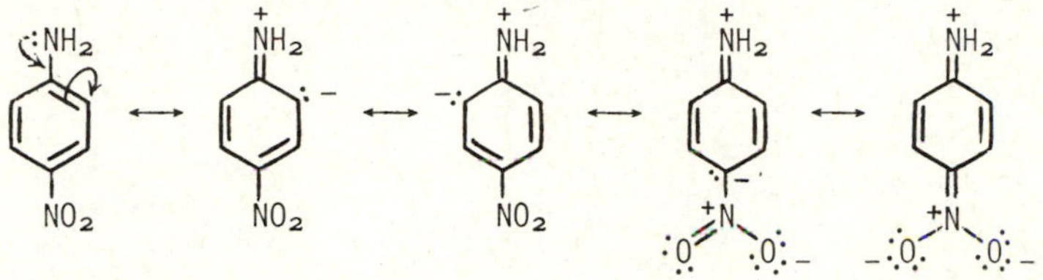

$\ddot{N}H_2$ … $\overset{+}{N}H_2$ … $\overset{+}{N}H_2$ … $\overset{+}{N}H_2$ … $\overset{+}{N}H_2$ (resonance forms of p-nitroaniline; NO_2, NO_2, NO_2, $^+N(O)O^-$, $^-O\text{–}^+N\text{–}O^-$)

When the nitro group is *para* to the amino group, conjugation of the amino group with the nitro group can also occur. *p*-Nitroaniline is thus even less basic than *m*-nitroaniline.

29.2 Least basic ⟶ most basic

a) NH_2/NO_2 < NH_2/$O{=}C\text{–}H$ < NH_2/Br

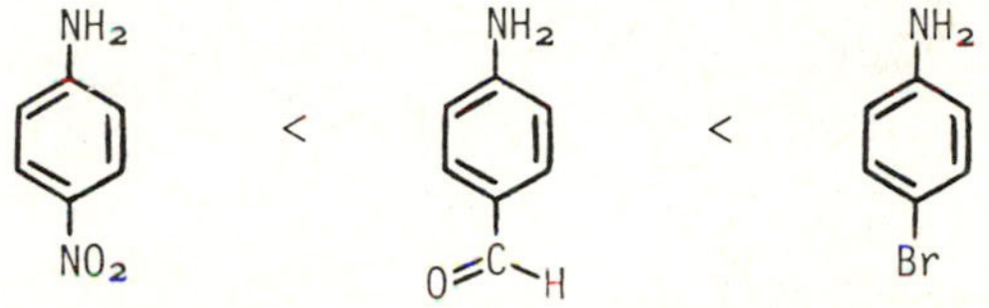

b) NH_2/$O{=}C\text{–}CH_3$ < NH_2/Cl < NH_2/CH_3

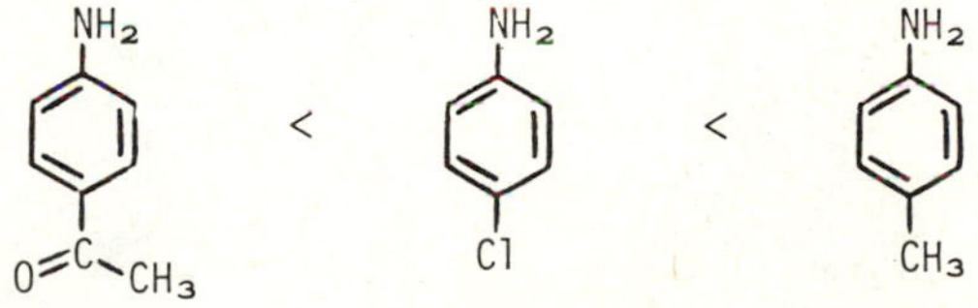

c) NH_2/CF_3 < NH_2/CH_2F < NH_2/CH_3

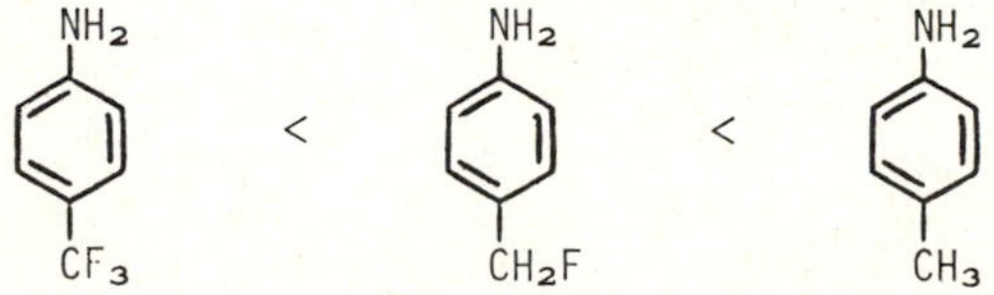

29.3

$C_6H_6 \xrightarrow[H_2SO_4]{HNO_3} O_2N\text{-}C_6H_5 \xrightarrow[C_2H_5OH]{H_2/Pt} H_2N\text{-}C_6H_5 \xrightarrow{(CH_3CO)_2O} CH_3CONH\text{-}C_6H_5 + CH_3COOH$

$\xrightarrow{HOSO_2Cl} CH_3CONH\text{-}C_6H_4\text{-}SO_2Cl \xrightarrow{H_2N\text{-thiazole}} CH_3CONH\text{-}C_6H_4\text{-}SO_2\text{-}NH\text{-thiazole}$

$\xrightarrow{NaOH,\ H_2O} H_2N\text{-}C_6H_4\text{-}SO_2\text{-}NH\text{-thiazole}$

sulfathiazole

29.4

In the last resonance form, the nitrogen lone pair is delocalized onto oxygen, rather than into the aromatic ring. Acetanilide is less activated toward electrophilic aromatic substitution than is aniline.

29.5 Reductions of nitro groups can also be carried out with $SnCl_2$ and H_3O^+, followed by $:OH^-$ and H_2O.

a) $C_6H_6 \xrightarrow[H_2SO_4]{HNO_3} C_6H_5NO_2 \xrightarrow[C_2H_5OH]{H_2/Pt} C_6H_5NH_2 \xrightarrow{2\ CH_3Br} C_6H_5N(CH_3)_2 + C_6H_5\overset{+}{N}(CH_3)_3\,Br^-$

b) $C_6H_5NHCOCH_3 \xrightarrow{Cl_2} p\text{-}Cl\text{-}C_6H_4NHCOCH_3 \xrightarrow[H_2O]{NaOH} p\text{-}Cl\text{-}C_6H_4NH_2$

(prob. 29.3)

p-chloroaniline

c) $C_6H_6 \xrightarrow[H_2SO_4]{HNO_3} C_6H_5NO_2 \xrightarrow[CF_3COOH]{Cl_2O} m\text{-}Cl\text{-}C_6H_4NO_2 \xrightarrow[2.\ NaOH,\ H_2O]{1.\ SnCl_2,\ H_3O^+} m\text{-}Cl\text{-}C_6H_4NH_2$

m-chloroaniline

d)

Benzene $\xrightarrow[AlCl_3]{CH_3Cl}$ toluene $\xrightarrow[H_2SO_4]{HNO_3}$ o-nitrotoluene (NO_2, CH_3) + p-nitrotoluene (NO_2, CH_3)

$\downarrow$ H_2/Pt, C_2H_5OH

o-toluidine (NH_2, CH_3) + p-toluidine (NH_2, CH_3)

$\downarrow$ $(CH_3CO)_2O$

$NHCOCH_3$, CH_3 (ortho) + $NHCOCH_3$, CH_3 (para) $\xrightarrow[AlCl_3]{CH_3Cl}$ $NHCOCH_3$, CH_3, CH_3 $\xrightarrow[H_2O]{NaOH}$ NH_2, CH_3, CH_3

2,4-dimethyl-
aniline

29.6 Aryldiazonium salts are more stable than alkyldiazonium salts for two reasons: (1) Overlap of the nitrogen pi electrons with the aromatic ring stabilizes an aryldiazonium salt relative to an alkyldiazonium salt.

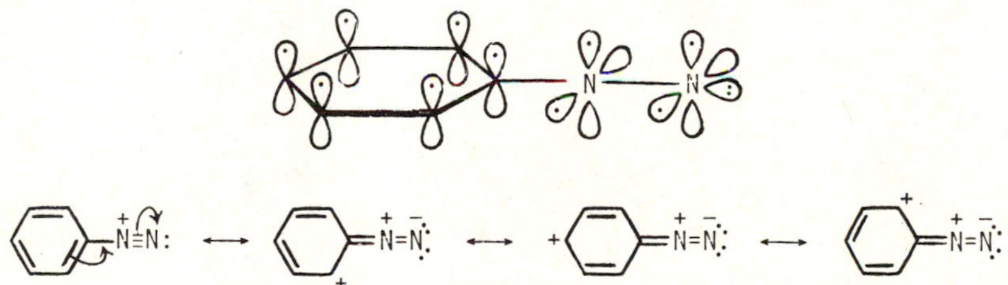

(2) Loss of nitrogen from an aryldiazonium salt produces a phenyl carbocation, which is less stable than an alkyl carbocation.

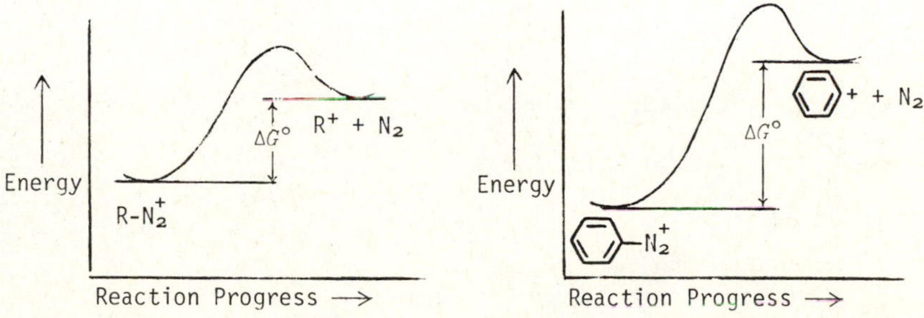

29.7 a)

Benzene $\xrightarrow[AlCl_3]{CH_3Cl}$ toluene $\xrightarrow[H_2SO_4]{HNO_3}$ p-nitrotoluene (CH_3, NO_2) $\xrightarrow[\text{2. NaOH, } H_2O]{\text{1. } SnCl_2,\ H_3O^+}$ p-toluidine (CH_3, NH_2) $\xrightarrow[H_2SO_4]{NaNO_2}$ (CH_3, $^+N_2\ HSO_4^-$) $\xrightarrow{CuBr}$ (CH_3, Br) $\xrightarrow[H_2O]{KMnO_4}$ (COOH, Br)

p-bromobenzoic acid

b)

Benzene $\xrightarrow[AlCl_3]{CH_3Cl}$ (CH_3) $\xrightarrow[H_2O]{KMnO_4}$ (COOH) $\xrightarrow[H_2SO_4]{HNO_3}$ (COOH, NO_2) $\xrightarrow[\text{2. NaOH, } H_2O]{\text{1. } SnCl_2,\ H_3O^+}$ (COOH, NH_2) $\xrightarrow[H_2SO_4]{NaNO_2}$ (COOH, $\overset{+}{N}_2\ HSO_4^-$) $\xrightarrow{CuBr}$ (COOH, Br)

m-bromobenzoic acid

c)

Benzene $\xrightarrow[H_2SO_4]{HNO_3}$ (NO_2) $\xrightarrow[CF_3COOH]{Cl_2O}$ (NO_2, Cl) $\xrightarrow[\text{2. NaOH, } H_2O]{\text{1. } SnCl_2,\ H_3O^+}$ (NH_2, Cl) $\xrightarrow[H_2SO_4]{NaNO_2}$ ($^+N_2\ HSO_4^-$, Cl) $\xrightarrow{CuBr}$ (Br, Cl)

m-bromochlorobenzene

Reduction of nitro groups can also be achieved by catalytic hydrogenation.

d)

(CH_3, $^+N_2\ HSO_4^-$) [from 29.7a] $\xrightarrow{CuCN}$ (CH_3, CN) $\xrightarrow{H_3O^+}$ (CH_3, COOH)

p-methylbenzoic acid

e) $C_6H_5NHCOCH_3$ (Prob. 29.3) $\xrightarrow{2\ Br_2}$ 2,4-dibromoacetanilide $\xrightarrow[H_2O]{NaOH}$ 2,4-dibromoaniline $\xrightarrow[H_2SO_4]{NaNO_2}$ 2,4-dibromobenzenediazonium HSO_4^- $\xrightarrow{CuBr}$ 1,2,4-tribromobenzene

1,2,4-tribromobenzene

<u>29.8</u>

$C_6H_5-\overset{+}{N}{\equiv}N:\ HSO_4^-$ + $C_6H_5-\ddot{N}(CH_3)_2$ $\longrightarrow$ [$C_6H_5-N{=}N-C_6H_5(H)={\overset{+}{N}}(CH_3)_2$] $\xrightarrow{-H^+}$ $C_6H_5-N{=}N-C_6H_4-\ddot{N}(CH_3)_2$

p-(*N*,*N*-dimethylamino)azobenzene

<u>29.9</u>

C_6H_6 $\xrightarrow[H_2SO_4]{HNO_3}$ $C_6H_5NO_2$ $\xrightarrow[C_2H_5OH]{H_2/Pt}$ $C_6H_5NH_2$ $\xrightarrow{(CH_3CO)_2O}$ $C_6H_5NHCOCH_3$ $\xrightarrow{SO_3,\ H_2SO_4}$ $p\text{-}HO_3S-C_6H_4-NHCOCH_3$ $\xrightarrow[H_2O]{NaOH}$ $p\text{-}Na^+\ {}^-O_3S-C_6H_4-NH_2$ $\xrightarrow[H_2SO_4]{NaNO_2}$ $p\text{-}Na^+\ {}^-O_3S-C_6H_4-{}^+N_2\ HSO_4^-$

$Na^+\ {}^-O_3S-C_6H_4-\overset{+}{N}_2\ HSO_4^-$ + $C_6H_5-\ddot{N}(CH_3)_2$ (prob. 29.5) $\xrightarrow{-H^+}$ $Na^+\ {}^-O_3S-C_6H_4-N{=}N-C_6H_4-\ddot{N}(CH_3)_2$

methyl orange

29.10

Phenol (OH) $\xrightarrow{2\ Cl_2}$ 2,4-dichlorophenol (OH, Cl, Cl) $\xrightarrow[2.\ BrCH_2COOC_2H_5]{1.\ NaOH}$ $OCH_2COOC_2H_5$ (2,4-dichloro) $\xrightarrow[2.\ H_3O^+]{1.\ NaOH,\ H_2O}$ OCH_2COOH (2,4-dichloro)

2,4-dichlorophenoxyacetic acid

29.11

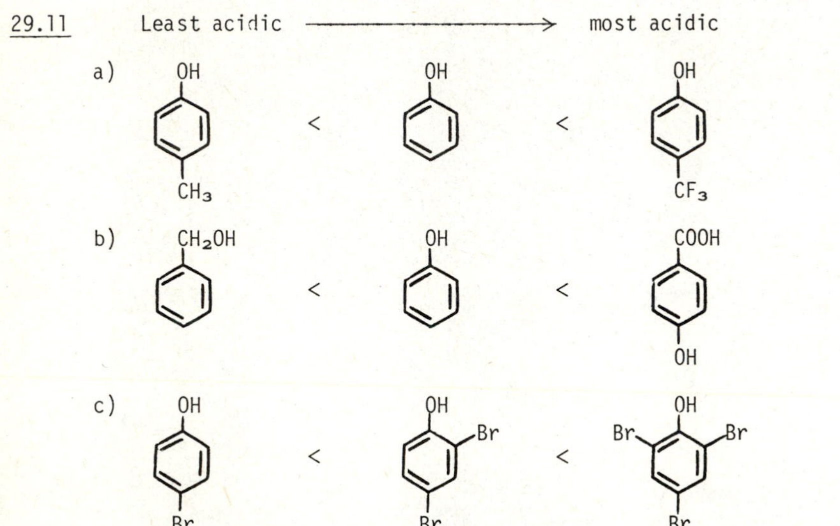

29.12

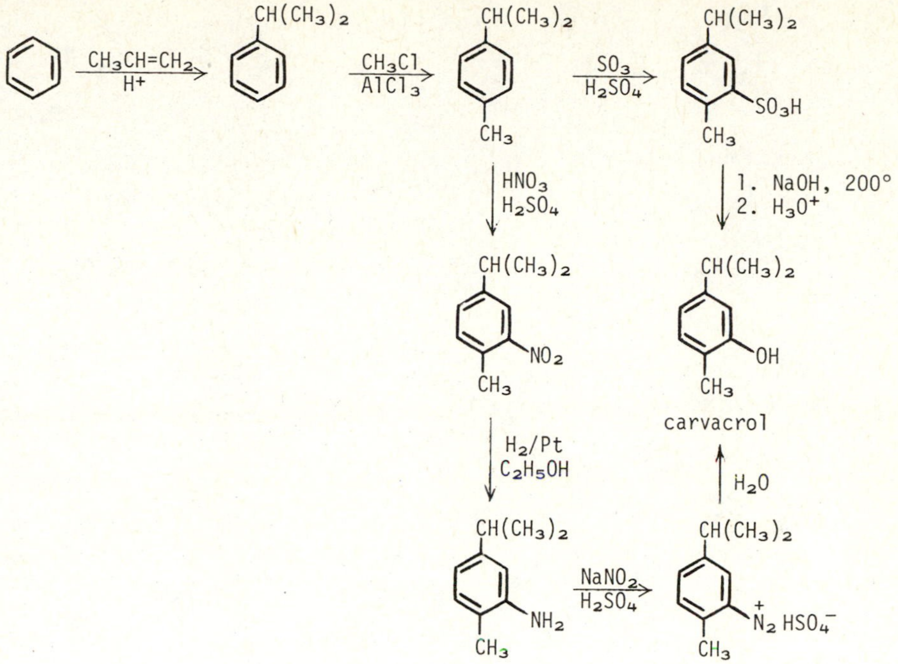

Benzene $\xrightarrow[H^+]{CH_3CH{=}CH_2}$ $C_6H_5CH(CH_3)_2$ $\xrightarrow[AlCl_3]{CH_3Cl}$ p-$CH_3C_6H_4CH(CH_3)_2$ $\xrightarrow[H_2SO_4]{SO_3}$ (2-SO_3H, 1-CH_3, 4-$CH(CH_3)_2$) $\xrightarrow[2.\ H_3O^+]{1.\ NaOH,\ 200°}$ (2-OH, 1-CH_3, 4-$CH(CH_3)_2$) carvacrol

p-$CH_3C_6H_4CH(CH_3)_2$ $\xrightarrow[H_2SO_4]{HNO_3}$ (2-NO_2) $\xrightarrow[C_2H_5OH]{H_2/Pt}$ (2-NH_2) $\xrightarrow[H_2SO_4]{NaNO_2}$ (2-$\overset{+}{N}_2\ HSO_4^-$) $\xrightarrow{H_2O}$ carvacrol

29.13

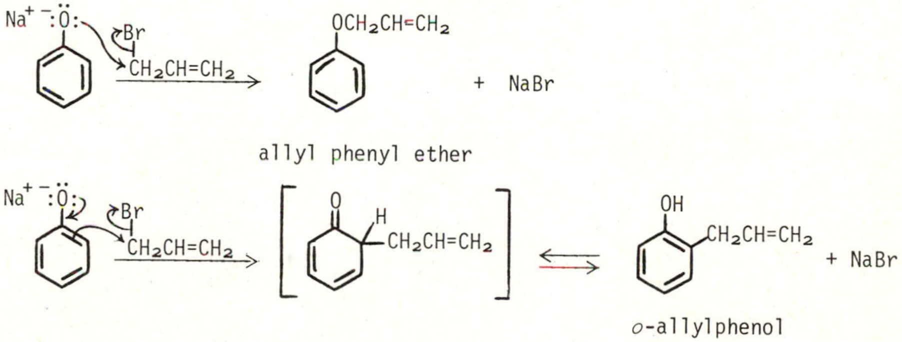

$Na^+\ ^-:\ddot{O}:$–C_6H_5 + $BrCH_2CH{=}CH_2$ $\longrightarrow$ $C_6H_5OCH_2CH{=}CH_2$ + NaBr

allyl phenyl ether

$Na^+\ ^-:\ddot{O}:$–C_6H_5 + $BrCH_2CH{=}CH_2$ $\longrightarrow$ [6-H, 6-$CH_2CH{=}CH_2$-cyclohexa-2,4-dienone] $\rightleftarrows$ OH, $CH_2CH{=}CH_2$ (ortho) + NaBr

o-allylphenol

29.14

$CH_3CH_2O:^-$ + 2,3-dimethoxy-5-methyl-6-R-1,4-benzoquinone $\rightleftarrows$ [CH_3CH_2O, CH_3O adduct enolate ($:\ddot{O}:^-$)] $\rightleftarrows$ 2-CH_3CH_2O, 3-CH_3O, 5-CH_3, 6-R-quinone + $CH_3O:^-$

$CH_3CH_2O:^-$ addition $\rightleftarrows$ [CH_3O, CH_3CH_2O adduct enolate] $\rightleftarrows$ $CH_3O:^-$ + 2,3-diethoxy (CH_3CH_2O, CH_3CH_2O), 5-CH_3, 6-R-1,4-benzoquinone

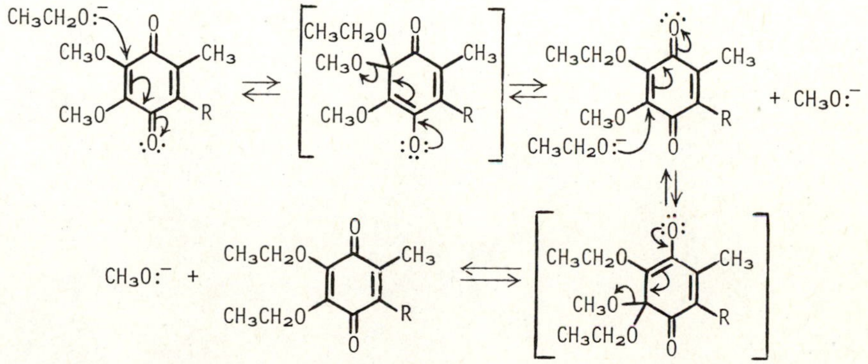

Conjugate addition of ethoxide to the double bond produces an intermediate that eliminates methoxide anion. The diethoxyl product predominates because of the large excess of ethoxide.

29.15

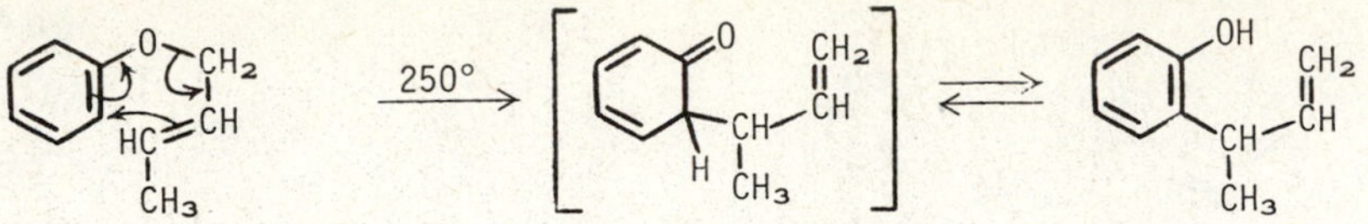

2-butenyl phenyl ether $\xrightarrow{250°}$ [ ] $\rightleftharpoons$

29.16 a) *p*-bromophenol b) 2,3-dichloro-*N*-methylaniline

c) 2-methyl-1,4-benzenediamine d) 3-methoxy-5-methylphenol

e) 1,3,5-benzenetriol f) *N*,3-diethyl-*N*,5-dimethylaniline

29.17

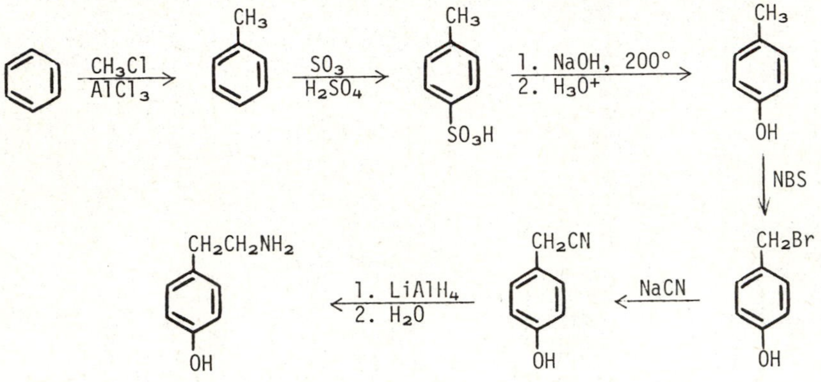

$\xrightarrow[AlCl_3]{CH_3Cl}$ $\xrightarrow[H_2SO_4]{SO_3}$ $\xrightarrow[2.\ H_3O^+]{1.\ NaOH,\ 200°}$ NBS; NaCN; $\xleftarrow[2.\ H_2O]{1.\ LiAlH_4}$

tyramine

29.18 a) $\xrightarrow[H_2SO_4]{HNO_3}$ $\xrightarrow[C_2H_5OH]{H_2/Pt}$

b) $\xrightarrow[NH_3]{KNH_2}$

c) $\xrightarrow{SOCl_2}$ $\xrightarrow{2\,NH_3}$ $\xrightarrow[H_2O,\ \Delta]{^-:OH,\ Br_2}$

$KMnO_4$, H_2O

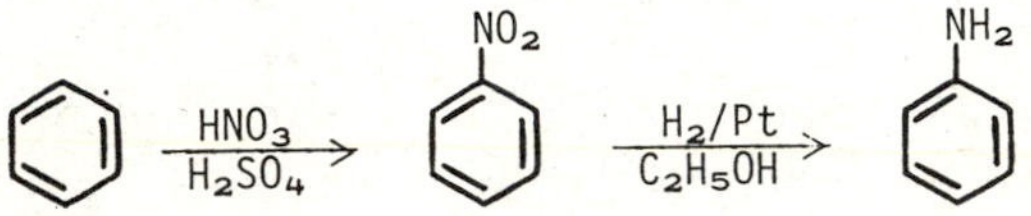

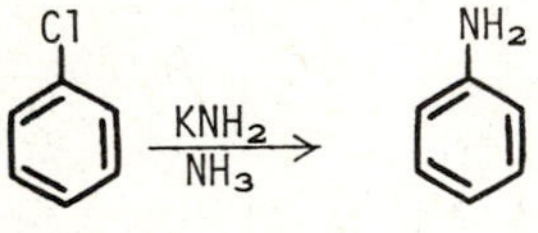

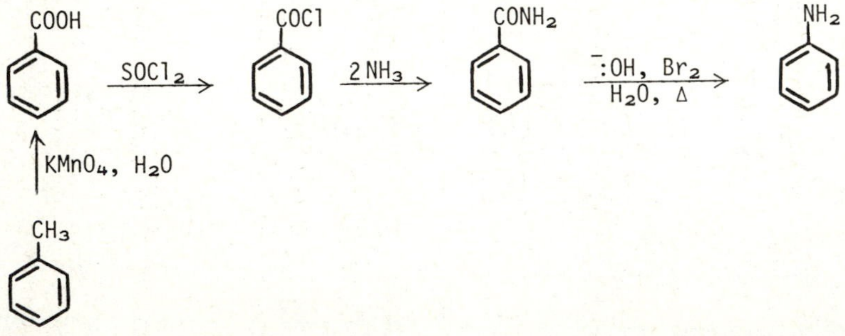

d)

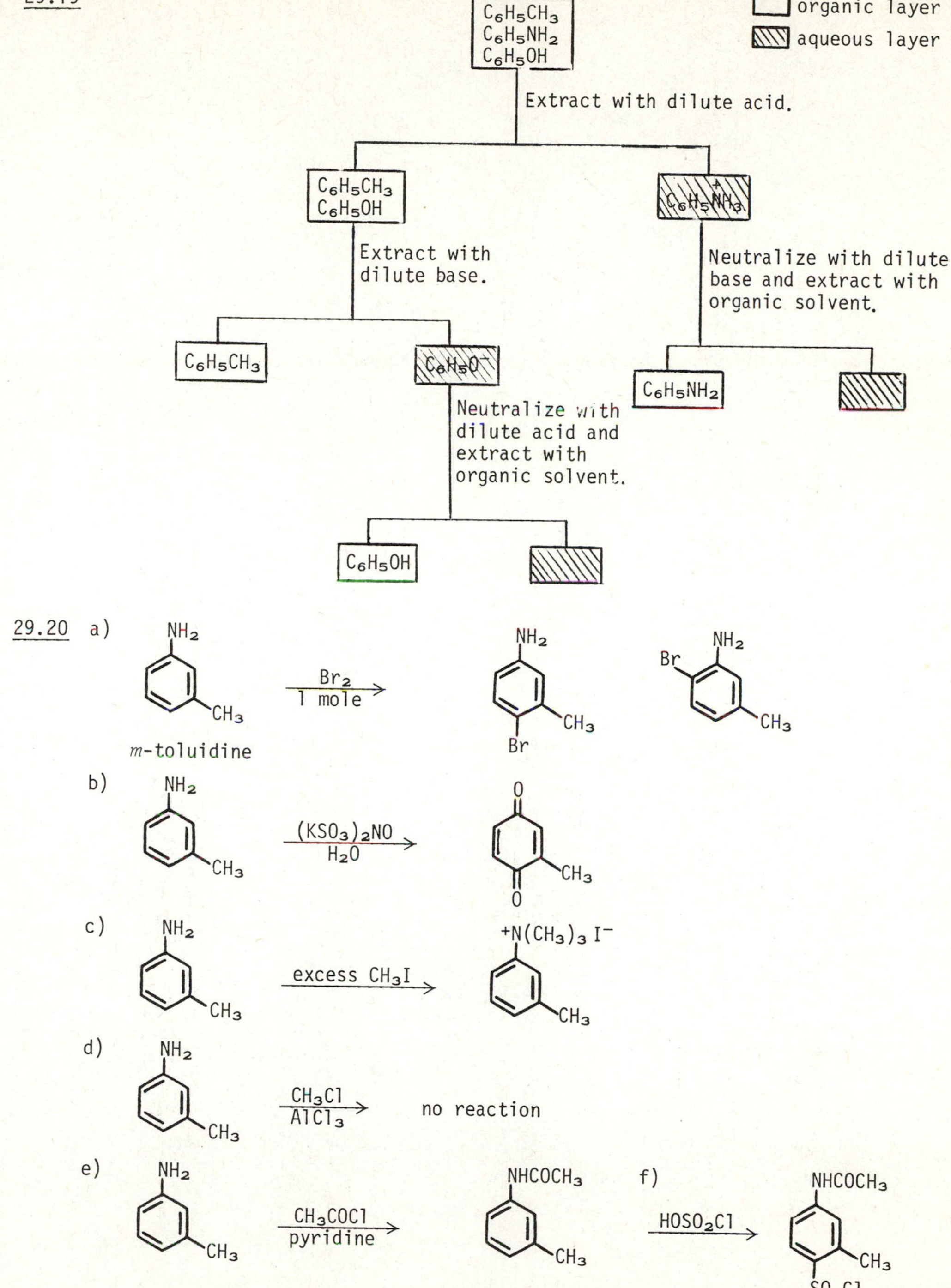
29.19
organic layer
aqueous layer
C6H5CH3
C6H5NH2
C6H5OH
Extract with dilute acid.
C6H5CH3
C6H5OH
C6H5NH3+
Extract with dilute base.
Neutralize with dilute base and extract with organic solvent.
C6H5CH3
C6H5O−
Neutralize with dilute acid and extract with organic solvent.
C6H5NH2
C6H5OH
29.20
a)
NH2
CH3
m-toluidine
Br2
1 mole
NH2
CH3
Br
NH2
Br
CH3
b)
NH2
CH3
(KSO3)2NO
H2O
O
CH3
O
c)
NH2
CH3
excess CH3I
+N(CH3)3 I−
CH3
d)
NH2
CH3
CH3Cl
AlCl3
no reaction
e)
NH2
CH3
CH3COCl
pyridine
NHCOCH3
CH3
f)
HOSO2Cl
NHCOCH3
CH3
SO2Cl

29.21

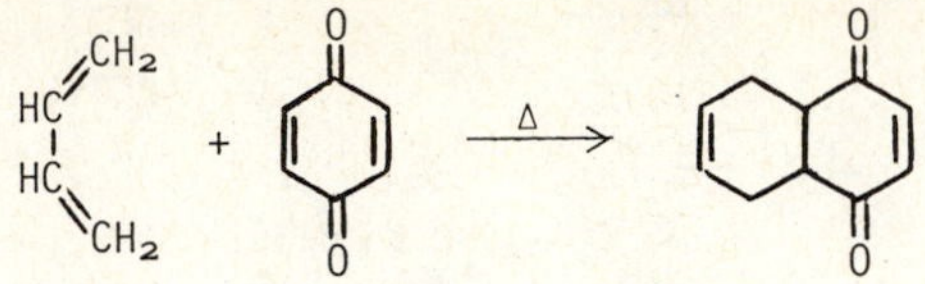

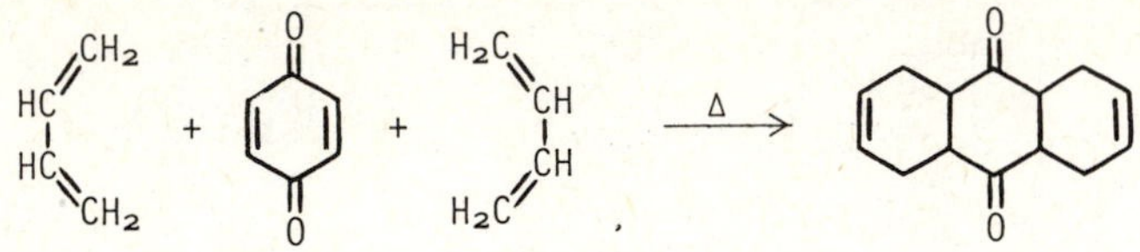

$$CH_2{=}CH{-}CH{=}CH_2 + \text{p-benzoquinone} \xrightarrow{\Delta} \text{(Diels–Alder adduct)}$$

$$CH_2{=}CH{-}CH{=}CH_2 + \text{p-benzoquinone} + CH_2{=}CH{-}CH{=}CH_2 \xrightarrow{\Delta} \text{(bis-adduct)}$$

29.22

$$\underline{A} \xrightarrow{H_3O^+ \text{ or } OH^-} \underline{B}$$

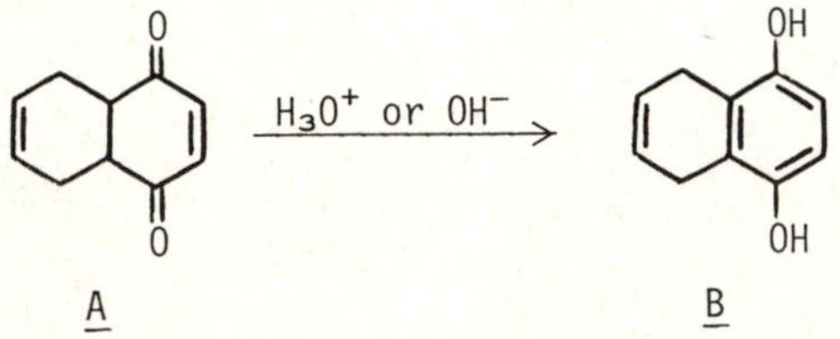

The two-proton singlet at 6.7 δ is due to the aromatic ring protons.

29.23 $CH_3C{\equiv}N$ + HCl + $ZnCl_2$ $\rightleftharpoons$ $CH_3C{\equiv}\overset{+}{N}H\ ZnCl_3^-$

$$\text{HO-C}_6\text{H}_4\text{-OH} + CH_3C{\equiv}\overset{+}{N}H\ ZnCl_3^- \longrightarrow [\ \ ZnCl_3^-\ ] \longrightarrow \ \overset{+}{N}H_2\ Cl^- + ZnCl_2$$

$$H_2\ddot{O}: \ \ {}^{+}NH_2 \rightleftharpoons [\ H_2\overset{+}{O}:\ :NH_2\ ] \rightleftharpoons [\ HO\ \overset{+}{N}H_3\ ]$$

$$+ NH_4^+ \rightleftharpoons [\ :\overset{+}{O}H + NH_3\ ]$$

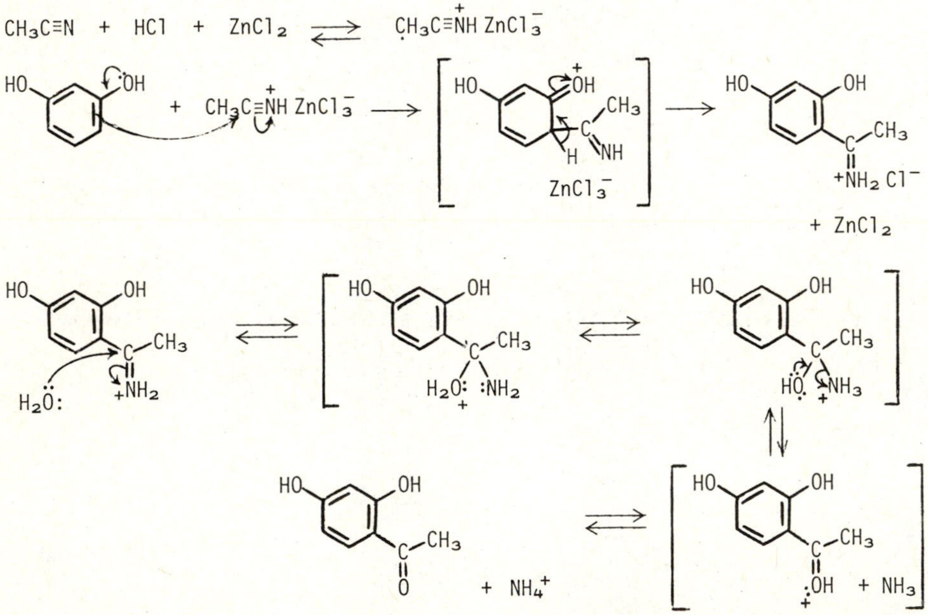

The Hoesch reaction is mechanistically similar to Friedel-Crafts acylation.

29.24 The nitrogen lone pair of electrons of diphenylamine can overlap the pi electron system of either ring. Electron delocalization occurs to an even greater extent for diphenylamine than for aniline. Because the energy difference between non-protonated and protonated amine is much greater for diphenylamine than for aniline, diphenylamine is non-basic.

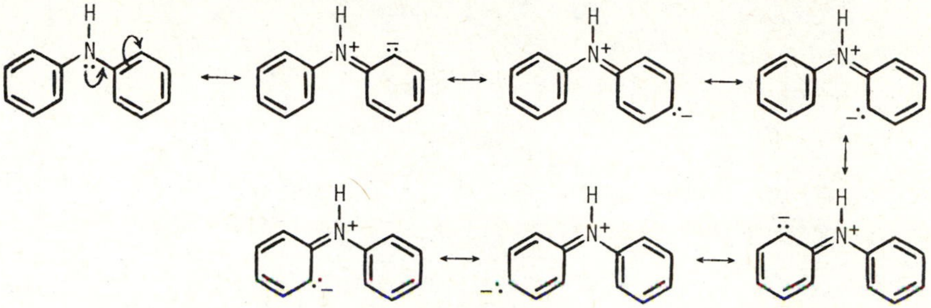

H H H H
N N+ N+ N+

H H H
N+ N+ N+

29.25

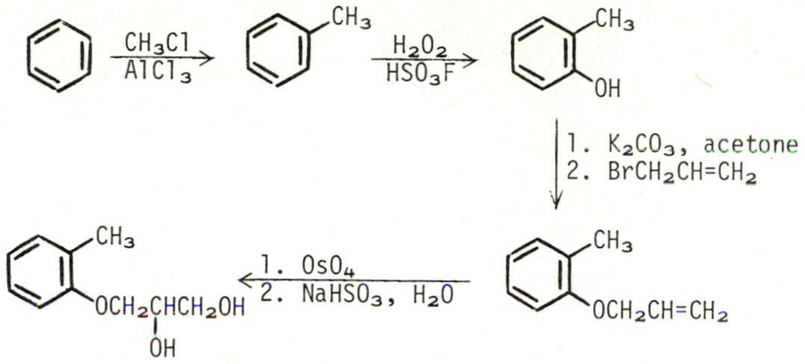

$\xrightarrow[AlCl_3]{CH_3Cl}$ CH_3 $\xrightarrow[HSO_3F]{H_2O_2}$ CH_3 OH

1. K_2CO_3, acetone
2. $BrCH_2CH=CH_2$

CH_3 OCH_2CHCH_2OH OH $\xleftarrow[2.\ NaHSO_3,\ H_2O]{1.\ OsO_4}$ CH_3 $OCH_2CH=CH_2$

mephenesin

29.26

$\xrightarrow[HSO_3F]{H_2O_2}$ OH $\xrightarrow[H_2O]{(KSO_3)_2NO}$ O O $\xrightarrow{SnCl_2}$ OH OH

1. NaOH
2. CO_2, pressure
3. H^+

OH COOH OH

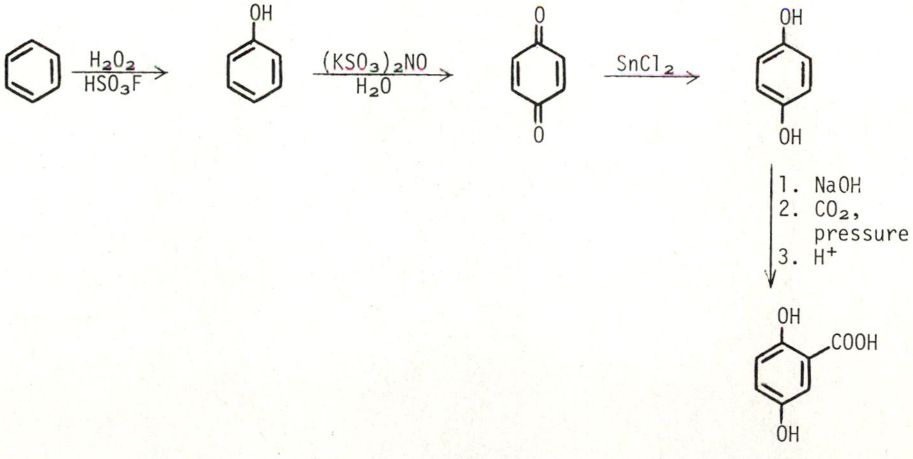

gentisic acid

29.27

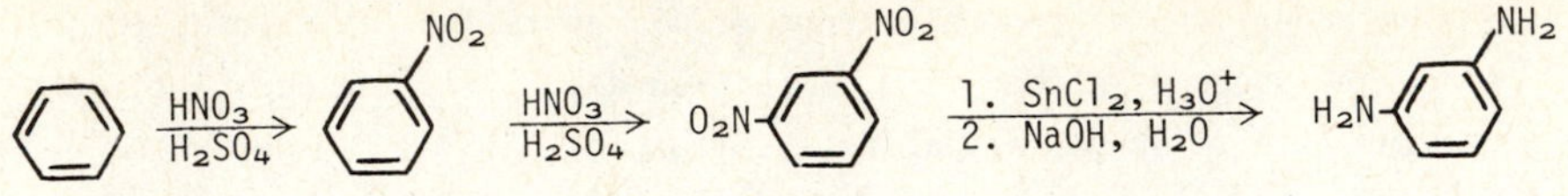

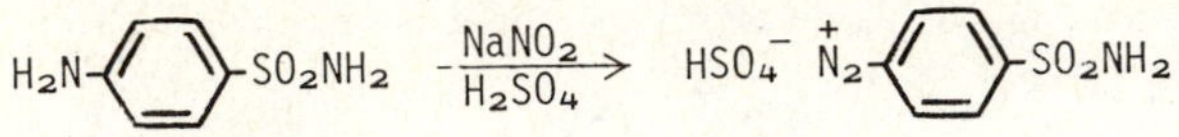

Benzene $\xrightarrow[H_2SO_4]{HNO_3}$ nitrobenzene $\xrightarrow[H_2SO_4]{HNO_3}$ m-dinitrobenzene (O_2N, NO_2) $\xrightarrow[\text{2. NaOH, } H_2O]{\text{1. } SnCl_2,\ H_3O^+}$ m-diaminobenzene (H_2N, NH_2)

$H_2N-C_6H_4-SO_2NH_2 \xrightarrow[H_2SO_4]{NaNO_2} HSO_4^-\ \overset{+}{N_2}-C_6H_4-SO_2NH_2$

sulfanilimide
(from text)

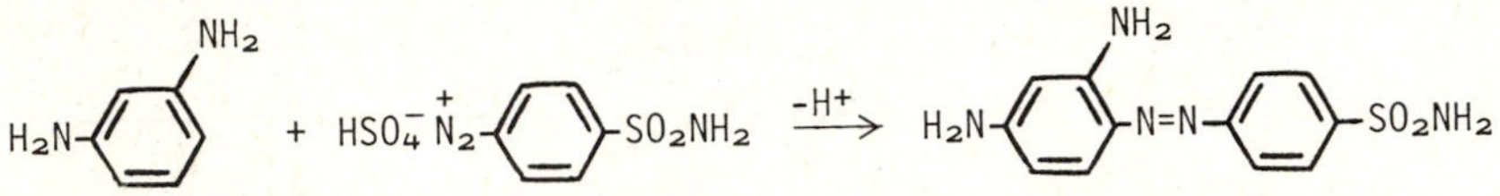

$H_2N-C_6H_4-NH_2$ (meta) + $HSO_4^-\ \overset{+}{N_2}-C_6H_4-SO_2NH_2 \xrightarrow{-H^+} H_2N-C_6H_3(NH_2)-N{=}N-C_6H_4-SO_2NH_2$

prontosil

29.28

$\overset{+}{Na}\ {}^-O_3S-C_6H_4-\overset{+}{N_2}\ HSO_4^-$ + β-naphthol (HO) $\longrightarrow$ $\overset{+}{Na}\ {}^-O_3S-C_6H_4-N{=}N-$(2-hydroxynaphthyl, HO)

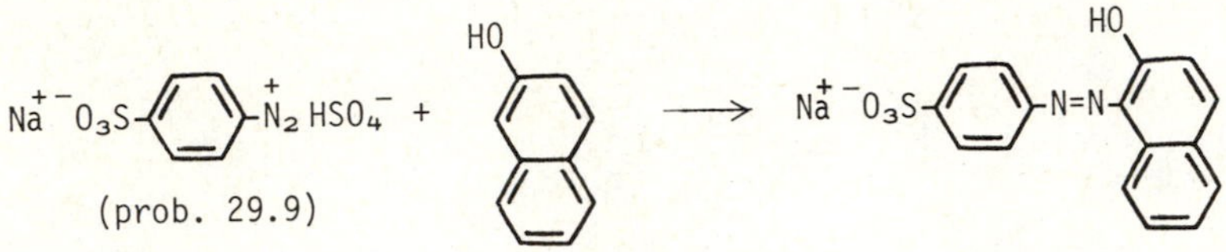

(prob. 29.9)

Orange II

Electrophilic substitution on β-naphthol occurs between the hydroxyl group and the fused ring.

29.29

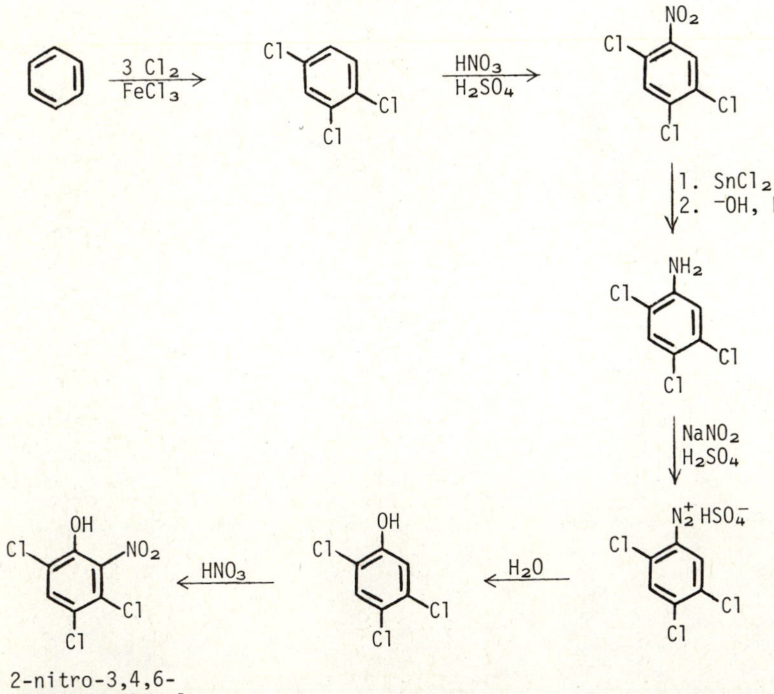

Benzene $\xrightarrow[FeCl_3]{3\ Cl_2}$ 1,2,4-trichlorobenzene (Cl, Cl, Cl) $\xrightarrow[H_2SO_4]{HNO_3}$ 2,4,5-trichloronitrobenzene (NO_2, Cl, Cl, Cl)

$\downarrow$ 1. $SnCl_2$, H_3O^+ 2. ^-OH, H_2O

2,4,5-trichloroaniline (NH_2, Cl, Cl, Cl)

$\downarrow$ $NaNO_2$, H_2SO_4

$\overset{+}{N_2}\ HSO_4^-$ (Cl, Cl, Cl) $\xrightarrow{H_2O}$ 2,4,5-trichlorophenol (OH, Cl, Cl, Cl) $\xrightarrow{HNO_3}$ (OH, Cl, NO_2, Cl, Cl)

2-nitro-3,4,6-
trichlorophenol

29.30

Cl Cl Cl $\ddot{O}H$ + $CH_2=\overset{+}{O}H$ ⟶ [Cl Cl Cl $\overset{+}{O}H$ H CH_2OH] ⟶ OH Cl Cl Cl CH_2OH + H^+

[OH Cl Cl Cl CH_2 H $\overset{+}{O}H$ Cl Cl Cl] + H_2O ⟵ [OH Cl Cl Cl $CH_2\overset{+}{O}H_2$ $\ddot{O}H$ Cl Cl Cl]

OH OH Cl CH_2 Cl Cl Cl Cl Cl + H_3O^+

hexachlorophene

29.31

Benzene $\xrightarrow[FeCl_3]{Cl_2}$ Cl $\xrightarrow[H_2SO_4]{HNO_3}$ NO_2 Cl $\xrightarrow[CF_3COOH]{Cl_2O}$ NO_2 Cl Cl

1. $SnCl_2$, H_3O^+
2. NaOH, H_2O

NH_2 Cl Cl

Benzene $\xrightarrow[SO_3]{H_2SO_4}$ SO_3H $\xrightarrow[2.\ H_3O^+]{1.\ NaOH,\ 200°}$ OH $\xrightarrow[2.\ CO_2 \quad 3.\ H^+]{1.\ NaOH}$ OH COOH

Cl_2

OH COCl Cl $\xleftarrow{SOCl_2}$ OH COOH Cl

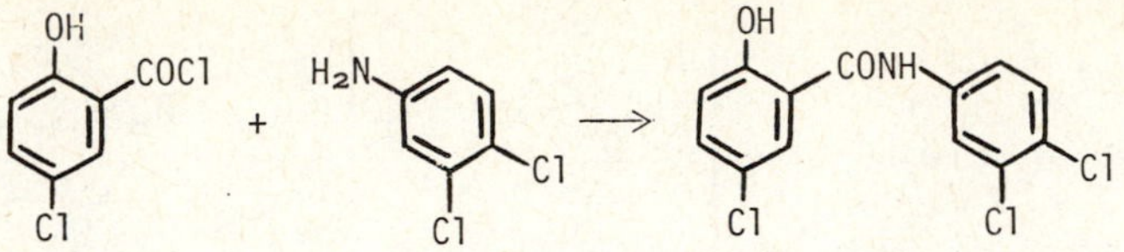

OH COCl + H_2N Cl Cl → OH CONH Cl Cl Cl

Cl

trichlorosalicylanilide

29.32 Structural formula: $C_8H_{10}O$ contains 4 multiple bonds and/or rings.

Infrared: The broad band at 3500 cm^{-1} indicates the presence of a hydroxyl group. The absorbances at 1500 cm^{-1} and 1600 cm^{-1}, as well as at 830 cm^{-1}, are due to an aromatic ring. Compound A is probably a phenol.

^1H NMR: The triplet at 1.16 δ (3H) is coupled with the quartet at 2.54 δ (2H). These two absorptions are due to an ethyl group.

The peaks at 6.80 δ (4H) are due to an aromatic ring. The symmetrical splitting pattern of these peaks indicate that the aromatic ring is *p*-disubstituted.

The absorption at 5.50 δ (1H) is due to an -OH proton.

Compound A

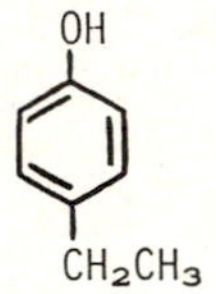

OH

CH_2CH_3

p-ethylphenol

29.33 $NHCOCH_3$ / OCH_2CH_3 $\xrightarrow[H_2O]{^-:OH}$ NH_2 / OCH_2CH_3 + CH_3COO^- $\xrightarrow{HI}$ NH_2 / OH + CH_3CH_2I

$C_{10}H_{13}NO_2$ phenacetin — $C_8H_{11}NO$ *p*-ethoxyaniline — C_6H_7NO *p*-aminophenol

$\downarrow$ $(KSO_3)_2NO$, H_2O

O / O

What you should know:

After doing these problems, you should be able to:

1. Predict the effects of substituents on the acidity and basicity of aromatic amines and phenols;
2. Synthesize aromatic amines and phenols by several methods;
3. Use diazonium salts in the synthesis of substituted aromatic compounds;
4. Synthesize simple dyes using diazo coupling reactions;
5. Predict the products of reactions involving phenols;
6. Identify aromatic amines and phenols from spectroscopic data.

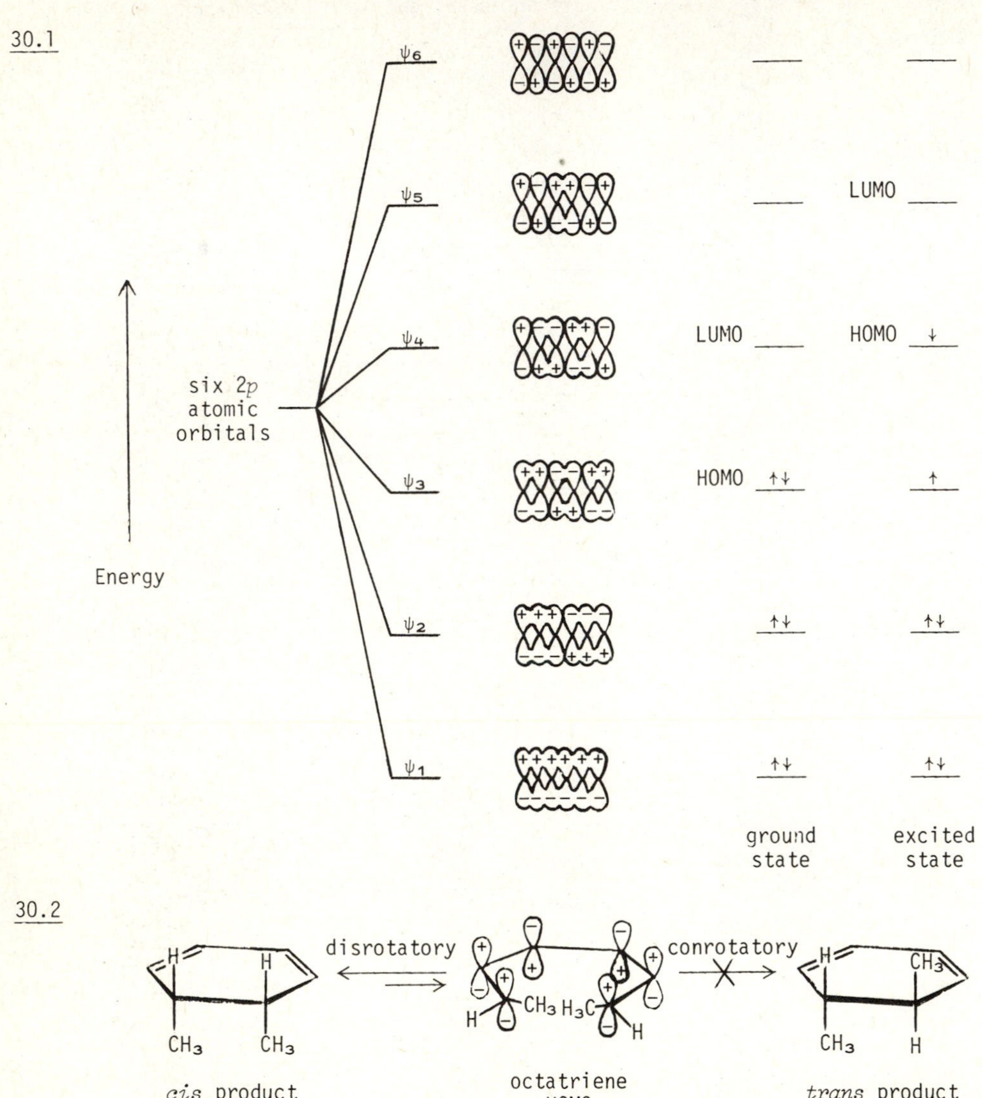

The symmetry of the octatriene HOMO predicts that ring closure will occur by a disrotatory path and that only *cis* product will be formed.

30.3 Note: *Trans*-3,4-dimethylcyclobutene is chiral; the *S*,*S* enantiomer will be used for this argument.

A

E,*E*

B

Z,*Z*

Conrotatory ring opening of *trans*-3,4-dimethylcyclobutene can occur in either a clockwise or a counterclockwise manner. Clockwise opening (path A) yields the *E*,*E* isomer; counterclockwise opening (path B) yields the *Z*,*Z* isomer. Production of (2*Z*,4*Z*)-hexadiene is disfavored because of unfavorable steric interactions between the methyl groups in the transition state leading to ring-opened product.

30.4

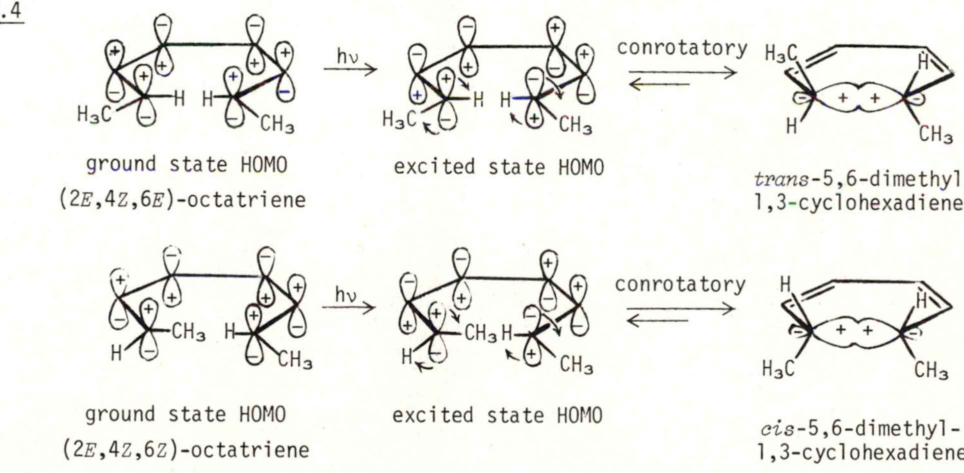

Photochemical electrocylic reactions of 6 pi electron systems always occur in a conrotatory manner.

30.5

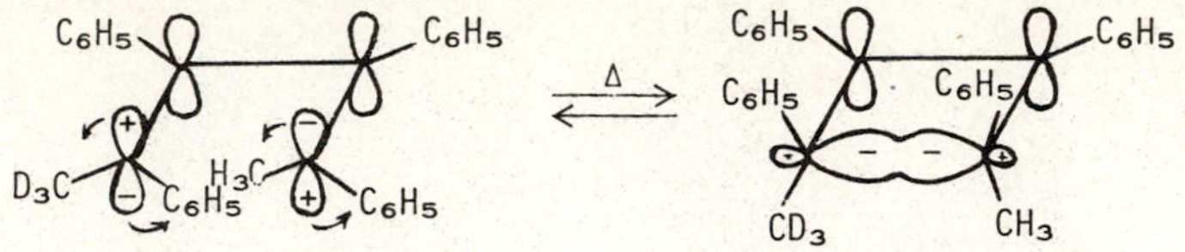

The diene can cyclize by either of two conrotatory paths to form cyclobutenes A and B.

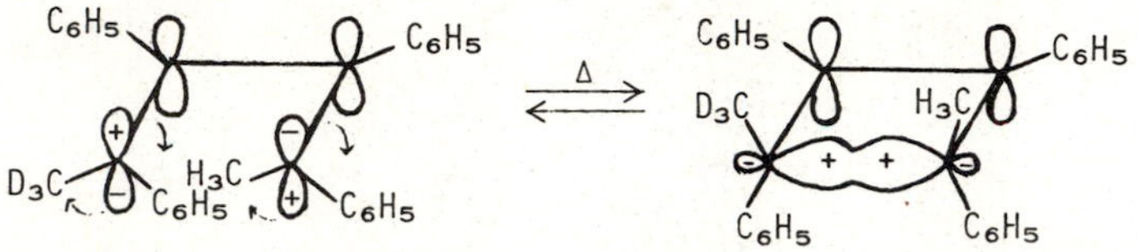

Using B as an example:

Opening of each cyclobutene ring can occur by either of two conrotatory routes to yield the isomeric dienes.

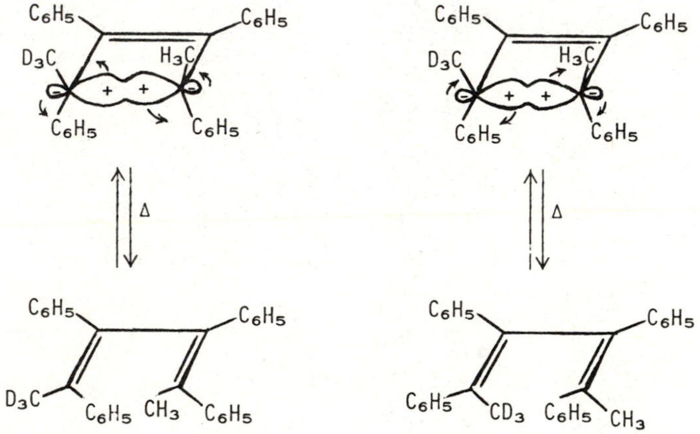

30.6 A photochemical electrocyclic reaction involving two electron pairs proceeds in a *disrotatory* manner (Table 30.1).

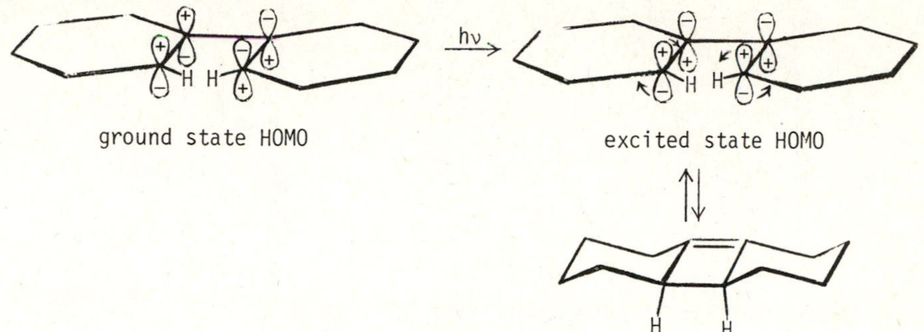

The two hydrogen atoms in the four-membered ring are *cis* to each other in the cyclobutene product.

30.7

(2*E*,4*E*)-hexadiene (2*E*,4*Z*)-hexadiene

Diene LUMO

Alkene HOMO

The Diels-Alder reaction is a thermal [4+2] cycloaddition, which occurs with suprafacial geometry. The stereochemistry of the diene is maintained in the product.

30.8

Diene HOMO

Triene LUMO

The reaction of cyclopentadiene and cycloheptatrienone is a [6+4] cycloaddition. This thermal cycloaddition proceeds with suprafacial geometry since five electron pairs are involved in the concerted process. The pi electrons of the carbonyl group do not take part in the reaction.

30.9

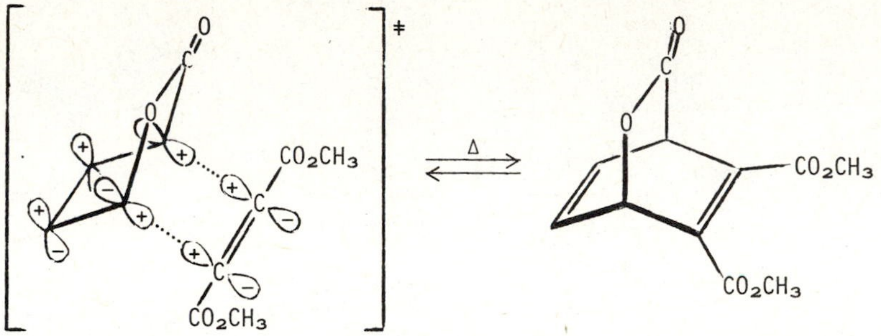

Formation of the bicyclic ring system occurs by a suprafacial [4+2] Diels-Alder cycloaddition process. Only one pair of pi electrons from the alkyne is involved in the reaction; the carbonyl pi electrons are not involved.

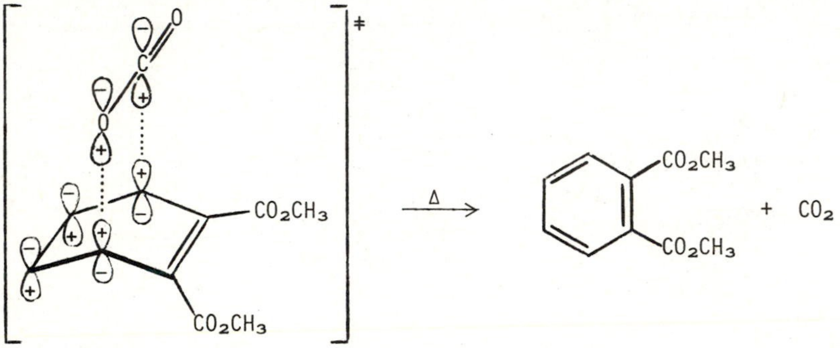

Loss of CO_2 is a reverse Diels-Alder [4+2] cycloaddition reaction.

30.10

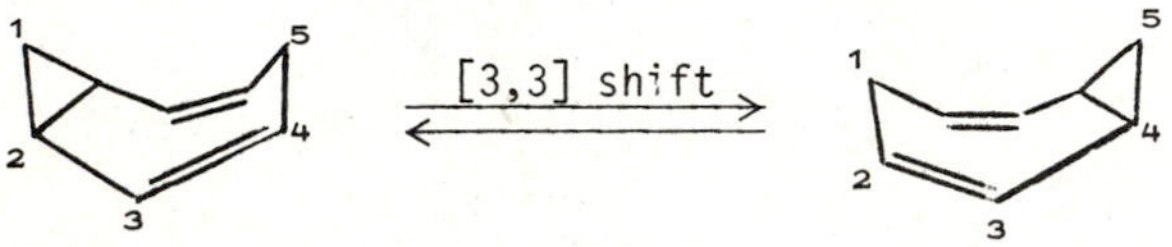

The ^{13}C NMR spectrum of homotropilidene would show five peaks if rearrangement were slow. In fact, rearrangement occurs at a rate that is too fast for NMR to detect. The ^{13}C NMR spectrum taken at room temperature is an average of the two equilibrating forms, in which positions 1 and 5 are equivalent, as are positions 2 and 4. Thus, only three distinct types of carbons are visible in the ^{13}C NMR spectrum of homotropilidene.

30.11 Scrambling of the deuterium label of 1-deuterioindene occurs by a series of [1,5] sigmatropic rearrangements. This thermal reaction involves three electron pairs -- one pair of pi electrons from the six-membered ring, the pi electrons from the five membered ring, and two electrons from a carbon-deuterium (or hydrogen) single bond -- and proceeds with suprafacial geometry.

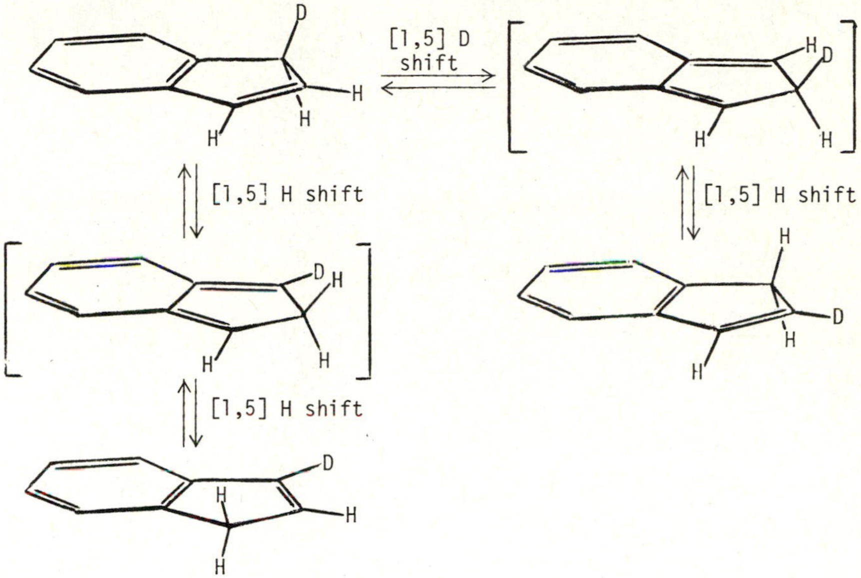

[1,5] D shift

[1,5] H shift

[1,5] H shift

[1,5] H shift

30.12 This [1,7] sigmatropic reaction proceeds with antarafacial geometry since four electron pairs are involved in the rearrangement.

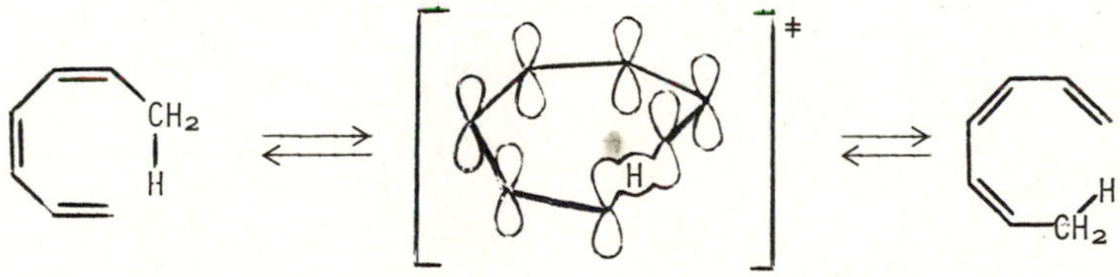

30.13

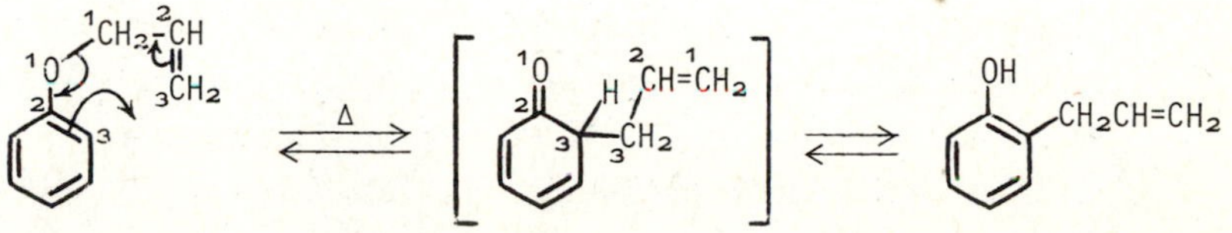

The Claisen rearrangement of an unsubstituted allyl phenyl ether is a [3,3] sigmatropic rearrangement in which the allyl group usually ends up in the position *ortho* to oxygen. In this problem both *ortho* positions are occupied by methyl groups. The Claisen intermediate undergoes a second [3,3] rearrangement, and the final product is *p*-allyl phenol.

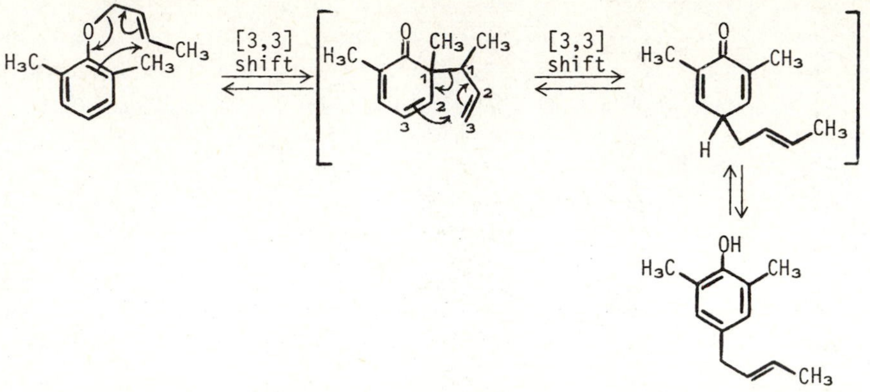

30.14

	Type of reaction	Number of electron pairs	Stereochemistry
a)	thermal electrocyclic	four	conrotatory
b)	photochemical electrocyclic	four	disrotatory
c)	photochemical cycloaddition	four	suprafacial
d)	thermal cycloaddition	four	antarafacial
e)	photochemical sigmatropic rearrangement	four	suprafacial

30.15 a) An *electrocyclic reaction* is a reversible, pericyclic process in which a ring is formed by the reorganization of the pi electrons of a conjugated polyene.

b) *Conrotatory* motion in a pericyclic reaction occurs when the lobes of two orbitals both rotate in either a clockwise or a counterclockwise fashion.

c) A *suprafacial* pericyclic reaction occurs between orbital lobes on the same face of one component and orbital lobes on the same face of the other component.

d) An *antarafacial* pericyclic reaction occurs between orbital lobes on the same face of one component and orbital lobes on opposite faces of the other component.

e) *Disrotatory motion* occurs when the lobes of one orbital in an electrocyclic reaction rotate clockwise and the lobes of the other orbital rotate counterclockwise.

f) A *sigmatropic rearrangement* is a pericyclic process in which a sigma-bonded group migrates across a pi electron system.

30.16 a)

Rotation of the orbitals in the 10 electron system occurs in a disrotatory fashion. According to the rules in Table 30.1, the reaction should be carried out under thermal conditions.

b)

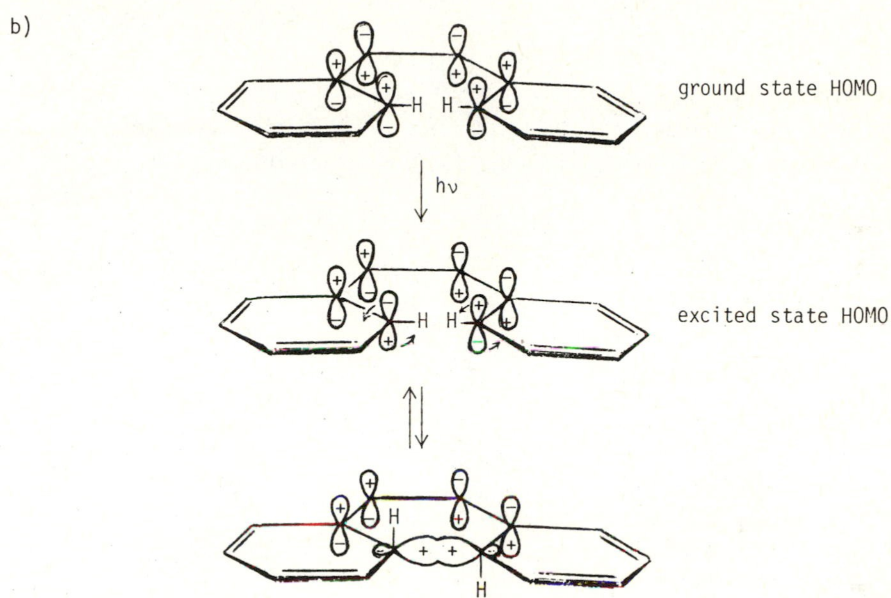

In order for the hydrogens to be *trans* in the product, rotation must occur in a conrotatory manner. This can happen only if the HOMO has the symmetry pictured. For a 6 pi electron system, this HOMO must arise from photochemical excitation of a pi electron. To obtain a product having the correct stereochemistry, the reaction must be carried out under photochemical conditions.

30.17

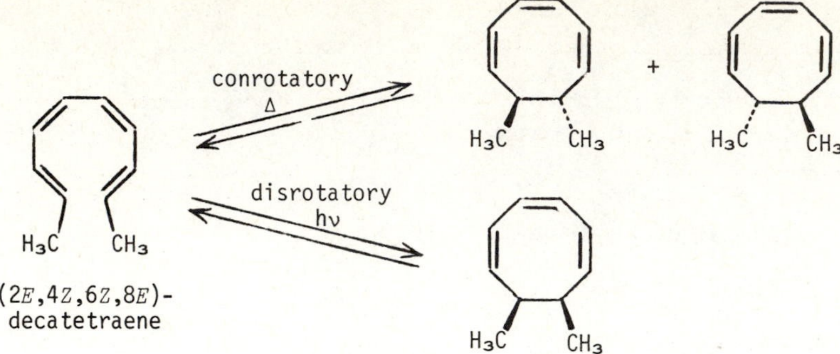

Four electron pairs undergo reorganization in this electrocyclic reaction. The thermal reaction occurs with conrotatory motion to yield a pair of enantiomeric *trans*-7,8-dimethyl-1,3,5-cyclooctatrienes. The photochemical cyclization occurs with disrotatory motion to yield the *cis*-7,8-dimethyl isomer.

30.18

conrotatory Δ

disrotatory hν

(2*E*,4*Z*,6*Z*,8*Z*)-decatetraene

30.19

thermal reaction — disrotatory

HOMO

photochemical reaction — conrotatory

HOMO

Two electrocyclic reactions, involving three electron pairs each, occur in this isomerization. The thermal reaction is a disrotatory process that yields two *cis*-fused six-membered rings. The photochemical reaction yields the *trans*-fused isomer. The two pairs of pi electrons in the eight-membered ring do not take part in the electrocyclic reaction.

30.20

	Type of reaction	Number of electron pairs	Stereochemistry
a)	photochemical [1,5] sigmatropic rearrangement	3	antarafacial
b)	thermal [4+6] cycloaddition	5	suprafacial
c)	thermal [1,7] sigmatropic rearrangement	4	antarafacial
d)	photochemical [2+6] cycloaddition	4	suprafacial

30.21

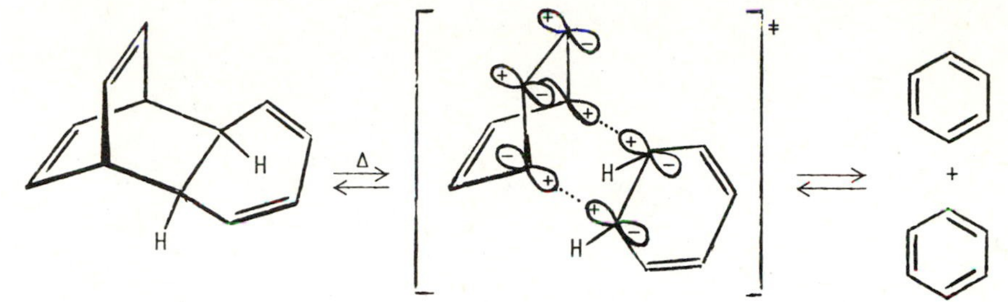

This reaction is a reverse [4+2] cycloaddition. The reacting orbitals have the correct symmetry for the reaction to take place by a favorable suprafacial process.

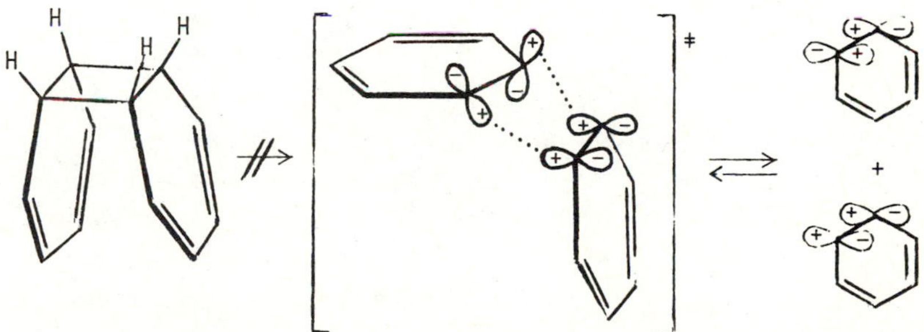

This [2+2] reverse cycloaddition is not likely to occur as a concerted process because the required antarafacial geometry for the thermal reaction is not possible for a four-electron system.

30.22

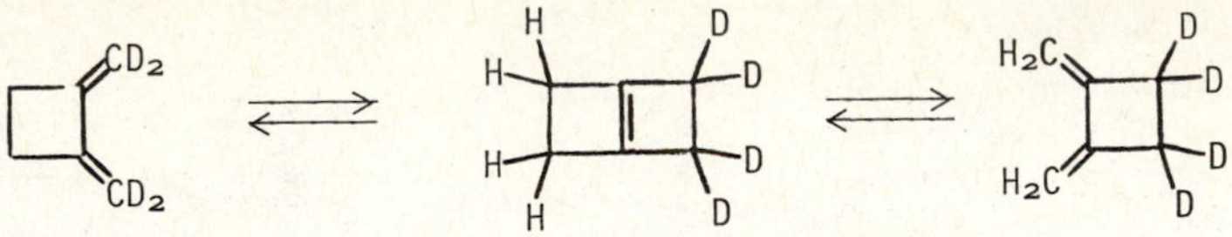

Each electrocyclic reaction involves two pairs of electrons and proceeds in a conrotatory manner.

30.23

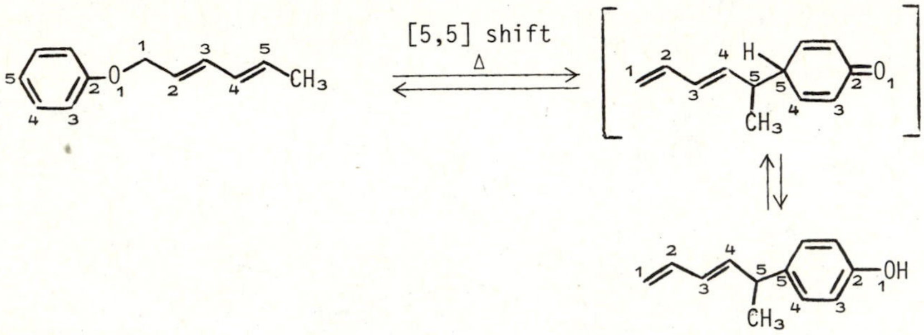

[5,5] shift

Δ

This thermal sigmatropic rearrangement is a suprafacial process since five electron pairs are involved in the reaction.

30.24

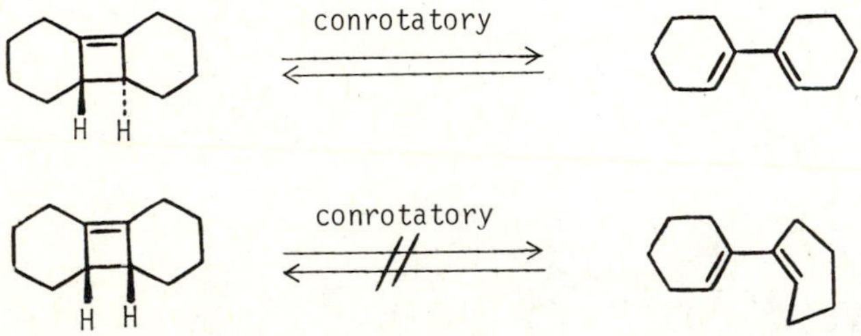

conrotatory

conrotatory

The observed product can be formed by a four-electron *pericyclic* process only if the four-membered ring geometry is *trans*. Ring-opening of the *cis* isomer by a concerted process would form a severely strained six-membered ring containing a *trans* double bond. Reaction of the *cis* isomer to yield the observed product occurs instead by a higher energy, non-concerted path.

30.25

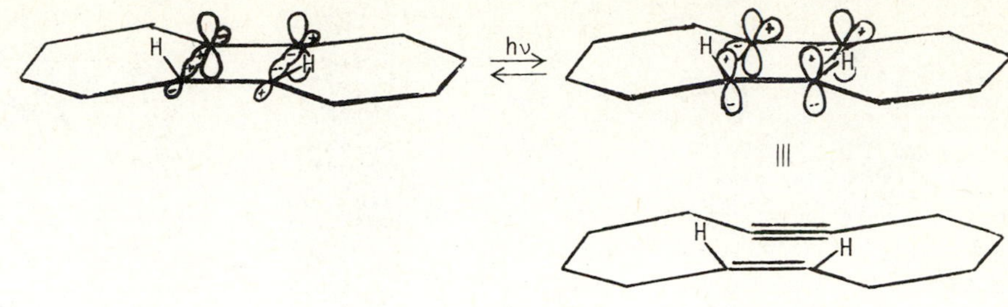

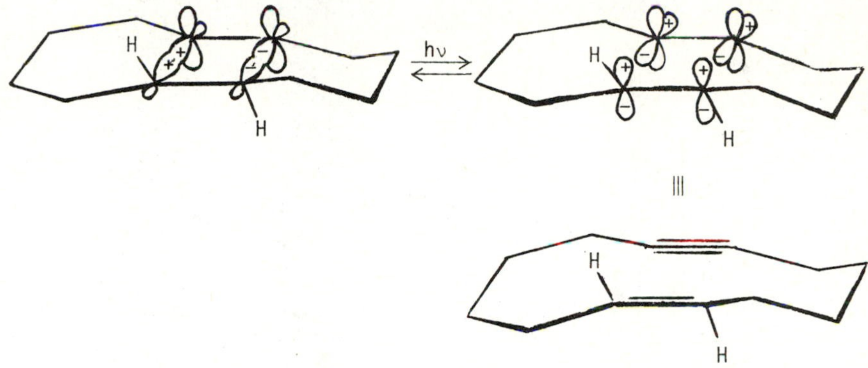

Both of these reactions are reverse [2+2] photochemical cycloadditions, which occur with suprafacial geometry.

30.26

C_6H_5 N O OHC C C CH_3 ‡ ⇄ C_6H_5 N O CH_3 CHO

The first reaction is a Diels-Alder [4+2] cycloaddition, which proceeds with suprafacial geometry.

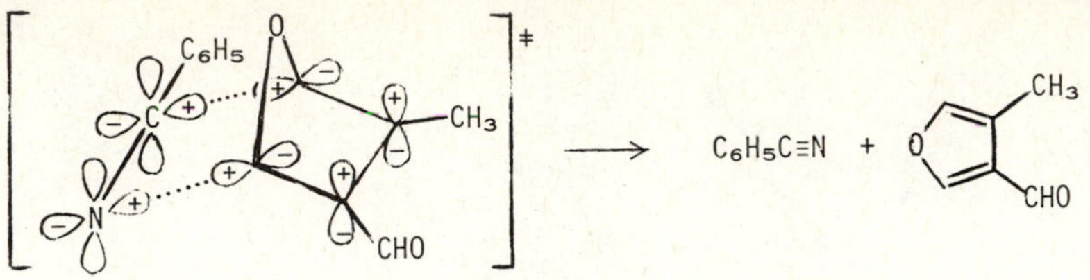

The second reaction is a reverse Diels-Alder [4+2] cycloaddition.

30.27

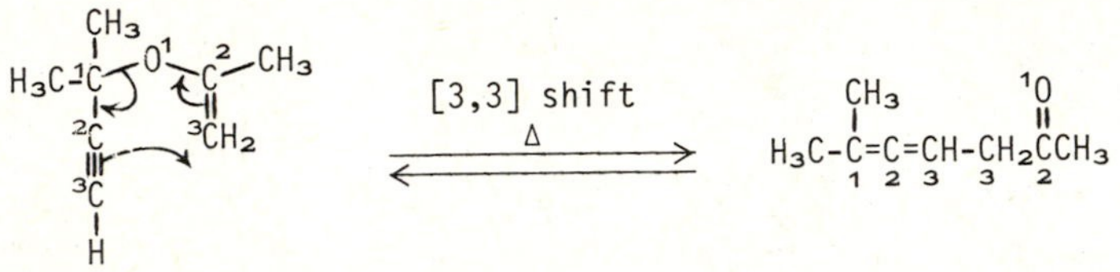

An allene is formed by a [3,3] sigmatropic rearrangement.

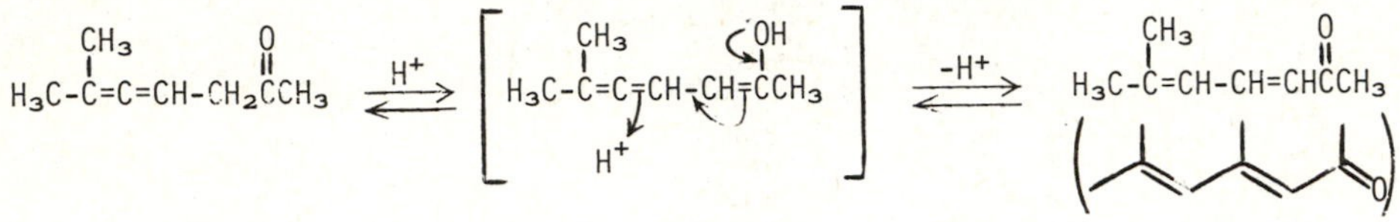

Acid catalyzes isomerization of the allene to a conjugated dienone *via* an intermediate enol.

30.28

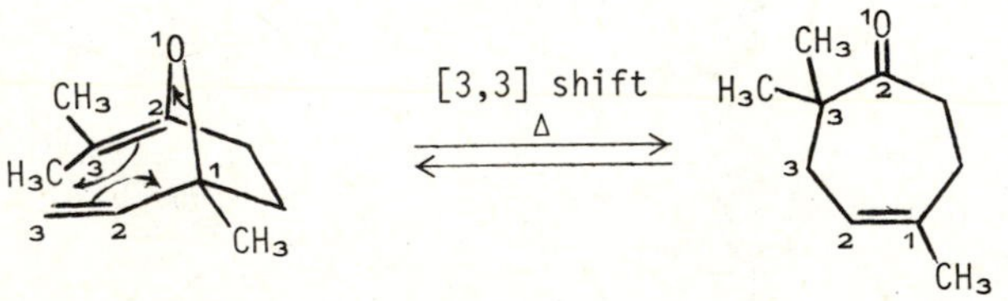

Karahanaenone is formed by a [3,3] sigmatropic rearrangement (Claisen rearrangement).

30.29

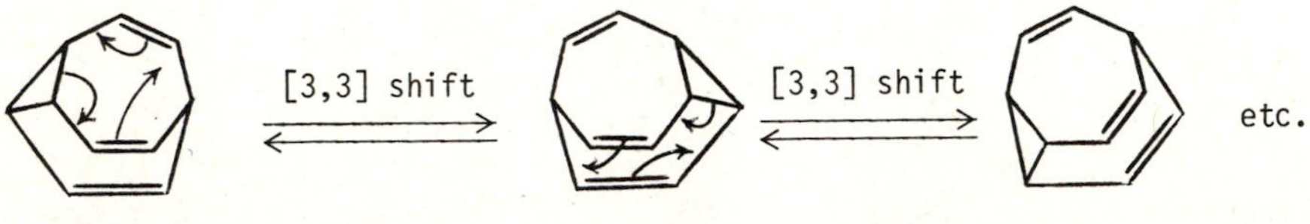

Bullvalene can undergo [3,3] sigmatropic rearrangements in all directions. At 100°, the rate of rearrangement is fast enough to make all hydrogen atoms equivalent, and only one signal is seen in the ^1H NMR spectrum.

30.30 Suprafacial shift:

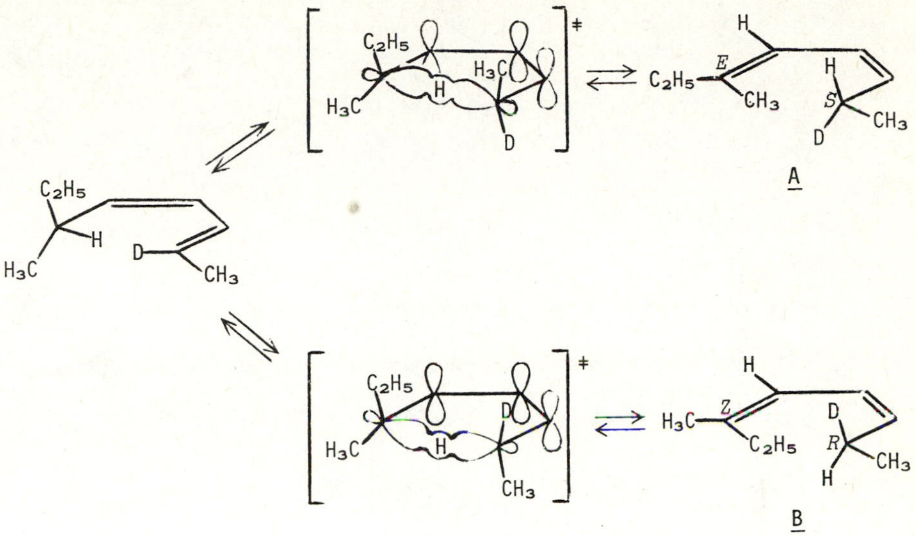

A

B

Antarafacial shift:

C
not formed

D
not formed

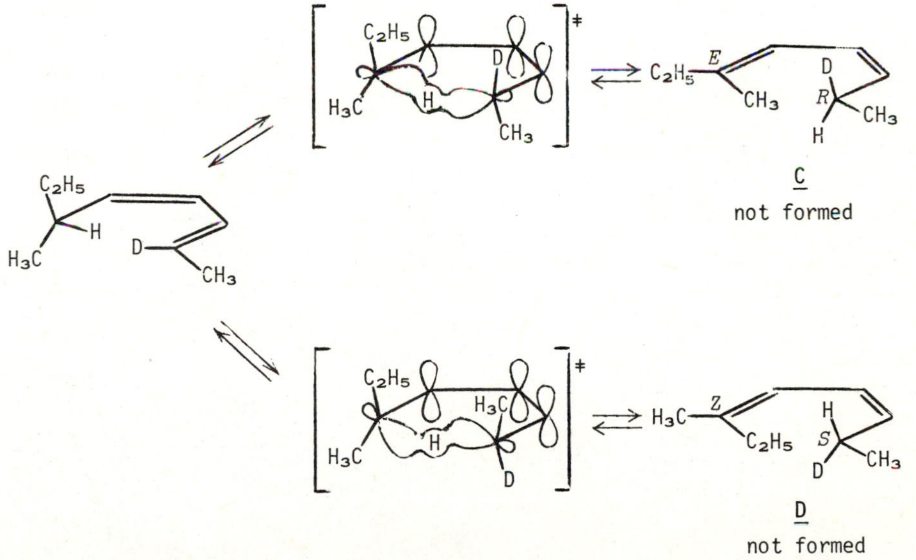

The observed products A and B result from a [1,5] sigmatropic hydrogen shift with suprafacial geometry, and they confirm the predictions of orbital symmetry. C and D are not formed.

30.31

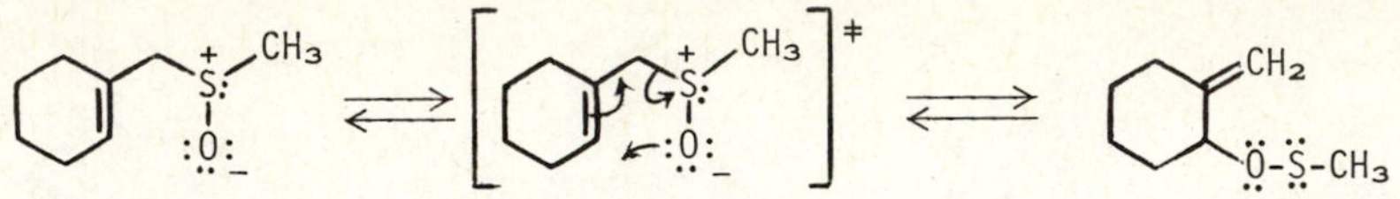

This [2,3] sigmatropic rearrangement involves three electron pairs and should occur with suprafacial geometry.

30.32 1.

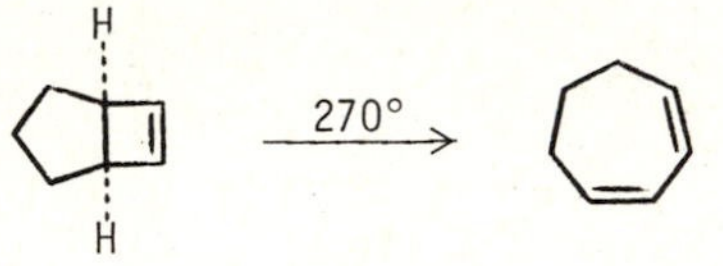

270°

2.

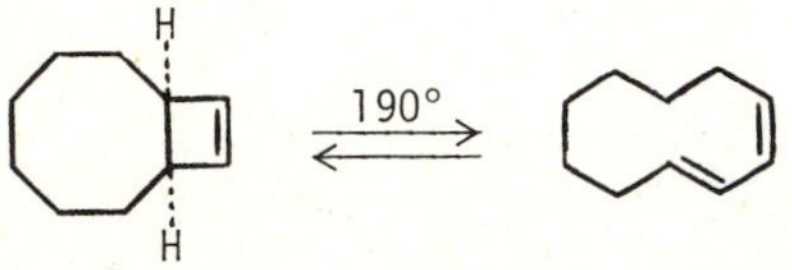

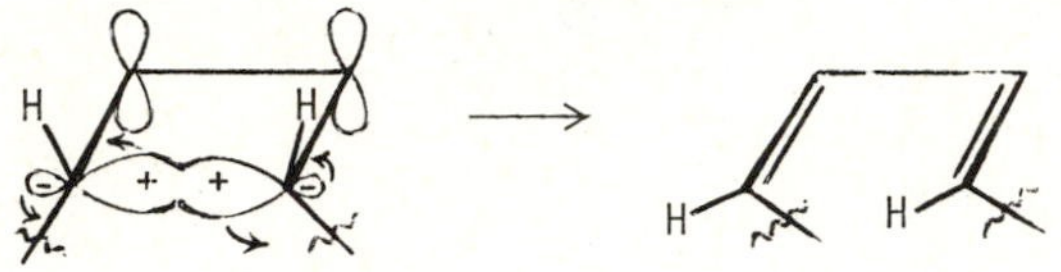

190°

Concerted thermal ring opening of a *cis* fused cyclobutene ring yields a product having one *cis* and one *trans* double bond. The ten-membered ring product of reaction 2 is large enough to accommodate a *trans* double bond, but a seven-membered ring containing a *trans* double bond is highly strained. Opening of the cyclobutene ring in reaction 1 occurs by a higher energy non-pericyclic process to yield a seven-membered ring having two *cis* double bonds.

30.33

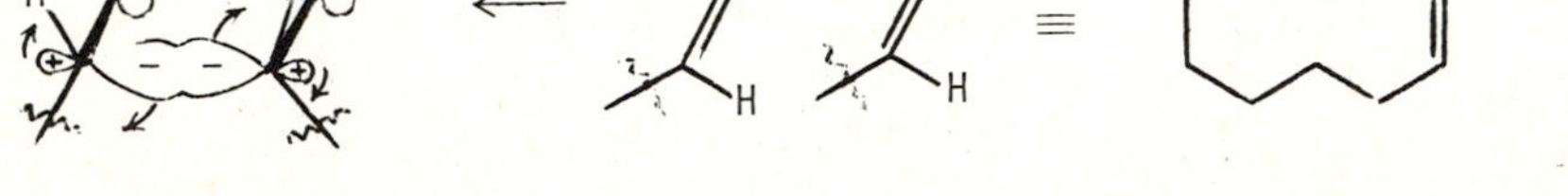

Thermal ring opening of the methylcyclobutene ring can occur by either of two symmetry-allowed conrotatory paths to yield the observed product mixture.

30.34

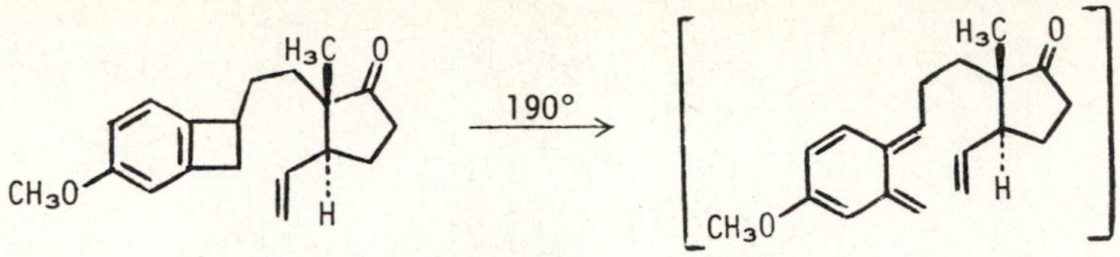

The first reaction is an electrocyclic opening of a cyclobutene ring.

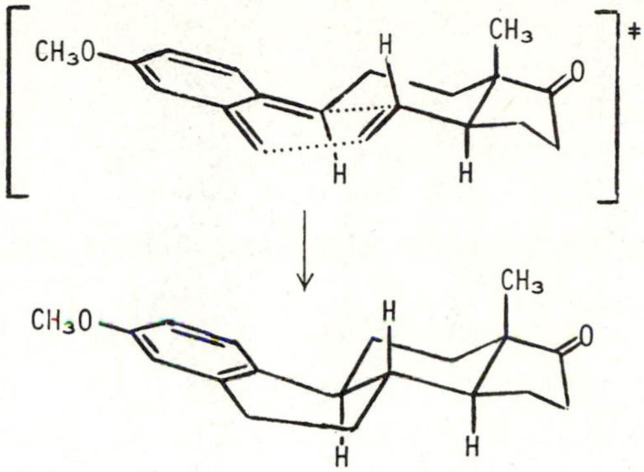

Formation of estrone methyl ether occurs by a Diels Alder [4+2] cycloaddition.

30.35

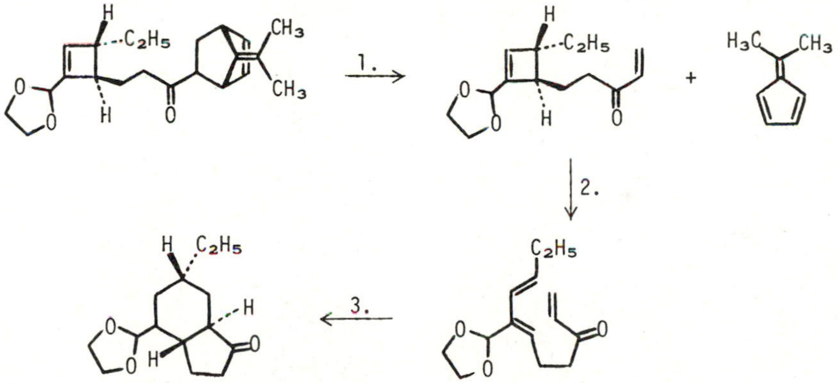

Reaction 1: Reverse Diels-Alder [4+2] cycloaddition;

Reaction 2: Conrotatory electrocyclic opening of a cyclobutene ring;

Reaction 3: Diels-Alder [4+2] cycloaddition.

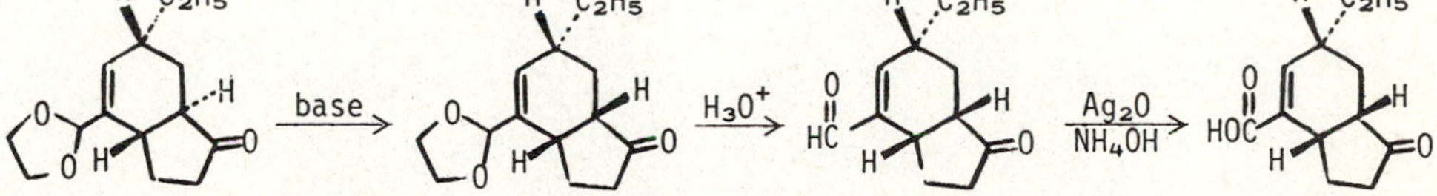

Treatment with base enolizes the ketone and changes the ring junction from *trans* to *cis*. A *cis* ring fusion is less strained when a six-membered ring is fused to a five-membered ring.

30.36

$$\text{C}_6\text{H}_5\text{-}\overset{+}{\text{N}}(\text{CH}_3)_2\text{CH}_2\text{CH=CH}_2 \xrightarrow[\Delta]{[3,3]\ \text{shift}} \left[\text{cyclohexadienylidene-}\overset{+}{\text{N}}(\text{CH}_3)_2,\ \text{H},\ \text{CH}_2\text{CH=CH}_2 \right] \rightleftarrows o\text{-}(\text{CH}_2\text{CH=CH}_2)\text{C}_6\text{H}_4\overset{+}{\text{N}}\text{H}(\text{CH}_3)_2$$

What you should know:

After doing these problems, you should be able to:

1. Understand the general principle of molecular orbitals;
2. Be able to define the terms HOMO, LUMO, conrotatory, disrotatory, suprafacial, antarafacial, electrocyclic reaction, cycloaddition reaction, sigmatropic rearrangement;
3. Determine the HOMO and LUMO of a conjugated pi system;
4. Predict the stereochemistry of thermal and photochemical electrocyclic reactions;
5. Know the stereochemical requirements for cycloaddition reactions and predict if these reactions are likely to occur;
6. Classify sigmatropic rearrangements by order and predict their products;
7. Know the stereochemical rules for pericyclic reactions.

CHAPTER 31 AMINO ACIDS, PEPTIDES, AND PROTEINS

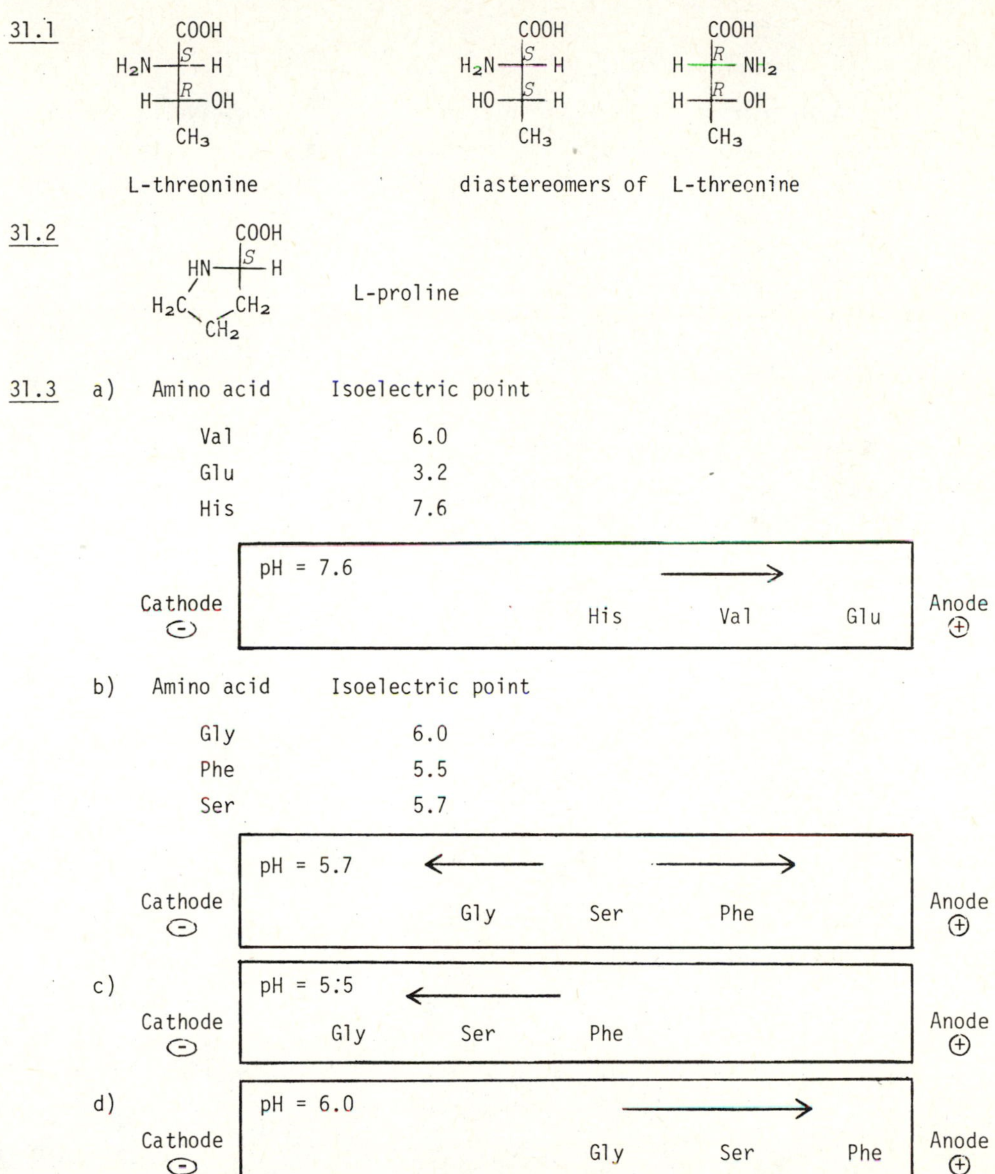

31.4 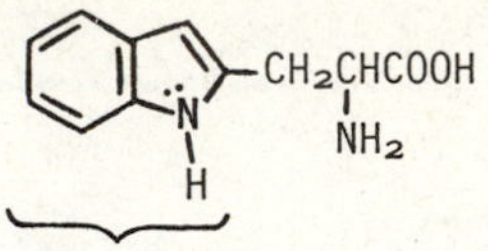 tryptophan

indole ring system

The indole nitrogen lone pair of electrons is part of the pi electron system of the two rings. Protonation of the indole nitrogen is unlikely to occur because it would disrupt the aromaticity of the indole ring system. Tryptophan is not basic, and its isoelectric point is near neutral pH.

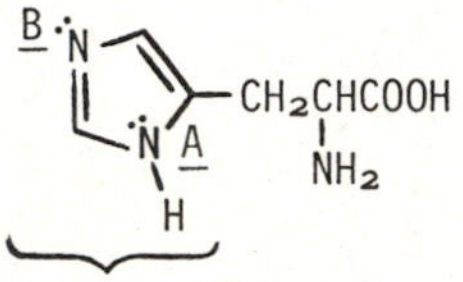

histidine

imidazole ring

Because the lone pair of electrons of nitrogen A is part of the imidazole ring pi system, nitrogen A is not basic. The sp^2 orbital containing the lone pair electrons of nitrogen B lies in the plane of the imidazole ring. Although nitrogen B can be protonated, it is less basic than an aliphatic amine nitrogen. Histidine is weakly basic, and its isoelectric point is slightly higher than neutral.

31.5 The first step of the Strecker synthesis can proceed by two different routes.

a) 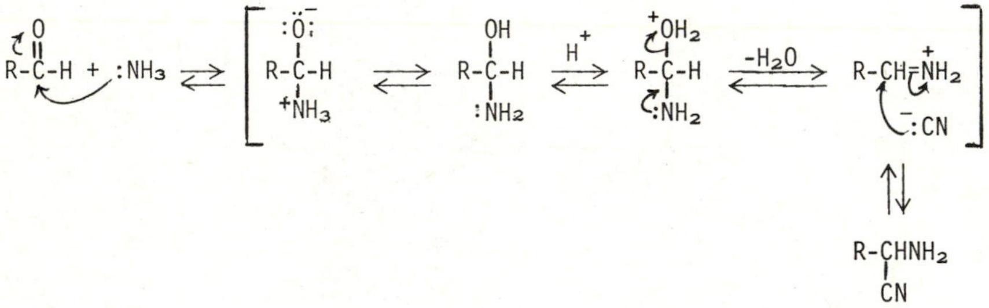 R-C(=O)-H + :NH3 ⇌ [R-C(-O:⁻)(-H)(-⁺NH3) ⇌ R-C(OH)(H)(:NH2) ⇌(H⁺) R-C(⁺OH2)(H)(:NH2) ⇌(-H2O) R-CH=⁺NH2 + ⁻:CN] ⇌ R-CH(NH2)(CN)

In this series of steps, a *nucleophilic acyl substitution* reaction forms an imine from an aldehyde. *Nucleophilic addition* of cyanide to the imine yields the α-amino nitrile.

b) 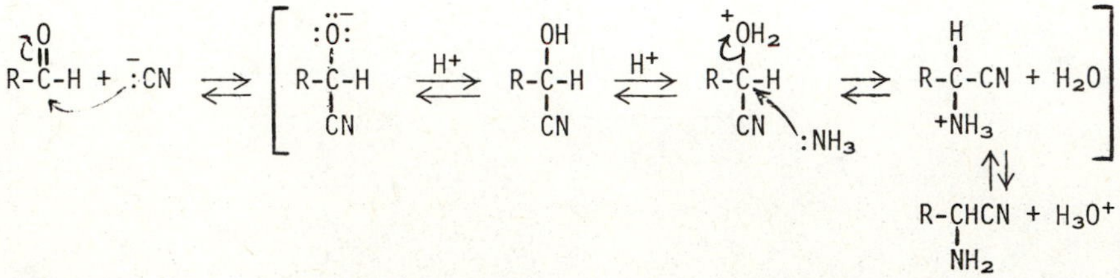 R-C(=O)-H + ⁻:CN ⇌ [R-C(-O:⁻)(-H)(CN) ⇌(H⁺) R-C(OH)(H)(CN) ⇌(H⁺) R-C(⁺OH2)(H)(CN) + :NH3 ⇌ R-C(H)(CN)(⁺NH3) + H2O] ⇌ R-CH(CN)(NH2) + H3O⁺

Nucleophilic addition of HCN to the aldehyde forms the cyanohydrin. *S_N2 displacement* of H_2O by NH_3 produces the α-amino nitrile.

31.6

$$\text{HO-}\underset{\text{HO}}{\text{C}_6\text{H}_3}\text{-CH}_2\text{CHO} \xrightarrow[H_2O]{NH_4Cl,\ KCN} \text{HO-}\underset{\text{HO}}{\text{C}_6\text{H}_3}\text{-CH}_2\underset{\text{NH}_2}{\text{CH}}\text{C}\equiv\text{N}$$

3,4-dihydroxyphenyl-acetaldehyde

$$\xrightarrow{H_3O^+} \text{HO-}\underset{\text{HO}}{\text{C}_6\text{H}_3}\text{-CH}_2\underset{^{+}\text{NH}_3}{\text{CH}}\text{COOH}$$

L-dopa

31.7

$$CH_3CH_2\overset{CH_3}{C}HCH_2COOH \xrightarrow[2.\ C_2H_5OH]{1.\ Br_2,\ PBr_3} CH_3CH_2\overset{CH_3}{C}H\underset{Br}{C}HCO_2C_2H_5$$

$$\xrightarrow{\text{potassium phthalimide } (C_6H_4(C{=}O)_2N^{:-}\ K^+)} C_6H_4(C{=}O)_2N\text{-}\underset{CO_2C_2H_5}{C}H\text{-}\overset{}{C}H(CH_3)CH_2CH_3$$

$$\xrightarrow{^{-}OH,\ H_2O} H_2N\text{-}\underset{CO_2^-}{C}H\overset{CH_3}{C}HCH_2CH_3$$

isoleucine

31.8

	Amino acid	Halide
a)	$(CH_3)_2CHCH_2\underset{NH_2}{C}HCOOH$ leucine	$(CH_3)_2CHCH_2Br$
b)	(imidazol-4-yl, N-H)$-CH_2\underset{NH_2}{C}HCOOH$ histidine	(imidazol-4-yl, N-H)$-CH_2Br$
c)	(indol-2-yl, N-H)$-CH_2\underset{NH_2}{C}HCOOH$ tryptophan	(indol-2-yl, N-H)$-CH_2Br$
d)	$CH_3SCH_2CH_2\underset{NH_2}{C}HCOOH$ methionine	$CH_3SCH_2CH_2Br$

31.9

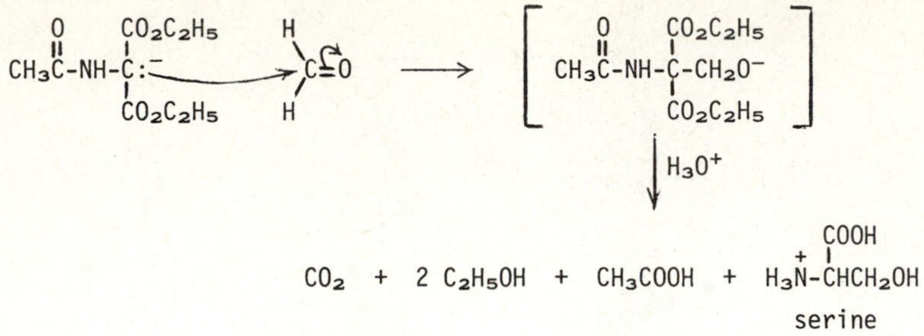

$$CH_3\overset{O}{\overset{\|}{C}}\text{-NH-}\underset{CO_2C_2H_5}{\overset{CO_2C_2H_5}{C}}:^- + \underset{H}{\overset{H}{}}\!\!>C{=}O \longrightarrow \left[CH_3\overset{O}{\overset{\|}{C}}\text{-NH-}\underset{CO_2C_2H_5}{\overset{CO_2C_2H_5}{C}}\text{-}CH_2O^- \right]$$

$$\xrightarrow{H_3O^+} CO_2 + 2\ C_2H_5OH + CH_3COOH + H_3\overset{+}{N}\text{-}\overset{COOH}{CH}CH_2OH$$

serine

The synthesis of serine is a variation of the Knoevenagel reaction (Sec. 26.7).

31.10 H-Val-Tyr-Gly-OH H-Tyr-Gly-Val-OH H-Gly-Val-Tyr-OH
H-Val-Gly-Tyr-OH H-Tyr-Val-Gly-OH H-Gly-Tyr-Val-OH

31.11

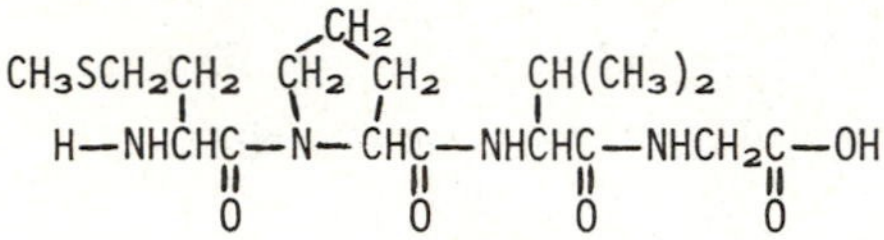

$$H\text{—}NH\overset{CH_3SCH_2CH_2}{CH}\overset{O}{\overset{\|}{C}}\text{—}N\text{—}\overset{CH_2CH_2CH_2}{CH}\overset{O}{\overset{\|}{C}}\text{—}NH\overset{CH(CH_3)_2}{CH}\overset{O}{\overset{\|}{C}}\text{—}NHCH_2\overset{O}{\overset{\|}{C}}\text{—}OH$$

H— Met — Pro — Val — Gly —OH

31.12 Only primary amines can form the extensively conjugated purple ninhydrin product. A secondary amine such as proline yields a product containing a shorter system of conjugated bonds, which absorbs at a shorter wavelength (440 nm vs. 570 nm).

31.13 a) H-Arg-Pro-OH
H-Pro-Leu-Gly-OH
H-Gly-Ile-Val-OH

The complete sequence:
H-Arg-Pro-Leu-Gly-Ile-Val-OH

b) H-Val-Met-Trp-OH
H-Trp-Asp-Val-OH
H-Val-Leu-OH

The complete sequence:
H-Val-Met-Trp-Asp-Val-Leu-OH

31.14 Trypsin cleaves peptide bonds at the carboxyl side of *lysine* and *arginine*.

Chymotrypsin cleaves peptide bonds at the carboxyl side of *phenylalanine*, *tyrosine and tryptophan*.

H-Asp-Arg-Val-Tyr-Ile-His-Pro-Phe-OH —trypsin→ H-Asp-Arg-OH + H-Val-Tyr-Ile-His-Pro-Phe-OH

H-Asp-Arg-Val-Tyr-Ile-His-Pro-Phe-OH —chymotrypsin→ H-Asp-Arg-Val-Tyr-OH + H-Ile-His-Pro-Phe-OH

31.15 H-Pro-Leu-Gly-OH

H-Gly-Pro-Arg-OH

H-Arg-Pro-OH

The complete sequence:

H-Pro-Leu-Gly-Pro-Arg-Pro-OH

31.16 The tripeptide is cyclic.

cyclo(Ala—Leu—Phe) (Leu at top, Ala and Phe at base)   cyclo(Ala—Phe—Leu) (Phe at top, Ala and Leu at base)

31.17

Leu = $H_2N\text{-}CH(CH_2CH(CH_3)_2)\text{-}COOH$ R = $-CH_2CH(CH_3)_2$

1. Protect the amino group of leucine.

$$(CH_3)_3COC(=O)OC(=O)OC(CH_3)_2 + \underset{\text{Leu}}{H_2NCH(R)COOH} \xrightarrow{(C_2H_5)_3N} (CH_3)_3COC(=O)\text{-NH-}CH(R)COOH + CO_2 + (CH_3)_3COH$$

2. Protect the carboxylic acid group of alanine.

$$\underset{\text{Ala}}{H_2NCH(CH_3)COOH} + CH_3OH \xrightarrow{H^+} H_2NCH(CH_3)COOCH_3 + H_2O$$

3. Couple the protected acids with DCC.

$$(CH_3)_3COC(=O)NHCH(R)COOH + H_2N\text{-}CH(CH_3)COOCH_3 + C_6H_{11}N{=}C{=}NC_6H_{11}$$

$$\downarrow$$

$$(CH_3)_3COC(=O)NHCH(R)C(=O)\text{-NHCH}(CH_3)COOCH_3 + C_6H_{11}NHC(=O)NHC_6H_{11}$$

4. Remove the leucine protecting group.

$$(CH_3)_3COC(=O)NHCH(R)C(=O)\text{-NHCH}(CH_3)COOCH_3 \xrightarrow{CF_3COOH} H_3\overset{+}{N}CH(R)C(=O)\text{-NHCH}(CH_3)COOCH_3 + (CH_3)_2C{=}CH_2 + CO_2$$

5. Remove the alanine protecting group.

$$H_3\overset{+}{N}CH(R)C(=O)\text{-NHCH}(CH_3)COOCH_3 \xrightarrow[2.\ H_3O^+]{1.\ ^-OH,\ H_2O} H_3\overset{+}{N}CH(CH_2CH(CH_3)_2)C(=O)\text{—NHCH}(CH_3)COOH + CH_3OH$$

H-Leu — Ala—OH

31.18 a)

1. H-Leu-Ala-OH + $(CH_3)_3COC(=O)OC(=O)OC(CH_3)_3$ $\xrightarrow{(CH_3CH_2)_3N}$ BOC-Leu-Ala-OH
(Prob. 31.17)

2. H-Gly-OH + CH_3OH $\xrightarrow{H^+}$ H-Gly-OCH_3

3. BOC-Leu-Ala-OH + H-Gly-OCH_3 $\xrightarrow{DCC}$ BOC-Leu-Ala-Gly-OCH_3

4. BOC-Leu-Ala-Gly-OCH_3 $\xrightarrow{CF_3COOH}$ H-Leu-Ala-Gly-OCH_3

5. H-Leu-Ala-Gly-OCH_3 $\xrightarrow[2.\ H_3O^+]{1.\ ^-OH,\ H_2O}$ H-Leu-Ala-Gly-OH

b)

1. H-Gly-OH + $(CH_3)_3COC(=O)OC(=O)OC(CH_3)_3$ $\xrightarrow{(CH_3CH_2)_3N}$ BOC-Gly-OH

2. BOC-Gly-OH + H-Leu-Ala-OCH_3 $\xrightarrow{DCC}$ BOC-Gly-Leu-Ala-OCH_3
(Prob. 31.17, part 4)

3. BOC-Gly-Leu-Ala-OCH_3 $\xrightarrow{CF_3COOH}$ H-Gly-Leu-Ala-OCH_3

4. H-Gly-Leu-Ala-OCH_3 $\xrightarrow[2.\ H_3O^+]{1.\ ^-OH,\ H_2O}$ H-Gly-Leu-Ala-OH

31.19 The presence of proline in a polypeptide chain interrupts α-helix formation. The amide nitrogen of proline has no hydrogen that can contribute to the hydrogen-bonded structure of an α-helix. In addition, the pyrrolidine ring of proline restricts rotation about the C-N bond and reduces flexibility in the polypeptide chain.

31.20

```
   COOH            COOH
H──┼──NH2       H──┼──NH2
   CH2OH           CH3
 D-serine       D-alanine
```

31.21 Water is a polar solvent; chloroform is a non-polar solvent. Since charged species are less stable in non-polar solvents than in polar solvents, amino acids exist as the non-ionic amino carboxylic acid form in chloroform.

31.22

amino acid	isoelectric point
histidine	7.6
serine	5.7
glutamic acid	3.2

The optimum pH for the electrophoresis of three amino acid occurs at the isoelectric point of the amino acid intermediate in acidity. At this pH one amino acid migrates toward the cathode (the least acidic), one migrates toward the anode (the most acidic), and the amino acid intermediate in acidity does not migrate. In this example, electrophoresis at pH = 5.7 allows the maximum separation of the three amino acids.

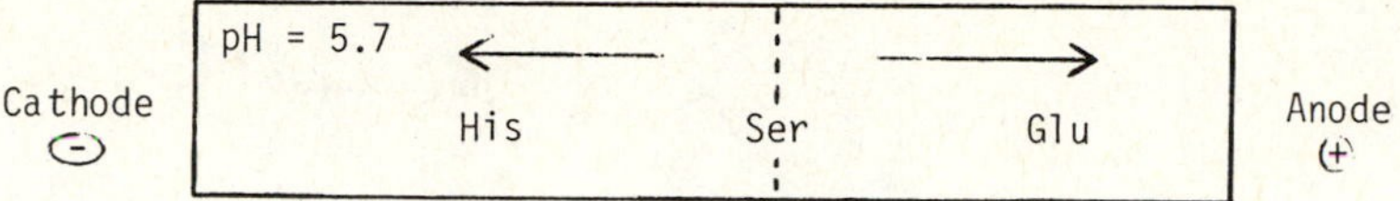

31.23 100 g of cytochrome C contains 0.43 g iron.

100 g of cytochrome C contains:

$$\frac{0.43 \text{ g Fe}}{55.8 \text{ g/mol Fe}} = 0.0077 \text{ moles Fe}$$

$$\frac{100 \text{ g cytochrome C}}{0.0077 \text{ moles Fe}} = \frac{X \text{ g cytochrome C}}{1 \text{ mole Fe}}$$

$$13{,}000 \text{ g/mol Fe} = X$$

Cytochrome C has a *minimum* molecular weight of 13,000.

31.24 a) $(CH_3)_2CHCH(\overset{+}{N}H_3)\text{-}COO^-$ (L-valine) $\xrightarrow{CH_3CH_2OH,\ H^+}$ $(CH_3)_2CHCH(\overset{+}{N}H_3)COOCH_2CH_3$

b) $(CH_3)_2CHCH(\overset{+}{N}H_3)\text{-}COO^-$ $\xrightarrow[(CH_3CH_2)_3N]{(CH_3)_3OC(=O)OC(=O)OC(CH_3)_3}$ $(CH_3)_2CHCH(NHC(=O)OC(CH_3)_3)COO^-$

c) $(CH_3)_3CHCH(\overset{+}{N}H_3)\text{-}COO^-$ $\xrightarrow{KOH,\ H_2O}$ $(CH_3)_2CHCH(NH_2)COO^-$

31.25 a)

```
        CH3     C6H5    SH
        |       |       |
        CHCH3   CH2     CH2     CH3
        |       |       |       |
H—NHCHC—NHCHC—NHCHC—NHCHC—OH
        ||      ||      ||      ||
        O       O       O       O
```

H— Val — Phe — Cys — Ala —OH

b)

```
   HOOCCH2      CH2         CH2CH3  CH(CH3)2
        |      /   \        |       |
        CH2  CH2   CH2      CHCH3   CH2
        |      \   /        |       |
H—NHCHC—N——CHC—NHCHC—NHCHC—OH
        ||       ||      ||      ||
        O        O       O       O
```

H— Glu — Pro — Ile — Leu —OH

31.26

```
   ..NH                 +NH2          ..NH2          ..NH2          ..NH2
     ||                  ||          +  |              |              |
H2N̈-C-N̈HR  --H+-->  [ H2N̈-C-N̈HR <--> H2N=C-N̈HR <--> H2N̈-C=NHR <--> H2N̈-C-N̈HR ]
                                                         +              +
```

The protonated guanidino group can be stabilized by resonance.

31.27 CH_3OCH_2Cl + $SnCl_4$ $\rightleftarrows$ $CH_3\overset{+}{O}=CH_2\ SnCl_5^-$

[aryl–polymer] + $CH_2=\overset{+}{O}CH_3$ $SnCl_5^-$ $\longrightarrow$ [arenium ion: +, H, CH_2OCH_3, $SnCl_5^-$] $\longrightarrow$ [aryl–polymer]$-CH_2OCH_3$ + HCl + $SnCl_4$

[aryl–polymer]$-CH_2OCH_3$ $\overset{H^+}{\rightleftarrows}$ [[aryl–polymer]$-CH_2-\overset{+}{O}(H)CH_3$, $:Cl^-$] $\longrightarrow$ [aryl–polymer]$-CH_2Cl$ + $HOCH_3$

31.28 1. $(CH_3)_3CO\overset{O}{\overset{\|}{C}}O\overset{O}{\overset{\|}{C}}OC(CH_3)_3$ + H-Val-OH $\xrightarrow{(CH_3CH_2)_3N}$ $(CH_3)_3CO\overset{O}{\overset{\|}{C}}$-Val-OH (BOC-Val-OH)

2. BOC-Val-OH + Cl-CH_2-(Polymer) $\longrightarrow$ BOC-Val-OCH_2-(Polymer)

3. BOC-Val-OCH_2-(Polymer) $\xrightarrow{\text{1. wash; 2. } CF_3COOH}$ H-Val-OCH_2-(Polymer)

4. BOC-Ala-OH + H-Val-OCH_2-(Polymer) $\xrightarrow{\text{1. DCC; 2. wash}}$ BOC-Ala-Val-OCH_2-(Polymer)

5. BOC-Ala-Val-OCH_2-(Polymer) $\xrightarrow{CF_3COOH}$ H-Ala-Val-OCH_2-(Polymer)

6. BOC-Phe-OH + H-Ala-Val-OCH_2-(Polymer)

$\downarrow$ 1. DCC 2. wash

BOC-Phe-Ala-Val-OCH_2-(Polymer)

7. BOC-Phe-Ala-Val-OCH_2-(Polymer)

$\downarrow$ CF_3COOH, HBr

H-Phe-Ala-Val-OH + $HOCH_2$-(Polymer)

31.29 a) Tyr + Gly $\longrightarrow$ Tyr-Gly yield: 90%

Tyr-Gly + Gly $\longrightarrow$ Tyr-Gly-Gly yield: 90% x 90% = 81%

Tyr-Gly-Gly + Phe $\longrightarrow$ Tyr-Gly-Gly-Phe yield: 81% x 90% = 73%

Tyr-Gly-Gly-Phe + Met $\longrightarrow$ Tyr-Gly-Gly-Phe-Met yield: 73% x 90% = 66%

Overall yield: 66%

b) Tyr + Gly $\longrightarrow$ Tyr-Gly yield: 90%

Tyr-Gly + Gly $\longrightarrow$ Tyr-Gly-Gly yield: 90% x 90% = 81%

Phe + Met $\longrightarrow$ Phe-Met yield: 90%

Tyr-Gly-Gly + Phe-Met $\longrightarrow$ Tyr-Gly-Gly-Phe-Met yield: 81% x 90% = 73%

Overall yield: 73%

31.30

$$O_2N\text{-}C_6H_3(NO_2)\text{-}F + H_2\ddot{N}\text{-}CHR\text{-}C(=O)\text{-NH-Peptide} \rightleftharpoons \left[O_2N\text{-}C_6H_3(F)(=N^+O^-O^-)\text{-}\overset{+}{N}H_2\text{-}CHR\text{-}C(=O)\text{-NH-Peptide} \right]$$

$$\rightleftharpoons \left[O_2N\text{-}C_6H_3(F)(=N^+O^-O^-)\text{-NH-}CHR\text{-}C(=O)\text{-NH-Peptide} + H^+ \right] \longrightarrow O_2N\text{-}C_6H_3(NO_2)\text{-NHCHR-}C(=O)\text{-NH-Peptide} + :F^-$$

This reaction proceeds via a nucleophilic aromatic substitution mechanism.

31.31 2,4-Dinitrofluorobenzene (Sanger's reagent) reacts not only with the *N*-terminal amino group of a protein, but also with the terminal $-NH_2$ group of amino acids such as lysine or arginine. In order to use Sanger's reagent for end-group analysis in a protein containing lysine or arginine, these amino acids must first have their basic groups protected.

31.32 a)

$$H_2NCHRCOOH + H_2NCHRCOOH \xrightarrow{DCC} H_2NCHRC(=O)\text{-}NHCHRCOOH$$

In this sequence of steps a dipeptide is formed from two equivalents of $H_2NCHRCOOH$. The mechanism is pictured in Fig. 31.5.

b)

DCC (C_6H_{11}-N=C=N-C_6H_{11}) + H^+ + dipeptide H_2N-RCH-C(=O)-NH-CHR-C(=O)OH $\longrightarrow$ [C_6H_{11}-N=C(NH-C_6H_{11})-$\overset{+}{O}$H-C(=O)-CHR-NH-C(=O)-RCH-$\ddot{N}H_2$ $\rightleftharpoons$ C_6H_{11}-N=C(NH-C_6H_{11})-O-C(=$\overset{+}{O}$H)-CHR-NH-C(=O)-RCH-$\ddot{N}H_2$]

$\downarrow$

[C_6H_{11}-N=C(NH-C_6H_{11})-O-C(OH)(cyclic: $H_2\overset{+}{N}$-RCH-C(=O)-NH-CHR) $\rightleftharpoons$ C_6H_{11}-N=C(NH-C_6H_{11})-O-C(O-H)(cyclic: HN-RCH-C(=O)-NH-CHR), H^+] $\longleftarrow$

C_6H_{11}-NH-C(=O)-NH-C_6H_{11} + 2,5-diketopiperazine (3,6-di-R) + H^+

DCC couples the carboxylic acid end of the dipeptide to the amino end to yield the 2,5-diketopiperazine.

31.33 ^1H NMR shows that the two methyl groups of *N*,*N*-dimethylformamide are non-equivalent at room temperature. If rotation around the CO-N bond were unrestricted, the methyl groups would be interconvertible, and their ^1H NMR absorptions would coalesce into a single signal.

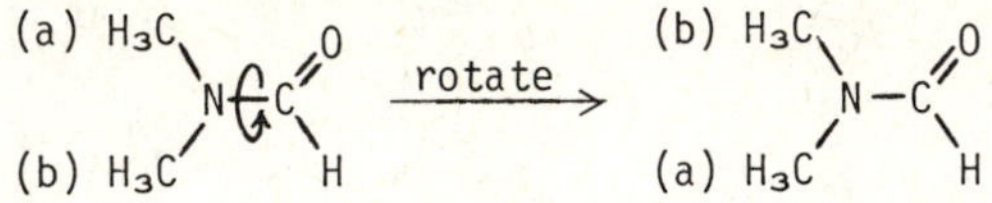

(a) H_3C, (b) H_3C — N—C(=O)H —rotate→ (b) H_3C, (a) H_3C — N—C(=O)H

The presence of two absorptions shows that there is a barrier to rotation around the CO-N bond. This barrier is due to the partial double bond character of the CO-N bond, as indicated by the two resonance forms. Heating to 180°C supplies enough energy to allow rapid rotation and to cause the two NMR absorptions to merge.

$(H_3C)_2\ddot{N}-CH=O \longleftrightarrow (H_3C)_2\overset{+}{N}=CH-O^-$

31.34 H-Gly-Gly-Asp-Phe-Pro-Val-Pro-Leu-OH

31.35

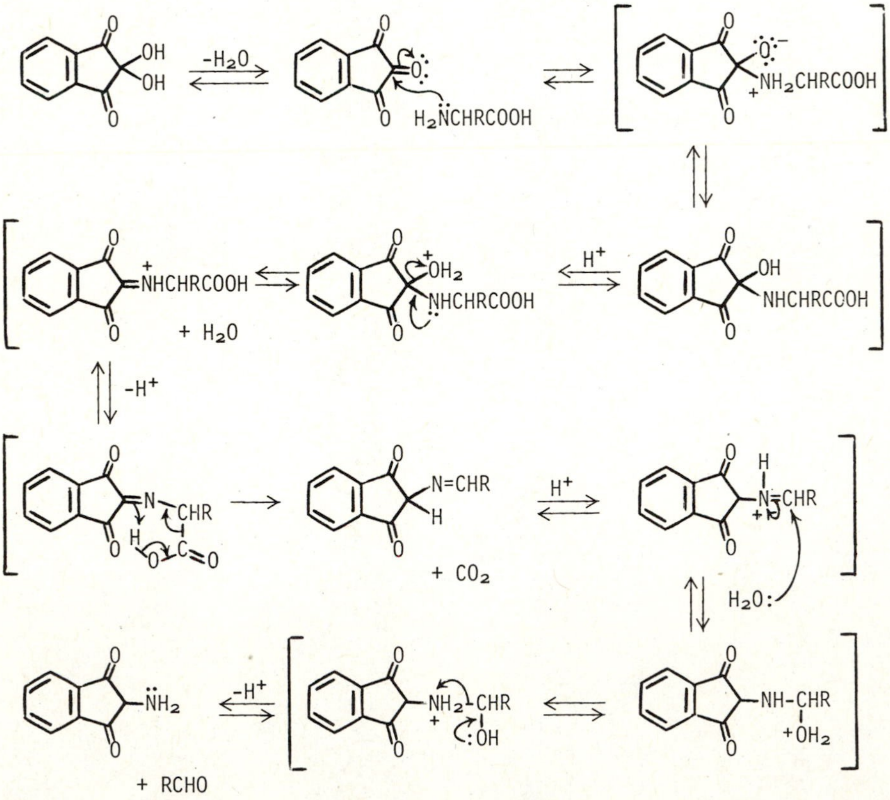

Ninhydrin (OH, OH) ⇌ ($-H_2O$) triketone + $H_2\ddot{N}CHRCOOH$ ⇌ [O^-, $\overset{+}{N}H_2CHRCOOH$] ⇌ [OH, NHCHRCOOH] ⇌ (H^+) $\overset{+}{O}H_2$, NHCHRCOOH ⇌ [$=\overset{+}{N}HCHRCOOH$ + H_2O] ⇌ ($-H^+$) [$=N-CHR$, H, O, C=O] → N=CHR, H + CO_2 ⇌ (H^+) [H, $\overset{+}{N}=CHR$, H_2O:] ⇌ [NH-CHR, $^+OH_2$] ⇌ [$\overset{+}{N}H_2$-CHR, :OH] ⇌ ($-H^+$) $\ddot{N}H_2$ + RCHO

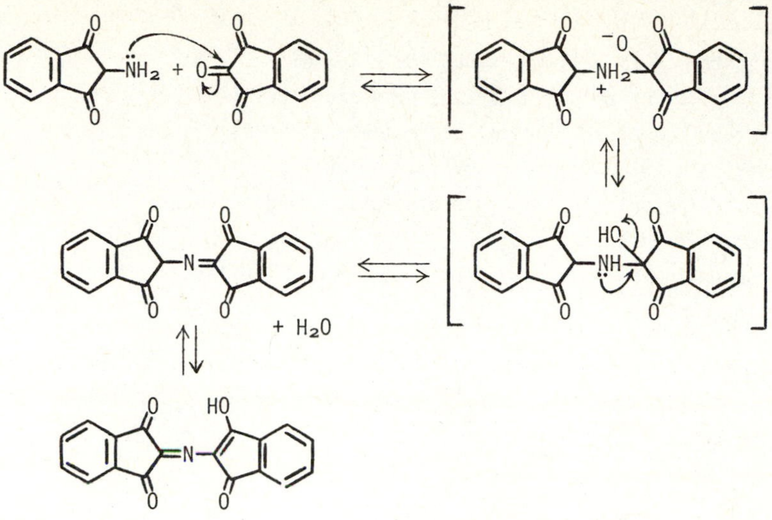

+ H_2O

31.36

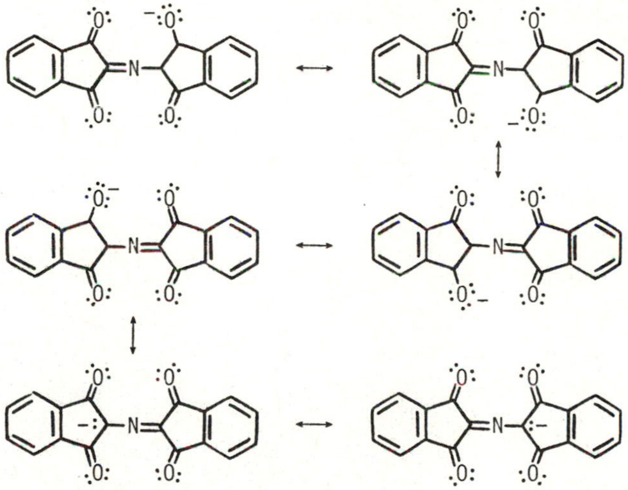

31.37 H-Cys-Tyr-Ile-Gln-Asn-Cys-Pro-Leu-Gly-OH
(NH_2)

reduced oxytocin

```
  ,Asn-Cys-Pro-Leu-Gly-OH
 /      |
Gln     S             (NH2)
 |      |
Ile     S
 \      |
  `Tyr-Cys-H
```

oxidized oxytocin

The C-terminal end of oxytocin is an amide, but this cannot be determined from the information given.

What you should know:

After doing these problems you should be able to:

1. Draw from memory the structure of each amino acid;
2. Understand the stereochemical and acid-base behavior of amino acids;
3. Synthesize α-amino acids by several routes;
4. Determine the primary structure of peptides and proteins;
5. Synthesize simple peptides;
6. Understand the classification of proteins and the levels of structure of proteins.

CHAPTER 32 LIPIDS

32.1

$$CH_3CH_2CH_2CH_2-CH{=}CH-CH{=}CH-CH{=}CH-CH_2CH_2CH_2CH_2CH_2CH_2CH_2COOH$$

(9Z,11E,13E)-octadecatrienoic acid

(eleostearic acid)

1. O_3
2. Zn, CH_3COOH

$$CH_3CH_2CH_2CH_2\overset{O}{\overset{\|}{C}}H + H\overset{O}{\overset{\|}{C}}-\overset{O}{\overset{\|}{C}}H + H\overset{O}{\overset{\|}{C}}-\overset{O}{\overset{\|}{C}}H + H\overset{O}{\overset{\|}{C}}CH_2CH_2CH_2CH_2CH_2CH_2CH_2COOH$$

The stereochemistry of the double bonds can't be determined from the information given.

32.2 $CH_3(CH_2)_7C{\equiv}C(CH_2)_7COOH \xrightarrow[\text{2. Zn, } CH_3COOH]{\text{1. } O_3} CH_3(CH_2)_7COOH + HOOC(CH_2)_7COOH$

stearolic acid — nonanoic acid — nonanedioic acid

32.3

$$CH_3\overset{O}{\overset{\|}{C}}-S-CoA + H-S-ACP \rightleftharpoons \left[CH_3\overset{O^-}{\overset{|}{C}}(-S-CoA)(-\overset{+}{S}H-ACP) \rightleftharpoons CH_3\overset{OH}{\overset{|}{C}}(-S-CoA)(-S-ACP) \rightleftharpoons CH_3\overset{O^-}{\overset{|}{C}}(-\overset{+}{S}H-CoA)(-S-ACP) \right]$$

$$\rightleftharpoons CH_3\overset{O}{\overset{\|}{C}}-S-ACP + H-S-CoA$$

This is an example of a nucleophilic acyl substitution reaction.

32.4

$$CH_3\overset{O}{\overset{\|}{C}}-\underset{COOH}{C}H\overset{O}{\overset{\|}{C}}-S-ACP \longrightarrow \left[\begin{matrix} H_3C \\ HO \end{matrix} \!\!>\! C{=}CH\overset{O}{\overset{\|}{C}}-S-ACP + CO_2 \right] \longrightarrow CH_3\underset{O}{\underset{\|}{C}}CH_2\overset{O}{\overset{\|}{C}}-S-ACP$$

32.5 A fatty acid synthesized from $^{14}CH_3COOH$ will have an alternating labeled and unlabeled carbon chain. The carboxylic acid carbon will be unlabeled.

$$\overset{*}{C}H_3CH_2\overset{*}{C}H_2CH_2\overset{*}{C}H_2CH_2\overset{*}{C}H_2CH_2\overset{*}{C}H_2CH_2\overset{*}{C}H_2CH_2\overset{*}{C}H_2COOH$$

32.6 a)

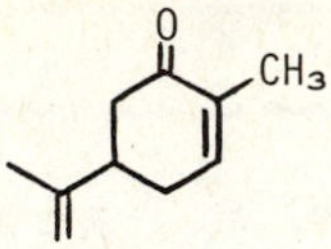

carvone

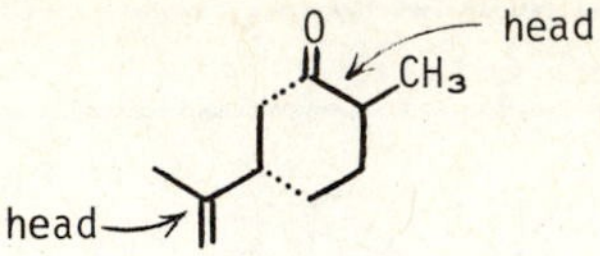

head

head

b)

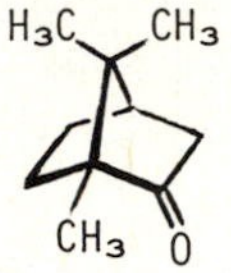

camphor

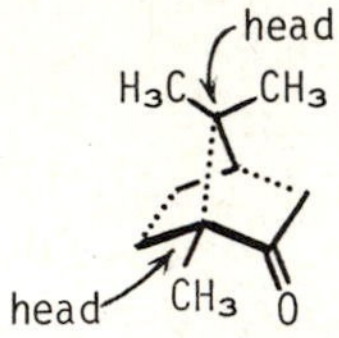

head

head

c)

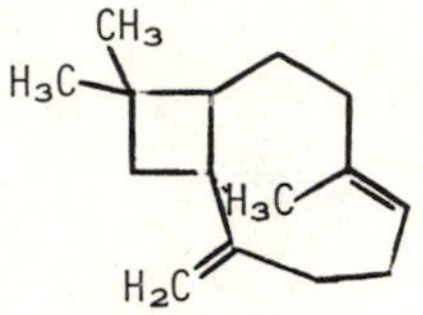

caryophyllene
(a sesquiterpene)

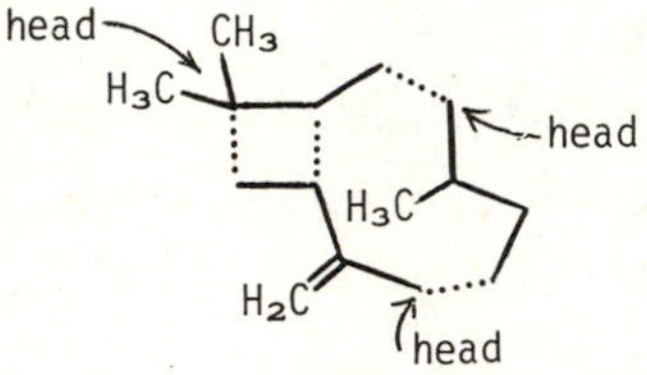

head

head

head

32.7

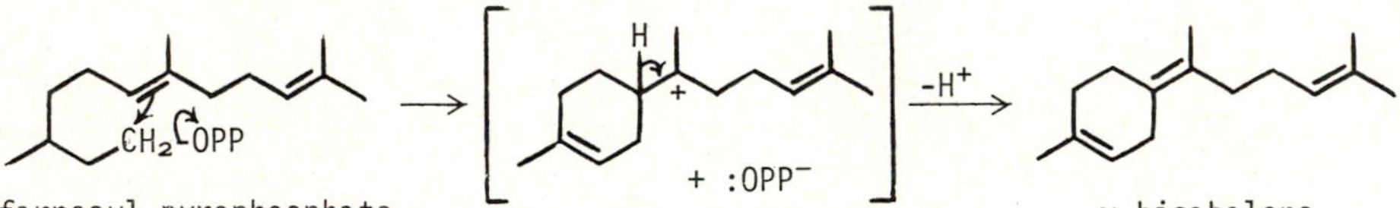

farnesyl pyrophosphate

+ :OPP$^-$

$-H^+$

γ-bisabolene

32.8 a)

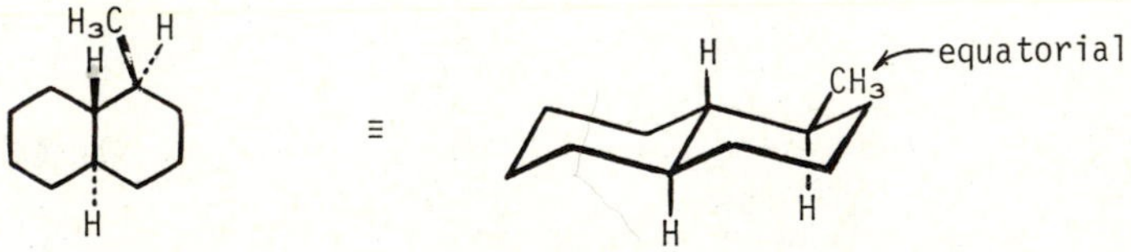

equatorial

b)

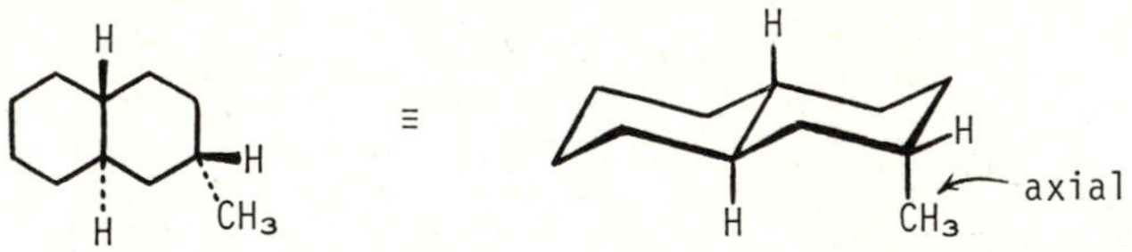

axial

32.9

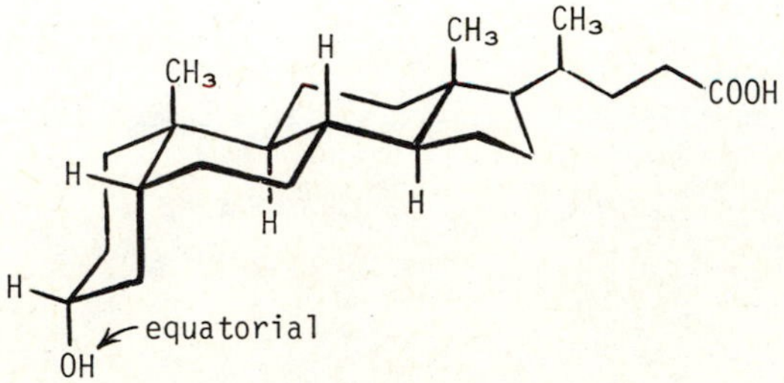

equatorial

Lithocholic acid

32.10

$-H^+$

lanosterol

32.11

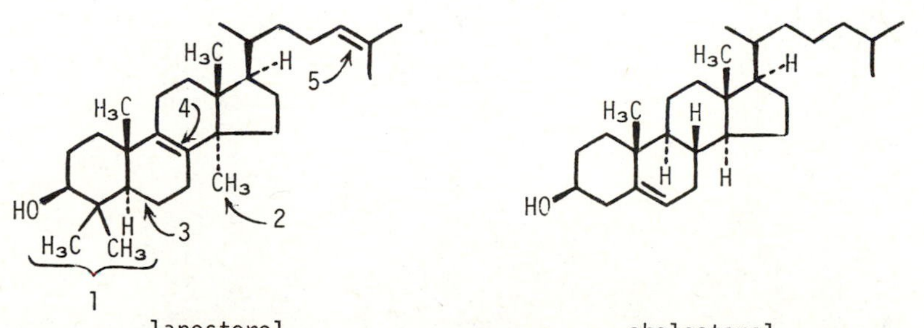

lanosterol cholesterol

Differences between lanosterol and cholesterol:

lanosterol	cholesterol
1. Two methyl groups at C4.	1. Two hydrogens at C4.
2. One methyl group at C14.	2. One hydrogen at C14.
3. C5-C6 single bond.	3. C5-C6 double bond.
4. C8-C9 double bond.	4. C8-C9 single bond.
5. Double bond in side chain.	5. Saturated side chain.

32.12

$$\begin{array}{l} \overset{O}{\overset{\|}{CH_2OC(CH_2)_{16}CH_3}} \\ | \\ \overset{O}{\overset{\|}{CHOC(CH_2)_7CH{=}CH(CH_2)_7CH_3}} \\ | \\ \overset{O}{\overset{\|}{CH_2OC(CH_2)_{16}CH_3}} \end{array} \qquad \begin{array}{l} \overset{O}{\overset{\|}{CH_2OC(CH_2)_{16}CH_3}} \\ | \\ {}^{*}\overset{O}{\overset{\|}{CHOC(CH_2)_{16}CH_3}} \\ | \\ \overset{O}{\overset{\|}{CH_2OC(CH_2)_7CH{=}CH(CH_2)_7CH_3}} \end{array}$$

optically inactive — optically active

1. ^{-}OH, H_2O
2. H_3O^+

$$\begin{array}{l} CH_2OH \\ | \\ CHOH \\ | \\ CH_2OH \end{array} + 2\ CH_3(CH_2)_{16}COOH + CH_3(CH_2)_7CH{=}CH(CH_2)_7COOH$$

stearic acid — oleic acid

Four different groups are bonded to the central glycerol carbon atom in the optically active fat.

32.13

$$\begin{array}{l} \overset{O}{\overset{\|}{CH_2OC(CH_2)_7CH{=}CH(CH_2)_7CH_3}} \\ | \\ \overset{O}{\overset{\|}{CHOC(CH_2)_7CH{=}CH(CH_2)_7CH_3}} \\ | \\ \overset{O}{\overset{\|}{CH_2OC(CH_2)_7CH{=}CH(CH_2)_7CH_3}} \end{array}$$

glyceryl trioleate

a)

$$\text{glyceryl trioleate} \xrightarrow[CCl_4]{Br_2} \begin{array}{l} \overset{O}{\overset{\|}{CH_2OC(CH_2)_7CHBrCHBr(CH_2)_7CH_3}} \\ | \\ \overset{O}{\overset{\|}{CHOC(CH_2)_7CHBrCHBr(CH_2)_7CH_3}} \\ | \\ \overset{O}{\overset{\|}{CH_2OC(CH_2)_7CHBrCHBr(CH_2)_7CH_3}} \end{array}$$

b)

$$\text{glyceryl trioleate} \xrightarrow{H_2/Pd} \begin{array}{l} \overset{O}{\overset{\|}{CH_2OC(CH_2)_{16}CH_3}} \\ | \\ \overset{O}{\overset{\|}{CHOC(CH_2)_{16}CH_3}} \\ | \\ \overset{O}{\overset{\|}{CH_2OC(CH_2)_{16}CH_3}} \end{array}$$

c)

$$\text{glyceryl trioleate} \xrightarrow{NaOH,\ H_2O} \begin{array}{l} CH_2OH \\ | \\ CHOH \\ | \\ CH_2OH \end{array} + 3\ CH_3(CH_2)_7CH{=}CH(CH_2)_7COO^-\ Na^+$$

d)

$$\text{glyceryl trioleate} \xrightarrow[\text{2. Zn, } CH_3COOH]{\text{1. } O_3} \begin{array}{l} CH_2O\overset{O}{\overset{\|}{C}}(CH_2)_7\overset{O}{\overset{\|}{C}}H \\ | \\ CHO\overset{O}{\overset{\|}{C}}(CH_2)_7\overset{O}{\overset{\|}{C}}H \\ | \\ CH_2O\overset{O}{\overset{\|}{C}}(CH_2)_7\overset{O}{\overset{\|}{C}}H \end{array} + 3\ CH_3(CH_2)_7\overset{O}{\overset{\|}{C}}H$$

e)

$$\text{glyceryl trioleate} \xrightarrow[\text{2. } H_3O^+]{\text{1. } LiAlH_4} \begin{array}{l} CH_2OH \\ | \\ CHOH \\ | \\ CH_2OH \end{array} + 3\ CH_3(CH_2)_7CH{=}CH(CH_2)_7CH_2OH$$

f)

$$\text{glyceryl trioleate} \xrightarrow[\text{2. } H_3O^+]{\text{1. } CH_3MgBr} \begin{array}{l} CH_2OH \\ | \\ CHOH \\ | \\ CH_2OH \end{array} + 3\ CH_3(CH_2)_7CH{=}CH(CH_2)_7\overset{OH}{\overset{|}{C}}(CH_3)_2$$

g)

$$\text{glyceryl trioleate} \xrightarrow[\text{2. } CH_2N_2]{\text{1. NaOH, } H_2O} \begin{array}{l} CH_2OH \\ | \\ CHOH \\ | \\ CH_2OH \end{array} + 3\ CH_3(CH_2)_7CH{=}CH(CH_2)_7\overset{O}{\overset{\|}{C}}OCH_3$$

32.14

$$\text{vaccenic acid: } \begin{array}{ccc} CH_3(CH_2)_4CH_2 & & CH_2(CH_2)_8COOH \\ & C{=}C & \\ H & & H \end{array}$$

$$\text{vaccenic acid} \xrightarrow[\text{2. Zn, } CH_3COOH]{\text{1. } O_3} \underset{\text{heptanal}}{CH_3(CH_2)_5\overset{O}{\overset{\|}{C}}H} + \underset{\text{11-oxoundecanoic acid}}{H\overset{O}{\overset{\|}{C}}(CH_2)_9COOH}$$

$$\text{vaccenic acid} \xrightarrow[\text{Zn/Cu}]{CH_2I_2} \begin{array}{ccc} CH_3(CH_2)_4CH_2 & & CH_2(CH_2)_8COOH \\ H\diagdown C & \text{—} & C\diagup H \\ & \diagdown CH_2 \diagup & \end{array}$$

lactobacillic acid

32.15

$$cis\text{-}CH_3(CH_2)_7CH{=}CH(CH_2)_7COOH$$

oleic acid

a) oleic acid $\xrightarrow{CH_2N_2 \textit{ or } CH_3OH,\ H^+}$ $CH_3(CH_2)_7CH{=}CH(CH_2)_7COOCH_3$

methyl oleate

b) methyl oleate (part a) $\xrightarrow[Pd/C]{H_2}$ $CH_3(CH_2)_{16}COOCH_3$

methyl stearate

c) oleic acid $\xrightarrow[\text{2. Zn, } CH_3COOH]{\text{1. } O_3}$ $\underset{\text{nonanal}}{CH_3(CH_2)_7CHO}$ + $\underset{\text{9-oxononanoic acid}}{OHC(CH_2)_7COOH}$

d) 9-oxononanoic acid (part c) $\xrightarrow[NH_4OH]{Ag_2O}$ $HOOC(CH_2)_7COOH$ nonanedioic acid

e) oleic acid $\xrightarrow[CCl_4]{Br_2}$ $CH_3(CH_2)_7CHBrCHBr(CH_2)_7COOH$

$\downarrow$ 1. 3 $NaNH_2/NH_3$ 2. H_3O^+

$CH_3(CH_2)_7C{\equiv}C(CH_2)_7COOH$
stearolic acid

f) oleic acid $\xrightarrow[Pd/C]{H_2}$ $CH_3(CH_2)_{15}CH_2COOH$ (stearic acid) $\xrightarrow[2.\ H_2O]{1.\ Br_2,\ PBr_3}$ $CH_3(CH_2)_{15}CHBrCOOH$ (2-bromostearic acid)

g) methyl stearate (part b) $\xrightarrow[2.\ H_3O^+]{1.\ DIBAH}$ $CH_3(CH_2)_{16}CHO$

stearic acid $\xrightarrow[Br_2]{HgO}$ $CH_3(CH_2)_{15}CH_2Br$ $\xrightarrow{Mg}$ $CH_3(CH_2)_{15}CH_2MgBr$

$CH_3(CH_2)_{16}CHO$ $\xrightarrow[2.\ H_3O^+]{1.\ CH_3(CH_2)_{15}CH_2MgBr}$ $CH_3(CH_2)_{16}CH(OH)(CH_2)_{16}CH_3$

$\downarrow$ Jones' (CrO_3, H_2O, H_2SO_4)

$CH_3(CH_2)_{16}C(=O)(CH_2)_{16}CH_3$
18-pentatriacontanone

an alternate method:

2 $CH_3(CH_2)_{16}COOCH_3$ (methyl stearate (part b)) $\xrightarrow[2.\ H_3O^+]{1.\ NaOCH_3}$ $CH_3(CH_2)_{16}C(=O)CH(COOCH_3)(CH_2)_{15}CH_3$ + $HOCH_3$

$\downarrow$ H_3O^+, Δ

$CH_3(CH_2)_{16}C(=O)CH_2(CH_2)_{15}CH_3$ + CO_2 + CH_3OH

This synthesis uses a Claisen condensation, followed by a β-ketoester decarboxylation.

32.16

$CH_3(CH_2)_7C{\equiv}CH$ $\xrightarrow{NaNH_2}$ $CH_3(CH_2)_7C{\equiv}C{:}^-\ Na^+$ + NH_3

$\downarrow$ $I{-}CH_2(CH_2)_5CH_2Cl$

$CH_3(CH_2)_7C{\equiv}C(CH_2)_6CH_2CN$ $\xleftarrow{NaCN}$ $CH_3(CH_2)_7C{\equiv}C(CH_2)_6CH_2Cl$ + NaI

$\downarrow$ H_3O^+

$CH_3(CH_2)_7C{\equiv}C(CH_2)_7COOH$
stearolic acid

S_N2 displacement by acetylide occurs at iodine, rather than at chlorine, because $:I^-$ is a better leaving group than $:Cl^-$.

32.17 - 32.19

Only a fraction of the possible stereoisomers of these compounds are found in nature or can be synthesized. Some stereoisomers have highly strained ring fusions; others contain high energy 1,3-diaxial interactions.

a)

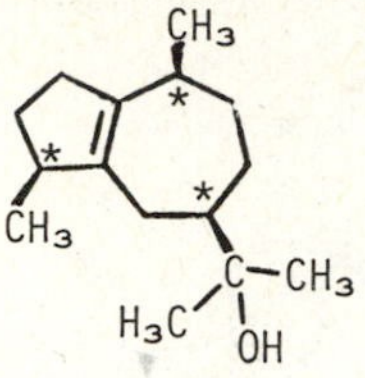

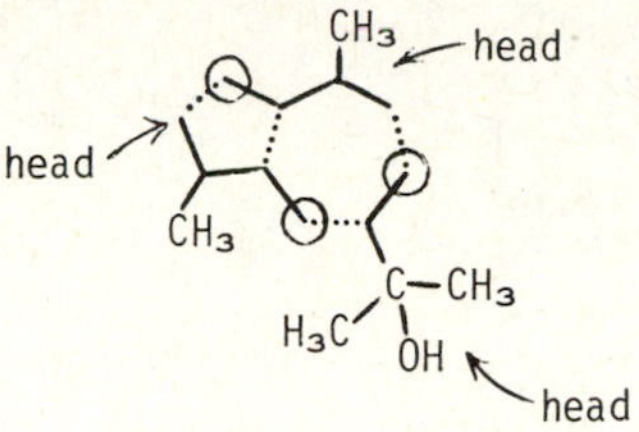

CH_3 head head CH_3 $C-CH_3$ H_3C OH head

guaiol
(8 possible stereoisomers)

b)

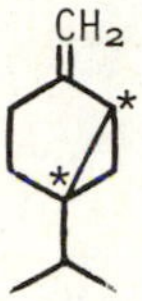

CH_2 head head

sabinene
(4 possible stereoisomers)

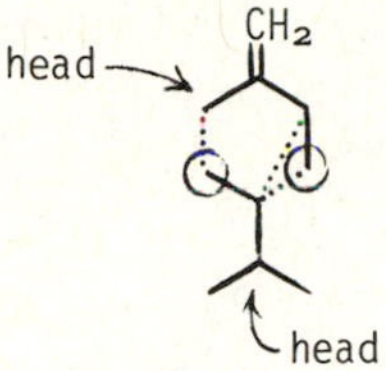

c)

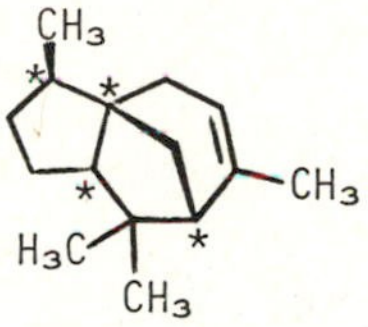

head CH_3 head CH_3 H_3C CH_3 head

cedrene
(16 possible stereoisomers)

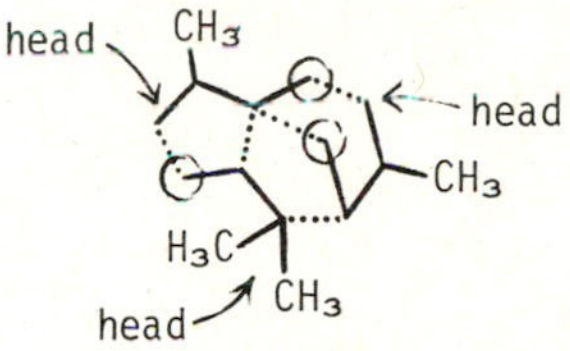

If carbon 1 of each pyrophospate were isotopically labeled, the labels would appear at the circled positions of the terpenes.

32.20

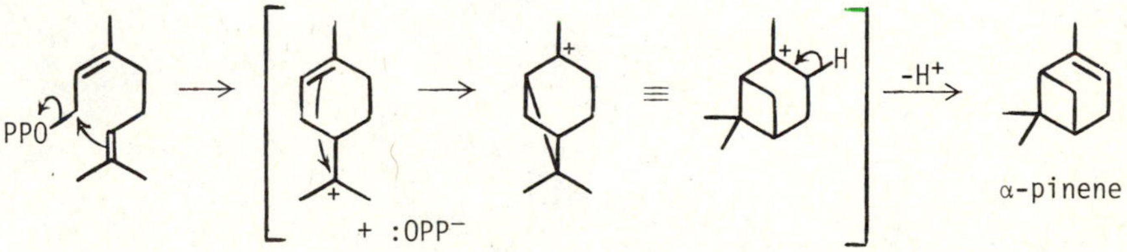

PPO → [+ :OPP⁻ → ≡] $\xrightarrow{-H^+}$ α-pinene

32.21

$\xrightarrow{H^+}$ [→ H] $\downarrow -H^+$

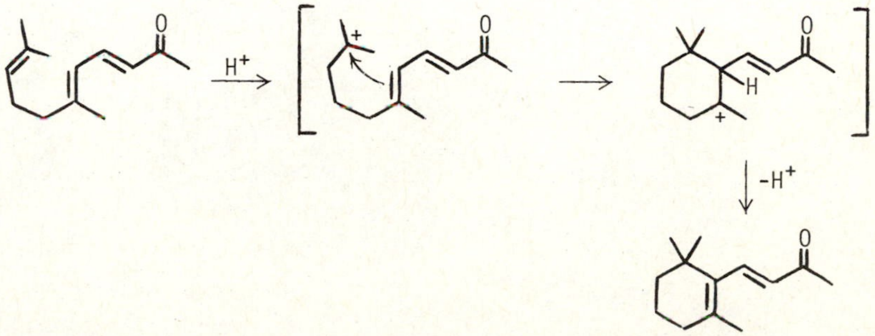

32.22

trans-decalin *cis*-decalin

Three 1,3-diaxial interactions cause *cis*-decalin to be less stable than *trans*-decalin.

32.23

32.24

32.25 - 32.26

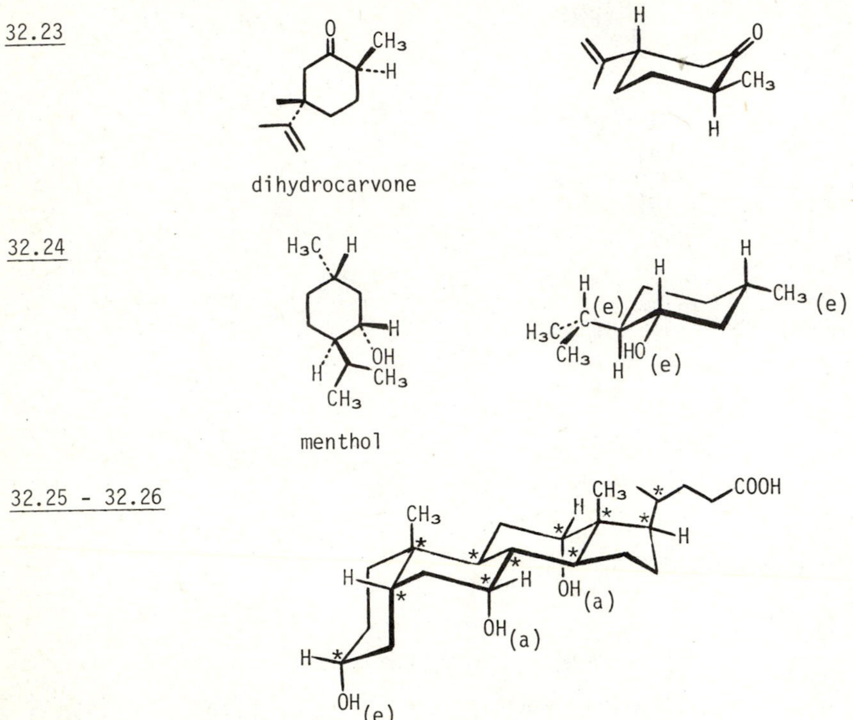

Cholic acid has eleven chiral centers and $2^{11} = 2048$ possible stereoisomers (some of these are very strained).

32.27 a)

cholic acid $\xrightarrow[H^+]{C_2H_5OH}$

b)

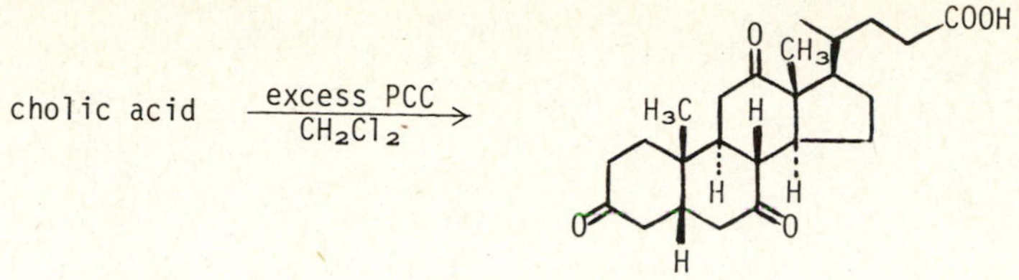

cholic acid $\xrightarrow[CH_2Cl_2]{\text{excess PCC}}$

COOH, O, CH_3, H_3C, H, H, H, O, O, H

c)

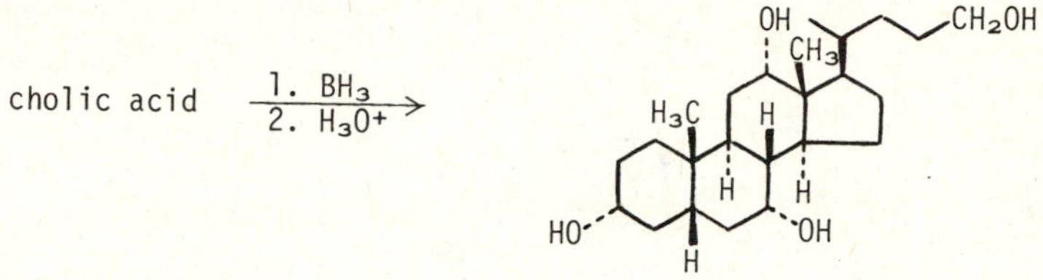

cholic acid $\xrightarrow[2.\ H_3O^+]{1.\ BH_3}$

OH, CH_2OH, CH_3, H_3C, H, H, H, HO, OH, H

32.28 a)

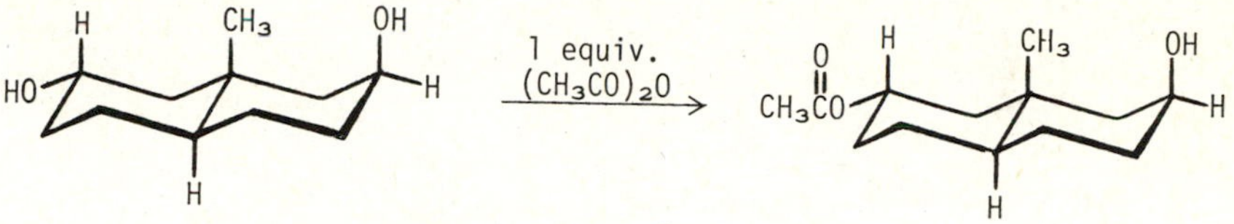

H, CH_3, OH, HO, H, H $\xrightarrow[(CH_3CO)_2O]{\text{1 equiv.}}$ $CH_3\overset{O}{\overset{\|}{C}}O$, H, CH_3, OH, H, H

b)

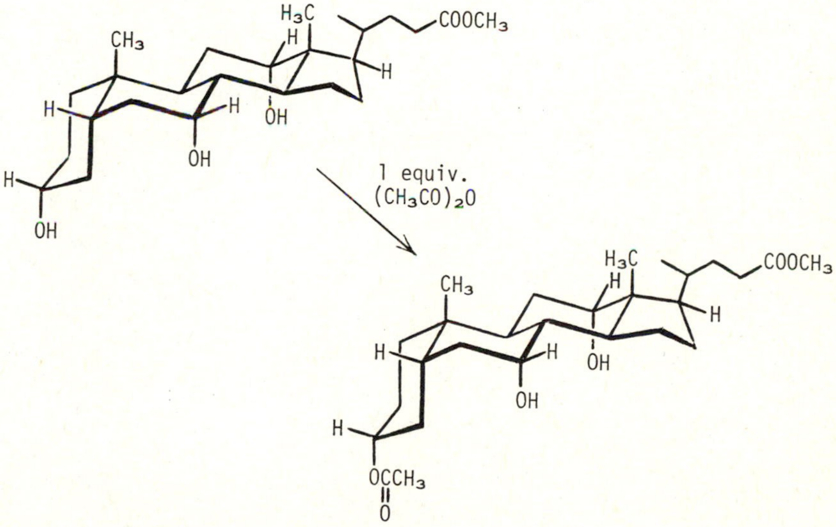

H_3C, $COOCH_3$, CH_3, H, H, H, H, OH, OH, H, OH

1 equiv.
$(CH_3CO)_2O$

H_3C, $COOCH_3$, CH_3, H, H, H, H, OH, OH, H, $O\underset{\underset{O}{\|}}{C}CH_3$

32.29

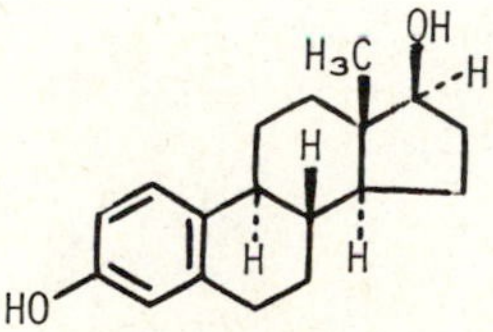

OH, H_3C, H, H, H, H, HO

estradiol

CH_3, OH, HO, H_3C

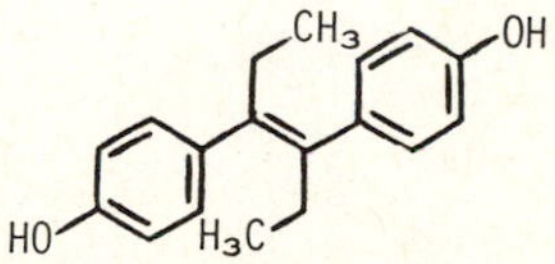

diethylstilbestrol

32.30

HO–C6H5 → (1. K_2CO_3; 2. CH_3Br) → CH_3O–C6H5 → ($CH_3CH_2\overset{O}{\overset{\|}{C}}Cl$, $AlCl_3$) → CH_3O–C6H4–$\overset{O}{\overset{\|}{C}}CH_2CH_3$

→ (1. $NaBH_4$; 2. H_3O^+) → CH_3O–C6H4–$\overset{OH}{\overset{|}{C}}HCH_2CH_3$

→ (PBr_3) → CH_3O–C6H4–$\overset{Br}{\overset{|}{C}}HCH_2CH_3$

→ (Mg) → CH_3O–C6H4–$\overset{MgBr}{\overset{|}{C}}HCH_2CH_3$

→ ($CH_3CH_2\overset{O}{\overset{\|}{C}}$–C6H4–$OCH_3$) → CH_3O–C6H4–C(H)(CH_3CH_2)–C(HO)(CH_2CH_3)–C6H4–OCH_3

→ (HI) → CH_3O–C6H4–C(CH_3CH_2)=C(CH_2CH_3)–C6H4–OCH_3

diethylstilbestrol

32.31

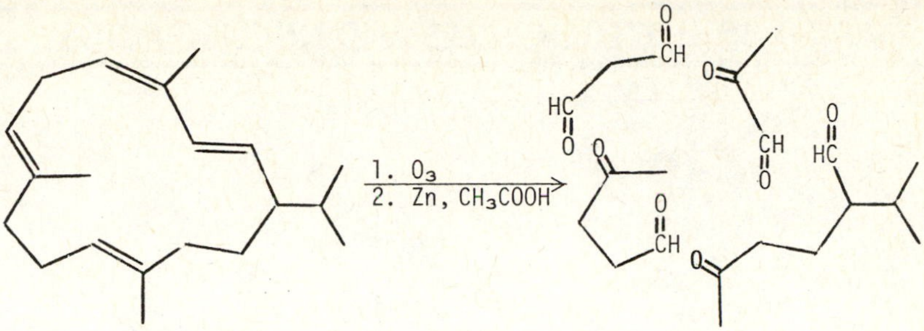

cembrene

1. O_3
2. Zn, CH_3COOH

1 equiv. H_2

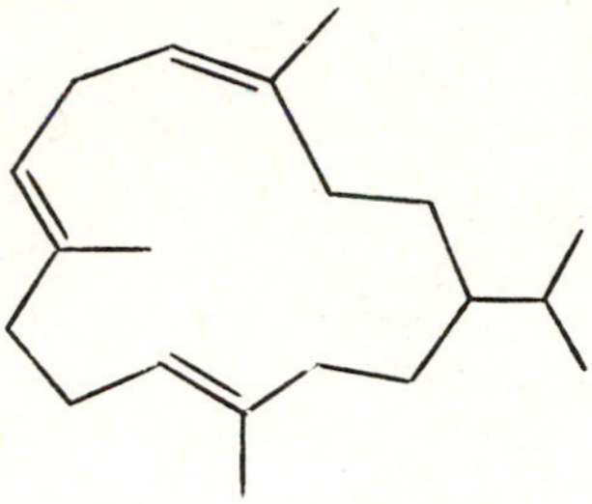

dihydrocembrene

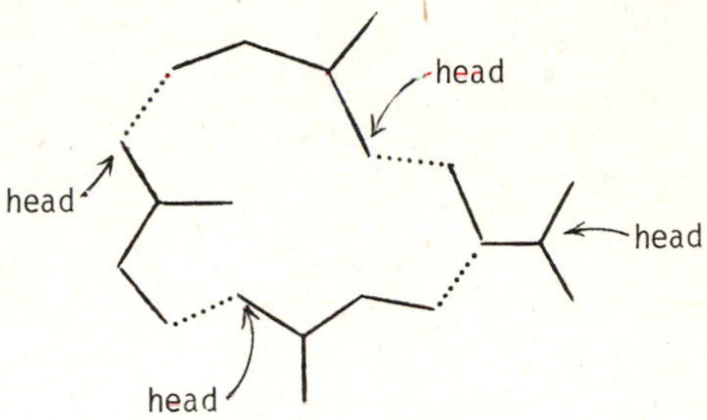

head

head

head

head

One equivalent of H_2 hydrogenates the least substituted double bond. Dihydrocembrene exhibits no ultraviolet absorption because it is not conjugated.

What you should know:

After doing these problems, you should be able to:

1. Deduce the structure of a fat;
2. Predict the products of reactions of fats;
3. Understand the principles of fatty acid, terpene, and steroid biosynthesis;
4. Locate the isoprene units in a terpene;
5. Draw the structures and conformations of steroids and other fused-ring systems.

33.1

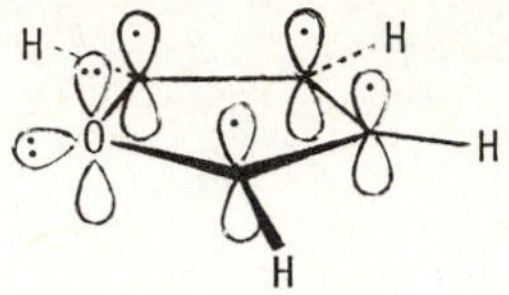

furan

One oxygen lone pair is in a *p* orbital that is part of the pi electron system of furan. The other oxygen lone pair is in an sp^2 orbital that lies in the plane of the furan ring.

33.2

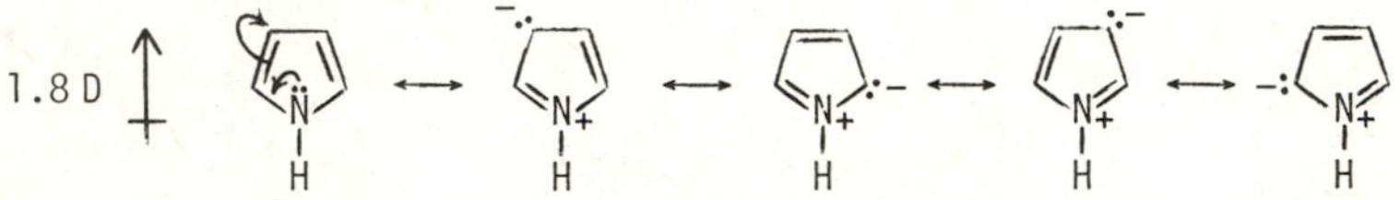

1.8 D

The dipole moment of pyrrole points in the direction indicated because the ring carbons are more electron-rich than nitrogen.

33.3

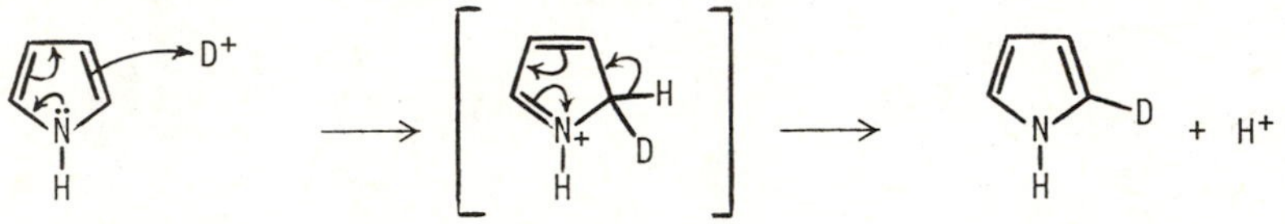

D^+

$+ H^+$

2-deuteriopyrrole

33.4

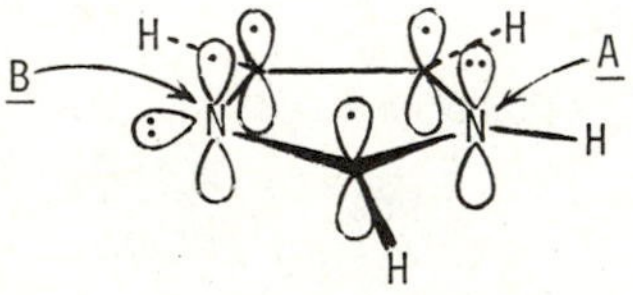

imidazole

Nitrogen atom B is more basic because its lone pair of electrons lies in an sp^2 orbital and is more available for donation to a Lewis acid than the lone pair of electrons of nitrogen A, which is part of the ring pi system.

33.5 Attack at C2:

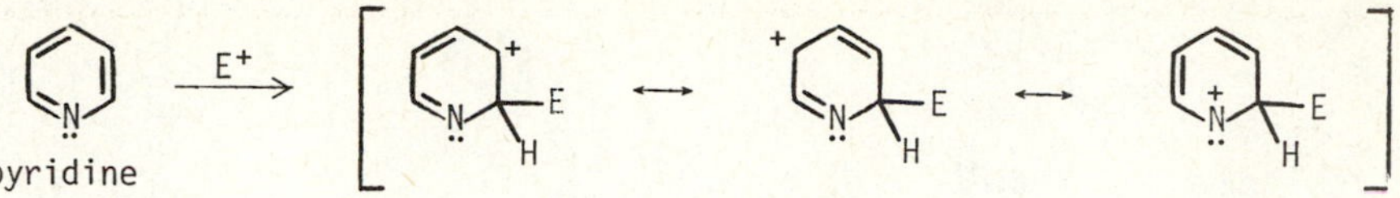

pyridine

Attack at C3:

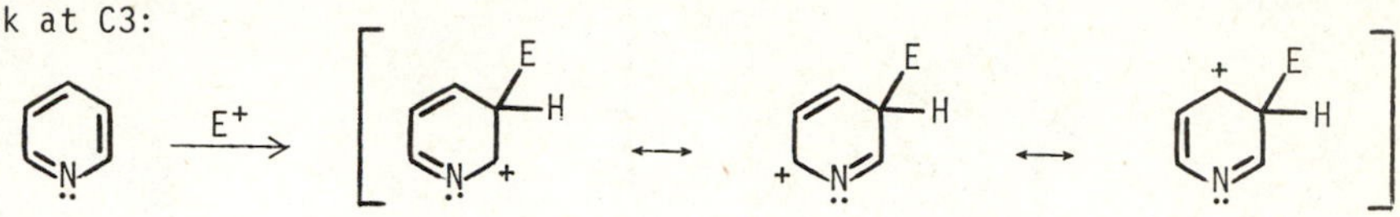

Attack at C4:

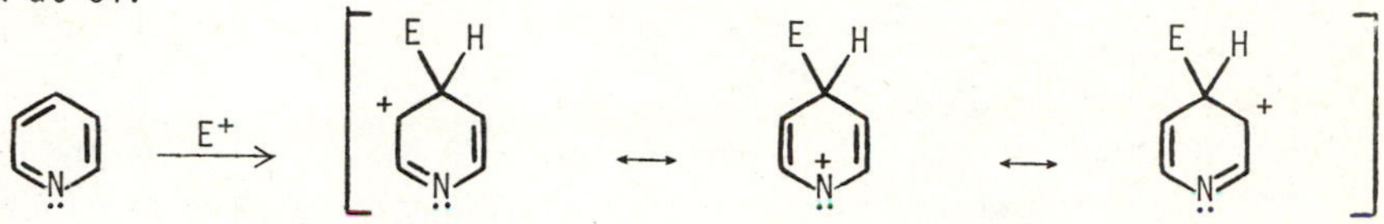

C3 attack is favored over C2 or C4 attack. The positive charge of the cationic intermediate of C3 attack is delocalized onto three carbon atoms, rather than onto two carbons and the electronegative pyridine nitrogen. Electrophilc attack at nitrogen is also possible.

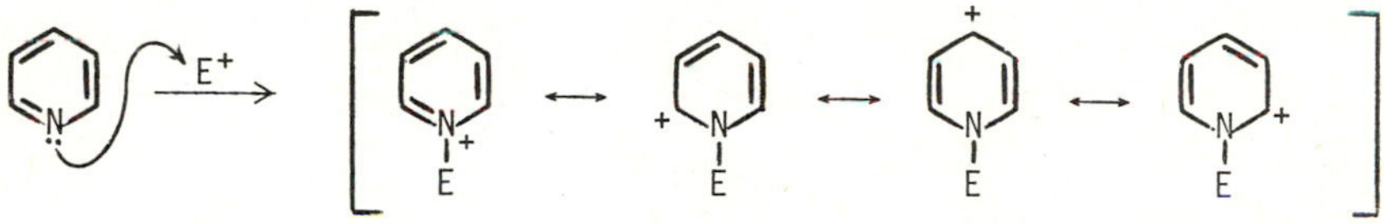

33.6 Attack at C3:

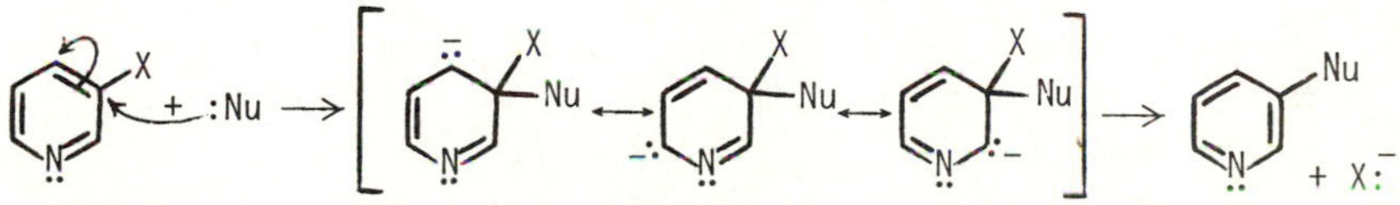

Attack at C4:

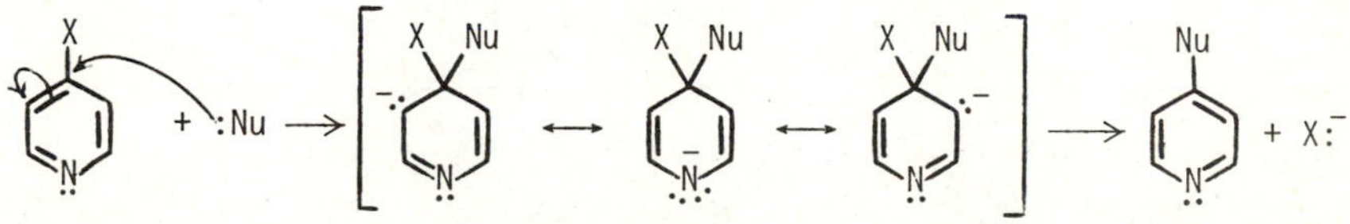

The negative charge resulting from C4 attack can be stabilized by nitrogen. Since no such stabilization is possible for C3 attack, reaction at C3 does not occur.

33.7 3-bromopyridine + NH_3 4-amino-pyridine 3-amino-pyridine

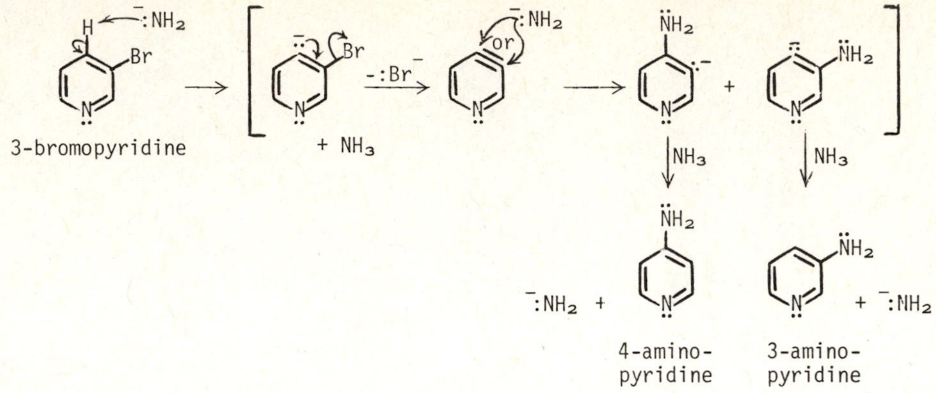

Reaction of 3-bromopyridine with $NaNH_2$ occurs by a benzyne mechanism. Since $^{-}:NH_2$ can add to either end of the triple bond of the benzyne intermediate, a mixture of products is formed.

33.8 $CH_2CH_2\ddot{N}(CH_3)_2$

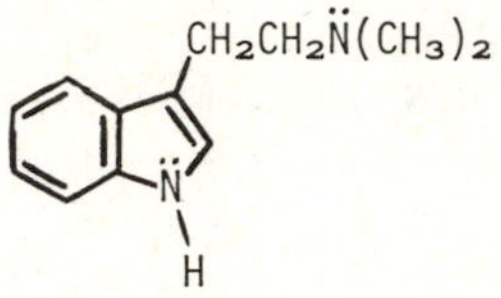

The aliphatic nitrogen atom of *N*,*N*-dimethyltryptamine is more basic because its lone electron pair is more available for donation to a Lewis acid. The aromatic nitrogen electron lone pair is part of the ring pi electron system.

33.9 C2 attack:

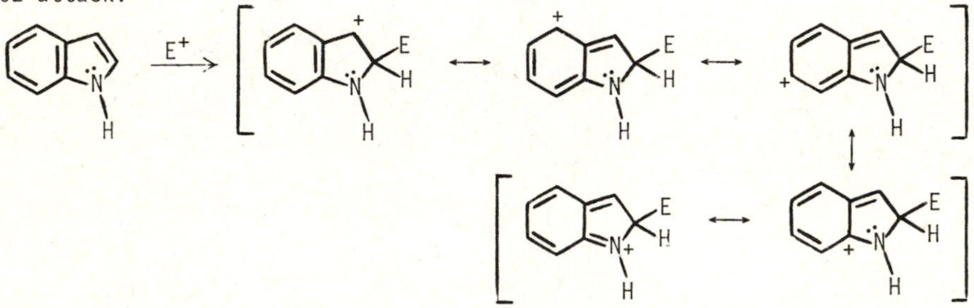

C3 attack:

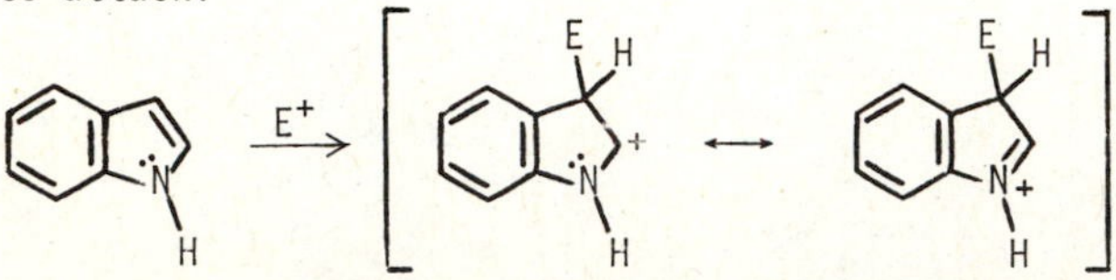

Positive charge can be stabilized by the nitrogen lone pair electrons in both C2 and C3 attack. In C2 attack, however, stabilization by nitrogen also destroys the aromaticity of the fused benzene ring. Reaction at C3 is favored, even though the cationic intermediate has fewer resonance forms, because the aromaticity of the six-membered ring is preserved in the most favored resonance form.

33.10

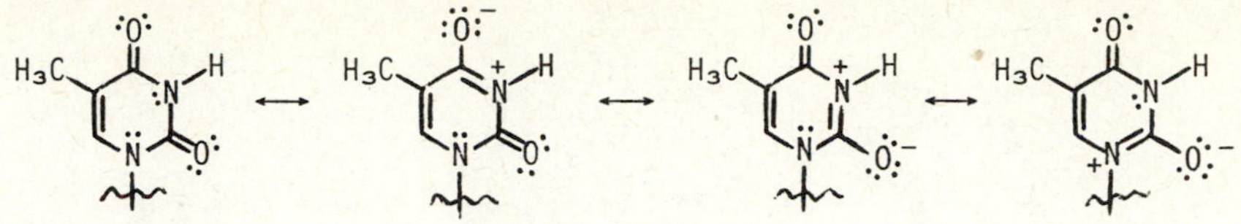

lactam

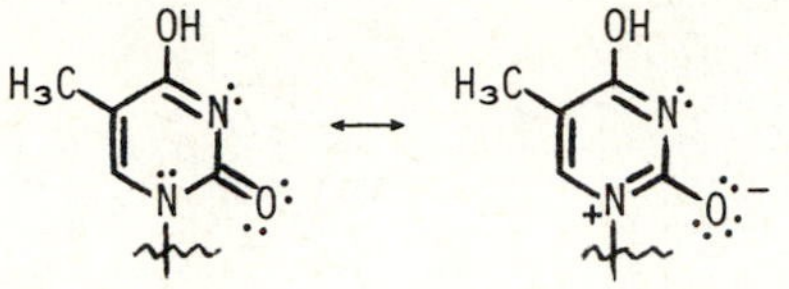

lactim

The lactam form of 2'-deoxythimidine has greater resonance stabilization than the lactim form.

33.11 DNA G-G-C-T-A-A-T-C-C-G-T is complementary to
DNA C-C-G-A-T-T-A-G-G-C-A

33.12

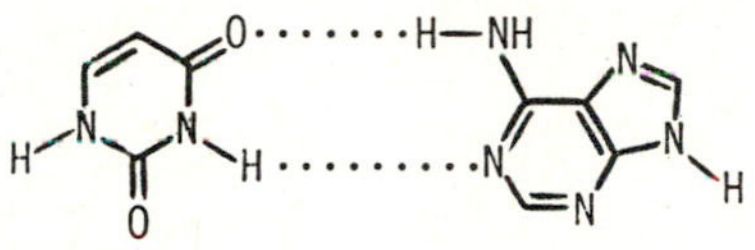

uracil adenine

33.13 DNA G-A-T-T-A-C-C-G-T-A is complementary to
RNA C-U-A-A-U-G-G-C-A-U

33.14 - 33.15

The mRNA base sequence:	CUU-AUG-GCU-UGG-CCC-UAA
The amino acid sequence:	Leu-Met-Ala-Trp-Pro-OH (stop)
The tRNA sequence:	GAA UAC CGA ACC GGG AUU
The DNA sequence:	GAA-TAC-CGA-ACC-GGG-ATT

33.16 Remember:

1. Only a few of the many possible splittings occur in each reaction.
2. Cleavage occurs at both sides of the reacting nucleotide.
3. In this problem, only the major products are given for A > G and G > A.

^{32}P-A-A-C-A-T-G-G-C-G-C-T-T-A-T-G-A-C-G-A

reaction	fragments
a) A > G	^{32}P ^{32}P-A ^{32}P-A-A-C ^{32}P-A-A-C-A-T-G-G-C-G-C-T-T ^{32}P-A-A-C-A-T-G-G-C-G-C-T-T-A-T-G ^{32}P-A-A-C-A-T-G-G-C-G-C-T-T-A-T-G-A-C-G ^{32}P-A-A-C-A-T-G-G-C-G-C-T-T-A-T-G-A-C-G-A

reaction	fragments
b) G > A	^{32}P-A-A-C-A-T ^{32}P-A-A-C-A-T-G ^{32}P-A-A-C-A-T-G-G-C ^{32}P-A-A-C-A-T-G-G-C-G-C-T-T-A-T ^{32}P-A-A-C-A-T-G-G-C-G-C-T-T-A-T-G-A-C ^{32}P-A-A-C-A-T-G-G-C-G-C-T-T-A-T-G-A-C-G-A
c) C	^{32}P-A-A ^{32}P-A-A-C-A-T-G-G ^{32}P-A-A-C-A-T-G-G-C-G ^{32}P-A-A-C-A-T-G-G-C-G-C-T-T-A-T-G-A ^{32}P-A-A-C-A-T-G-G-C-G-C-T-T-A-T-G-A-C-G-A
d) C + T	^{32}P-A-A ^{32}P-A-A-C-A ^{32}P-A-A-C-A-T-G-G ^{32}P-A-A-C-A-T-G-G-C-G ^{32}P-A-A-C-A-T-G-G-C-G-C ^{32}P-A-A-C-A-T-G-G-C-G-C-T ^{32}P-A-A-C-A-T-G-G-C-G-C-T-T-A ^{32}P-A-A-C-A-T-G-G-C-G-C-T-T-A-T-G-A ^{32}P-A-A-C-A-T-G-G-C-G-C-T-T-A-T-G-A-C-G-A

33.17 X-T-C-A-G-T-A-C-C-G-A-T-T-C-G-G-T-A-C

33.18

OCH_3 CH_3O C Polymer CO O H CH_2O: + $ZnBr_2$ O H Base → [OCH_3 CH_3O C Polymer CO O H $CH_2\overset{+}{O}$:$\overset{-}{Z}$nBr Br O H Base] ↓ [OCH_3 CH_3O C + + Polymer CO O H $CH_2O:ZnBr_2^-$ O H Base] $\xleftarrow{H_2O}$ OCH_3 CH_3O C OH + Polymer CO O H CH_2OH O H Base

Cleavage of DMT ethers proceeds by an S_N1 mechanism and is rapid because the DMT cation formed is unusually stable.

33.19 The pyrrole anion, $C_4H_4N:^-$, is a 6 pi electron species that is isoelectronic with the cyclopentadienyl anion. Both of these anions possess the stability of 6 pi electron systems.

33.20

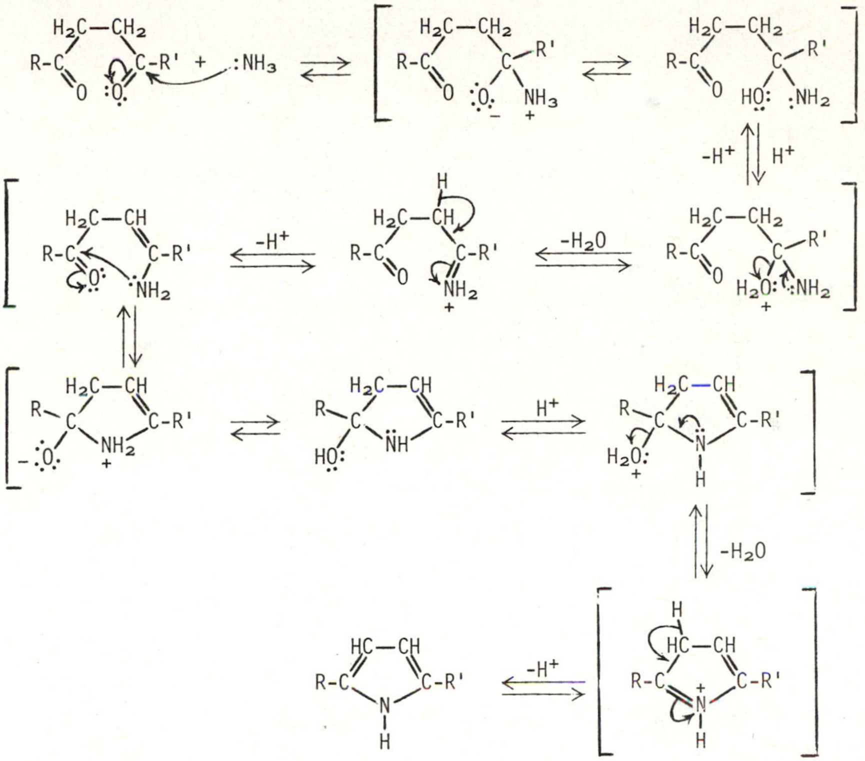

33.21

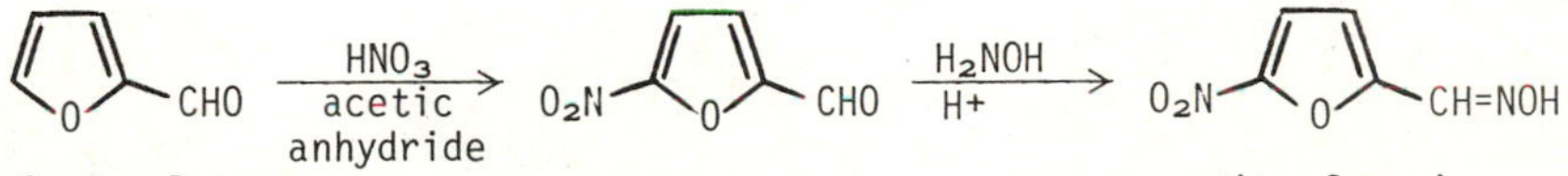

HNO$_3$, acetic anhydride; H$_2$NOH, H+

furfural → nitrofuroxime

33.22

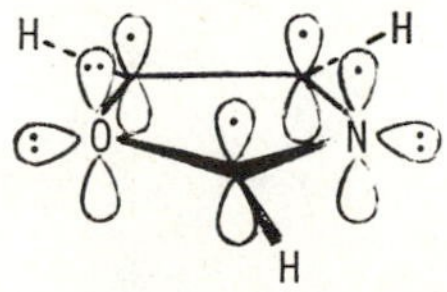

oxazole

Oxazole is an aromatic 6 pi electron heterocycle. Two oxygen electrons and one nitrogen electron are in *p* orbitals that are part of the pi electron system of the ring, along with one electron from each carbon. An oxygen lone pair and a nitrogen lone pair are in sp^2 orbitals that lie in the plane of the ring. Since the nitrogen lone pair is available for donation to acids, oxazole (pK_b = 12.7) is more basic than pyrrole (pK_b = 14).

33.23

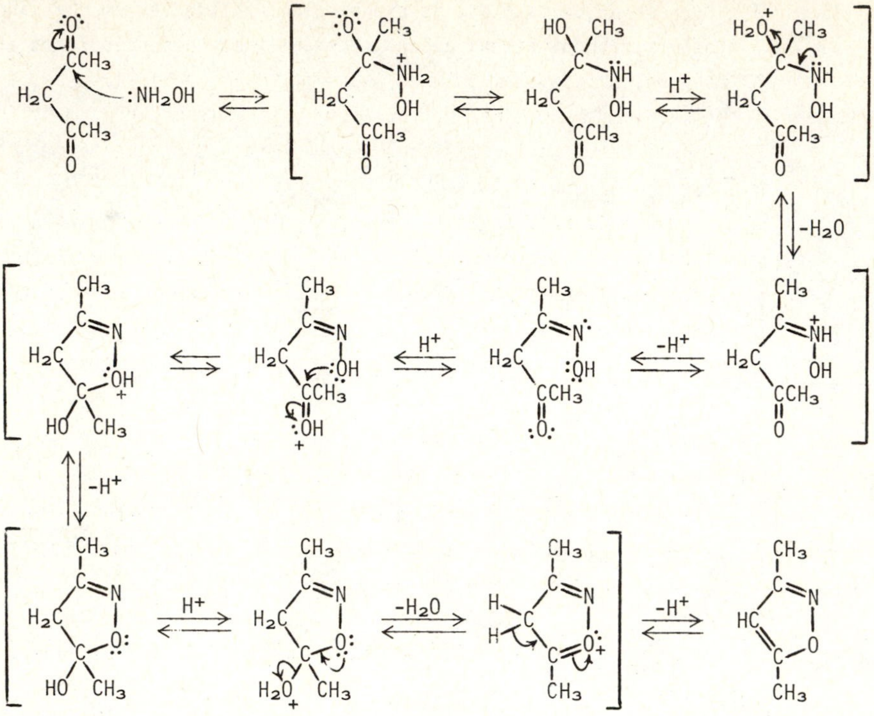

3,5-dimethyl-
isoxazole

33.24 a)

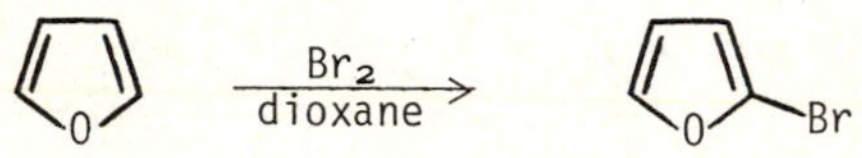

b)

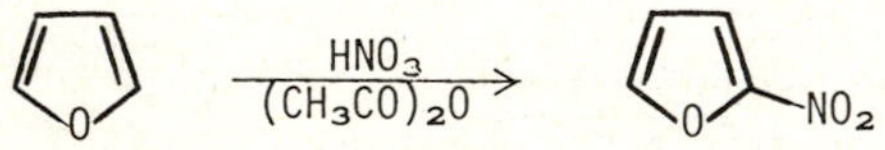

c)

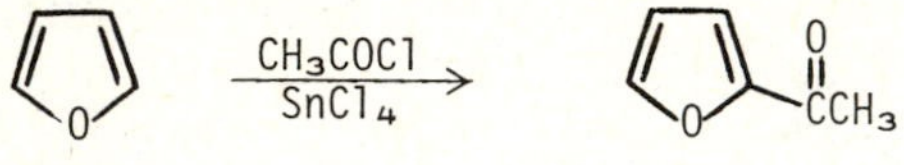

d)

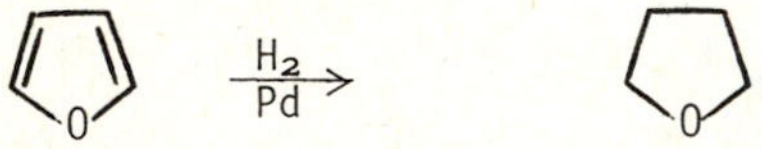

e)

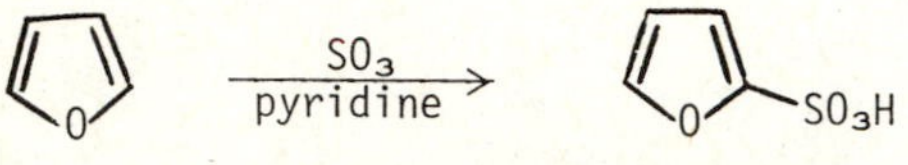

33.25 a) A *heterocycle* is a cyclic compound whose ring framework is composed of one or more different elements in addition to carbon.

b) *DNA* is a biological polymer whose monomeric units are nucleotides. A nucleotide is composed of a heterocyclic amine base, deoxyribose, and a phosphate group. DNA is the transmitter of the genetic code of all complex living organisms.

c) A *base pair* is a specific pair of heterocyclic amine bases that hydrogen-bond to each other in a DNA double helix and during protein synthesis.

d) *Transcription* is the process by which the genetic message contained in DNA is read by RNA and carried from the nucleus to the ribosomes.

e) *Translation* is the process by which RNA decodes the genetic message and uses the information to synthesize proteins.

f) *Replication* is the enzyme-catalyzed process by which a new molecule of DNA is produced. A DNA double helix separates into two strands, and nucleotide monomers line up and base-pair with each strand. Two new DNA double helices are produced, each one identical to the original DNA helix.

g) A *codon* is a sequence of three mRNA bases that specifies a particular amino acid to be used in protein synthesis.

h) An *anticodon* is a sequence of three tRNA bases that is complementary to the codon sequence. Each tRNA covalently bonds to a specific amino acid. The anticodon brings the amino acid into correct position for protein synthesis by hydrogen-bonding to its particular codon.

33.26 - 33.27

UAC is the codon for tyrosine. It was transcribed from ATG of the DNA chain.

mRNA codon DNA

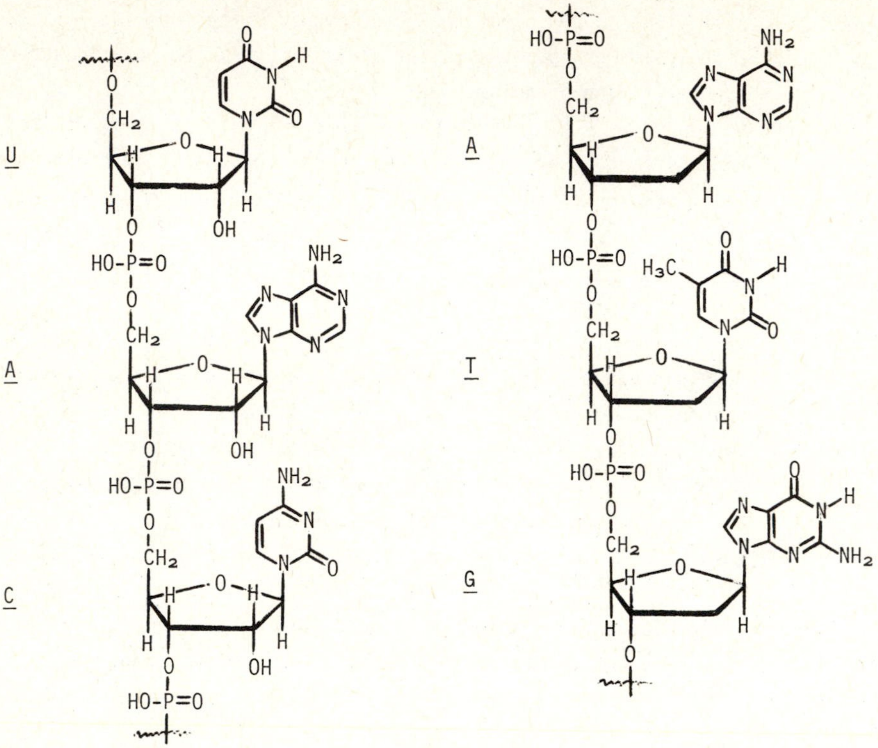

U A C A T G

33.28 H - Tyr - Gly - Gly - Phe - Met - OH is coded by

(UAC)	(GGU)	(GGU)	(UUU)	(AUG)	(UAA)
(UAU)	(GGC)	(GGC)	(UUC)		(UAG)
	(GGA)	(GGA)			(UGA)
	(GGG)	(GGG)			

33.29

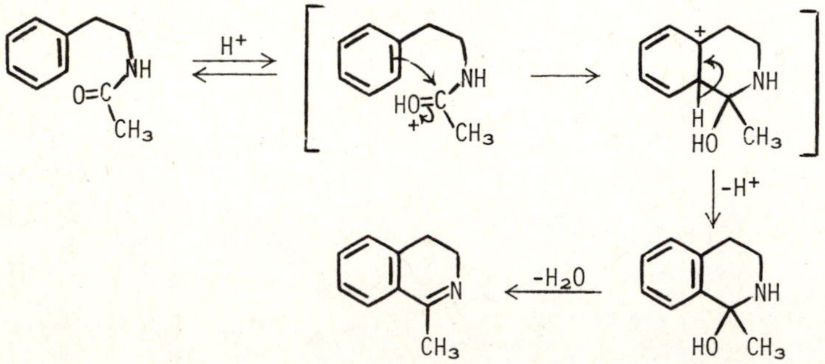

H^+ $-H^+$ $-H_2O$

The cyclization is an electrophilic aromatic substitution.

33.30

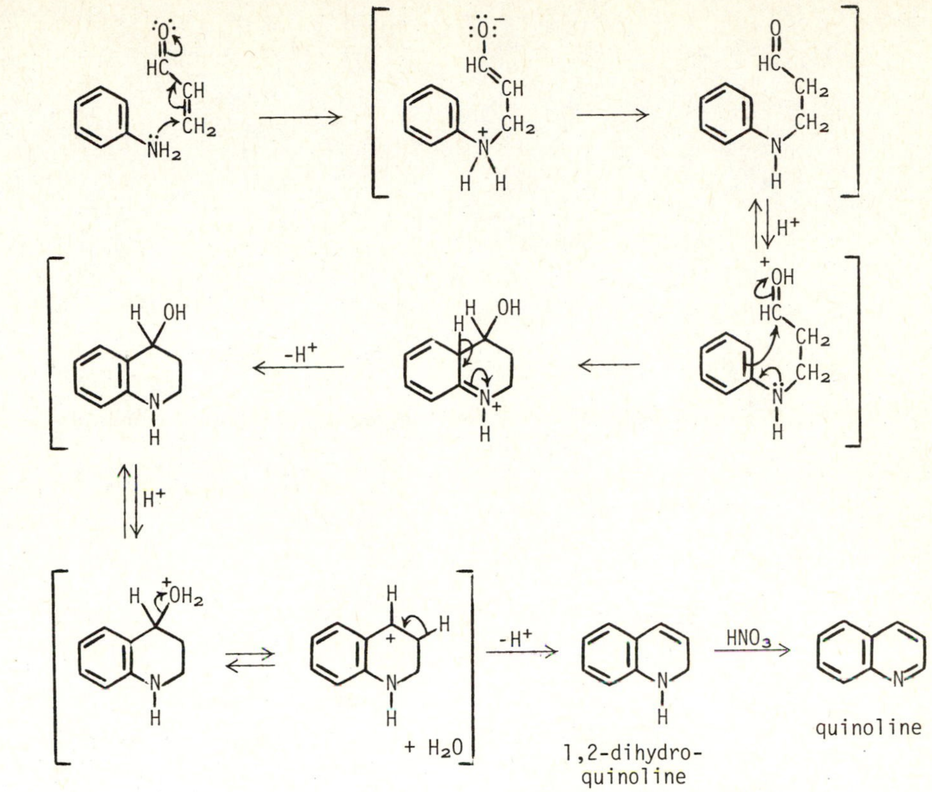

1,2-dihydro-quinoline

quinoline

Conjugate addition of aniline is followed by electrophilic aromatic substitution.

33.31 1. First, protect the nucleotides.

a) Bases are protected by amide formation.

adenine $\xrightarrow{C_6H_5COCl}$ (NHCOC$_6$H$_5$)

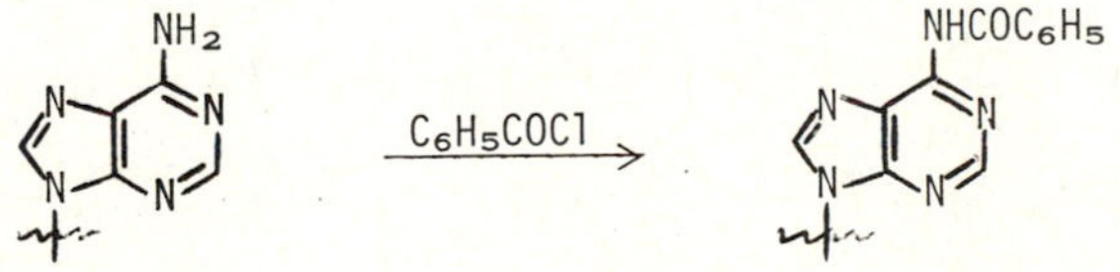

cytosine $\xrightarrow{C_6H_5COCl}$ (NHCOC$_6$H$_5$)

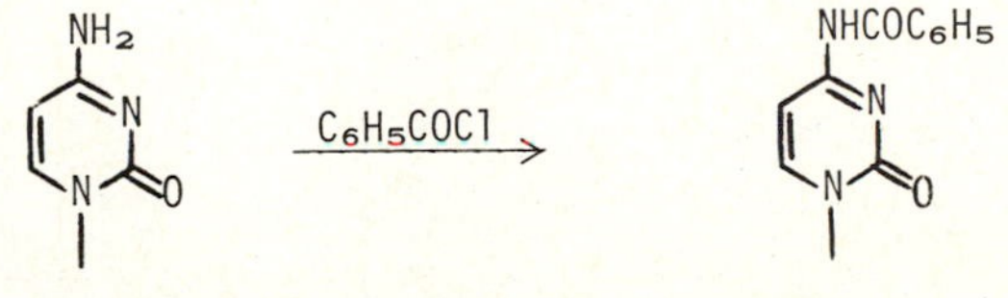

guanine $\xrightarrow{(CH_3)_2CHCOCl}$ (NHCOCH(CH$_3$)$_2$)

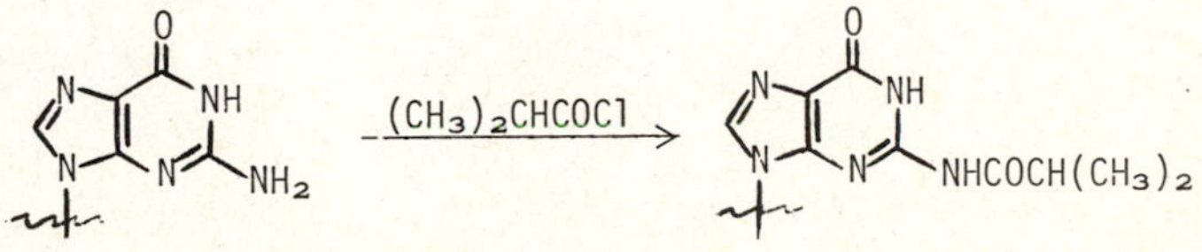

Thymine does not need to be protected.

b) The 5' hydroxyl group is protected as its *p*-dimethoxytrityl (DMT) ether.

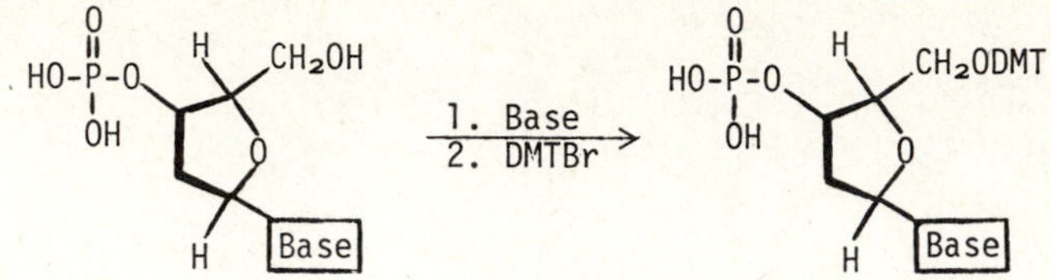

1. Base
2. DMTBr

c) The phosphate group is protected as its mono-*o*-chlorophenyl ester.

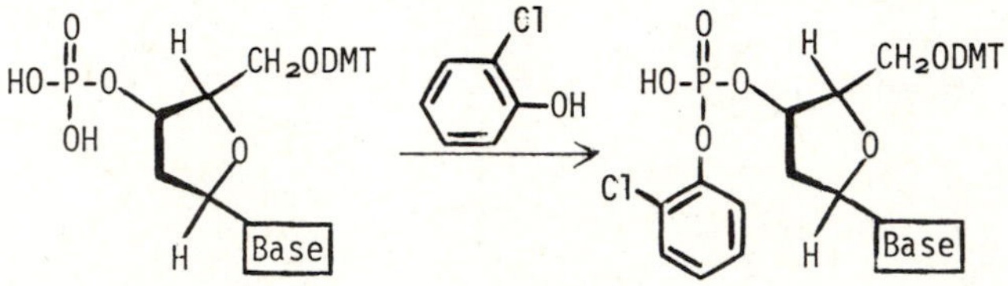

2. Attach a protected 2'-deoxycytidine nucleo*side* to the polymer carboxylic acid group.

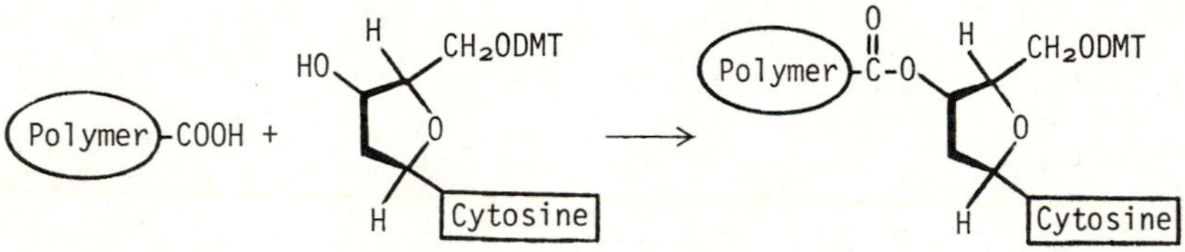

Cleave the DMT ether.

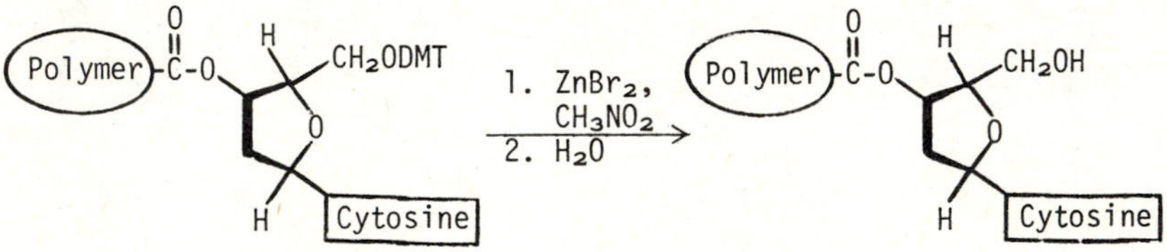

1. $ZnBr_2$, CH_3NO_2
2. H_2O

3. Couple the protected nucleotides to the polymer-2'-deoxycytidine, using TPST.
   a) Couple protected 2'-deoxythymidine to 2'-deoxycytidine, using TPST.

Polymer C(=O)–O–[2'-deoxycytidine, CH_2OH] + HO–P(=O)(O–C$_6$H$_4$Cl)–O–[2'-deoxythymidine, CH_2ODMT]

↓ TPST

Polymer C(=O)–O–[Cytosine]–CH_2–O–P(=O)(O–C$_6$H$_4$Cl)–O–[Thymine, CH_2ODMT]

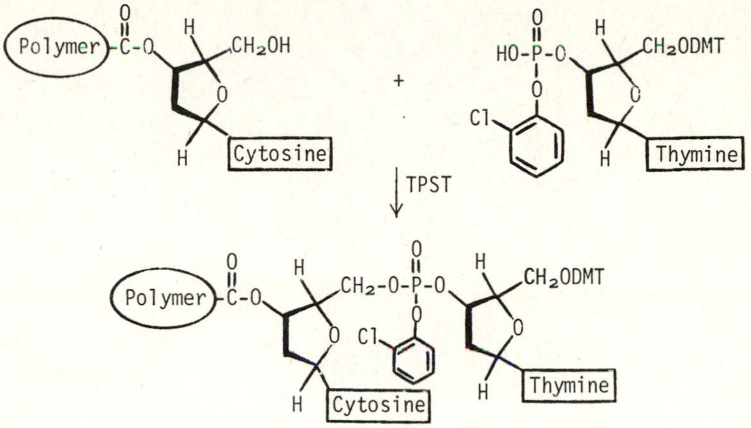

   Cleave the DMT ether.

   b) Couple protected 2'-deoxyadenosine to polymer-2'-deoxycytidine-2'-deoxythymidine using TPST, then cleave the DMT ether.
   c) Couple protected 2'-deoxyguanosine to polymer-trinucleotide.

4. Cleave all protecting groups.
   a) Cleave the DMT ether using $ZnBr_2$, H_2O.
   b) Cleave the chlorophenylphosphates, base protecting groups, and polymer ester linkage with NH_3, H_2O:

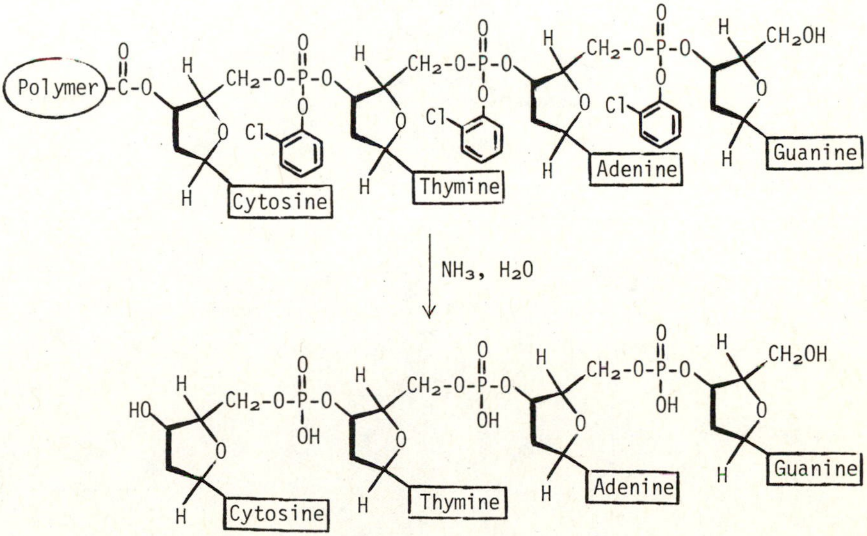

Polymer C(=O)–O–[Cytosine]–CH_2–O–P(=O)(O–C$_6$H$_4$Cl)–O–[Thymine]–CH_2–O–P(=O)(O–C$_6$H$_4$Cl)–O–[Adenine]–CH_2–O–P(=O)(O–C$_6$H$_4$Cl)–O–[Guanine, CH_2OH]

↓ NH_3, H_2O

HO–[Cytosine]–CH_2–O–P(=O)(OH)–O–[Thymine]–CH_2–O–P(=O)(OH)–O–[Adenine]–CH_2–O–P(=O)(OH)–O–[Guanine, CH_2OH]

33.32

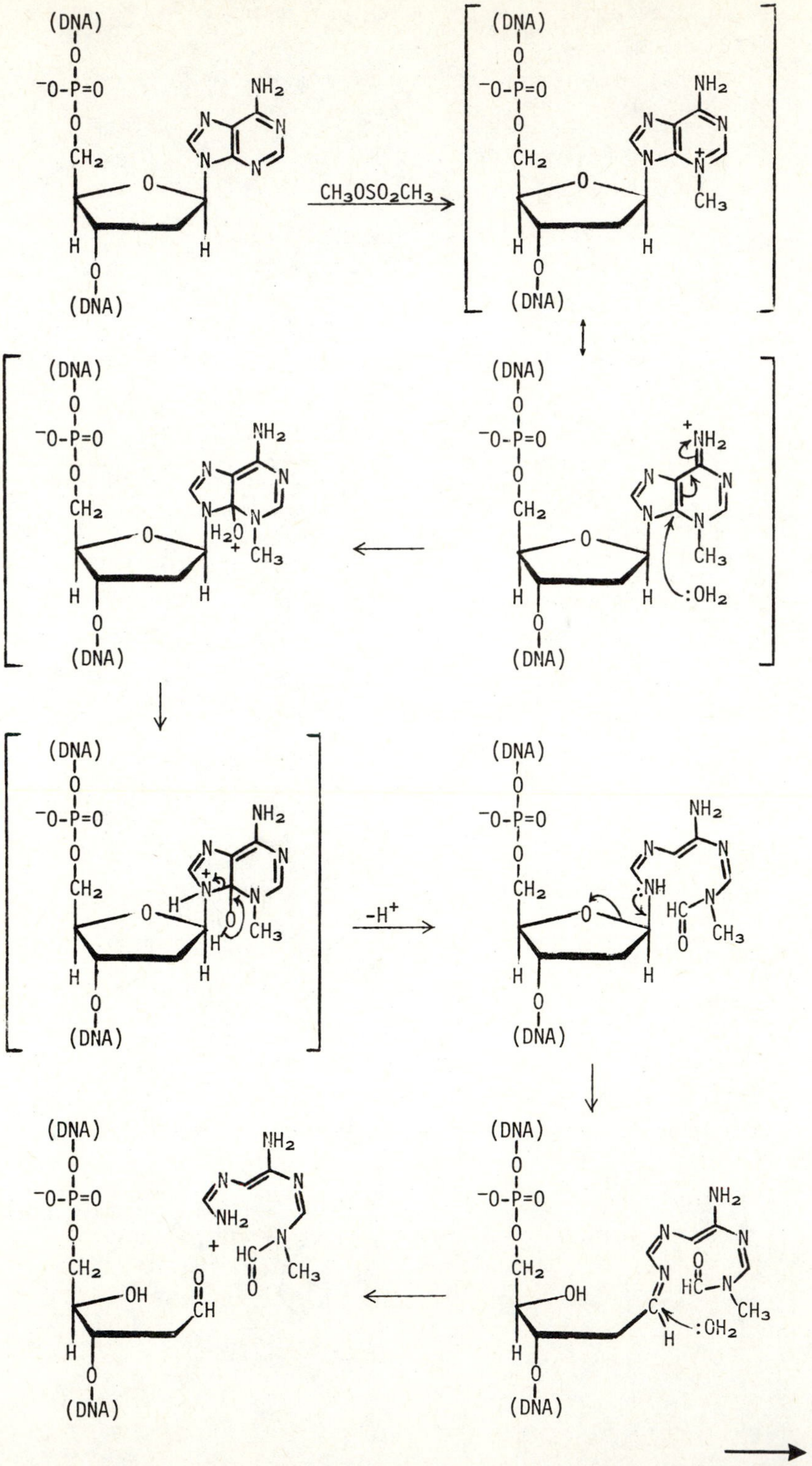

$CH_3OSO_2CH_3$

$-H^+$

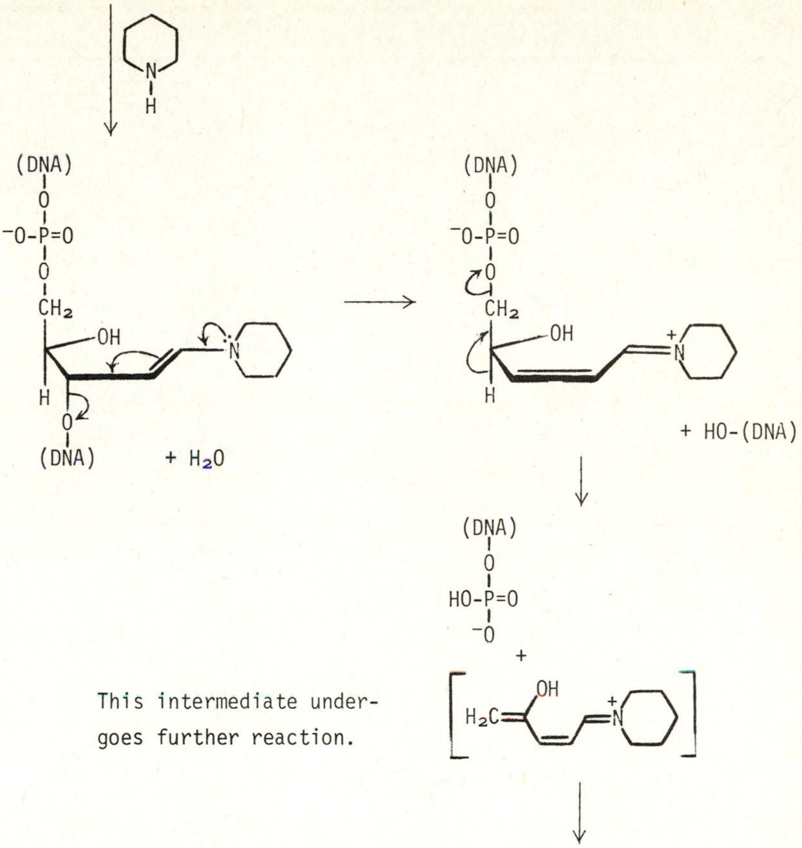

This intermediate undergoes further reaction.

<u>What you should know:</u>

After doing these problems, you should be able to:

1. Explain the electronic and acid-base properties of heterocyclic compounds;
2. Understand the chemical reactivity of heterocycles and predict the products of reactions involving heterocyclic compounds;
3. Know the structure of nucleosides and nucleotides and how they hydrogen-bond;
4. Understand the processes of DNA replication, transcription, translation, and protein synthesis;
5. Outline the method of laboratory DNA synthesis;
6. Outline the process of DNA sequencing and deduce the sequence of a nucleic acid fragment.

34.1 monomer polymer

a) CH_2=CHOCH_3 methyl vinyl ether — polymer: ~CH_2-CH(OCH_3)-CH_2-CH(OCH_3)-CH_2-CH(OCH_3)~

b) CH_2=(cyclohexadiene)=CH_2 — polymer: ~CH_2-C_6H_4-CH_2-CH_2-C_6H_4-CH_2~

c) CHCl=CHCl 1,2-dichloroethylene — polymer: ~CH(Cl)-CH(Cl)-CH(Cl)-CH(Cl)-CH(Cl)-CH(Cl)~

34.2 Addition of an initiator to a styrene double bond produces a radical that can be stabilized by the phenyl ring.

C_6H_5CH=CH_2 —In·→ $C_6H_5\dot{C}$HCH_2In ⟷ (ring radical) CHCH$_2$In etc.

If this radical adds to the CH_2= end of another styrene double bond, the new radical can also be stabilized by a phenyl ring.

$C_6H_5\dot{C}$HCH_2In + C_6H_5CH=CH_2 ⟶ $C_6H_5\dot{C}$HCH_2-CH(C_6H_5)CH_2In ⟷ (ring radical) etc.

If addition occurs at the PhCH= end of the double bond, the product radical cannot be stabilized. The product having phenyl groups on neighboring carbons is not formed.

$C_6H_5\dot{C}$HCH_2In + C_6H_5CH=CH_2 ⟶ ·CH_2CH(C_6H_5)-CH(C_6H_5)CH_2In phenyl groups on adjacent carbons

34.3 most reactive ⟶ least reactive

CH_2=CHC_6H_5 > CH_2=$CHCH_3$ > CH_2=CHCl > CH_2=$CHCO_2CH_3$

The alkenes most reactive to cationic polymerization contain electron-donating functional groups that can stabilize the carbocation intermediate. The reactivity order of substituents in cationic polymerization is similar to the reactivity order of substituted benzenes in electrophilic aromatic substitution.

34.4 most reactive ⟶ least reactive

$CH_2{=}CHCN > CH_2{=}CF_2 > CH_2{=}CHC_6H_5 > CH_2{=}CHCH_3$

Anionic polymerization occurs most readily with alkenes having electron-withdrawing groups.

34.5

$$HO^- + \overset{O}{CH_2\text{-}CH_2} \longrightarrow \left[HOCH_2CH_2\text{-}\ddot{O}:^- \xrightarrow{\overset{O}{CH_2\text{-}CH_2}} HOCH_2CH_2\text{-}O\text{-}CH_2CH_2\ddot{O}:^- \right]$$

repeat many times (with $\overset{O}{CH_2\text{-}CH_2}$) ↓

$$HOCH_2CH_2\text{-}O\text{-}(CH_2CH_2\text{-}O\text{-}CH_2CH_2\text{-}O)\text{-}$$

34.6

$$n\ CH_2{=}CCl_2 \xrightarrow{In\cdot} \text{-}\overset{H}{\underset{H}{C}}\text{—}\overset{Cl}{\underset{Cl}{C}}\text{—}\overset{H}{\underset{H}{C}}\text{—}\overset{Cl}{\underset{Cl}{C}}\text{—}\overset{H}{\underset{H}{C}}\text{—}\overset{Cl}{\underset{Cl}{C}}\text{-}$$

Vinylidene chloride does not polymerize in isotactic, syndiotactic, or atactic forms because no asymmetric centers are formed during polymerization.

34.7 None of the polypropylenes rotate plane-polarized light. If an optically inactive reagent and an achiral compound react, the product must be optically inactive. For every chiral center generated, an enantiomeric chiral center is also generated, and the resulting polymer mixture is racemic.

34.8 A vinyl branch in a diene polymer is the result of an occasional 1,2-double bond addition to the polymer chain.

$$\text{-}CH_2CH{=}CH\dot{C}H_2 + \underset{CH=CH_2}{CH{=}CH_2} \longrightarrow \text{-}CH_2CH{=}CHCH_2\dot{C}H_2\underset{CH=CH_2}{CH} \xrightarrow{\text{etc.}}$$

34.9

$$\underset{\text{-}CH_2}{\overset{H_3C}{}}C{=}C\underset{CH_2\text{-}}{\overset{H}{}} \xrightarrow{O_3} \underset{\text{-}CH_2}{\overset{H_3C}{}}C{=}O + O{=}C\underset{CH_2\text{-}}{\overset{H}{}}$$

Ozone causes oxidative cleavage of the double bonds in rubber and breaks the polymer chain.

34.10

$$CH_2{=}\overset{CH_3}{C}CH{=}CH_2 + \underset{CH_3}{\overset{CH_3}{C}}{=}CH_2 \xrightarrow{H^+} \text{-}CH_2\overset{CH_3}{C}{=}CHCH_2\underset{CH_3}{\overset{CH_3}{C}}CH_2CH_2\overset{CH_3}{C}{=}CHCH_2\underset{CH_3}{\overset{CH_3}{C}}CH_2\text{-}$$

2-methyl-1,3-butadiene   2-methyl-propene

34.11 $\sim CH_2CH{=}CHCH_2CH_2CH{=}CHCH_2 \sim \xrightarrow{h\nu} \sim CH_2CH{=}CH\dot{C}HCH_2CH{=}CHCH_2 \sim$

$\downarrow C_6H_5CH{=}CH_2$

$\sim CH_2CH{=}CHCHCH_2CH{=}CHCH_2 \sim$ with side chain $CH_2\dot{C}HC_6H_5$

$\xleftarrow[\text{repeat many times}]{C_6H_5CH=CH_2}$

$\sim CH_2CH{=}CHCHCH_2CH{=}CHCH_2 \sim$ with side chain $(CH_2CHC_6H_5)_n CH_2\dot{C}HC_6H_5$

Irradiation homolytically cleaves an allylic C-H bond because it has the lowest bond energy. The resulting radical adds to styrene to produce a polystyrene graft.

34.12 Either aqueous acid or aqueous base hydrolyze the amide bonds of nylon. The bonds of orlon and polyethylene are inert to acid and to base.

acid $\sim (CH_2)_5NHC(=O)(CH_2)_5 \sim \xrightarrow{H_3O^+} \left[\sim (CH_2)_5NHC(OH)(\overset{+}{O}H_2)(CH_2)_5 \sim\right] \longrightarrow \sim (CH_2)_5\overset{+}{N}H_3 + HOOC(CH_2)_5 \sim$

base $\sim (CH_2)_5NHC(=O)(CH_2)_5 \sim \xrightarrow{^-OH} \left[\sim (CH_2)_5NHC(O^-)(OH)(CH_2)_5 \sim\right] \xrightarrow{H_2O} \sim (CH_2)_5NH_2 + {}^-OOC(CH_2)_5 \sim$

34.13 a) $BrCH_2CH_2CH_2Br + HOCH_2CH_2CH_2OH \xrightarrow{\text{base}} \sim CH_2CH_2CH_2OCH_2CH_2CH_2O \sim$

b) $HOCH_2CH_2OH + HOC(=O)(CH_2)_6C(=O)OH \xrightarrow{H^+} \sim OCH_2CH_2OC(=O)(CH_2)_6C(=O)OCH_2CH_2OC(=O)(CH_2)_6C(=O) \sim$

c) $H_2N(CH_2)_6NH_2 + ClC(=O)(CH_2)_4C(=O)Cl \longrightarrow \sim NH(CH_2)_6NHC(=O)(CH_2)_4C(=O)NH(CH_2)_6NHC(=O)(CH_2)_4C(=O) \sim$

34.14 $HOC(=O){-}C_6H_4{-}C(=O)OH$ + $H_2N{-}C_6H_4{-}NH_2$

1,4-benzenedicarboxylic acid   1,4-benzenediamine

$\downarrow$

$\sim C(=O){-}C_6H_4{-}C(=O)NH{-}C_6H_4{-}NHC(=O){-}C_6H_4{-}C(=O)NH{-}C_6H_4{-}NH \sim$

Kevlar

34.15

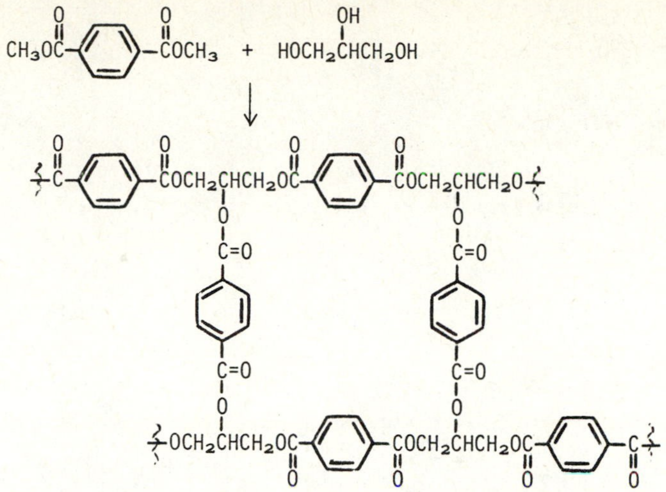

$CH_3OC(=O)\text{-}C_6H_4\text{-}C(=O)OCH_3 + HOCH_2CH(OH)CH_2OH \longrightarrow$ cross-linked polyester

The product of the reaction of dimethylterephthalate with glycerol has a high degree of cross-linking. This polymer is more rigid than Dacron.

34.16

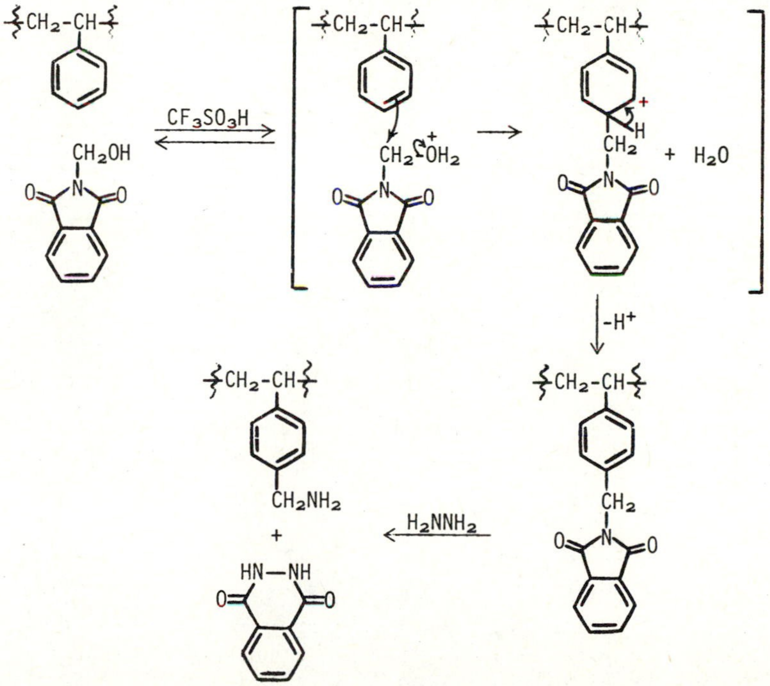

CF_3SO_3H $-H^+$ H_2NNH_2 $+ H_2O$

The mechanism for hydrolysis of the phthalimide is illustrated in problem 28.9.

34.17

natural rubber $\xrightarrow[Pd/C]{H_2}$ $-CH_2CH(CH_3)CH_2CH_2CH_2CH(CH_3)CH_2CH_2-$

The product of catalytic hydrogenation of natural rubber is atactic. This product also results from the radical copolymerization of propene with ethylene.

$$CH_3CH{=}CH_2 + CH_2{=}CH_2 \xrightarrow{In\cdot} -CH(CH_3)CH_2CH_2CH_2CH(CH_3)CH_2CH_2CH_2-$$

34.18

$$H_2C{=}O \underset{}{\overset{H^+}{\rightleftharpoons}} [H_2C{=}\overset{+}{O}H] \rightarrow \xrightarrow{-H^+} HOC_6H_4CH_2OH \overset{H^+}{\rightleftharpoons} [HOC_6H_4CH_2\overset{+}{O}H_2] \rightarrow [\ldots] + H_2O \xrightarrow{-H^+} HOC_6H_4CH_2C_6H_4OH \xrightarrow{} etc.$$

This product can react many times with additional formaldehyde and phenol to yield Bakelite. Reaction occurs at both *ortho* and *para* positions of phenol.

34.19

	polymer	monomer unit	type of polymer
a)	$-CH_2CH(NO_2)-CH_2CH(NO_2)-$	$CH_2{=}CH(NO_2)$	chain-growth
b)	$-CFCl-CF_2-CFCl-CF_2-$	$CFCl{=}CF_2$	chain-growth
c)	$-CH_2-O-CH_2-O-CH_2-O-$	$CH_2{=}O$	chain-growth
d)	$-O-C_6H_4-\overset{O}{\overset{\|}{C}}O-C_6H_4-\overset{O}{\overset{\|}{C}}-$	$HO-C_6H_4-\overset{O}{\overset{\|}{C}}OH$	step-growth
e)	$=CH(CH_2)_4CH{=}N-(CH_2)_6-N=$	$H_2N(CH_2)_6NH_2$ and $OHC(CH_2)_4CHO$	step-growth

34.20

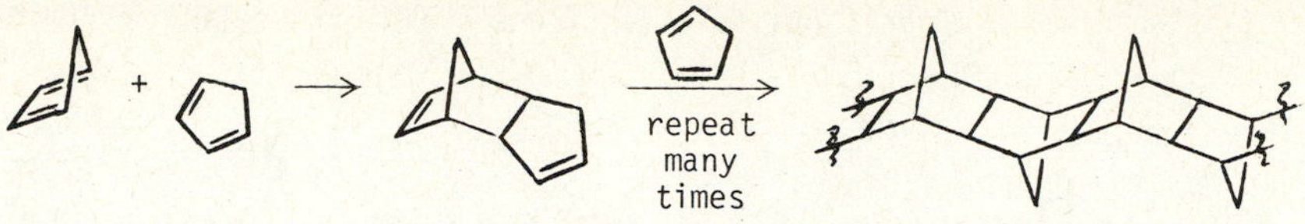

repeat many times

polycyclopentadiene

Polycyclopentadiene is the product of successive Diels-Alder additions of cyclopentadiene to a growing polymer chain. Strong heat causes depolymerization of the chain and reversion to cyclopentadiene monomer units.

34.21 a)

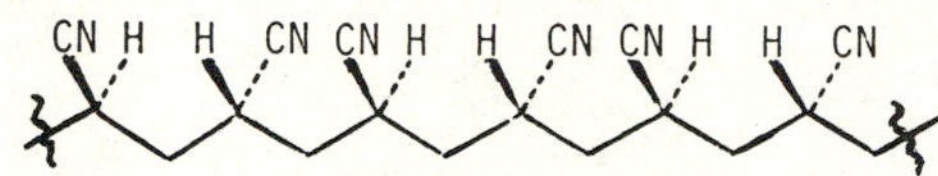

syndiotactic polyacrylonitrile

b)

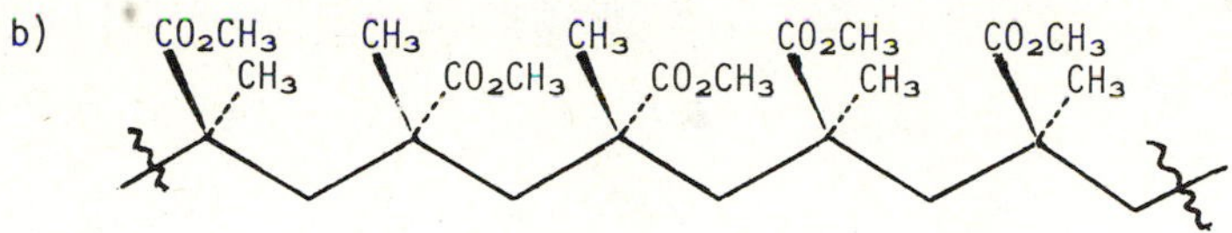

atactic poly(methyl methylacrylate)

c)

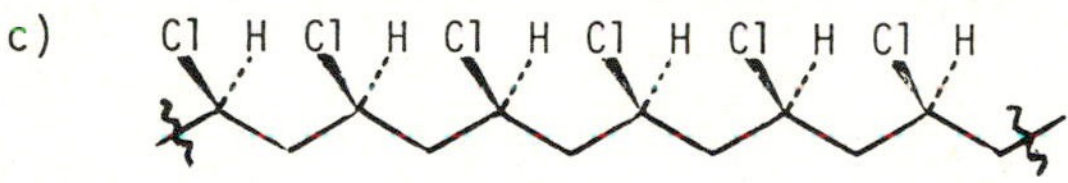

isotactic poly(vinyl chloride)

34.22 1. *p*-Divinylbenzene is incorporated into the growing polystyrene chain.

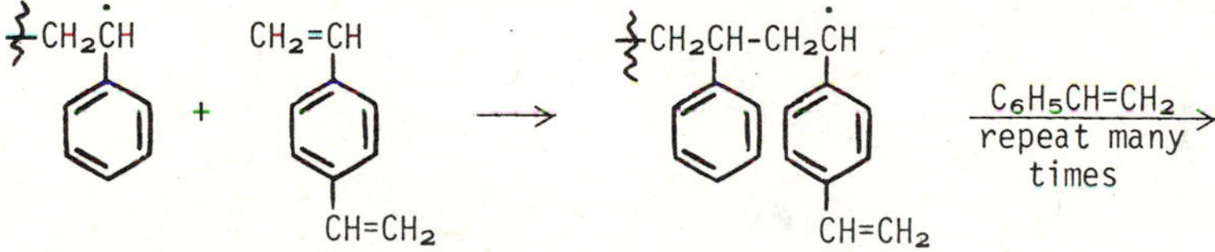

2. Another growing polystyrene chain reacts with the second double bond of *p*-divinylbenzene.

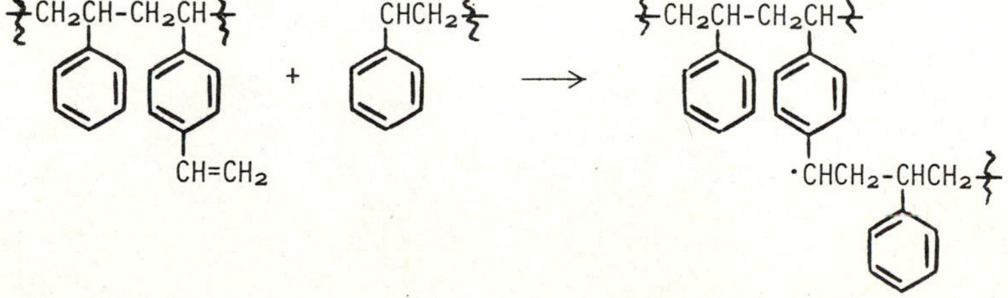

CH_2CH-CH_2CH $CH=CH_2$ + $\dot{C}HCH_2$ → CH_2CH-CH_2CH $\cdot CHCH_2-CHCH_2$

3. The final product contains polystyrene chains cross-linked by *p*-divinylbenzene units.

$$\text{-CH}_2\text{CH(C}_6\text{H}_5\text{)-CH}_2\text{CH(C}_6\text{H}_4\text{-}\dot{\text{C}}\text{HCH}_2\text{-CH(C}_6\text{H}_5\text{)CH}_2\text{-)-} \xrightarrow{C_6H_5CH=CH_2} \text{-CH}_2\text{CH(C}_6\text{H}_5\text{)-CH}_2\text{CH(C}_6\text{H}_4\text{-}\dot{\text{C}}\text{H(C}_6\text{H}_5\text{)CH}_2\text{-CHCH}_2\text{-CH(C}_6\text{H}_5\text{)CH}_2\text{-)-}$$

34.23

$$\underset{\text{1,3-butadiene}}{CH_2{=}CHCH{=}CH_2} \xrightarrow[\text{1,4-addition}]{Br_2,\ CCl_4} BrCH_2CH{=}CHCH_2Br \xrightarrow{2\ NaCN} NCCH_2CH{=}CHCH_2CN + 2\ NaBr \xrightarrow[Pd]{H_2} \underset{\text{1,6-hexanediamine}}{H_2NCH_2CH_2CH_2CH_2CH_2CH_2NH_2}$$

34.24 The white coating on the distillation flask is caused by the thermal polymerization of nitroethylene.

$$n\ CH_2{=}CH(NO_2) \xrightarrow{\text{heat}} \text{-}[CH_2CH(NO_2)]\text{-}$$

34.25

$$n\ CH_2{=}CH(OCCH_3{=}O) \xrightarrow{In\cdot} \text{-CH}_2\text{CH(OCOCH}_3\text{)-CH}_2\text{CH(OCOCH}_3\text{)-} \xrightarrow[H_2O]{OH^-} \text{-CH}_2\text{CH(OH)-CH}_2\text{CH(OH)-}$$

Poly(vinyl alcohol) is formed by radical polymerization of vinyl acetate, followed by hydrolysis of the acetate groups.

$$\text{-CH}_2\text{CH(OH)-CH}_2\text{CH(OH)-} + H\text{-}\overset{O}{\overset{\|}{C}}\text{-}n\text{-}C_3H_7 \xrightarrow{H^+} \text{-CH}_2\text{CH-CH}_2\text{CH-}\ (\text{O-C(H)}(n\text{-}C_3H_7)\text{-O ring})$$

Reaction of poly(vinyl alcohol) with *n*-butanal produces poly(vinyl butyral).

34.26 The polyamide pictured is a step-growth polymer of a benzene tetracarboxylic acid and an aromatic diamine.

$$\underset{\text{1,2,4,5-benzenetetracarboxylic acid}}{C_6H_2(COOH)_4} + \underset{\text{1,4-benzenediamine}}{H_2N\text{-}C_6H_4\text{-}NH_2} \xrightarrow{\text{heat}} \underset{\text{a polyamide}}{\text{-}[N(CO)_2C_6H_2(CO)_2N\text{-}C_6H_4]\text{-}}$$

1,2,4,5-benzenetetracarboxylic acid

1,4-benzenediamine

a polyamide

34.27

$$HO\overset{O}{\overset{\|}{C}}(CH_2)_6\overset{O}{\overset{\|}{C}}OH + H_2N\text{-}C_6H_{10}\text{-}CH_2\text{-}C_6H_{10}\text{-}NH_2 \longrightarrow \text{-}\!\!\!\{\overset{O}{\overset{\|}{C}}(CH_2)_6\overset{O}{\overset{\|}{C}}\text{-}NH\text{-}C_6H_{10}\text{-}CH_2\text{-}C_6H_{10}\text{-}NH\}\!\!\!\text{-}$$

Qiana

To synthesize the diamine:

$$2\ H_2N\text{-}C_6H_5 \xrightarrow[H^+]{CH_2O} H_2N\text{-}C_6H_4\text{-}CH_2\text{-}C_6H_4\text{-}NH_2 \xrightarrow[Pd/C]{H_2} H_2N\text{-}C_6H_{10}\text{-}CH_2\text{-}C_6H_{10}\text{-}NH_2$$

34.28

$$HO\!:^- + \text{(β-propiolactone)} \longrightarrow HO\overset{O}{\overset{\|}{C}}CH_2CH_2\ddot{\underset{..}{O}}\!:^- \xrightarrow{\text{β-propiolactone}} HO\overset{O}{\overset{\|}{C}}CH_2CH_2O\overset{O}{\overset{\|}{C}}CH_2CH_2\ddot{\underset{..}{O}}\!:^-$$

repeat many times

$$\text{-}\!\!\!\{\overset{O}{\overset{\|}{C}}CH_2CH_2O\overset{O}{\overset{\|}{C}}CH_2CH_2O\}\!\!\!\text{-}$$

This polymer is a polyester.

34.29

$$\text{phthalic anhydride} + HOCH_2\overset{OH}{\overset{|}{C}}HCH_2OH\ (\text{glycerol}) \longrightarrow \text{glyptal}$$

phthalic anhydride

glycerol

glyptal

This polymer and the one in problem 34.15 are closely related.

34.30 Melamine + CH_2O → [intermediate] → (+ melamine) → [intermediate dimer]; Melamine, CH_2O, repeat many times → Melmac

H2N NH2 N N N NH2 + CH2O → [H2N NHCH2OH N N N NH2]
H2N NH2 N N N NH2
Melamine

[H2N N NH-CH2-NH N NH2 N N N N NH2 NH2]
Melamine, CH_2O
repeat many times

Melmac

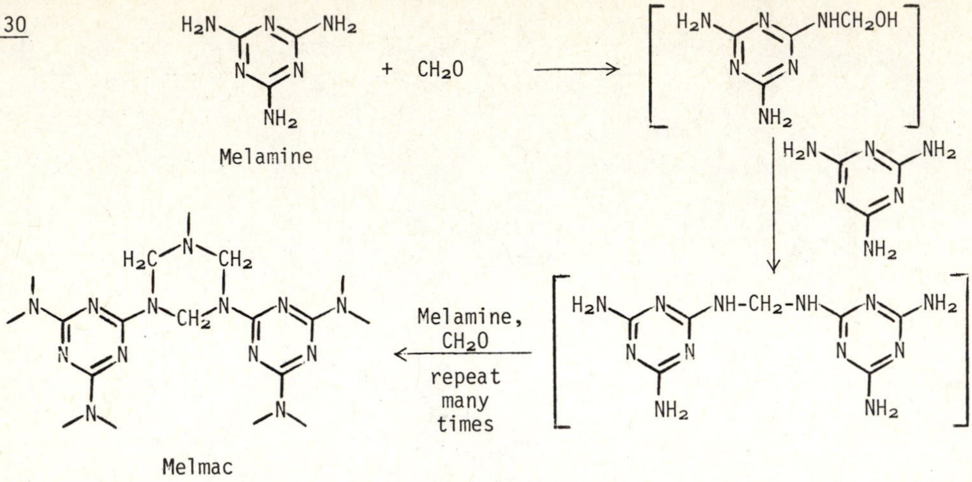

34.31

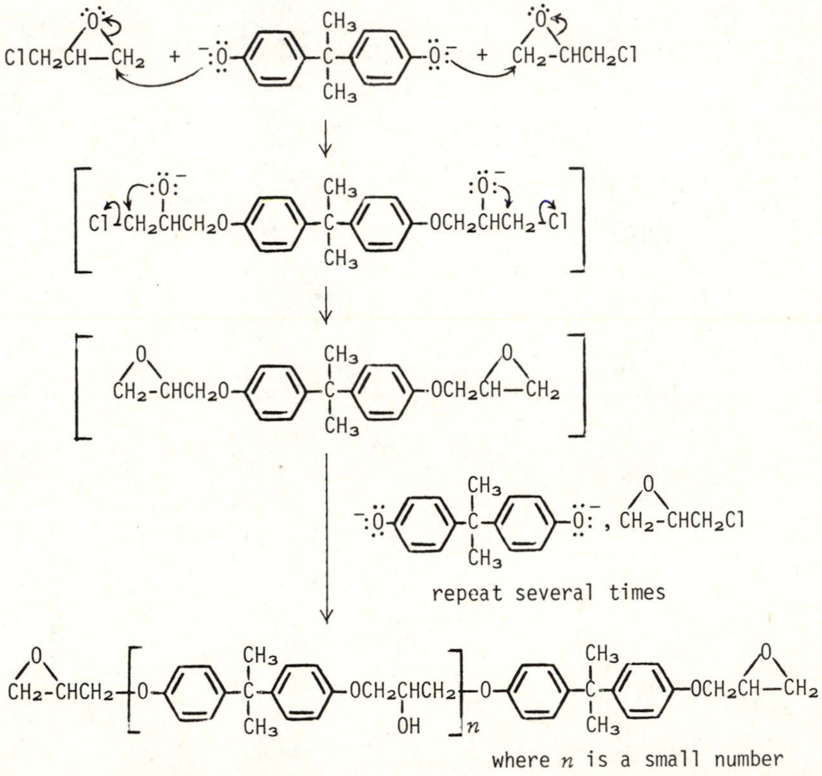

$ClCH_2CH—CH_2$ (epoxide) + $^-:\ddot{O}$–C6H4–C(CH_3)2–C6H4–$\ddot{O}:^-$ + CH_2-$CHCH_2Cl$ (epoxide)

↓

[Cl-CH_2CHCH_2O(–$:\ddot{O}:^-$)–C6H4–C(CH_3)2–C6H4–OCH_2CHCH_2(–$:\ddot{O}:^-$)-Cl]

↓

[CH_2-$CHCH_2O$(epoxide)–C6H4–C(CH_3)2–C6H4–$OCH_2CH—CH_2$(epoxide)]

↓ $^-:\ddot{O}$–C6H4–C(CH_3)2–C6H4–$\ddot{O}:^-$, CH_2-$CHCH_2Cl$ (epoxide)
repeat several times

CH_2-$CHCH_2$[–O–C6H4–C(CH_3)2–C6H4–OCH_2CHCH_2(OH)–]$_n$–O–C6H4–C(CH_3)2–C6H4–$OCH_2CH—CH_2$ (epoxide ends)

where n is a small number

The prepolymer contains epoxide rings and hydroxyl groups. Copolymerization with a triamine occurs at the epoxide ends of the prepolymer.

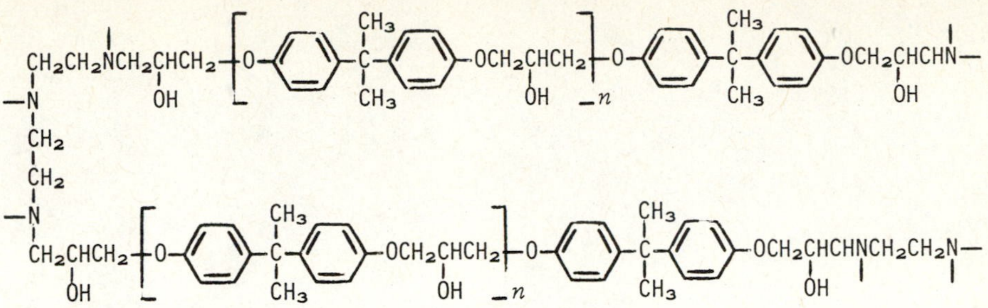

34.32 1. Formation of the diamine.

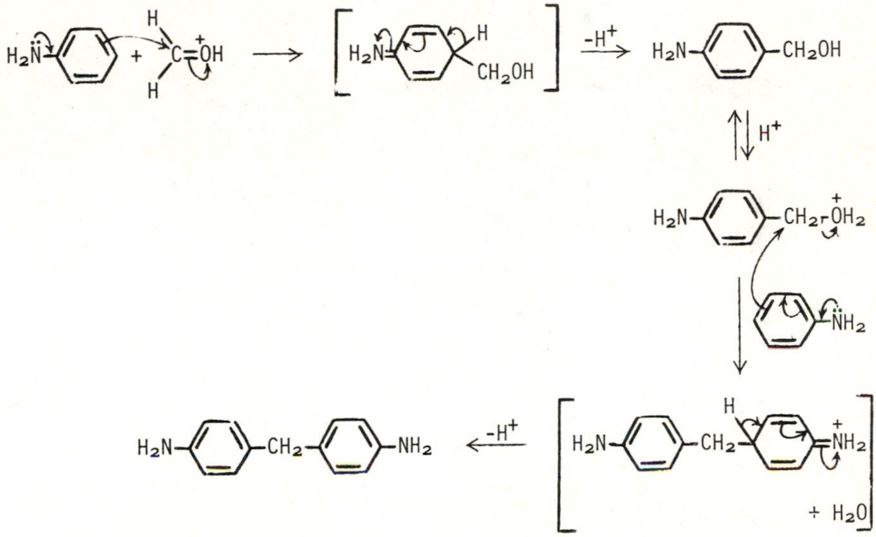

$H_2N\text{-}C_6H_4\text{-}H + H_2C{=}\overset{+}{O}H \longrightarrow [H_2\overset{+}{N}{=}C_6H_4(H)CH_2OH] \xrightarrow{-H^+} H_2N\text{-}C_6H_4\text{-}CH_2OH$

$\underset{H^+}{\rightleftharpoons} H_2N\text{-}C_6H_4\text{-}CH_2\text{-}\overset{+}{O}H_2 \xrightarrow{C_6H_5NH_2} [H_2N\text{-}C_6H_4\text{-}CH_2\text{-}C_6H_4(H){=}\overset{+}{N}H_2 + H_2O] \xrightarrow{-H^+} H_2N\text{-}C_6H_4\text{-}CH_2\text{-}C_6H_4\text{-}NH_2$

2. Reaction of the diamine with two equivalents of phosgene.

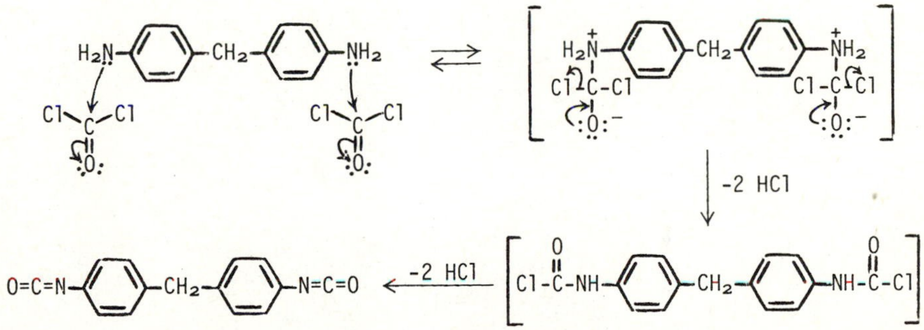

$H_2N\text{-}C_6H_4\text{-}CH_2\text{-}C_6H_4\text{-}NH_2 + 2\,Cl_2C{=}O \rightleftharpoons [H_2\overset{+}{N}(CCl_2O^-)\text{-}C_6H_4\text{-}CH_2\text{-}C_6H_4\text{-}\overset{+}{N}H_2(CCl_2O^-)] \xrightarrow{-2\,HCl} [Cl\text{-}C(O)\text{-}NH\text{-}C_6H_4\text{-}CH_2\text{-}C_6H_4\text{-}NH\text{-}C(O)\text{-}Cl] \xrightarrow{-2\,HCl} O{=}C{=}N\text{-}C_6H_4\text{-}CH_2\text{-}C_6H_4\text{-}N{=}C{=}O$

MDI

34.33

$$H_2N\text{-}\overset{O}{\overset{\|}{C}}\text{-}NH_2 + CH_2O \xrightarrow{\Delta}$$

What you should know:

After doing these problems, you should be able to:

1. Locate the monomer units of a polymer and deduce the structure of a polymer, given its monomer units;
2. Write the mechanisms of radical, cationic, and anionic polymerization, and understand the formation of step-growth polymers;
3. Understand the stereochemistry of polymerization and the structures of atactic, isotactic, and syndiotactic polymers;
4. Know the structures of copolymers, block polymers, and graft polymers.

ADDITIONAL READING

Chapters 1 and 2 - Structure and Bonding.

J. R. Partington, "A History of Chemistry," Volumes I - IV, MacMillan, London, 1961 - 1964.

W. L. Masterton and E. L. Slowinski, "Chemical Principles," 4th ed., Saunders, Philadelphia 1977.

R. E. Dickerson, H. B. Gray, and G. P. Haight, "Chemical Principles," 3nd ed., Benjamin, Menlo Park, 1979.

M. J. Sienko and R. A. Plane, "Chemical Principles and Properties," 2nd ed., McGraw-Hill, New York, 1974.

L. Salem, "A Faithful Couple: The Electron Pair," **J. Chem. Educ., 55,** 344 (1978).

R. J. Gillespie, "The Electron-Pair Repulsion Model for Molecular Geometry," **J. Chem. Educ., 47,** 18 (1970).

R. H. Maybury, "The Language of Quantum Mechanics," **J. Chem. Educ., 39,** 367 (1962).

D. Kolb, "Acids and Bases," **J. Chem. Educ., 55,** 459 (1978).

D. Kolb, "The pH Concept," **J. Chem. Educ., 56,** 49 (1979).

D. Kolb, "The Chemical Formula, Part II: Determination," **J. Chem Educ., 55,** 109 (1978).

Chapter 3 - Nature of Organic Compounds: Alkanes.

O. T. Benfey, "The Names and Structures of Organic Compounds," Wiley, New York, 1966.

J. H. Fletcher, O. C. Dermer, and R. B. Fox, "Nomenclature of Organic Compounds: Principles and Practice," Advances in Chemistry Series No. 126, American Chemical Society, Washington, D. C., 1974.

"Nomenclature of Organic Chemistry, Sections A, B, C, D, E, F, and H," International Union of Pure and Applied Chemistry, Pergamon Press, Oxford, 1979.

C. A. Kingsbury, "Conformations of Substituted Ethanes," **J. Chem. Educ., 56,** 431 (1979).

D. Kolb and K. E. Kolb, "Petroleum Chemistry," **J. Chem. Educ., 56,** 465 (1979).

E. L. Eliel, "Stereochemistry of Carbon Compounds," McGraw-Hill, New York, 1962.

Chapter 4 - Organic Reactions.

J. March, "Advanced Organic Chemistry," 2nd ed., McGraw-Hill, New York, 1977, Chapter 6.

W. H. Saunders, Jr., "Ionic Aliphatic Reactions," Prentice-Hall, Englewood Cliffs, 1965.

P. Sykes, "A Guidebook to Mechanism in Organic Chemistry," 5th ed., Longman, Green, New York 1981.

R. Breslow, "Organic Reaction Mechanisms," 2nd ed., Benjamin, Menlo Park, 1969.

Chapters 5 and 6 - Alkenes.

H. Salzman, "Arthur Lapworth: The Genesis of Reaction Mechanism," **J. Chem. Educ., 49**, 750 (1972).

W. R. Dolbier Jr., "Electrophilic Additions to Alkenes," **J. Chem. Educ., 46**, 342 (1969).

J. March, "Advanced Organic Chemistry," 2nd ed., McGraw-Hill, New York, 1977, Chapters 15 and 17.

N. Isenberg and M. Grdinic, "A Modern Look at Markovnikov's Rule and the Peroxide Effect," **J. Chem. Educ., 46**, 601 (1969).

P. Wiseman, "Ethylene by Naphtha Cracking: Free Radicals in Action," **J. Chem. Educ., 54**, 154 (1977).

S. Patai (ed.), "Chemistry of the Alkenes, Part I" Wiley Interscience, New York, 1964.

J. Zabicky (ed.), "Chemistry of the Alkenes, Part II," Wiley Interscience, New York, 1970.

Chapter 7 - Alkynes.

S. Patai (ed.), "The Chemistry of the Carbon-Carbon Triple Bond, Parts I and II," Wiley Interscience, New York, 1978.

Chapter 8 - Stereochemistry.

D. F. Mowery, Jr., "Criteria for Optical Activity in Organic Molecules," **J. Chem. Educ., 46**, 269 (1969).

R. S. Cahn, "An Introduction to the Sequence Rule," **J. Chem. Educ., 41**, 116 (1964).

E. L. Eliel, "Elements of Stereochemistry," Wiley, New York, 1969.

E. L. Eliel, "Stereochemistry of Carbon Compounds," McGraw-Hill, New York, 1962.

Chapter 9 - Alkyl Halides.

W. A. Pryor, "Introduction to Free Radical Chemistry," Prentice-Hall, Englewood Cliffs, 1966.

S. Patai (ed.), "Chemistry of the Carbon-Halogen Bond, Parts I and II," Wiley Interscience, New York, 1973.

Chapter 10 - Nucleophilic Substitution Reactions.

J. March, "Advanced Organic Chemistry," 2nd ed., McGraw-Hill, New York, 1977, Chapters 10 and 17.

A. Streitwieser, Jr., "Solvolytic Displacement Reactions," McGraw-Hill, New York, 1962.

Chapter 11 - Mass Spectroscopy and Infrared Spectroscopy.

F. W. McLafferty, "Interpretation of Mass Spectroscopy," 3rd ed., Benjamin, Menlo Park, 1981.

J. R. Dyer, "Applications of Absorption Spectroscopy of Organic Compounds," Prentice-Hall, Englewood Cliffs, 1965.

J. W. Cooper, "Spectroscopic Techniques for Organic Chemists," Wiley Interscience, New York, 1980.

R. M. Silverstein and G. C. Bassler, "Spectrometric Identification of Organic Compounds," 3rd ed., Prentice-Hall, Englewood Cliffs., 1974.

Chapter 12 - Nuclear Magnetic Resonance Spectroscopy.

J. R. Dyer, "Applications of Absorption Spectroscopy of Organic Compounds," Prentice-Hall, Englewood Cliffs, 1965.

J. W. Cooper, "Spectroscopic Techniques for Organic Chemists," Wiley Interscience, New York, 1980.

R. M. Silverstein and G. C. Bassler, "Spectrometric Identification of Organic Compounds," 3rd ed., Prentice-Hall, Englewood Cliffs., 1974.

F. W. Wehrli and T. Wirthlin, "Interpretation of Carbon-13 NMR Spectra," Heyden, Philadelphia, 1978.

Chapter 13 - Conjugated Dienes and Ultraviolet Spectroscopy.

H. H. Jaffe and M. Orchin, "Theory and Application of Ultraviolet Spectroscopy," Wiley, New York, 1962.

J. R. Dyer, "Applications of Absorption Spectroscopy of Organic Compounds," Prentice-Hall, Englewood Cliffs, 1965.

J. W. Cooper, "Spectroscopic Techniques for Organic Chemists," Wiley Interscience, New York, 1980.

Chapters 14, 15 and 16 - Aromatic Compounds.

R. Breslow, "The Nature of Aromatic Molecules," **Scientific American**, August, 1972, p. 32.

D. Kolb, "The Aromatic Ring," **J. Chem. Educ., 56**, 334 (1979).

M. Blummer, "Polycyclic Aromatic Compounds in Nature," **Scientific American**, March, 1976, p. 34.

G. M. Badger, "Aromatic Character and Aromaticity," Cambridge University Press, London, 1969.

G. A. Olah, "Friedel-Crafts Chemistry," Wiley, New York, 1973.

Chapter 18 - Alicyclic Molecules.

L. N. Ferguson, "Ring Strain and Reactivity of Alicycles," **J. Chem. Educ., 47**, 46 (1970).

J. B. Lambert, "The Shapes of Organic Molecules," **Scientific American**, January, 1970, p. 58.

E. L. Eliel, "Stereochemistry of Carbon Compounds," McGraw-Hill, New York, 1962.

M. Jones, Jr., "Carbenes," **Scientific American**, February, 1976, p. 101.

W. Kirmse, "Carbene Chemistry," Academic Press, New York, 1964.

Chapter 19 - Ethers and Epoxides.

S. Patai (ed.), "The Chemistry of the Ether Linkage", Wiley Interscience, New York, 1967.

S. Patai (ed.), "The Chemistry of Ethers, Crown Ethers, Hydroxyl Groups, and Their Sulfur Analogues, Parts I and II", Wiley Interscience, New York, 1980.

Chapter 20 - Alcohols.

S. Patai (ed.), "The Chemistry of the Hydroxyl Group, Parts I and II," Wiley Interscience, New York, 1971.

S. Patai (ed.), "The Chemistry of Ethers, Crown Ethers, Hydroxyl Groups, and Their Sulfur Analogues, Parts I and II", Wiley Interscience, New York, 1980.

K. L. Rinehart, "Oxidation and Reduction of Organic Compounds," Prentice-Hall, Englewood Cliffs, 1973.

Chapter 21 - Overview of Carbonyl Compounds.

C. D. Gutsche, "The Chemistry of Carbonyl Compounds," Prentice-Hall, Englewood Cliffs, 1967.

Chapter 22 - Aldehydes and Ketones.

H. Hart and M. Sasaoka, "Simple Enols: How Rare Are They?," **J. Chem. Educ., 57**, 685 (1980).

S. Patai (ed.), "The Chemistry of the Carbonyl Group, Part I." Wiley Interscience, New York, 1966.

J. Zabicky (ed.), "The Chemistry of the Carbonyl Group, Part II," Wiley Interscience, New York, 1970.

C. D. Gutsche, "The Chemistry of Carbonyl Compounds," Prentice-Hall, Englewood Cliffs, 1967.

Chapters 23 and 24 - Carboxylic Acids and Their Derivatives.

S. Patai (ed.), "The Chemistry of Carboxylic Acids and Esters," Wiley Interscience, New York, 1969.

J. Zabicky (ed.), "The Chemistry of Amides," Wiley Interscience, New York, 1970.

H. O. House, "Modern Synthetic Reactions," 2nd ed., Benjamin, Menlo Park, 1972, Chapter 11.

C. D. Gutsche, "The Chemistry of Carbonyl Compounds," Prentice-Hall, Englewood Cliffs, 1967.

Chapter 25 - Carbonyl Alpha-Substitution Reactions.

H. O. House, "Modern Synthetic Reactions," 2nd ed., Benjamin, Menlo Park, 1972, Chapter 9.

C. D. Gutsche, "The Chemistry of Carbonyl Compounds," Prentice-Hall, Englewood Cliffs, 1967.

Chapter 26 - Carbonyl Condensation Reactions.

H. O. House, "Modern Synthetic Reactions," 2nd ed., Benjamin, Menlo Park, 1972, Chapter 10.

C. D. Gutsche, "The Chemistry of Carbonyl Compounds," Prentice-Hall, Englewood Cliffs, 1967.

Chapter 27 - Carbohydrates.

C. S. Hudson, "Emil Fischer's Discovery of the Configuration of Glucose," **J. Chem. Educ., 18,** 353 (1941).

W. W. Pigman and D. Horton (eds.), "The Carbohydrates: Chemistry and Biochemistry," continuing series, Academic Press, New York, 1972 -.

Chapters 28 and 29 - Amines and Phenols.

J. M. McIntosh, "Phase-Transfer Catalysis Using Quaternary 'Onium Salts," **J. Chem. Educ., 55,** 235 (1978).

I. T. Miller and H. D. Springall, "Sidgwick's Organic Chemistry of Nitrogen," 3rd ed., Oxford Press, London, 1966.

P. A. S. Smith, "The Chemistry of Open-Chain Organic Nitrogen Compounds, Vols. I and II," Benjamin, Menlo Park, 1966.

S. Patai, "The Chemistry of Quinonoid Compounds," Wiley Interscience, New York, 1974.

S. Patai, "The Chemistry of the Amino Group," Wiley Interscience, New York, 1968.

H. Zollinger, "Diazo and Azo Chemistry," Wiley, New York, 1961.

Chapter 30 - Pericyclic Reactions.

R. H. Wollenberg and R. Belloli, "Woodward-Hoffmann Made Easy," **Chem. Brit., 10,** 95 (1974).

T. L. Gilchrist and R. C. Storr, "Organic Reactions and Orbital Symmetry," Cambridge University Press, London, 1972.

R. E. Lehr and A. P. Marchand, "Orbital Symmetry: A Problem-Solving Approach," Academic Press, Inc., New York, 1972.

Chapter 31 - Amino Acids, Peptides, and Proteins.

R. E. Dickerson and I. Geis, "The Structure and Action of Proteins," Harper and Row, New York, 1969.

M. Calvin and M. J. Jorgenson (eds.), "Bio-Organic Chemistry: Readings from Scientific American," Freeman, San Francisco, 1968.

A. L. Lehninger, "Biochemistry," 2nd ed., Worth, New York, 1975, Chapters 1 - 9.

L. Stryer, "Biochemistry," 2nd ed., Freeman, San Francisco, 1981, Chapters 2 - 7.

Chapter 32 - Lipids.

M. Calvin and M. J. Jorgenson (eds.), "Bio-Organic Chemistry: Readings from Scientific American," Freeman, San Francisco, 1968.

A. L. Lehninger, "Biochemistry," 2nd ed., Worth, New York, 1975, Chapter 11.

L. Stryer, "Biochemistry," 2nd ed., Freeman, San Francisco, 1981, Chapter 20.

J. H. Richards and J. B. Hendrickson, "Biosynthesis of Steroids, Terpenes, and Acetogenins," Benjamin, Menlo Park, 1964.

Chapter 33 - Heterocycles and Nucleic Acids.

L. A. Paquette, "Principles of Modern Heterocyclic Chemistry," Benjamin, Menlo Park, 1968.

A. L. Lehninger, "Biochemistry," 2nd ed., Worth, New York, 1975, Chapters 12, 31 - 34.

L. Stryer, "Biochemistry," 2nd ed., Freeman, San Francisco, 1981, Chapters 22, 24 - 27.

Chapter 34 - Synthetic Polymers.

F. W. Harris, et.al., "State of the Art: Polymer Chemistry," **J. Chem. Educ., 58**, 836-955 (1981).

F. W. Billmeyer, Jr., "Textbook of Polymer Science," 2nd ed., Wiley, New York, 1971.

L. Mandelkern, "An Introduction to Macromolecules," Springer-Verlag, New York, 1972.

J. K. Stille, "Industrial Organic Chemistry," Prentice-Hall, Englewood Cliffs, 1968.

GLOSSARY

Absolute configuration (Section 8.7): the actual three-dimensional structure of a chiral molecule. Absolute configurations are specified verbally by the Cahn-Ingold-Prelog R,S convention and are represented on paper by Fischer projections.

Absorption spectrum (Section 11.8): a plot of wavelength of incident light versus amount of light absorbed. Organic molecules show absorption spectra in both the infrared and ultraviolet regions of the electromagnetic spectrum. By interpreting these spectra, useful structural information about the sample can be obtained. (See: infrared spectrum, ultraviolet spectrum)

Acetal (Section 22.11): a functional group consisting of two ether-type oxygen atoms bound to the same carbon, $R_2C(OR')_2$. Acetals are often used as protecting groups for ketones and aldehydes since they are stable to basic and nucleophilic reagents but can be easily removed by acidic hydrolysis.

Achiral (Section 8.6): having a lack of handedness. A molecule is achiral if it has a plane of symmetry and is thus superimposable on its mirror image. (See chiral)

Activating group (Section 15.8): an electron-donating group such as hydroxyl (-OH) or amino ($-NH_2$) that increases the reactivity of an aromatic ring toward electrophilic aromatic substitution. All activating groups are ortho/para directing.

Activation energy (Section 4.8): the difference in energy levels between ground state and transition state. The amount of activation energy required by a reaction determines the rate at which the reaction proceeds. The majority of organic reactions have activation energies of 10 - 25 kcal/mol.

Acylation (Section 16.6): the introduction of an acyl group, -COR, onto a molecule. For example, acylation of an alcohol yields an ester ($R'OH \rightarrow R'OCOR$), acylation of an amine yields an amide ($R'NH_2 \rightarrow R'NHCOR$), and acylation of an aromatic ring yields an alkyl aryl ketone ($ArH \rightarrow ArCOR$).

Acylium ion (Section 16.6): a resonance-stabilized carbocation in which the positive charge is located at a carbonyl-group carbon, $R\text{-}\overset{+}{C}\text{=}O \leftrightarrow R\text{-}C\equiv\overset{+}{O}$. Acylium ions are strongly electrophilic and are involved as intermediates in Friedel-Crafts acylation reactions.

1,4-Addition (Sections 13.5 and 22.15): addition of a reagent to the ends of a conjugated pi system. Conjugated dienes yield 1,4 adducts when treated with electrophiles such as HCl. Conjugated enones yield 1,4 adducts when treated with nucleophiles such as cyanide ion.

Aldaric acid (Section 27.8): the dicarboxylic acid resulting from oxidation of an aldose.

Alditol (Section 27.8): the polyalcohol resulting from reduction of the carbonyl group of a sugar.

Aldonic acid (Section 27.8): the monocarboxylic acid resulting from mild oxidation of an aldose.

Alicyclic (Section 3.10): referring to an aliphatic cyclic hydrocarbon such as a cycloalkane or cycloalkene.

Aliphatic (Section 3.2): referring to a nonaromatic hydrocarbon such as a simple alkane, alkene, or alkyne.

Alkaloid (Section 28.10): a naturally occurring compound that contains a basic amine functional group. Morphine is an example of an alkaloid.

Alkylation (Sections 16.2, 19.4, and 25.9): introduction of an alkyl group onto a molecule. For example, aromatic rings can be alkylated to yield arenes (ArH $\longrightarrow$ ArR), alkoxide anions can be alkylated to yield ethers (R'O $\longrightarrow$ R'OR), and enolate anions can be alkylated to yield new carbonyl compounds ($R'_2C{=}C(R')O \longrightarrow RR'_2C{-}COR'$).

Allylic (Section 9.5): used to refer to the position next to a double bond. For example, $CH_2{=}CHCH_2Br$ is an allylic bromide, and an allylic radical is a conjugated, resonance-stabilized species in which the unpaired electron is in a *p* orbital next to a double bond ($C{=}C{-}\dot{C} \longleftrightarrow \dot{C}{-}C{=}C$).

Angle strain (Section 18.2): the strain introduced into a molecule when a bond angle is deformed from its ideal value. Angle strain is particularly important in small-ring cycloalkanes where it results from compression of bond angles to less than their ideal tetrahedral values. For example, cyclopropane has approximately 22 kcal/mol angle strain owing to bond deformations from the 109° tetrahedral angle to 60°.

Anomers (Section 27.6): cyclic stereoisomers of sugars that differ only in their configurations at the hemiacetal (anomeric) carbon.

Antarafacial (Section 30.8): a word used to describe the geometry of pericyclic reactions. An antarafacial reaction is one that takes place on opposite faces of the two ends of a pi electron system. (See suprafacial)

Anti conformation (Section 3.9): the geometric arrangement around a carbon-carbon single bond, in which the two largest substituents are 180° apart as viewed in a Newman projection.

Anti stereochemistry (Section 6.1): referring to opposite sides of a double bond or molecule. An anti addition reaction is one in which the two ends of the double bond are attacked from different sides. For example, addition of Br_2 to cyclohexene yields trans-1,2-dibromocyclohexane, the product of anti addition. An anti elimination reaction is one in which the two groups leave from opposite sides of the molecule. (See syn stereochemistry)

Antibonding orbital (Section 1.7): a molecular orbital that is higher in energy than the atomic orbitals from which it is formed.

Anticodon (Section 33.14): a sequence of three bases on tRNA that read the codons on mRNA and bring the correct amino acids into position for protein synthesis.

Aromaticity (Chapter 14): the special characteristics of cyclic conjugated pi electron systems that result from their electronic structures. These characteristics include unusual stability, the presence of a ring current in the ^1H NMR spectrum, and a tendency to undergo substitution reactions rather than addition reactions on treatment with electrophiles. Aromatic molecules must be planar, cyclic, conjugated species that have 4n + 2 pi electrons.

Asymmetric center (Section 8.6): see chiral center.

Atactic polymers (Section 34.6): chain-growth polymers that have a random stereochemical arrangement of substituents on the polymer backbone. These polymers result from high-temperature radical-initiated polymerization of alkene monomers.

Aufbau principle (Section 1.3): a guide for determining the ground-state electronic configuration of elements by filling the lowest energy orbitals first.

Axial bond (Section 18.10): a bond to chair cyclohexane that lies along the ring axis perpendicular to the rough plane of the ring. (See equatorial bond)

Axial bonds

Base peak (Section 11.5): the most intense peak in a mass spectrum.

Bent bonds (Section 18.5): the bonds in small rings such as cyclopropane that bend away from the internuclear line and overlap at a slight angle, rather than head-on. Bent bonds are highly strained and highly reactive.

Benzylic (Section 16.8): referring to the position next to an aromatic ring. For example, a benzylic cation is a resonance-stabilized, conjugated carbocation having its positive charge located on a carbon atom next to the benzene ring in a pi orbital that overlaps the aromatic pi system.

Benzyne (Section 15.4): an unstable intermediate having a triple bond in a benzene ring. Benzynes are implicated as intermediates in certain nucleophilic aromatic substitution reactions of aryl halides with strong bases.

Betaine (Section 22.13): a neutral dipolar molecule that has nonadjacent positive and negative charges. For example, the initial adducts of Wittig reagents with carbonyl compounds are betaines.

```
O⁻     +  R'
 \   |   /
R-C-C---P-R'
 /   |   \
R         R'
```

A Wittig betaine

Bimolecular reaction (Section 10.4): a reaction that occurs between two reagents.

Block copolymer (Section 34.8): a polymer consisting alternating homopolymer blocks. Block copolymers are usually prepared by initiating chain-growth polymerization of one monomer, followed by addition of an excess of a second monomer.

Boat cyclohexane (Section 18.15): a three-dimensional conformation of cyclohexane that bears a slight resemblance to a boat. Boat cyclohexane has no angle strain, but has a large number of eclipsing interactions that make it less stable than chair cyclohexane.

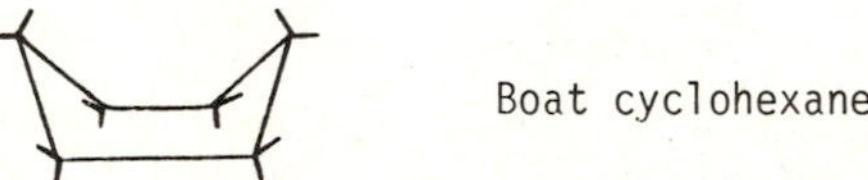

Boat cyclohexane

Bond angle (Section 1.8): the angle formed between two adjacent bonds.

Bond-dissociation energy (Section 4.7): the amount of energy needed to homolytically break a bond to produce two radical fragments.

Bond length (Section 1.7): the equilibrium distance between the nuclei of two atoms that are bonded to each other.

Bond strength (Section 1.7): see Bond-dissociation energy.

Bonding orbital (Section 1.7): a molecular orbital that is lower in energy than the atomic orbitals from which it is formed.

Bronsted acid (Section 2.6): a substance that donates a hydrogen ion (proton) to a base.

Carbanion (Section 9.8): a carbon-anion, or substance that contains a trivalent, negatively charged carbon atom ($R_3C:^-$). Carbanions are sp^3 hybridized and have eight electrons in the outer shell of the negatively charged carbon.

Carbene (Section 18.6): a neutral substance that contains a divalent carbon atom having only six electrons in its outer shell ($R_2C:$).

Carbinolamine (Section 22.9): a molecule that contains the $R_2C(OH)NH_2$ functional group. Carbinolamines are produced as unstable intermediates during the nucleophilic addition of amines to carbonyl groups.

Carbocation (Section 5.8): a carbon-cation, or substance that contains a trivalent, positively charged carbon atom having six electrons in its outer shell (R_3C^+). Carbocations are planar and sp^2 hybridized.

Carbocycle (Section 14.8): a cyclic molecule that has only carbon atoms in the ring. (See heterocycle)

Carbohydrate (Chapter 27): a polyhydroxy aldehyde or polyhydroxy ketone. The name derives from the fact that glucose, the most abundant carbohydrate, has the formula $C_6H_{12}O_6$ and was originally thought to be a hydrate of carbon. Carbohydrates can be either simple sugars such as glucose or complex sugars such as cellulose. Simple sugars are those that cannot be hydrolyzed to yield smaller molecules, whereas complex sugars are those that can be hydrolyzed to yield simpler sugars.

Chain reaction (Section 4.4): a reaction that, once initiated, sustains itself in an endlessly repeating cycle of propagation steps. The radical chlorination of alkanes in an example of a chain reaction that is initiated by irradiation with light and that then continues in a series of propagation steps.

Step 1.	Initiation:	$Cl_2 \longrightarrow 2\ Cl\cdot$
Steps 2 and 3.	Propagation:	$Cl\cdot + CH_4 \longrightarrow HCl + \cdot CH_3$ $\cdot CH_3 + Cl_2 \longrightarrow CH_3Cl + Cl\cdot$
Step 4.	Termination:	$R\cdot + R\cdot \longrightarrow R\text{-}R$

Chain-growth polymer (Section 34.1): a polymer produced by a chain reaction procedure in which an initiator adds to a carbon-carbon double bond to yield a reactive intermediate. The chain is then built as more monomers add successively to the reactive end of the growing chain.

Chair cyclohexane (Section 18.9): a three-dimensional conformation of cyclohexane that resembles the rough shape of a chair. The chair form of cyclohexane, which has neither angle strain nor eclipsing strain, represents the lowest energy conformation of the molecule.

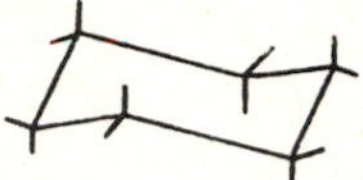

Chair cyclohexane

Chemical shift (Section 12.3): the position on the NMR chart where a nucleus absorbs. By convention, the chemical shift of tetramethylsilane is arbitrarily set at zero, and all other absorptions usually occur downfield (to the left on the chart). Chemical shifts are expressed in delta units, δ, where one delta equals one part per million of the spectrometer operating frequency. For example, one delta on a 60 megahertz instrument equals 60 Hertz. The chemical shift of a given nucleus is related to the chemical environment of that nucleus in the molecule, thus allowing one to obtain structural information by interpreting the NMR spectrum.

Chiral (Section 8.6): having handedness. Chiral molecules are those that do not have a plane of symmetry and are therefore not superimposable on their mirror image. A chiral molecule thus exists in two forms, one right handed and one left handed. The most common (though not the only) cause of chirality in a molecule is the presence of a carbon atom that is bonded to four different substituents. (See achiral)

Chiral center (Section 8.6): an atom (usually carbon) that is bonded to four different groups and is therefore chiral. (See chiral)

Chromatography (Section 11.1): a technique for separating a mixture of compounds into pure components. Chromatography operates on a principle of differential adsorption whereby different compounds adsorb to a stationary support phase and are then carried along at different rates by a mobile phase.

Cis-trans isomers (Section 3.10 and 5.3): special kinds of stereoisomers that differ in their stereochemistry about a double bond or on a ring. Cis-trans isomers are also called geometric isomers.

Codon (Section 33.14): a three-base sequence on the messenger RNA chain that encodes the genetic information necessary to cause specific amino acids to be incorporated into proteins. Codons on mRNA are read by complementary anticodons on tRNA.

Concerted (Section 13.7): referring to a reaction that takes place in a single step without intermediates. For example, the Diels-Alder cycloaddition reaction is a concerted process.

Configuration (Section 8.7): the three-dimensional arrangement of atoms bonded to a chiral center relative to the stereochemistry of other chiral centers in the same molecule.

Conformation (Section 3.7): the exact three-dimensional shape of a molecule at any given instant, assuming that rotation around single bonds is frozen.

Conformational analysis (Section 18.13): a means of assessing the minimum-energy conformation of a substituted cycloalkane by totalling the steric interactions present in the molecule. Conformational analysis is particularly useful in assessing the relative stabilities of different conformations of substituted cyclohexane rings.

Conjugate addition (Section 22.15): addition of a nucleophile to the β-carbon atom of an α,β-unsaturated carbonyl compound. (See 1,4-addition).

Conjugate base (Section 2.6): the anion that results from dissociation of a Bronsted acid.

Conjugation (Section 13.1): a series of alternating single and multiple bonds with overlapping *p* orbitals. For example, 1,3-butadiene is a conjugated diene, 3-buten-2-one is a conjugated enone, and benzene is a cyclic conjugated triene.

Conrotatory (Section 30.5): a term used to indicate the fact that *p* orbitals must rotate in the same direction during electrocyclic ring opening or ring closure. (See disrotatory)

Constitutional isomers (Section 3.2 and 8.3): isomers that have their atoms connected in a different order. For example, butane and 2-methylpropane are constitutional isomers.

Copolymer (Section 34.8): a polymer formed by chain-growth polymerization of a mixture of two or more different monomer units.

Coupling constant (Section 12.12): the magnitude (expressed in Hertz) of the spin-spin splitting interaction between nuclei whose spins are coupled. Coupling constants are denoted *J*.

Covalent bond (Section 1.6): a bond formed by sharing electrons between two nuclei. (See ionic bond)

Cracking (Section 5.1): a process used in petroleum refining in which large alkanes are thermally cracked into smaller fragments.

Cycloaddition (Sections 4.5 and 30.1): a pericyclic reaction in which two reactants add together in a single step to yield a cyclic product. The Diels-Alder reaction between a diene and a dienophile to give a cyclohexene is the best-known example of a cycloaddition.

Deactivating group (Section 15.8): an electron-withdrawing substituent that decreases the reactivity of an aromatic ring towards electrophilic aromatic substitution. Most deactivating groups, such as nitro, cyano, and carbonyl are meta-directors, but halogen substituents are ortho/para directors.

Decarboxylation (Section 23.9): a reaction that involves loss of carbon dioxide from the starting material. β-Keto acids decarboxylate particularly readily on heating.

Degenerate orbitals (Section 14.5): two or more orbitals that have the same energy level.

Dehydration (Section 6.11): a reaction that involves loss of water from the starting material. Most alcohols can be dehydrated to yield alkenes, but aldol condensation products (β-hydroxy ketones) dehydrate particularly readily.

Dehydrohalogenation (Section 6.11): a reaction that involves loss of HX from the starting material. Alkyl halides undergo dehydrohalogenation to yield alkenes on treatment with strong base.

Delocalization (Section 9.6): a spreading out of electron density over a conjugated pi electron system. For example allylic cations and allylic anions are delocalized because their charges are spread out by resonance stabilization over the entire pi-electron system.

Denaturation (Section 31.15): the physical changes that occur in proteins when secondary and tertiary structures are disrupted. Denaturation is usually brouught about by heat treatment or by a change in pH and is accompanied by a loss of biological activity.

Deshielding (Section 12.2): an effect observed in NMR that causes a nucleus to absorb downfield (to the left) of tetramethylsilane standard. Deshielding is caused by a withdrawal of electron density from the nucleus and is responsible for the observed chemical shifts of vinylic and aromatic protons.

Deuterium isotope effect (Section 10.11): a tool for use in mechanistic investigations to establish whether or not a C-H bond is broken in the rate-limiting step of a reaction. Since carbon--deuterium bonds are stronger and less easily broken than carbon--protium bonds, one can measure the reaction rates of both protium- and deuterium-substituted substrates and see if they are the same or different. If they are different, a deuterium isotope effect is present, indicating that C-H bond breakage is rate-limiting.

Dextrorotatory (Section 8.1): a word used to describe an optically active substance that rotates the plane of polarization of plane-polarized light in a right-handed (clockwise) direction. The direction of rotation is not, however, related to the absolute configuration of the molecule. (See levorotatory)

Diastereomer (Section 8.8): a term that indicates the relationship between non-mirror-image stereoisomers. Diastereomers are stereoisomers that have the same configuration at one or more chiral centers, but differ at other chiral centers.

Diazotization (Section 29.4): the conversion of a primary amine, RNH_2 into a diazonium salt, RN_2^+ by treatment with nitrous acid. Aryl diazonium salts are stable, but alkyl diazonium salts are extremely reactive and are rarely isolable.

Dielectric constant (Section 10.9): a measure of the ability of a solvent to act as an insulator of electric charge. Solvents that have high dielectric constants are highly polar and are particularly valuable in S_N1 reactions because of their ability to stabilize the developing positive charge of the intermediate carbocation.

Dienophile (Section 13.8): a compound containing a double bond that can take part in the Diels-Alder cycloaddition reaction. The most reactive dienophiles are those that have electron-withdrawing groups such as nitro, cyano, or carbonyl on the double bond.

Dipolar aprotic solvent (Section 10.5): a dipolar solvent that cannot function as a hydrogen ion donor. Dipolar aprotic solvents such as dimethyl sulfoxide (DMSO), hexamethylphosphoramide (HMPA), and dimethylformamide (DMF) are particularly useful in S_N2 reactions because of their ability to solvate cations.

Dipole moment, (Section 2.5): a measure of the polarity of a molecule. A dipole moment arises when the centers of gravity of positive and negative charges within a molecule do not coincide.

Disrotatory (Section 30.5): a term used to indicate the fact that *p* orbitals rotate in opposite directions during electrocyclic ring opening or ring closing. (See conrotatory)

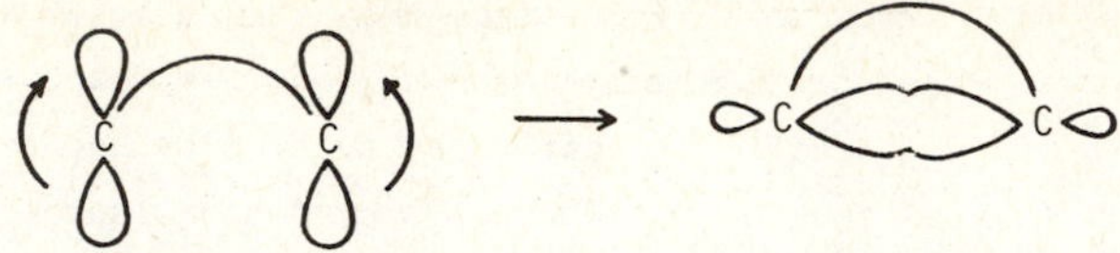

Dissociation constant (Section 23.3): a measure of the extent to which a molecule dissociates into ions. The dissociation constant, K_{diss}, for the reaction $AB \rightleftarrows A^+ + B^-$ is given by the expression

$$K_{diss} = \frac{[A^+][B^-]}{[AB]}$$

where the quantities in brackets represent the molar concentrations of the reactant and products.

d,l form (Section 27.3): a shorthand way of indicating the racemic modification of a compound. (See racemic)

DNA (Section 33.9): deoxyribonucleic acid, the biopolymer consisting of deoxyribonucleotide units linked together through phosphate-sugar bonds. DNA is found in the nucleus of cells and contains an organism's genetic information.

Doublet (Section 12.6): a two-line NMR absorption caused by spin-spin splitting when the spin of the nucleus under observation couples with the spin of a neighboring magnetic nucleus.

Downfield (Section 12.3): used to refer to the left hand portion of the NMR chart. (See deshielding)

Eclipsed conformation (Section 3.7): the geometric arrangement around a carbon-carbon single bond in which the bonds to substituents on one carbon are parallel to the bonds to substituents on the neighboring carbon as viewed in a Newman projection. For example, the eclipsed conformation of ethane has the C-H bonds on one carbon lined up with the C-H bonds on the neighboring carbon.

Eclipsed conformation

Eclipsing strain (Section 18.4): the strain energy in a molecule caused by electron repulsions between eclipsed bonds. Eclipsing strain is also called torsional strain.

Elastomers (Section 34.13): amorphous polymers that have the ability to stretch out and then return to their previous shape. These polymers have irregular shapes that prevent crystallite formation and have little cross-linking between chains.

Electrocyclic reaction (Section 30.1): a unimolecular pericyclic reaction in which a ring is formed or broken by a concerted reorganization of electrons through a cyclic transition state. For example the cyclization of 1,3,5-hexatriene to yield 1,3-cyclohexadiene is an electrocyclic reaction,

Electromagnetic spectrum (Section 11.8): the range of electromagnetic energy, including infrared, ultraviolet and visible radiation.

Electron affinity (Section 1.5): the measure of an atom's tendency to gain an electron and form an anion. Elements on the right side of the periodic table such as the halogens have higher electron affinities than elements on the left side.

Electronegativity (Section 2.4): the ability of an atom to attract electrons and thereby polarize a bond. As a general rule, electronegativity increases in going across the periodic table from right to left and in going from bottom to top.

Electrophile (Section 4.1): an electron-lover, or substance that accepts an electron pair from a nucleophile in a polar bond-forming reaction.

Electrophoresis (Section 31.3): a technique used for separating charged organic molecules, particularly proteins and amino acids. The mixture to be separated is placed on a buffered gel or paper and an electric potential is applied across the ends of the apparatus. Negatively charged molecules migrate towards the positive electrode and positively charged molecules migrate towards the negative electrode.

Elution (Section 11.2): the removal of a substance from a chromatography column.

Empirical formula (Section 2.8): a formula that gives the relative proportions of elements in a compound in smallest whole numbers.

Enantiomers (Section 8.3): stereoisomers of a chiral substance that have a mirror image relationship. Enantiomers must have opposite configurations at all chiral centers in the molecule.

Endothermic (Section 4.6): a term used to describe reactions that absorb energy and that therefore have positive enthalpy changes. In reaction energy diagrams, the products of endothermic reactions have higher energy levels than the starting materials.

Entgegen (E) (Section 5.6): a term used to describe the stereochemistry of a carbon-carbon double bond. The two groups on each carbon are first assigned priorities according to the Cahn-Ingold-Prelog sequence rules, and the two carbons are then compared. If the high priority groups on each carbon are on opposite sides of the double bond, the bond has E geometry. (See Zusammen).

Enthalpy change, ΔH (Section 4.6): the heat of reaction. The enthalpy change that occurs during a reaction is a measure of the difference in total bond energy between reactants and products.

Entropy change, ΔS (Section 4.6): the amount of disorder. The entropy change that occurs during a reaction is a measure of the difference in disorder between reactants and products.

Enzyme (Section 6.10): a biological catalyst. Enzymes are large proteins that catalyze specific biochemical reactions.

Epoxide (Section 19.7): a three-membered ring ether functional group.

Equatorial bond (Section 18.10): a bond to cyclohexane that lies along the rough equator of the ring. (See axial bond)

Equatorial bonds

Equilibrium constant (Section 2.6): a measure of the equilibrium position for a reaction. The equilibrium constant, K_{eq}, for the reaction A + B $\rightleftarrows$ C + D is given by the expression

$$K_{eq} = \frac{[C][D]}{[A][B]}$$

where the numbers in brackets refer to the molar concentrations of the reactants and products.

Essential oil (Section 32.6): the volatile oil that is obtained by steam distillation of a plant extract.

Excited-state configuration (Section 1.8): an electronic configuration having a higher energy level than the ground state. Excited states are normally obtained by excitation of an electron from a bonding orbital to an antibonding one, such as occurs during irradiation of a molecule with light of the proper frequency.

Exothermic (Section 4.6): a term used to describe reactions that release energy and that therefore have negative enthalpy changes. On reaction energy diagrams, the products of exothermic reactions have energy levels lower than those of starting materials.

Fat (Section 32.1): a solid triacylglycerol derived from animal sources.

Fibers (Section 34.13): thin threads produced by extruding a molten polymer through small holes in a die.

Fibrous protein (Section 31.13): proteins that consist of polypeptide chains arranged side by side in long threads. These proteins are tough, insoluble in water, and are used in nature for structural materials such as hair, hooves, and fingernails.

Fingerprint region (Section 11.10): the complex region of the infrared spectrum from 1500 cm^{-1} to 400 cm^{-1}. If two substances have identical absorption patterns in the fingerprint region of the IR, they are almost certainly identical.

Fischer projection (Section 8.7): a means of depicting the absolute configuration of chiral molecules on a flat page. A Fischer projection employs a cross to represent the chiral center; the horizontal arms of the cross represent bonds coming out of the plane of the page, whereas the vertical arms of the cross represent bonds going back into the plane of the page.

Fischer projection

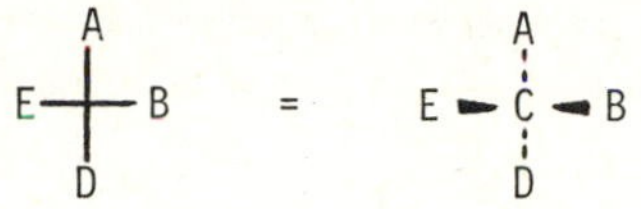

```
  A              A
  |              :
E-+-B    =   E ▬ C ▬ B
  |              :
  D              D
```

Formal charge (Section 2.3): the difference in the number of electrons owned by an atom in a molecule and by the same atom in its elemental state. The formal charge on an atom is given by the formula:

$$\text{Formal charge} = \begin{pmatrix}\text{\# of outer shell electrons} \\ \text{in a free atom}\end{pmatrix} - \begin{pmatrix}\text{\# of outer shell electrons} \\ \text{in a bound atom}\end{pmatrix}$$

Frequency (Section 11.8): the number of electromagnetic wave cycles that travel past a fixed point in a given unit of time. Frequencies are usually expressed in units of cycles per second, or Hertz.

Functional group (Section 3.1): an atom or group of atoms that is part of a larger molecule and that has a characteristic chemical reactivity. Functional groups display the same chemistry in all molecules of which they are a part.

Gated-decoupled mode (Section 12.5): a mode of ^{13}C NMR spectrometer operation in which all one-carbon resonances are of equal intensity. Operating in this mode allows one to integrate the spectrum to find out how many of each kind of carbon atom is present.

Gauche conformation (Section 3.9): the conformation of butane in which the two methyl groups lie 60° apart as viewed in a Newman projection. This conformation has 0.9 kcal/mol steric strain.

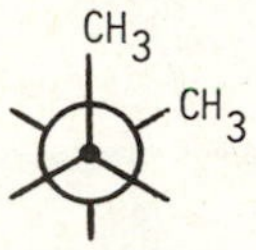

CH_3 CH_3

Gauche conformation

Geometric isomers (Sections 3.10 and 5.3): an old term for cis-trans isomers.

Gibbs free energy change (Section 4.6): the total amount of free energy change, both enthalpy and entropy, that occurs during a reaction. The standard Gibb's free energy change for a reaction is given by the formula

$$\Delta G^\circ = \Delta H^\circ - T\Delta S^\circ$$

Globular protein (Section 31.13): proteins that are coiled into compact, nearly spherical shapes. These proteins are generally water soluble, are mobile within the cell, and are the structural class to which enzymes belong.

Glycol (Section 19.8): a 1,2-diol such as ethylene glycol, $HOCH_2CH_2OH$.

Glycoside (Section 27.8): a cyclic acetal formed by reaction of a sugar with another alcohol.

Graft copolymer (Section 34.8): a copolymer that consists of homopolymer chains grafted onto a different homopolymer backbone. Graft copolymers are prepared by x-ray irradiation of a homopolymer to generate radical sites along the chain, followed by addition of a second monomer.

Ground state (Section 1.3): the most stable, lowest energy electronic configuration of a molecule.

Halohydrin (Section 6.2): a 1,2-disubstituted haloalcohol such as is obtained on addition of HOBr to an alkene.

Halonium ion (Section 6.1): a species containing a positively charged, divalent halogen. Three-membered-ring bromonium ions are implicated as intermediates in the elctrophilic addition of bromine to alkenes.

Hammond postulate (Section 5.11): a postulate stating that we can get a picture of what a given transition state looks like by looking at the structure of the nearest stable species. Exothermic reactions have transition states that resemble starting material, whereas endothermic reactions have transition states that resemble products.

Haworth projection (Section 27.5): a means of viewing stereochemistry in cyclic hemiacetal forms of sugars. Haworth projections are drawn so that the ring is flat and is viewed from an oblique angle with the hemiacetal oxygen at the upper right.

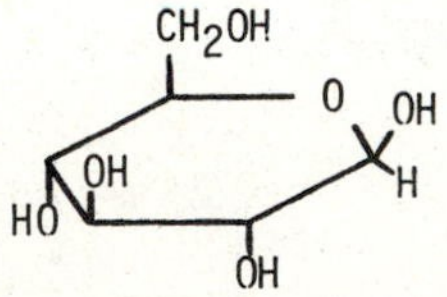

CH_2OH O OH OH H HO OH

Haworth projection of glucose

Heat of combustion (Section 18.3): the amount of heat released when a compound is burned in a calorimeter according to the equation

$$C_nH_m + O_2 \longrightarrow n\ CO_2 + m/2\ H_2O$$

Heat of hydrogenation (Section 5.4): the amount of heat released when a carbon-carbon double bond is hydrogenated. Comparison of heats of hydrogenation for different alkenes allows one to determine the stability of the different double bonds.

Heterocycle (Section 14.8 and Chapter 33): a cyclic molecule whose ring contains more than one kind of atom. For example, pyridine is a heterocycle that contains five carbon atoms and one nitrogen atom in its ring.

Heterogenic bond formation (Section 4.1): what occurs when one partner donates both electrons in forming a new bond. Polar reactions always involve heterogenic bond formation:

$$A^+ + B:^- \longrightarrow A:B$$

Heterolytic bond breakage (Section 4.1): the kind of bond breaking that occurs in polar reactions when one fragment leaves with both of the bonding electrons, as in the equation:

$$A:B \longrightarrow A^+ + B:^-$$

HOMO (Section 30.6): an acronym for the highest occupied molecular orbital. The symmetries of the HOMO and LUMO are important in pericyclic reactions. (See LUMO)

Homogenic bond formation (Section 4.3): what occurs in radical reactions when each partner donates one electron to the new bond:

$$A\cdot + B\cdot \longrightarrow A:B$$

Homolytic bond breakage (Section 4.3): the kind of bond breaking that occurs in radical reactions when each fragment leaves with one bonding electron according to the equation:

$$A:B \longrightarrow A\cdot + B\cdot$$

Homopolymer (Section 34.8): a polymer made by chain-growth polymerization of a single monomer unit.

Huckel's rule (Section 14.6): a rule stating that monocyclic conjugated molecules having 4n + 2 pi electrons (n = an integer) show the unusual stability associated with aromaticity.

Hybrid orbital (Section 1.8): an orbital that is mathematically derived from a combination of ground-state (*s*, *p*, *d*) atomic orbitals. Hybrid orbitals, such as the *sp*^{3}, *sp*^{2}, and *sp* hybrids of carbon, are strongly directed and form stronger bonds than ground-state atomic orbitals.

Hydration (Section 6.3): addition of water to a molecule, such as occurs when alkenes are treated with strong sulfuric acid.

Hydroboration (Section 6.4): addition of borane (BH_3) or an alkyl borane to an alkene. The resultant trialkylborane products are useful synthetic intermediates than can be oxidized to yield alcohols.

Hydrogenation (Section 6.6): addition of hydrogen to a double or triple bond to yield the saturated product.

Hydrogen bond (Section 20.2): a weak (5 kcal/mol) attraction between a hydrogen atom bonded to an electronegative element and an electron lone pair on another atom. Hydrogen bonding plays an important role in determining the secondary structure of proteins and in stabilizing the DNA double helix.

Hyperconjugation (Sections 5.4 and 5.10): a weak stabilizing interaction that results from overlap of a p orbital with a neighboring sigma bond. Hyperconjugation is important in stabilizing carbocations and in stabilizing substituted alkenes.

Inductive effect (Sections 2.4 and 15.9): the electron attracting or electron withdrawing effect that is transmitted through sigma bonds as the result of a nearby dipole. Electronegative elements have an electron-withdrawing inductive effect, whereas electropositive elements have an electron-donating inductive effect.

Infrared spectroscopy (Section 11.8): a kind of optical spectroscopy that uses infrared energy. IR spectroscopy is particularly useful in organic chemistry for determining the kinds of functional groups present in molecules.

Initiator (Section 4.4): a substance with an easily broken bond that is used to initiate radical chain reactions. For example, radical chlorination of alkanes is initiated when light energy breaks the weak chlorine-chlorine bond to form chlorine radicals.

Intermediate (Section 4.9): a species that is formed during the course of a multi-step reaction but is not the final product. Intermediates are more stable than transition states, but may or may not be stable enough to isolate.

Intramolecular, intermolecular (Section 26.7): Reactions that occur within the same molecule are intramolecular, whereas reactions that occur between two molecules are intermolecular.

Ion pair (Section 10.8): a loose complex between two ions in solution. Ion pairs are implicated as intermediates in S_N1 reactions in order to account for the partial retention of stereochemistry that is often observed.

Ionic bond (Section 1.5): a bond between two ions due to the electrical attraction of unlike charges. Ionic bonds are formed between strongly electronegative elements (such as the halogens) and strongly electropositive elements (such as the alkali metals).

Ionization energy (Section 1.5): the amount of energy required to remove an electron from an atom. Elements on the far right of the periodic table have high ionization energies, and elements on the far left of the periodic table have low ionization energies.

Isoelectric point (Section 31.3): the pH at which the number of positive charges and the number of negative charges on a protein or amino acid are exactly balanced.

Isomers (Section 3.2): compounds that have the same molecular formula but have different structures.

Isoprene rule (Section 32.6): an observation to the effect that terpenoids appear to be made up of isoprene (2-methyl-1,3-butadiene) units connected in a head-to-tail fashion. Monoterpenes have two isoprene units, sesquiterpenes have three isoprene units, diterpenes have four isoprene units, and so on.

Isotactic polymer (Section 34.6): a chain-growth polymer in which all substituents on the polymer backbone have the same three-dimensional orientation.

Kekule structure (Section 1.6): a representation of molecules in which a line between atoms is used to represent a bond. (See line-bond structure)

Kinetic control (Section 13.6): Reactions that follow the lowest activation-energy pathway are said to be kinetically controlled. The product formed in a kinetically controlled reaction is the one that is formed most rapidly, but is not necessarily the most stable. (See thermodynamic control)

Kinetics (Section 10.3): referring to rates of reactions. Kinetics measurements can be extremely important in helping to determine reaction mechanisms.

Leaving group (Section 10.5): the group that is replaced in a substitution reaction. The best leaving groups in nucleophilic substitution reactions are those that form the most stable, least basic, anions.

Levorotatory (Section 8.1): used to describe an optically active substance that rotates the plane of polarization of plane-polarized light in a left-handed (counterclockwise) direction. (See dextrorotatory)

Lewis acid (Section 2.7): a substance having a vacant low-energy orbital that can accept an electron pair from a base. All electrophiles are Lewis acids, but transition metal salts such as $AlCl_3$ and $ZnCl_2$ are particularly good ones. (See Lewis base)

Lewis base (Section 2.7): a substance that donates an electron lone pair to an acid. All nucleophiles are Lewis bases. (See Lewis acid)

Lewis structure (Section 1.6): a representation of a molecule showing covalent bonds as a pair of electron dots between atoms.

Line-bond structure (Section 2.1): a representation of a molecule showing covalent bonds as lines between atoms. (See Kekule structure):

Lipid (Chapter 32): a naturally occurring substance isolated from cells and tissues by extraction with nonpolar solvents. Lipids belong to many different structural classes, including fats, terpenes, prostaglandins, and steroids.

Lipophilic (Section 32.2): fat-loving. Long non-polar hydrocarbon chains tend to cluster together in polar solvents because of their lipophilic properties.

Lone-pair electrons (Section 1.12): non-bonding electron pairs that occupy valence orbitals. It is the lone-pair electrons that are used by nucleophiles in their reactions with electrophiles.

LUMO (Section 30.6): an acronym for lowest unoccupied molecular orbital. The symmetries of LUMO and HOMO are important in determining the stereochemistry of pericyclic reactions. (See HOMO)

Magnetic equivalence (Section 12.9): used to describe nuclei that have identical chemical and magnetic environments, and that therefore absorb at the same place in the NMR spectrum. For example, the six hydrogens in benzene are magnetically equivalent, as are the six carbons.

Markovnikov's rule (Section 5.9): a guide for determining the regiochemistry (orientation) of electrophilic addition reactions. In the addition of HX to an alkene, the hydrogen atom becomes bonded to the alkene carbon that has fewer alkyl substituents. A modern statement of this same rule is that electrophilic addition reactions proceed via the most stable carbocation intermediate.

Mechanism (Section 4.6): a complete description of how a reaction occurs. A mechanism must account for all starting materials and all products, and must describe the details of each individual step in the overall reaction process.

Meso (Section 8.9): A meso compound is one that contains chiral centers but is nevertheless achiral by virtue of a symmetry plane. For example, (2R,3S)-butanediol has two chiral carbon atoms, but is achiral because of a symmetry plane between carbons 2 and 3.

Micelle (Section 32.2): a spherical cluster of soap-like molecules that aggregate in aqueous solution. The ionic heads of the molecules lie on the outside where they are solvated by water, and the organic tails bunch together on the inside of the micelle.

Mobile phase (Section 11.1): the solvent (either gas or liquid) used in chromatography to move material along the solid adsorbent phase. (See chromatography)

Molecular formula (Section 2.8): an expression of the total numbers of each kind of atom present in a molecule. The molecular formula must be a whole-number multiple of the empirical formula.

Molecular ion (Section 11.5): the cation produced in the mass spectrometer by loss of an electron from the parent molecule. The mass of the molecular ion corresponds to the molecular weight of the sample.

Molecular orbital (Section 1.7): an orbital that is the property of the entire molecule rather than of an individual atom. Molecular orbitals result from overlap of two or more atomic orbitals when bonds are formed, and may be either bonding, non-bonding, or antibonding. Bonding molecular orbitals are lower in energy than the starting atomic orbitals, non-bonding M.O.s are equal in energy to the starting orbitals, and antibonding orbitals are higher in energy.

Monomer (Section 34): the simple starting units from which polymers are made.

Multiplet (Section 12.6): a symmetrical pattern of peaks in an nmr spectrum that arises by spin-spin splitting of a single absorption because of coupling between neighboring magnetic nuclei.

Mutarotation (Section 27.6): the spontaneous change in optical rotation observed when a pure anomer of a sugar is dissolved in water. Mutarotation is caused by the reversible opening and closing of the acetal linkage, which yields an equilibrium mixture of anomers.

Neighboring-group effect (Section 27.8): the effect on a reaction of a nearby functional group.

Newman projection (Section 3.7): a means of indicating stereochemical relationships between substituent groups on neighboring carbons. The carbon-carbon bond is viewed end-on, and the carbons are indicated by a circle.

Bonds radiating from the center of the circle are attached to the front carbon, and bonds radiating from the edge of the circle are attached to the rear carbon.

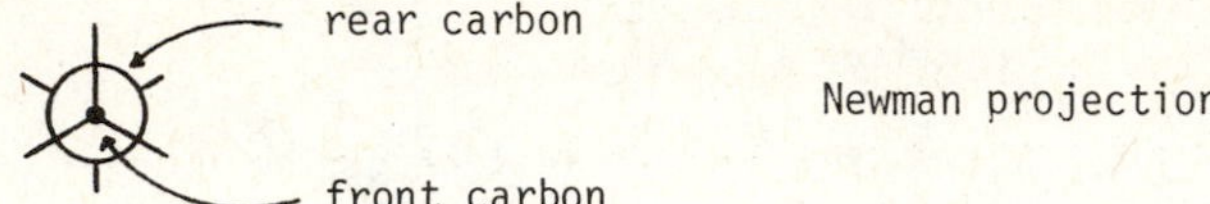

Nitrogen rule (Section 28.11): a rule stating that compounds having an odd number of nitrogens give rise to an odd-numbered molecular ion in the mass spectrum. Conversely, compounds with an even number of nitrogens give rise to even-numbered molecular ions.

Node (Sections 1.2 and 13.3): the surface of zero electron density between lobes of orbitals. For example, a p orbital has a nodal plane passing through the center of the nucleus, perpendicular to the line of the orbital.

Normal alkane (Section 3.2): a straight-chain alkane, as opposed to a branched alkane. Normal alkanes are denoted by the suffix n, as in n-C_4H_{10} (n-butane).

Nuclear magnetic resonance, NMR (Chapter 12): a spectroscopic technique that provides information about the carbon-hydrogen framework of a molecule. NMR works by detecting the energy absorption accompanying the transition between nuclear spin states that occurs when a molecule is placed in a strong magnetic field and irradiated with radio-frequency waves. Different nuclei within a molecule are in slightly different magnetic environments and therefore show absorptions at slightly different frequencies.

Nucleophile (Section 4.1): a nucleus-lover, or species that donates an electron pair to an electrophile in a polar bond-forming reaction. Nucleophiles are also Lewis bases. (See electrophile)

Nucleoside (Section 33.9): a nucleic acid constituent, consisting of a sugar residue bonded to a heterocyclic purine or pyrimidine base.

Nucleotide (Section 33.9): a nucleic acid constituent, consisting of a sugar residue bonded both to a heterocyclic purine or pyrimidine base and to a phosphoric acid. Nucleotides are the monomer units from which DNA and RNA are constructed.

Off-resonance mode (Section 12.6): a mode of ^{13}C NMR spectrometer operation that allows for the observation of spin-spin splitting between carbons and their attached hydrogens. Carbons bonded to one hydrogen show a doublet, carbons attached to two hydrogens show a triplet, and carbons attached to three hydrogens show a quartet in the off-resonance NMR.

Olefin: an alternative name for an alkene.

Optical isomers (Section 8.3): enantiomers. Optical isomers are isomers that have a mirror image relationship.

Optically active (Section 8.1): a substance that rotates the plane of polarization of plane-polarized light. Note that an optically active sample must contain chiral molecules, but that all samples with chiral molecules are not optically active. Thus, a racemic sample is optically inactive even though the individual molecules are chiral. (See chiral)

Orbital (Section 1.2) the volume of space in which an electron is most likely to be found. Orbitals are described mathematically by wavefunctions, which delineate the behavior of electrons around nuclei.

Ozonide (Section 6.8) the product formed by addition of ozone to a carbon-carbon double bond. Ozonides are usually treated with a reducing agent such as zinc in acetic acid to produce carbonyl compounds.

Paraffins (Section 3.6): a trivial name for alkanes.

Pauli exclusion principle (Section 1.3): a statement of the fact that no more than two electrons can occupy the same orbital, and that those two must have spins of opposite sign.

Peptides (Section 31.6): amino acid polymers in which the individual amino acid residues are linked by amide bonds. (See proteins)

Pericyclic reaction (Section 4.5 and Chapter 30): a reaction that occurs by a concerted reorganization of bonding electrons in a cyclic transition state.

Periplanar (Section 10.10): a conformation in which bonds to neighboring atoms have a parallel arrangement. In an eclipsed conformation, the neighboring bonds are syn-periplanar; in a staggered conformation, the bonds are anti-periplanar.

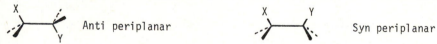

Peroxide (Section 19.2): a molecule containing an oxygen-oxygen bond functional group, R-O-O-R' or R-O-O-H. The peroxides present as explosive impurities in ether solvents are usually of the latter type. Since the oxygen-oxygen bond is weak and easily broken, peroxides are often used to initiate radical chain reactions.

Phase-transfer catalysts (Section 28.9): agents that cause the transfer of ionic reagents between phases, thus catalyzing reactions. Tetraalkylammonium salts, $R_4N^+ \; X^-$, are often used to transport inorganic anions from the aqueous phase to the organic phase where the desired reaction then occurs. For example, permanganate ion is solubilized in benzene in the presence of tetraalkylammonium ions.

Phospholipid (Section 32.3): lipids that contain a phosphate residue. For example, phosphoglycerides contain a glycerol backbone linked to two fatty acids and a phosphoric acid.

Pi bond (Section 1.7): the covalent bond formed by sideways overlap of atomic orbitals. For example, carbon-carbon double bonds contain a pi bond formed by sideways overlap of two *p* orbitals.

Plane of symmetry (Section 8.6): an imaginary plane that bisects a molecule such that one half of the molecule is the mirror image of the other half. Molecules containing a plane of symmetry are achiral.

Plane-polarized light (Section 8.1): ordinary light that has its electric vectors in a single plane rather than in random planes. The plane of polarization is rotated when the light is passed through a solution of a chiral substance.

Polar reaction (Section 4.1): a reaction in which bonds are made when a nucleophile donates two electrons to an electrophile, and in which bonds are broken when one fragment leaves with both electrons from the bond. Polar reactions are the most common class of reactions. (See heterogenic and heterolytic reactions)

Polarity (Sections 2.4 and 4.1): the unsymmetrical distribution of electrons in molecules that results when one atom attracts electrons more strongly than another.

Polarizability (Section 4.1): the measure of the change in a molecule's electron distribution in response to changing electric interactions with solvents or ionic reagents.

Polymer (Chapter 34): a large molecule made up of repeating smaller units. For example, polyethylene is a synthetic polymer made from repeating ethylene units, and DNA is a biopolymer made of repeating deoxyribonucleotide units.

Primary, secondary, tertiary, quaternary (Section 3.4): terms used to describe the substitution pattern at a specific site. A primary site has one organic substituent attached to it, a secondary site has two organic substituents, a tertiary site has three, and a quaternary site has four.

	Carbon	**Hydrogen**	**Alcohol**	**Amine**
primary	$R-\underline{C}H_3$	$R-C\underline{H}_3$	RCH_2OH	$R-NH_2$
secondary	$R_2\underline{C}H_2$	$R_2C\underline{H}_2$	R_2CHOH	R_2NH
tertiary	$R_3\underline{C}H$	$R_3C\underline{H}$	R_3COH	R_3N
quaternary	$R_4\underline{C}$			

R = any organic substituent

Primary structure (Section 31.14): the amino acid sequence in a protein. (See secondary structure, tertiary structure)

Principle of maximum overlap (Section 1.7): The strongest bonds are formed when overlap between orbitals is greatest.

Propagation step (Section 4.4): the step or series of steps in a radical chain reaction that carry on the chain. The propagation steps must yield both product and a reactive intermediate to carry on the chain.

Propargylic position (Section 7.9): the position next to a carbon-carbon triple bond.

Prostaglandin (Section 32.5): a member of the class of lipids with the general carbon skeleton:

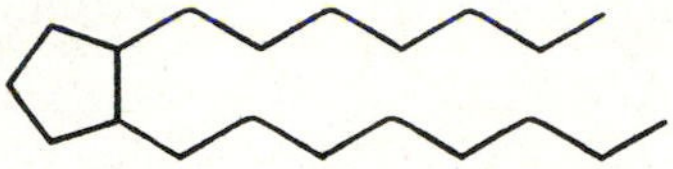

Prostaglandins are present in nearly all body tissues and fluids, where they serve a large number of important hormonal functions.

Protecting group (Section 20.11): a group that is introduced to protect a sensitive functional group towards reaction elsewhere in the molecule. After serving its protective function, the group is then removed. For example, ketones and aldehydes are often protected as acetals by reaction with ethylene glycol, and alcohols are often protected as tetrahydropyranyl ethers.

Protein (Section 31.13): a large peptide containing fifty or more amino acid residues. Proteins serve both as structural materials (hair, horns, fingernails) and as enzymes that control an organism's chemistry. (See peptide)

Protic solvent (Section 10.9): a solvent such as water or alcohol that can serve as a proton donor. Protic solvents are particularly good at stabilizing anions by hydrogen bonding, thereby lowering their reactivity. (See dipolar aprotic solvent)

Proton-noise-decoupled mode (Section 12.4): the most common manner of ^{13}C NMR spectrometer operation, in which all non-equivalent carbon atoms in the sample show a single, unsplit resonance. Operating the spectrometer in this mode allows one to count the number of chemically different carbon atoms present in the sample molecule.

Quartet (Section 12.6): a set of four peaks in the NMR, caused by spin-spin splitting of a signal by three adjacent nuclear spins.

Quaternary (See primary)

Quaternary structure (Section 31.14): the highest level of protein structure, involving a specific aggregation of individual proteins into a larger cluster.

R,S convention (Section 8.7): a method for defining the absolute configuration around chiral centers. The Cahn-Ingold-Prelog sequence rules are used to assign relative priorities to the four substituents on the chiral center and the center is oriented such that the group of lowest (fourth) priority faces directly away from the viewer. If the three remaining substituents have a right-handed or clockwise relationship in going from first to second to third priority, then the chiral center is denoted *R* (rectus, right). If the three remaining substituents have a left-handed or counterclockwise relationship, the chiral center is denoted *S* (sinister, left). (see sequence rules)

Racemic mixture (Section 8.11): a mixture consisting of equal parts (+) and (-) enantiomers of a chiral substance. Even though the individual molecules are chiral, racemic mixtures are optically inactive.

Racemization (Section 8.11): the process whereby one enantiomer of a chiral molecule becomes converted into a 50:50 mixture of enantiomers, thus losing its optical activity. For example, this might happen during an S_N1 reaction of a chiral alkyl halide.

Radical (Section 4.3): When used in organic nomenclature, the word radical refers to a part of a molecule that appears in its name - for example the phenyl in phenyl acetate. Chemically, however, a radical is a species that has an odd number of electrons, such as the chlorine radical, Cl· .

Radical reaction (Section 4.3): a reaction in which bonds are made by donation of one electron from each of two reagents, and in which bonds are broken when each fragments leaves with one electron. (See homogenic, homolytic)

Rate-limiting step (Section 10.7): the slowest step in a multi-step reaction sequence. It is the rate-limiting step that acts as a kind of bottleneck in multi-step reactions and that is observed by kinetics measurements.

Reaction energy diagram (Section 4.8): a pictorial representation of the course of a reaction, in which potential energy is plotted as a function of reaction progress. Starting materials, transition states, intermediates, and final products are all represented, and their appropriate energy levels are indicated.

Reducing sugar (Section 27.8): any sugar that reduces silver ion in the Tollens' test or cupric ion in the Fehling's or Benedict's tests. All sugars that are aldehydes or that can be readily converted into aldehydes are reducing. Glycosides, however, are not reducing sugars.

Refining (Section 3.5): the process by which petroleum is converted into gasoline and other useful products.

Regiochemistry (Section 5.9): a term describing the orientation of a reaction that occurs on an unsymmetrical substrate. Markovnikov's rule, for example, predicts the regiochemistry of electrophilic addition reactions.

Regiospecific (Section 5.9): a term describing a reaction that occurs with a specific regiochemistry to give a single product, rather than a mixture of products.

Replication (Section 33.12): the process by which double-stranded DNA uncoils and is replicated to produce two new copies.

Resolution (Section 28.5): the process by which a racemic mixture is separated into its two pure enantiomers. For example, a racemic carboxylic acid might be converted by reaction with a chiral amine base into a diastereomeric mixture of salts, which could be separated by fractional crystallization. Regeneration of the free acids would then yield the two pure enantiomeric acids.

Resonance effect (Sections 13.10 and 15.9): the effect by which substituents donate or withdraw electron density through orbital overlap with neighboring pi bonds. For example, an oxygen or nitrogen substituent donates electron density to an aromatic ring by overlap of the O or N orbital with the aromatic ring *p* orbitals. A carbonyl substituent, however, withdraws electron density from an aromatic ring by *p*-orbital overlap. These effects are particularly important in determining whether a given group is meta-directing or ortho,para-directing in electrophilic aromatic substitution reactions.

Resonance hybrid (Section 9.6): a molecule, such as benzene, that cannot be represented adequately by a single Kekule structure but must instead be considered as an average of two or more resonance structures. The resonance structures themselves differ only in the positions of their electrons, not of their nuclei.

Ring current (Section 14.11): the circulation of pi electrons induced in aromatic rings by an external magnetic field. This effect accounts for the pronounced downfield shift of aromatic-ring protons in the ^{1}H NMR.

Ring-flip (Section 18.11): the molecular motion that converts one chair conformation of cyclohexane into another chair conformation. The effect of a ring-flip is to convert an axial substituent into an equatorial substituent.

RNA (Section 33.9): ribonucleic acid, the biopolymer found in cells that serves to transcribe the genetic information found in DNA and use that information to direct the synthesis of proteins.

Saccharide (Section 27.1): a sugar.

Saponification (Section 24.7): an old term for the base-induced hydrolysis of an ester to yield a carboxylic acid salt.

Saturated (Section 5.2): A saturated molecule is one that has only single bonds and thus cannot undergo addition reactions. Alkanes, for example, are saturated, but alkenes are unsaturated.

Sawhorse structure (Section 3.7): a stereochemical manner of representation that portrays a molecule using a stick drawing and gives a perspective view of the conformation around single bonds.

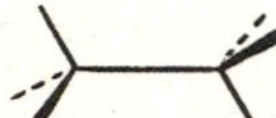

Sawhorse structure

Second-order reaction (Section 10.3): a reaction whose rate-limiting step is bimolecular and whose kinetics are therefore dependent on the concentration of two reagents.

Secondary (See primary)

Secondary structure (Section 31.14): the level of protein substructure that involves organization of chain sections into ordered arrangements such as β-pleated sheets or α-helices.

Sequence rules (Sections 5.6 and 8.7): a series or rules devised by Cahn, Ingold, and Prelog for assigning relative priorities to substituent groups on a double-bond carbon atom or on a chiral center. Once priorities have been established, _E_,_Z_-double bond geometry and _R_,_S_-configurational assignments can be made. (See entgegen, _R_,_S_ convention, zusammen)

Shielding (Section 12.2): an effect observed in NMR that causes a nucleus to absorb toward the right (upfield) side of the chart. Shielding is caused by donation of electron density to the nucleus. (See deshielding)

Sigma bond (Section 1.7): a covalent bond formed by head-on overlap of atomic orbitals.

Sigmatropic reaction (Section 30.1): a pericyclic reaction that involves the migration of a group from one end of a pi electron system to the other. For example, the [1.5] sigmatropic rearrangement of a hydrogen atom in cyclopentadiene is such a reaction.

Skew conformation (Section 3.7): any conformation about a single bond that is intermediate between staggered and eclipsed. (See staggered conformation; eclipsed conformation)

Soap (Section 32.2): the mixture of long-chain fatty acid salts obtained on base hydrolysis of animal fat.

Solid-phase synthesis (Section 31.2): a technique of synthesis whereby the starting material is covalently bound to a solid polymer bead and reactions are carried out on the bound substrate. After the desired transformations have been effected, the product is cleaved from the polymer and is isolated. This technique is particularly useful in peptide synthesis (Merrifield method).

Solvolysis (Section 10.9): an S_N1 reaction in which the intermediate cation reacts with a nucleophilic solvent. Most solvolyses are carried out with polar, non-basic solvents such as water or acetic acid.

sp Orbital (Section 1.11): a hybrid orbital mathematically derived from the combination of an s and a p atomic orbital. The two sp orbitals that result from hybridization are oriented at an angle of 180° to each other.

sp^2 Orbital (Section 1.10): a hybrid orbital mathematically derived by combination of an s atomic orbital with two p atomic orbitals. The three sp^2 hybrid orbitals that result lie in a plane at angles of 120° to each other.

sp^3 Orbital (Section 1.9): a hybrid orbital mathematically derived by combination of an s atomic orbital with three p atomic orbitals. The four sp^3 hybrid orbitals that result are directed towards the corners of a tetrahedron at angles of 109° to each other.

Specific rotation, $[\alpha]_D$ (Section 8.2): The specific rotation of a chiral compound is a physical constant that is defined by the equation:

$$[\alpha]_D = \frac{\text{observed rotation}}{\text{path length} \times \text{concentration}} = \frac{\alpha}{l \times c}$$

where l is the pathlength of the sample solution expressed in decimeters, and c is the concentration of the sample solution expressed in g/mL.

Spin-spin splitting (Section 12.6): the splitting of an NMR signal into a multiplet caused by an interaction between nearby magnetic nuclei whose spins are coupled. The magnitude of spin-spin splitting is given by the coupling constant, *J*.

Staggered conformation (Section 3.7): the three-dimensional arrangement of atoms around a carbon-carbon single bond in which the bonds on one carbon exactly bisect the bond angles on the second carbon as viewed end-on. (See eclipsed conformation)

Staggered conformation

Stationary phase (Section 11.1): the solid support used in chromatography. The molecules to be chromatographically separated adsorb to the stationary phase and are moved along by the mobile phase. Silica gel (hydrated SiO_2) and alumina (Al_2O_3) are often used as stationary phases in column chromatography of organic mixtures. (See mobile phase)

Step-growth polymer (Section 34.1): a polymer produced by a series of polar reactions between two difunctional monomers. The polymer normally has the two monomer units in alternating order and usually has other atoms in addition to carbon in the polymer backbone. Nylon, a polyamide produced by reaction between a diacid and a diamine, is an example of such a polymer.

Stereochemistry (Chapter 8): the branch of chemistry concerned with the three-dimensional arrangement of atoms in molecules.

Stereoisomers (Section 8.3): isomers that have their atoms connected in the same order but that have different three-dimensional arrangements. The term stereoisomer includes both enantiomers and diastereomers but does not include constitutional isomers.

Stereospecific (Section 13.8): a term indicating that only a single stereoisomer is produced in a given reaction, rather than a mixture.

Steric strain (Section 3.9): the strain imposed on a molecule when two groups are too close together and try to occupy the same space. Steric strain is responsible both for the greater stability of trans versus cis alkenes, and for the greater stability of equatorially substituted versus axially substituted cyclohexanes.

Steroid (Section 32.8): a lipid whose structure is based on the tetracyclic carbon skeleton:

Steroids occur in both plants and animals and have a variety of important hormonal functions.

Suprafacial (Section 30.8): a word used to describe the geometry of pericyclic reactions. Suprafacial reactions take place on the same side of the two ends of a pi electron system. (See antarafacial)

Symmetry-allowed, symmetry-disallowed (Section 30.4): A symmetry-allowed reaction is a pericyclic process that has a favorable orbital symmetry for reaction through a concerted pathway. A symmetry-disallowed reaction is one that does not have favorable orbital symmetry for reaction through a concerted pathway.

Syn stereochemistry (Section 6.1): A syn addition reaction is one in which the two ends of the double bond are attacked from the same side. For example, OsO_4 induced hydroxylation of cyclohexene yields *cis*-1,2-cyclohexanediol, the product of syn addition. A syn elimination is one in which the two groups leave from the same side of the molecule.

Syndiotactic polymer (Section 34.6): a chain-growth polymer in which the substituents on the polymer backbone have a regular alternating stereochemistry.

Tautomers (Section 7.4): isomers that are rapidly interconverted. For example, enols and ketones are tautomers since they are rapidly interconverted on treatment with either acid or base catalysts.

Terpenes (Section 32.6): lipids that are formally derived by head-to-tail polymerization of isoprene units. (See isoprene rule)

Tertiary (Section 3.4): (see primary)

Tertiary structure (Section 31.14): the level of protein structure that involves the manner in which the entire protein chain is folded into a specific three-dimensional arrangement.

Thermodynamic control (Section 13.6): Equilibrium reactions that yield the lowest-energy, most stable product are said to be thermodynamically controlled. Although most stable, the product of a thermodynamically controlled reaction is not necessarily formed fastest. (See kinetic control)

Thermoplastic (Section 34.13): a polymer that is hard at room temperature but that becomes soft and pliable when heated. Thermoplastics are used for the manufacture of a variety of molded objects.

Thermosetting resin (Section 34.13): a polymer that is highly cross-linked and that sets into a hard, insoluble mass when heated. Bakelite is the best-known example of such a polymer.

Torsional strain (Section 3.7): the strain in a molecule caused by electron repulsion between eclipsed bonds. Torsional strain plays a major role in destabilizing boat cyclohexane relative to chair cyclohexane. (See eclipsing strain)

Transcription (Section 33.13): the process by which the genetic information encoded in DNA is read and used to synthesize RNA in the nucleus of the cell. A small portion of double-stranded DNA uncoils, and complementary ribonucleotides line up in the correct sequence for RNA synthesis.

Transition state (Section 4.8): an imaginary activated complex between reagents, representing the highest energy point on a reaction curve. Transition states are unstable complexes that cannot be isolated.

Translation (Section 33.14): the process by which the genetic information transcribed from DNA onto mRNA is read by tRNA and used to direct protein synthesis.

Tree diagram (Section 12.13): a diagram used in NMR to help sort out the complicated splitting patterns that can arise from multiple couplings.

Triacylglycerol (Section 32.1): lipids such as animal fat and vegetable oil consisting chemically of triesters of glycerol with long-chain fatty acids.

Triplet (Section 12.6): a symmetrical three-line splitting pattern observed in the ^{1}H NMR when a proton has two equivalent neighbor protons or in the ^{13}C NMR when a carbon is bonded to two hydrogens.

Ultraviolet (UV) spectroscopy (Section 13.11): an optical spectroscopy employing ultraviolet irradiation. UV spectroscopy provides structural information about the extent of pi electron conjugation in organic molecules.

Unimolecular reaction (Section 10.7): a reaction that occurs by spontaneous transformation of the starting material without the intervention of other reagents. For example, the dissociation of a tertiary alkyl halide in the S_N1 reaction is a unimolecular process.

Unsaturated (Section 5.2): An unsaturated molecule is one that has multiple bonds and can undergo addition reactions. Alkenes and alkynes, for example, are unsaturated. (See saturated)

Upfield (Section 12.3): used to refer to the right-hand portion of the NMR chart. (See shielding)

Van der Waals forces (Section 3.6): the attractive forces between molecules that are caused by dipole-dipole interactions. Van der Waals forces are one of the primary forces responsible for holding molecules together in the liquid state.

Vicinal (Section 7.11): a term used to refer to a 1,2-disubstitution pattern. For example, 1,2-dibromoethane is a vicinal dibromide.

Vinylic (Section 7.3): a term that refers to a substituent at a double-bond carbon atom. For example, chloroethylene is a vinylic chloride, and enols are vinylic alcohols.

Vulcanization (Section 34.7): a process for hardening rubber by heating in the presence of elemental sulfur. The sulfur functions by forming cross links between polymer chains.

Wave function (Section 1.1): the mathematical expression that defines the behavior of an electron. The square of the wave function is the probability function that defines the shapes of orbitals.

Wavelength (Section 11.8): the length of a wave from peak to peak. The wavelength of electromagnetic radiation is inversely proportional to frequency and inversely proportional to energy. (See frequency)

Wavenumber (Section 11.9): The wavenumber is the reciprocal of the wavelength in centimeters. Thus, wavenumbers are expressed in cm^{-1}.

Ylide (Section 22.13): a neutral dipolar molecule in which the positive and negative charges are adjacent. For example, the phosphoranes used in Wittig reactions are ylides.

Zaitsev's rule (Section 6.12): a rule stating that E2 elimination reactions normally yield the more highly substituted alkene as major product.

Zusammen (<u>Z</u>) (Section 5.6): a term used to describe the stereochemistry of a carbon-carbon double bond. The two groups on each carbon are assigned priorities according to the Cahn-Ingold-Prelog sequence rules, and the two carbons are compared. If the high priority groups on each carbon are on the same side of the double bond, the bond has <u>Z</u> geometry. (See Entgegen, sequence rules)

Zwitterion (Section 31.2): a neutral dipolar molecule in which the positive and negative charges are not adjacent. For example, amino acids exist as zwitterions. (Zwitterions are also called betaines.)

$H_3\overset{+}{N}\text{-CHR-COO}^-$ A zwitterion

NOMENCLATURE OF POLYFUNCTIONAL ORGANIC COMPOUNDS

Judging from the number of incorrect names that appear in the chemical literature, it is probably safe to say that relatively few practicing organic chemists are fully conversant with the rules of organic nomenclature. Simple hydrocarbons and monofunctional compounds present few problems; the basic rules governing the naming of such compounds are logical and easy to understand. Problems, however, are often encountered with polyfunctional compounds. Whereas most chemists could probably correctly identify hydrocarbon **1** shown below as 3-ethyl-2,5-dimethylheptane, rather few could correctly identify polyfunctional compound **2**. Should we consider **2** as an ether? as an ethyl ester? as a ketone? as an alkene? It is, of course, all four, but it has only one correct name: ethyl 3-(4-methoxy-2-oxo-3-cyclohexenyl)propanoate.

1 3-ethyl-2,5-dimethylheptane

CH_3CHCH_3 (branch)
$CH_3CH_2CHCH_2CHCH_2CH_3$
CH_3 (branch)

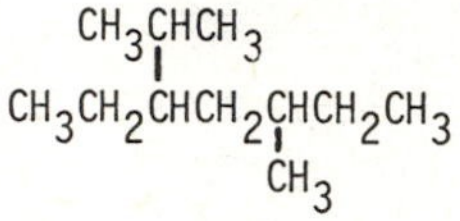

2 ethyl 3-(4-methoxy-2-oxo-3-cyclo-hexenyl)propanoate

O, $COOCH_2CH_3$, CH_3O

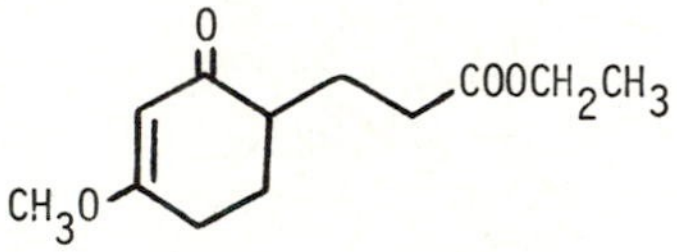

Naming polyfunctional organic compounds is really not much harder than naming monofunctional compounds. All that is required is a prior knowledge of monofunctional-compound nomenclature and rigid application of a set of additional rules. In the following discussion, it is assumed that you have a good command of the rules of monofunctional-compound nomenclature that were given throughout the text as each new functional group was introduced. A list of where these rules can be found is shown in Table 1.

Functional Group	Text Section	Functional Group	Text Section
Alkanes	3.3	Aldehydes	22.2
Cycloalkanes	18.1	Carboxylic Acids	23.1
Alkenes	5.5	Acid Halides	24.1
Alkynes	7.2	Acid Anhydrides	24.1
Alkyl Halides	9.1	Amides	24.1
Aromatic Compounds	14.1	Esters	24.1
Ethers	19.1	Nitriles	24.1
Alcohols	20.1	Amines	28.1
Ketones	22.2		

Table 1. Where to Find Nomenclature Rules for Simple Functional Groups.

The name of a polyfunctional organic molecule has four parts:

1. **Suffix** - the part that identifies the principal functional-group class to which the molecule belongs.
2. **Parent** - the part that identifies the size of the main chain or ring.
3. **Substituent Prefixes** - parts that identify what substituents are located on the main chain or ring.
4. **Locants** - numbers that tell where substituents are located on the main chain or ring.

To arrive at the correct name for a complex molecule, the above four name parts must be identified and then expressed in the proper order and format. Let's look at the four parts.

The Suffix - Functional-Group Precedence

A polyfunctional organic molecule may contain many different kinds of functional groups, but, for nomenclature purposes, we must choose just one suffix. It is not correct to use two suffixes; thus, keto ester **3** shown below would have to be named either as a ketone with an -one suffix or as an ester with an -oate suffix, but could not be named as an -onoate. Similarly, amino alcohol **4** would have to be named either as an alcohol (-ol) or as an amine (-amine) but could not properly be named as an -olamine. The only exception to this rule is in naming compounds that have double or triple bonds. For example, $H_2C{=}CHCH_2COOH$ is 3-butenoic acid, and $HC{\equiv}CCH_2CH_2CH_2CH_2OH$ is 5-hexyn-1-ol.

3 named as an ester with a keto (oxo) substituent [methyl 4-oxopentanoate]

$$CH_3\overset{O}{\overset{\|}{C}}-CH_2CH_2COOCH_3$$

4 named as an alcohol with an amino substituent [5-amino-2-pentanol]

$$CH_3\overset{OH}{\overset{|}{C}}HCH_2CH_2CH_2NH_2$$

How do we choose which suffix to use? Functional groups are divided into two classes, **principal groups** and **subordinate groups,** as shown in Table 2. Principal groups are those that may be cited either as prefixes or as suffixes, whereas subordinate groups are those that may be cited only as prefixes. Within the principal groups, an order of precedence has been established. The proper suffix for a given compound is determined by identifying all of the functional groups present and then choosing the principal group of highest priority. For example, Table 2 indicates that keto ester **3** must be named as an ester rather than as a ketone, since an ester functional group is higher in priority than a ketone. Similarly, amino alcohol **4** must be named as an alcohol rather than as an amine. The correct name of **3** is methyl 4-oxopentanoate, and the correct name of **4** is 5-amino-2-pentanol. Further examples are shown below:

5 named as a cyclohexanecarboxylic acid with an oxo substituent [4-oxocyclohexanecarboxylic acid]

COOH

O

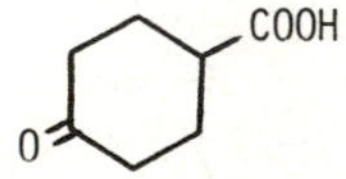

6 named as a carboxylic acid with a chlorocarbonyl substituent [5-chlorocarbonyl-2,2-dimethyl-pentanoic acid]

$$HOOC-\underset{CH_3}{\underset{|}{\overset{CH_3}{\overset{|}{C}}}}-CH_2CH_2CH_2COCl$$

7 named as an ester with an oxo substituent [methyl 5-methyl-6-oxohexanoate]

$$CH_3\overset{CHO}{\overset{|}{C}}HCH_2CH_2CH_2COOCH_3$$

Functional Group Class	Structure	Name When Used as Suffix	Name When Used as Prefix
Principal Groups			
carboxylic acids	-COOH	-oic acid -carboxylic acid	carboxy
carboxylic anhydrides	-C-O-C-	-oic anhydride -carboxylic anhydride	
carboxylic esters	-COOR	-oate -carboxylate	alkoxycarbonyl
acyl halides	-COCl	-oyl halide -carbonyl halide	halocarbonyl (haloformyl)
amides	$-CONH_2$	-amide -carboxamide	amido
nitriles	-C≡N	-nitrile -carbonitrile	cyano
aldehydes	-CHO	-al -carbaldehyde	formyl
	=O		oxo (either aldehyde or ketone)
ketones	=O	-one	oxo
alcohols	-OH	-ol	hydroxy
phenols	-OH	-ol	hydroxy
thiols	-SH	-thiol	mercapto, sulfhydryl
amines	$-NH_2$	-amine	amino
imines	=NH	-imine	imino
alkenes	C=C	-ene	
alkynes	C≡C	-yne	
alkanes	C-C	-ane	

(Table 2 continued on next page)

(Table 2 continued)

Functional Group Class	Structure	Name When Used as Suffix	Name When Used as Prefix
Subordinate Groups			
ethers	-OR	——	alkoxy
sulfides	-SR	——	alkylthio
halides	-F,Cl,Br,I	——	halo
nitro	$-NO_2$	——	nitro
azides	$-N{=}\overset{+}{N}{=}\overset{-}{N}$	——	azido
diazo	$=\overset{+}{N}{=}\overset{-}{N}$	——	diazo

Table 2. Classification of Functional Groups for Purposes of Nomenclature. The principal functional groups are listed in order of decreasing priority, but the subordinate functional groups have no established priority order. Principal functional groups may be cited either as prefixes or as suffixes; subordinate functional groups may be cited only as prefixes.

The Parent - Selecting the Main Chain or Ring

The parent or base name of a polyfunctional organic compound is usually quite easy to identify. If the group of highest priority is part of an open chain, we simply select the longest chain that contains the largest number of principal functional groups. If the highest priority group is attached to a ring, we use the name of that ring system as the parent. For example, compounds **8** and **9** are isomeric aldehydo acids, and both must be named as acids rather than as aldehydes according to Table 2. The longest chain in compound 8 has seven carbons, and the substance is therefore named 6-methyl-7-oxoheptanoic acid. Compound **9** also has a chain of seven carbons, but the longest chain that contains both of the principal functional groups has only three carbons; the correct name of this compound is 3-oxo-2-pentylpropanoic acid.

8 named as a substituted heptanoic acid
[6-methyl-7-oxoheptanoic acid]

$$CH_3\underset{}{\overset{CHO}{\overset{|}{C}}}HCH_2CH_2CH_2CH_2COOH$$

9 named as a substituted propanoic acid
[3-oxo-2-pentylpropanoic acid]

$$CH_3CH_2CH_2CH_2CH_2\overset{CHO}{\overset{|}{C}}HCOOH$$

Similar rules apply for compounds **10 - 13**, which contain rings. Compounds **10** and **11** are isomeric keto nitriles, and both must be named as nitriles according to Table 2. Substance **10** is named as a benzonitrile since the -CN functional group is a substituent on the aromatic ring, but substance **11** is named as an acetonitrile since the -CN functional group is on an open chain. The correct names are 2-acetyl-4-methylbenzonitrile (**10**) and (2-acetyl-4-methylphenyl)acetonitrile (**11**). Compounds **12** and **13** are both keto acids and must be named as acids. The correct names are 3-(2-oxocyclohexyl)propanoic acid (**12**) and 2-(3-oxopropyl)cyclohexanecarboxylic acid (**13**).

10 named as a substituted benzonitrile
[2-acetyl-4-methylbenzonitrile]

CN, H_3C, $COCH_3$

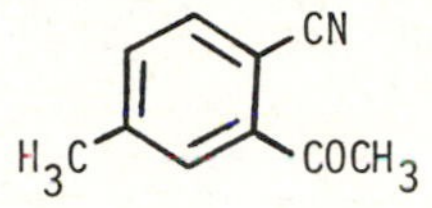

11 named as a substituted acetonitrile
[(2-acetyl-4-methylphenyl)acetonitrile]

CH_2CN, H_3C, $COCH_3$

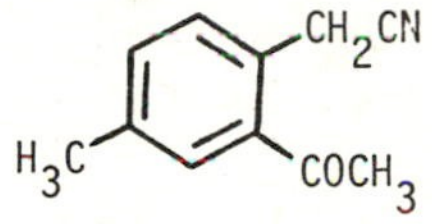

12 named as a carboxylic acid
[3-(2-oxocyclohexyl)propanoic acid]

O, COOH

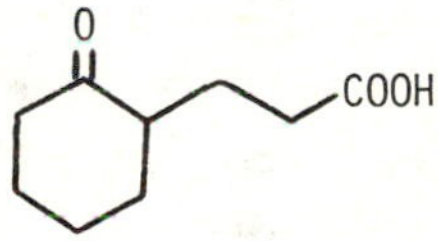

13 named as a carboxylic acid
[2-(3-oxopropyl)cyclohexanecarboxylic acid]

COOH, CHO

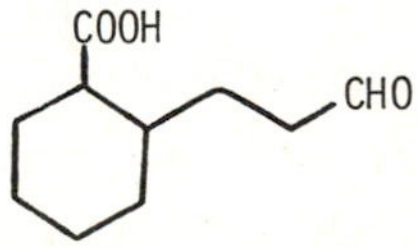

The Prefixes and Locants

With the suffix and parent name established, the next step is to identify and number all substituents on the parent chain or ring. These substituents include all alkyl groups and all functional groups other than the one cited in the suffix. For example, compound **14** contains three different functional groups (carboxyl, keto, and double bond). Since the carboxyl group is highest in priority, and since the longest chain containing the functional groups is seven carbons long, **14** is a heptenoic acid. In addition, the main chain has an oxo (keto) substituent and three methyl groups. Numbering from the end nearer the highest-priority functional group, we find that **14** is 2,5,5-trimethyl-4-oxo-2-heptenoic acid. Note that the final "e" of heptene is deleted in the word "heptenoic". This deletion only occurs when the name would have two adjacent vowels (thus, "heptenoic" has the final "e" deleted, but "heptenenitrile" retains the "e"). Look back at some of the other compounds we have considered to see other examples of how prefixes and locants are assigned.

14 named as a heptenoic acid [2,5,5-trimethyl-4-oxo-2-heptenoic acid]

$$\begin{array}{c} H_3C\ \ O\ \ \ \ \ \ \ \ CH_3 \\ CH_3CH_2C\text{-}C\text{-}CH{=}C\text{-}COOH \\ H_3C \end{array}$$

Writing the Name

Once the name parts have been established, the entire name is written out. Several additional rules apply:

1. Order of Prefixes.
   When the substituents have been identified, the main chain has been numbered, and the proper multipliers such as di- and tri- have been assigned, the name is written with the substituents listed in alphabetical, rather than numerical order. Multipliers such as di- and tri- are not used for alphabetization purposes, but the prefixes iso- and tert- are used.

   15 5-amino-3-methyl-2-pentanol (not 3-methyl-5-amino-2-pentanol)

$$\begin{array}{c} CH_3 \\ H_2NCH_2CH_2CHCHCH_3 \\ OH \end{array}$$

2. Use of Hyphens. Single- and Multiple-Word Names.
   One difficulty that often arises in constructing a name is whether to write the compound as a single word or as multiple words. Should we write methylbenzene or methyl benzene? 2-butanone or 2 butanone? diethylether or diethyl ether? The general rule in such cases is to determine whether the principal functional group is itself an element or compound. If it is, then the name is written as a single word; if it is not, then the name is written as multiple words. For example, methylbenzene (one word) is correct because the parent, benzene, is itself a compound. Diethyl ether, however, is written as two words because the parent, ether, is a class name rather than a compound name. Some further examples are shown:

16 dimethylmagnesium (one word, since magnesium is an element)

$H_3C\text{-}Mg\text{-}CH_3$

17 2-bromopropanoic acid (two words, since "acid" is not a compound)

$CH_3CHBrCOOH$

18 4-(dimethylamino)pyridine (one word, since pyridine is a compound)

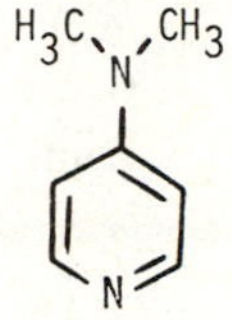

H3C
CH3
N
N

19 methyl cyclopentanecarboxylate

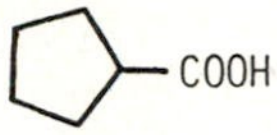

COOH

3. Parentheses.

Parentheses are used to denote complex substituents when ambiguity would otherwise arise. For example, chloromethylbenzene has two substituents on a benzene ring, but (chloromethyl)benzene has only one complex substituent. Note that the expression in parentheses is not set off by hyphens from the rest of the name.

20 chloromethylbenzene (two substituents)

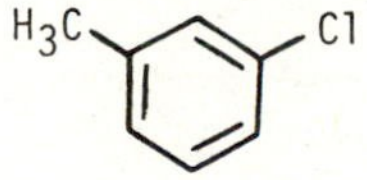

H3C
Cl

21 (chloromethyl)benzene (one complex substituent)

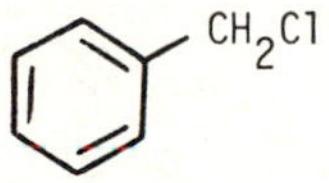

CH2Cl

22 2-(1-methylpropyl)pentanedioic acid (The 1-methylpropyl group is a complex substituent on C-2 of the main chain.)

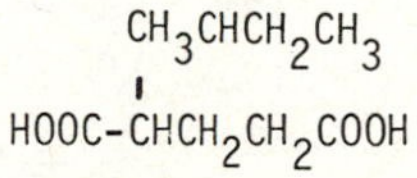

CH3CHCH2CH3
HOOC-CHCH2CH2COOH

Additional Reading:

Further explanations of the rules of organic nomenclature can be found in the following references:

O. T. Benfey, "The Names and Structures of Organic Compounds," Wiley, New York, 1966.

J. H. Fletcher, O. C. Dermer, and R. B. Fox, "Nomenclature of Organic Compounds: Principles and Practice," Advances in Chemistry Series No. 126, American Chemical Society, Washington, D. C. 1974.

"Nomenclature of Organic Chemistry, Sections A, B, C, D, E, F, and H," International Union of Pure and Applied Chemistry, Pergamon Press, Oxford, 1979.

SYMBOLS AND ABBREVIATIONS

A	symbol for Angstrom unit (10^{-8} cm)
Ac-	acetyl group, CH_3CO-
Ar-	aryl group
at. no.	atomic number
at. wt.	atomic weight
$[\alpha]_D$	specific rotation
BOC	*tert*-butoxycarbonyl group, $(CH_3)_3COCO$-
bp	boiling point
n-Bu	*n*-butyl group, $CH_3CH_2CH_2CH_2$-
sec-Bu	*sec*-butyl group, $CH_3CH_2C(CH_3)$-
t-Bu	*tert* -butyl group, $(CH_3)_3C$-
cm	centimeter
cm^{-1}	wavenumber or reciprocal centimeter
D	Debye units (10^{-18} esu cm)
D	stereochemical designation of carbohydrates and amino acids
DBN	diazabicyclononene
DCC	dicyclohexylcarbodiimide, C_6H_{11}-N=C=N-C_6H_{11}
δ	chemical shift in ppm downfield from TMS
Δ	symbol for heat; also symbol for change
$\Delta G°$	standard Gibbs free energy of a reaction
$\Delta G^{\ddagger}$	activation energy
$\Delta H°$	standard enthalpy of a reaction
$\Delta H^{\ddagger}$	enthalpy of activation
$\Delta S°$	standard entropy of a reaction
$\Delta S^{\ddagger}$	entropy of activation
DIBAH	diisobutylaluminum hydride, $[(CH_3)_2CHCH_2]_2AlH$
dl	racemic mixture
dm	decimeter
DMF	dimethylformamide, $(CH_3)_2NCHO$
DMSO	dimethyl sulfoxide, $(CH_3)_2SO$
DMT	dimethoxytrityl
DNA	deoxyribonucleic acid

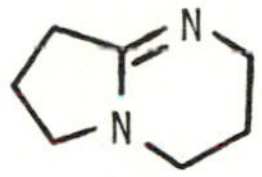

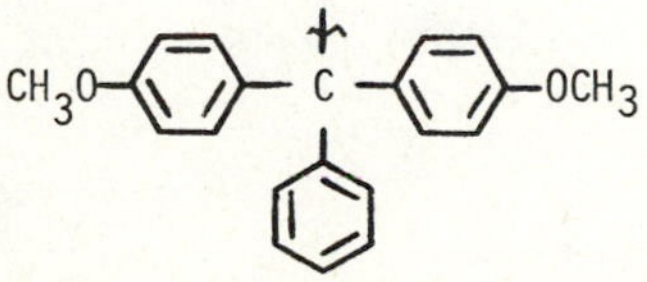

DNP	dinitrophenyl group, as in 2,4-DNP (2,4-dinitrophenylhydrazone)
(*E*)	entgegen, stereochemical designation of double bond geometry
E1	unimolecular elimination reaction
E2	bimolecular elimination reaction
EA	electron affinity
Et	ethyl group, CH_3CH_2-
eu	entropy unit
g	gram
HMPA	hexamethylphosphoramide, $[(CH_3)_2N]_3PO$
HOMO	highest occupied molecular orbital
$h\nu$	symbol for light
Hz	Hertz, or cycles per second
i	iso
IP	ionization potential
IR	infrared
J	symbol for coupling constant
K	symbol for equilibrium constant
K_a	symbol for acid dissociation constant
Kcal	kilocalories
L	stereochemical designation of carbohydrates and amino acids
LAH	lithium aluminum hydride, $LiAlH_4$
LDA	lithium diisopropylamide, $Li\text{-}N[CH(CH_3)_2]_2$
ln	natural logarithm
LUMO	lowest unoccupied molecular orbital
Me	methyl group, CH_3-
mg	milligram
MHz	megahertz (10^6 cycles per second)
mL	milliliter
mm	millimeter
MO	molecular orbital
mol wt.	molecular weight
mp	melting point
m/z	symbol for mass-to-charge ratio in mass spectrometry
μ	symbol for dipole moment
μg	microgram (10^{-6} gram)
mμ	millimicron (nanometer, 10^{-9} meter)
n	normal, straight chain alkane or alkyl group
ng	nanogram (10^{-9} gram)
nm	nanometer (10^{-9} meter)
NMR	nuclear magnetic resonance
-OAc	acetate group, CH_3COO-
PCC	pyridinium chlorochromate

Ph phenyl group, C_6H_5-

pH measure of acidity of aqueous solution

pK_a measure of acid strength (= $-\log K_a$)

ppm parts per million

n-Pr n-propyl group, $CH_3CH_2CH_2$-

i-Pr isopropyl group, $(CH_3)_2CH$-

prim primary

R- symbol for a generalized alkyl group

(R) rectus, stereochemical designation of chiral centers

RNA ribonucleic acid

(S) sinister, stereochemical designation of chiral centers

sec- secondary

S_N1 unimolecular substitution reaction

S_N2 bimolecular substitution reaction

tert- tertiary

THF tetrahydrofuran

THP tetrahydropyranyl group

TMS tetramethylsilane nmr standard, $(CH_3)_4Si$

Tos tosylate group

TPST 2,4,6-triisopropylbenzenesulfonyltetrazole

UV ultraviolet

yr year

X- halogen group

(Z) zusammen, stereochemical designation of double bond geometry

Symbol	Meaning
$\longrightarrow$	chemical reaction in direction indicated
$\rightleftarrows$	reversible chemical reaction
$\longleftrightarrow$	resonance symbol indicating that structures on both sides of the arrow contribute to a resonance hybrid
$\Longrightarrow$	arrow indicating a thought process, rather than an actual chemical reaction
$\curvearrowright$	curved arrow indicating direction of electron flow
$+\!\!\!\longrightarrow$	direction of dipole moment
$\equiv$	is equivalent to
$>$	greater than
$<$	less than
$\sim$	approximately equal to
$-\!\!\!\wr\!\!-$	indicates that the organic fragment shown is a part of a larger molecule
▬	single bond coming out of the plane of the paper
- - -	single bond receding into the plane of the paper
......	partial bond
δ^+, δ^-	partial charge
*	isotopically labeled atom
‡	denoting the transition state

BOND DISSOCIATION ENERGIES

(A-B ⟶ A• + •B)

Bond	ΔH° (kcal/mol)	Bond	ΔH° (kcal/mol)	Bond	ΔH° (kcal/mol)
H-H	104	$H_2C{=}CH$-H	108	$CH_3\overset{O}{\overset{\|}{C}}$-H	86
H-F	136	$H_2C{=}CH$-Cl	88	$CH_3\overset{O}{\overset{\|}{C}}-CH_3$	77
H-Cl	103	$H_2C{=}CHCH_2$-H	87	HO-H	119
H-Br	88	$H_2C{=}CHCH_2$-Cl	69	HO-OH	51
H-I	71	C_6H_5-H	112	CH_3O-H	102
Cl-Cl	58	C_6H_5-Cl	97	CH_3S-H	88
Br-Br	46	C_6H_5-Br	82	CH_3CH_2O-H	103
I-I	36	C_6H_5-OH	112	$CH_3CH_2O-CH_3$	81
CH_3-H	104	$C_6H_5-CH_3$	102	H-CN	130
CH_3-F	109	$C_6H_5CH_2$-H	85	CH_3-CN	112
CH_3-Cl	84	$C_6H_5CH_2$-Cl	70	NH_2-H	103
CH_3-Br	70	$C_6H_5CH_2-CH_3$	72		
CH_3-I	56	HC≡C-H	125		
CH_3-OH	91	CH_3-CH_3	88		
CH_3-NH_2	80	$CH_3CH_2-CH_3$	85		
CH_3CH_2-H	98	$(CH_3)_2CH-CH_3$	84		
CH_3CH_2-Cl	81	$(CH_3)_3C-CH_3$	81		
CH_3CH_2-Br	68	$H_2C{=}CH-CH_3$	97		
CH_3CH_2-I	53	$H_2C{=}CHCH_2-CH_3$	74		
CH_3CH_2-OH	91				
$(CH_3)_2CH$-H	95				
$(CH_3)_2CH$-Cl	80				
$(CH_3)_2CH$-Br	68				
$(CH_3)_3C$-H	91				
$(CH_3)_3C$-Cl	79				
$(CH_3)_3C$-Br	65				
$(CH_3)_3C$-I	50				

ACIDITIES OF SOME ORGANIC COMPOUNDS

Compound	pK_a
CH_3SO_3H	- 1.8
$CH(NO_2)_3$	0.1
O_2N–⌬–OH (with NO_2, NO_2)	0.3
CCl_3COOH	0.5
CF_3COOH	0.5
CBr_3COOH	0.7
HOOCC≡CCOOH	1.2; 2.5
HOOCCOOH	1.2; 3.7
$CHCl_2COOH$	1.3
$CH_2(NO_2)COOH$	1.3
HC≡CCOOH	1.9
Z HOOCCH=CHCOOH	1.9; 6.3
⌬ (COOH, NO_2)	4.2
$CH_3COCOOH$	2.4
$NCCH_2COOH$	2.5
$CH_3C≡CCOOH$	2.6
CH_2FCOOH	2.7
$CH_2ClCOOH$	2.8
$HOOCCH_2COOH$	2.8; 5.6
$CH_2BrCOOH$	2.9
⌬ (COOH, Cl)	3.0
⌬ (COOH, OH)	3.0
E HOOCCH=CHCOOH	3.1; 4.6
CH_2ICOOH	3.2
CHOCOOH	3.2
O_2N–⌬–COOH	3.4
O_2N, O_2N–⌬–COOH	3.5
$HSCH_2COOH$	3.5; 10.2
$CH_2(NO_2)_2$	3.6
CH_3OCH_2COOH	3.6
CH_3COCH_2COOH	3.6; 7.8
$HOCH_2COOH$	3.7
HCOOH	3.7
Cl, ⌬–COOH	3.8
Cl–⌬–COOH	4.0
CH_2BrCH_2COOH	4.0
O_2N–⌬ (NO_2, OH)	4.1
⌬–COOH	4.2
H_2C=CHCOOH	4.2
$HOOCCH_2CH_2COOH$	4.2; 5.7
$HOOCCH_2CH_2CH_2COOH$	4.3; 5.4
Cl, Cl, Cl–⌬–OH, Cl, Cl	4.5
CH_3COOH	4.7
$H_2C=C(CH_3)COOH$	4.7
CH_3CH_2COOH	4.8
$(CH_3)_3CCOOH$	5.0
$CH_3COCH_2NO_2$	5.1
ring (O, O)	5.3
$O_2NCH_2COOCH_3$	5.8
ring (O, CHO)	5.8
Cl, Cl–⌬–OH, Cl	6.2
⌬–SH	6.6
HCO_3H	7.1
⌬ (NO_2, OH)	7.2
$(CH_3)_2CHNO_2$	7.7
Cl–⌬–OH, Cl	7.8
CH_3CO_3H	8.2

Compound	pK_a	Compound	pK_a	Compound	pK_a
OH, Cl	8.5	$CH_2(CN)_2$	11.2	CH_3COCH_3	20.0
$CH_3CH_2NO_2$	8.5	CCl_3CH_2OH	12.2		23
F_3C–OH	8.7	Glucose	12.3	$CH_3COOCH_2CH_3$	25
$CH_3COCH_2COCH_3$	9.0	$(CH_3)_2C{=}NOH$	12.4	HC≡CH	25
HO, OH	9.3; 11.1	$CH_2(COOCH_3)_2$	12.9	CH_3CN	25
OH, OH	9.3; 12.6	$CHCl_2CH_2OH$	12.9	$CH_3SO_2CH_3$	28
CH_2SH	9.4	$CH_2(OH)_2$	13.3	$(C_6H_5)_3CH$	32
HO, OH	9.9; 11.5	CH_3CHO	13.5	$(C_6H_5)_2CH_2$	34
$CH_3COCH_2SOCH_3$	10.0	$(CH_3)_2CHCHO$	13.8	CH_3SOCH_3	35
OH	10.0	$HOCH_2CH(OH)CH_2OH$	14.1	NH_3	35
OH, CH_3	10.3	CH_2ClCH_2OH	14.3	$CH_3CH_2NH_2$	35
CH_3NO_2	10.3		15.0	$(CH_3CH_2)_2NH$	36
CH_3SH	10.3	CH_3OH	15.1	CH_3	41
$CH_3COCH_2COOCH_3$	10.6	CH_2OH	15.4		43
CH_3COCHO	11.0	$H_2C{=}CHCH_2OH$	15.5	$H_2C{=}CH_2$	44
		CH_3CH_2OH	15.9	CH_4	49
		$CH_3CH_2CH_2OH$	16.1		
		CH_3COCH_2Br	16.1		
		=O	16.7		
		$(CH_3)_2CHOH$	17.1		
		$(CH_3)_3COH$	19.2		

An acidity list covering more than 5000 organic compounds has been published: E. P. Serjeant and B. Dempsey (eds.), "Ionization Constants of Organic Acids in Aqueous Solution," IUPAC Chemical Data Series No. 23, Pergamon Press, Oxford, England, 1979.

INFRARED ABSORPTION FREQUENCIES

Functional Group Class		Frequency (cm^{-1})	Text Section
Alcohol	-O-H	3300 - 3600 (s)	20.12
	>C-O-	1050 (s)	20.12
Aldehyde	-CO-H	2720, 2820 (m)	22.17
aliphatic	>C=O	1725 (s)	22.17
aromatic	>C=O	1705 (s)	22.17
Alkane	>C-H	2850 - 2960 (s)	11.11
	>C-C<	800 - 1300 (m)	11.11
Alkene	=C-H	3020 - 3100 (m)	11.11
	>C=C<	1650 - 1670 (m)	11.11
	$RCH=CH_2$	910, 990 (m)	11.11
	$R_2C=CH_2$	890 (m)	11.11
Alkyne	≡C-H	3300 (s)	11.11
	-C≡C-	2100 - 2260 (m)	11.11
Alkyl bromide	>C-Br	500 - 600 (s)	11.11
Alkyl chloride	>C-Cl	600 - 800 (s)	11.11
Amine, primary	H-N-H	3400, 3500 (s)	28.11
secondary	>N-H	3350 (s)	28.11
Ammonium salt	$\geq\overset{+}{N}$-H	2200 - 3000 (broad)	28.11
Aromatic ring	Ar-H	3030 (m)	14.11
monosubstituted	Ar-R	690 - 710 (s)	14.11
		730 - 770 (s)	
o-disubstituted		735 - 770 (s)	14.11
m-disubstituted		690 - 710 (s)	14.11
		810 - 850 (s)	
p-disubstituted		810 - 840 (s)	14.11
Carboxylic acid	-O-H	2500 - 3300 (broad)	23.10
associated	>C=O	1710 (s)	23.10
free	>C=O	1760 (s)	23.10
Acid anhydride	>C=O	1820, 1760 (s)	24.11

Acid chloride			
aliphatic	$>C=O$	1810 (s)	24.11
aromatic	$>C=O$	1770 (s)	24.11
Amide, aliphatic	$>C=O$	1690 (s)	24.11
aromatic	$>C=O$	1675 (s)	24.11
N-substituted	$>C=O$	1680 (s)	24.11
N,N-disubstituted	$>C=O$	1650 (s)	24.11
Ester, aliphatic	$>C=O$	1735 (s)	24.11
aromatic	$>C=O$	1720 (s)	24.11
Ether	$-O-C\lessgtr$	1050 - 1150 (s)	19.10
Ketone, aliphatic	$>C=O$	1715 (s)	22.17
aromatic	$>C=O$	1690 (s)	22.17
6-memb. ring	$>C=O$	1715 (s)	22.17
5-memb. ring	$>C=O$	1750 (s)	22.17
Nitrile, aliphatic	$-C\equiv N$	2250 (m)	24.11
aromatic	$-C\equiv N$	2230 (m)	24.11
Phenol	$-O-H$	3500 (s)	29.10

(s) = strong; (m) = medium intensity

PROTON NMR CHEMICAL SHIFTS

Type of Proton		Chemical shift (ppm)	Text Section
Alkyl, primary	$R-CH_3$	0.7 - 1.3	12.10
Alkyl, secondary	$R-CH_2-R$	1.2 - 1.4	12.10
Alkyl, tertiary	R_3C-H	1.4 - 1.7	12.10
Allylic	$>C=C-C-H$	1.6 - 1.9	12.10
α to carbonyl	$-C(=O)-C-H$	2.0 - 2.3	22.17
Benzylic	Ar-C-H	2.3 - 3.0	14.11
Acetylenic	R-C≡C-H	2.5 - 2.7	12.10
Alkyl chloride	Cl-C-H	3.0 - 4.0	12.10
Alkyl bromide	Br-C-H	2.5 - 4.0	12.10
Alkyl iodide	I-C-H	2.0 - 4.0	12.10
Amine	>N-C-H	2.2 - 2.6	28.11
Epoxide	-C—C-H (O bridging)	2.5 - 3.5	19.10
Alcohol	HO-C-H	3.5 - 4.5	20.12
Ether	R-O-C-H	3.5 - 4.5	19.10
Vinylic	>C=C-H	5.0 - 6.5	12.10
Aromatic	Ar-H	6.5 - 8.0	14.11
Aldehyde	R-C(=O)—H	9.7 - 10.0	22.17
Carboxylic acid	R-C(=O)-O-H	11.0 - 12.0	23.10
Alcohol	R-O-H	3.5 - 4.5	20.12
Phenol	Ar-O-H	2.5 - 6.0	29.10

REAGENTS USED IN ORGANIC SYNTHESIS

The following table summarizes the uses of some important reagents in organic chemistry. The reagents are listed alphabetically, followed by a brief description of the uses of each and references to the appropriate text sections.

Acetic acid: Reacts with vinylic organoboranes to yield alkenes. The net effect of alkyne hydroboration followed by protonolysis with acetic acid is reduction of the alkyne to a cis alkene (Section 7.5).

Acetic anhydride: Reacts with alcohols to yield acetate esters (Sections 24.6, 27.8).

Aluminum chloride: Acts as a Lewis acid catalyst in Friedel-Crafts alkylation and acylation reactions of aromatic-ring compounds (Sections 16.2, 16.6).

Borane: Adds to alkenes, giving alkylboranes that can be oxidized with alkaline hydrogen peroxide to yield alcohols (Section 6.4).

- Adds to alkynes, giving vinylic organoboranes that can either be treated with acetic acid to yield cis alkenes (Section 7.5) or can be oxidized with alkaline hydrogen peroxide to yield ketones (Section 7.5).

- Reduces carboxylic acids to yield primary alcohols (Section 23.9).

Boron trichloride: Cleaves ethers under mild conditions, yielding alcohols and alkyl chlorides (Section 19.6).

Bromine: Adds to alkenes yielding 1,2-dibromides (Sections 6.1, 13.5).

- Adds to alkynes yielding either 1,2-dibromoalkenes or 1,1,2,2-tetrabromoalkanes (Section 7.3).

- Reacts with arenes in the presence of ferric bromide catalyst to yield bromoarenes (Section 15.1).

- Reacts with ketones in acetic acid solvent to yield α-bromo ketones (Section 25.3).

- Reacts with carboxylic acids in the presence of phosphorus tribromide to yield α-bromo carboxylic acids (Hell-Volhard-Zelinskii reaction; Section 25.4).

- Reacts with methyl ketones in the presence of sodium hydroxide to yield carboxylic acids and bromoform (Haloform reaction; Section 25.7).

- Oxidizes aldoses to yield aldonic acids (Section 27. 8).

_N_-Bromosuccinimide: Reacts with alkenes in the presence of aqueous dimethylsulfoxide to yield bromohydrins (Section 6.2).

- Reacts with alkenes in the presence of light to yield allylic bromides (Wohl-Ziegler reaction; Section 9.5).

- Reacts with alkylbenzenes in the presence of light to yield benzylic bromides (Section 16.8).

Di-_tert_-butoxy dicarbonate: Reacts with amino acids to give t-BOC protected amino acids suitable for use in peptide synthesis (Section 31.11).

Butyllithium: Reacts with alkynes to yield alkyne dianions that can be alkylated (Section 7.9).

- Reacts with dialkylamines to yield lithium dialkylamide bases such as LDA [lithium diisopropylamide] (Section 25.5).

Carbon dioxide: Reacts with Grignard reagents to yield carboxylic acids (Section 23.6).

- Reacts with phenoxide anions to yield _o_-hydroxybenzoic acids (Kolbe-Schmitt carboxylation reaction; Section 29.9).

Chlorine: Adds to alkenes to yield 1,2-dichlorides (Section 6.1, 13.5).

- Reacts with alkanes in the presence of light to yield chloroalkanes by a radical chain reaction pathway (Section 9.4).

- Reacts with arenes in the presence of ferric chloride catalyst to yield chloroarenes (Section 15.2).

Chlorine monoxide: Reacts with arenes in the presence of trifluoroacetic acid to yield chloroarenes. The reaction is particularly effective with rings that have electron withdrawing substituents (Section 15.2).

_m_-Chloroperoxybenzoic acid: Reacts with alkenes to yield epoxides (Section 19.7).

Chromium trioxide: Oxidizes alcohols in aqueous sulfuric acid (Jones reagent) to yield carbonyl-containing products. Primary alcohols yield carboxylic acids and secondary alcohols yield ketones (Sections 20.10, 22.5, 23.6).

Cuprous bromide: Reacts with arenediazonium salts to yield bromoarenes (Sandmeyer reaction; Section 29.4).

Cuprous chloride: Reacts with arenediazonium salts to yield chloroarenes (Sandmeyer reaction; Section 29.4).

Cuprous cyanide: Reacts with arenediazonium salts to yield substituted benzonitriles (Sandmeyer reaction; Section 29.4).

Cuprous iodide: Reacts with organolithiums to yield lithium diorganocopper reagents (Gilman reagents; Section 9.9).

Diazabicyclononene (DBN): Reacts with alkyl halides to yield alkenes in an elimination reaction (Section 6.12).

Diazomethane: Reacts with carboxylic acids to yield methyl esters (Section 24.4).

Dicyclohexylcarbodiimide (DCC): Couples an amine with a carboxylic acid to yield an amide. DCC is often used in peptide synthesis (Section 31.11).

Diethyl acetamidomalonate: Reacts with alkyl halides in a common method of α-amino acid synthesis (Section 31.4).

Diethylaluminum cyanide: Reacts with α,β-unsaturated ketones to yield β-ketonitriles (Section 22.15).

Dihydropyran: Reacts with alcohols in the presence of an acid catalyst to yield tetrahydropyranyl ethers that serve as useful hydroxyl protecting groups (Section 20.11).

Diiodomethane: Reacts with alkenes in the presence of zinc-copper couple to yield cyclopropanes (Simmons-Smith reaction; Section 18.5).

Diisobutylaluminum hydride (DIBAH): Reduces esters to yield aldehydes (Sections 22.3, 24.6).

- Reduces nitriles to yield aldehydes (Section 24.9).

2,4-Dinitrophenylhydrazine: Reacts with ketones and aldehydes to yield 2,4-DNPs that serve as useful crystalline derivatives (Section 22.9).

Disiamylborane: A hindered dialkylborane that adds to terminal alkynes, giving trialkylboranes that can be oxidized with alkaline hydrogen peroxide to yield aldehydes (Section 7.5).

1,2-Ethanedithiol: Reacts with ketones or aldehydes in the presence of an acid catalyst to yield dithioacetals that can be reduced with Raney nickel to yield alkanes (Section 22.12).

Ethylene glycol: Reacts with ketones or aldehydes in the presence of a acid catalyst to yield acetals that serve as useful carbonyl-protecting groups (Section 22.11).

Ferric bromide: Acts as a catalyst for the reaction of arenes with bromine to yield bromoarenes (Section 15.1).

Ferric chloride: Acts as a catalyst for the reaction of arenes with chlorine to yield chloroarenes (Section 15.2).

Grignard reagent: Adds to carbonyl-containing compounds (ketones, aldehydes, esters) to yield alcohols (Section 20.7).

Hydrazine: Reacts with ketones or aldehydes in the presence of potassium hydroxide to yield the corresponding alkanes (Wolff-Kishner reaction; Section 22.10).

Hydrogen bromide: Adds to alkenes to yield alkyl bromides. Markovnikov regiochemistry is observed (Sections 5.7, 13.5).

- Adds to alkenes in the presence of a peroxide catalyst to yield alkyl bromides. Non-Markovnikov regiochemistry is observed (Section 6.5).

- Adds to alkynes to yield either bromoalkenes or 1,1-dibromoalkanes (Section 7.3).

- Reacts with alcohols to yield alkyl bromides (Sections 9.7, 19.6).

Hydrogen chloride: Adds to alkenes to yield alkyl chlorides. Markovnikov regiochemistry is observed (Sections 5.7, 13.5).

- Adds to alkynes to yield either chloroalkenes or 1,1-dichloroalkanes (Section 7.3).

- Reacts with alcohols to yield alkyl chlorides (Section 9.7).

Hydrogen cyanide: Adds to ketones and aldehydes to yield cyanohydrins (Section 22.8).

Hydrogen iodide: Reacts with alcohols to yield alkyl iodides (Section 19.6).

Hydrogen peroxide: Oxidizes organoboranes to yield alcohols. Used in conjunction with addition of borane to alkenes, the overall transformation effects syn Markovnikov addition of water to an alkene (Section 6.4).

- Oxidizes vinylic boranes to yield aldehydes. Since the vinylic borane starting materials are prepared by hydroboration of a terminal alkyne with disiamyl borane, the overall transformation is the hydration of a terminal alkyne to yield an aldehyde (Section 7.5).

- Reacts with arenes in the presence of trifluoromethanesulfonic acid to yield phenols (Section 15.5).

- Oxidizes sulfides to yield sulfoxides (Section 19.11).

- Reacts with α-phenylselenenyl ketones to yield α,β-unsaturated ketones (Section 25.8).

Hydroxylamine: Reacts with ketones and aldehydes to yield oximes (Section 22.9).

- Reacts with aldoses to yield oximes as the first step in the Wohl degradation of aldoses (Section 27.8).

Hypophosphorous acid: Reacts with arenediazonium salts to yield arenes (Section 29.4).

Iodine: Reacts with arenes in the presence of cupric chloride or hydrogen peroxide to yield iodoarenes (Section 15.2).

- Reacts with carboxylic acids in the presence of lead tetraacetate to yield alkyl iodides (Hunsdiecker reaction; Section 23.9).

- Reacts with methyl ketones in the presence of aqueous sodium hydroxide to yield carboxylic acids and iodoform (Section 25.7).

Iodomethane: Reacts with alkoxide anions to yield methyl ethers (Section 19.4).

- Reacts with carboxylate anions to yield methyl esters (Section 24.7).

- Reacts with enolate ions to yield α-methylated carbonyl compounds (Section 25.9).

- Reacts with amines to yield methylated amines (Section 28.7).

Iron: Reacts with nitroarenes in the presence of mineral acid to yield anilines (Section 29.3).

Lead tetraacetate: Reacts with carboxylic acids in the presence of iodine to yield alkyl iodides and carbon dioxide (Hunsdiecker reaction; Section 23.9).

Lindlar catalyst: Acts as a catalyst for the hydrogenation of alkynes to yield cis alkenes (Section 7.6).

Lithium: Reduces alkynes in liquid ammonia solvent to yield trans alkenes (Section 7.6).

- Reacts with organohalides to yield organolithium compounds (Section 9.9).

Lithium aluminum hydride: Reduces primary alkyl halides to yield alkanes (Section 9.8).

- Reduces ketones, aldehydes, esters, and carboxylic acids to yield alcohols (Section 20.5).

- Reduces amides to yield amines (Section 24.8).

- Reduces alkyl azides to yield amines (Section 28.7).

- Reduces nitriles to yield amines (Section 28.7).

Lithium diisopropylamide (LDA): Reacts with carbonyl compounds (aldehydes, ketones, esters) to yield enolate ions (Sections 25.8, 25.9).

Lithium diorganocopper reagent (Gilman reagent): Couples with alkyl halides to yield alkanes (Section 9.9).

- Adds to α,β-unsaturated ketones to give 1,4-addition products (Section 22.15).

Lithium tri-tert-butoxyaluminum hydride: Reduces acid chlorides to yield aldehydes (Section 24.5).

Lithium triethylborohydride: Reduces primary and secondary alkyl halides to yield alkanes (Section 9.8).

Magnesium: Reacts with organohalides to yield Grignard reagents (Section 9.8).

Mercuric acetate: Adds to alkenes in the presence of water, giving β-hydroxy organomercury compounds that can be reduced with sodium borohydride to yield alcohols. The overall reaction effects the Markovnikov hydration of an alkene (Section 6.3).

- Adds to alkenes in the presence of acetic acid, giving β-acetoxy organomercury compounds that can be reduced with sodium borohydride to yield esters. The overall reaction effects a net addition of acetic acid to the alkene (Section 6.4).

Mercuric oxide: Reacts with carboxylic acids in the presence of bromine to yield alkyl bromides and carbon dioxide (Hunsdiecker reaction; Section 23.9).

Mercuric sulfate: Acts as a catalyst for the addition of water to alkynes in the presence of aqueous sulfuric acid, yielding ketones (Section 7.4).

Mercuric trifluoroacetate: Adds to alkenes in the presence of alcohol, giving β-alkoxy organomercury compounds that can be reduced with sodium borohydride to yield ethers. The overall rection effects a net addition of an alcohol to an alkene (Section 19.5).

Methyl sulfate: A reagent used to methylate heterocyclic amine bases during Maxam-Gilbert DNA sequencing (Section 33.15).

Nitric acid: Reacts with arenes in the presence of sulfuric acid to yield nitroarenes (Section 15.3).

- Oxidizes aldoses to yield aldaric acids (Section 27.8).

Nitronium tetrafluoroborate: Reacts with arenes to yield nitroarenes (Section 15.3).

Nitrous acid: Reacts with amines to yield diazonium salts (Section 29.4).

Osmium tetroxide: Adds to alkenes to yield 1,2-diols (Sections 6.7, 20.4).

- Reacts with alkenes in the presence of periodic acid to cleave the carbon-carbon double bond, yielding ketone or aldehyde fragments (Section 6.9).

- Reacts with alkenes in the presence of potassium permanganate to cleave the carbon-carbon double bond, yielding carboxylic acid fragments (Section 6.9).

Oxalyl chloride: Reacts with carboxylic acids yielding acid chlorides (Sections 16.6, 24.4).

Ozone: Adds to alkenes to cleave the carbon-carbon double bond and give ozonides. The ozonides can then be reduced either with sodium borohydride to yield alcohols, or with zinc in acetic acid to yield carbonyl compounds (Sections 6.8, 22.4, 22.5).

Palladium on barium sulfate: Acts as a hydrogenation catalyst in the Rosenmund reduction of acid chlorides to yield aldehydes (Sections 22.3, 24.5).

Palladium on carbon: Acts as a hydrogenation catalyst in the reduction of carbon-carbon multiple bonds. Alkenes and alkynes are reduced to yield alkanes (Sections 6.6, 7.6).

- Acts as a hydrogenation catalyst in the reduction of aryl ketones to yield alkylbenzenes (Section 16.7).

- Acts as a hydrogenation catalyst in the reduction of nitroarenes to yield anilines (Section 29.3).

Periodic acid: Reacts with 1,2-diols to yield carbonyl-containing cleavage products (Section 6.9).

Peroxyacetic acid: Oxidizes sulfoxides to yield sulfones (Section 19.11)

Phenyl isothiocyanate: A reagent used in the Edman degradation of peptides to identify N-terminal amino acids (Section 31.9).

Phenylselenenyl bromide: Reacts with enolate ions to yield α-phenylselenenyl ketones. On oxidation of the product with hydrogen peroxide, an α,β-unsaturated ketone is produced (Section 25.8).

Phosphorus oxychloride: Reacts with secondary and tertiary and alcohols to yield alkene dehydration products (Sections 6.12, 20.9).

Phosphorus tribromide: Reacts with alcohols to yield alkyl bromides (Section 9.7).

- Reacts with carboxylic acids in the presence of bromine to yield α-bromo carboxylic acids (Hell-Volhard-Zelinskii reaction; Section 25.4).

Phosphorus trichloride: Reacts with carboxylic acids to yield acid chlorides (Section 24.4).

Platinum oxide (Adam's catalyst): Acts as a hydrogenation catalyst in the reduction of alkenes and alkynes to yield alkanes (Section 6.6).

Potassium hydroxide: Reacts with alkyl halides to yield alkenes by an elimination reaction (Section 6.12).

- Reacts with 1,1- or 1,2-dihaloalkanes to yield alkynes by a twofold elimination reaction (Section 7.11).

Potassium nitrosodisulfonate (Fremy's salt): Oxidizes phenols and anilines to yield quinones (Section 29.9).

Potassium permanganate: oxidizes alkenes under alkaline conditions to yield 1,2-diols (Section 6.7).

- Oxidizes alkenes under neutral or acidic conditions to give carboxylic acid double-bond cleavage products (Sections 6.9).
- Oxidizes alkynes to give carboxylic acid triple-bond cleavage products (Section 7.10).
- Oxidizes arenes to yield benzoic acids (Section 16.9).
- Oxidizes alkenes in the presence of sodium periodate to yield double-bond cleavage products (Section 23.6).

Potassium phthalimide: Reacts with alkyl halides to yield an N-alkylphthalimide that is hydrolyzed by aqueous sodium hydroxide to yield an amine (Gabriel amine synthesis; Section 28.7).

Potassium *tert*-butoxide: Reacts with allylic halides to yield conjugated dienes in an elimination reaction (Section 13.1).

- Reacts with chloroform in the presence of an alkene to yield a dichlorocyclopropane (Section 18.5).

Pyridine: Reacts with α-bromo ketones to yield α,β-unsaturated ketones (Section 25.3).

- Acts as a catalyst for the reaction of alcohols with acid chlorides to yield esters (Section 24.5).
- Acts as a catalyst for the reaction of alcohols with acetic anhydride to yield acetate esters (Section 24.6).

Pyridinium chlorochromate: Oxidizes primary alcohols to yield aldehydes and secondary alcohols to yield ketones (Sections 20.10, 22.3, 22.4).

Pyrrolidine: Reacts with ketones to yield enamines for use in the Stork enamine reaction (Section 22.9).

Raney nickel: Reduces dithioacetals to yield alkanes by a desulfurization reaction (Section 22.12).

Rhodium on carbon: Acts as a hydrogenation catalyst in the reduction of benzene rings to yield cyclohexanes (Section 16.10).

Silver oxide: Oxidizes primary alcohols in aqueous ammonia solution to yield aldhydes (Tollens oxidation; Sections 22.5, 23.6).

- Acts as a catalyst for the reaction of alcohols with alkyl halides to yield ethers (Section 27.8).

Sodium amide: Reacts with terminal alkynes to yield acetylide anions (Section 7.7).

- Reacts with 1,1- or 1,2-dihalides to yield alkynes by a twofold elimination reaction (Section 7.11).

- Reacts with aryl halides to yield anilines by a benzyne aromatic substitution mechanism (Section 15.14).

Sodium azide: Reacts with alkyl halides to yield alkyl azides (Section 28.7).

- Reacts with acid chlorides to yield acyl azides. On heating in the presence of water, acyl azides yield amines and carbon dioxide (Section 28.7).

Sodium bisulfite: Reduces osmate esters, prepared by treatment of an alkene with osmium tetroxide, to yield 1,2-diols (Sections 6.7, 20.4).

Sodium borohydride: Reduces organomercury compounds, prepared by oxymercuration of alkenes, to convert the C-Hg bond to C-H (Section 6.3).

- Reduces ozonides, prepared by treatment of an alkene with ozone, to yield alcohols (Section 6.8).

- Reduces ketones and aldehydes to yield alcohols (Section 20.5).

- Reduces quinones to yield hydroquinones (Section 29.9).

Sodium cyanide: Reacts with alkyl halides to yield alkanenitriles (Section 23.6).

Sodium cyanoborohydride: Reacts with ketones and aldehydes in the presence of ammonia to yield an amine by a reductive amination process (Section 28.7).

Sodium dichromate: Oxidizes primary alcohols to yield carboxylic acids and secondary alcohols to yield ketones (Sections 20.10, 22.5).

- Oxidizes alkylbenzenes to yield benzoic acids (Section 16.9).

Sodium hydride: Reacts with alcohols to yield alkoxide anions (Section 20.9).

Sodium hydroxide: Reacts with arenesulfonic acids at high temperature to yield phenols (Section 15.4).

- Reacts with aryl halides to yield phenols by a benzyne aromatic substitution mechanism (Sections 15.13, 15.14).

- Reacts with methyl ketones in the presence of iodine to yield carboxylic acids and iodoform (Section 25.7).

Sodium iodide: Reacts with arenediazonium salts to yield aryl iodides (Section 29.4).

Sodium periodate: Oxidizes alkenes in the presence of potassium permanganate to yield double-bond cleavage products (Section 23.6).

Stannous chloride: Reduces nitroarenes to yield anilines (Sections 15.14, 29.3).

- Reduces quinones to yield hydroquinones (Section 29.9).

Sulfur trioxide: Reacts with arenes in sulfuric acid solution to yield arenesulfonic acids (Section 15.4).

Sulfuric acid: Reacts with alcohols to yield alkenes (Section 6.12).

- Reacts with alkynes in the presence of water and mercuric acetate to yield ketones (Section 7.4).

Thionyl chloride: Reacts with primary and secondary alcohols to yield alkyl chlorides (Section 9.7).

- Reacts with carboxylic acids to yield acid chlorides (Sections 16.6, 24.4).

Thiourea: Reacts with primary alkyl halides to yield thiols (Section 20.13).

***p*-Toluenesulfonyl chloride:** Reacts with alcohols to yield tosylates (Section 20.9).

Trifluoroacetic acid: Acts as a catalyst for the reaction of arenes with Cl_2O to yield aryl chlorides (Section 15.2).

- Acts as a catalyst for cleaving *tert*-butyl ethers, yielding alcohols and 2-methylpropene (Section 19.6).

- Acts as a catalyst for cleaving the t-BOC protecting group from amino acids in peptide synthesis (Section 31.11).

Trifluoromethanesulfonic acid: Acts as a catalyst for the reaction of arenes with hydrogen peroxide to yield phenols (Section 15.5).

Triisopropylsulfonyltetrazole (TPST): Acts as a coupling reagent for use in DNA synthesis (Section 33.16).

Triphenylphosphine: Reacts with primary alkyl halides to yield the alkyltriphenylphosphonium salts used in Wittig reactions (Section 22.13).

Zinc bromide: Acts as a Lewis acid catalyst to cleave DMT ethers in DNA synthesis (Section 33.16).

Zinc: Reduces ozonides, produced by addition of ozone to alkenes, to yield ketones and aldehydes (Section 6.8).

- Reduces disulfides to yield thiols (Section 20.13).

Zinc-copper couple: Reacts with diiodomethane in the presence of alkenes to yield cyclopropanes (Simmons-Smith reaction; Section 18.5).

SUMMARY OF FUNCTIONAL GROUP PREPARATIONS

The following table summarizes the synthetic methods by which important functional groups can be prepared. The functional groups are listed alphabetically, followed by reference to the appropriate text section and a brief description of each synthetic method.

Acetals	(Sec. 22.11)	from ketones and aldehydes by acid-catalyzed reaction with alcohols
Acid anhydrides	(Sec. 24.4)	from dicarboxylic acids by heating
	(Sec. 24.6)	from acid chlorides by reaction with carboxylate salts
Acid bromides	(Sec. 24.4)	from carboxylic acids by reaction with PBr_3
Acid chlorides	(Sec. 16.6, 24.4)	from carboxylic acids by reaction with either $SOCl_2$, PCl_3, or oxalyl chloride
Alcohols	(Sec. 6.3)	from alkenes by oxymercuration/demercuration
	(Sec. 6.4)	from alkenes by hydroboration/oxidation
	(Sec. 6.7)	from alkenes by hydroxylation with either OsO_4 or $KMnO_4$
	(Sec. 10.4, 10.5)	from alkyl halides and tosylates by S_N2 reaction with hydroxide ion
	(Sec. 19.6)	from ethers by acid-induced cleavage
	(Sec. 19.8)	from epoxides by acid-catalyzed ring opening with either H_2O or HX
	(Sec. 19.8)	from epoxides by base-induced ring opening
	(Sec. 20.5)	from ketones and aldehydes by reduction with $NaBH_4$ or $LiAlH_4$
	(Sec. 20.7)	from ketones and aldehydes by addition of Grignard reagents
	(Sec. 23.8)	from carboxylic acids by reduction with either $LiAlH_4$ or BH_3
	(Sec. 24.5)	from acid chlorides by reduction with $LiAlH_4$
	(Sec. 24.5)	from acid chlorides by reaction with Grignard reagents
	(Sec. 24.6)	from acid anhydrides by reduction with $LiAlH_4$
	(Sec. 20.5, 24.7)	from esters by reduction with $LiAlH_4$

	(Sec. 20.7, 24.7)	from esters by reaction with Grignard reagents
Aldehydes	(Sec. 6.8)	from disubstituted alkenes by ozonolysis
	(Sec. 6.9)	from 1,2-diols by cleavage with sodium periodate
	(Sec. 7.5)	from terminal alkynes by hydroboration with disiamylborane followed by oxidation
	(Sec. 20.10, 22.3)	from primary alcohols by oxidation
	(Sec. 22.3, 24.5)	from acid chlorides by catalytic hydrogenation (Rosenmund reduction)
	(Sec. 24.5)	from acid chlorides by partial reduction with $LiAl(O\text{-}\underline{t}\text{-}Bu)_3H$
	(Sec. 22.3, 24.7)	from esters by reduction with DIBAH $[H\text{-}Al(\underline{i}\text{-}Bu)_2]$
	(Sec. 24.9)	from nitriles by partial reduction with DIBAH
Alkanes	(Sec. 6.6)	from alkenes by catalytic hydrogenation
	(Sec. 9.8)	from alkyl halides by synthesis and protonolysis of Grignard reagents
	(Sec. 9.8, 10.4)	from primary alkyl halides by reduction with $LiBH(Et)_3$
	(Sec. 9.9)	from alkyl halides by coupling with Gilman reagents
	(Sec. 18.7)	from cyclopropanes by catalytic hydrogenation
	(Sec. 22.10)	from ketones and aldehydes by Wolff-Kishner reaction
Alkenes	(Sec. 6.12, 10.12)	from alkyl halides by treatment with strong base (E2 reaction)
	(Sec. 6.13)	from alcohols by dehydration
	(Sec. 7.5)	from alkynes by hydroboration followed by protonolysis
	(Sec. 7.6)	from alkynes by catalytic hydrogenation using the Lindlar catalyst
	(Sec. 7.6)	from alkynes by reduction with lithium in liquid ammonia
	(Sec. 22.13)	from ketones and aldehydes by treatment with alkylidenetriphenylphosphoranes (Wittig reaction)
	(Sec. 25.3)	from alpha-bromo ketones by heating with pyridine
	(Sec. 25.8)	from saturated ketones by phenylselenenylation, followed by oxidative elimination
	(Sec. 28.8)	from amines by methylation and Hofmann elimination
Alkynes	(Sec. 7.7, 7.8)	from terminal alkynes by alkylation of acetylide anions

	(Sec. 7.9)	from terminal alkynes by alkylation of the dianion
	(Sec. 7.11)	from dihalides by base-induced double dehydrohalogenation
Amides	(Sec. 24.5)	from carboxylic acids by heating with ammonia
	(Sec. 24.5, 28.8)	from acid chlorides by treatment with an amine or ammonia
	(Sec. 24.6)	from acid anhydrides by treatment with an amine or ammonia
	(Sec. 24.7)	from esters by treatment with an amine or ammonia
	(Sec. 24.9)	from nitriles by partial hydrolysis with either acid or base
	(Sec. 31.11)	from a carboxylic acid and an amine by treatment with dicyclohexylcarbodiimide (DCC)
Amines	(Sec. 10.4, 28.7)	from primary alkyl halides by treatment with ammonia
	(Sec. 22.15)	from conjugated enones by addition of primary or secondary amines
	(Sec. 24.8, 28.7)	from amides by reduction with $LiAlH_4$
	(Sec. 24.9, 28.7)	from nitriles by reduction with $LiAlH_4$
	(Sec. 28.7)	from primary alkyl halides by Gabriel synthesis
	(Sec. 28.7)	from acid chlorides by Curtius rearrangement of acyl azides
	(Sec. 28.7)	from primary amides by Hofmann rearrangement
	(Sec. 28.7)	from primary alkyl azides by reduction with $LiAlH_4$
	(Sec. 28.7)	from ketones and aldehydes by reductive amination with an amine and $NaBH_3CN$
Amino Acids	(Sec. 31.4)	from alpha-bromo acids by S_N2 reaction with ammonia
	(Sec. 31.4)	from aldehydes by reaction with KCN and ammonia (Strecker synthesis)
	(Sec. 31.4)	from alpha-keto acids by reductive amination
	(Sec. 31.4)	from primary alkyl halides by alkylation with diethyl acetamidomalonate
Arenes	(Sec. 16.2 - 16.5)	from arenes by Friedel-Crafts alkylation with a primary alkyl halide
	(Sec. 16.7)	from aryl alkyl ketones by catalytic reduction of the keto group
	(Sec. 29.4)	from arenediazonium salts by treatment with phosphorous acid
Arylamines	(Sec. 15.4, 29.3)	from nitroarenes by reduction with either Fe, Sn, or H_2/Pd.

Arenediazonium salts	(Sec. 29.4)	from arylamines by reaction with nitrous acid
Arenesulfonic acids	(Sec. 15.4)	from arenes by electrophilic aromatic substitution with SO_3/H_2SO_4
Azides	(Sec. 10.4, 18.7)	from primary alkyl halides by S_N2 reaction with azide ion
Carboxylic acids	(Sec. 6.8)	from mono- and 1,2-disubstituted alkenes by ozonolysis
	(Sec. 6.9)	from 1,2-diols by cleavage with sodium periodate
	(Sec. 16.9)	from arenes by side-chain oxidation with $Na_2Cr_2O_7$ or $KMnO_4$
	(Sec. 22.5)	from aldehydes by oxidation
	(Sec. 23.6)	from alkyl halides by conversion into Grignard reagents followed by reaction with CO_2
	(Sec. 23.6, 24.9)	from nitriles by vigorous acid or base hydrolysis
	(Sec. 24.5)	from acid chlorides by reaction with aqueous base
	(Sec. 24.6)	from acid anhydrides by reaction with aqueous base
	(Sec. 24.7)	from esters by hydrolysis with aqueous base
	(Sec. 24.8)	from amides by hydrolysis with aqueous base
	(Sec. 25.7)	from methyl ketones by reaction with halogen and base (haloform reaction)
	(Sec. 29.9)	from phenols by treatment with CO_2 and base (Kolbe carboxylation)
Cyanohydrins	(Sec. 22.8)	from aldehydes and ketones by reaction with HCN
Cycloalkanes	(Sec. 16.10)	from arenes by rhodium-catalyzed hydrogenation
	(Sec. 18.6)	from alkenes by addition of dichlorocarbene
	(Sec. 18.6)	from alkenes by reaction with CH_2I_2 and Zn/Cu (Simmons-Smith reaction)
Disulfides	(Sec. 20.13)	from thiols by oxidation with bromine
Enamines	(Sec. 22.9)	from ketones or aldehydes by reaction with secondary amines
Epoxides	(Sec. 19.7)	from alkenes by treatment with a peroxyacid
	(Sec. 19.7)	from halohydrins by treatment with base

Esters	(Sec. 24.4, 10.4)	from carboxylic acid salts by S_N2 reaction with primary alkyl halides
	(Sec. 24.4)	from carboxylic acids by acid-catalyzed reaction with an alcohol (Fischer esterification)
	(Sec. 24.4)	from carboxylic acids by reaction with diazomethane
	(Sec. 24.5)	from acid chlorides by base-induced reaction with an alcohol
	(Sec. 24.6)	from acid anhydrides by base-induced reaction with an alcohol
	(Sec. 25.9)	from alkyl halides by alkylation with diethyl malonate
	(Sec. 25.9)	from esters by treatment of their enolate ions with alkyl halides
	(Sec. 29.9)	from phenols by base-induced reaction with an acid chloride
Ethers	(Sec. 6.3, 19.5)	from alkenes by alkoxymercuration/demercuration
	(Sec. 10.4, 19.4)	from primary alkyl halides by S_N2 reaction with alkoxide ions (Williamson ether synthesis)
	(Sec. 15.13)	from activated haloarenes by reaction with alkoxide ions
	(Sec. 15.14)	from unactivated haloarenes by reaction with alkoxide ions via benzyne intermediates
	(Sec. 19.7)	from alkenes by epoxidation with peroxyacids
	(Sec. 29.9)	from phenols by reaction of phenoxide ions with primary alkyl halides
Halides, alkyl	(Sec. 5.8, 5.9)	from alkenes by electrophilic addition of HX
	(Sec. 6.1)	from alkenes by addition of halogen
	(Sec. 6.2)	from alkenes by electrophilic addition of hypohalous acid (HOX) to yield halohydrins
	(Sec. 6.5)	from alkenes by radical-catalyzed addition of HBr
	(Sec. 7.3)	from alkynes by addition of halogen
	(Sec. 7.3)	from alkynes by addition of HX
	(Sec. 9.5)	from alkenes by allylic bromination with N-bromosuccinimide (NBS)
	(Sec. 9.7)	from alcohols by reaction with HX
	(Sec. 9.7)	from alcohols by reaction with $SOCl_2$

	(Sec. 9.7)	from alcohols by reaction with PBr_3
	(Sec. 10.4, 10.5)	from alkyl tosylates by S_N2 reaction with halide ions
	(Sec. 16.8)	from arenes by benzylic bromination with N-bromosuccinimide (NBS)
	(Sec. 18.7)	from cyclopropanes by electrophilic ring opening with HBr
	(Sec. 19.6)	from ethers by cleavage with either HX or BBr_3
	(Sec. 23.9)	from carboxylic acids by treatment with Ag_2O and bromine (Hunsdiecker reaction)
	(Sec. 25.3)	from ketones by alpha-halogenation with bromine
	(Sec. 25.4)	from carboxylic acids by alpha-halogenation with phosphorus and bromine (Hell-Volhard-Zelinskii reaction)
Halides, aryl	(Sec. 15.1, 15.2)	from arenes by electrophilic aromatic substitution with halogen
	(Sec. 29.4)	from arenediazonium salts by reaction with cuprous halides (Sandmeyer reaction)
	(Sec. 33.3)	from aromatic heterocycles by electrophilic aromatic substitution with halogen
Halohydrins	(Sec. 6.2)	from alkenes by electrophilic addition of hypohalous acid (HOX)
	(Sec. 19.8)	from epoxides by acid-induced ring opening with HX
Imines	(Sec. 22.9)	from ketones or aldehydes by reaction with primary amines
Ketones	(Sec. 6.8)	from alkenes by ozonolysis
	(Sec. 6.9)	from 1,2-diols by cleavage reaction with sodium periodate
	(Sec. 7.4)	from alkynes by mercuric-ion-catalyzed hydration
	(Sec. 7.5)	from alkynes by hydroboration/oxidation
	(Sec. 16.6)	from arenes by Lewis-acid-catalyzed reaction with an acid chloride (Friedel-Crafts acylation)
	(Sec. 20.10, 22.4)	from secondary alcohols by oxidation
	(Sec. 22.4, 24.5)	from acid chlorides by reaction with lithium diorganocopper (Gilman) reagents
	(Sec. 22.15)	from conjugated enones by addition of lithium diorganocopper reagents
	(Sec. 24.9)	from nitriles by reaction with Grignard reagents
	(Sec. 25.9)	from primary alkyl halides by alkylation with ethyl acetoacetate

	(Sec. 25.9)	from ketones by alkylation of their enolate ions with primary alkyl halides
Nitriles	(Sec. 10.4, 10.5)	from primary alkyl halides by S_N2 reaction with cyanide ion
	(Sec. 22.15)	from conjugated enones by addition of HCN
	(Sec. 24.9)	from primary amides by dehydration with $SOCl_2$
	(Sec. 25.9)	from nitriles by alkylation of their alpha-anions with primary alkyl halides
	(Sec. 29.4)	from arenediazonium ions by treatment with CuCN
Nitroarenes	(Sec. 15.3)	from arenes by electrophilic aromatic substitution with nitric/sulfuric acids
Organometallics	(Sec. 9.8)	formation of Grignard reagents from organohalides by treatment with magnesium
	(Sec. 9.9)	formation of organolithium reagents from organohalides by treatment with lithium
	(Sec. 9.9)	formation of lithium diorganocopper reagents (Gilman reagents) from organolithium reagents by treatment with cuprous halides
Phenols	(Sec. 15.4, 29.4)	from arenes by electrophilic aromatic substitution with H_2O_2 and FSO_3H
	(Sec. 15.4, 29.8)	from arenesulfonic acids by fusion with KOH
	(Sec. 29.4)	from arenediazonium salts by reaction with aqueous acid
	(Sec. 29.8)	from aryl halides by nucleophilic aromatic substitution with hydroxide ion
Quinones	(Sec. 29.9)	from phenols by oxidation with Fremy's salt $[(KSO_3)_2NO]$
	(Sec. 29.9)	from arylamines by oxidation with Fremy's salt
Sulfides	(Sec. 19.11)	from thiols by S_N2 reaction of thiolate ions with primary alkyl halides
Sulfones	(Sec. 19.11)	from sulfides or sulfoxides by oxidation with peroxyacids
Sulfoxides	(Sec. 19.11)	from sulfides by oxidation with H_2O_2
Thioacetals	(Sec. 22.12)	from ketones and aldehydes by acid-catalyzed reaction with thiols
Thiols	(Sec. 10.4, 10.5)	from primary alkyl halides by S_N2 reaction with hydrosulfide anion
	(Sec. 20.13)	from primary alkyl halides by S_N2 reaction with thiourea followed by hydrolysis

ORGANIC NAME REACTIONS

Acetoacetic ester synthesis (Section 25.9): a multi-step reaction sequence for converting a primary alkyl halide into a methyl ketone having three more carbon atoms in the chain.

$$R\text{-}CH_2X \quad + \quad CH_3CO\bar{C}HCOOEt \xrightarrow[2.\ H_3O^+,\ \Delta]{1.\ \Delta} RCH_2\text{-}CH_2COCH_3$$

Adams' catalyst (Section 6.6): PtO_2, a catalyst used for the hydrogenation of carbon-carbon double bonds.

Aldol condensation reaction (Section 26.2): the nucleophilic addition of an enol or enolate ion to a ketone or aldehyde, yielding a β-hydroxy ketone.

$$2\ R\text{-}\overset{O}{\overset{\|}{C}}\text{-}\overset{|}{\underset{|}{C}}\text{-}H \xrightarrow{\text{Base}} R\text{-}\overset{O}{\overset{\|}{C}}\text{-}\overset{|}{\underset{|}{C}}\text{-}\overset{OH}{\overset{|}{\underset{R}{\underset{|}{C}}}}\text{-}\overset{|}{\underset{|}{C}}\text{-}H$$

Amidomalonate amino acid synthesis (Section 31.4): a multi-step reaction sequence, similar to the malonic ester synthesis, for converting a primary alkyl halide into an amino acid.

$$R\text{-}CH_2X \quad + \quad :\bar{C}(NHAc)(COOEt)_2 \xrightarrow[2.\ H_3O^+,\ \Delta]{1.\ \Delta} R\text{-}CH_2\text{-}\underset{NH_2}{\underset{|}{C}}HCOOH$$

Cannizzaro reaction (Section 22.14): the disproportionation reaction that occurs when a nonenolizable aldehyde is treated with base.

$$2\ R_3C\text{-}CHO \xrightarrow[2.\ H_3O^+]{1.\ HO^-} R_3C\text{-}COOH \quad + \quad R_3C\text{-}CH_2OH$$

Claisen condensation reaction (Section 26.9): a nucleophilic acyl substitution reaction that occurs when an ester enolate ion attacks the carbonyl group of a second ester molecule. The product is a β-keto ester.

$$2\ R\text{-}CH_2COOEt \xrightarrow[2.\ H_3O^+]{1.\ NaOEt} R\text{-}CH_2\text{-}\overset{O}{\overset{\|}{C}}\text{-}\underset{R}{\underset{|}{C}}HCOOEt$$

Claisen rearrangement (Sections 29.9 and 30.12): the thermal [3.3] sigmatropic rearrangement of an allyl vinyl ether or an allyl phenyl ether.

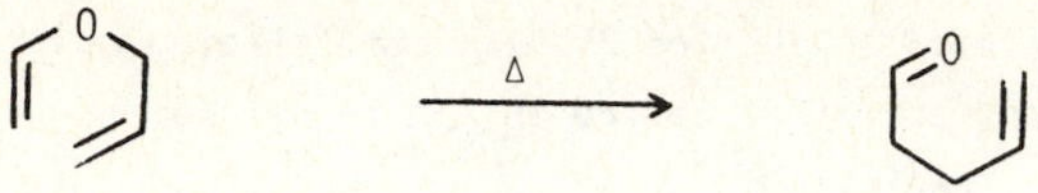

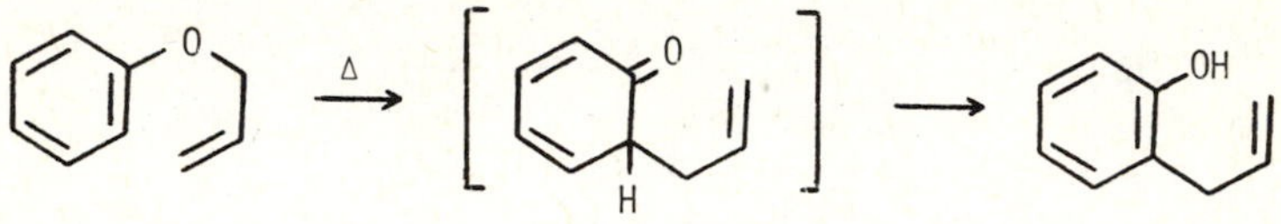

Cope rearrangement (Section 30.12): the thermal [3.3] sigmatropic rearrangement of a 1,5-diene to a new 1,5-diene.

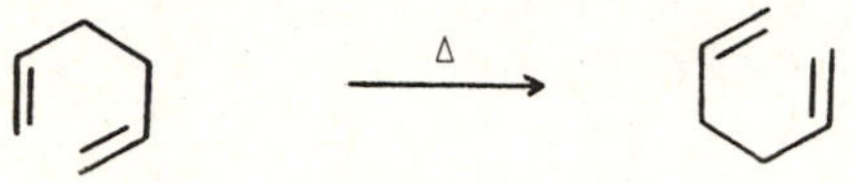

Curtius rearrangement (Section 28.7): the thermal rearrangement of an acyl azide to an isocyanate, followed by hydrolysis to yield an amine.

$$R-\overset{O}{\overset{\|}{C}}-N=\overset{+}{N}=\overset{-}{N} \xrightarrow[2.\ H_2O]{1.\ \Delta} R-NH_2 + CO_2$$

Diazonium coupling reaction (Section 29.4): the coupling reaction between an aromatic diazonium salt and a phenol or aniline.

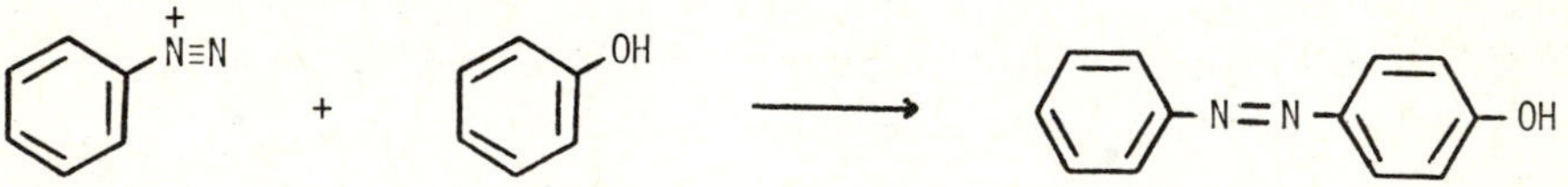

Dieckmann reaction (Section 26.11): the intramolecular Claisen condensation reaction of a 1,6- or 1,7-diester, yielding a cyclic β-keto ester.

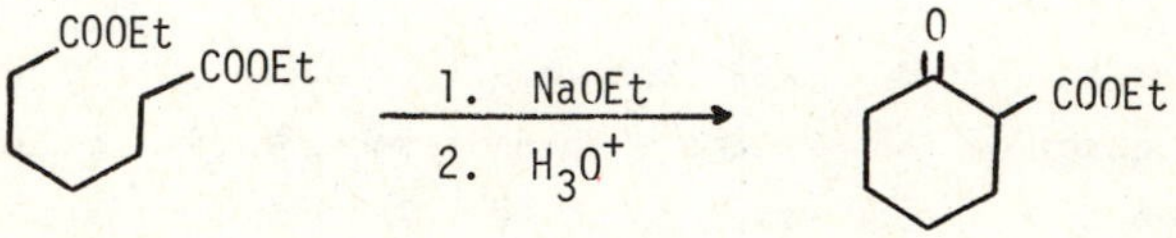

Diels-Alder cycloaddition reaction (Sections 13.7 and 30.8): the thermal reaction between a diene and a dienophile to yield a cyclohexene ring.

$$\text{diene} + \text{dienophile} \xrightarrow{\Delta} \text{cyclohexene}$$

Edman degradation (Section 31.9): a method for cleaving the N-terminal amino acid from a peptide by treatment of the peptide with N-phenylisothiocyanate.

$$Ph\text{-}N{=}C{=}S + H_2N\text{-}CH(R)\text{-}C(=O)\text{-}NH\text{-} \longrightarrow \text{(phenylthiohydantoin, Ph–N, S, NH, R, H, O)} + H_2N\text{-}$$

Fehling's test (Section 27.8): a chemical test for aldehydes, involving treatment with cupric ion.

Fischer esterification reaction (Section 24.4): the acid-catalyzed reaction between a carboxylic acid and an alcohol, yielding the ester.

$$R\text{-}COOH + R'\text{-}OH \xrightarrow{H^+,\ \Delta} R\text{-}COOR'$$

Friedel-Crafts reaction (Section 16.2): the alkylation or acylation of an aromatic ring by treatment with an alkyl- or acyl chloride in the presence of a Lewis-acid catalyst.

$$C_6H_6 \xrightarrow[AlCl_3]{R\text{-}Cl\ \ (RCOCl)} C_6H_5\text{-}R\ \ (\text{-}COR)$$

Gabriel amine synthesis (Section 28.7): a multi-step sequence for converting a primary alkyl halide into a primary amine by alkylation with potassium phthalimide, followed by hydrolysis.

$$R\text{-}CH_2X + \text{phthalimide anion } (:\bar{N}) \xrightarrow[2.\ NaOH]{1.\ \Delta} R\text{-}CH_2\text{-}NH_2$$

Gilman reagent (Section 9.9): a lithium dialkylcopper reagent, R_2CuLi, prepared by treatment of a cuprous salt with an alkyllithium. Gilman reagents undergo a coupling reaction with alkyl halides, and undergo a 1,4-addition reaction with α,β-unsaturated ketones.

Grignard reaction (Section 20.7): the nucleophilic addition reaction of an alkylmagnesium halide to a ketone, aldehyde, or ester carbonyl group.

$$\text{R-Mg-X} + \text{C=O} \xrightarrow[2.\ H_3O^+]{1.\ \Delta} \text{R-C(OH)}$$

Grignard reagent (Section 9.8): an organomagnesium halide, RMgX, prepared by reaction between an organohalide and magnesium metal.

Haloform reaction (Section 25.7): the conversion of a methyl ketone to a carboxylic acid and haloform by treatment with halogen and base.

$$\text{R-C(=O)-CH}_3 \xrightarrow[2.\ H_3O^+]{1.\ \text{NaOH},\ X_2} \text{R-COOH} + \text{CHX}_3$$

Hell-Volhard-Zelinskii reaction (Section 25.4): the α-bromination of carboxylic acids by treatment with bromine and phosphorus.

$$\text{R-CH}_2\text{-COOH} \xrightarrow[2.\ H_2O]{1.\ Br_2,\ P} \text{R-CH(Br)-COOH}$$

Hinsberg test (Section 28.8): a chemical means of distinguishing among primary, secondary, and tertiary amines by observing their reactions with benzenesulfonyl chloride.

Hofmann elimination (Section 28.8): a method for effecting the elimination reaction of an amine to yield an alkene. The amine is first treated with excess iodomethane, and the resultant quaternary ammonium salt is heated with silver oxide.

$$R_2N\text{-C-C-H} \xrightarrow[2.\ Ag_2O,\ \Delta]{1.\ CH_3I} \text{C=C}$$

Hofmann rearrangement (Section 28.7): the rearrangement of an N-bromoamide to a primary amine by treatment with aqueous base.

$$R\text{-}\overset{O}{\overset{\|}{C}}\text{-}NHBr \xrightarrow{NaOH,\ \Delta} R\text{-}NH_2 + CO_2$$

Hunsdiecker reaction (Section 23.9): a reaction for converting a carboxylic acid into an alkyl halide by treatment with mercuric oxide and halogen.

$$R\text{-}COOH \xrightarrow{HgO,\ X_2} R\text{-}X + CO_2$$

Jones reagent (Section 20.10): a solution of CrO_3 in acetone/aqueous sulfuric acid. This reagent oxidizes primary and secondary alcohols to carbonyl compounds under mild conditions.

Kiliani-Fischer synthesis (Section 27.8): a multi-step sequence for chain-lengthening an aldose into the next higher homolog.

$$\begin{matrix} CHO \\ | \\ R \end{matrix} \xrightarrow[3.\ Na\text{-}Hg]{1.\ HCN \quad 2.\ \Delta} \begin{matrix} CHO \\ | \\ CHOH \\ | \\ R \end{matrix}$$

Koenigs-Knorr reaction (Section 27.8): a method for synthesizing glycosides by reaction between an alcohol and a bromo-substituted carbohydrate.

$$\text{(CH}_2\text{OAc; OAc; AcO; AcO; Br pyranose)} \xrightarrow{ROH,\ Ag_2O} \text{(CH}_2\text{OAc; OAc; AcO; AcO; OR pyranose)}$$

Kolbe-Schmitt carboxylation reaction (Section 29.9): a method for introducing a carboxyl group in the ortho position of a phenol by treatment of the phenoxide anion with CO_2.

$$C_6H_5O^- \xrightarrow[2.\ H_3O^+]{1.\ CO_2} o\text{-}HOC_6H_4COOH$$

Malonic ester synthesis (Section 25.9): a multi-step sequence for converting an alkyl halide into a carboxylic acid with the addition of two carbon atoms to the chain.

$$R\text{-}CH_2X \quad + \quad \bar{:}CH(COOEt)_2 \xrightarrow[2.\ H_3O^+,\ \Delta]{1.\ \Delta} R\text{-}CH_2\text{-}CH_2COOEt$$

Maxam-Gilbert DNA sequencing (Section 33.15): a rapid and efficient method for sequencing long chains of DNA by employing selective cleavage reactions.

McLafferty rearrangement (Section 22.17): a general mass spectral fragmentation pathway for carbonyl compounds having a hydrogen three carbon atoms away from the carbonyl carbon.

H O e⁻ → + H O+·

Meisenheimer complex (Section 15.13): an intermediate formed in the nucleophilic aryl substitution reaction of a base with a nitro-substituted aromatic ring.

Cl O_2N NO_2 $\xrightarrow{HO^-}$ Cl OH O_2N NO_2

Merrifield solid-phase peptide synthesis (Section 31.12): a rapid and efficient means of peptide synthesis in which the growing peptide chain is attached to an insoluble polymer support.

Michael reaction (Section 26.12): the 1,4-addition reaction of a stabilized enolate anion to an α,β-unsaturated carbonyl compound.

$$R\text{-}\overset{O}{\overset{\|}{C}}\text{-}CH_2\text{-}\overset{O}{\overset{\|}{C}}\text{-}R' \quad + \quad H_2C{=}CH\text{-}\overset{O}{\overset{\|}{C}}\text{-}R'' \xrightarrow{\text{Base}} (R\text{-}\overset{O}{\overset{\|}{C}})(R'\text{-}\overset{O}{\overset{\|}{C}})CH\text{-}CH_2CH_2\text{-}\overset{O}{\overset{\|}{C}}\text{-}R''$$

Nagata hydrocyanation reaction (Section 22.15): a reaction for effecting the conjugate (1,4)-addition of HCN to an α,β-unsaturated ketone by treatment with diethylaluminum cyanide.

$$\text{>C=C-C(=O)-R} \xrightarrow[\text{2. } H_3O^+]{\text{1. } Et_2AlCN} \text{N}\equiv\text{C-C-C-C(=O)-R}$$

Raney nickel (Section 22.12): a specially prepared form of nickel that is used for desulfurization of thioacetals.

$$(RS)(RS)C< \xrightarrow{\text{Raney Ni}} (H)(H)C<$$

Robinson annulation reaction (Section 26.14): a multi-step sequence for building a new cyclohexenone ring onto a ketone. The sequence involves an initial Michael reaction of the ketone followed by an internal aldol cyclization.

$$\text{methyl vinyl ketone} + \text{cyclic ketone} \xrightarrow{\text{Base}} \text{fused cyclohexenone}$$

Rosenmund reduction (Section 22.3): a method for conversion of an acid chloride into an aldehyde by catalytic hydrogenation.

$$\text{R-C(=O)-Cl} \xrightarrow{H_2/BaSO_4} \text{R-C(=O)-H}$$

Sandmeyer reaction (Section 29.4): a method for converting aryldiazonium salts into aryl halides by treatment with cuprous halide.

$$\text{Ar-}\overset{+}{\text{N}}\equiv\text{N} + \text{CuBr} \longrightarrow \text{Ar-Br}$$

Schotten-Baumann reaction (Section 24.5): a method for preparing esters by treatment of an acid chloride with an alcohol in the presence of aqueous base.

$$R\text{-}\overset{O}{\overset{\|}{C}}\text{-}Cl \quad + \quad R'\text{-}OH \xrightarrow{NaOH,\ H_2O} R\text{-}\overset{O}{\overset{\|}{C}}\text{-}OR'$$

Simmons-Smith reaction (Section 18.6): a method for preparing cyclopropanes by treating an alkene with CH_2I_2 and zinc-copper.

$$>C{=}C< \quad + \quad CH_2I_2 \xrightarrow{Zn\text{-}Cu} \text{cyclopropane } (CH_2)$$

Stork enamine reaction (Section 26.13): a multi-step sequence whereby ketones are converted into enamines by treatment with a secondary amine, and the enamines are then used in Michael reactions.

$$R\text{-}\overset{O}{\overset{\|}{C}}\text{-}\overset{|}{C}\text{-}H \xrightarrow{HNR_2} R\text{-}\overset{NR_2}{\overset{|}{C}}{=}C< \xrightarrow[2.\ H_3O^+]{1.\ H_2C{=}CHCOR'} R\text{-}\overset{O}{\overset{\|}{C}}\text{-}\overset{|}{\underset{|}{C}}\text{-}CH_2CH_2\overset{O}{\overset{\|}{C}}\text{-}R'$$

Strecker amino acid synthesis (Section 31.4): a multi-step sequence for converting an aldehyde into an amino acid by initial treatment with ammonium cyanide, followed by hydrolysis.

$$R\text{-}CHO \xrightarrow{NH_3,\ KCN} R\text{-}\overset{NH_2}{\overset{|}{C}H}\text{-}CN \xrightarrow{H_3O^+} R\text{-}\overset{NH_2}{\overset{|}{C}H}\text{-}COOH$$

Tollen's test (Section 22.5): a chemical test for detecting aldehydes by treatment with ammoniacal silver nitrate. A positive test is signaled by formation of a silver mirror on the walls of the reaction vessel.

Walden inversion (Section 10.1): the inversion of stereochemistry at a chiral center that occurs during S_N2 reactions.

$$Nu:^- \quad + \quad >C\text{-}X \longrightarrow Nu\text{-}C< \quad + \quad :X^-$$

Williamson ether synthesis (Section 19.4): a method for preparing ethers by treatment of a primary alkyl halide with an alkoxide ion.

$$R\text{-}O:^- \quad + \quad R'CH_2X \longrightarrow R\text{-}O\text{-}CH_2R'$$

Wittig reaction (Section 22.12): a general method of alkene synthesis by treatment of a ketone or aldehyde with an alkylidenetriphenylphosphorane.

$$R\text{-}\overset{O}{\overset{\|}{C}}\text{-}R' \quad + \quad \rangle C{=}P(Ph)_3 \longrightarrow \begin{matrix} R \\ 'R \end{matrix}\rangle C{=}C\langle$$

Wohl degradation (Section 27.8): a multi-step reaction sequence for degrading an aldose into the next lower homolog.

$$\begin{matrix} CHO \\ H\text{-}\!\!+\!\!\text{-}OH \\ R \end{matrix} \xrightarrow[\text{3. } NaOCH_3]{\text{1. } NH_2OH \quad \text{2. } Ac_2O} \begin{matrix} CHO \\ | \\ R \end{matrix}$$

Wohl-Ziegler reaction (Section 9.5): a reaction for effecting allylic bromination by treatment of an alkene with N-bromosuccinimide.

$$\rangle C{=}\overset{|}{C}\text{-}\underset{|}{\overset{|}{C}}\text{-}H \xrightarrow{NBS,\ CCl_4} \rangle C{=}\overset{|}{C}\text{-}\underset{|}{\overset{|}{C}}\text{-}Br$$

Wolff-Kishner reaction (Section 22.10): a method for converting a ketone or aldehyde into the corresponding hydrocarbon by treatment with hydrazine and strong base.

$$\diagup\overset{O}{\overset{\|}{C}}\diagdown \xrightarrow{N_2H_4,\ KOH} \diagup\overset{H\ \ H}{C}\diagdown$$

Woodward-Hoffmann orbital symmetry rules (Section 30.13): a series of rules for predicting the stereochemistry of pericyclic reactions. Even-electron species react thermally through either antarafacial or conrotatory pathways, whereas odd-electron species react thermally through either suprafacial or disrotatory pathways (even-antara-con; odd-supra-dis).

Ziegler-Natta polymerization (Section 34.6): a method for carrying out the stereoregular polymerization of alkenes by using titanium-aluminum catalysts.

NOBEL PRIZE WINNERS IN CHEMISTRY

1901 Jacobus H. van't Hoff (Dutch):
"for the discovery of laws of chemical dynamics and of osmotic pressure"

1902 Emil Fischer (German):
"for syntheses in the groups of sugars and purines"

1903 Svante A. Arrhenius (Swedish):
"for his theory of electrolytic dissociation"

1904 Sir William Ramsey (British):
"for the discovery of gases in different elements in the air and for the determination of their place in the periodic system"

1905 Adolf von Baeyer (German):
"for his researches on organic dyestuffs and hydroaromatic compounds"

1906 Henri Moissan (French):
"for his research on the isolation of the element fluorine and for placing at the service of science the electric furnace that bears his name"

1907 Eduard Buchner (German):
"for his biochemical researches and his discovery of cell-less formation"

1908 Ernest Rutherford (British):
"for his investigation into the disintegration of the elements and the chemistry of radioactive substances"

1909 Wilhelm Ostwald (German):
"for his work on catalysis and on the conditions of chemical equilibrium and velocities of chemical reactions"

1910 Otto Wallach (German):
"for his initiative work in the field of alicyclic substances"

1911 Marie Curie (French):
"for her services to the advancement of chemistry by the discovery of the elements radium and polonium"

1912 Victor Grignard (French):
"for the discovery of the so-called Grignard reagent, which has greatly helped in the development of organic chemistry"
Paul Sabatier (French):
"for his method of hydrogenating organic compounds in the presence of finely divided metals"

1913 Alfred Werner (Swiss):
"in recognition of his work on the linkage of atoms in molecules"

1914 Theodore W. Richards (U.S.):
"for his exact determinations of the atomic weights of a great number of chemical elements"

1915 Richard M. Willstatter (German):
"for his research on coloring matter in the vegetable kingdom, principally on chlorophyll"

1916 No award

1917 No award

1918 Fritz Haber (German):
"for the synthesis of ammonia from its elements, nitrogen and hydrogen"

1919 No award

1920 Walther H. Nernst (German):
"for his thermochemical work"

1921 Frederick Soddy (British):
"for his contributions to the chemistry of radioactive substances and his investigations into the origin and nature of isotopes"

1922 Francis W. Aston (British):
"for his discovery, by means of his mass spectrograph, of the isotopes of a large number of nonradioactive elements, as well as for his discovery of the whole-number rule"

1923 Fritz Pregl (Austrian):
"for his invention of the method of microanalysis of organic substances"

1924 No award

1925 Richard A. Zsigmondy (German):
"for his elucidation of the heterogeneous nature of colloid solutions"

1926 Theodor Svedberg (Swedish):
"for his work on disperse systems"

1927 Heinrich O. Wieland (German):
"for his research on bile acids and analogous substances"

1928 Adolf O. R. Windaus (German):
"for his studies on the constitution of the sterols and their connection with the vitamins"

1929 Arthur Harden (British):
Hans von Euler-Chelpin (Swedish):
"for their investigation on the fermentation of sugar and of fermentative enzymes"

1930 Hans Fischer (German):
"for his researches into the constitution of hemin and chlorophyll"

1931 Frederich Bergius (German):
Carl Bosch (German):
"for their contributions to the invention and development of chemical high-pressure methods"

1932 Irving Langmuir (U.S.):
"for his discoveries and investigations in surface chemistry"

1933 No award

1934 Harold C. Urey (U.S.):
"for his discovery of heavy hydrogen"

1935 Frederic Joliot (French):
Irene Joliot-Curie (French):
"for their synthesis of new radioactive elements"

1936 Peter J. W. Debye (Dutch/U.S.):
"for his contributions to the study of molecular structure through his investigations on dipole moments and on the diffraction of X-rays and electrons in gases"

1937 Walter N. Haworth (British):
"for his researches into the constitution of carbohydrates and vitamin C"
Paul Karrer (Swiss):
"for his researches into the constitution of carotenoids, flavins, and vitamins A and B"

1938 Richard Kuhn (German):
"for his work on carotenoids and vitamins"

1939 Adolf F. J. Butenandt (German):
"for his work on sex hormones"
Leopold Ruzicka (Swiss):
"for his work on polymethylenes and higher terpenes"

1940 No award

1941 No award

1942 No award

1943 Georg de Hevesy (Hungarian):
"for his work on the use of isotopes as tracer elements in researches on chemical processes"

1944 Otto Hahn (German):
"for his discovery of the fission of heavy nuclei"

1945 Artturi I. Virtanen (Finnish):
"for his researches and inventions in agricultural and nutritive chemistry"

1946 James B. Sumner (U.S.):
"for his discovery that enzymes can be crystallized"
John H. Northrop (U.S.):
Wendell M. Stanley (U.S.):
"for their preparation of enzymes and virus proteins in a pure form"

1947 Sir Robert Robinson (British):
"for his research on certain vegetable products of great biological importance, particularly alkaloids"

1948 Arne W. K. Tiselius (Swedish):
"for his researches on electrophoresis and adsorption analysis, especially for his discoveries concerning the complex nature of the serum proteins"

1949 William F. Giauque (U.S.):
"for his work in the field of chemical thermodynamics, particularly concerning the behavior of substances at extremely low temperatures"

1950 Kurt Alder (German):

Otto P. H. Diels (German):

"for the development of the diene synthesis"

1951 Edwin M. McMillan (U.S.):

Glenn T. Seaborg (U.S.):

"for their discoveries in the chemistry of the transuranium elements"

1952 Archer J. P. Martin (British):

Richard L. M. Synge (British):

"for their development of partition chromatography

1953 Hermann Staudinger (German):

"for his work in developing the chemistry of high polymers"

1954 Linus C. Pauling (U.S.):

"for his discoveries of the structures of proteins"

1955 Vincent du Vigneaud (U.S.):

"for his work on biochemically important sulfur compounds"

1956 Sir Cyril N. Hinshelwood (British):

Nikolai N. Semenov (U.S.S.R.):

"for their research in clarifying the mechanisms of chemical reactions in gases"

1957 Sir Alexander R. Todd (British):

"for his work on nucleotide and nucleotide coenzymes"

1958 Frederick Sanger (British):

"for his work on the structure of proteins, particularly insulin"

1959 Jaroslav Heyrovsky (Czechoslovakian):

"for his important work with the polarograph"

1960 Willard F. Libby (U.S.):

"for his work in nuclear chemistry and its application to geochronology"

1961 Melvin Calvin (U.S.):

"for his partial explanation of photosynthesis"

1962 John C. Kendrew (British):

Max F. Perutz (British):

"for their studies of the structures of globular proteins"

1963 Giulio Natta (Italian):

Karl Ziegler (German):

"for their work in the controlled polymerization of hydrocarbons through the use of organometallic catalysts"

1964 Dorothy C. Hodgkin (British):

"for her work in determining the structures of biochemical substances, particularly vitamin B-12 and penicillin, by X-ray techniques"

1965 Robert B. Woodward (U.S.):

"for his meritorious contributions to the 'art' of organic synthesis"

1966 Robert S. Mulliken (U.S.):
"for his fundamental work concerning chemical bonds and the electronic structure of molecules by the molecular orbital method"

1967 Manfred Eigen (German):
Ronald G. W. Norrish (British):
George Porter (British):
"for their studies of extremely fast chemical reactions, effected by disturbing the equilibrium with very short pulses of energy"

1968 Lars Onsager (U.S.):
"for his discovery of the reciprocal relations bearing his name, which are fundamental for the thermodynamics of irreversible processes"

1969 Derek H. R. Barton (British):
Odd Hassel (Norwegian):
"for their work in conformational theory"

1970 Luis F. Leloir (Argentinian):
"for his discovery of sugar nucleotides and their role in the biosynthesis of carbohydrates"

1971 Gerhard Herzberg (Canadian):
"for his contributions to knowledge of electronic structure and geometry of molecules, particularly free radicals"

1972 Christian B. Anfinsen (U.S.):
Stanford Moore (U.S.):
William H. Stein (U.S.):
"for their fundamental contributions to enzyme chemistry"

1973 Ernst Otto Fischer (German):
Geoffrey Wilkinson (British):
"for their pioneering work, performed independently, on the chemistry of the organometallic sandwich compounds"

1974 Paul J. Flory (U.S.):
"for his work in polymer chemistry"

1975 John Cornforth (Australian/British):
"for his work in enzyme-catalyzed reactions"
Vladimir Prelog (Yugoslavian/Swiss):
"for his work in organic molecules and reactions"

1976 William N. Lipscomb (U.S.):
"for his studies on the structures of boranes illuminating problems of chemical bonding"

1977 Ilya Pregogine (Belgian):
"for his researches in nonlinear irreversible thermodynamics"

1978 Peter Mitchell (British):
"for his contribution to the understanding of biological energy transfer through the formulation of the chemiosmotic theory"

1979 Herbert C. Brown (U.S.):

"for his application of boron compounds to synthetic organic chemistry"

Georg Wittig (German):

"for developing phosphorus reagents, presently bearing his name"

1980 Paul Berg (U.S.):

"for his fundamental studies of the biochemistry of nucleic acids"

Walter Gilbert (U.S.)

Frederick Sanger (British):

"for their contributions concerning the determination of base sequences in nucleic acids"

1981 Kenichi Fukui (Japanese)

Roald Hoffmann (U.S.):

"for their development of the principles of orbital symmetry conservation"

1982 Aaron Klug (British):

"for his development of crystallographic electron microscopy and his structural elucidation of biologically important nucleic acid - protein complexes"

TOP 40 ORGANIC CHEMICALS (U.S. CHEMICAL INDUSTRY, 1981)

Ethylene (14,434,000 tons/yr):

prepared by thermal cracking of ethane and propane during petroleum refining; used as starting material for manufacture of polyethylene, ethylene oxide, ethylene glycol, ethylbenzene, 1,2-dichloroethane, and other bulk chemicals.

Propylene (7,008,000 tons/yr):

prepared by steam cracking of light hydrocarbon fractions during petroleum refining; used as starting material for the manufacture of polypropylene, acrylonitrile, propylene oxide, and isopropyl alcohol.

1,2-Dichloroethane (4,586,000 tons/yr):

prepared by addition of chlorine to ethylene in the presence of $FeCl_3$ catalyst at 50 °C; used as a chlorinated solvent and as starting material for the manufacture of vinyl chloride.

Methanol (4,204,000 tons/yr):

prepared by high temperature reaction of a mixture of H_2, CO, and CO_2 ("synthesis gas") over a catalyst at 100 atmospheres presssure; used as a solvent and as starting material for the manufacture of formaldehyde, acetic acid, and methyl *tert*-butyl ether.

Ethylbenzene (3,924,000 tons/yr):

prepared during catalytic reforming in petroleum refining and by an acid-catalyzed Friedel-Crafts alkylation of benzene with ethylene; used almost exclusively for production of styrene.

Vinyl chloride (3,360,000 tons/yr):

prepared by addition of chlorine to ethylene followed by elimination of HCl; used as starting material for preparation of poly(vinyl chloride) polymers (hoses, pipe, molded objects).

Styrene (3,306,000 tons/yr):

prepared by high-temperature catalytic dehydrogenation of ethylbenzene; used in the manufacture of polystyrene polymers (thermoplastics, packaging materials).

Dimethyl terephthalate (3,175,000 tons/yr):

prepared from *p*-xylene by oxidation and esterification; used in the manufacture of polyester polymers (textiles, upholstery, recording tape, and film).

Formaldehyde (2,928,000 tons/yr):

prepared by air oxidation of methanol over a silver or metal oxide catalyst; used in the manufacture of phenolic resins, melamine resins, and plywood adhesives.

Ethylene oxide (2,555,000 tons/yr):

prepared by high-temperature air oxidation of ethylene over a silver catalyst; used as starting material for the preparation of ethylene glycol and poly(ethylene glycol).

Ethylene glycol (2,028,000 tons/yr):

prepared by high-temperature reaction between water and ethylene oxide at neutral pH; used as antifreeze and as a starting material for polymers and latex paints.

***p*-Xylene** (1,848,000 tons/yr):

prepared by separation from the mixed xylenes that result during catalytic reforming in gasoline refining; used as starting material for manufacture of the dimethyl terephthalate needed for polyester synthesis.

Cumene (1,653,000 tons/yr):

prepared by a phosphoric-acid-catalyzed Friedel-Crafts reaction between benzene and propylene; used primarily for conversion into phenol and acetone.

1,3-Butadiene (1,523,000 tons/yr):

prepared by steam cracking of gas oil during petroleum refining and by dehydrogenation of butane and butene; used primarily as a monomer component in the manufacture of styrene-butadiene rubber (SBR), polybutadiene rubber, and acrylonitrile-butadiene-styrene (ABS) copolymers.

Acetic acid (1,353,000 tons/yr):

prepared by metal-catalyzed air oxidation of acetaldehyde under pressure at 80 °C and by reaction of methanol with carbon monoxide; used to make vinyl acetate polymers, ethyl acetate solvent, and cellulose acetate polymers.

Phenol (1,276,000 tons/yr):

prepared from cumene by air oxidation to cumene hydroperoxide, followed by acid-catalyzed decomposition; used as starting material for preparing phenolic resins, epoxy resins, and caprolactam.

Acetone (1,084,000 tons/yr):

prepared by acid-catalyzed decomposition of cumene hydroperoxide and by air oxidation of isopropyl alcohol at 300 °C over a metal oxide catalyst; used as a solvent and as starting material for synthesizing bisphenolA and methyl methacrylate.

Acrylonitrile (1,003,000 tons,yr):

prepared by the Sohio ammoxidation process in which propylene, ammonia, and air are passed over a catalyst at 500 °C; used in the preparation of acrylic fibers, nitrile rubber, and acrylonitrile-butadiene-styrene (ABS) copolymer.

Propylene oxide (1,000,000 tons/yr):

prepared from propylene by high-temperature oxidation with air and by formation of propylene chlorohydrin followed by loss of HCl; used in the manufacture of propylene glycol, polyurethanes, and polyesters.

Vinyl acetate (967,000 tons/yr):

prepared from reaction of acetic acid, ethylene, and oxygen; used for manufacture of poly(vinyl acetate) (paint emulsions, plywood adhesives, textiles).

Cyclohexane (875,000 tons/yr):

prepared by catalytic hydrogenation of benzene; used as starting material for synthesis of the caprolactam and adipic acid needed for nylon.

Isopropyl alcohol (822,000 tons/yr):

prepared by direct high-temperature addition of water to propylene; used in cosmetics formulations, as a solvent and deicer, and as starting material for manufacture of acetone.

Benzene (674,000 tons/yr):

obtained from petroleum by catalytic reforming of hexane and cyclohexane over a platinum catalyst; used as starting material for the synthesis of ethylbenzene, cumene, cyclohexane, and aniline.

Ethanol (579,000 tons/yr):

prepared by direct vapor phase hydration of ethylene at 300 °C over an acidic catalyst; used as a solvent, as a constituent of cleaning preparations, and as starting material for ester synthesis.

Toluene (500,000 tons/yr):

prepared during catalytic reforming of petroleum; used as a gasoline additive and as a degreasing solvent.

***o*-Xylene** (474,000 tons/yr):

obtained by separation from the mixed xylenes that result during catalytic reforming in petroleum refining; used as starting material for preparation of phthalic acid and phthalic anhydride.

Caprolactam (463,000 tons/yr):

prepared from phenol by conversion into cyclohexanone, followed by formation and acid-catalyzed rearrangement of cyclohexanone oxime; used as starting material for the manufacture of nylon-6.

Butanol (412,000 tons/yr):

prepared in the oxo process by reaction of propylene with carbon monoxide; used as solvent and as a starting material for synthesis of butyl acetate and dibutyl phthalate.

Phthalic anhydride (406,000 tons/yr):

prepared by oxidation of o-xylene at 400 °C and by oxidation of naphthalene obtained from coal tar; used for the synthesis of polyesters and plasticizers.

Tetrachloromethane (359,000 tons/yr):

prepared by reaction of carbon disulfide with chlorine and by radical chlorination of methane; used as starting material for the synthesis of the chlorofluorocarbon gases (Freons) used as aerosol propellants.

Tetrachloroethylene (349,000 tons/yr):

prepared by high-temperature chlorination of ethane, followed by loss of HCl; used as a solvent for metal degreasing and for dry cleaning.

Aniline (318,000 tons/yr):

prepared by catalytic reduction of nitrobenzene with hydrogen at 350 °C; used as starting material for preparing toluene diisocyanate and for the synthesis of dyes and pharmaceuticals.

2-Butanone (313,000 tons/yr):

prepared by oxidation of 2-butanol over a ZnO catalyst at 400 °C; used as a solvent for vinyl coatings, lacquers, rubbers, and paint removers.

1,1,1-Trichloroethane (302,000 tons/yr):

prepared by addition of HCl to vinyl chloride to give 1,1-dichloroethane, followed by radical chlorination with Cl_2; used as a solvent, industrial cleaner, and metal degreaser.

Dichloromethane (280,000 tons/yr):

prepared by radical chlorination of methane at 500 °C; used in decaffeinating coffee and as a solvent and paint stripper.

Dodecylbenzene (257,000 tons/yr):

prepared by Friedel-Crafts alkylation of benzene with dodecene; used in the manufacture of detergents.

Bisphenol A (255,000 tons/yr):

prepared by reaction of phenol with acetone; used in the manufacture of epoxy resins and adhesives, polycarbonates, and polysulfones.

Propylene glycol (240,000 tons/yr):

prepared by high temperature reaction of propylene oxide with water; used in the preparation of polyesters and as an additive in the food industry.

Ethanolamines (204,000 tons/yr):

prepared by reaction of ammonia with ethylene oxide at 100 °C; used in soaps, detergents, cosmetics, and corrosion inhibitors.

Carbon disulfide (200,000 tons/yr):

prepared by reaction of sulfur and methane at 700 °C in the presence of a catalyst; used in the synthesis of viscose rayon and cellophane film, and as starting material for the preparation of tetrachloromethane.